Nonhuman Primate Welfare

Lauren M. Robinson • Alexander Weiss
Editors

Nonhuman Primate Welfare

From History, Science, and Ethics to Practice

Editors
Lauren M. Robinson
Wolf Science Center
University of Veterinary Medicine Vienna
Vienna, Austria

Language Research Center
Georgia State University
Atlanta, GA, USA

Alexander Weiss
School of Philosophy, Psychology and
Language Sciences, Department of Psychology
University of Edinburgh
Edinburgh, UK

Kyoto University, Wildlife Research Center
Kyoto, Japan

ISBN 978-3-030-82707-6   ISBN 978-3-030-82708-3 (eBook)
https://doi.org/10.1007/978-3-030-82708-3

© Springer Nature Switzerland AG 2023

This work is subject to copyright. All rights are reserved by the Publisher, whether the whole or part of the material is concerned, specifically the rights of translation, reprinting, reuse of illustrations, recitation, broadcasting, reproduction on microfilms or in any other physical way, and transmission or information storage and retrieval, electronic adaptation, computer software, or by similar or dissimilar methodology now known or hereafter developed.

The use of general descriptive names, registered names, trademarks, service marks, etc. in this publication does not imply, even in the absence of a specific statement, that such names are exempt from the relevant protective laws and regulations and therefore free for general use.

The publisher, the authors, and the editors are safe to assume that the advice and information in this book are believed to be true and accurate at the date of publication. Neither the publisher nor the authors or the editors give a warranty, expressed or implied, with respect to the material contained herein or for any errors or omissions that may have been made. The publisher remains neutral with regard to jurisdictional claims in published maps and institutional affiliations.

"The book cover photo shows a Japanese macaque with her infant. It was taken by Lauren M. Robinson with kind permission from and within the Royal Zoological Society of Scotland's Highland Wildlife Park."

This Springer imprint is published by the registered company Springer Nature Switzerland AG.
The registered company address is: Gewerbestrasse 11, 6330 Cham, Switzerland

To Charlie and Danny
—Alexander Weiss

To Imogene, Sammie, and Terri
—Lauren M. Robinson

Preface

> Both animals and caretakers are likely to benefit from positive feedback from one another. Engaging in interactions that are pleasant for both caretakers and the animals are likely to create a better work environment for caretakers and to help instill positive attitudes regarding the animals they care for.[1]
> —Corri D. Waitt (1975–2014)

This book originated in a photo contest that was held by Springer and that the first editor had entered into. During conversations with Springer that took place after the contest was over, it emerged that an edited book on nonhuman primate welfare was lacking from their collection. Soon after these conversations with Springer, the first editor contacted potential authors and found that there were enough people willing to contribute chapters. A proposal was written, and Springer accepted the proposal.

The topic of animal welfare is broad and cuts across disciplines. Therefore, two goals guided our choice of authors and areas to cover. The first goal was to provide a resource for people who work with nonhuman primates. As such, we sought out a wide range of experts who could write about topics related to the welfare of primates in different settings, such as in laboratories, zoos, and sanctuaries, and primates from different backgrounds, including those kept as pets or used in the entertainment industry. The second, related, goal was to provide a resource for people who guide or shape policy on, or engender debates about, the proper use and care of primates (note that we fully expect that these two audiences will overlap!). To these ends, we sought out chapters from philosophers, ethicists, legal scholars, and both advocates and adversaries of the use of nonhuman primates in research. Finally, in support of these goals, we recruited authors whose chapters would provide historical overviews for these other chapters and tried to find authors willing to write about the situation in countries other than the USA, member nations of the European Union, and other countries in Europe, such as the UK. We hope that the readers of this book, including those that we did not mention, and especially students, will agree that we have been modestly successful in achieving our goals.

---

[1] Taken from Waitt C. et al. (2002). The effects of caretaker-primate relationships on primates in the laboratory. J Appl Anim Welf Sci 5(4), 309–319. https://doi.org/10.1207/S15327604JAWS0504_05

Of course, as with Top 10 lists, we expect that opinions will differ with regard to what we did and did not include in this book. In particular, we expect that some chapters and/or their authors will raise some readers' hackles. We therefore feel compelled to explain our reasons for including these chapters and/or inviting these authors. We decided early on that the readers of this book would be best served if we provided a broad range of viewpoints and so we gave all of the authors the freedom to present their views and stepped in only to provide guidance and feedback so that their chapters are the best representations of those views. We think that readers, upon closer consideration, will agree that the potential for these uncensored views to spark debate will engage non-experts and ultimately benefit captive nonhuman primates.

A few notes of thanks are in order. This book could not have been completed without the contributions of the authors. Their hard work and patience are much appreciated. We also gratefully acknowledge the assistance from the editors at Springer. They were always on hand to answer our questions. In addition, a considerable portion of Alex's work on the book and chapter was completed during a Visiting Professorship at the Kyoto University Wildlife Research Center. He thanks his hosts, and especially Professor Miho Inoue-Murayama, for their help and for making this wonderful opportunity possible.

Since humans began to keep primates in captivity, humans have made vast strides in understanding their cognitive abilities, social bonds, and personalities. We hope that the science, practice, information, and wisdom presented in this book will go some way in ensuring that, whether they are in our care or not, these evolutionary cousins of ours will lead good lives.

Finally, a personal note on the epigraph. Corri Waitt was an American who had come to the UK in 1998 to study primate behavior and welfare at the University of Stirling. Corri and I (AW) were introduced by a mutual friend, and Corri was one of the first people outside of Edinburgh that I met after I moved to the UK in 2005.

Every so often Corri had an early morning flight to catch. On these occasions, she would ask if she could stay in the spare bedroom in my flat on 11 Johnston Terrace, where I lived at the time. I was always happy to have her over. On most of those visits, we stayed in my sitting room, from which one can see Edinburgh Castle, and we would chat. I would drink tea with lemon and sugar, and I do not recall whether she joined me or not (it is funny and tragic how minor details such as these fade so quickly, especially as they are what we end up treasuring). Corri was terrific company. I learned a lot about life in the UK and many other things. I also learned that she was bright and warm and had a terrific sense of humor and that she was kind, sweet, and dedicated to being decent to everybody, human and nonhuman alike.

Corri lost her life to an aggressive form of cancer in 2014. Everybody who knew her, including some of the contributors to this book, knows what a loss her death was. I did not have Corri in mind when I agreed to co-edit this book, but if she was alive, I would have invited Corri to contribute a chapter. Unfortunately, that was not to be. All I can do now is to hope that this is the book that Corri would have enjoyed reading and that, by sharing these memories, I am keeping her spirit alive.

Vienna, Austria Lauren M. Robinson
Kyoto, Japan Alexander Weiss

Contents

Part I History of Nonhuman Primates in Captivity and Primate Welfare in Different Settings

The History of Primates in Zoos 3
Geoff Hosey

The History of Chimpanzees in Biomedical Research 31
Patricia V. Turner

Using Primates in Captivity: Research, Conservation, and Education 57
Mark J. Prescott

The Welfare of Primates in Zoos 79
Kathy R. Baker and Holly L. Farmer

Welfare of Primates in Laboratories: Opportunities for Refinement 97
Hannah M. Buchanan-Smith, Lou Tasker, Hayley Ash, and Melanie L. Graham

The Welfare of Primates Kept as Pets and Entertainers 121
Rachel Hevesi

Primates Under Human Care in Developing Countries: Examples From Latin America 145
R. G. Ferreira, C. Ruiz-Miranda, S. Sita, S. Sánchez-López, A. Pissinatti, S. Corte, L. Jerusalinsky, P. G. Wagner, and C. Maas

Part II Assessing Nonhuman Primate Welfare

Using Behavior to Assess Primate Welfare 171
Corrine K. Lutz and Kate C. Baker

Cognitive Bias Tasks: A New Set of Approaches to Assess Welfare in Nonhuman Primates 207
Emily J. Bethell and Dana Pfefferle

Physiological Measures of Welfare 231
John P. Capitanio, Jessica Vandeleest, and Darcy L. Hannibal

Questionnaires and Their Use in Primate Welfare 255
Marieke Cassia Gartner

Part III Nonhuman Primate Housing and Husbandry

Meeting Cognitive, Behavioral, and Social Needs of Primates in Captivity 267
Catherine F. Talbot, Lisa A. Reamer, Susan P. Lambeth, Steven J. Schapiro, and Sarah F. Brosnan

Primate Breeding Colonies: Colony Management and Welfare 307
James C. Ha and Adrienne F. Sussman

Common Husbandry, Housing, and Animal Care Practices 323
Kristine Coleman, Gregory Timmel, Kamm Prongay, and Kate C. Baker

Housing and Husbandry for Primates in Zoos 355
H. L. Farmer, K. R. Baker, and F. Cabana

Humane Endpoints and End of Life in Primates Used in Laboratories 375
Sarah Wolfensohn

Part IV Individual Differences, Application, and Improvement of Nonhuman Primate Welfare

Primate Personality and Welfare 395
Lauren M. Robinson and Alexander Weiss

Sociality, Health, and Welfare in Nonhuman Primates 413
Brianne A. Beisner, Darcy L. Hannibal, Jessica J. Vandeleest, and Brenda McCowan

Research Benefits of Improving Welfare in Captive Primates 445
Steven J. Schapiro and Jann Hau

Enrichment 463
Caralyn Kemp

Challenging Cognitive Enrichment: Examples from Caring for the Chimpanzees in the Kumamoto Sanctuary, Japan and Bossou, Guinea 501
Naruki Morimura, Satoshi Hirata, and Tetsuro Matsuzawa

Training Research Primates 529
Mollie Bloomsmith, Jaine Perlman, Andrea Franklin, and Allison L. Martin

Part V Biomedical Research, Ethics, and Legislation Surrounding Nonhuman Primate Welfare

Arguments Against Using Nonhuman Primates in Research 559
Jarrod Bailey

The Indispensable Contribution of Nonhuman Primates to Biomedical Research 589
Stefan Treue and Roger Lemon

An Unexpected Symbiosis of Animal Welfare and Clinical Relevance in a Refined Nonhuman Primate Model of Human Autoimmune Disease 605
Bert A. 't Hart, Jon D. Laman, and Yolanda S. Kap

Animal Welfare, Animal Rights, and a Sanctuary Ethos 627
Lori Gruen and Erika Fleury

The Welfare Impact of Regulations, Policies, Guidelines, and Directives and Nonhuman Primate Welfare 643
Kathryn Bayne, Jann Hau, and Timothy Morris

Index 661

Part I

History of Nonhuman Primates in Captivity and Primate Welfare in Different Settings

The History of Primates in Zoos

Geoff Hosey

Abstract

Primates have been kept in captivity for at least 5000 years, but only in the last 200 years they have been maintained in facilities that we would regard as zoos. During those 200 years, many important advances have been made both in zoo practice and philosophy and in the housing and husbandry of primates. Initially, zoos attempted to display many different species in a way that followed taxonomic principles. Modern zoos concentrate on a smaller number of species, many of which are of conservation importance, housed according to habitat or ecological principles. Housing has changed, too, from relatively barren cages with little furniture to naturalistic and sometimes semi-free-ranging designs. Husbandry has moved from being based on ideas of what has worked in the past to a more evidence-based scientific approach, and this has resulted in better health and longevity, and improved breeding success in the animals. At the same time, zoos have become a significant resource for researchers interested in primate biology. This chapter surveys these changes through the recent history of primates in zoos.

Keywords

Zoo · Primate · History · Housing · Husbandry · Human–animal interactions · Conservation · Research

G. Hosey (✉)
University of Bolton, Bolton, UK

© Springer Nature Switzerland AG 2023
L. M. Robinson, A. Weiss (eds.), *Nonhuman Primate Welfare*,
https://doi.org/10.1007/978-3-030-82708-3_1

1 Introduction

Primates, both nonhuman and occasionally human, have been kept and displayed in captivity for a large part of our recorded history. Most of the facilities in which these animals were kept, certainly prior to the start of the nineteenth century, would not be regarded as zoos by any modern definition, as they were usually not open to the public and rarely included research, conservation, or education among their priorities. Nevertheless, they can be regarded as the forerunners of the modern zoo, and continuities can be seen between these old menageries and the early zoos (Rothfels 2002). In the transition from menagerie to modern zoo, there have been a great many changes: a move away from showing individuals of as many species as possible toward larger self-sustaining groups of a smaller number of species; improvements in housing and husbandry and consequently better animal welfare and breeding success; an increase in research to provide an evidence base for animal management; and more systematic education to promote positive attitudes toward and enhanced support for conservation among zoo visitors. This chapter will review some of these changes, mostly exemplified by changes in European and North American zoos, since it is these that have driven most of these changes, and for which most of the documentary evidence is available.

2 Development of the Modern Zoo

Archeological evidence tells us that primates have been kept captive since at least the Bronze Age. Skeletons of 112 animals, including 11 baboons, have been excavated at the ancient Egyptian capital of Hierakonpolis. The presence of healed bone fractures, as well as evidence that some animals were eating cultivated plants, suggests that the animals were part of a menagerie, dating from around 3500 BC (Rose 2010). During the first millennium BC, menageries were kept elsewhere in Egypt and also in Babylon and Assyria (Baratay and Hardouin-Fugier 2002) and it is likely that these contained monkeys (Morris and Morris 1966). Hamadryas baboons *Papio hamadryas*, probably lived wild in Egypt at that time (Groves 2008) so these were hardly exotics, and they were probably kept in captivity because of their religious significance and as a demonstration of the ruler's power (Rose 2010). But more exotic species can also be seen in the archeological record. A Minoan fresco from Akrotiri, for example, portrays some "blue monkeys" (Fig. 1) which Groves (2008) has identified as Tantalus monkeys, *Chlorocebus tantalus*, and which probably came to Crete via Egypt.

Throughout medieval and Renaissance times, menageries, "ferocious" animals like bears and big cats, as well as ungulates, were kept by kings, emperors, and the powerful. These animals were often gifts from other rulers and might be used for games, fights, and hunting (Baratay and Hardouin-Fugier 2002). Primates occasionally found their way into these collections. For example, monkeys were present in the court menagerie in Milan, where Leonardo da Vinci used them as living subjects for his anatomical observations (Hahn 2004). Some famous examples include the

Fig. 1 Blue monkeys on a Minoan fresco in Akrotiri, dated to before 1600 BC; Groves (2008) has identified these as Tantalus monkeys *C. tantalus*. Picture: Wikimedia Commons

Fig. 2 *The Monkey Room in the Tower* by Thomas Rowlandson (1799). Picture: National Gallery of Art, Washington DC

monkeys at the Versailles menagerie, set up by Louis XIV in 1664 (Morris and Morris 1966), and the eighteenth-century "monkey room" in the Tower of London (Fig. 2). Within this room, visitors and monkeys were free to intermingle, but this only lasted a short time before the room was closed due to there being too much aggression between monkeys and visitors (Hahn 2004).

Institutions that we would now recognize as zoos differed from the previous menageries in that they were open to the public (or at least some sectors of the public) and they were at least nominally interested in furthering science, which usually meant anatomy and taxonomy. They started to appear toward the end of the eighteenth century; the earliest was at Schönbrunn (now Vienna Zoo, opened in 1752), followed by the Jardin des Plantes in Paris (1793) and London Zoo (1828). In North America, the first zoo was in Philadelphia (1874), followed by Cincinnati in 1875. The number of zoos has grown enormously in the past two centuries; it is difficult to say just how many but the major accrediting zoo associations list 377 member institutions in Europe and the Middle East (European Association of Zoos and Aquariums [EAZA] 2016), 233 in North America (Association of Zoos and Aquariums [AZA] 2016), and 100 in Australasia (Zoo and Aquarium Association [ZAA] 2016). There are other, smaller zoo associations in other areas of the world, and an unknown number of nonaccredited zoos. Thus, the number of primates held in zoos is also likely to be large.

3   Species Kept by Zoos

The last 200 years have seen an increase in the taxonomic diversity of primates kept in zoos, but a reduction in the overall number of species. Nevertheless, during 200 years of modern zoo history we would expect that most species of primates have been kept somewhere at some time. It may seem a simple question to ask what has been kept where and when, but as with many apparently simple questions, finding an answer is not so easy. There are at least three ways we could find out what species zoos have, and have had, in the past.

Firstly, we could ask them. This essentially is what the editors of *International Zoo Yearbook* (IZY) did for each issue between 1961 and 1998. These editors asked zoos to give information on the species of mammals bred in the previous year and the number of rare species (i.e., species on an International Union for the Conservation of Nature [IUCN] list) currently held. Table 1 lists the number of taxa reported in those two years for those two categories. In this table, "taxa" means species or subspecies; hybrids between taxa were listed by IZY, but are not included in this table. In the table, we can see that, firstly, in 1961 there were some noticeable gaps across the families, for both numbers of births and rare species, particularly across the strepsirrhines. Secondly, numbers for both births and rare species are generally higher in 1998 than in 1961, reflecting an increase in zoos, better husbandry, and an increased emphasis on establishing breeding populations of endangered species. Finally, views of what constitutes a "rare" species have changed. Thus, in 1961 ring-tailed lemur *Lemur catta*, Sulawesi macaque *Macaca nigra,* and siamang *Symphalangus syndactylus* were listed among the rare species, but not in 1998. We can also trawl through notes, publications, reminiscences, and historic documents to construct species holdings for different collections. This onerous

Table 1 Numbers of taxa in which births were reported (births), and numbers of rare taxa (rare taxa) listed in *International Zoo Yearbook* in 1961 and 1998. Taxa may be species or subspecies

Family	Births 1961	Births 1998	Rare taxa 1961	Rare taxa 1998
Cheirogaleidae	0	3	3	6
Lepilemuridae	0	0	0	0
Lemuridae	3	16	10	17
Indriidae	0	3	0	3
Daubentoniidae	0	1	0	1
Galagidae	4	4	0	0
Lorisidae	0	4	1	0
Tarsiidae	0	0	0	2
Callitrichidae	5	22	2	13
Cebidae	6	8	0	1
Aotidae	1	3	0	0
Pitheciidae	1	3	1	7
Atelidae	3	13	0	13
Cercopithecidae	45	68	11	23
Hylobatidae	3	9	1	7
Hominidae	4	6	4	6

Table 2 Number of taxa held in zoos within the European Association of Zoos and Aquaria (EAZA) area (current taxa), and number of taxa that were previously held but are not held now (previous taxa), according to data contained within the Zootierliste (2016) website

Family	Current taxa	Previous taxa
Cheirogaleidae	4	7
Lepilemuridae	0	1
Lemuridae	18	1
Indriidae	1	3
Daubentoniidae	1	0
Galagidae	4	12
Lorisidae	6	3
Tarsiidae	1	2
Callitrichidae	23	15
Cebidae	13	8
Aotidae	5	2
Pitheciidae	3	15
Atelidae	14	17
Cercopithecidae	74	78
Hylobatidae	12	10
Hominidae	9	2

task has, as far as I am aware, only been done for zoos in the EAZA area (i.e., Europe and the Middle East; though the list is not restricted to EAZA zoos) and is maintained and kept up to date by Zootierliste (2016). The number of taxa of each primate family currently and formerly held in these zoos is shown in Table 2. When interpreting these numbers, we must note that for much of zoo history many subspecies were not recognized, and even those that were recognized were often not distinguished, so a species with, say, two subspecies might be listed as three taxa, i.e., the two subspecies and species not identified at subspecies level. This accounts for how, for example, zoos managed to find eleven different hominid taxa to exhibit.

Some of the species in these historic data are not found in zoos today. The indri *Indri indri*, for instance, does not occur captive in any collection anywhere in the world. A way of maintaining these animals in captivity has never been found. Historically, the Jardin des Plantes in Paris imported eight indris in 1939, which all died within a month, but no other European zoo has attempted to keep them. Another species that is, perhaps surprisingly, absent is the mountain gorilla *Gorilla beringei beringei*. No zoo anywhere keeps these animals today, but the Zootierliste records show that Rome Zoo had a male and female in 1969, London had a male in 1961–1962, Cologne had two females between 1969 and 1978, and Dusseldorf (which closed in 1944) had a female from 1929 to 1930. None of these gorillas lived very long, which suggests that, unlike western lowland gorillas, which are housed successfully in zoos, mountain gorillas need housing and husbandry, which we either cannot or do not know how to provide. Regardless, Table 2 shows that there are many primate taxa that have been kept in the past but are not, for whatever reason, kept now.

Reinterpretation of the historic records yields surprising information. For example, the bonobo *Pan paniscus* was only formally described in 1929 as a subspecies of the chimpanzee on the basis of museum skins and skeletons and was elevated to species status in 1933 (Mittermeier et al. 2013). Yet, the zoo records show that a male bonobo lived at Artis Zoo (Amsterdam) between 1911 and 1916, but was assumed to be a chimpanzee (Hediger 1970). Indeed, it has been suggested that Sally, a chimpanzee who lived at London Zoo between 1883 and 1891, was actually a bonobo (Barrington-Johnson 2005). In both of these cases, it was recognized at the time that the animals were different from the "usual" sort of chimpanzee, but taxonomy had not caught up with these differences. This example raises interesting questions about what other species may have gone unrecognized in zoos over the years, simply because they were acquired before formal scientific description. In any case, for the person interested in seeing as many primate taxa as possible a trip to the past would be a fascinating experience, but by modern standards we would find it depressing that so many of these taxa were represented by just ones and twos of individuals, as was common at the time, rather than viable groups or populations, as is more customary now.

Neither of these methods answer the question of what primates are currently held worldwide by zoos. To do this, we can consult the database maintained by Species360 (formally the International Species Information System (ISIS) 2016). Species360 provides an online record-keeping system that is used by most of the accredited zoos in the world to maintain records on their animals, such as births, deaths, transfers between zoos, and veterinary treatments, and in some cases, historic records as well; and also contains an inventory of which animals in each taxon are kept in zoos, and in which zoos these animals are located. However, as not all zoos subscribe to Species360, these inventories will be underestimates. Table 3 lists the number of animals in each family located in Species360-registered zoos (as of May 27, 2016), along with the number of species recognized by Species360 in that family. These two variables are strongly correlated ($r(14) = 0.803$, $p < 0.001$),

Table 3 Number of animals currently (as of May 25, 2016) held in Species360-listed zoos across the different primate families

Family	No of species in Species360 database	No of animals in Species360 zoos
Cheirogaleidae	23	274
Lepilemuridae	8	0
Lemuridae	20	6309
Indriidae	12	81
Daubentoniidae	1	51
Galagidae	6	381
Lorisidae	10	376
Tarsiidae	7	17
Callitrichidae	43	6514
Cebidae	22	4078
Aotidae	8	195
Pitheciidae	41	659
Atelidae	8	1496
Cercopithecidae	124	10,986
Hylobatidae	14	1582
Hominidae	6	3063

showing that the most speciose families are represented by the greatest number of animals.

Nevertheless, two families (Lemuridae and Hominidae) appear to be overrepresented in captivity based on the number of species in those families. In the case of Lemuridae, this is partly because the ring-tailed lemur, *Lemur catta*, does particularly well in captivity (with 282 births recorded in the 12 months up to May 2016), and partly because of increased attention by zoos toward the endangered species in this family. For Hominidae, there has been a great deal of conservation attention on all three genera (gorillas *Gorilla*: 790 animals in 122 zoos; chimpanzees and bonobos *Pan*: 1612 animals in 188 zoos; orangutans *Pongo*: 661 animals in 143 zoos), and of course, these are very popular animals with zoo visitors. Conversely, Pitheciidae is represented by rather fewer animals than we would expect for the number of species, with only sakis *Pithecia* (mostly the white-faced saki *Pithecia pithecia*) having a substantial zoo representation (465 animals in 139 zoos). This perhaps reflects the fact that the IUCN, which assesses the vulnerability of species to extinction, lists many of the species in this family to be of "Least Concern" for conservation.

Nine genera are not held by any Species360 member zoos: hairy-eared dwarf lemur *Allocebus*, sportive lemurs *Lepilemur*, fork-marked lemurs *Phaner*, woolly lemurs *Avahi*, indris *Indri*, needle-clawed galagos *Euoticus*, angwantibos *Arctocebus*, false potto *Pseudopotto*, and yellow-tailed woolly monkey *Oreonax*. Some of these (e.g., *Euoticus* and *Arctocebus*) are of IUCN Least Concern, and others (e.g., *Indri*) are of greater conservation importance, but may be species that are very difficult to keep in captivity. Nevertheless, Table 3 shows that there are currently 36,062 primates held in Species360 member zoos across the world. Comparable historical data are not available, but limited comparison with the IZY data shows that more animals of a wider range of taxa are now maintained in zoos. These numbers reflect the drive since the 1960s to establish self-sustaining zoo populations, and a greater emphasis on building populations of endangered species.

4   Iconic Apes

Iconic animals are those who become famous among the public as individuals, sometimes because they have a distinctive appearance and often because they have idiosyncratic behaviors or have had life experiences that make them famous. In principle, a mouse lemur or a marmoset could become iconic, but usually it is the great apes, and gorillas and chimpanzees in particular, who show the differences in appearance, behavior, and personality that lead them to become iconic. Nowadays, iconic animals might be used to help promote the education and attitude raising activities of the zoo, but in the past their main importance has been to draw more visitors to the zoo.

The first chimpanzee at London Zoo, Tommy, arrived in 1835 at Bristol and was transported to London in a stagecoach, accompanied by a keeper (Barrington-Johnson 2005). Tommy only survived for 6 months, and it would be close on 50 years before London exhibited another chimpanzee. Chimpanzees in zoos were few and far between in the nineteenth century, so one chimpanzee, Consul, at Belle Vue Zoo in Manchester, in 1893 (Fig. 3), caused a sensation and became famous despite only surviving for fifteen months before dying of dysentery (Morris and Morris 1966). It was claimed at the time that Consul could put on his own hat and coat, eat with a knife and fork, put coals on the fire, smoke a pipe, and engage in various other human-like behaviors (Morris and Morris 1966). Consequently, "Consul" became a popular name for other chimpanzees, including one at London Zoo some years later, and the name lives on in the fossil genus *Proconsul*, meaning "before Consul." Other famous chimpanzees include Cholmondeley (or Chumley) at London Zoo, made famous in the writings of Gerald Durrell (1960), and Congo (1954–64), another London Zoo chimpanzee whose claims to fame were the numerous paintings he produced (Fig. 4) and the publicity he received from appearing in the television series *Zoo Time* (Barrington-Johnson 2005).

When that first chimpanzee arrived at London, the gorilla was still something of a mythical animal, which was not formally known to science until 1847. Intriguingly, the first live gorilla to reach Europe was exhibited in a traveling menagerie in 1855 but was mistakenly thought to be a chimpanzee (Morris and Morris 1966). London Zoo's first gorilla was Mumbo in 1887, who only lived for a few months, possibly because he was fed on sausages, beer, and cheese sandwiches as well as fruit (Barrington-Johnson 2005). London's most famous gorilla was Guy, who arrived in 1947 at 18 months of age, and lived until 1978, when he died under anesthetic during treatment for dental problems, the result of years of eating sweets given to him by visitors (Barrington-Johnson 2005), a practice that is now generally not

Fig. 3 Consul, the chimpanzee at Belle Vue Zoo (Manchester) in 1893, wearing his hat and coat and smoking his pipe. Picture: Wikimedia Commons

Fig. 4 A painting by Congo, a chimpanzee who lived in London Zoo from 1954 until 1964, and produced about 400 such paintings in his lifetime. Picture: Wikimedia Commons

Fig. 5 Snowflake, the white gorilla who lived for 37 years at Barcelona Zoo. Picture: Wikimedia Commons

permitted in zoos. But before Guy, Bristol Zoo had an equally famous gorilla, Albert, who arrived at the zoo in 1930 and lived until 1948, the only gorilla in the UK at the time of his arrival (Warin and Warin 1985). Prior to Albert, no zoo-housed gorillas had survived for very long, and the species was considered too delicate to be successfully kept, certainly in the British climate (Schomberg 1957). Perhaps the most famous gorilla in the USA was Willie B, who lived alone in an enclosure of concrete and iron bars at Atlanta Zoo from 1961 until 1988, when he was released for the first time into a new outdoor enclosure. Here, he integrated with other gorillas, became a father, and lived apparently happily until his death in 2000, following which more than seven thousand people attended a memorial service in his honor (Hanson 2002). Finally, we can mention the albino gorilla, Snowflake (Fig. 5), who lived at the Barcelona Zoo from 1966 to 2003 (Jonch 1968). During his life, Snowflake fathered 22 offspring by three different mates, and none of them showed albinism. Eventually, he was euthanized after developing skin cancer. Our attitudes toward great apes can often be ambiguous (Morris and Morris 1966), but our fascination with iconic individuals is still strong and constitutes an important part of zoo history.

5 Human Primates on Exhibition in the Zoo

Strange though it might seem to modern sensibilities, humans were sometimes exhibited at zoos in the past. During the period of world exploration in Medieval and Renaissance times, examples of peoples from newly discovered or newly conquered areas of the world were often exhibited in royal courts or menageries. The practice of exhibiting people persisted in different guises into the start of the twentieth century (Baratay and Hardouin-Fugier 2002). During the nineteenth century, these displays of people often took the form of "ethnographical exhibitions," where people from what were then seen as exotic parts of the world were exhibited in their own costume and artifacts. Much of the time, these exhibitions were part of touring fairs or international exhibitions, but sometimes took place in zoos, as for example with the exhibition of people from various African tribes at Vienna (1897–98) and Berlin (1907) zoos (Baratay and Hardouin-Fugier 2002). A major influence on this was the animal dealer Carl Hagenbeck (1844–1913; more on him later), who developed ethnographical exhibitions largely in response to a downturn in the animal trade as a result of war in the Sudan. When he moved his own zoo in Hamburg to Stellingen in 1907, his ethnographical exhibitions went with it and partly paved the way for the open enclosure style of animal exhibition for which he became famous (Baratay and Hardouin-Fugier 2002).

Exhibition of people appears to have been less of a tradition in North American zoos, although Bronx Zoo in New York famously exhibited an African Pygmy, Ota Benga (Fig. 6), in 1904 (Hanson 2002). Ota Benga had reportedly been purchased from slave traders for an exhibition in St Louis before transfer to the Bronx, where he

Fig. 6 Ota Benga at the Bronx Zoo in New York. Picture: Wikimedia Commons

had the run of the zoo, but for a time lived in the monkey house. The zoo's director, William T. Hornaday, apparently had strong views on white superiority (Baratay and Hardouin-Fugier 2002), but exhibition of Ota Benga was ended after protest from New York's black community (Hanson 2002). After release from the zoo, Ota Benga was initially cared for in an orphan asylum, but tragically he took his own life in 1916 at the age of 32. The ethnographical exhibitions of Europe and the display of Ota Benga were products of the prevailing world view of white Europeans and Americans at the time. From the perspective of modern attitudes, the underlying racism is obvious.

6 Changes in Housing

Enormous changes in the housing of primates (indeed, of all zoo animals) have taken place since the establishment of the first modern-type zoos two hundred years ago. Nowadays, enclosure design attempts to satisfy the welfare and biological needs of the animals, the viewing needs of the public, and the management needs of the keepers (Hosey et al. 2013; Farmer et al. 2022). This is in contrast to nineteenth-century cages, which were little more than holding facilities for the animals. The basic model for nineteenth-century primate cages initially was the Jardin des Plantes in Paris, where the first monkey house was constructed in 1801 (Baratay and Hardouin-Fugier 2002). At London Zoo in 1829, the monkeys were at first tethered to poles, each of which had a small shelter at the top. The monkeys had leather belts around them attached by a chain to a ring which could slide up and down the pole, allowing the animals to wander at will within chain range, but also climb up to the shelter at the top of the pole, often with objects stolen from visitors. One animal, a lion-tailed macaque *Macaca silenus* (called "wanderoos" at that time) named Jack, stole so much from visitors that eventually all the monkeys were confined to the monkey house (Barrington-Johnson 2005).

The Monkey House at London Zoo was typical of zoo housing at the time and had outdoor enclosures (Fig. 7). It was replaced in 1864 with a structure, which had no outdoor enclosures, as the damp London climate was not deemed suitable for the animals (Barrington-Johnson 2005). This basic design, with or without outdoor enclosures, became the general pattern for much of the nineteenth century (Fig. 8). As can be seen in the illustrations, the cages contained branches for the animals to climb on, but otherwise were, by today's standards, relatively barren, with sparse cage furniture and few opportunities for the animals to escape public view. Modifications to this design were often motivated by what was perceived as architectural attractiveness rather than usefulness for the animals, which is reflected in zoo buildings erected in the style of Egyptian or Greek temples, or Eastern palaces (e.g., the monkey house at Cologne Zoo; Hancocks 2010). Some of these buildings still exist in zoos, usually because the zoo is required to protect them as architectural heritage, even though they are, by modern standards, unsuitable for housing animals.

By the end of the nineteenth century, it was becoming increasingly clear that improvements were needed, not least because the animals were just not surviving for

The History of Primates in Zoos

Fig. 7 First Monkey House at London Zoo in 1835, showing the outdoor enclosures. Picture: Museum of London

Fig. 8 Monkey House at Vienna Zoo in 1898, showing little advance over similar buildings from the start of the century. Picture: Wikimedia Commons

Fig. 9 Baboons *Papio papio* at Paris Zoo, Vincennes, in a Hagenbeck-styled enclosure built in the 1930s, seen here still in use in 1966. Picture: Geoff Hosey

very long (Baratay and Hardouin-Fugier 2002). For example, analysis of the records of 3780 monkeys held at the Jardin des Plantes between 1830 and 1939 shows that the average length of stay for each monkey was just 18.6 months and 17.7 months for great apes (Baratay and Hardouin-Fugier 2002). As most of these animals would have been very young when caught (a widespread method of catching animals was to kill parents and take the babies), it is likely that very few primates reached maturity in the zoo.

A major advance came with a new style of housing, pioneered by Carl Hagenbeck in Hamburg, involving open enclosures where the animals were separated from the public by ditches, moats, and low walls. These enclosures could be landscaped with artificial rocks and trees, and they gave the illusion of animals in semi-liberty. This style of housing was incorporated into Hagenbeck's new zoo at Stellingen near Hamburg, which opened in 1907, and it was copied by many zoos around the world (e.g., at Antwerp, Rome, Budapest, Detroit, Cincinnati: Baratay and Hardouin-Fugier 2002) for many decades afterward. In this tradition, monkeys could be displayed on an artificial hill, as at Monkey Hill at London Zoo, which was constructed in the 1920s (Barrington-Johnson 2005), and the "Big Rock" at Paris Zoo, dating from the 1930s (Fig. 9). These enclosures looked, from a human point of view, to be far better for the animals than the older style of cages, but of course no systematic studies were done on them at the time. They were, however, not without their problems, which included occasional escapes and drowning of nonswimming

Fig. 10 Squirrel monkeys *Saimiri sciureus* in a "hygienic" cage at Flamingo Park Zoo in 1969. Picture: Geoff Hosey

primates in moats (Hediger 1970); nevertheless, they represented an advance in design and paved the way for the more naturalistic enclosures of the 1970s and 80s.

There was an unfortunate but fortunately brief episode in the 1920s and 1930s when some zoos allowed architects to design housing where the priority appeared to be architectural aspirations rather than animal needs. Famous examples include the penguin pool at London Zoo and the polar bear enclosure at Dudley Zoo. These were both products of the designer Berthold Lubetkin of the firm Tekton and built in modernist, functional design principles. Primates did not escape this trend. Similar architecture is seen in the Round House for gorillas at London Zoo (another Tekton) and Monkey Island at Melbourne Zoo, the latter built by Percy Everett in 1938 and demolished in 1971 (Grow 2009). Even without modernism, the 1920s and 30s saw increasing emphasis on cages and enclosures that were easy to clean. This period, which became known as the "hygiene" or "disinfectant" era (Hancocks 2001), extended well into modern times, and examples could be seen into the 1980s. These enclosures had bare walls, little if any floor covering, and sparse cage furniture (Fig. 10), and frequently had viewing windows for the public. Interestingly, this "hygiene" trend has continued in laboratories, but these too are moving away from this (Coleman et al. 2022).

Since the 1970s, there has been a shift in emphasis toward housing that attempts to satisfy the perceived needs of the animals. This shift led to the development of a range of (sometimes overlapping) styles, such as naturalistic, immersion, and even free-range designs. Naturalistic enclosures, unlike "hygienic" enclosures, had earth

Fig. 11 Chimpanzee enclosure at Arnhem Zoo, seen here in 1989, an early example of a naturalistic primate enclosure. Picture: Geoff Hosey

and plants, and were intended to be a little bit like the animals' natural habitats. Early examples include the chimpanzee enclosure at Arnhem Zoo from 1970, which actually resembled parkland (Coe and Lee 1996) and therefore was not really like chimpanzee African habitat at all (Fig. 11). Another example is the gorilla enclosure at Woodland Park Zoo in Seattle, WA. From 1978, this enclosure was designed using information from field studies and attempted to replicate the animals' natural environment (Hancocks 1980, 2010; Coe and Lee 1996). It is a short step from this to an "immersion" enclosure, which aims not only to provide a naturalistic habitat for the animals, but also to draw the public into the animals' world, so they have an experience like that of seeing the animal in the wild (Fig. 12).

Naturalistic and immersion exhibits for zoo primates have now become widespread (Coe 1989). To human eyes, these enclosures look good, and many zoo visitors regard them as good for the welfare of the animals, though these perceptions are not necessarily based on any welfare knowledge (Melfi et al. 2004). Furthermore, they appear to raise visitors' positive attitudes toward the animals, and possibly their concerns for the animals, too (Lukas and Ross 2014). Importantly, empirical studies on captive primates transferred from traditional style to naturalistic enclosures show changes in behavior. These include increases in positive behaviors, such as grooming, and decreases in negative behaviors, such as aggression and stereotypies, across a range of species (sifakas *Propithecus verreauxi*: Macedonia 1987; mandrills *Mandrillus sphinx*: Chang et al. 1999; Hanuman langurs *Semnopithecus entellus*: Little and Sommer 2002; chimpanzees and gorillas: Ross et al. 2011). It is likely that

Fig. 12 Tropical world at Brookfield Zoo (Chicago), home to gorillas and guenons in an immersion-type enclosure, 2013. Picture: Geoff Hosey

the success of these enclosures is not particularly because of how closely they resemble the animals' natural habitats, but more about the fact that they are complex, giving the animals increased choice and control, and providing opportunities to perform the species-typical behaviors which are important to them (Kurtycz et al. 2014; Wilson 1982).

Some zoos have extended these ideas still further by exhibiting some of their primates in semi-free-range enclosures, where the animals either have the run of the whole zoo or are confined to an area where the limits of the confinement are not always obvious. A pioneer of this approach, Monkey Jungle in Goulds, Florida, maintains small neotropical monkeys (squirrel monkeys, howler monkeys) in a 15-acre patch of subtropical hardwood jungle (Dumond 1967), where human visitors are restricted to enclosed walkways (Fig. 13), separate from the monkeys. In most zoos that have enclosures in which primates are semi-free-ranging, however, visitors are allowed to walk among the animals. One of the earliest zoos to do this was Apenheul Primate Park in the Netherlands (Mager and Griede 1986), where lemurs, callitrichids, squirrel monkeys, and Barbary macaques live in what are essentially large walk-through enclosures (Fig. 14).

Few empirical studies have been undertaken on the welfare aspects of semi-free-range enclosures, though studies at Jersey Zoo show that cotton-top tamarins *Saguinus oedipus* adapt well to this lifestyle (Price 1992), and visitors spend longer looking at them and think they learn more from them than from tamarins in cages (Price et al. 1994). Despite the apparent benefits to animal welfare and visitor education, there is clearly scope for negative human–animal interactions to take place (Mun et al. 2013). This is, perhaps, more of a concern now that many zoos encourage close human–animal proximity through enclosure design, using carefully

Fig. 13 Visitors in an enclosed walkway at Goulds Monkey Jungle, while the monkeys are in semi-free-range in a forest outside, 2002. Picture: Geoff Hosey

Fig. 14 Large walk-through lemur area at Apenheul Primate Park, 2007. Picture: Geoff Hosey

placed heated pads or comfortable areas close to public viewing areas (Vicky Melfi, personal communication). Other unexpected welfare concerns may also occur. It has, for instance, been reported that free-ranging lemurs develop frostbite when they fail to come indoors in cold weather (Vicky Melfi, personal communication). At the moment, we know too little about the welfare consequences of semi free-range and walk-through enclosures.

7 Changes in Husbandry

For many years, the way zoos have looked after their animals has been largely determined by trial and error, and the passing on of knowledge of "what works." For most species coming into zoos, even the most basic understanding of their ecology in the wild was lacking, so virtually nothing was known of what they ate, what sorts of social structure they had, or what mating and infant-rearing behaviors they showed. Consequently, until at least the 1950s, animals were often maintained singly or in pairs, breeding performance was poor, and longevity was low. Since the 1960s, two developments have led to huge improvements in zoo husbandry. Firstly, there has been an enormous growth in field studies, and our knowledge of the basic biology and life history of many primate species has benefitted as a result. Secondly, there has been much greater application of science in the zoo, particularly in nutrition and animal health, but increasingly in animal welfare as well (Melfi 2009), so that zoo animal husbandry is becoming more evidence-based and less tradition-led.

As a result, the longevity of many zoo animals has increased. Fifty years ago, New World primates in zoos rarely lived longer than 10 years, Old World primates had a normal life span in captivity of 10–20 years, and apes might live to 30 or more (Jones 1968). Thirty years later, a database containing ages of several hundred zoo-housed primates (Hakeem et al. 1996) showed greater ages achieved by the animals; nine zoo-housed spider monkeys *Ateles* sp., for example, had a mean age at death of 31.1 years. Over a 30-year period up to the end of the last century, the maximum longevities of zoo-housed gorillas increased by 62% to 54 years, those of orangutans by 92% to 59 years, pygmy marmosets *Cebuella pygmaea* by 267% to 17.9 years, and Goeldi's monkeys *Callimico goeldii* by 678% to 18 years (Kitchener and Macdonald 2002). While these increases in lifespan are largely to be welcomed, and testify to the advances in husbandry and welfare that have been made by zoos, they nevertheless produce new challenges, from knowing how to deal with the pains and discomforts of geriatric animals to trying to manage an increasing population of postreproductive animals.

Improvements in husbandry have also led to higher birth rates. Consider, for example, the case of the gorilla. As noted previously, for years gorillas were regarded as a delicate species that did not take well to captivity, and they certainly did not breed in captivity. The first captive-born gorilla was Colo at Columbus Zoo in 1956 (Fig. 15), followed by Goma at Basel Zoo in 1959. But it was not until 1961 that the second American gorilla was born (Tomoko at Washington National Zoo: Carmichael et al. 1962), and in the same year, the second Basel gorilla, Jambo, was

Fig. 15 Colo, the first ever zoo-born gorilla, photographed here at Columbus Zoo in 2009. Picture: Wikimedia Commons

born (Lang 1962). By 1976, Martin (1981) was reporting the first birth of a gorilla at London Zoo, while also pointing out that between 1956 and 1975 only 136 gorilla births had been reported by zoos, and only 105 of those infants survived. Martin was mostly concerned with showing how scientific advances, and in particular the monitoring of hormones, contributed to London Zoo's success. By 1998, however, *International Zoo Yearbook* reported 33 gorilla births in 16 different zoos in 1995 and 37 in 29 zoos in 1996.

These developments have been due largely to advances in knowledge and practice in animal nutrition and health, but in the last 50 years the "well-being" or psychological health of zoo animals has become a priority as well. Ensuring this might involve long-term changes to the animals' physical (enclosure design) and social (appropriate group composition) environment, and can also be brought about by promoting behavioral change through shorter-term interventions, such as environmental enrichment (Hosey 2005). These sorts of interventions have been undertaken in zoos for a long time, but in a relatively unsystematic way (Mellen and Sevenich MacPhee 2001); since the 1970s, there has been a more systematic approach to undertaking interventions, using better knowledge of the biological needs of the animals (Shepherdson 2003). Some of the earliest of these, pioneered by Markowitz (1982), were operant devices aimed at raising levels of animal activity, and some of them permitted visitor participation, such as a coin-operated food dispenser at Portland Zoo which signaled its activation to gibbons, who then raced across the enclosure to receive food (Markowitz and Woodworth 1978). This approach was criticized at the time, notably by Hutchins et al. (1978), for being too artificial, but the use of enrichment has evolved considerably since then, and implements features from both the "behavioral engineering" and the "naturalistic" approaches to enrichment (see Kemp 2022; Morimura et al. 2022).

8   Human–Animal Interaction

In the early days of zoos, it was often possible, as we have seen (Fig. 2), for visitors to walk among, and interact directly with captive monkeys. This appears to have frequently led to bad behavior on the part of both the human and nonhuman primates, leading to the animals being confined to cages. The recent trend toward walk-through enclosures could, perhaps, be seen as a return to this way of displaying animals, though clearly, walk-through enclosures are only appropriate for certain species. In most zoos that have primate walk-throughs, direct interaction or touching the animals is forbidden, though some visitors still try (Fig. 16). There is also a long history of visitors being able to feed zoo animals, including primates (Hediger 1970), a practice that has generally been discouraged or banned only in the past few decades. Again, some visitors try to feed the animals anyway, and this can lead to undesirable changes in the animals' behavior, such as increases in aggression (Fa 1989) and, if unsuitable food or objects are given, this practice can lead to animal deaths. Hediger (1970), for instance, recounts the case of a healthy woolly monkey at Zurich Zoo dying from an intestinal blockage after eating a hazelnut-sized piece of gravel offered by a visitor. There are potential risks to humans too. At Chester Zoo in the 1960s, the chimpanzees apparently became proficient at throwing

Fig. 16 A visitor attempting to touch a ruffed lemur at Apenheul Primate Park, 1999. Picture: Geoff Hosey

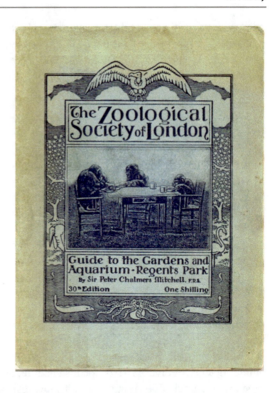

Fig. 17 London Zoo guidebook cover from 1933, showing the chimpanzee "tea party"

objects across the moats at human visitors (Morris and Morris 1966). More recently, a chimpanzee at Furuvik Zoo in Sweden reportedly collected and hoarded stones in anticipation of throwing them at visitors (Osvath 2009).

Sometimes encounters between visitors and zoo-housed primates have been through stage shows, of which chimpanzee "tea parties" have been among the best known. These were put on for entertainment, but also to give an educational message about the cognitive skills of the apes, and their similarity to humans. London Zoo started their chimpanzee tea parties in the 1920s (Morris and Morris 1966) and did not end them until the 1960s (Fig. 17). Similar "tea parties" were run at Bronx Zoo, using orangutans (Kreger and Mench 1995), and other kinds of shows with chimpanzees and orangutans, presented as educational shows, were common across North American zoos until the 1960s (Kreger and Mench 1995). Shows are becoming popular again, particularly for conservation education, but are generally less anthropomorphic than these earlier shows.

9 Zoos and Primate Research

Research was one of the guiding principles of many of the zoos founded in the nineteenth century; however, it did not become common practice until much later (Baratay and Hardouin-Fugier 2002). When zoo animals were used for research, it was usually in anatomy and species description, for example on the differences between chimpanzees and gorillas (Hochadel 2011). Indeed, the director of Berlin Zoo, Alfred Brehm, gave informal lectures on anatomy in front of the monkey house at Berlin in the 1860s (Hochadel 2005). Thus, the animals were only of scientific value to much of the scientific community once they were dead, and scientific observation of living animals in zoos was not considered worthwhile or was only reported anecdotally (Baratay and Hardouin-Fugier 2002; Hochadel 2011). There were, of course, some notable exceptions to this, such as Darwin's observations of emotional expressions in primates, which were carried out at London Zoo (Black 2002).

The scientific study of primates in zoos started to change in the 1930s. At London Zoo, Solly Zuckerman undertook a behavioral study of the hamadryas baboon colony, which had been set up on Monkey Hill. Zuckerman (2013) described the sexual and dominance fighting occurring in the group and established in many people's minds the belief, held until surprisingly recently, that zoo primate colonies were hyperaggressive. However, the colony consisted of 39 adult males and only nine females. As Morris and Morris (1966) later commented, it was a testament to the flexibility of baboon behavior "that there was not an immediate and total bloodbath." Ironically, although it would be another 50 years before the next large-scale study of social behavior in a zoo primate group, that later study (de Waal 1982) would demonstrate the mechanisms by which primates maintain social relationships in their societies (in this case chimpanzees at Arnhem Zoo).

Zoo primates were also proving to be valuable for research in comparative psychology in the USA. Here, Abraham Maslow, better known for his "hierarchy of needs," undertook research on social dominance in monkeys at Vilas Park Zoo in Madison (Maslow 1936). With Harry Harlow, he also experimented on the learning abilities of a range of primate species at the Bronx Zoo (Harlow et al. 1932; Maslow 1936). Other workers subsequently conducted similar research at a variety of zoos, but the behavior of zoo animals was widely seen as so artificial as to render any research results coming from zoos as meaningless (Rumbaugh 1972). It has only really been in the past few decades that the enormous potential of zoo-housed primates as subjects for comparative research has been realized through the construction of dedicated research facilities at zoos. Examples of these research facilities include the Wolfgang Kohler Primate Research Centre at Leipzig Zoo (Pennisi 2001) and the Living Links to Human Evolution Research Centre at Edinburgh Zoo (MacDonald and Whiten 2011). Zoos now undertake and host a significant amount of primate research (e.g., Gartner and Weiss 2018; Behringer et al. 2018), and although there is evidence of a taxonomic bias in much of this research in favor of great apes (Melfi 2005), zoo collections nevertheless represent a considerable resource for the primate researcher.

10   Primates in Zoos, Then and Now

The modern zoo is about 200 years old, but those 200 years, and particularly the last 50 of them, have seen huge changes, and a visitor from 1816 would be astounded now not just at the physical transformations zoos have gone through, but also the revolution in underlying philosophy. A primatologist visiting a zoo in the mid-nineteenth century might well see an array of different species, though there would be few great apes or lemurs. Most of the monkeys they would see would be in a Monkey House, housed singly or in small groups, in sparsely furnished cages. A century on and that primatologist might still see an array of species in a Monkey House, and they might still be housed singly or in small groups, but the enclosures would probably be better furnished, and some of the larger terrestrial species (macaques, baboons) could be observable in open, landscaped, moated enclosures. A primatologist visiting a zoo now will notice that they are seeing many of the species that they have already seen in other zoos, but often they will be looking at active groups in naturalistically furnished enclosures. The zoos have changed from what was effectively a kind of "living museum," exhibiting examples of lots of different species, to a "conservation and education center," where exhibits are parts of larger interzoo population management programs. There are, of course, challenges, which are being addressed now, and will continue to be challenges in the future. These include ensuring the best welfare for the animals, genetic management of dispersed populations in light of dwindling wild populations, maintenance of wild-type behaviors and avoidance of "domestication," and the management of surplus animals. Nevertheless, for the primatologist, a good zoo is certainly an exciting place to visit!

Acknowledgements The author is extremely grateful to Sam Ward of Nottingham Trent University for obtaining the ISIS (Species360) data for him. The author also thanks, as always, Vicky Melfi of Hartpury University for her critical reading of this chapter and suggestions to improve it.

References

AZA (Association of Zoos and Aquariums) (2016). www.aza.org/current-accreditation-list/. Accessed 23 May 2016
Baratay E, Hardouin-Fugier E (2002) Zoo: a history of zoological gardens in the West. Reaktion Books, London
Barrington-Johnson J (2005) The zoo: the story of London Zoo. Robert Hale, London
Behringer V, Stevens J, Deschner T, Hohmann G (2018) Getting closer: contributions of zoo studies to research on the physiology and development of bonobos *Pan paniscus*, chimpanzees *Pan troglodytes* and other primates. Int Zoo Yearb 52(1):34–47. https://doi.org/10.1111/izy.12176
Black J (2002) Darwin in the world of emotions. J R Soc Med 95(6):311–313. https://doi.org/10.1258%2Fjrsm.95.6.311
Carmichael L, Kraus MB, Reed T (1962) The Washington National Zoological Park gorilla infant, Tomoko. Int Zoo Yearb 3(1):88–93. https://doi.org/10.1111/j.1748-1090.1962.tb03407.x

Chang TR, Forthman DL, Maple TL (1999) Comparison of confined mandrill (*Mandrillus sphinx*) behavior in traditional and "ecologically representative" exhibits. Zoo Biol 18(3):163–176. https://doi.org/10.1002/(SICI)1098-2361(1999)18:3%3C163::AID-ZOO1%3E3.0.CO;2-T

Coe JC (1989) Naturalizing habitats for captive primates. Zoo Biol 8(S1):117–125. https://doi.org/10.1002/zoo.1430080512

Coe JC, Lee GH (1996) One hundred years of evolution in great ape facilities in American zoos 1986–1996. Proceedings of the AZA 1995 Western Regional conference, AZA, Bethesda, MD

Coleman K, Timmel G, Prongay K, Baker KC (2022) Common husbandry, housing, and animal care practices. In: Robinson LM, Weiss A (eds) Nonhuman primate welfare: from history, science, and ethics to practice. Springer, Cham, pp 317–348

de Waal F (1982) Chimpanzee politics. Jonathan Cape, London

Dumond FV (1967) Semi-free-ranging colonies of monkeys at Goulds Monkey Jungle. Int Zoo Yearb 7(1):202–207. https://doi.org/10.1111/j.1748-1090.1967.tb00394.x

Durrell G (1960) A zoo in my luggage. Rupert Hart-Davis, London

EAZA (European Association of Zoos and Aquaria) (2016). www.aza.org/current-accreditation-list/. Accessed 23 May 2016

Fa J (1989) Influence of people on the behavior of display primates. In: Segal E (ed) Housing, care and psychological well-being of captive and laboratory primates. Noyes, Park Ridge, NJ, pp 270–290

Farmer HL, Baker KR, Cabana F (2022) Housing and husbandry for primates in zoos. In: Robinson LM, Weiss A (eds) Nonhuman primate welfare: from history, science, and ethics to practice. Springer, Cham, pp 349–368

Gartner MC, Weiss A (2018) Studying primate personality in zoos: Implications for the management, welfare and conservation of great apes. Int Zoo Yearb 52(1):79–91. https://doi.org/10.1111/izy.12187

Groves CP (2008) Extended family: long lost cousins: a personal look at the history of primatology. Conservation International, Arlington, VA

Grow R (2009) Modernism and the zoo: London penguins and Melbourne monkeys. Spirit Progress 10(3):11–12. https://doi.org/10.3316/ielapa.811142 68525

Hahn D (2004) The tower menagerie: The amazing 600-year history of the royal collection of wild and ferocious beasts kept at the Tower of London. Jeremy P. Tarcher/Penguin, London

Hakeem A, Sandoval GR, Jones M, Allman J (1996) Brain and life span in primates. In: Handbook of the psychology of aging, 4th edn. Academic Press, New York, pp 78–104

Hancocks D (1980) Bringing nature into the zoo: inexpensive solutions for zoo environments. Int J Stud Anim Prob 1(4):170–177

Hancocks D (2001) A different nature: the paradoxical world of zoos and their uncertain future. University of California Press, Berkeley

Hancocks D (2010) The history and principles of zoo exhibition. In: Kleiman D, Thompson K, Baer C (eds) Wild mammals in captivity: principles and techniques for zoo management. Chicago University Press, Chicago, pp 121–136

Hanson E (2002) Animal attractions: nature on display in American zoos. Princeton University Press, Princeton

Harlow HF, Uehling H, Maslow AH (1932) Comparative behavior of primates. I. Delayed reaction tests on primates from the lemur to the orang-outan. J Comp Psychol 13(3):313–343. https://doi.org/10.1037/h0073864

Hediger H (1970) Man and animal in the zoo. Routledge & Kegan Paul, London

Hochadel O (2005) Science in the 19th-century zoo. Endeavour 29(1):38–42. https://doi.org/10.1016/j.endeavour.2004.11.002

Hochadel O (2011) Watching exotic animals next door: "Scientific" observations at the zoo (ca. 1870–1910). Sci Context 24(2):183–214. http://doi.org/10.1017/S0269889711000068

Hosey GR (2005) How does the zoo environment affect the behaviour of captive primates? Appl Anim Behav Sci 90(2):107–129. https://doi.org/10.1016/j.applanim.2004.08.015

Hosey G, Melfi V, Pankhurst S (2013) Zoo animals: behaviour, management, and welfare, 2nd edn. Oxford University Press, Oxford

Hutchins M, Hancocks D, Calip T (1978) Behavioral engineering in the zoo: a critique. Int Zoo News Part I 25(7):18–23

International Species Information System (ISIS) (2016). www2.isis.org. Accessed 25–27 May 2016

Jonch A (1968) The white lowland gorilla *Gorilla g. gorilla* at Barcelona Zoo. Int Zoo Yearb 8 (1):196–197. https://doi.org/10.1111/j.1748-1090.1968.tb00482.x

Jones ML (1968) Longevity of primates in captivity. Int Zoo Yearb 8(1):183–192. https://doi.org/10.1111/j.1748-1090.1968.tb00479.x

Kemp C (2022) Enrichment. In: Robinson LM, Weiss A (eds) Nonhuman primate welfare: from history, science, and ethics to practice. Springer, Cham, pp 451–488

Kitchener A, MacDonald AA (2002) The longevity legacy: the problem of old animals in zoos. Adv Ethol 37:7–10

Kreger MD, Mench JA (1995) Visitor—animal interactions at the zoo. Anthrozoös 8(3):143–158. https://doi.org/10.2752/089279395787156301

Kurtycz LM, Wagner KE, Ross SR (2014) The choice to access outdoor areas affects the behavior of great apes. J Appl Anim Welf Sci 17(3):185–197. https://doi.org/10.1080/10888705.2014.896213

Lang EM (1962) Jambo, the second gorilla born at Basel Zoo. Int Zoo Yearb 3(1):84–88. https://doi.org/10.1111/j.1748-1090.1962.tb03406.x

Little KA, Sommer V (2002) Change of enclosure in langur monkeys: implications for the evaluation of environmental enrichment. Zoo Biol 21(6):549–559. https://doi.org/10.1002/zoo.10058

Lukas KE, Ross SR (2014) Naturalistic exhibits may be more effective than traditional exhibits at improving zoo-visitor attitudes toward African apes. Anthrozoös 27(3):435–455. https://doi.org/10.2752/175303714X14023922797904

MacDonald C, Whiten A (2011) The 'Living Links to Human Evolution' research centre in Edinburgh Zoo: a new endeavour in collaboration. Int Zoo Yearb 45(1):7–17. https://doi.org/10.1111/j.1748-1090.2010.00120.x

Macedonia JM (1987) Effects of housing differences upon activity budgets in captive sifakas (*Propithecus verreauxi*). Zoo Biol 6(1):55–67. https://doi.org/10.1002/zoo.1430060107

Mager WB, Griede T (1986) Using outside areas for tropical primates in the northern hemisphere: *Callitrichidae*, *Saimiri*, and *Gorilla*. In: Benirschke K (ed) Primates: the road to self-sustaining populations. Springer, New York, pp 471–477

Markowitz H (1982) Behavioral enrichment in the zoo. Van Nostrand Reinhold, New York

Markowitz H, Woodworth G (1978) Experimental analysis and control of group behavior. In: Markowitz H, Stevens VJ (eds) Behavior of captive wild animals. BMJ Publishing Group, Chicago, pp 107–131

Martin R (1981) Breeding great apes in captivity. New Sci 72(1022):100–102

Maslow AH (1936) The role of dominance in the social and sexual behavior of infra-human primates: I. Observations at Vilas Park Zoo. Pedagog Semin J Genet Psychol 48(2):261–277. https://doi.org/10.1080/08856559.1936.10533730

Melfi V (2005) The appliance of science to zoo-housed primates. Appl Anim Behav Sci 90 (2):97–106. https://doi.org/10.1016/j.applanim.2004.08.017

Melfi V (2009) There are big gaps in our knowledge, and thus approach, to zoo animal welfare: a case for evidence-based zoo animal management. Zoo Biol 28(6):574–588. https://doi.org/10.1002/zoo.20288

Melfi VA, McCormick W, Gibbs A (2004) A preliminary assessment of how zoo visitors evaluate animal welfare according to enclosure style and the expression of behavior. Anthrozoös 17 (2):98–108. https://doi.org/10.2752/089279304786991792

Mellen J, Sevenich MacPhee M (2001) Philosophy of environmental enrichment: past, present, and future. Zoo Biol 20(3):211–226. https://doi.org/10.1002/zoo.1021

Mittermeier RA, Wilson DE, Rylands AB (eds) (2013) Handbook of the mammals of the world: primates, vol 3. Lynx Edicions, Barcelona

Morimura N, Hirata S, Matsuzawa T (2022) Challenging cognitive enrichment: examples from caring for the chimpanzees in the Kumamoto Sanctuary, Japan and Bossou, Guinea. In: Robinson LM, Weiss A (eds) Nonhuman primate welfare: from history, science, and ethics to practice. Springer, Cham, pp 489–516

Morris R, Morris D (1966) Men and apes. Hutchinson, London

Mun JSC, Kabilan B, Alagappasamy S, Guha B (2013) Benefits of naturalistic free-ranging primate displays and implications for increased human–primate interactions. Anthrozoös 26(1):13–26. https://doi.org/10.2752/175303713X13534238631353

Osvath M (2009) Spontaneous planning for future stone throwing by a male chimpanzee. Curr Biol 19(5):R190–R191. https://doi.org/10.1016/j.cub.2009.01.010

Pennisi E (2001) Zoo's new primate exhibit to double as research lab. Science 293(5533):1247. https://doi.org/10.1126/science.293.5533.1247

Price EC (1992) Adaptation of captive-bred cotton-top tamarins (*Saguinus oedipus*) to a natural environment. Zoo Biol 11(2):107–120. https://doi.org/10.1002/zoo.1430110206

Price EC, Ashmore LA, McGivern AM (1994) Reactions of zoo visitors to free-ranging monkeys. Zoo Biol 13(4):355–373. https://doi.org/10.1002/zoo.1430130409

Rose M (2010) World's first zoo-Hierakonpolis, Egypt. Archaeol Arch 63(1):25–32

Ross SR, Wagner KE, Schapiro SJ, Hau J, Lukas KE (2011) Transfer and acclimatization effects on the behavior of two species of African great ape (*Pan troglodytes* and *Gorilla gorilla gorilla*) moved to a novel and naturalistic zoo environment. Int J Primatol 32(1):99–117. https://doi.org/10.1007/s10764-010-9441-3

Rothfels N (2002) Savages and beasts: The birth of the modern zoo. Johns Hopkins University Press, Baltimore

Rumbaugh DM (1972) Zoos: valuable adjuncts for instruction and research in primate behavior. Bioscience 22(1):26–29. https://doi.org/10.2307/1296181

Schomberg G (1957) British zoos: a study of animals in captivity. Allen Wingate, London

Shepherdson DJ (2003) Environmental enrichment: past, present and future. Int Zoo Yearb 38(1):118–124. https://doi.org/10.1111/j.1748-1090.2003.tb02071.x

Warin RP, Warin A (1985) Portrait of a zoo: Bristol Zoological Gardens 1835–1985. Redcliffe, Bristol

Wilson SF (1982) Environmental influences on the activity of captive apes. Zoo Biol 1(3):201–209. https://doi.org/10.1002/zoo.1430010304

ZAA (Zoo and Aquarium Association) (2016). www.zooaquarium.org.au/index.php/member-location-map. 23 May 2016

Zootierliste (2016). www.zootierliste.de/. 24 May 2016

Zuckerman S (2013) The social life of monkeys and apes. Kegan Paul, London

The History of Chimpanzees in Biomedical Research

Patricia V. Turner

Abstract

Chimpanzees have been used in biomedical research for over a century. With the removal of chimpanzees from biomedical research in the European Union, Japan, and more recently, the USA, this topic is of public and scientific interest. This chapter will cover the use of chimpanzees in biomedical research from the early 1900s, including consideration of the impact of captive breeding and housing on welfare, the debate over chimpanzee use in research and the decision leading to their removal from mainstream biomedical research, and the potential repercussions this will have for both chimpanzees and humans.

Keywords

Chimpanzee · Animal welfare · Toxicology · Ape

1 Introduction

The aim of this chapter is to recall the history of how and why chimpanzees came to be used in biomedical research and the subsequent impact on chimpanzee welfare. It is always important when considering events that occurred outside of the immediate contemporary era to remember the historical context during which they occurred, including the very different societal perceptions and ethical values that were prevalent in the early to mid-twentieth century. The use of chimpanzees in research began at a time when it was permissible to display a willing pygmy human, originally a member of the Mbuti tribe from the Congo, in the 1904 St Louis World's Fair and then later within an ape exhibit at the Bronx Zoo (Bradford and Blume 1992).

P. V. Turner (✉)
Department of Pathobiology, University of Guelph, Guelph, ON, Canada
e-mail: pvturner@uoguelph.ca

© Springer Nature Switzerland AG 2023
L. M. Robinson, A. Weiss (eds.), *Nonhuman Primate Welfare*,
https://doi.org/10.1007/978-3-030-82708-3_2

Similarly, at the turn of the last century and throughout the 1950s, medical studies were routinely conducted on psychiatric and cancer patients (both adults and children), prison inmates, African Americans, and indigenous people of North America, Africa, and elsewhere without informed consent or other forms of approval (Gamble 1997; see the following for examples: Rice 2008; Mukherjee 2010). Animal welfare was not a consideration for biomedical research subjects in the early 1900s, and few, if any, research review or ethical oversight processes were in place for human or animal studies throughout North America or Europe (Gamble 1997; Elkeles 2004; Layman 2009). Scientists were limited only by their imaginations, and, sometimes, their pocketbooks, in that national extramural research funding programs were rare and sparsely funded by today's standards (IOM 1990). The public, at least in North America, fueled by post-World War I medical successes and new life-saving medicines, including disinfectants, newer anesthetic agents, and rudimentary antimicrobial therapeutics, was keen to see medical success continue and contribute to a thriving economy and increasing standard of living (Block 2001; Aminov 2010; Hampton 2017). Many forms of scientific investigation, including direct infection of humans with incurable diseases, were considered reasonable in the cause of advancing science and medicine (Elkeles 2004).

2 Early Chimpanzee Use in Research and the Road to the Yerkes Primate Research Center

The use of chimpanzees in biomedical research began in multiple Western countries, including Prussia, Belgium, France, the United Kingdom, Russia, and the USA on a sporadic and haphazard basis toward the end of the nineteenth century and into the early twentieth century. Research was initially based on chance availability of animals or their cadavers from zoos or other private collections (Fridman and Nadler 2002). In the mid- to late nineteenth century, apes were of significant interest to scientists, particularly within the context of evolutionary discussions (Honess 2016). At the time, observational field studies were viewed dimly by the scientific community as being unscientific and uncontrolled, but the controlled study of chimpanzees under laboratory conditions was hindered since specimens were expensive and difficult to procure (de Waal 2005).

Apes were so rare as research subjects that chimpanzees and bonobos were not recognized as different species until 1933. Prior to this, both were referred to as "chimpanzees" and studied together (Coolidge 1933). All apes were initially wild-sourced for research, and young chimpanzees were plucked from their mothers in Africa (the mothers and several other family members often being killed in the process) and traded to sailors in African ports and markets in the late nineteenth century. Welfare was poor and many animals died from malnutrition and exposure during the long ship journeys between Africa and Europe (Fridman and Nadler 2002). Even for chimpanzees that survived the voyage, life was often short as little was known about appropriate management and captive care of chimpanzees. This included the intensive and prolonged altricial needs of infants, as well as specifics of diet, housing, and hygiene needed for successfully keeping animals of all ages. Up to the mid-1950s, infectious disease containment and hygiene practices were minimal for captive research chimpanzees, and it was not uncommon for juveniles and adults to die suddenly and spontaneously of dysentery or respiratory disease (http://first100chimps.wesleyan.edu/; Yerkes 1943). In 1915, the first recorded captive-bred chimpanzee was born within a private Cuban menagerie managed by an idiosyncratic and wealthy amateur naturalist and animal lover, Madame Rosalia Abreu (Wynne 2008). However, it would still be many years until chimpanzees were bred for commercial sale for research use.

Early scientific research on chimpanzees included descriptive studies on comparative physiology, anatomy, pathology, neuroanatomy, and behavior, but sufficient numbers were generally unavailable for controlled experimental studies. One notable exception to this was Elie Metchnikoff's studies in Paris in 1903, in which he and his co-workers purchased 50–60 chimpanzees and orangutans for syphilis experiments. This research resulted in production of the first reliable animal models for the study of the disease and progress in finding early treatments for syphilis (Krause 1996; Rossiianov 2002). There are few details available about the care or fate of these animals, but the development of a reliable animal model for studying syphilis and the advancement of efficacious treatments were considered major medical breakthroughs. At the time, it was estimated that 16% of Paris's population was infected with syphilis, accounting for 11% of all annual recorded deaths for the city (Krause 1996).

The first organized chimpanzee research center, the Anthropoid Research Station, was established on Tenerife, one of the Canary Islands, by the Prussian Academy of Sciences in 1912. Today, it is accepted that the site was selected, in part, to spy on nearby shipping lanes, and because it had an appropriate ambient climate and ready food sources for chimpanzees brought in from nearby Cameroon (Ley 1990). Before World War I, there was minimal state budget for primate or other research. The construction and operation of the station were made possible by a generous donation (approximately $5.8 million U.S. dollars in 2018) from a Prussian banker, who was interested in the ontogeny of individual and group morality (Lück and Jaeger 1988). Wolfgang Köhler, a German psychologist, directed the center from 1913 and studied the development of "insight" using controlled experimental trials in nine chimpanzees, as well as in many other animal species, until research ended in 1917 and the center was abandoned. Primate research centers that included chimpanzees were subsequently established in the 1920s by France (Pastoria in Guinea by the Pasteur Institute in 1923; more of a staging site that supplied Paris laboratories with several hundred wild-caught juvenile chimpanzees), Russia (Sukhum Station on the Black Sea in 1927), and other European countries, and used to study infectious and tropical diseases, behavior, and reproductive biology (Haraway 1989; Fridman and Nadler 2002; Rossiianov 2002). As early as the 1920s, there were significant concerns within the scientific community about depletion of wild animal populations. This was because of the juvenile capture/adult depopulation methods used by African hunters to acquire chimpanzees for research, and because of sporadic but devastating infectious gastroenteritis outbreaks that periodically killed hundreds of chimpanzee juveniles waiting to be transported out of local holding centers (Rossiianov 2002).

Although the goal of a large-scale, funded American ape research center was not realized until 1930, Robert Yerkes, an American psychologist considered by many to be the father of modern primatology, had written about the need for a U.S.-based center for anthropoid study as early as 1916. Lobbying for an ape research center occurred during the contextual background of the Nature Movement, in which there was tremendous interest in establishing public collections of animal remains within museums, and furthering education and scientific knowledge about the natural world. This was part of an ongoing attempt to try to understand the place of humans in the natural order of primates and the scientific underpinnings of primate personality (Haraway 1989). In Yerkes' vision, the center would have stable, long-term funding and facilities for maintaining, breeding, and experimenting on chimpanzees and other primate species. In addition, he envisioned that the center would be located in a climate suitable for indoor–outdoor living of the animals, while at the same time, avoiding the enervating effects on humans of a hot, tropical climate (Yerkes 1916). Further, Yerkes proposed that all forms of scientific study, including behavioral, biomedical, fundamental science, and comparative biology, would occur together at this site to make best use of the animals.

So committed was Yerkes to this concept of a national ape research resource that he indicated a willingness to devote his life to its inception (Yerkes 1916). Subsequently, to enhance his understanding of primate care and to initiate psychobiology research on chimpanzees, Yerkes purchased two wild-caught juvenile apes using personal funds from an importer associated with the Bronx Zoo, a chimpanzee and a bonobo, although he did not know that the two animals were of different species at the time. He housed and studied the two apes, and subsequently others, in a renovated barn situated on his farm in New Hampshire (Yerkes 1977). The first two chimpanzees died within the year from different infectious diseases, a reminder again of how little was known about captive care of these animals at the time. Through this experience and visits to other sites that held chimpanzees, including that of Abreu, Yerkes gained a great deal of knowledge and practical experience about captive chimpanzee management and breeding. This also formed the basis of his training approach for subsequent research on chimpanzees by research team members at Yale (Yerkes 1925, 1943). Although seemingly basic by current standards, Yerkes' recipe for success in housing chimpanzees was to socially house animals in large, clean spaces with a choice of shade or sunlight, together with lots of fresh air, a varied diet, regular feeding, and, where possible, indoor–outdoor access for exercise. Some measure of hygiene was afforded by removing uneaten food on a regular basis.

Yerkes subsequently transitioned to Yale from Harvard and established the Primate Laboratory in 1924 (Hilgard 1965). The cramped laboratory space soon became inadequate for the needs of the chimpanzees and other primates, and the cold, damp climate of Cambridge, MA, proved to be unsuitable for chimpanzee breeding. The location served as a temporary site to initiate chimpanzee research until additional funding and a more suitable property could be obtained. Eventually, a site in Orange County, Florida, near Jacksonville, was purchased and the Yale Anthropoid Laboratory first opened its doors in 1930. The laboratory was funded initially by a $500,000 grant from the Rockefeller and Carnegie Foundations, and from Yale University (Haraway 1989). Because of funding constraints, the site was not the densely forested, multi-acre property that Yerkes had envisioned to permit free-roaming of captive research chimpanzees, but it did provide a private, fenced compound with an outdoor exercise area of just under an acre (0.4 ha) (Riesen 1977). The site was populated with captive chimpanzees from various private and research collections in the USA and Europe and supplemented later by wild-caught juveniles and chimpanzees born at the research station.

The first recorded American captive chimpanzee birth occurred in September 1930, marking the start of systematic breeding of chimpanzees in the USA for research purposes (Fridman and Nadler 2002). Scrupulous attention was given to documenting detailed life histories of each chimpanzee housed at the Yale Anthropoid Laboratory, work that was generally given over to graduate students, who also provided much of the care and feeding of the animals. Primate veterinarians did not exist at the time, and local physicians and pediatricians were prevailed upon to provide medical care to the chimpanzees (Yerkes and Learned 1926; Riesen 1977). Robert Yerkes' primary research at the time revolved around studies of personality and comparative psychology in chimpanzees (Yerkes 1943). He was adamant that, whenever possible, animals should be trained to cooperate in experimental procedures, for both humane and scientific reasons (Yerkes 1943; Haraway 1989). This principle was also adopted by and expected of his graduate students (Haslerud 1977). That said, positive punishment was used at times when chimpanzees resisted participation in daily experimental trials (Yerkes 1943).

The nature of the comparative psychology studies conducted by Yerkes, and his associates at the Orange Park facility were wide-ranging and sometimes quite invasive. There were no animal research ethics committees, and graduate research projects and experimental approaches were generally agreed upon by the primary advisor or a small graduate research advisory committee. Examples of research conducted at the Orange Park facility included studies on social dynamics, chimpanzee sex, vision deprivation experiments, in which infant chimpanzees were raised in the dark to 16 months of age (Riesen 1947), isolation rearing of infants (Yerkes 1940; Haslerud 1977), avoidance and frustration studies (Haslerud 1977), and morphine addiction studies (Spragg 1977).

Following Yerkes' death in 1956, Emory University assumed ownership of the Orange Park facility. The chimpanzees were eventually moved to Emory University in Georgia in 1965 after the U.S. National Institutes of Health (NIH) designated Emory University as an official Regional Primate Research Center. An outdoor research field station was built nearby in 1966 to house and breed various primate species, including chimpanzees.

3 Use and Care of Captive Chimpanzees in Research

Although the types of research that have been conducted with captive chimpanzees (e.g., behavioral, aerospace, biomedical, and linguistic) are considered separately in the following section, they are all part of a greater continuum. This is because the history and development of each are linked, the housing and husbandry provided to animals for each of these purposes have been comparable throughout history, and surplus or "spent" animals from one branch of research frequently were transferred and subsequently used in others. Yerkes referred to the chimpanzee as the "servant of science" (Yerkes 1943, p 1) and encouraged the widespread use and exchange of chimpanzees as a replacement for forms of research that could not be conducted on humans (Yerkes 1943).

3.1 Chimpanzees in Aerospace Studies

Primates became involved in American military aerospace studies in the 1940s, and young chimpanzees were used for some of these studies in the early 1950s. The use of primates paved the way for subsequent human tests and space flights. Primates, specifically chimpanzees, were deemed more similar to humans than the dogs that were used in the Soviet space programs, and thus more relevant as a research species, and there was a race for space supremacy during the Cold War. Some early studies involved rapid acceleration and decompression testing, which resulted in the instantaneous death of the animals. Chimpanzees were generally injected with a sedative before trials (Zinser 2014). Later trials involved training and preparation of chimpanzees for the Mercury space flights in anticipation of human flights, which meant that the animals would have been more prepared for physical effects of the flights, such as weightlessness, restraint in-flight sleds, and various noises and vibrations. Initially, the infant chimpanzees needed for the aerospace tests were wild-caught in Cameroon and group-housed in indoor facilities in the USA. Later, a successful breeding colony was established at the Holloman Aero-Medical (HAM) Research Laboratory in Alamogordo, New Mexico. This facility had animals and the necessary personnel, including veterinarians and veterinary technicians, to care for the young chimpanzees; however, before the development of the Mercury space program, the personnel lacked training in behavioral management of animals (Angelo 2007). Behavioral management was later recognized to be essential in training animals to perform various maneuvers during flights when no humans were around. Subsequently, chimpanzees, and later, veterinary personnel, were sent to the Wenner-Gren Aeronautical Research Laboratory at the University of Kentucky for behavioral and methodology training, respectively (Henry and Mosley 1963).

Numerous chimpanzees were used for developing and optimizing flight training techniques and simulations, and as is the case for human astronauts, many animals underwent rigorous training for space flight with the expectation that only a few of the highest performing animals of a specific body size would be taking part in flight

Fig. 1 Examining young chimpanzees in an outdoor run at the Holloman Air Force Base. Photo courtesy of Dr. W.E. Britz

missions. Photographs from this period show close physical relationships between the young chimpanzees and their human caregivers (Fig. 1), with animals being carried like infants by military personnel or sitting in chairs or on a table in the company of personnel (Henry and Mosley 1963). As part of their training, animals were habituated to long periods of restraint in the flight capsules. Further, positive reinforcement and shock-avoidance techniques were used to train animals to respond with high discrimination to complex instrument signals that the animal would need to monitor during actual flight. One chimpanzee, Ham, received 219 h of training over 15 months in addition to centrifugal launch simulations (Henry and Mosley 1963). Ham undertook the first suborbital ballistic flight from Cape Canaveral, Florida, on January 31, 1961, and captured the hearts and minds of millions of Americans, later appearing on the cover of Life magazine on February 10, 1961. After similar rigorous training, Enos, another young male chimpanzee, was sent into an orbital flight on November 29, 1961 (Henry and Mosley 1963), becoming the first being to orbit the earth. During the flight, because of an inadvertent crossing of wires, Enos received repeated shocks despite correctly performing functions throughout (Conlee and Boysen 2005).

Ham was retired from research in 1963 and singly housed on display for the next 17 years at the National Zoo in Washington, DC. He was then moved to a group setting at the North Carolina Zoo in 1981, where he died prematurely, 2 years later, at approximately 24 years of age. Enos was returned to the Air Force colony and died of bacterial enteritis 11 months following his orbit of the Earth (Angelo 2007), prior to his fifth birthday. Shortly after these flights, the U.S. Air Force research program with chimpanzees ended, and by 1963 most of the remaining chimpanzees were either leased to Fred Coulston, a toxicologist and researcher based out of the Albany Medical College in New York, for private and public biomedical research use or maintained for breeding purposes by New Mexico State University in Alamogordo (Conlee and Boysen 2005). Breeding and research occurred at the original Holloman Air Force Base in New Mexico, which was equipped with solid but barren concrete and metal enclosures and small metal cages for holding the chimpanzees. These were typical of enclosures built at the time for chimpanzees in all settings, including zoos (Hosey 2022). While good for disease control, little consideration was afforded to captive animals in terms of their psychological or behavioral well-being, resulting in poor welfare.

3.2 Chimpanzees in Linguistic Studies

There was longstanding interest within the psychobiology community in determining whether apes could learn to communicate through the use of language. Several poorly standardized experiments of the 1940s and later demonstrated that even when chimpanzees or other apes were brought into the home at an early age and raised similarly to a human child, they were incapable of vocalizing words (Rumbaugh 1977; Haraway 1989). In 1966, Beatrix and Allen Gardner, a husband and wife scientist team working at the University of Nevada in Reno, purchased a wild-caught infant chimpanzee, Washoe, and brought her to their home. Their goal was to test whether chimpanzees could be taught to express themselves using American Sign Language (AMESLAN) (Gardner and Gardner 1969). Washoe had no difficulty learning sign language, but she quickly outgrew her trailer and backyard accommodations, necessitating a move to the Institute for Primate Studies in Norman, Oklahoma, and later to Central Washington University in Ellensburg, Washington. Washoe went on to teach her adopted chimpanzee son, Loulis, AMESLAN, and remained a signing chimpanzee until her death in 2007 at 42 years (Fouts 1997). David and Ann Premack, working at the University of California in Santa Barbara, conducted similar experiments on a chimpanzee named Sarah using plastic linguistic markers and found that Sarah mastered 130 symbols with up to 80% reliability (Rumbaugh 1977). Publications describing the sensational results from both groups sparked discussions and debates as to possible investigator bias in subjectively interpreting signing, and fueled interest in determining whether chimpanzees could actually communicate complex thoughts using sign language or by other means. This interest also led to almost two decades of linguistic research on chimpanzee and gorilla infants in the USA in search of finding a common language (Haraway 1989). The question of language capabilities of apes also forced reinterpretation of man's place in the animal continuum, since these results together with the chimpanzee tool use observations of Jane Goodall indicated that two human qualities, language and toolmaking, were not exclusive to humans (Haraway 1989).

Juvenile animals were thought to be optimal for language research, both because they were easier to handle than adults and because their brains at that age were felt to be more receptive for learning, meaning that infant chimpanzees were taken from their mothers when they were as young as 2 weeks, and placed under human foster care (Fouts 1997). This resulted in a further need for captive chimpanzee breeding with the rise of new sites, such as the Institute for Primate Studies in Oklahoma (Hampton 2016a). This institute was directed by psychotherapist William Lemmon, a keen proponent of domination and fear training techniques, who had a fondness for using a cattle prod and restraint chains with his research chimpanzees (Fouts 1997). Lemmon also commonly "prescribed" infant chimpanzee foster rearing to many of his patients in Norman, Oklahoma (Fouts 1997). Most of the infant chimpanzees from language training programs eventually outgrew their quarters, became too dangerous to work with, or were made redundant by lack of interest and a reduction in federal funding for this line of language research in the mid-1980s (Hampton 2016b). Although a few university-based chimpanzee language research centers in the USA persisted through the decades, in the 1980s many of these animals were transferred or sold to biomedical research centers, such as the Laboratory for Experimental Medicine and Surgery in Primates (LEMSIP) in New York, to be used for biomedical studies (Hess 2008).

3.3   Chimpanzee Use in Experimental Medicine and Biology

Because of their rarity, chimpanzees were not used much in early biomedical research. Other than some rather bizarre and failed Soviet attempts to produce a chimpanzee–human hybrid in 1927 by Il'ya Ivanovich Ivanov (Rossiianov 2002; Etkind 2008) and exploration of chimpanzee testicular slice transplants into aging men (purportedly to restore virility) by Serge Voronoff, a Russian surgeon working in Paris in 1919 (Kahn 2005; Kozminski and Bloom 2012), early research use of chimpanzees was largely confined to infectious disease and behavioral studies. The use of chimpanzees in experimental biology and medical research expanded markedly in the late 1950s (Fridman and Nadler 2002) and was commensurate with the development of in-country national primate research centers and breeding colonies in the USA, Japan, UK, the Netherlands, Austria, and elsewhere. A marked increase in the number of scientific publications related to primates occurred simultaneously, and between 1968 and 1976, an estimated 2800 research institutes were using primates in research worldwide (Fridman and Nadler 2002). Chimpanzee use tended to be restricted to purpose-built nationally funded or supported primate research centers or private institutions, such as pharmaceutical companies, because of the specialized knowledge, handling facilities, and high cost required to care for and use them.

For over 40 years, chimpanzees were considered essential for research on hepatitis A, hepatitis B, and hepatitis C infections and vaccine development, since chimpanzees readily develop persistent infections to human hepatitis agents, but do not develop clinical signs of disease for hepatitis B and hepatitis C (Hillis 1961; Deinhardt et al. 1962; Buynak et al. 1976; Prince and Brotman 2001; Bukh 2004; Catanese and Dorner 2015).

Hepatitis B and hepatitis C viruses and the human immunodeficiency virus (HIV) were known to be transferred through blood products, and there was interest in developing rapid screening assays to assure a safe national blood supply for humans. The development of novel assays used samples from infected chimpanzees (Tobler and Busch 1997). Not-for-profit research centers, such as LEMSIP and the New York Blood Center (NYBC), came into being between 1965 and the early 1970s, and developed on-site (LEMSIP: New York state) and offshore (NYBC: Vilab II in Liberia) chimpanzee breeding colonies to support research and development of tests of and vaccines against hepatitis B and hepatitis C viruses.

While resulting in tremendous health gains for humans, there was a significant cost to chimpanzees from the hepatitis research. Any infectious disease study using animals that may result in a major risk for serious human disease must be conducted under enhanced biosecurity conditions, including more barren housing and less freedom of movement. In the case of hepatitis research, this necessitated single housing of chimpanzees in stainless steel cages continuously for months to years to minimize the potential for bites and scratches of human caregivers and researchers, and to enhance ease of handling and access for sampling and observation. For hepatitis studies, blood was generally collected from conscious animals using squeeze-back cages (a cage with a false back that can be pulled forward to forcibly press animals against the front wall of the cage) on a weekly basis, often for 4–6 months or beyond, depending on the nature of the study (Prince et al. 2005). To track the course of infection and the extent of disease, multiple percutaneous liver biopsies were collected from immobilized or sedated chimpanzees at least four to twelve times over the course of a study at semi-weekly (Sakai et al. 2003), weekly, or monthly intervals (Su et al. 2002). No information was provided in most papers published at the time regarding peri-operative analgesic use despite the discomfort and pain known to be associated with percutaneous liver biopsy. It is thus unknown whether analgesics were administered and not reported or not used. Studies required chronic infections to mimic human disease and were commonly 6 months to 2 years in duration, although one study mentions testing animals over 5 years following the initial infection (Mizukoshi et al. 2002), the entire duration for which most of these animals would be singly housed in an indoor cage. Following the conclusion of a study, chimpanzees could never be returned to social housing conditions because of the risk of transmitting infections to uninfected chimpanzees and caregivers. Thus, infected chimpanzees had to be housed apart from their conspecifics, something that was noted as being highly unnatural and psychologically damaging by Robert Yerkes (1943). When possible, cohorts of similarly infected chimpanzees were formed and group-housed, and these animals became available for other studies. In all, it is thought that just over 500 chimpanzees were used globally in hepatitis C research between 1998 and 2007 (Bettauer 2010).

Chimpanzees were also used for HIV research in the late 1980s and throughout the 1990s and at least 150 animals were infected in the USA for this work (van Akker et al. 1994). In 2000 alone, there were 23 ongoing Public Health Service-funded grants for HIV-related research (including breeding and maintenance programs) in captive chimpanzees (Conlee and Boysen 2005). Chimpanzees were eventually recognized to be a poor model for HIV infection, because, in general, their $CD4^+$ T cells do not become severely depleted postinfection and they almost never develop immunodeficiency-related disease, such as is seen in HIV-positive humans with acquired immunodeficiency syndrome (AIDS) (van Akker et al. 1994). One chimpanzee chronically infected with HIV did demonstrate AIDS-like disease 10 years after initial infection (Novembre et al. 1997). However, the cost and unreliability of chimpanzees as a model argued strongly against their return to use for testing candidate HIV vaccines, particularly when more promising primate models, such as SIV-infected rhesus macaques, had been developed (Prince 1999). Throughout this era and those preceding, there was less consideration for appropriate chimpanzee behavioral management with an emphasis instead on maximizing the data extracted from each animal.

Chimpanzees have been used extensively for testing novel pharmaceuticals and environmental toxins, although because of the expense of chimpanzee studies, their use in these studies was restricted to work deemed most critical or necessary for federal regulatory approval (Hobson et al. 1976; Mueller et al. 1985). From 1963 to 2002, over 1000 chimpanzees at the Coulston Foundation in Alamogordo, New Mexico, were used for toxicology and infectious disease contract research. While some of the research at the Coulston Foundation was conducted through funding from NIH contracts, other work was conducted for private industry, the results of which were never published, making it difficult to estimate the number of chimpanzees used in studies and the scope and nature of studies in which animals were used. Animal numbers at the Coulston Foundation varied over the years as the facility accumulated animals when other chimpanzee research or breeding facilities closed, such as LEMSIP in New York and the New Mexico State University breeding colony in Alamogordo (Macilwain 1997). The Coulston facility was never assessed and accredited by AAALAC, International, a voluntary not-for-profit accrediting organization that assesses animal research programs using high standards for animal care and welfare. Further, the facility received numerous citations and fines from U.S. Department of Agriculture inspectors for Animal Welfare Act violations over its years of operation for issues such as poor hygiene, inadequate cage size, and inadequate veterinary care that resulted in chimpanzee deaths (Wadman 1999). The NIH eventually withdrew support for breeding chimpanzees at the Coulston Foundation and failed to renew key research contracts in the late 1990s, and in 2001, the U.S. Food and Drug Administration withdrew site registration, which precluded the site from conducting further regulatory studies. The care of the animals (chimpanzees and other primate species) was turned over to the NIH when the Coulston Foundation closed in 2002.

Other lines of biomedical research using chimpanzees have included studies of photosensitive epilepsy (Naquet et al. 1967), respiratory virus infection and vaccine development (Bem et al. 2011), poliovirus vaccine development (Chen et al. 2011), kidney and heart xenotransplantation to humans (Cooper et al. 2015), effects of prefrontal lobotomies on task performance (Jacobsen et al. 1935; Rosvold et al. 1961), malaria, and a host of other applications (see Stephens 1995 for further examples).

3.4 Large-scale Sourcing of Chimpanzees for Experimental Studies

Wild, infant chimpanzees were the primary source of animals used for all types of captive chimpanzee research with over 8000 animals estimated to have been exported from Sierra Leone alone between 1928 and 1979 (Kabasawa 2011). It has been suggested that five to ten adult animals were killed for each infant collected, leading to a significant and rapid depopulation of chimpanzees in the West African countries from which chimpanzees were being harvested (Kabasawa 2011). In the 1970s, approximately 60% of all wild-captured chimpanzees were exported to the USA and Japan for research use, the rest going to research centers in Austria, Germany, the Netherlands, France, and elsewhere, as well as sporadic animals being exported for use in public and private zoos, entertainment, and personal collections (Kabasawa 2011). Chimpanzee export and depletion of wild populations became an issue of significant concern to ecologists, politicians, animal rights activists, and primatologists during the 1960s. When the Convention on International Trade in Endangered Species of Flora and Fauna (CITES) was enacted in 1975, international chimpanzee exportation theoretically became highly controlled. However, Sierra Leone, one of the largest exporters of chimpanzees globally, did not become a CITES signatory country until 1995 (Kabasawa 2011; Turner 2017). Thus, instituting CITES protection for chimpanzees slowed but did not immediately quell wild animal export for research use, as animals continued to be exported for research use in Europe and elsewhere, albeit in smaller numbers, throughout the 1980s and beyond (Cherfas 1986). Further, to reconcile the use of chimpanzees in research and in captive zoo and entertainment settings despite their international threatened status (reinforced in the U.S. Endangered Species Act), in 1976, the U.S. Fish and Wildlife Service introduced a split listing designation for chimpanzees, allowing for the continued use of captive animals in research.

Prior to the establishment of captive breeding programs, chimpanzees purchased as pets or for private animal collections, those used by the entertainment industry in advertising, television, and movies, animals purchased for zoo performances in the early twentieth century (see Hevesi 2022, for a review of primates in these industries), and even some chimpanzees with difficult temperaments from zoo settings were sporadically transferred to research facilities (Goodall 2003). This occurred after animals entered adolescence and became difficult to handle because of their physical strength and when they lost some of their physical attractiveness and "cuteness" to humans with the onset of sexual maturity. There were few perceived options for disposing of these unwanted chimpanzees, since most zoos were unable to accommodate extra animals and animal sanctuaries that were capable of caring for mature chimpanzees were few and far between (Goodall 2003). These animals frequently came with a host of abnormal behaviors, such as hair plucking, rocking, and coprophagy, and were often difficult to integrate into social settings in research colonies because of years of social isolation, cramped housing conditions, and adverse training techniques that they had been exposed to as juveniles. This also limited the types of studies that these animals could be used for in research settings. Thus, despite the fact that several hundred chimpanzees were transitioned from pet, entertainment, or zoo facilities to research institutions over several decades, a more reliable source of chimpanzees was required to fulfill research needs following the onset of international animal import/export bans.

Some research centers were successful in breeding chimpanzees, particularly in the USA and Japan (Graham 1981; Morimura et al. 2011). Despite this, the long intergeneration interval of chimpanzees meant that it took many years for animal numbers to increase to a point that a country became self-sufficient and no longer needed to import chimpanzees. In addition, chimpanzee infant mortality rates remained relatively high, further contributing to slow growth in colony numbers (Graham 1981). In terms of estimating numbers of chimpanzees needed for research, hundreds of animals were initially anticipated for study of HIV in the 1980s and many available breeding age chimpanzees were considered to be unsuitable for reproductive or other research projects because they were chronically infected with various diseases from previous research, such as hepatitis B virus. Following a national census of research chimpanzees, a detailed review of animal genetics and health histories, and extensive expert consultations, the U.S. government concluded that there was a need for coordinated action to protect national health interests. In 1986, the NIH funded five breeding centers for chimpanzees, charging these centers with communicating and harmonizing best practices between themselves to optimize animal care (Graham 1981; Hobson et al. 1991; NRC 1997). The U.S. chimpanzee breeding programs were highly successful, and within less than a decade, the U.S. government population of chimpanzees rose to almost 1000 animals. Because research needs in HIV studies were less than anticipated during this same interval, this created a surplus of chimpanzees that were being maintained in research centers at significant cost to the American taxpayer.

Discussions were initiated within the newly formed National Chimpanzee Management Program Advisory Committee regarding how to care for and manage the U.S. federal government's captive population of research animals. Despite the great expense of caring for chimpanzees within institutional settings, and the long life spans of chimpanzees (40–50 years or more), the Committee recommended against euthanizing surplus animals. This consideration was made in recognition of their special connection to humans, but also because of an anticipated public outcry at such an action. To keep expenses in check, however, the NIH instituted a 10-year chimpanzee breeding moratorium in 1995 (IOM 2011). In addition, because future chimpanzee needs could not be accurately predicted, a recommendation was made for the U.S. federal government to maintain a population of about 1000 research chimpanzees and for the government to consider developing a sanctuary for future potential retirement of chimpanzees from research (NRC 1997). The care for these animals would be overseen by a newly formed U.S. Chimpanzee Management Program (NRC 1997). The issue of whether the breeding moratorium on federally owned research chimpanzees should continue was reviewed in 2007 by a panel of nongovernment experts (the Chimpanzee Management Plan Working Group); however, this group was unable to reach a decision on the issue (NCRR 2007). Subsequently, the NIH's National Advisory Research Resources Council determined that an indefinite ban on chimpanzee breeding should be maintained (IOM 2011).

3.5 Captive Research Chimpanzee Housing and Care

Until the mid-1960s, care of research chimpanzees was largely in the hands of scientific investigators and the institutions they worked for. Minimum standards for housing and husbandry of research chimpanzees were not established in the USA until the Animal Welfare Act was passed in 1966 (U.S. Department of Agriculture 2017) and much later in the European Union (1986) and Japan (1983) (IOM 2011; Morimura et al. 2011). The conditions of the 1966 Animal Welfare Act are provided for basic care and sanitation practices, minimum cage sizes, and unannounced veterinary inspections for specific covered species, but the Act did little to address aspects of social, behavioral, physical, psychological, and environmental management that are now known to be critical to captive animal health and well-being. Similar to many zoos at the time, cage sizes were not based on scientific evidence, but were adopted from cages being used at the time that seemed to be adequate, and materials used for building cages and enclosures tended to be those that could be readily sanitized and disinfected, such as stainless steel and concrete. This resulted in indoor housing environments that were sterile and barren in appearance, since minimal consideration was given to environmental enrichment until the 1990s (Wolfle 1999). Depending on the location and climate, some facilities also had large outdoor enclosures or corrals that could be used to house small colonies of animals used for breeding or that were not on active research protocols (Fig. 2).

The U.S. Animal Welfare Act was an important first step in providing regulation for research animal care, but it was not until amendments to the Act in 1971 and 1986 that the need for written research protocols with a priori animal ethics committee review was required (Turner 2017). Similar requirements were imposed by the U.S. Public Health Policy (1979) on holders of federally funded grants. Changes to the Animal Welfare Act in 1985 included recommendations for improving the environment of captive primates to enhance the animals' psychological well-being and a requirement for documenting the plan, but few specifics were provided, and so few consistent practices were adopted across research facilities (U.S. Department of Agriculture 1985; Wolfle 1999). A significant number of studies on primate environment and social housing were published in the 1990s, leading to scientific debate amongst primatologists, ethicists, and veterinarians regarding what was needed to optimize the housing of nonhuman primates. Publication of the National Research Council's report on "Psychological Well-Being of Nonhuman Primates" helped to

Fig. 2 Large outdoor exercise structure at the Southwest Foundation for Biomedical Research, San Antonio, Texas in 1989. Photo courtesy of Dr. T.M. Butler

provide the research community with a more fulsome understanding of environmental enrichment and behavioral well-being (NRC 1998). The report emphasized the need for social housing as the most important attribute of primate care, as well as provision of food foraging resources, cage furnishings, such as perches and visual hiding places, objects that could be manipulated, and other resources, such as opportunities for exercise, to ensure that captive research primates could perform at least some natural behaviors and had sufficient stimulation to promote mental well-being and minimize the development of abnormal behaviors (NRC 1998). While special provisions were permitted to deal with program exceptions, such as single housing of animals when scientifically justified, these guidelines were broad-ranging and applicable to all nonhuman primate species held in research settings, including chimpanzees, and had a significant impact on improving the captive research primate environment.

4 Ethical Concerns Regarding Captive Chimpanzee Welfare and the Development of a Research Chimpanzee Sanctuary

Following the moratorium on breeding of chimpanzees for research purposes in the late 1990s, arguments were made for increased use of chimpanzees in research in a range of applications (VandeBerg and Zola 2005). However, the larger issues

Fig. 3 Chimpanzees seen foraging in a Chimp Haven habitat. Photo courtesy of R. Smith

surrounding the ethics of using such intelligent and sentient animals in invasive biomedical research projects combined with the high federal cost of maintaining chimpanzees in research centers, estimated to be $300,000 to $500,000 USD per animal over the course of a 30- to 45-year lifespan, led to a reduction in support for their use, although overall numbers of U.S. federal grants funding chimpanzee research remained relatively stable during this time (Cohen 2007; Knight 2008; de Waal 2012). In 2008, the Great Ape Protection Act was introduced into the U.S. Congress in an attempt to ban the use of great apes of any species in research, but the proposal was never enacted in this or subsequent attempts (U.S. Congress 2011).

The mounting costs of maintaining chimpanzees in research facilities had not gone unnoticed within the U.S. federal government, and in 2000, resulted in a directive from Congress to retire unused federally owned chimpanzees to a sanctuary largely as a cost-saving measure (US Chimpanzee Health Improvement, Maintenance and Protection Act 2000). Chimp Haven, a chimpanzee sanctuary in Keithville, Louisiana, that had been established in 1995, was awarded the federal contract in 2000 for providing care for any future retired government chimpanzees, although the first two federally owned chimpanzees did not arrive until 2005 (Figs. 3 and 4). In 2007, a loophole was closed in an existing public service act, permanently preventing retired chimpanzees from being transferred back to research facilities or otherwise being used for invasive experimentation (U.S. Chimp Haven is Home Act 2007). Other private or not-for-profit companies that had been using chimpanzees in research, such as NYBC and various pharmaceutical companies, subsequently

Fig. 4 Chimpanzees in a corral at Chimp Haven. Photo courtesy of R. Smith

phased out chimpanzee use between 2005 and 2014 on a voluntary basis. The NYBC phase-out was highly controversial since the annual financial provisions that the organization provided for supporting the retired colony of research chimpanzees living on small islands off the coast of Liberia were withdrawn abruptly in 2015, as an institutional cost-saving measure. To prevent these hepatitis-infected chimpanzees from starving to death, immediate daily maintenance and veterinary care were taken on by international animal protection groups, but the controversy continues to highlight the need for defining ethical institutional responsibility when long-lived animals are used for research purposes (Gorman 2015).

Elsewhere, there was a significant movement away from the use of chimpanzees in research in the 1990s and early 2000s, created in part by heavy government lobbying from animal protection and public interest groups. The UK had significantly restricted all research on great apes in 1986 but committed to an outright Home Office prohibition of granting new licenses for great ape research in 1997 (Knight 2008). The Dutch government issued a similar prohibition on great ape experimentation in 2002, although ongoing studies did not conclude until 2004, following which all remaining chimpanzees housed for biomedical experimentation were transferred to zoos or sanctuaries. Several other European countries and countries elsewhere around the world instituted similar research bans, regardless of whether they had ever used chimpanzees in biomedical research. In Japan, the sole pharmaceutical company using chimpanzees for testing voluntarily agreed to stop experimentation in 2006, and subsequently supported development of the Kumamoto Sanctuary, in conjunction with Kyoto University, to care for the retired animals (Cohen 2007; Knight 2008; Morimura et al. 2022). In response to public pressure, the European Union in 2010 instituted an outright ban on the use of great apes in research under Directive 2010/63/EU, which went into effect in 2013 (European Commission 2010).

5 Research Chimpanzees Under the Microscope: Reports of the Institute of Medicine (IOM) and the NIH Council of Councils Working Group

The chimpanzees held at the Alamogordo Primate Facility in New Mexico under the responsibility of the NIH and following the closing of the Coulston Foundation in 2002 had never been available for research since the facilities were considered inadequate. In 2010, discussions occurred within the U.S. federal government regarding the possibility of transferring these animals to the Southwest National Primate Research Center in San Antonio, TX. This was seen as a means of making these chimpanzees available once more for research studies, and as a cost-saving consolidation measure, since the outsourced care of these animals was costing the government approximately $2 million annually (IOM 2011). This sparked an outcry from animal activists, primatologists, and the New Mexico government, the latter being concerned about the possible loss of employment that would be brought about by the facility's closure (IOM 2011). Subsequently, the U.S. National Academies of Sciences, Engineering, and Medicine was charged with conducting an in-depth analysis of the need for chimpanzee research, a task eventually delegated to the IOM. The IOM ad hoc expert committee, composed of scientists, bioethicists, veterinarians, and consultants, was tasked with determining the necessity of chimpanzee use for current or future biomedical or behavioral research. The committee was specifically advised not to include cost or ethical considerations surrounding chimpanzee use as part of their deliberations (IOM 2011).

Following a year of reviewing the use of chimpanzees in research and numerous peer-reviewed scientific publications, availability of animals, international guidelines covering the use of great apes in research, and public comment, the committee submitted their report (IOM 2011). Through their efforts, the committee developed and published core principles to define acceptable conditions for biomedical and behavioral research with chimpanzees and defining a set of criteria as to when chimpanzees could be used in research (IOM 2011). The specific criteria were subdivided into biomedical versus behavioral research requirements. For biomedical research, the committee proposed that chimpanzees could be used only if no other comparable models existed, the research could not be conducted on humans, and life-threatening or debilitating results could be expected in humans without occurrence of the research. For behavioral research, the committee stipulated that research on chimpanzees should only occur if the information is unattainable by other means and if the animals voluntarily acquiesced to participate in study procedures through training, in an effort to reduce potential pain and distress. The report also emphasized (but did not define) that chimpanzees should be held in ethologically appropriate environments and recommended that another committee be established to review current and ongoing chimpanzee research protocols using NIH-owned chimpanzees to ensure that they met these criteria.

The NIH accepted the IOM report in its entirety shortly after it was released. Subsequently, another working group was formed and was charged with implementing the report, including evaluating federally funded research grants with chimpanzees against the IOM criteria, determining the size and placement of a colony of chimpanzees for future research, and developing a process for reviewing future requests to use chimpanzees in NIH-supported research. This new Council of Councils' Working Group was composed of biomedical and animal behavioral scientists, veterinarians, a bioethicist, and a primatologist. In considering future colony placement of active vs. inactive animals, the Working Group quickly realized that defining an "ethologically appropriate physical and social environment" (IOM 2011) would be essential to completing the task. As part of its effort to define housing needs, the Working Group conducted site visits to all U.S. federally funded national research centers housing chimpanzees and two chimpanzee sanctuaries (Chimp Haven and Save the Chimps) and solicited expert national and international scientific opinion as to requirements for appropriate housing of chimpanzees. A subcommittee for emerging infectious disease research needs was also convened to assist with addressing the current and future need for chimpanzees in biomedical research.

The Council of Councils' Working Group submitted their report to the NIH in 2013. The report made extensive recommendations for appropriate environments for retired and working research chimpanzees, and, based on the review of federal grants, determined that the majority of research chimpanzees should be retired to the federal sanctuary, since most of the currently funded grants did not meet the IOM criteria and the sanctuary provided the most "species-appropriate environment" for the animals (NIH 2013). Up to 50 chimpanzees could be maintained for future potential biomedical research use, but these animals were to be held in an appropriate environment and continuation of the colony should be reviewed every 5 years to determine whether there was an ongoing need. The Working Group further suggested that noninvasive behavioral research could be conducted at the federal sanctuary, if needed. Finally, the Working Group developed the concept for an independent oversight process for future NIH-supported research proposals with chimpanzees, separate from the regular extramural NIH scientific program review processes, in which investigators would have to justify how their proposals met the IOM criteria (NIH 2013).

6   NIH Decision and Post-Decision Impact

The NIH accepted the Council of Councils' Working Group report with the exception that research facilities housing chimpanzees would not have to meet the space requirements suggested within the Working Group report (i.e., 93 m^2/animal). Subsequently, the Chimp Amendments of 2013 (S. 1561) were passed removing the $30 million USD cap that had been placed on sanctuary spending as this cap would have prevented the transfer of research chimpanzees to the federal sanctuary at Chimp Haven. This was followed by removal of the split listing status for captive versus wild chimpanzees by the U.S. Fish and Wildlife services, such that all chimpanzees would be considered as endangered species (U.S. FWS 2015). In November 2015, the NIH issued a statement indicating that all chimpanzees, including the potential colony of 50 reserve animals, would also be retired to the federal sanctuary (Reardon 2015), paving the way for the end of all publicly funded chimpanzee use for biomedical research in the USA.

Following the NIH's announcement, concerns were expressed within the scientific community that removing chimpanzees from biomedical research would result in a perilous threat to humankind in the event of a future pandemic. The global COVID-19 pandemic demonstrated that many other animal models were available for studying that virus and developing efficacious therapeutics. However, a key point made by the NIH Working Group in their 2013 report was that none of the highest level containment laboratories in the USA were capable of safely holding chimpanzees, and chimpanzee use had never been envisioned for highly pathogenic infectious disease studies when these institutions were constructed in the early 2000s. This is largely because smaller, more easily contained and cared for, purpose-bred nonhuman primate models, such as rhesus and cynomolgus macaques, are readily available, and these models have proven to be as susceptible to many of the same serious infectious diseases and conditions as humans.

By early 2018, 232 of the total NIH-owned or NIH-supported research chimpanzees had been transferred to the federal sanctuary at Chimp Haven with another 272 still held in three research facilities (NIH 2018). Concerns had been voiced by the research facilities housing and maintaining these chimpanzees about the safety of transporting up to 177 of the remaining animals, many of whom had chronic degenerative diseases or other adverse behavioral or medical conditions (NIH 2018). To address this, the NIH's Council of Councils assembled another Working Group that was charged with assessing the safety of relocating these at-risk chimpanzees (NIH 2018). This Working Group visited the three sites still holding retired research chimpanzees and Chimp Haven and developed recommendations to guide relocation decisions for the remaining chimpanzees. The Working Group recommended that all chimpanzees be relocated to the federal sanctuary at Chimp Haven unless the move was associated with a high safety risk for a particular animal, that the facilities caring for U.S. federally owned chimpanzees work collectively to develop and use a common rubric for assessing the health and well-being of the chimpanzees in their care, that the facilities develop shared standard operating procedures to optimize the relocation of chimpanzees and work together to ensure successful relocation of animals, and that any disagreements arising concerning relocation of any chimpanzee would be decided by an independent NIH-determined veterinary panel (NIH 2018). A harmonized chimpanzee health categorization framework was subsequently developed and approved in 2019 and is being utilized to make decisions about relocation of the remaining NIH-owned or NIH-supported chimpanzees (NIH 2019). Under this framework, any chimpanzee deemed medically unfit to be transported to the federal sanctuary will be retired in situ at the research facility in which it is currently held.

7   Conclusion

With the 2015 U.S. federal rulings and a decision the same year by the Franceville International Center for Medical Research (CIRMF) in Gabon to retire their colony of 50 chimpanzees (Sarabian et al. 2017), the use of chimpanzees for biomedical research has largely ended on a global scale. Intensive study of chimpanzees over the past 100 years has certainly helped to increase human knowledge about the natural world and advance science and medicine in many important ways. But, at the same time, it has raised significant ethical issues associated with keeping these large, highly intelligent animals with complex physical, psychological, social, and behavioral needs in restrictive housing environments in research settings. While it has been difficult for some in the research community to accept, the near-complete end of all biomedical use of chimpanzees represents refinement of animal use. As newer, validated, and easier to care for animal models are developed, such as humanized mice, these models can replace more challenging and difficult to manage models, such as chimpanzees.

References

Aminov RI (2010) A brief history of the antibiotic era: Lessons learned and challenges for the future. Front Microbiol 1:1–7. https://doi.org/10.3389/fmicb.2010.00134

Angelo JA (2007) Frontiers in space: human space flight. Fact on File, Inc., New York

Bem RA, Domachowske JB, Rosenberg HF (2011) Animal models of human respiratory syncytial virus disease. Am J Phys Lung Cell Mol Phys 301(2):L148–L156. https://doi.org/10.1152/ajplung.00065.2011

Bettauer RH (2010) Chimpanzees in hepatitis C virus research: 1998–2007. J Med Primatol 39(1):9–23. https://doi.org/10.1111/j.1600-0684.2009.00390.x

Block SS (2001) Disinfection, sterilization, and preservation. Lippincott Williams & Wilkins, Philadelphia

Bradford PV, Blume H (1992) Ota Benga: the pygmy in the zoo. St. Martin's Press, New York

Bukh J (2004) A critical role for the chimpanzee model in the study of hepatitis C. Hepatology 39(6):1469–1475. https://doi.org/10.1002/hep.20268

Buynak EB, Roehm RR, Tytell AA et al (1976) Development and chimpanzee testing of a vaccine against human hepatitis B. Exp Biol Med 151(4):694–700. https://doi.org/10.3181/00379727-151-39288

Catanese MT, Dorner M (2015) Advances in experimental systems to study hepatitis C virus in vitro and in vivo. Virology 479–480:221–233. https://doi.org/10.1016/j.virol.2015.03.014

Chen Z, Chumakov K, Dragunsky E et al (2011) Chimpanzee-human monoclonal antibodies for treatment of chronic poliovirus excretors and emergency postexposure prophylaxis. J Virol 85(9):4354–4362. https://doi.org/10.1128/JVI.02553-10

Cherfas J (1986) Drugs firm accused over chimpanzee cages. New Sci 111:16–17

Chimp Act Amendments of 2013. https://www.congress.gov/bill/113th-congress/senate-bill/1561. Accessed 27 February 2017

Cohen J (2007) The endangered lab chimp. Science 315(5811):450–452. https://doi.org/10.1126/science.315.5811.450

Committee on Long-Term Care of Chimpanzees, Institute for Laboratory Animal Research, Commission on Life Sciences, National Research Council (NRC) (1997) Chimpanzees in research: strategies for their ethical care, management, and use. National Academies Press, Washington, DC

Conlee KM, Boysen ST (2005) Chimpanzees in research: past, present, and future. In: Salem DJ, Rowan AN (eds) The state of the animals III. Humane Society Press, Washington, DC, pp 119–133

Coolidge HJ (1933) *Pan paniscus*. Pigmy chimpanzee from South of the Congo River. Am J Phys Anthropol 18:1–59. https://doi.org/10.1002/ajpa.1330180113

Cooper DKC, Ekser B, Tector AJ (2015) A brief history of clinical xenotransplantation. Int J Surg 23B:205–210. https://doi.org/10.1016/j.ijsu.2015.06.060

de Waal FBM (2005) A century of getting to know the chimpanzee. Nature 437(7055):56–59. https://doi.org/10.1038/nature03999

de Waal FBM (2012) Research chimpanzees may get a break. PLoS Biol 10(3):e1001291. https://doi.org/10.1371/journal.pbio.1001291

Deinhardt F, Courtois G, Dherte P et al (1962) Studies of liver function tests in chimpanzees after inoculation with human infectious hepatitis virus. Am J Hyg 75(3):311–321. https://doi.org/10.1093/oxfordjournals.aje.a120252

Elkeles B (2004) The German debate on human experimentation between 1880 and 1914. In: Roelke V, Maio G (eds) Twentieth century ethics of human subjects research. Franz Steiner, Stuttgart, pp 19–33

Etkind A (2008) Beyond eugenics: the forgotten scandal of hybridizing humans and apes. Stud Hist Phil Biol Biomed Sci 39(2):205–210. https://doi.org/10.1016/j.shpsc.2008.03.004

European Commission (2010) Directive 2010/63/EU of the European Parliament and of the Council of 22 September 2010 on the protection of animals used for scientific purposes OJL276/33. Off J Eur Union L276:33

Fouts RS (1997) Next of kin. Harper Collins, New York

Fridman EP, Nadler RD (2002) Medical primatology: history, biological foundations and applications. Taylor & Francis, New York

Gamble VN (1997) Under the shadow of Tuskegee: African Americans and health care. Am J Public Health 87(11):1773–1778. https://doi.org/10.2105/AJPH.87.11.1773

Gardner RA, Gardner BT (1969) Teaching sign language to a chimpanzee. Science 165(3894):664–672. https://doi.org/10.1126/science.165.3894.664

Goodall J (2003) Problems faced by wild and captive chimpanzees: finding solutions. In: Armstrong SJ, Botzler RC (eds) The animal ethics reader. Routledge, New York, pp 145–152

Gorman J (2015) Chimpanzees in Liberia, used in New York Blood Center Research, face uncertain future. New York Times, May 28, 2015. https://www.nytimes.com/2015/05/29/science/chimpanzees-liberia-new-york-blood-center.html?_r=0. Accessed January 15, 2017

Graham CE (1981) Editorial report: a national chimpanzee breeding plan. Am J Primatol 1(1):99–101. https://doi.org/10.1002/ajp.1350010113

Hampton J (2016a) Remembering Norman's Chimps. Part II: Norman psychology professor rears chimps. The Norman Transcript, Norman, Oklahoma. https://www.normantranscript.com/news/university_of_oklahoma/part-norman-psychology-professor-rears-chimps/article_90febf6b-4250-544b-882f-e2c50e82472e.html. Accessed July 13, 2019

Hampton J (2016b) Remembering Norman's Chimps. Part IV: Lack of funding leads to end of chimp program at OU. The Norman Transcript, Norman, Oklahoma. https://www.normantranscript.com/news/university_of_oklahoma/part-lack-of-funding-leads-to-end-of-chimp-program/article_899e45f7-a9f1-5edc-8efc-21c625a1a280.html. Accessed July 13, 2019

Hampton E (2017) How World War I revolutionized medicine. The Atlantic, Feb 24, 2017. https://www.theatlantic.com/health/archive/2017/02/world-war-i-medicine/517656/. Accessed July 14, 2019

Haraway D (1989) Primate visions: gender, race, and nature in the world of modern science. Routledge, London

Haslerud GM (1977) Reminiscences of the Yerkes Orange Park Laboratory in the mid-1930s. In: Bourne GH (ed) Progress in ape research. Elsevier, New York, pp 21–27

Henry JP, Mosley JD (1963) Results of the Project Mercury ballistic and orbital chimpanzee flights. NASA, SP-39, 71

Hess E (2008) Nim Chimpsky: the chimp who would be human. Bantam Dell, New York

Hevesi R (2022) The welfare of primates kept as pets and entertainers. In: Robinson LM, Weiss A (eds) Nonhuman primate welfare: from history, science, and ethics to practice. Springer, Cham, pp 121–144

Hilgard ER (1965) Robert Mearns Yerkes, 1867–1956, A biographical memoir. Biographical Memoirs 36:385–425

Hillis WD (1961) An outbreak of infectious hepatitis among chimpanzee handlers at a United States Air Force base. Am J Epidemiol 73(3):316–328. https://doi.org/10.1093/oxfordjournals.aje.a120191

Hobson WC, Coulston F, Faiman C et al (1976) Reproductive endocrinology of female chimpanzees: a suitable model of humans. J Toxicol Environ Health 1(4):657–668. https://doi.org/10.1080/15287397609529364

Hobson WC, Graham CE, Rowell TJ (1991) National chimpanzee breeding program: Primate Research Institute. Am J Primatol 24(3-4):257–263. https://doi.org/10.1002/ajp.1350240311

Honess P (2016) A brief history of primate research: Global health improvements and ethical challenges. Arch Med Biomed Res 2(4):151–157. https://doi.org/10.4314/ambr.v2i4.7

Hosey G (2022) The history of primates in Zoos. In: Robinson LM, Weiss A (eds) Nonhuman primate welfare: from history, science, and ethics to practice. Springer, Cham, pp 3–30

Institute of Medicine (IOM) (1990) Funding health sciences research: a strategy to restore balance. Washington, DC

Institute of Medicine (IOM) (2011) Chimpanzees in biomedical and behavioral research: assessing the necessity. National Academic Press, Washington, DC

Jacobsen CF, Wolfe JB, Jackson TA (1935) An experimental analysis of the functions of the frontal association areas in primates. J Nerv Ment Dis 82(1):1–14. https://doi.org/10.1097/00005053-193507000-00001

Kabasawa A (2011) The chimpanzees of West Africa: From "man-like beast" to "our endangered cousin". In: Matsuzawa T, Humle T, Sugiyama Y (eds) The chimpanzees of Bossou and Nimba. Springer, New York, pp 45–57

Kahn A (2005) Regaining lost youth: the controversial and colorful beginnings of hormone replacement therapy in aging. J Gerontol Ser A Biol Med Sci 60(2):142–147. https://doi.org/10.1093/gerona/60.2.142

Knight A (2008) The beginning of the end for chimpanzee experiments? Philos Ethics Humanit Med 3:16. https://doi.org/10.1186/1747-5341-3-16

Kozminski MA, Bloom DA (2012) A brief history of rejuvenation operations. J Urol 187(3):1130–1134. https://doi.org/10.1016/j.juro.2011.10.134

Krause RM (1996) Metchnikoff and syphilis research during a decade of discovery, 1900-1910. ASM News 62(6):307–310

Layman EJ (2009) Human experimentation: historical perspective of breaches of ethics in US health care. Health Care Manag 28(4):354–374. https://doi.org/10.1097/HCM.0b013e3181bddbc2

Ley R (1990) A whisper of espionage: Wolfgang Köhler and the apes of Tenerife. Avery Publishing Group, Inc, New York

Lück H, Jaeger S (1988) Wolfgang Köhler and the Anthropoid Research Station on Tenerife Island. Revista de Historia de la Psicología 9(2–3):295–308

Macilwain C (1997) Closure of primate lab angers both researchers and critics. Nature 390(6658): 321. https://doi.org/10.1038/36921

Mizukoshi E, Nascimbeni M, Blaustein JB et al (2002) Molecular and immunological significance of chimpanzee major histocompatibility complex haplotypes for hepatitis C virus immune response and vaccination studies. J Virol 76(12):6093–6103. https://doi.org/10.1128/JVI.76.12.6093-6103.2002

Morimura N, Idani G, Matsuzawa T (2011) The first chimpanzee sanctuary in Japan: an attempt to care for the "surplus" of biomedical research. Am J Primatol 73(3):226–232. https://doi.org/10.1002/ajp.20887

Morimura N, Hirata S, Matsuzawa T (2022) Challenging cognitive enrichment: examples from caring for the chimpanzees in the Kumamoto Sanctuary, Japan and Bossou, Guinea. In: Robinson LM, Weiss A (eds) Nonhuman primate welfare: from history, science, and ethics to practice. Springer, Cham, pp 489–516

Mueller WF, Coulston F, Korte F (1985) The role of the chimpanzee in the evaluation of the risk of foreign chemicals to man. Regul Toxicol Pharmacol 5(2):182–189. https://doi.org/10.1016/0273-2300(85)90031-5

Mukherjee S (2010) The emperor of all maladies: a biography of cancer. Scribner, New York

Naquet R, Killam KF, Rhodes JM (1967) Flicker stimulation with chimpanzees. Life Sci 6(15):1575–1578. https://doi.org/10.1016/0024-3205(67)90166-X

National Center for Research Resources (NCRR) (2007) Report of the Chimpanzee Management Plan Working Group. Washington, DC

National Institutes of Health (NIH) (2013) Council of Councils Working Group on the use of chimpanzees in NIH-supported research. Washington, DC. https://dpcpsi.nih.gov/council/pdf/FNL_Report_WG_Chimpanzees.pdf. Accessed July 14, 2019

National Institutes of Health (NIH) (2018) Council of Councils Working Group on assessing the safety of relocating at-risk chimpanzees. Washington, DC. https://dpcpsi.nih.gov/council/chimpanzee_advice. Accessed July 14, 2019

National Institutes of Health (NIH) (2019) Chimpanzee health categorization framework: Harmonized across NIH-supported facilities. Washington, DC. Available at: https://dpcpsi.nih.gov/council/chimpanzee_advice. Accessed July 14, 2019

National Research Council (NRC) (1998) The psychological well-being of nonhuman primates. National Academy Press, Washington, DC https://dpcpsi.nih.gov/sites/default/files/orip/document/Report%20of%20the%20Chimpanzee%20Management%20Plan%20Working%20Group%20.pdf. Accessed January 12, 2017

Novembre FJ, Saucier M, Anderson DC et al (1997) Development of AIDS in a chimpanzee infected with human immunodeficiency virus type 1. J Virol 71(5):4086–4091. https://doi.org/10.1128/jvi.71.5.4086-4091.1997

Prince AM (1999) Virulent HIV strains, chimpanzees, and trial vaccines. Science 283(5405):1117. https://doi.org/10.1126/science.283.5405.1115e

Prince AM, Brotman B (2001) Perspectives on hepatitis B studies with chimpanzees. ILAR J 42(2):85–88. https://doi.org/10.1093/ilar.42.2.85

Prince AM, Brotman B, Lee D et al (2005) Protection against chronic hepatitis C virus infection after rechallenge with homologous, but not heterologous, genotypes in a chimpanzee model. J Infect Dis 192(10):1701–1709. https://doi.org/10.1086/496889

Reardon S (2015) NIH to retire all research chimpanzees. Nature. https://doi.org/10.1038/nature.2015.18817

Rice TW (2008) The historical, ethical, and legal background of human-subjects research. Respir Care 53(10):1325–1329

Riesen AH (1947) The development of visual perception in man and chimpanzee. Science 106(2744):107–108. https://doi.org/10.1126/science.106.2744.107

Riesen A (1977) Introduction. In: Bourne G (ed) Progress in ape research. Academic Press, New York

Rossiianov K (2002) Beyond species: Il'ya Ivanov and his experiments on cross-breeding humans with anthropoid apes. Sci Context 15(2):277–316. https://doi.org/10.1017/S0269889702000455

Rosvold HE, Szwarcbart MK, Mirsky AF, Mishkin M (1961) The effect of frontal-lobe damage on delayed-response performance in chimpanzees. J Comp Physiol Psychol 54(4):368–374. https://doi.org/10.1037/h0043331

Rumbaugh DM (1977) The emergence and state of ape language research. In: Bourne G (ed) Progress in ape research. Academic Press, New York, pp 75–83

Sakai A, Claire MS, Faulk K et al (2003) The p7 polypeptide of hepatitis C virus is critical for infectivity and contains functionally important genotype-specific sequences. Proc Natl Acad Sci U S A 100(20):11646–11651. https://doi.org/10.1073/pnas.1834545100

Sarabian C, Ngoubangoye B, MacIntosh AJJ (2017) Avoidance of biological contaminants through sight, smell and touch in chimpanzees. R Soc Open Sci 4(11):170968. https://doi.org/10.1098/rsos.170968

Spragg SDS (1977) Reminiscences of early days in New Haven and Orange Park. In: Bourne G (ed) Progress in ape research. Elsevier, New York, pp 39–42

Stephens ML (1995) Chimpanzees in laboratories: distribution and types of research. ATLA-Altern Lab Anim 23(5):579–583

Su AI, Pezacki JP, Wodicka L et al (2002) Genomic analysis of the host response to hepatitis C virus infection. Proc Natl Acad Sci U S A 99(24):15669–15674. https://doi.org/10.1073/pnas.202608199

Tobler LH, Busch MP (1997) History of posttransfusion hepatitis. Clin Chem 43(8):1487–1493. https://doi.org/10.1093/clinchem/43.8.1487

Turner PV (2017) International regulations and legislation for captive primate care. In: Fuentes A (ed) International encyclopedia of primatology, vol 1. Wiley-Blackwell, Ames, IA, pp 148–157

U.S. Chimp Haven is Home Act (2007) Public Law 110-170. https://www.congress.gov/bill/110th-congress/senate-bill/1916?r=65. Accessed 15 January 2017

U.S. Chimpanzee Health Improvement, Maintenance and Protection Act (2000) Public Law 106-551. https://www.gpo.gov/fdsys/pkg/PLAW-106publ551/html/PLAW-106publ551.htm. Accessed January 15, 2017

U.S. Congress (2011) The Great Ape Protection and Cost Savings Act (H.R.1513/S.810). Reintroduced on April 18, 2011, to the 112th Congress in both the U.S. House of Representatives and the Senate

U.S. Department of Agriculture (1985) Animal Welfare Act

U.S. Department of Agriculture (2017) Animal Welfare Act and Animal Welfare Regulations. National Agricultural Library. http://awic.nal.usda.gov/government-and-professional-resources/federal-laws/animal-welfare-act

U.S. Fish and Wildlife Service (U.S. FWS) (2015) Endangered and threatened wildlife and plants; Listing all chimpanzees as endangered species. 50 CFR Part 17, Federal Register 80(15):34500

van Akker R, Balls M, Eichberg JW et al (1994) Chimpanzees in AIDS research: a biomedical and bioethical perspective. J Med Primatol 23(1):49–51. https://doi.org/10.1111/j.1600-0684.1994.tb00095.x

VandeBerg JL, Zola SM (2005) A unique biomedical resource at risk. Nature 437(7055):30–32. https://doi.org/10.1038/437030a

Wadman M (1999) Financial doubts over future of chimp lab. Nature 398(6729):644. https://doi.org/10.1038/19353

Wolfle TL (1999) Psychological well-being of nonhuman primates: a brief history. J Appl Anim Welf Sci 2(4):297–302. https://doi.org/10.1207/s15327604jaws0204_4

Wynne CDL (2008) Rosalià Abreu and the apes of Havana. Int J Primatol 29:289–302. https://doi.org/10.1007/s10764-008-9242-0

Yerkes RM (1916) Provision for the study of monkeys and apes. Science 43(1103):231–234. https://doi.org/10.1126/science.43.1103.231

Yerkes RM (1925) Almost human. Century/Random House UK, London

Yerkes RM (1940) Laboratory chimpanzees. Science 91(2362):336–337. https://doi.org/10.1126/science.91.2362.336

Yerkes RM (1943) Chimpanzees: a laboratory colony. Yale University Press, New Haven

Yerkes DN (1977) Home life with chimpanzees. Part I. In: Bourne G (ed) Progress in ape research. Academic Press, New York, pp 5–7

Yerkes RM, Learned BW (1926) Chimpanzee intelligence and its vocal expressions. Williams and Wilkins Co, Baltimore

Zinser LM (2014) Pilot logbook lies and more. XLibris, Bloomington, IN

Using Primates in Captivity: Research, Conservation, and Education

Mark J. Prescott

Abstract

Nonhuman primates (henceforth, primates) are among the most extensively studied animal species on the planet. In this chapter I provide a brief overview of the reasons why primatology is a popular and thriving science, why primates are valuable research subjects, the scientific disciplines in which they are used, and the species and numbers involved. Globally, an estimated 100,000 to 200,000 primates are used in research every year, the majority (an estimated two-thirds) being long-tailed and rhesus macaques used mainly for pharmaceutical development, neuroscience, and infectious disease studies. I give examples of what primates may experience as part of their involvement in regulated and unregulated scientific procedures and outline how the associated ethical and welfare issues are typically addressed. Although primate research projects are conducted in a variety of settings, special attention is given in this chapter to laboratory- and zoo-based research. The role of zoos in primate conservation and education is also discussed. I conclude with some broad principles for good practice in the design, conduct, and reporting of primate research, aimed principally at students and early career scientists. Adoption of high scientific and ethical standards is important for continued funding and public support for primate research, and for garnering maximum value from it.

Keywords

Ethics · Experimental design · Primatology · Research methods · Welfare · 3Rs

M. J. Prescott (✉)
National Centre for the Replacement, Refinement and Reduction of Animals in Research (NC3Rs), London, UK
e-mail: mark.prescott@nc3rs.org.uk

1 Why Study Primates?

> Then join our leaping lines that scumfish through the pines,
> That rocket by where, light and high, the wild-grape swings,
> By the rubbish in our wake, and the noble noise we make,
> Be sure, be sure, we're going to do some splendid things!
> —Extract from Road-Song of the Bandar-Log [monkey-people]
>
> The Jungle Book (Kipling 1894)

Primates are among the most extensively studied animals on the planet. For many students and scientists studying these fascinating animals, their motivation is to advance knowledge about the biology, behavior, and evolution of their study species because this information has intrinsic interest. Other researchers conduct more applied research of direct relevance to the management and welfare of primates in captivity, or the conservation of threatened and endangered primate species. However, the most extensive use of primates is in biomedical research and testing, where these animals act as models of humans, owing to similarities in our anatomy, genetics, biochemistry, physiology, and behavior. Primates have played a crucial role in improving our understanding of normal and abnormal body function and in the development of treatments and therapies for disease, where they often serve as the final test system for assessment of the safety and efficacy of potential medicines before human clinical trials (Phillips et al. 2014; Weatheall et al. 2015; Scientific Committee on Health, Environmental and Emerging Risks 2017). I have been fortunate in my career to work in all three of these domains, which has helped to inform the perspectives outlined in this chapter. My experience in the zoo, field, and laboratory, from basic research in behavior and ecology, through conservation, to animal welfare, illustrates the multidisciplinary nature of primatology and the opportunities this provides for collaboration and career development.

There are many reasons for specializing in the study of primates. The order Primates is a diverse one, including over 500 species from the tiny 30 g Berthe's mouse lemur (*Microcebus berthae*) to the huge 100–250 kg Eastern gorilla (*Gorilla beringei*). Most primate species are highly social, active, and dexterous, with extensive behavioral repertoires, making them engaging animals to watch and study. The rich behavior, intelligence, adaptability, and complex social relationships of these relatively large-brained mammals provide fertile ground for varied research questions in ethology, psychology, and anthropology. To quote Allison Jolly (1985, p. 230) *"primates are animals that love and hate and think."*

As our closest relatives, the study of primates, both living and extinct, can tell us a great deal about ourselves and our evolution from ancestral primates. This close phylogenetic relationship, and its implications for the way in which primates experience the world, also raises important questions about whether they should be used in research and how we should treat them in captivity. Hence, the use of primates in invasive biomedical research and their confinement and breeding in zoos are contentious issues (Bailey 2022; Gruen and Fleury 2022; Kreger and Mench 1995; Regan 1996; Boyd Group 2002; Weatheall et al. 2015)

More than half of the world's apes, monkeys, lemurs, and lorises are now threatened with extinction as agriculture and industrial activities destroy forest habitats and the animals' populations are hit by hunting and trade (Estrada et al. 2017). The more knowledge we have about the ecology and behavior of these species, and of the anthropogenic challenges that they face, the better able we will be to conserve them.

Being tropical or subtropical animals, and predominantly arboreal, the study of primates in their natural habitats involves, for many scientists, travel to exotic forested environments, bringing a sense of adventure. This attracts many early career scientists to primatology. The short time I spent censusing New World monkey populations in the Bolivian rainforest was an incredibly stimulating sensory and zoological experience. Being in their world as opposed to vice versa profoundly changed my understanding of the New World monkeys that hitherto I had studied only in the zoo environment (Buchanan-Smith et al. 2000). Without this fieldwork, I could not have gained a proper appreciation of their ecological niche and adaptations to it, such as their superb large-scale navigational abilities in the dense, three-dimensional forest environment, so far in excess of my own. A greater understanding of New World monkey ecology subsequently enabled me to provide more appropriate environments for these animals in captivity, designed to meet species-specific needs and support good welfare.

Within biomedical research, primates are generally used where no other nonhuman species is suitable (e.g., where only they are susceptible to the disease under study), or where they possess the biological characteristics that allow investigation of a scientific question (e.g., in basic neuroscience research, where the parts of the brain under investigation, and the relevant cognitive or behavioral abilities, are absent or less well developed in other species). However, that is not to say that primates are necessarily always the best animal model or always predictive of responses in humans. For example, consider the phase 1 clinical trial of the humanized monoclonal antibody TGN1412, which induced a severe inflammatory reaction ("cytokine storm") and systemic organ failure in the healthy human volunteers; effects that were not predicted by the preclinical safety testing in cynomolgus macaques using doses around 500 times larger (Eastwood et al. 2010).

2 Primate Research Settings

Primates are used in a variety of research settings. Some readers may dislike the word "used" in this context, but I consider it appropriate because the individual animals involved cannot consent to the research, and generally do not voluntarily participate in, or benefit from, it (though there are exceptions, e.g., Matsuzawa 2007; Whitehouse et al. 2013; Hopper et al. 2016). Studies are conducted in the field with primate populations in their natural habitats, in the laboratory under highly controlled conditions and limited space, in zoos and sanctuaries characterized by often large numbers of unfamiliar human visitors, and in semi-free ranging conditions where habitats and social structures are replicated in a managed setting. Each setting has strengths and limitations from a scientific perspective, and all raise ethical and welfare issues. Some of the ways in which these issues can be addressed are presented later in this chapter and in more detail in other chapters in this volume (Baker and Farmer 2022; Bayne et al. 2022; Buchanan-Smith et al. 2022).

A great deal of primate research is conducted in zoos, which have a long history of keeping primates for exhibition (Hosey 2022; Gippoliti 2006). Zoo-based research has contributed to improvements in primate husbandry, nutrition, health, welfare, breeding, and conservation. It has also played a role in improving understanding of primate behavior and cognition. Zoos provide the opportunity to address research questions not easily studied in the wild, permitting controlled experiments to test hypotheses, such as in my own research on the adaptive value of mixed species associations and color vision polymorphism in tamarins (*Saguinus* spp.) (Prescott and Buchanan-Smith 1999, 2002; Prescott et al. 2005; Smith et al. 2012). In most cases, however, zoo-based ethologists must consider the possibility that the behavior of their research subjects may be influenced by, for example, the presence, density, and activity of zoo visitors or the enclosure design and husbandry practices, and take this into account during the design of their experiments and when drawing conclusions (Hosey 2005). Both captive and field studies have their limitations and advantages. They provide complementary insights into the function and mechanisms of primate behavior, and ideally information gained from one should inform the other (Hardie et al. 2003; Janson and Brosnan 2013).

Many zoo-based primate research projects are collaborations between zoo staff, academics from local universities, and these academics' students. Some of the most productive collaborations have arisen from university research and public engagement centers situated within zoos (Bowler et al. 2012; Waller et al. 2012). This includes the Max Planck Institute's Wolfgang Köhler Primate Research Center at Leipzig Zoo (Wolfgang Köhler Primate Research Center 2017) and the Living Links to Human Evolution Research Centre at Edinburgh Zoo, used by scientists at the Universities of St Andrews, Stirling, Edinburgh, and Abertay (Living Links 2021). Developments such as these reflect the research, education, and conservation objectives of modern, professionally accredited zoos. Indeed, there is a legislative requirement in some jurisdictions for zoos to engage in such activities (European Commission 2015).

Modern zoos have an essential role in the conservation of endangered primates through their involvement in captive breeding programs and the preservation of genetic diversity. Increasingly these *ex situ* conservation efforts are supplemented by crucial work *in situ* to protect or replenish natural habitats, monitor species in the wild, and engage local communities (Minteer and Collins 2013). The Association of Zoos and Aquariums (AZA) (2016) has recommended that zoos allocate at least 3% of their budget to field conservation. Although the long-term impact of zoo-based re-introduction projects is still to be determined, there have been some notable successes with primate species (e.g., golden lion tamarin *Leontopithecus rosolia*: Kierulff et al. 2012; Ferreira et al. 2022 Western lowland gorilla *G. gorilla gorilla*: King et al. 2012, 2014). The public engagement and educational activities of zoos can also contribute to conservation by direct fund raising and by improving awareness of the threats to biodiversity and how to best support conservation efforts (Moss et al. 2015). The role of urban zoos in enabling people to get face-to-face with animal species and in fostering personal connections with the animal kingdom should not be underestimated. My own scientific interest in animal behavior and welfare, and ultimately my career in the animal sciences, was forged during childhood visits to the local zoo, and many colleagues would say the same. It is important, however, to acknowledge the potential negative impact of the visiting public on the welfare of zoo animals (Hosey 2005; Farmer et al. 2022).

3 Disciplines, Species, Origins, and Numbers

Primates are used across a wide variety of scientific disciplines, including ethology, ecology, conservation, anthropology, psychology, anatomy, neuroscience, neurology, pharmacology, toxicology, microbiology, immunology, genetics, biochemistry, reproduction, endocrinology, ophthalmology, dentistry, surgery, transplantation, veterinary medicine, and animal welfare science. Carlsson et al. (2004) assessed the global use in research of primates and primate-derived biological material by reviewing 4411 studies reported in 2937 peer-reviewed journal articles in 2001. The most common areas of primate research were microbiology (including HIV/AIDS research; 26%), neuroscience (19%), and biochemistry/chemistry (12%). Most (84%) of this research was conducted in North America, Europe, and Japan. Literature reviews have also been undertaken to survey primate use in Asia and Sweden (Hagelin 2004, 2005). One limitation of this bibliometric approach is that industry studies using primates are far less likely to be published than those funded by public sources (Lexchin et al. 2003).

More recently, the Association of Primate Veterinarians (APV) surveyed its members regarding biomedical research involving primates in the USA and Canada. The most common uses for primates included pharmaceutical research and development, and neuroscience, neurology, or neuromuscular disease research (Lankau et al. 2014). This picture is broadly in accordance with that for the European Union, where the largest fraction of primate use (73%) is for safety assessment (toxicity testing) of potential new pharmaceuticals to meet regulatory requirements (Scientific Committee on Health, Environmental and Emerging Risks 2017). The National Centre for the Replacement, Reduction and Refinement of Animals in Research (NC3Rs) reviews all primate research proposals submitted to the UK's major public funders of bioscience research, to help ensure that the 3Rs (i.e., the replacement, reduction, and refinement of animal use) are implemented in the proposed research. Neuroscience, drug development, and vaccinology are the disciplines which receive the largest number of proposals and funding (Fig. 1).

The most commonly used species in biomedical research worldwide are the long-tailed macaque, also known as the crab-eating or cynomolgus macaque (*Macaca fascicularis*) and rhesus macaque (*Macaca mulatta*) (Carlsson et al. 2004; Lankau et al. 2014). Eighty percent of the regulated procedures using primates in the European Union in 2014 were performed on long-tailed macaques (Scientific

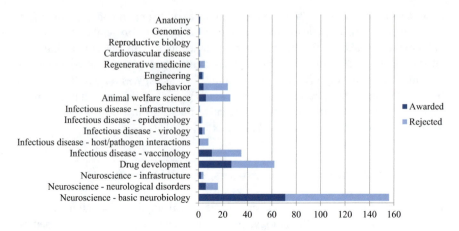

Fig. 1 Number, discipline, and outcome of research proposals involving the use of primates submitted to the major UK public funders of bioscience research and reviewed by the NC3Rs under its peer review service to date. (370 proposals reviewed from September 2004 to September 2016, 169 Wellcome Trust, 102 Medical Research Council, 66 Biotechnology and Biological Sciences Research Council, 22 NC3Rs, 11 other public funders. Classification conducted by the author based on the proposal titles and abstracts)

Committee on Health, Environmental and Emerging Risks 2017). Neither of these macaque species is endangered, being listed as "least concern" on the IUCN Red List (International Union for Conservation of Nature 2021). Other macaque species, African green or vervet monkeys (*Chlorocebus aethiops*), baboons (*Papio* spp.), and New World monkey species, including common marmosets (*Callithrix jacchus*), are also used. Most of these animals are purpose-bred for research. Macaques used in the pharmaceutical industry are most often imported, predominantly from breeding centers in Asia, whereas academia tends to use domestic sources (Lankau et al. 2014; Home Office 2016). Common chimpanzees (*Pan troglodytes*) and other apes are these days rarely used in regulated scientific procedures, largely due to changes in legislation (Turner 2022; European Commission 2010; Kaiser 2015). None have been used in the European Union since 1999 and invasive research with chimpanzees in the USA has been winding down since 2015, following policy change by the National Institutes of Health and Fish and Wildlife Service. Of course, they and other types of primate continue to be used in unregulated projects aimed at understanding their biology and behavior, improving their health and welfare, and species preservation.

Data on the total number of primates used annually in research and testing is incomplete. Based on their survey of the published literature, Carlsson et al. (2004) estimated between 100,000 and 200,000 primates are used globally per year. Lankau et al. (2014) report that, according to US Department of Agriculture records, primate use in biomedical research in the USA increased during 2000 to 2010, peaking at more than 71,000 primates during the 2010 fiscal year, and has since decreased (declining from 2010 levels by 9% in 2011 and 11% in 2012). China has emerged as a major region for primate research (Zhang et al. 2014; Cyranoski 2016), but figures on the number of primates used in China are not available. According to the latest available statistics for the European Union, primate use decreased from around 10,000 in 2008 to around 6000 in 2011 (Scientific Committee on Health, Environmental and Emerging Risks 2017), but numbers fluctuate year-on-year. Many factors contribute to national and regional trends in animal use, including changes in the research priorities of major funding bodies, the development of alternatives to animal experiments, and economic factors, such as the outsourcing of research studies overseas. The preponderance of biopharmaceuticals in the drug development pipeline, and their species- and target-specificity, may lead to a future increase in industry use of primates (Buckley et al. 2011). However, considerable effort is being put into stemming this increase by questioning the scientific rationale for primate use on a case-by-case basis and via the use of more efficient study designs. For example, working with the pharmaceutical industry and regulatory bodies, the NC3Rs has identified opportunities to reduce by up to 64% the number of monkeys required per monoclonal antibody in development (Chapman et al. 2012).

4 Welfare, Ethics, and Legislation

The experience of primates used in research, and the impact on their welfare, can be very different because of the large diversity of disciplines, procedures, and settings in which they are used. Differences in legislation and in awareness of the available opportunities to refine primate use and care also contribute to variation in research and husbandry practices and hence animal welfare. Within the European Union, scientific procedures likely to cause pain, suffering, distress, or lasting harm equivalent to, or higher than, that caused by the introduction of a needle in accordance with good veterinary practice, are regulated under Directive 2010/63/EU (European Commission 2010). Procedures above this threshold are classified as mild, moderate, severe, or non-recovery. The vast majority of primates are used in mild and moderate severity procedures (93–99%, Carlsson et al. 2004; Home Office 2016).

Under the Directive, primates may be re-used if the previous procedure was of mild or moderate severity, the animal's state of health and well-being has been fully restored, and the further procedure is classified as mild, moderate, or non-recovery. However, in many biomedical experiments, the animals are euthanized as an integral part of the experiment (e.g., for analysis of clinical pathology, or confirmation of recording/stimulation sites) or because humane endpoints have been reached (Wolfensohn 2022; Lankau et al. 2014). In circumstances where euthanasia is not required, and primates are not reused, a minority are retired to sanctuaries, where this is judged to be in the best welfare interests of the individual animals concerned (Prescott 2006; Kerwin 2006). However, this practice is not common, partly due to the paucity of new homes with suitable conditions.

Toxicity tests of pharmaceuticals are among the most common studies involving primates. Such tests are typically 1–4 weeks (acute) or 6–9 months (chronic) in duration and involve dosing of the test substance (once or repeatedly, usually by the

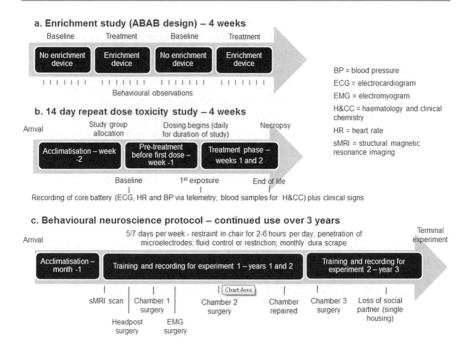

Fig. 2 Schematic representation of events in three common primate study types: (**a**) behavioral study to evaluate the impact on animal welfare of a novel environmental enrichment device; (**b**) 14-day repeat dose toxicity study to characterize the toxicological effects of a test compound following repeated administration (adapted from Tasker 2012); (**c**) long-term neuroscience study in the awake, behaving state to understand the causal relationship between neuronal activity and cognitive function

intended route of clinical administration), sampling of blood and possibly other body fluids at regular intervals for hematology and clinical chemistry, and observations of clinical signs in group housed animals (Walker et al. 2007) (Fig. 2). There may also be recording of parameters such as electrocardiogram, heart rate, and blood pressure, generally via external telemetry, and sometimes involving temporary separation from the group (Prior et al. 2016). Contingent harms may arise from, for example, international supply and transport (Prescott and Jennings 2004). Such protocols are generally classified as mild or moderate depending on the dosing and sampling regimen and the expected clinical effects of the test substance (Animal Procedures Committee 2002; Expert Working Group on Severity Classification Criteria 2009).

In contrast, most neuroscience studies involving electrophysiological recording of brain activity in the awake, behaving state—another common study type—typically last several years. The basic paradigm involves surgeries under anesthesia (with analgesia and post-operative care) to implant a post into the skull for head fixation and one or more recording chambers for access to the brain (head implant), daily penetration of electrodes while being restrained for 2–6 h in a specifically-designed chair, and control of food and/or fluid to motivate the animals to work reliably and for extended periods for small food or fluid rewards in behavioral or cognitive tasks (Prescott et al. 2010; McMillan et al. 2017) (Fig. 2). In the UK, such long-term protocols have a prospective severity classification of severe. This reflects the seriousness of the adverse effects or complications that can occur in a minority of animals, and the requirement under Directive 2010/63/EU for severity classification to take into account the lifetime experience of animals, the duration frequency and multiplicity of harmful techniques, the potential of cumulative suffering within a procedure, and the application of refinement techniques (European Commission 2010; Pickard et al. 2013).

Perspectives on the moral acceptability of invasive primate experiments vary considerably within the scientific community (t'Hart et al. 2022; Gruen and Fleury 2022; Prescott 2010) and there is undoubtedly major concern about this use among the public (European Commission 2006; von Roten 2013; Clemence and Leaman 2016). Regardless of the ethical perspective one favors, there is strong agreement that primates and other animals matter morally, and therefore as a minimum they should only be used where there are no alternatives, and their welfare should be optimized (Boyd Group 2002; Scientific Committee on Health, Environmental and Emerging Risks 2017). This position is reflected in national legislation on the protection of animals used in science which mandates that the 3Rs be applied in the design and conduct of primate research to minimize both animal use and suffering, and that the human interest in obtaining some benefit for mankind (e.g., alleviating or preventing human suffering, furthering scientific knowledge) must be balanced against the interests of the primates in avoiding harm (Bayne et al. 2022; Bayne and Morris 2012; Chapman et al. 2015; Graham and Prescott 2015; Prescott 2010, 2017). This so-called harm-benefit assessment, based on utilitarian ethics, is usually performed case-by-case for each proposed research project by the regulatory authority and/or institutional ethics committee before deciding whether to authorize the project and on what terms (Home Office 2015; Brønstad et al. 2016; Laber et al. 2016). However, some ethicists have questioned whether the utilitarian approach is the most appropriate one for primate research, given the capacities of these animals (Quigley 2007; Rossi 2009). Just because there is a scientific rationale and the potential for medical benefit, and because harm will be minimized, this still does not necessarily make it right to proceed. Regulatory authorities in some regions have gone further to protect primates, establishing limits and controls on their use. For example, Directive 2010/63/EU places an effective ban on the use of great apes in scientific procedures (save for exceptional circumstances), prohibits the use of wild-caught primates, and requires more frequent inspections of establishments keeping or using primates than for other species (European Commission 2010).

Most zoo-based primate research falls below the threshold required for a license under legislation on the protection of animals used in science and is focused on observation of behavior (e.g., studies aimed at evaluating improvements in enclosure design, the effects of visitors, or cognitive capabilities such as theory of mind). However, this depends on the jurisdiction; for example, in Australia even observational research is required to be licensed. Where physiological variables are measured to assess health, welfare, or reproductive status, the methods used are often non-invasive (e.g., analysis of salivary, fecal, or hair cortisol as an index of stress), or else the research capitalizes on invasive procedures performed during routine health checks and other recognized veterinary practices. In addition, most projects are non-terminal and aimed at benefiting the welfare and/or survival of captive and wild populations of the species. Nonetheless, there are instances in which research use may conflict with the welfare interests of the animals involved and, as in the biomedical field, zoos have established institutional ethics committees to review such situations and decide on an appropriate course of action. Such committees are recommended in the UK under Secretary of State's Standards of Modern Zoo Practice (Department for Environment, Food and Rural Affairs 2012) and typically also consider ethical and welfare issues raised by the transport, relocation, maintenance, breeding, and disposal of zoo-housed primates. For further information on criteria for the evaluation of zoo research projects and the regulatory framework for zoos, see Kleiman (1985) and Hosey et al. (2013).

Aside from scientific procedures, the housing, husbandry, and social conditions provided for captive primates are key determinants of their welfare (Coleman et al. 2022; Talbot et al. 2022). There is an increased interest from regulatory, funding, and accrediting agencies to ensure improved conditions for captive primates, and these are helping to raise standards (NC3Rs 2006; Coleman et al. 2011; Whittaker and Laule 2012). For example, the proportion of primates housed socially in laboratories has increased in recent years (Chapman et al. 2015; Bennett 2016; Baker 2016). The welfare of zoo-housed primates tends to be better than those in laboratories, but caregivers for primates in laboratories (including breeding/supplying centers) and zoos are dealing with many of the same animal management challenges (e.g., providing adequate nutrition; environmental enrichment to support physical and psychological well-being; training for cooperation with husbandry using positive reinforcement techniques; minimization of inbreeding) (McCann et al. 2007; Prescott and Buchanan-Smith 2003, 2004, 2007). It is frustrating, therefore, that there is not greater information sharing between laboratories and zoos. National and international primatological societies could do more to encourage this, for example by organizing cross-sector scientific seminars and/or exchange visits.

Welfare and ethics are also important in field primatology. Making trails/transects through habitats to access, observe, and census primate populations can cause disturbance and open habitats for hunting for bushmeat. Studies involving capture, handling, sampling of tissues, and radio-tracking have obvious welfare implications. Even habituation to human researchers can put the animals in danger, for example by exposing the primates to infectious diseases, hunters, or trappers (Fedigan 2010; Gruen et al. 2013). The overarching principle should be to minimize harm to the animals, ecosystem, and people involved (Riley et al. 2014).

5   Good Research Practice

For primate research to be legitimate, ethical, and high quality, it is important not only to minimize harm to the animals involved, but also to ensure that the research is methodologically sound and that there is effective dissemination of the findings. This is particularly important in today's competitive funding environment and given increasing emphasis by funding bodies on issues such as application of the 3Rs, scientific rigor, and achieving impact (National Institutes of Health 2016; NC3Rs 2006; Bateson et al. 2011; Research Councils UK 2015). Below I offer some broad principles for good practice in the design, conduct, and reporting of primate research. It goes without saying that researchers should ensure they have the necessary licenses and permissions, and adequate funds in place, before commencing their projects.

1. *Consider the need to use primates, and search the published literature for alternatives.*
   Researchers should carefully consider at the outset the need to use primates to address their research question or questions. Clearly, there are instances where there can be no substitute, for example, where the research aims to understand the behavior of a particular primate species, or to learn about the ecology of a primate population in the field. In other cases, alternative approaches may be available (whether animal or non-animal), which can be advantageous from an ethical or scientific perspective, or both. Hence, there should always be fair-minded appraisal of potential alternatives to primate use (Burm et al. 2014; Scientific Committee on Health Environmental and Emerging Risks 2017). Biomedical researchers should be wary of the often disingenuous and misleading information from both pro- and anti-vivisection organizations about the availability or lack of alternatives to primates, and instead conduct a thorough search of the literature, including mainstream biomedical (e.g., PubMed) and specialist 3Rs-specific databases (e.g., DB-ALM, ALTWEB). This will also help to avoid unnecessary duplication of primate studies (as opposed to circumstances where replication of previous work is scientifically important). Project evaluators should also be sensitive to areas where replacement opportunities have been identified and to take these opportunities into account in their work. Where use of primates is required, conducting a formal systematic review of the literature can improve the value of the planned work and enable researchers to apply the 3Rs, for example by avoiding the use of uninformative models (e.g., those that are poorly predictive of responses in man) or supporting a reduction in animal numbers and/or suffering by helping to inform sample size calculations and choice of outcome measures (Petticrew and Davey Smith 2012; Irvine et al. 2015). For advice and guidance on conducting systematic reviews and meta-analyses of preclinical studies, researchers may wish to consult the Systematic Review Facility (CAMARADES/NC3Rs 2017), which includes a free app to help researchers to utilize these methodologies.

2. *Possess or gain a sound understanding of the biology and behavior of the study species.*
   A sound knowledge of the biology and behavior of the study species underpins good science, animal management, and welfare, and can facilitate data collection and interpretation, regardless of the setting. For example, a good grasp of species-specific behavior will enable more successful dawn-to-dusk follows in the field, training for cooperation with scientific procedures in the laboratory, and environmental enrichment in the zoo and laboratory. Published literature, online resources, and scientific events are good sources of information (Jennings and Prescott 2009; Prescott 2016). For refreshing knowledge about the natural history and behavior of long-tailed and rhesus macaques, and of common marmosets, I recommend The Macaque Website (NC3Rs 2015), and the Common Marmoset Care site (University of Stirling/Primate Society of Great Britain/NC3Rs 2011). Both websites contain videos, images, and sound files to help users interpret the facial expressions, body postures, and vocalizations of these primate species. Primate veterinary associations also have useful information online, such as drug doses and normal physiological values for commonly used species (Association of Primate Veterinarians 2017). For some studies, it is important not only to have a good knowledge about the biology of the study species, but also knowledge about genetic (or other) differences within that species so that the populations and individuals most appropriate to the research questions will be chosen for research.
3. *Optimize the welfare of the animals used.*
   Optimizing the welfare of primates in our care, and effective implementation of refinement techniques, depend on the ability of staff to recognize signs of pain and distress, as well as signs that indicate positive welfare. Advice on welfare indicators and assessment is given in Part II of this volume. Animals free from unnecessary pain and distress yield better quality, reproducible data (Poole 1997). Evidence of the link between good welfare and good science is growing within primatology (e.g., Schnell and Gerber 1997; Prescott and Buchanan-Smith 1999; Reinhardt 2004; Graham et al. 2012). Therefore, there is a scientific as well as ethical imperative to minimize harm and promote good welfare (Buchanan-Smith et al. 2022; Ross et al. 2010; Graham and Prescott 2015). While most researchers take this responsibility seriously, they may not be aware of all of the available opportunities to refine primate use and care (e.g., Rennie and Buchanan-Smith 2006a, b, c; Jennings and Prescott 2009; NC3Rs 2010). Researchers who give inadequate attention to the 3Rs, persist with outdated practices and/or employ spurious arguments to defend the status quo, should be challenged by colleagues within their institution, not least because of the reputational risk (Brown et al. 2013). A good culture of care within the research institution is key to ensuring appropriate attitudes toward animals and the adoption of good practice (Coleman 2011).
4. *Ensure all staff members are adequately trained and competent, and that there are sufficient resources and infrastructure to enable best practice.*
   Appropriate training for staff working with primates is essential to ensure compliance with legislation, good quality science, and implementation of the 3Rs.

Students and early career researchers should ensure they have opportunities to acquire the scientific and practical skills required to carry out their research to a high standard. Scientists at all levels should undertake continuing professional development to keep abreast of the latest developments in primate use, care, and welfare, and follow best (not just locally accepted) practice. Publications, scientific symposia, training courses, and visits to other establishments all provide opportunities to expand knowledge and skills. Another requirement for best practice is access to sufficient resources and infrastructure. The UK bioscience funding bodies will consider requests in grant proposals for resources for implementing the 3Rs, and for attending events relevant to funded research and the 3Rs (NC3Rs/AMRC/BBSRC/Defra/EPSRC/MRC/NERC/Wellcome Trust 2008).

There is guidance available within the literature on established research methods used in primate studies, for example, on recording animal behavior (Martin and Bateson 2007), field methods such as census and survey techniques for assessing population density (Setchell and Curtis 2011), and the conduct of regulated procedures, such as dosing, sampling, anesthesia, and imaging (Wolfe-Coote 2005). Comprehensive guidance on conducting zoo-based science is available from the British and Irish Association of Zoos and Aquariums (BIAZA 2013). Its *Handbook of Zoo Research* covers study planning through to publication, with sections on behavior observation methods, ecological methods, surveys and questionnaires, and analysis of zoo records.

5. *Ensure robust experimental design, with appropriate sample sizes for adequate statistical power, and the use of measures to reduce subjective bias.*

Recent years have seen growing concern about the reliability and reproducibility of *in vivo* research (see articles in the *Nature* specials archive: *Challenges in irreproducible research*; Nature 2018). Poor experimental design and reporting have been implicated as major contributing factors (Academy of Medical Sciences/Biotechnology and Biological Sciences Research Council/Medical Research Council/Wellcome Trust 2015; Institute for Laboratory Animal Research 2015; Freedman et al. 2015). There is no reason to suspect that primate research is any better in this regard than other fields of research. Researchers using primates should ensure that they apply principles of good experimental design, statistical analysis, and reporting in their research, in order to increase the robustness and validity of the results, maximize the knowledge gained from the work, and avoid any wastage of animals, effort, resources, and public funds.

The high cost of acquiring and maintaining primates for laboratory research can lead to the use of small sample sizes in primate studies, but this is a false economy. Experiments should be appropriately powered to detect biologically meaningful effects. Use of efficient experimental designs (e.g., crossover designs) can help to reduce animal numbers (Bate and Clark 2014), as can collaborating with other research groups to utilize the same cohort of study animals. Care should be taken to use methods to minimize bias, such as randomization in the allocation of animals to experimental groups and blinding in the assessment of the outcomes. Failure to do so can lead to overestimation of treatment effects, leading to poor decision making and problems with reproducibility later down the line (Vesterinen et al. 2010). Statistical analysis methods should be used. Small sample sizes are also frequently used in zoo-based studies. In such cases, zoo-based researchers should take care to use appropriate statistical tests and to be realistic when extrapolating the findings to the population (Plowman 2008). The BIAZA (2013) *Handbook of Zoo Research* is a good source of advice in these matters.

To guide *in vivo* researchers in the design of their experiments, the NC3Rs has developed the Experimental Design Assistant (EDA) (NC3Rs 2016). The EDA consists of a website with comprehensive guidance on experimental design and a web application that uses computer-based reasoning to provide tailored feedback and advice on experimental plans. The system also includes dedicated support for randomization, blinding, and sample size calculation, helping researchers to design robust, well-powered experiments that meet the requirements of various funding bodies.

6. *Report studies comprehensively and transparently using the ARRIVE Guidelines.* Comprehensive reporting of primate studies allows the study findings to be properly evaluated and utilized, thus maximizing the value of the information published and minimizing the number of unnecessary studies in the future. Unfortunately, the reporting of primate studies is not always exemplary. The NC3Rs and NIH Office of Laboratory Animal Welfare jointly funded a systematic survey of the quality of reporting, experimental design, and statistical analysis in 271 randomly chosen papers describing research on live rats, mice, and primates carried out in UK and US publicly funded research institutions. The survey found major omissions in the reporting of study hypotheses and objectives, the number and characteristics of the animals used, and the use of randomization and blinding to reduce bias in animal selection and outcome assessment (Kilkenny et al. 2009). In addition, only 70% of the publications that used statistics described the statistical tests used and presented the results with a measure of variability or error (e.g., standard deviation or confidence interval). In their review of 2937 articles reporting primate research, Carlsson et al. (2004) found that animal characteristics (e.g., sex, age, and weight), and the conditions under which the animals were housed and used, were infrequently described (none of these parameters were described in more than 50% of the articles analyzed). They called for editors to require authors to provide comprehensive information concerning the research subjects (e.g., their origin), treatment conditions, and experimental procedures in the studies they publish to facilitate replication of studies and assessment of animal welfare.

The ARRIVE Guidelines (NC3Rs 2020) were developed by a working group of the NC3Rs to improve the reporting of animal-based studies (Kilkenny et al. 2010; Percie du Sert et al. 2020a, b). The guidelines lay out the items that should be included in *in vivo* research papers in order that their results and conclusions can be properly evaluated and utilized by readers. They are endorsed by well over 1000 journals internationally, including major bioscience titles like *Nature*, *Cell*, and the *Public Library of Science* family of journals (e.g., *PLoS One*), as well as specialist journals, such as *International Journal of Primatology*, and all major UK research funding bodies. All researchers using primates should ensure that they take account of the guidelines in the design and reporting of their studies, and journal editors and peer reviewers should ensure that published studies fulfill the ARRIVE criteria.

7. *Publish all research findings, including negative results, and exploit other mechanisms of knowledge transfer for maximum impact.*
Most publicly funded research organizations have policies in place requiring grant holders to make their peer-reviewed articles available open access, preferably immediately on publication. Although publication in a peer-reviewed journal is a basic, vital step in the dissemination of scientific research findings, it does not alone constitute an adequately reliable means of ensuring knowledge transfer and impact. Researchers should therefore utilize all available routes for effective sharing of data and technology (e.g., data repositories, social media, training workshops, collaboration with industry or other stakeholders) to enable translation of their research for maximum benefit for animals and society. This should include the publication of negative and null results (e.g., in journals such as *PLoS One* and *Journal of Negative Results in Biomedicine*, or online preprint servers such as *bioRxiv*), since if these are not published then other researchers may undertake the same work that led nowhere. Publicly registering the protocol before a study is conducted can also help to reduce publication bias by preventing *p*-hacking (also known as data dredging or inflation bias), where data sets are repeatedly searched or alternative analyses tried until a significant result is found (Academy of Medical Sciences/Biotechnology and Biological Sciences Research Council/Medical Research Council/Wellcome Trust 2015). Although these issues apply to any area of science, they are of particular concern where highly sentient animals, such as primates, have been used (Bateson et al. 2011).

6   Conclusions

Primatology is a popular, thriving, and diverse science, addressing big questions such as what it means to be human, understanding the brain, animal consciousness, and how to conserve biodiversity in the face of global change. There is still much to learn about our closest relatives and continued demand for their use in biomedical research. As scientists working with these precious animals, we have an obligation to do the best and most humane science we can, regardless of discipline and research setting. The NC3Rs, zoological associations, and primatological societies have a variety of resources available to support scientists in this goal.

Acknowledgements   Thanks to Dr. Nathalie Percie du Sert, NC3Rs for helpful comments on this chapter.

References

Academy of Medical Sciences/Biotechnology and Biological Sciences Research Council/Medical Research Council/Wellcome Trust (2015) Reproducibility and reliability of biomedical research: improving research practise. In: The Academy of Medical Sciences, Symposium report. Academy of Medical Sciences. https://acmedsci.ac.uk/file-download/38189-56531416e2949.pdf. Accessed 16 Feb 2017

Animal Procedures Committee (2002) The use of primates under the Animals (Scientific Procedures) Act (1986): analysis of current trends with particular reference to regulatory toxicology. London. https://www.gov.uk/government/uploads/system/uploads/attachment_data/file/118994/primates.pdf. Accessed 16 Feb 2017

Association of Primate Veterinarians (2017) Education and resources. https://www.primatevets.org/education%2D%2Dresources. Accessed 16 Jun 2021

Association of Zoos and Aquariums (2016) A toolkit to engage in field conservation: created by the AZA Field Conservation Committee, Edition 2.0

Bailey J (2022) Arguments against using nonhuman primates in research. In: Robinson LM, Weiss A (eds) Nonhuman primate welfare: from history, science, and ethics to practice. Springer, Cham, pp 547–574

Baker KC (2016) Survey of 2014 behavioral management programs for laboratory primates in the United States. Am J Primatol 78(7):780–796. https://doi.org/10.1002/ajp.22543

Baker KR, Farmer HL (2022) The welfare of primates in zoo. In: Robinson LM, Weiss A (eds) Nonhuman primate welfare: from history, science, and ethics to practice. Springer, Cham, pp 79–96

Bate ST, Clark RA (2014) The design and statistical analysis of animal experiments. Cambridge University Press, Cambridge

Bateson P, Johansen-Berg H, Jones DK, et al. (2011) Review of research using non-human primates: report of a panel chaired by Professor Sir Patrick Bateson FRS. London

Bayne KAL, Morris TH (2012) Laws, regulations and policies relating to the care and use of nonhuman primates in biomedical research. In: Abee C, Mansfield K, Tardif S, Morris T (eds) Nonhuman primates in biomedical research: biology and management, 2nd edn, vol 1. Academic Press, London, pp 35–56

Bayne K, Hau J, Morris T (2022) The welfare impact of regulations, policies, guidelines and directives and nonhuman primate welfare. In: Robinson LM, Weiss A (eds) Nonhuman primate welfare: from history, science, and ethics to practice. Springer, Cham, pp 629–646

Bennett BT (2016) Association of primate veterinarians 2014 nonhuman primate housing survey. JAALAS 55(2):172–174

BIAZA (2013) Handbook of zoo research, guidelines for conducting research in zoos. http://www.biaza.org.uk/uploads/Research/BIAZA%20Handbook%20of%20Zoo%20Research%202014%20FINAL.pdf. Accessed 16 Feb 2017

Bowler MT, Buchanan-Smith HM, Whiten A (2012) Assessing public engagement with science in a university primate research centre in a national zoo. PLoS One 7(4):e34505. https://doi.org/10.1371/journal.pone.0034505

Boyd Group (2002) The use of non-human primates in research and testing. The British Psychological Society, Leicester

Brønstad A, Newcomer CE, Decelle T et al (2016) Current concepts of harm–benefit analysis of animal experiments–report from the AALAS–FELASA working group on harm–benefit analysis - Part 1. Lab Anim 50(1S):1–20. https://doi.org/10.1177/0023677216642398

Brown S, Flecknell PA, Jackson I, et al (2013) Independent investigation into animal research at Imperial College London. http://brownreport.info/. Accessed 16 Feb 2017

Buchanan-Smith HM, Hardie SM, Caceres C, Prescott MJ (2000) Distribution and forest utilization of *Saguinus* and other primates of the Pando Department, Northern Bolivia. Int J Primatol 21:353–379. https://doi.org/10.1023/A:1005483601403

Buchanan-Smith HM, Tasker L, Ash H, Graham ML (2022) Welfare of primates in laboratories: opportunities for improvement. In: Robinson LM, Weiss A (eds) Nonhuman primate welfare: from history, science, and ethics to practice. Springer, Cham, pp 97–120

Buckley LA, Chapman KL, Burns-Naas LA et al (2011) Considerations regarding nonhuman primate use in safety assessment of biopharmaceuticals. Int J Toxicol 30(5):583–590. https://doi.org/10.1177/1091581811415875

Burm SM, Prins J-B, Langermans J, Bajramovic JJ (2014) Alternative methods for the use of non-human primates in biomedical research. ALTEX-Altern Anim Ex 31(4):520–529. https://doi.org/10.14573/altex.1406231

CAMARADES/NC3Rs (2017) Systematic review facility. https://syrf.org.uk/ Accessed 16 Jun 2021

Carlsson H, Schapiro SJ, Farah I, Hau J (2004) Use of primates in research: a global overview. Am J Primatol 63(4):225–237. https://doi.org/10.1002/ajp.20054

Chapman KL, Andrews L, Bajramovic JJ et al (2012) The design of chronic toxicology studies of monoclonal antibodies: implications for the reduction in use of non-human primates. Regul Toxicol Pharmacol 62(2):347–354. https://doi.org/10.1016/j.yrtph.2011.10.016

Chapman KL, Bayne KAL, Couch J et al (2015) Opportunities for implementing the 3Rs in drug development and safety assessment studies using nonhuman primates. In: Bluemel J (ed) The nonhuman primate in nonclinical drug development and safety assessment. Elsevier, Amsterdam, pp 281–301

Clemence M, Leaman J (2016) Public attitudes to animal research in 2016. Ipsos MORI Social Research Institute/Department for Business, Energy and Industrial Strategy. www.ipsos-mori.com/Assets/Docs/Publications/sri-public-attitudes-to-animal-research-2016.pdf Accessed 16 Feb 2017

Coleman K (2011) Caring for nonhuman primates in biomedical research facilities: scientific, moral and emotional considerations. Am J Primatol 73(3):220–225. https://doi.org/10.1002/ajp.20855

Coleman K, Robertson ND, Bethea CL (2011) Long-term ovariectomy alters social and anxious behaviors in semi-free ranging Japanese macaques. Behav Brain Res 225(1):317–327. https://doi.org/10.1016/j.bbr.2011.07.046

Coleman K, Timmel G, Prongay K, Baker KC (2022) Common husbandry, housing, and animal care practices. In: Robinson LM, Weiss A (eds) Nonhuman primate welfare: from history, science, and ethics to practice. Springer, Cham, pp 317–348

Cyranoski D (2016) Monkey kingdom. Nature 532(7599):300–302. https://doi.org/10.1038/532300a

Department for Environment Food and Rural Affairs (2012) Zoos expert committee handbook, November 2012. https://www.gov.uk/government/uploads/system/uploads/attachment_data/file/69611/pb13815-zoos-expert-committee-handbook1.pdf. Accessed 16 Feb 2017

Eastwood D, Findlay L, Poole S et al (2010) Monoclonal antibody TGN1412 trial failure explained by species differences in CD28 expression on CD4+ effector memory T-cells. Br J Pharmacol 161(3):512–526. https://doi.org/10.1111/j.1476-5381.2010.00922.x

Estrada A, Garber PA, Rylands AB et al (2017) Impending extinction crisis of the world's primates: Why primates matter. Sci Adv 3(1):e1600946. https://doi.org/10.1126/sciadv.1600946

European Commission (2006) Results of questionnaire for the general public on the revision of Directive 86/609/EEC on the protection of animals used for experimental and other scientific purposes. http://ec.europa.eu/environment/chemicals/lab_animals/pdf/results_citizens.pdf. Accessed 16 Feb 2017

European Commission (2010) Directive 2010/63/EU of the European Parliament and of the Council of 22 September 2010 on the protection of animals used for scientific purposes OJL276/33. Official Journal of the European Union L276:33. http://eur-lex.europa.eu/LexUriServ/LexUriServ.do?uri=OJ:L:2010:276:0033:0079:en:PDF. Accessed 16 Feb 2017

European Commission (2015) EU Zoos directive. Good practices document. https://ec.europa.eu/environment/nature/pdf/EU_Zoos_Directive_Good_Practices.pdf. Accessed 16 Jun 2021

Expert Working Group on Severity Classification Criteria (2009) Expert working group on severity classification of scientific procedures performed on animals, Final Report. European Commission, Brussels. http://ec.europa.eu/environment/chemicals/lab_animals/pdf/report_ewg.pdf. Accessed 16 Feb 2017

Farmer HL, Baker KR, Cabana F (2022) Housing and husbandry for primates in zoos. In: Robinson LM, Weiss A (eds) Nonhuman primate welfare: from history, science, and ethics to practice. Springer, Cham, pp 349–368

Fedigan LM (2010) Ethical issues faced by field primatologists: asking the relevant questions. Am J Primatol 72(9):754–771. https://doi.org/10.1002/ajp.20814

Ferreira RG, Ruiz-Miranda C, Sita S, Sánchez-López S, Pissinatti A, Corte S, Jerusalinsky L, Wagner PG, Maas C (2022) Primates under human care in developing countries: examples from Latin America. In: Robinson LM, Weiss A (eds) Nonhuman primate welfare: from history, science, and ethics to practice. Springer, Cham, pp 145–170

Freedman LP, Cockburn IM, Simcoe TS (2015) The economics of reproducibility in preclinical research. PLoS Biol 13(6):e1002165. https://doi.org/10.1371/journal.pbio.1002165

Gippoliti S (2006) Applied primatology in zoos: History and prospects in the field of wildlife conservation, public awareness and animal welfare. Primate Rep 73:57–71

Graham ML, Prescott MJ (2015) The multifactorial role of the 3Rs in shifting the harm-benefit analysis in animal models of disease. Eur J Pharmacol 759:19–29. https://doi.org/10.1016/j.ejphar.2015.03.040

Graham ML, Rieke EF, Mutch LA et al (2012) Successful implementation of cooperative handling eliminates the need for restraint in a complex non-human primate disease model. J Med Primatol 41(2):89–106. https://doi.org/10.1111/j.1600-0684.2011.00525.x

Gruen L, Fleury E (2022) Animal welfare, animal rights, and a sanctuary ethos. In: Robinson LM, Weiss A (eds) Nonhuman primate welfare: from history, science, and ethics to practice. Springer, Cham, pp 613–628

Gruen L, Fultz A, Pruetz J (2013) Ethical issues in African great ape field studies. ILAR J 54 (1):24–32. https://doi.org/10.1093/ilar/ilt016

Hagelin J (2004) Use of live nonhuman primates in research in Asia. J Postgrad Med 50(4):253–256

Hagelin J (2005) Use of nonhuman primates in research in Sweden: 25 year longitudinal survey. ALTEX-Altern Anim Ex 22(1):13–18

Hardie SM, Prescott MJ, Buchanan-Smith HM (2003) Ten years of mixed-species troops at Belfast Zoological Gardens. Primate Rep 65:21–38

Home Office (2015) The harm-benefit analysis process: new project licence applications. Advice Note: 05/2015. Animals in Science Regulation Unit, Home Office, London. https://www.gov.uk/government/uploads/system/uploads/attachment_data/file/487914/Harm_Benefit_Analysis__2_.pdf. Accessed 16 Feb 2017

Home Office (2016) Annual statistics of scientific procedures on living animals Great Britain 2015. The Stationery Office, London. https://www.gov.uk/government/uploads/system/uploads/attachment_data/file/537708/scientific-procedures-living-animals-2015.pdf. Accessed 16 Feb 2017

Hopper LM, Shender MA, Ross SR (2016) Behavioral research as physical enrichment for captive chimpanzees. Zoo Biol 35(4):293–297. https://doi.org/10.1002/zoo.21297

Hosey G (2005) How does the zoo environment affect the behaviour of captive primates? Appl Anim Behav Sci 90(2):107–129. https://doi.org/10.1016/j.applanim.2004.08.015

Hosey G (2022) The history of primates in zoos. In: Robinson LM, Weiss A (eds) Nonhuman primate welfare: from history, science, and ethics to practice. Springer, Cham, pp 3–30

Hosey G, Melfi V, Pankhurst S (2013) Zoo animals: behaviour, management, and welfare, 2nd edn. Oxford University Press, Oxford

Institute for Laboratory Animal Research (2015) Reproducibility issues in research with animals and animal models: Workshop in brief. National Academy of Sciences, Washington, DC. https://www.nap.edu/download/21835. Accessed 16 Feb 2017

International Union for Conservation of Nature (2021) The IUCN red list of threatened species. https://www.iucnredlist.org/ Accessed 16 Jun 2021

Irvine C, Egan KJ, Shubber Z et al (2015) Efficacy of HIV postexposure prophylaxis: systematic review and meta-analysis of nonhuman primate studies. Clin Infect Dis 60(Suppl 3):S165–S169. https://doi.org/10.1093/cid/civ069

Janson CH, Brosnan SF (2013) Experiments in primatology: from the lab to the field and back again. In: Sterling EJ, Bynum N, Blair ME (eds) Primate ecology and conservation. Oxford University Press, Oxford, pp 177–194

Jennings M, Prescott MJ (2009) Refinements in husbandry, care and common procedures for non-human primates. Lab Anim 43(1S):1–47. https://doi.org/10.1258/la.2008.007143

Jolly A (1985) The evolution of primate behavior: a survey of the primate order traces the progressive development of intelligence as a way of life. Am Sci 73(3):230–239

Kaiser J (2015) NIH to end all support for chimpanzee research. Science November 1. http://www.sciencemag.org/news/2015/11/nih-end-all-support-chimpanzee-research. Accessed 16 Feb 2017

Kerwin A (2006) Overcoming the barriers to the retirement of Old and New World monkeys from research facilities. J Appl Anim Welf Sci 9(4):337–347. https://doi.org/10.1207/s15327604jaws0904_9

Kierulff MCM, Ruiz-Miranda CR, Oliveira PP et al (2012) The golden lion tamarin *Leontopithecus rosalia*: a conservation success story. Int Zoo Yearb 46(1):36–45. https://doi.org/10.1111/j.1748-1090.2012.00170.x

Kilkenny C, Parsons N, Kadyszewski E et al (2009) Survey of the quality of experimental design, statistical analysis and reporting of research using animals. PLoS One 4(11):e7824. https://doi.org/10.1371/journal.pone.0007824

Kilkenny C, Browne WJ, Cuthill IC et al (2010) Improving bioscience research reporting: The ARRIVE guidelines for reporting animal research. PLoS Biol 8(6):e1000412. https://doi.org/10.1371/journal.pbio.1000412

King T, Chamberlan C, Courage A (2012) Assessing initial reintroduction success in long-lived primates by quantifying survival, reproduction, and dispersal parameters: Western lowland gorillas (*Gorilla gorilla gorilla*) in Congo and Gabon. Int J Primatol 33:134–149. https://doi.org/10.1007/s10764-011-9563-2

King T, Chamberlan C, Courage A (2014) Assessing reintroduction success in long-lived primates through population viability analysis: Western lowland gorillas *Gorilla gorilla gorilla* in Central Africa. Oryx 48(2):294–303. https://doi.org/10.1017/S0030605312001391

Kipling R (1894) The jungle book. Macmillan and Co., London

Kleiman DG (1985) Criteria for the evaluation of zoo research projects. Zoo Biol 4(2):93–98. https://doi.org/10.1002/zoo.1430040202

Kreger MD, Mench JA (1995) Visitor—animal interactions at the zoo. Anthrozoös 8(3):143–158. https://doi.org/10.2752/089279395787156301

Laber K, Newcomer CE, Decelle T et al (2016) Recommendations for addressing harm–benefit analysis and implementation in ethical evaluation – Report from the AALAS–FELASA working group on harm–benefit analysis – Part 2. Lab Anim 50(1S):21–42. https://doi.org/10.1177/0023677216642397

Lankau EW, Turner PV, Mullan RJ, Galland GG (2014) Use of nonhuman primates in research in North America. Lab Anim Sci 53(3):278–282

Lexchin J, LA B, Djulbegovi B, et al (2003) Pharmaceutical industry sponsorship and research outcome and quality: Systematic review. BMJ 326:1167–1170. https://doi.org/10.1136/bmj.326.7400.1167

Living Links (2021) Publications. https://living-links.org/about/publications/. Accessed 16 Jun 2021

Martin P, Bateson P (2007) Measuring behaviour: an introductory guide, 3rd edn. Cambridge University Press, Cambridge

Matsuzawa T (2007) Comparative cognitive development. Dev Sci 10(1):97–103. https://doi.org/10.1111/j.1467-7687.2007.00570.x

McCann C, Buchanan-Smith HM, Jones-Engel L, et al (2007) IPS International guidelines for the acquisition, care and breeding of nonhuman primates, 2nd edn. Primate Care Committee, International Primatological Society. http://www.internationalprimatologicalsociety.org/docs/ips_international_guidelines_for_the_acquisition_care_and_breeding_of_nonhuman_primates_second_edition_2007.pdf. Accessed 16 Feb 2017

McMillan JL, Bloomsmith MA, Prescott MJ (2017) An international survey of approaches to chair restraint of nonhuman primates. Comp Med 67(5):442–451

Minteer BA, Collins JP (2013) Ecological ethics in captivity: balancing values and responsibilities in zoo and aquarium research under rapid global change. ILAR J 54(1):41–51. https://doi.org/10.1093/ilar/ilt009

Moss A, Jensen E, Gusset M (2015) Evaluating the contribution of zoos and aquariums to Aichi Biodiversity Target 1. Conserv Biol 29(2):537–544. https://doi.org/10.1111/cobi.12383

National Institutes of Health (2016) Rigor and reproducibility. https://grants.nih.gov/reproducibility/index.htm. Accessed 16 Feb 2017

Nature (2018) Challenges in irreproducible research. https://www.nature.com/collections/prbfkwmwvz/. Accessed 16 Jun 2021

NC3Rs (2006) Non-human primate accommodation, care and use, 1st edition. NC3Rs, London. https://www.nc3rs.org.uk/non-human-primate-accommodation-care-and-use. Accessed 16 Feb 2017

NC3Rs (2010) The welfare of non-human primates. https://nc3rs.org.uk/welfare-non-human-primates. Accessed 16 Jun 2021

NC3Rs (2015) The Macaque Website. https://www.nc3rs.org.uk/macaques/. Accessed 16 Jun 2021

NC3Rs (2016) Experimental design assistant. https://eda.nc3rs.org.uk/. Accessed 16 Jun 2021

NC3Rs (2020) ARRIVE guidelines. https://arriveguidelines.org/. Accessed 16 Jun 2021

NC3Rs/AMRC/BBSRC/Defra/EPSRC/MRC/NERC/Wellcome Trust (2008) Responsibility in the use of animals in bioscience research: expectations of the major research councils and charitable funding bodies. NC3Rs, London. https://www.nc3rs.org.uk/responsibility-use-animals-bioscience-research. Accessed 16 Feb 2017

Percie du Sert N, Hurst V, Ahluwalia A et al (2020a) The ARRIVE guidelines 2.0: updated guidelines for reporting animal research. PLoS Biol 18(7):e3000410. https://doi.org/10.1371/journal.pbio.3000410

Percie du Sert N, Ahluwalia A, Alam S et al (2020b) Reporting animal research: explanation and elaboration for the ARRIVE guidelines 2.0. PLoS Biol 18(7):e3000411. https://doi.org/10.1371/journal.pbio.3000411

Petticrew M, Davey Smith G (2012) The monkey puzzle: a systematic review of studies of stress, social hierarchies, and heart disease in monkeys. PLoS One 7(3):e27939. https://doi.org/10.1371/journal.pone.0027939

Phillips KA, Bales KL, Capitanio JP et al (2014) Why primate models matter. Am J Primatol 76(9):801–827. https://doi.org/10.1002/ajp.22281

Pickard J, Buchanan-Smith HM, Dennis MB, et al (2013) Review of the assessment of cumulative severity and lifetime experience in non-human primates used in neuroscience research. London

Plowman AB (2008) BIAZA statistics guidelines: toward a common application of statistical tests for zoo research. Zoo Biol 27(3):226–233. https://doi.org/10.1002/zoo.20184

Poole T (1997) Happy animals make good science. Lab Anim 31(2):116–124. https://doi.org/10.1258/002367797780600198

Prescott MJ (2006) Finding new homes for ex-laboratory and surplus zoo primates. Lab Prim Newsl 45:5–8. https://www.brown.edu/Research/Primate/lpn45-3.html#homing

Prescott MJ (2010) Ethics of primate use. Adv Sci Res 5(1):11–22. https://doi.org/10.5194/asr-5-11-2010

Prescott MJ (2016) Online resources for improving the care and use of non-human primates in research. Prim Biol 3(2):33–40. https://doi.org/10.5194/pb-3-33-2016

Prescott MJ (2017) The three Rs. In: The international encyclopedia of primatology. Wiley, London, pp 1–5

Prescott MJ, Buchanan-Smith HM (1999) Intra- and inter-specific social learning of a novel food task in two species of tamarin. Int J Comp Psychol 12(2):71–92

Prescott MJ, Buchanan-Smith HM (2002) Predation sensitive foraging in captive tamarins. In: Miller LE (ed) Eat or be eaten: predator sensitive foraging among primates. Cambridge University Press, Cambridge, pp 44–57

Prescott MJ, Buchanan-Smith HM (2003) Training nonhuman primates using positive reinforcement techniques. J Appl Anim Welf Sci 6(3):157–161. https://doi.org/10.1207/S15327604JAWS0603_01

Prescott MJ, Buchanan-Smith HM (2004) Cage sizes for tamarins in the laboratory. Anim Welf 13 (2):151–158

Prescott MJ, Buchanan-Smith HM (2007) Training laboratory-housed non-human primates, part I: A UK survey. Anim Welf 16(1):21–26

Prescott MJ, Jennings M (2004) Ethical and welfare implications of the acquisition and transport of non-human primates for use in research and testing. ATLA-Altern Lab Anim 32(S1):323–327. https://doi.org/10.1177/026119290403201s53

Prescott MJ, Buchanan-Smith HM, Smith AC (2005) Social interaction with non-averse groupmates modifies a learned food aversion in single- and mixed-species groups of tamarins (*Saguinus fuscicollis* and *S. labiatus*). Am J Primatol 65(4):313–326. https://doi.org/10.1002/ajp.20118

Prescott MJ, Brown VJ, Flecknell PA et al (2010) Refinement of the use of food and fluid control as motivational tools for macaques used in behavioural neuroscience research: report of a Working Group of the NC3Rs. J Neurosci Methods 193(2):167–188. https://doi.org/10.1016/j.jneumeth.2010.09.003

Prior H, Bottomley A, Champéroux P et al (2016) Social housing of non-rodents during cardiovascular recordings in safety pharmacology and toxicology studies. J Pharmacol Toxicol Methods 81:75–87. https://doi.org/10.1016/j.vascn.2016.03.004

Quigley M (2007) Non-human primates: the appropriate subjects of biomedical research? J Med Ethics 33(11):655–658. https://doi.org/10.1136/jme.2007.020784

Regan T (1996) Are zoos morally defensible? In: Norton BG (ed) Ethics on the ark: Zoos, animal welfare, and wildlife conservation. Smithsonian Institution Press, Washington, DC, pp 38–51

Reinhardt V (2004) Common husbandry-related variables in biomedical research with animals. Lab Anim 38(3):213–235. https://doi.org/10.1258/002367704323133600

Rennie AE, Buchanan-Smith HM (2006a) Refinement of the use of non-human primates in scientific research. Part III: Refinement of procedures. Anim Welf 15(3):239–261

Rennie AE, Buchanan-Smith HM (2006b) Refinement of the use of non-human primates in scientific research. Part II: Housing, husbandry and acquisition. Anim Welf 15(3):215–238

Rennie AE, Buchanan-Smith HM (2006c) Refinement of the use of non-human primates in scientific research. Part I: The influence of humans. Anim Welf 15(3):203–213

Research Councils UK (2015) Updated RCUK guidance for funding applications involving animal research

Riley EP, MacKinnon KC, Fernandez-Duque E, et al (2014) Code of best practices for field primatology. https://doi.org/10.13140/2.1.2889.1847

Ross SR, Wagner KE, Schapiro SJ, Hau J (2010) Ape behavior in two alternating environments: comparing exhibit and short-term holding areas. Am J Primatol 72(11):951–959. https://doi.org/10.1002/ajp.20857

Rossi J (2009) Nonhuman primate research: the wrong way to understand needs and necessity. Am J Bioethics 9(5):21–23. https://doi.org/10.1080/15265160902788728

Schnell CR, Gerber P (1997) Training and remote monitoring of cardiovascular parameters in non-human primates. Prim Rep 49:61–70

Scientific Committee on Health Environmental and Emerging Risks (2017) Preliminary Opinion on the need for non-human primates in biomedical research, production and testing of products and devices (update 2017). https://ec.europa.eu/health/sites/health/files/scientific_committees/scheer/docs/scheer_o_004.pdf. Accessed 16 Feb 2017

Setchell JM, Curtis DJ (2011) Field and laboratory methods in primatology: a practical guide. Cambridge University Press, Cambridge

Smith AC, Surridge AK, Prescott MJ et al (2012) Effect of colour vision status on insect prey capture efficiency of captive and wild tamarins (*Saguinus spp.*). Anim Behav 83(2):479–486. https://doi.org/10.1016/j.anbehav.2011.11.023

t'Hart BA, Laman JD, Kap YS (2022) An unexpected symbiosis of animal welfare and clinical relevance in a refined nonhuman primate model of human autoimmune disease. In: Robinson LM, Weiss A (eds) Nonhuman primate welfare: from history, science, and ethics to practice. Springer, Cham, pp 591–612

Talbot CF, Reamer LA, Lambeth SP, Schapiro SJ, Brosnan SF (2022) Meeting cognitive, behavioral, and social needs of primates in captivity. In: Robinson LM, Weiss A (eds) Nonhuman primate welfare: from history, science, and ethics to practice. Springer, Cham, pp 267–302

Tasker L (2012) Linking welfare and quality of scientific output in cynomolgus macaques (*Macaca fascicularis*) used for regulatory toxicology. PhD thesis, University of Stirling, Stirling. https://dspace.stir.ac.uk/bitstream/1893/9801/1/Lou%20Tasker%20Thesis%202012.pdf. Accessed 16 Feb 2017

Turner PV (2022) The history of chimpanzees in biomedical research. In: Robinson LM, Weiss A (eds) Nonhuman primate welfare: from history, science, and ethics to practice. Springer, Cham, pp 31–56

University of Stirling, Primate Society of Great Britain, NC3Rs (2011) Common marmoset care. www.marmosetcare.com. Accessed 16 Jun 2021

Vesterinen HM, Sena ES, ffrench-Constant C et al (2010) Improving the translational hit of experimental treatments in multiple sclerosis. Mult Scler 16(9):1044–1055. https://doi.org/10.1177/1352458510379612

von Roten FC (2013) Public perceptions of animal experimentation across Europe. Public Underst Sci 22(6):691–703. https://doi.org/10.1177/0963662511428045

Walker MD, Nelson JK, Bernal JC (2007) Toxicology (primates). In: Gad S (ed) Animal models in toxicology, 2nd edn. CRC Press, Boca Raton, FL, pp 797–799

Waller BM, Peirce K, Mitchell H, Micheletta J (2012) Evidence of public engagement with science: visitor learning at a zoo-housed primate research centre. PLoS One 7(9):e44680. https://doi.org/10.1371/journal.pone.0044680

Weatheall D, Goodfellow P, Harris J, et al (2015) The use of non-human primates in research: a working group report chaired by Sir David Weatherall. 2006. London

Whitehouse J, Micheletta J, Powell LE et al (2013) The impact of cognitive testing on the welfare of group housed primates. PLoS One 8(11):e78308. https://doi.org/10.1371/journal.pone.0078308

Whittaker M, Laule G (2012) Training techniques to enhance the care and welfare of nonhuman primates. Vet Clin North Am Exot Anim Pract 15(3):445–454. https://doi.org/10.1016/j.cvex.2012.06.004

Wolfe-Coote S (2005) The laboratory primate. Academic, London

Wolfensohn S (2022) Humane end points and end of life in primates used in laboratories. In: Robinson LM, Weiss A (eds) Nonhuman primate welfare: from history, science, and ethics to practice. Springer, Cham, pp 369–386

Wolfgang Köhler Primate Research Center (2017) Publications. http://wkprc.eva.mpg.de/english/files/public.htm. Accessed 16 Jun 2021

Zhang XL, Pang W, Hu XT et al (2014) Experimental primates and non-human primate (NHP) models of human diseases in China: current status and progress. Dongwuxue Yanjiu 35(6):447–464. https://doi.org/10.13918/j.issn.2095-8137.2014.6.447

The Welfare of Primates in Zoos

Kathy R. Baker and Holly L. Farmer

Abstract One challenge facing zoos is balancing welfare needs with other primary goals, which include conservation, education, research, and entertainment. Managing primates in zoos involves similar welfare challenges faced by primates in other environments, which are covered elsewhere in this volume. In this chapter we identify and discuss welfare challenges that are unique to zoo-housed primates. All captive primates experience the presence of familiar humans (animal care staff), however the presence of unfamiliar humans (visitors) is common in zoo environments. In addition to providing a resource to zoo visitors, zoo primates also have an important conservation role that may involve intensive social management to facilitate captive breeding. We first discuss the influence of both familiar and unfamiliar humans on the welfare of zoo primates. We then examine the impact of different methods of social management on primate welfare.

Keywords Zoo management · Human–animal interactions · Social management

K. R. Baker (✉)
Wild Planet Trust, Newquay Zoo, Cornwall, UK
e-mail: kathy.baker@wildplanettrust.org.uk

H. L. Farmer
Wild Planet Trust, Paignton Zoo Environmental Park, Devon, UK

© Springer Nature Switzerland AG 2023
L. M. Robinson, A. Weiss (eds.), *Nonhuman Primate Welfare*,
https://doi.org/10.1007/978-3-030-82708-3_4

1 Introduction

Modern zoological collections have five main responsibilities: animal welfare, conservation, education, research, and entertainment (Godinez and Fernandez 2019). When considering zoo primate welfare, it is important to be aware of these goals because there is a need to evaluate different welfare challenges for different species within a collection. For example, if the primary role of an individual animal is to act as an ambassador for their wild counterparts—an educational role—then one welfare challenge they would face would relate to the effects of exposure to large visitor numbers. On the other hand, if the main role of a species is to maintain a genetically diverse population for captive breeding purposes—a conservation role—then welfare challenges may include the animals' movement to another zoo or the introduction of animals for breeding purposes. With careful enclosure design and planning these goals need not clash. For example, research can occur in front of the public and provide entertainment and conservation education opportunities as well (Farmer et al. 2022).

Zoo design has recently moved toward larger and more naturalistic enclosures (Coe 2003; Hosey 2022). This has brought about improvements in meeting welfare needs as well as providing better educational experiences for visitors. The welfare impacts of enclosure design and management practices are discussed elsewhere in this volume (Coleman et al. 2022; Farmer et al. 2022; Kemp 2022), therefore this chapter focuses on three welfare challenges that we believe are pertinent to zoo-housed primates. The first relates to regular contact with familiar humans. This challenge is not unique to zoo-housed primates (Buchanan-Smith et al. 2022), but the interactions may be different in zoo environments. The second challenge is the presence of large numbers of visitors, which is common in zoos (Hosey 2005). The third challenge is that of social management, which is the manipulation of social groupings based on breeding requirements and/or housing and husbandry constraints. While this third challenge is not unique to zoos, its end goal, that is, maintaining self-sustaining genetically diverse (i.e., breeding) populations, may be different to the end goal of other captive facilities, such as laboratories and sanctuaries for which reproduction may not be a primary goal.

2 Presence of Familiar Humans

The welfare implications of human–animal interactions in zoos resemble those in laboratories and other captive environments (Buchanan-Smith et al. 2022). When human–animal interactions are consistent, human-animal relationships and human-animal bonds may develop (Hosey and Melfi 2012). Positive human-animal relationships develop when humans talk calmly and stroke/groom the animals, for example; negative human-animal relationships develop when humans shout at and roughly handle the animals, for example (Ward and Melfi 2015). The development of positive human-animal relationships can occur after even only small positive interventions. For instance, laboratory chimpanzees (*Pan troglodytes*) that spend

10 min per day engaging in positive interactions with their caregivers show an increase in play and grooming behaviors and a decrease in abnormal behaviors (Baker 2004).

Some evidence indicates that human-animal relationships may be further enhanced if caregivers use species-specific communication, such as gestures based on the species' natural mode of communication. Jensvold (2008) carried out a study on three male chimpanzees housed at Zoo Northwest Florida (now Gulf Breeze Zoo), in which caregivers communicated with chimpanzees using either chimpanzee or human means. For example, when grooming chimpanzees, the human caregiver would either lip-smack and make grooming noises or just examine the chimpanzee's hair without lip-smacking. The three chimpanzees differed in how they responded to the two types of caregiver communication style. For example, two chimpanzees spent more time grooming when caregivers used chimpanzee communication while the other chimpanzee spent more time grooming when the caregivers used human communication. This finding highlights the importance of considering species' natural mode of communication and individual differences in the effect that these interventions have on the human-animal relationship.

Human-animal bonds involve a relationship between a human and an animal that is reciprocal and persistent and that promotes a perceived increase in well-being for both parties (Hosey and Melfi 2012). These bonds are often reported by pet owners. Research on zoo-housed nonhuman primates has only recently examined these bonds. For instance, Hosey and Melfi (2012) explored the prevalence of human-animal bonds by surveying 130 zoo professionals at industry conferences. Irrespective of age, gender, or job role, 78 of the respondents reported that they had formed a bond with a zoo animal. Moreover, a quarter of respondents reported that they had formed a bond with a species of primate and, for over a half of these respondents, the primate was one of the apes. The respondents reported that the benefits that they themselves incurred included a sense of enjoyment and emotional attachment while they perceived benefits to animals as being related to improved husbandry and welfare. Other studies have found that human–animal interactions, such as positive reinforcement training, also enhance husbandry and welfare by making routine management situations, such as isolation (Spiezio et al. 2015) and medication administration (Melfi and Thomas 2005), less stressful for the animals.

It is important to add the caveat that some human–animal interactions may have adverse effects on animals. For example, two groups of chimpanzees and western lowland gorillas (*Gorilla gorilla gorilla*) at Lincoln Park Zoo, USA, were observed for 4 years as part of a continuous behavioral monitoring study. During this study, all occurrences of interactions with caretakers were recorded (provision of food, drink, enrichment, tactile contact, and friendly gestures). Both species showed higher levels of agonistic behaviors and lower levels of pro-social behaviors during observations when caretaker interactions occurred compared to sessions without caretaker interaction (Chelluri et al. 2013). Although the authors considered all interactions to be positive, the behaviors exhibited during the animals' interactions with caretakers suggest that the animals may compete for attention, and that this leads to stress-related behaviors.

Stockmanship may also play a part in human–animal interactions. Developed in the domestic and agriculture industry, good stockmanship refers to the extent to which animals are managed safely, effectively, and in a manner that is low stress for the animal and keeper (Ward and Melfi 2015). Ward and Melfi (2015) evaluated stockmanship of Sulawesi macaques (*Macaca nigra*), black rhinoceroses (*Diceros bicornis*), and Chapman's zebra (*Equus quagga chapman*) in a zoo context by evaluating animal responses to different keepers that delivered different cues when moving animals from one area of the exhibit to another. Ward and Melfi found that these species reacted differently to the different cues and that some keepers were able to initiate a quicker response than others. Based on these and other findings (Ward and Melfi 2013, 2015), the authors suggest that social species, such as macaques, respond more rapidly to general keeper cues and that solitary species, such as rhino, are more influenced by individual human–animal interactions and are more likely to form specific human-animal bonds due to their solitary nature.

Clearly, the human-animal relationships that develop between caregivers and primates can vary depending on many factors. To aid our understanding in the techniques used to evaluate the impact of these relationships, we can consult principles from the domestic, agriculture, and laboratory industries. In the next section, we evaluate the impact of unfamiliar humans on zoo primate welfare.

3   Presence of Unfamiliar Humans

Zoo primates are managed by a small team of familiar humans and exposed to a daily influx of unfamiliar humans: zoo visitors (Hosey and Melfi 2014). Without attracting and engaging visitors, zoos cannot pursue the other goals that are central to the mission of zoos. The presence of zoo visitors, however, may conflict with the goal of maintaining good welfare (Fernandez et al. 2009). Visitor effects can be negative, positive, or neutral (Hosey 2005), but studies on zoo primates overwhelmingly conclude that visitors have a negative impact on welfare (Hosey 2005). Aggression and abnormal behaviors, such as fur plucking, are the most commonly reported negative behaviors associated with increased visitor numbers. For example, Mallapur et al. (2005) observed lion-tailed macaques (*Macaca silenus*) at eight Indian zoos during days when visitors were present and days when visitors were absent. They found that the macaques engaged in more abnormal behaviors on days when visitors were present. The authors of this study highlighted that the level of disturbance by visitors in Indian zoos is relatively high due to the lack of well-established conservation and animal welfare awareness programs; behaviors such as shouting, teasing, feeding, and even physically harming animals are commonplace. If this is the case, it is reasonable to assume that, in some cases, it is the behavior, proximity, and/or type of contact that visitors have with the animals that adversely affect welfare and not simply the presence of visitors. Altering visitor behavior has been shown to reduce their adverse impact. In one study, Chamove et al. (1988) found that encouraging zoo visitors to behave submissively by crouching in front of primate exhibits resulted in less aggressive behavior from the primates.

It is difficult to assess which aspect of visitor interactions primates find most stressful. Is it visual contact with humans, the noise that human visitors make, or something else? One attempt to assess visitor-related disturbances involved imposing one-way screens at a black capped capuchin (*Cebus apella*) enclosure at Melbourne Zoo, Australia. These screens allowed visitors to view the capuchins, but from the capuchin side, the viewing window looked like a white screen. The experimenters watched the animals and took biological samples during the control (no modification to viewing windows) and the reduced visual contact condition. The reduced visual contact condition resulted in a reduction in group aggression and abnormal behaviors. In addition, fecal steroid metabolites, a measure of stress response (Capitanio et al. 2022), were lower in the reduced visual contact condition than in the control condition. The screens reduced the number of visitors present at the enclosures but not visitor behavior (e.g., banging on the viewing window) and noise levels. As the screens were not soundproof, the auditory stimuli were the same in each condition. The study therefore affirmed that it was visual signals such as direct eye contact with visitors that were the fear-eliciting stimuli (Sherwen et al. 2015).

Animal–visitor interactions are potentially enriching (Claxton 2011), but there is limited empirical evidence that visitors have a positive effect on the welfare of zoo-housed primates (Hosey 2005). Also, where positive effects of interactions are described, there are confounds, such as the presence of food rewards. For example, a study of visitor–chimpanzee interactions at Chester Zoo, UK, found that the longer chimpanzees interacted with humans, the more likely they were to receive food from visitors (Cook and Hosey 1995). Receiving this food in addition to their normal diet may pose a welfare issue if it results in nutritional imbalances and/or obesity.

The relationships that zoo animals have with familiar and unfamiliar humans are likely related. Hosey (2008) proposes a model where, if interactions with familiar and unfamiliar humans are positive, animals may learn to not fear humans, and this could lead to a greater likelihood that the animal will be enriched by humans and other environmental stimuli (Fig. 1). Therefore, future studies of human-animal relationships in zoo primates should consider the relationships that primates have with familiar and unfamiliar humans.

3.1 Assessing Human-Animal Relationships

It is difficult to draw firm conclusions about the impact of human-animal relationships on the welfare of zoo-housed primates due to the sheer variety of published studies. There is no standardized method for assessing the "visitor effect" across zoological collections.

One issue comes down to enclosure design. There are many ways that primate species are exhibited. For example, lemurs can be housed in traditional cages, island exhibits, and walk through (free-ranging) exhibits (Farmer et al. 2022). Enclosure design has a large influence on visitor pressure; visitors can get very close to animals

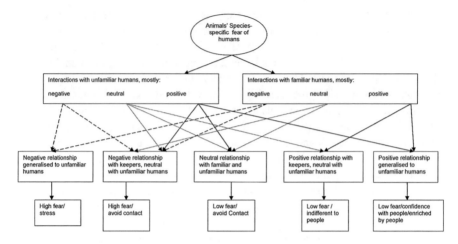

Fig. 1 A model of human–animal interactions and their consequences for human-animal relationships in zoo animals (from Hosey 2008)

in free-ranging exhibits but not when viewing animals in island enclosures or in traditional cages.

When evaluating visitor effects, one problem is how to quantify visitors. For example, researchers may count the number of visitors to an enclosure at a particular time and use these values or, as is more common, researchers may record the number of visitors as being "low" or "high." How these categories are quantified vary. For example, in their study of Diana monkey (*Cercopithecus diana*) behavior, Todd et al. (2007) coded density as being "low" if there were one to five visitors and as being "high" if there were more than five visitors. Bonnie et al.'s (2016) observations of gorilla behavior had similar category labels but different accompanying definitions: they considered 1–30 visitors "low" and more than 30 visitors "high." A different approach is to record total visitors to the zoo rather than at the species enclosure. For example, for her study of gorillas, Wells (2005) compared behavior on days of high visitor density, which they defined as weekends during the summer months (mean of 1288 visitors per day), with days of low visitor density, defined as weekdays during winter months (mean 6 visitors per day).

For most zoo research, cortisol is not used as a welfare indicator due to its cost and the inability to conduct the relevant analyses, there is also the added issue that cortisol could indicate excitement or arousal rather than stress. However, facilities that have collected cortisol data have demonstrated its potential utility in studying the visitor effect. At Chester Zoo, for example, Davis et al. (2005) collected urine samples from four female and three male spider monkeys (*Ateles geoffroyi rufiventris*) three to four times per week during opening hours and when the zoo was closed. They found that urinary cortisol was positively associated with visitor number, although this relationship was not strong and it was possibly nonlinear, and from this, concluded that other factors may have influenced the relationship.

Ultimately, the biggest issue when assessing the impact of human-animal relationships is the fact that welfare is an individual outcome: a stimulus perceived as stressful by one animal may be perceived as neutral or pleasant by another. As such, factors such as personality (Robinson and Weiss 2022) may contribute to these differences. For example, squirrel monkeys (*Saimiri sciureus*) housed at the Living Links Research Centre at the Edinburgh Zoo that were rated as being more playful and less cautious, depressed, and solitary were more likely to approach the viewing window when visitors were present (Polgár et al. 2017).

3.2 Reducing the Impact of Visitors on Welfare

Management practices have been developed to "dilute" the visitor effect. One practice is to reduce visual or auditory contact with visitors. A popular method involves the use of camouflage netting to reduce visual contact. For example, six gorillas at Belfast Zoo, Northern Ireland, UK, displayed significantly less aggression and abnormal behavior when camouflage netting was introduced (Blaney and Wells 2004). The authors noted that the barrier also encouraged quieter, more relaxed behavior on the part of visitors. Thus, the effect of the netting could be attributable to reduced visual contact, noise reduction, the change in visitor behavior, or a combination of these factors.

These techniques can introduce a conflict between visitor engagement and animal welfare. In short, although welfare may be improved by this buffering, visitor engagement may be negatively affected. For example, when screens were used to reduce visual contact between black capped capuchins and zoo visitors, the welfare of the animals was improved, but zoo visitors did not stay at the enclosure for as long as they did before the screens were introduced, potentially impacting on the opportunity to engage and/or educate visitors (Sherwen et al. 2015). However, in the study of camouflage netting in gorilla exhibits, when questioned by researchers, the public considered the animals to be more exciting and less aggressive when netting was present (Blaney and Wells 2004).

3.3 Direct Human-Animal Contact

All of the above examples deal with visitors having indirect contact with primates. However, visitor experiences, such as "keeper for a day" or "feed the animals," are becoming increasingly popular. These experiences offer the chance for visitors to get much closer to the animals—there may even be a chance for direct physical contact—and often take place at "off-show" areas that are normally only accessible by zoo staff. There have been few studies of these programs' impact on welfare.

Within the Wild Planet Trust (Paignton Zoo and Newquay Zoo) the authors of this chapter have been evaluating the implications of visitor experiences. Lemur species are a popular animal for feeding experiences due to their calm temperament. From November 2013 to February 2014, visitor feeding experiences with crowned

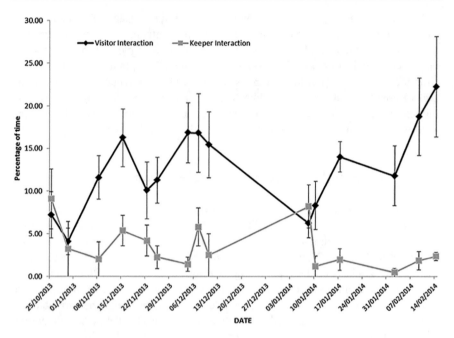

Fig. 2 Mean (± SE) percentage of time crowned lemurs ($N = 4$) spent interacting with keepers or visitors during visitor feed experiences at Newquay Zoo, UK (From Jones et al. 2016)

lemurs (*Eulemur coronatus*) at Newquay Zoo, UK were evaluated by means of behavioral observations during and immediately following either visitor feeds or keeper feeds (Jones et al. 2016). There were minor behavioral changes: during visitor feeds, lemurs spent more time interacting with keepers and less time engaged in aggressive behavior. The reduced aggression may indicate a positive effect of visitor feeds on welfare, but it should be noted that levels of aggression in all conditions were extremely low. The increased interaction with keepers during visitor feeds is probably an artifact of the feed condition, that is, during visitor feeds, keepers may have encouraged interactions for the benefit of visitors. This study also examined changes in the animal–keeper and animal–visitor interactions as the feeding experiences continued. Interestingly we found an increase in animal–visitor interaction and a decrease in animal–keeper interaction as feeding experiences progressed; however, when feeding experiences resumed after a 28-day break, interactions with keepers and visitors returned to levels seen during the first visitor feed experience (Fig. 2). This finding is interesting from two perspectives. First, it shows that the lemurs found the visitors enriching. Second, it suggests that making these events predictable, and thus allowing lemurs to habituate, enhances their well-being.

Regardless of whether visitors have direct or indirect interactions with zoo primates, it is important that the zoo industry continues to develop and standardize ways to evaluate the effects of visitors. Results of these studies, such as ours, can then be used to design programs that engage zoo visitors in such a way as to minimize any negative effects of these visitors.

4   Social Management

Many species housed in zoos are involved in captive breeding programs. These animals are bred and transferred between collections to maintain stable demographics and the genetic health of a population. The IUCN (International Union for Conservation of Nature) endorse captive breeding as an essential component in species conservation (IUCN 1987). Ex-situ species can be managed on both regional and global levels and the type of management varies (Hosey 2022; Prescott 2022).

According to the IUCN Guidelines on the Use of Ex situ Management for Species Conservation (2014), there are a number of roles for species managed through captive breeding programs. These roles are to function as an insurance population, temporary rescue, long-term *ex situ* population, demographic manipulation, source for population restoration, source for ecological replacement, source for assisted colonization, research, and/or training or for an education and awareness program (see Prescott 2022). Population restoration has only been realized for a small number of species and for primates, the main success story being the golden-lion tamarin *Leontopithecus rosalia* (see Ferreira et al. 2022). For similar projects to be successful, we need to maintain self-sustaining and genetically diverse captive populations. Therefore, there are coordinated efforts in the management of many zoo-housed primate species, which are based on genetics and demography.

In some cases, primates managed as part of captive breeding programs are housed in unnatural social groups. For species in which individuals emigrate from their natal group when they reach maturity, zoos may manipulate breeding opportunities or maintain animals in nonbreeding or in small groups, which can affect primate welfare.

Knowledge of social group composition and behavioral repertoires of wild members of a species are thus important for promoting positive welfare in zoo-housed primates. However, this information is limited for some species, because of a lack of research, the elusive nature of the species, or political and social conditions in range countries. Given the limited data on wild individuals, comparisons between zoos can be hugely beneficial. For example, following a series of male–male aggression incidents in a troop of spider monkeys, Chester Zoo, UK began an investigation into the social systems of captive groups. It is known that the social systems of wild *Ateles* involve a fission–fusion dynamic with groups comprising a variety of sex-age classes (Shimooka et al. 2010). To gather information on and to contextualize aggressive behavior in spider monkeys within zoos, Chester Zoo sent a survey to zoos that held the genus. The survey revealed that most zoo-housed spider monkeys were housed in small social groups and that, of the aggressive interactions that resulted in severe or fatal injuries, most were initiated by adult males. The results of the survey also led to several recommendations for ways to allow something like a fission–fusion system to operate, including providing larger, more complex enclosures and creating areas in which individuals can separate themselves from the group (Davis et al. 2009).

The Bornean orangutans (*Pongo pygmaeus*) at Apenheul Primate Park in Apeldoorn, Netherlands, are managed in a fission–fusion social system designed to mimic the social system of this species in the wild where male orangutans are semi-solitary and come together with females to breed (van Schaik et al. 2009). A study compared stress responses of orangutans housed in Apenheul to orangutans housed in permanent captive groups in other European zoos. The study did not find a significant difference in fecal glucocorticoid metabolites between these groups. However, individuals housed in Apenheul were more stressed by high visitor numbers. These findings suggest that, although providing a more naturalistic environment may reduce the influence of group size on social stress, there may be unanticipated welfare costs associated with these environments (Amrein et al. 2014).

Researchers often have access to studbook data for species that are managed in captivity. Studbook data allows researchers to assess the influence of social management practices on breeding success. Studbook data can also enable researchers to investigate the influence of species-specific behaviors and their effect on breeding or welfare. For example, howler monkeys (*Alouatta* spp.) are characterized by their vocalizations (Whitehead 1987, 1995), which in the wild serve many functions, including the regulation of the use of space, allowing neighbors to avoid one another, the demarcation of territory, opponent assessment, predator avoidance, and mate defense (see review by Da Cunha and Byrne 2006). In the wild, groups of black and gold howler monkey (*Alouatta caraya*) range to up to 19 animals and these groups contain adult males and females (for review see Antonio 2007). Analysis of European studbook data by Farmer et al. (2011) found that significantly more offspring were born (and survived to one year of age) to females and males housed in a family group than to pair-housed males and females. The same study found that males who had high rates of vocalizations had greater reproductive success than males who had low rates or males who did not vocalize; females housed with males who vocalized regularly also had higher rates of reproductive success. Four of the 12 males did not perform vocalizations. Based on these results, the authors recommended that zoos conduct playbacks of these vocalizations to encourage successful breeding in this species.

The performance of behaviors that may be deemed undesirable by zoo visitors, such as aggression and infanticide, are also important for maintaining social dynamics in many primate species. For example, in wild Sulawesi macaques (*Macaca nigra*), social aggression is a common and important behavior (Reed et al. 1997). The multi-male social system of macaque species leads to competition for access to receptive females. Males of similar rank engage in more aggressive interactions than those that differ in rank (Reed and O'Brien 1997). This suggests that aggressive behaviors should not be prevented in this species as these behaviors are used to maintain the social hierarchy. Moreover, by permitting these behaviors, the injuries that occur are less likely to be severe and animals can be monitored and treated quickly by veterinary staff.

4.1 Single-Sex Groups

In response to space limitations, zoos often need to manage the surplus of one sex or prevent breeding. The formation of single-sex groups is a common practice used to maintain or reduce population numbers.

The formation of single-sex groups can be difficult if such a social grouping is not common in the wild (Hosey, et al. 2009). A few primate species are reported to form bachelor groups in the wild (western lowland gorillas, Stoinski et al. 2001; chimpanzees, Fritz and Howell 1997; proboscis monkeys *Nasalis larvatus*, Sha et al. 2013; Murai 2004). With the increased success of captive breeding programs, the number of surplus animals is increasing. Consequently, single-sex groups are becoming an important management tool (Neier et al. 2013).

In zoos, western lowland gorilla bachelor groups are formed to manage the surplus of males and to socialize young males (Leeds et al. 2015) as these groups provide an environment where they can learn "appropriate" behaviors before being moved to a breeding situation. Comparable data on wild bachelor groups are mainly drawn from mountain gorillas (*Gorilla beringei beringei*) (Yamagiwa 1987; Stoinski et al. 2001; Robbins 1996). Members of wild mountain gorilla bachelor groups engage in relatively high levels of affiliative behaviors and the groups exhibit a high degree of social cohesion (Yamagiwa 1987; Robbins 1996). All captive bachelor gorilla groups are western gorillas, which, with respect to behavior, are distinct from mountain gorillas (Tutin 1996 as cited in Stoinski et al. 2001). Therefore, care must be taken in when comparing the social behaviors of the two subspecies.

Pullen (2005) compared social behaviors performed by bachelor-housed western gorillas at Paignton Zoo to males of the same species that were part of a breeding group at Belfast Zoo. Despite the small sample size ($n = 8$), the study showed that the two groups used different methods to manage social interactions and that these differences were influenced by the presence of females. Silverback males in both groups performed more aggressive behaviors than lower-ranking males. However, non-escalated aggression (chest beating with no contact) was performed more frequently by the silverback male in the breeding group. These findings suggest that the behavior of individuals in bachelor groups may differ to males in a breeding situation. More recently, Leeds et al. (2015) surveyed wounding in bachelor and mixed-sex groups of western lowland gorillas in 28 North American zoos. The study reported no differences between wounding rates in bachelor and mixed-sex groups when no young silverback males were present, but bachelor groups that contained a young silverback experienced higher rates than those without.

Because of difficulties in introducing unrelated young males into existing family groups (Johnstone-Scott 1988), Species Survival Plans (Association of Zoos and Aquariums, America) and EAZA (European Association of Zoos and Aquaria) ex situ programs for western lowland gorillas involve the formation of bachelor groups

(Pullen 2005). In North America, 23 zoos house at least one bachelor group (Neier et al. 2013) and of the 74 European zoos housing the species, 19 house one bachelor group and one houses two bachelor groups (Bemment 2016). As of 2015, half of the 22 male gorillas which had been moved from a bachelor group to a breeding situation had sired offspring (Bemment 2016). Guidelines for the successful formation and maintenance of bachelor groups includes a diversity in rearing history and a maximum group size of 3 or 4 animals, helping to ensure that bachelor groups may function as a long-term solution in managing surplus males (Stoinski et al. 2004). As it has only been commonplace to form bachelor groups over the last decade, only now are the animals involved in the initial formation of bachelor groups starting to mature. Therefore, the welfare of gorillas kept in permanent bachelor groups requires ongoing monitoring.

The white-faced saki monkey (*Pithecia pithecia*) is managed through an EAZA ex situ program. In response to a surplus of males, EAZA's management strategy involves forming bachelor groups (Webb 2017, personal communication). A study on male interactions in this species carried out in five European zoos compared social interactions between animals housed in breeding and bachelor groups. A bachelor group is considered socially compatible when low levels of aggressive interactions are observed (Fàbregas and Guillén-Salazar 2007). No significant differences in the performance of aggressive behaviors between the two saki monkey social groupings were reported. However, when the study separated aggression into physical and non-physical aggression, compared to males housed in breeding groups, males housed in bachelor groups spent more time engaged in non-physical aggression. The study suggests that bachelor group formation in white-faced saki monkeys is an effective management strategy as the males appear to have ways to avoid conflict (Prins 2015, unpublished data).

Successful bachelor group formation has been reported in other captive primate species. For instance, the formation of an all-male group of proboscis monkeys (*Nasalis larvatus*) at the Singapore Zoo resulted in less contact aggression compared to non-contact aggression, and by the sixth week of the introduction, almost all aggression had stopped (Sha et al. 2013). In wild proboscis monkeys, peripheral males form all male groups (Yeager 1990), thus this research suggests that the formation of bachelor groups may be a solution to managing surplus males in this species. Similarly, for white crowned mangabeys (*Cercocebus atys lunulatus*) housed at the Valencia Zoo, Spain, although non-contact aggression (facial threats) was performed at high rates, physical aggression was rare and mostly between animals in the same age-sex class. In addition, all animals were groomed by at least one other member, which suggests that the males were socially compatible (Fàbregas and Guillén-Salazar 2007). Similar findings have been reported in liontailed macaques (*Macaca silenus*; Stahl et al. 2001) and ruffed lemurs (*Varecia* spp.; Romano and Vermeer 2003). Even with the small sample sizes of these studies (most examined only one group), these findings suggest that bachelor groups can be an appropriate social grouping for many primate species. However, comparisons of multiple groups would provide stronger support for managing surplus males in this way; continued monitoring on the impact of non-contact aggression and the lack of reproductive opportunity, on the psychological welfare of these animals, is also essential.

4.2 Contraceptive Methods

Contraception is another means to manage surplus animals and to limit population growth. There is limited research into the long-term effects of contraception on the physiology and behavior of zoo-housed primates. Guidance on the implementation of contraception is available to EAZA member zoos through the EAZA Reproductive Management Group. For AZA institutions, the Reproductive Management Center has expanded to work toward improving reproductive management. Both organizations maintain databases containing over 30,000 records for the use of contraception for a range of species. Their aims are to provide zoos guidance on the use of contraceptives and to collate evidence on the effectiveness and reversibility of different types of contraceptives.

A range of contraceptive methods is available. However, most of the literature on contraception comes from laboratory studies with small sample sizes (see Wallace et al. 2016 for review).

Castration involves removing the testes and thus prevents testosterone production. There is limited work on the effects of castration in zoo-housed primates. In Javan langurs (*Trachypithecus auratus*), castration was used to maintain surplus males in social groups at two UK zoos (Port Lympne and Howletts Wild Animal Parks). The langurs were housed in seven groups: three bachelor pairs and four mixed-sex groups. All groups contained one intact male, with the remaining males being castrated, except for one pair of intact males and one mixed-sex group where all of the males were castrated. Bachelor pair males spent more time engaged in affiliative behaviors compared to males in mixed-sex groups. Moreover, the presence of females did not affect male–male interactions in the mixed-sex groups; all males showed a preference for females as social partners over males. Castrated males were more submissive than intact males, which suggested that castration may have influenced these males' social status (Dröscher and Waitt 2012).

Unlike castration, vasectomy involves blocking the vas deferens. This prevents the passage of sperm out of the penis. Vasectomy should not disrupt testosterone production and so is preferable to castration (Asa and Porton 2010). There is no published data concerning the impact of vasectomy on captive primate behavior. However, anecdotal evidence has been collected from two UK zoos (Paignton Zoo and Shaldon Wildlife Trust) on the use of vasectomy and contraceptive implants to prevent breeding in the white-faced saki monkey population. At Paignton Zoo, the adult male was vasectomized and no effect on behavior was reported initially (Silcocks 2014, unpublished data). However, over the first seven months after surgery, there was a decrease in the amount of social grooming this individual received from group members with grooming returning to original levels one year after the vasectomy (Thornton 2015, unpublished data). At Shaldon, the adult male of the pair was treated with a contraceptive implant Deslorelin. Social grooming rates after implantation increased over the year-long study period. The author concluded that, although both methods were effective for management and did not cause long-lasting changes in behavior, Deslorelin implants were less effective in preventing pregnancies (Thornton 2015, unpublished data).

The success of hormonal implants has been documented for a range of zoo-housed primate species, including chimpanzees (Bettinger et al. 1997; Bourry et al. 2005), western lowland gorillas (Sarfaty et al. 2012), hamadryas baboons (*Papio hamadryas*; Portugal and Asa 1995), white-faced saki monkeys (Savage et al. 2002), white-faced marmosets (*Callithrix geoffroyi*; Mustoe et al. 2012), and golden-headed lion tamarins (*Leontopithecus chrysomelas*; De Vleeschouwer et al. 2000). The contraceptive implant Norplant did not affect the duration of estrus cycles in female chimpanzees, but the duration of their sexual swellings and full-swelling phases were shorter than before implantation (Bettinger et al. 1997). Norplant was also used as a contraceptive in a troop of hamadryas baboons at Paignton Zoo. A study of this troop found no differences in self-directed behaviors or social interactions between implanted and non-implanted females (Plowman et al. 2005), although, later, many females removed their implants (Plowman, personal communication).

The long-term effect of keeping primates in a nonbreeding situation can impact their social competence and/or future breeding success, and ultimately their welfare. Preventing animals from breeding or delaying reproduction has led to reduced fertility in several mammalian and fish species (see Penfold et al. 2014 for a review), but there is only limited evidence that this occurs in nonhuman primates. A survey of zoo-raised chimpanzees that assessed the effect of rearing, age at which the animal was removed from the mother, sex, and participation in shows, revealed that there was no single aspect of rearing that influenced sexual competence; however, individuals that were reared alone with no exposure to conspecifics and individuals removed from their mother at less than 12 months were less likely to reproduce (King and Mellen 1994). By documenting the use of preventative methods in zoos (thought the AZA and EAZA reproductive management centers), we can monitor the long-term impact of captive management techniques and evaluate their success.

5 Conclusions

Primates in zoos are subject to many of the welfare challenges experienced by individuals in other captive situations, as covered elsewhere in this volume. However, zoo-housed primates face unique welfare challenges; being exposed to both familiar and unfamiliar humans on a regular basis and intensive social management to ensure self-sustaining captive populations. The modification of enclosures and management practices have been shown to mitigate the effects of visitor–primate interactions and should be considered in future enclosure designs. The welfare implications of social and genetic management of primates require ongoing monitoring in order to make future decisions evidence-based and to promote positive welfare.

References

Amrein M, Heistermann M, Weingrill T (2014) The effect of fission-fusion zoo housing on hormonal and behavioral indicators of stress in Bornean orangutans (*Pongo pygmaeus*). Int J Primatol 35:509–528. https://doi.org/10.1007/s10764-014-9765-5

Antonio AC (2007) Primate group size and abundance in the Caatinga dry forest, northeastern Brazil. Int J Primatol 28:1279–1297. https://doi.org/10.1007/s10764-007-9223-8

Asa CS, Porton IJ (2010) Contraception as a management tool for controlling surplus animals. In: Keilman DG, Thompson KV, Kirk Baer C (eds) Wild mammals in captivity: principles and techniques for zoo management, 2nd edn. University of Chicago Press, London, pp 469–482

Baker KC (2004) Benefits of positive human interaction for socially housed chimpanzees. Anim Welf 13(2):239–245

Bemment N (2016) EAZA Gorilla EEP: a review of 20 years of gorilla bachelor group management: 1995–2015. In: EAZA annual conference. EAZA, Belfast

Bettinger T, Cougar D, Lee DR et al (1997) Ovarian hormone concentrations and genital swelling patterns in female chimpanzees with Norplant implants. Zoo Biol 16(3):209–223. https://doi.org/10.1002/(SICI)1098-2361(1997)16:3<209::AID-ZOO2>3.0.CO;2-E

Blaney EC, Wells DL (2004) The influence of a camouflage net barrier on the behaviour, welfare and public perceptions of zoo-housed gorillas. Anim Welf 13:111–118

Bonnie KE, Ang MYL, Ross SR (2016) Effects of crowd size on exhibit use by and behavior of chimpanzees (*Pan troglodytes*) and Western lowland gorillas (*Gorilla gorilla*) at a zoo. Appl Anim Behav Sci 178:102–110. https://doi.org/10.1016/j.applanim.2016.03.003

Bourry O, Peignot P, Rouquet P (2005) Contraception in the chimpanzee: 12-year experience at the CIRMF Primate Centre, Gabon. J Med Primatol 34(1):25–34. https://doi.org/10.1111/j.1600-0684.2004.00088.x

Buchanan-Smith HM, Tasker L, Ash H, Graham ML (2022) Welfare of primates in laboratories: opportunities for improvement. In: Robinson LM, Weiss A (eds) Nonhuman primate welfare: from history, science, and ethics to practice. Springer, Cham, pp 97–120

Capitanio JP, Vandeleest J, Hannibal DL (2022) Physiological measures of welfare. In: Robinson LM, Weiss A (eds) Nonhuman primate welfare: from history, science, and ethics to practice. Springer, Cham, pp 231–254

Chamove AS, Hosey G, Schaetzel P (1988) Visitors excite primates in zoos. Zoo Biol 7(4):359–369. https://doi.org/10.1002/zoo.1430070407

Chelluri GI, Ross SR, Wagner KE (2013) Behavioral correlates and welfare implications of informal interactions between caretakers and zoo-housed chimpanzees and gorillas. Appl Anim Behav Sci 147(3-4):306–315. https://doi.org/10.1016/j.applanim.2012.06.008

Claxton AM (2011) The potential of the human-animal relationship as an environmental enrichment for the welfare of zoo-housed animals. Appl Anim Behav Sci 133(1-2):1–10. https://doi.org/10.1016/j.applanim.2011.03.002

Coe JC (2003) Steering the ark toward Eden: design for animal well-being. J Am Vet Med Assoc 223(7):977–980. https://doi.org/10.2460/javma.2003.223.977

Coleman K, Timmel G, Prongay K, Baker KC (2022) Common husbandry, housing, and animal care practices. In: Robinson LM, Weiss A (eds) Nonhuman primate welfare: from history, science, and ethics to practice. Springer, Cham, pp 317–348

Cook S, Hosey GR (1995) Interaction sequences between chimpanzees and human visitors at the zoo. Zoo Biol 14(5):431–440. https://doi.org/10.1002/zoo.1430140505

Da Cunha RGT, Byrne RW (2006) Roars of black howler monkeys (*Alouatta caraya*): evidence for a function in inter-group spacing. Behaviour 143(10):1169–1199. https://doi.org/10.1163/156853906778691568

Davis N, Schaffner CM, Smith TE (2005) Evidence that zoo visitors influence HPA activity in spider monkeys (*Ateles geoffroyii rufiventris*). Appl Anim Behav Sci 90(2):131–141. https://doi.org/10.1016/j.applanim.2004.08.020

Davis N, Schaffner CM, Wehnelt S (2009) Patterns of injury in zoo-housed spider monkeys: a problem with males? Appl Anim Behav Sci 116(2-4):250–259. https://doi.org/10.1016/j.applanim.2008.08.008

De Vleeschouwer K, Leus K, Van Elsacker L (2000) An evaluation of the suitability of contraceptive methods in golden-headed lion tamarins (*Leontopithecus chrysomelas*), with emphasis on melengestrol acetate (MGA) implants: (I) effectiveness, reversibility and medical side-effects. Anim Welf 9:251–271

Dröscher I, Waitt CD (2012) Social housing of surplus males of Javan langurs (*Trachypithecus auratus*): compatibility of intact and castrated males in different social settings. Appl Anim Behav Sci 141(3-4):184–190. https://doi.org/10.1016/j.applanim.2012.08.001

Fàbregas M, Guillén-Salazar F (2007) Social compatibility in a newly formed all-male group of white crowned mangabeys (*Cercocebus atys lunulatus*). Zoo Biol 26(1):63–69. https://doi.org/10.1002/zoo.20117

Farmer HL, Plowman AB, Leaver LA (2011) Role of vocalisations and social housing in breeding in captive howler monkeys (*Alouatta caraya*). Appl Anim Behav Sci 134(3-4):177–183. https://doi.org/10.1016/j.applanim.2011.07.005

Farmer HL, Baker KR, Cabana F (2022) Housing and husbandry for primates in zoos. In: Robinson LM, Weiss A (eds) Nonhuman primate welfare: from history, science, and ethics to practice. Springer, Cham, pp 349–368

Fernandez EJ, Tamborski MA, Pickens SR, Timberlake W (2009) Animal-visitor interactions in the modern zoo: conflicts and interventions. Appl Anim Behav Sci 120(1-2):1–8. https://doi.org/10.1016/j.applanim.2009.06.002

Ferreira RG, Ruiz-Miranda C, Sita S, Sánchez-López S, Pissinatti A, Corte S, Jerusalinsky L, Wagner PG, Maas C (2022) Primates under human care in developing countries: examples from Latin America. In: Robinson LM, Weiss A (eds) Nonhuman primate welfare: from history, science, and ethics to practice. Springer, Cham, pp 145–170

Fritz J, Howell S (1997) The behavior of captive male chimpanzees (*Pan troglodytes*) housed in multi-male bachelor versus mixed-sex social groups at the Primate Foundation of Arizona. Am J Primatol 49(1):54. https://doi.org/10.1002/(SICI)1098-2345(1999)49:1%3C39::AID-AJP3%3E3.0.CO;2-9

Godinez AM, Fernandez EJ (2019) What is the zoo experience? How zoos impact a visitor's behaviors, perceptions, and conservation efforts. Front Psychol 10. https://doi.org/10.3389/fpsyg.2019.01746

Hosey G (2005) How does the zoo environment affect the behaviour of captive primates? Appl Anim Behav Sci 90(2):107–129. https://doi.org/10.1016/j.applanim.2004.08.015

Hosey G (2008) A preliminary model of human-animal relationships in the zoo. Appl Anim Behav Sci 109(2-4):105–127. https://doi.org/10.1016/j.applanim.2007.04.013

Hosey G (2022) The history of primates in zoos. In: Robinson LM, Weiss A (eds) Nonhuman primate welfare: from history, science, and ethics to practice. Springer, Cham, pp 3–30

Hosey G, Melfi V (2012) Human-animal bonds between zoo professionals and the animals in their care. Zoo Biol 31(1):13–26. https://doi.org/10.1002/zoo.20359

Hosey G, Melfi V (2014) Are we ignoring neutral and negative human-animal relationships in zoos? Zoo Biol 34(1):1–8. https://doi.org/10.1002/zoo.21182

Hosey G, Melfi VA, Pankhurst S (2009) Captive breeding. In: Hosey G, Melfi VA, Pankhurst S (eds) Zoo animals: behaviour, management and welfare. Oxford University Press, New York, pp 292–345

IUCN (World Conservation Union) (1987) The IUCN policy statement on captive breeding. IUCN, Gland, Switzerland

Jensvold MLA (2008) Chimpanzee (*Pan troglodytes*) responses to caregiver use of chimpanzee behaviors. Zoo Biol 27(5):345–359. https://doi.org/10.1002/zoo.20194

Johnstone-Scott R (1988) The potential for establishing bachelor groups of western lowland gorillas *Gorilla g. gorilla*. Dodo 25:61–66

Jones H, McGregor PK, Farmer HLA, Baker KR (2016) The influence of visitor interaction on the behavior of captive crowned lemurs (*Eulemur coronatus*) and implications for welfare. Zoo Biol 35(3):222–227. https://doi.org/10.1002/zoo.21291

Kemp C (2022) Enrichment. In: Robinson LM, Weiss A (eds) Nonhuman primate welfare: from history, science, and ethics to practice. Springer, Cham, pp 451–488

King NE, Mellen JD (1994) The effects of early experience on adult copulatory behavior in zoo-born chimpanzees (*Pan troglodytes*). Zoo Biol 13(1):51–59. https://doi.org/10.1002/zoo.1430130107

Leeds A, Boyer D, Ross SR, Lukas KE (2015) The effects of group type and young silverbacks on wounding rates in western lowland gorilla (*Gorilla gorilla gorilla*) groups in North American zoos. Zoo Biol 34(4):296–304. https://doi.org/10.1002/zoo.21218

Mallapur A, Sinha A, Waran N (2005) Influence of visitor presence on the behaviour of captive lion-tailed macaques (*Macaca silenus*) housed in Indian zoos. Appl Anim Behav Sci 94(3-4):341–352. https://doi.org/10.1016/j.applanim.2005.02.012

Melfi VA, Thomas S (2005) Can training zoo-housed primates compromise their conservation? A case study using Abyssinian colobus monkeys (*Colobus guereza*). Anthrozoös 18(3):304–317. https://doi.org/10.2752/089279305785594063

Murai T (2004) Social behaviors of all-male proboscis monkeys when joined by females. Ecol Res 19(4):451–454. https://doi.org/10.1111/j.1440-1703.2004.00656.x

Mustoe AC, Jensen HA, French JA (2012) Describing ovarian cycles, pregnancy characteristics, and the use of contraception in female white-faced marmosets, *Callithrix geoffroyi*. Am J Primatol 74(11):1044–1053. https://doi.org/10.1002/ajp.22058

Neier, B., Boyer, D., Lukas, K., Ross S (2013) Wounding rates in bachelor and mixed sex groupings of lowland gorillas (*Gorilla gorilla gorilla*). Am J Primatol 75(S1):61. https://doi.org/10.1002/ajp.22188

Penfold LM, Powell D, Traylor-Holzer K, Asa CS (2014) "Use it or lose it": characterization, implications, and mitigation of female infertility in captive wildlife. Zoo Biol 33(1):20–28. https://doi.org/10.1002/zoo.21104

Plowman AB, Jordan NR, Anderson N et al (2005) Welfare implications of captive primate population management: Behavioural and psycho-social effects of female-based contraception, oestrus and male removal in hamadryas baboons (*Papio hamadryas*). Appl Anim Behav Sci 90(2):155–165. https://doi.org/10.1016/j.applanim.2004.08.014

Polgár Z, Wood L, Haskell MJ (2017) Individual differences in zoo-housed squirrel monkeys' (*Saimiri sciureus*) reactions to visitors, research participation, and personality ratings. Am J Primatol 79(5):1–10. https://doi.org/10.1002/ajp.22639

Portugal MM, Asa CS (1995) Effects of chronic melengestrol acetate contraceptive treatment on perineal tumescence, body weight, and sociosexual behavior of hamadryas baboons (*Papio hamadryas*). Zoo Biol 14(3):251–259. https://doi.org/10.1002/zoo.1430140306

Prescott MJ (2022) Using primates in captivity: research, conservation, and education. In: Robinson LM, Weiss A (eds) Nonhuman primate welfare: from history, science, and ethics to practice. Springer, Cham, pp 57–78

Prins EF (2015) Surplus management techniques for captive white-faced saki monkeys (*Pithecia pithecia*): contraception and bachelor group formation. University of Plymouth

Pullen PK (2005) Preliminary comparisons of male/male interactions within bachelor and breeding groups of western lowland gorillas (*Gorilla gorilla gorilla*). Appl Anim Behav Sci 90(2):143–153. https://doi.org/10.1016/j.applanim.2004.08.016

Reed C, O'Brien TG, Kinnaird M (1997) Male social behavior and dominance hierarchy in the Sulawesi crested black macaque (*Macaca nigra*). Int J Primatol 18:247–260. https://doi.org/10.1023/A:1026376720249

Robbins MM (1996) Male-male interactions in heterosexual and all-male wild mountain gorilla groups. Ethology 102(7):942–965. https://doi.org/10.1111/j.1439-0310.1996.tb01172.x

Robinson LM, Weiss A (2022) Primate personality and welfare. In: Robinson LM, Weiss A (eds) Nonhuman primate welfare: from history, science, and ethics to practice. Springer, Cham, pp 387–402

Romano G, Vermeer J (2003) Preliminary observations on a bachelor group of ruffed lemurs at La Vallee des Singes. Int Zoo News 50:5–8

Sarfaty A, Margulis SW, Atsalis S (2012) Effects of combination birth control on estrous behavior in captive western lowland gorillas, *Gorilla gorilla gorilla*. Zoo Biol 31(3):350–361. https://doi.org/10.1002/zoo.20401

Savage A, Zirofsky DS, Shideler SE et al (2002) Use of levonorgestrel as an effective means of contraception in the white-faced saki (*Pithecia pithecia*). Zoo Biol 21(1):49–57. https://doi.org/10.1002/zoo.10006

Sha JCM, Alagappasamy S, Chandran S et al (2013) Establishment of a captive all-male group of proboscis monkey (*Nasalis larvatus*) at the Singapore Zoo. Zoo Biol 32(3):281–290. https://doi.org/10.1002/zoo.21020

Sherwen SL, Harvey TJ, Magrath MJL et al (2015) Effects of visual contact with zoo visitors on black-capped capuchin welfare. Appl Anim Behav Sci 167:65–73. https://doi.org/10.1016/j.applanim.2015.03.004

Shimooka Y, Campbell CJ, Di Fiore A et al (2010) Demography and group composition of Ateles. In: Campbell CJ (ed) Spider monkeys: behavior, ecology and evolution of the genus Ateles. Cambridge University Press, Cambridge, pp 329–348

Spiezio C, Piva F, Regaiolli B, Vaglio S (2015) Positive reinforcement training: a tool for care and management of captive vervet monkeys (*Chlorocebus aethiops*). Anim Welf 24(3):283–290. https://doi.org/10.7120/09627286.24.3.283

Stahl D, Herrmann F, Kaumanns W (2001) Group formation of a captive all-male group of lion-tailed macaques (*Macaca silenus*). Primate Rep 59:93–108

Stoinski TS, Hoff MP, Lukas KE, Maple TL (2001) A preliminary behavioral comparison of two captive all-male gorilla groups. Zoo Biol 20(1):27–40. https://doi.org/10.1002/zoo.1003

Stoinski TS, Lukas KE, Kuhar CW, Maple TL (2004) Factors influencing the formation and maintenance of all-male gorilla groups in captivity. Zoo Biol 23(3):189–203. https://doi.org/10.1002/zoo.20005

Todd PA, Macdonald C, Coleman D (2007) Visitor-associated variation in captive Diana monkey (*Cercopithecus diana diana*) behaviour. Appl Anim Behav Sci 107(1-2):162–165. https://doi.org/10.1016/j.applanim.2006.09.010

van Schaik CP, Marshall AJ, Wich SA (2009) Geographic variation in orangutan behavior and biology: its functional interpretation and its mechanistic basis. In: Wich SA, Atmoko SSU, Setia TM, van Schaik CP (eds) Orangutans: geographic variation in behavioral ecology and conservation, Oxford University Press, New York, pp 351–361.

Wallace PY, Asa CS, Agnew M, Cheyne SM (2016) A review of population control methods in captive-housed primates. Anim Welf 25(1):7–20. https://doi.org/10.7120/09627286.25.1.007

Ward SJ, Melfi V (2013) The implications of husbandry training on zoo animal response rates. Appl Anim Behav Sci 147(1-2):179–185. https://doi.org/10.1016/j.applanim.2013.05.008

Ward SJ, Melfi V (2015) Keeper-animal interactions: differences between the behaviour of zoo animals affect stockmanship. PLoS One 10(10):e0140237. https://doi.org/10.1371/journal.pone.0140237

Wells DL (2005) A note on the influence of visitors on the behaviour and welfare of zoo-housed gorillas. Appl Anim Behav Sci 93(1-2):13–17. https://doi.org/10.1016/j.applanim.2005.06.019

Whitehead JM (1987) Vocally mediated reciprocity between neighbouring groups of mantled howling monkeys, *Alouatta palliata palliata*. Anim Behav 35(6):1615–1627. https://doi.org/10.1016/S0003-3472(87)80054-4

Whitehead JM (1995) *Vox alouattinae*: a preliminary survey of the acoustic characteristics of long-distance calls of howling monkeys. Int J Primatol 16:121–144. https://doi.org/10.1007/BF02700156

Yamagiwa J (1987) Intra- and inter-group interactions of an all-male group of Virunga mountain gorillas (*Gorilla gorilla beringei*). Primates 28:1–30. https://doi.org/10.1007/BF02382180

Yeager CP (1990) Proboscis monkey (*Nasalis larvatus*) social organization: group structure. Am J Primatol 20(2):95–106. https://doi.org/10.1002/ajp.1350200204

Welfare of Primates in Laboratories: Opportunities for Refinement

Hannah M. Buchanan-Smith, Lou Tasker, Hayley Ash, and Melanie L. Graham

Abstract

The use of primates in regulated research and testing means that they are intentionally subjected to scientific procedures that have the potential to cause pain, suffering, distress, or lasting harm. These harms, combined with keeping primates in restricted laboratory conditions, are balanced against the potential (primarily human) benefits gained from their use. In this chapter, we provide a brief overview of the use of primates in laboratories, the estimated number, and purpose of use, and summarize the evidence that primates are especially vulnerable and deserve special protection compared to other animals. The 3Rs (replacement, reduction, and refinement) framework, underpinning humane science, is described, and we emphasize both the ethical and scientific needs for refinement. Refinement refers to all approaches used (by humans responsible for their care) to minimize harms and improve welfare for those primates that are still used in research after the application of the replacement and reduction principles. There is a growing body of evidence demonstrating an interplay between animals' welfare and experimental parameters, and that this interplay affects the validity and reliability of scientific output. With this perspective, we argue that it is better to collect no data than to collect poor (e.g., invalid, unreliable) data. It is, after all, unacceptable for primates to suffer in vain and violates utilitarian principles underlying animal use. Furthermore, inconsistency in experimental approach may introduce conflicting results, increasing the likelihood of using more animals, and delaying delivery of promising therapies to the clinic. We focus

H. M. Buchanan-Smith (✉) · L. Tasker · H. Ash
Faculty of Natural Sciences, University of Stirling, Stirling, Scotland, UK
e-mail: h.m.buchanan-smith@stir.ac.uk

M. L. Graham
Department of Surgery, University of Minnesota, St. Paul, MN, USA

Veterinary Population Medicine Department, University of Minnesota, St. Paul, MN, USA

on mitigating the major welfare issues faced by primates housed in laboratories through coordinated refinements across their life spans. Drawing on examples from cynomolgus macaques (*Macaca fascicularis*), an Old World monkey commonly used during the development of medical products, we highlight the importance of understanding the critical role humans play in the laboratory, providing environments, performing husbandry, and undertaking procedures that promote welfare and decrease harms. Our theoretical premise is that if primates are to be "fit for purpose" (i.e., well suited for the designated role), we need a proactive, concerted approach for implementing refinement that spans their lifetime.

Keywords

3Rs · Fit for purpose · Regulated research · *Macaca fascicularis* · Reliable · Valid data

1 Introduction

> ...refinement is never enough, and we should always seek further for reduction and if possible replacement. (Russell and Burch 1959, p. 66)

Using animals for research and testing in laboratories has, by its nature, the potential to cause "pain, suffering, distress or lasting harm." This is precisely how regulated scientific procedures are defined (e.g., Home Office 1986), and as such, scientific research is strictly controlled through legislation in many (but not all) countries (Bayne et al. 2010). Intentionally conducting scientific procedures that have the potential to adversely affect the welfare of animals raises its own ethical issues, and the use of nonhuman primates (hereafter primates), as opposed to other animals, is also a special case. In this chapter, we focus on regulated laboratory studies (and not unlicensed behavioral or cognitive research on primates), describing the ethical framework of the 3Rs, the importance of promoting welfare given the link with quality of scientific output, and the major welfare issues affecting primates in laboratories. Our main emphasis lies with how we can improve the welfare of laboratory-housed primates through coordinated refinements across the lifespan, recognizing the critical role humans play in devising opportunities for reducing harms, and advancing primate welfare.

Animal welfare has been the focus of scientific study for many years yet constructing a single definition and approach to measurement has been difficult (reviewed in Fraser 2009). It is accepted that welfare is broad in concept, multidimensional in nature (Dawkins 2004), and lies on a continuum from poor to good (Broom 1999). In this chapter, we adopt an integrated approach to the concept and assessment of welfare that includes both physical and psychological aspects. Defined by Broom (1986, 2010), the welfare of an animal is its state as regard its attempts to cope with its environment, such that failure to cope leads to profound deviations in biological functioning. Thus, animal welfare, as a biological state within the animal, is relevant to scientists who use primates in biomedical research and testing to benefit humans in some way. When primate welfare is poor, primates are not "fit for purpose" as models of normal functioning.

2 Differences Between Laboratories and Other Captive Settings

Factors affecting the welfare of primates housed in laboratories differ in multiple ways from those in other captive settings (see Table 2). As described in this chapter, laboratory animals have regulated procedures conducted upon them to characterize a pathophysiologic process or intentionally model clinical disease that can result in pain, suffering, or distress similar to the target patient. Factors negatively affecting welfare of primates in zoos may be high visitor numbers (Hosey 2000, 2022), a stressor not present in laboratories. There are some rare cases where zoo-housed primates are released back into the wild, with concomitant stress (such as the golden lion tamarin, *Leontopithecus rosalia*, Teixeira et al. 2007). Nonetheless, compared to laboratory-housed primates, zoo-housed primates are likely to have better welfare. They are usually housed in more natural social groupings with conspecifics of both sexes and all (or nearly all) age classes. Their enclosures are also comparatively larger, more complex, and include access to outdoor runs providing more choice and control (Coleman et al. 2022). Primates kept as pets, however, have a host of welfare-related issues related to inappropriate rearing, housing, and husbandry (Hevesi 2022), as do those who live in sanctuaries (Brent 2007) given their life experiences.

3 Primate Use in Laboratories

The number of primates used in laboratories worldwide is not known exactly, but estimates over 16 years ago were in the region of one to two hundred thousand (Carlsson et al. 2004). Primates are used as models for humans because of their genetic, physiological, and psychological similarities, primarily in the fields of microbiology, immunology, neuroscience, biochemistry, pharmacology, and toxicology (see Hau et al. 2000; Carlsson et al. 2004; Weatherall et al. 2006; Chapman et al. 2010).

The most commonly used species of primate used for research and testing are rhesus macaques (*Macaca mulatta*) and cynomolgus macaques (*M. fascicularis*). A range of other Old World monkeys are used less frequently, such as pigtail macaques (*M. nemestrina*), baboons (*Papio* spp.), and vervet monkeys (*Chlorocebus pygerythrus*). Of the New World monkeys, common marmosets (*Callithrix jacchus*) are the most frequently used, but also others such as tamarins (*Saguinus* spp.), and squirrel (*Saimiri* spp.), capuchin (*Sapajus* spp.), and night (*Aotus* spp.) monkeys. Chimpanzees (*Pan troglodytes*) were the only great ape used in biomedical research, but European legislation (European Parliament and the Council of the European Union 2010 and Turner 2022), and other bans or National Institutes of Health funding limitations mean they are now no longer or rarely used (Kaiser 2013; Graham and Prescott 2015; Grimm 2015). However, Gabon continues to conduct biomedical research on chimpanzees (Kaiser 2013).

Primates are usually purpose-bred for use in research. The use of wild-caught animals is generally no longer accepted (McCann et al. 2007; European Commission 2010). This is because of the stress of capture and transport, and the associated morbidity and mortality (McCann et al. 2007), and presence of disease (Weber et al. 1999). In addition, they may be less suitable as models when their life histories are not known (Howard 2002) as this may introduce unwanted variation and bias experimental outcomes, resulting in studies lacking in experimental rigor and that are consequently less robust in prediction (Howard 2002).

Purpose-bred primates maintain their evolved capacities (i.e., adaptations to survive and reproduce in the wild), and so an understanding of natural history is critical if we are to provide environments that may help to promote welfare. The Jennings and Prescott (2009) report chapters in the Universities Federation for Animal Welfare (UFAW) Handbook (2010; 9th ed. in prep), and Marini et al. (2019) provide important information about the natural history, veterinary, and welfare aspects of the most commonly used primates in research.

Not all animals are protected by legislation, and this varies by country (Bayne et al. 2022). For example, in Europe all vertebrates and some invertebrates are protected, and certain animals, such as primates, cats, dogs, and Equidae, get special protection (European Commission 2010). It appears that this is due to public concern for animals that humans keep as pets or close companions, and our ability to empathize with these animals. There is, however, no robust scientific evidence that the animals with special protection are capable of suffering more than those without special protection (Buchanan-Smith 2010b; Hubrecht 2014).

What evidence is there to suggest that primates require special protection? Their phylogenetic closeness to humans is exactly the reason why they are used, and this similarity is also the basis for our apprehension concerning their use (i.e., they may suffer like humans). The brains of primates are larger in relation to body size than other mammals used in laboratories (Dunbar and Shultz 2007), and brain size is associated with mental capacities and cognitive complexity. However, cognitive complexity does not necessarily mean a greater potential to suffer pain (see Mendl and Paul 2004). The sensation of pain may be the same for an individual which *experiences* it as to one which is *consciously aware* and feels pain (Bekoff 2002). An animal therefore does not need to be self-aware to experience pain. Indeed, Broom (2010) has argued that cognitive complexity may reduce suffering as it helps individuals cope with adverse conditions and allows for more possibilities of pleasure. On the other hand, the most cognitively complex primates, the great apes, may also be able to empathize with the suffering of others, and dread future events, increasing their own ability to suffer (Smith and Boyd 2002; Mendl and Paul 2004). These arguments are not fully evaluated in relation to welfare, but suggest that primates, and certainly great apes, are indeed a special case.

In our view, the strongest scientific arguments that primates require special protection are (1) the intricacy of adverse effects resulting from inappropriate rearing (e.g., Parker and Maestripieri 2011) and (2) that their larger brains have evolved for dealing with the complexity of their social and physical worlds. Primates are long-lived compared with other laboratory animals (e.g., rodents), and this poses challenges for care staff who need to provide the opportunities for good welfare throughout their lives, including as their needs change. The provision of appropriate rearing, together with physical and social complexities in the laboratory environment, can be very challenging given the constraints of laboratory life and the requirements of studies. Table 2 outlines the key stages in a cynomolgus macaque life cycle, with potential negative welfare impacts and opportunities for refinement.

There is considerable debate about whether the suffering that primates experience in laboratory research is cumulative (see Honess and Wolfensohn 2010; Pickard et al. 2013; Wolfensohn et al. 2015). What is clear is that some individuals are unable to cope and are euthanized (Wolfensohn 2022). This may be due to additive stacking up when "the residual effects of repeated procedures may add up" or it may be due to additive potentiation when "suffering from earlier events may actually increase the negative impact on welfare of subsequent events" (Pickard et al. 2013, p. 6). In addition to suffering from direct scientific procedures (e.g., surgery, disease modeling, adverse effects from a test item), the intelligence of primates means they may suffer from boredom and fear. Therefore, the consequences of this, inadequate rearing histories, and environments, together with the scientific procedures conducted upon them, can lead to poor welfare (Buchanan-Smith 2010a, b) (Table 1).

4   Ethical Framework of 3Rs and Welfare

The utilitarian approach is adopted for dealing with the ethical dilemma of using animals in research and testing, is enshrined in legislation, and underpins many local ethical review processes (e.g., European Commission 2010; U.S. Department of Agriculture 2013; including China and India; see Graham and Prescott 2015). This pragmatic approach weighs the ethical importance of the individual and their capacity to suffer against the interests of the other parties concerned (Singer 1975; Sandøe et al. 1997). In practice, this approach is known as a harm-benefit assessment. It is currently applied to primate use prospectively by mandatory ethics boards (e.g., Institutional Animal Care and Use Committees, Animal Welfare and Ethical Review Body) and retrospectively where the actual costs to the animal are reviewed in light of the results of scientific study (European Commission 2010). The perceived harm to the animal in terms of its likely experience of pain, distress, or lasting harm, including intensity, duration, and frequency, is weighed against the anticipated benefits of the research for humans (or other animals or the environment) (Graham and Prescott 2015).

In addition to requiring assessment of the harms and benefits of the proposed research, the legislative framework requires the 3R principle be applied to the project

Table 1 Examples of issues that may compromise welfare of primates (and many other animals) housed and used in laboratory research and testing, and associated refinements

Housing and husbandry	Example of welfare compromise	Refinement opportunity
Individual identification.	Freeze branding, tattooing, microchip, and temporary methods may be painful and impact on behavior (reviewed in Rennie and Buchanan-Smith 2006b)	Sensitive placement of tattoos (never on face). Primates should be anesthetized for tattooing. Combined temporary and permanent methods to minimize frequency of intrusive handling, together with positive reinforcement training (PRT) to accept scanning of microchips, or accept temporary dyes (reviewed in Rennie and Buchanan-Smith 2006b).
Small enclosures lacking environmental complexity.	Space is restricted in laboratories, and few have outdoor areas, limiting the ability to perform species-typical behavioral repertoires. Primates may become bored (Buchanan-Smith 2010a, b).	Factors that should be taken into account when determining enclosure size and design include morphometric, ecological, locomotor, physiological, social, reproductive, and behavioral characteristics (see Buchanan-Smith et al. 2004). Increasing choice, complexity, and opportunities for control improve welfare and coping ability (see Buchanan-Smith and Badihi 2012).
Separation from family earlier than weaning would normally occur in the wild. Unnatural social groups and loss of social support.	Early weaning has a range of physical and psychological effects that negatively impact welfare and quality of scientific output, including behavioral disturbances (e.g., stereotypies and self-injury), growth, health and survival, and immune consequences (reviewed in Prescott et al. 2012).	Prescott et al. (2012) describe the range of factors that should determine weaning age in macaques, with a focus on behavioral, weight, and health criteria, as well as age (not normally less than 10–14 months specified). Marmosets and tamarins should remain in their natal groups until at least 8 months, and 12 months for those destined to breed, to gain experience with rearing 2 sets of younger siblings (Council of Europe Appendix A to Convention ETS 123).

(continued)

Table 1 (continued)

Housing and husbandry	Example of welfare compromise	Refinement opportunity
Regular room and cage changes, with changes in grouping. Noisy enclosures—often metal, rooms may be power-hosed.	Room and cage change (Crockett et al. 1995, 2000), and social regrouping (Shively et al. 1997) adversely impact macaques. Regular changes in rooms/social grouping may lead to instability. Cage cleaning is stressful and masks olfactory communication (Epple 1970).	Careful advance planning may increase stability of groups and rooms. For marmosets, where olfactory communication is important, ensure some continuity of familiar scents (e.g., cleaning half the cage, or keep one branch or enrichment device which has been scent marked) (Prescott 2006).
Regulated scientific procedure	Example of welfare compromise	Refinement opportunity
Use as disease models (e.g., Parkinson's disease (PD), arthritis), which may include genetic modification.	MPTP-treated monkeys used as models for PD show akinesia, rigidity, and postural abnormalities; some display a "climbing syndrome" or "obstinate progression syndrome" (reviewed in Vitale et al. 2009). The development of a transgenic model of Huntington's disease, a neurodegenerative disorder characterized in humans by motor impairment, cognitive deterioration, and psychiatric disturbances followed by death within 10–15 years of the onset of the symptoms. The transgenic rhesus macaque exhibits clinical features including dystonia and chorea (Yang et al. 2008).	Physical and social refinements for PD primates are comprehensively described by Vitale et al. (2009) at all stages, including preparation, injections, restraint, and at the various stages of disease progression, to humane end points. These include soft enclosures for individuals who climb and fall, to minimize injury. For genetically modified animals: appropriate treatment of conditions produced, restriction of gene expression to tissues of interest or to certain time periods, and clear criteria to remove primates from a study, or humanely end life to stop further suffering (Dennis 2002).
Toxicology testing	The test substance may cause sickness and health deterioration. Historically, primates used in toxicology studies have had limited social and physical environmental enrichment, given concerns about confounding or negating the study data (e.g., unstable groups, ingesting material) (Bayne 2003).	A list of refinements including social and physical enrichment, and refinements to capture, handling, restraint, and administration and sampling is provided in Rennie and Buchanan-Smith (2006a, b, Rennie and Buchanan-Smith 2006). Consideration should be given to providing comfortable, quiet areas to individuals suffering the effects of administered substances or surgery, and this impacts on types of environmental enrichment appropriate.

Regulated scientific procedure	Example of welfare compromise	Refinement opportunity
Surgery	Surgery is common, for example, in neuroscience and implanting internal telemetry devices for remote recording, but even with appropriate anesthesia and analgesics surgery may lead to complications (e.g., Rennie and Buchanan-Smith 2006b; Pickard et al. 2013).	Improvement in headposts includes biocompatible titanium, which is simpler to implant, more securely anchored, easier to maintain, and less obtrusive than devices attached with traditional acrylic (Adams et al. 2007). Morton et al. (2003) provide an account of refinements for all aspects of telemetry.
Capture, handling, and restraint to collect data (such as blood samples, electrocardiogram), administer substances, and provide medical care, which often has intrinsically aversive components for disease models.	There is extensive evidence that capture, handling, and restraint can be stressful (reviewed in Rennie and Buchanan-Smith 2006b). Drug Metabolism and Pharmacokinetics (DMPK) requires sampling at fixed intervals (e.g., waking animals at night), and sleep deprivation adversely alters biological functioning (McEwen 2006).	Human socialization and PRT are key to minimizing stress associated with restraint (see Prescott and Buchanan-Smith 2007). Careful planning of housing can minimize disruption of nonstudy animals and DMPK animals, without disruption to social groups. Methods of sampling and refinements, including PRT, long-term catheterization techniques to reduce painful catheter starts or more invasive approaches for blood collection/drug delivery (e.g., portal vascular access), and sonophoresis, are reviewed in Rennie and Buchanan-Smith (2006b). Refinement for administration of substances is reviewed in Rennie and Buchanan-Smith (2006b) and Morton et al. (2001). Caron et al. (2015) describe miniaturized blood sampling techniques to minimize volume required to be taken.
Single housing in metabolism cages, to allow collection of samples (e.g., urine).	It is widely recognized that single housing of primates is detrimental to psychological well-being (e.g., Hartner et al. 2001; McCann et al. 2007)	Primates can be trained to produce a urine sample on request (McKinley et al. 2003) or other mediums such as saliva may be collected (e.g., Lutz et al. 2000; Ash et al. 2018) obviating the need for single housing.

(continued)

Table 1 (continued)

Regulated scientific procedure	Example of welfare compromise	Refinement opportunity
Food and fluid restriction.	Prescott et al. (2010) describes the (understudied) effects of food and fluid restriction on physiological and behavioral responses in animals, which can potentially compromise their health and well-being.	Prescott et al. (2010) details refinements to food and fluid control as motivational tools for macaques used in behavioral neuroscience research, including alternatives and type of reward (e.g., appetitive rewards), level of control, and breaks in regimen.
Stress in anticipation of event.	The order in which samples are taken are known to affect blood cell counts and plasma cortisol (Capitanio et al. 1996; Flow and Jaques 1997).	Individuals should not be restrained and dosed or sampled in view of others. Reliably signaling a stressful event for individual animals may reduce stress (see Bassett and Buchanan-Smith 2007).
Intentional death—this is often required as part of the experiment, to allow postmortem analysis, or when primates are no longer required and cannot be reused.	Rennie and Buchanan-Smith (2006b) describe issues related to welfare leading up to euthanasia, and humane end points.	Extremely competent staff and use of PRT are important refinements in euthanasia—administration of the euthanizing agent must result in rapid loss of consciousness before death ensues (Rennie and Buchanan-Smith 2006b). Rehoming potential is also discussed. The OECD (2000) describes refinements, including validation, use of earlier end points, and avoidance of using death and moribundity as end points.

from its experimental design to its execution (Home Office 1986; European Commission 2010). The 3Rs are replacement, reduction, and refinement. Replacement is concerned with the absolute or relative replacement of animals for scientific use. Reduction emphasizes the need to reduce to a minimum the number of animals through good experimental design, the sharing of data and/or resources, or by using modern techniques to obtain more information from the same number of animals (thereby reducing future use of animals). Refinement has been defined as "any approach which avoids or minimises the actual or potential pain, distress and other adverse effects experienced at any time during the life of the animals involved, and which enhances their well-being" (Buchanan-Smith et al. 2005, pp. 379–380). This definition highlights the need to consider all stages of an animal's life, from birth to death, including the promotion of good welfare of breeding animals not used in research, and involves not just minimization of harms, but takes a proactive stance to enhance welfare through to death with the use of humane end points as required (Wolfensohn 2022). It should also be noted that, although each R is often considered separately, they have a complex interplay (de Boo et al. 2005). For example, reuse may reduce the number of individuals used, but increase the suffering of individual animals. Table 2 highlights some of the main welfare issues that a macaque used in toxicology may experience across their lifespan and describes the opportunities for refinement.

The 3Rs provide a platform for uniting welfare together with scientific merit (e.g., refinement: Richter et al. 2010; Tasker 2012; Hall et al. 2015). They can also help increase public support for animal research by highlighting that alternatives are being sought and that animal welfare is prioritized (Leaman et al. 2014). However, despite the widespread scientific support of the 3Rs, there are barriers to uptake of refinements, including staff time, motivation, knowledge, skills, and resources. Laboratories need to have ongoing programs to critically appraise practice in light of new evidence and resources so that the most up-to-date refinements are used (Lloyd et al. 2008). Several publications provide detailed and comprehensive accounts of refinements for primates (e.g., Rennie and Buchanan-Smith 2006a, b; Rensnie and Buchanan-Smith 2006; Jennings and Prescott 2009). To implement refinement successfully requires understanding what welfare is, how it can be assessed, and having a strategy for rapid implementation of changes and their evaluation. Underlying this process should be an acceptance that refinement is a necessary and continuous process—it is a permanent challenge for care staff and scientists (Tasker 2012).

While welfare in the laboratory is formally considered in terms of refinement, one of the greatest influences on the development of animals and their resilience (i.e., their coping ability) as adults is their early rearing environment (Parker and Maestripieri 2011). Hence, the welfare of primates under study may be profoundly affected by the conditions in which they are born, reared, and kept prior to their use in a study.

In macaques, natural weaning from the mother's milk is usually seen at 10–14 months of age (Harvey et al. 1987); it is a gradual process involving withdrawal of milk and dependence on the mother for caregiving over a period of weeks or months (Lee 1996). Offspring remain with their mother beyond weaning for up to 24 months of age (Ross 1992). In captive breeding, infants are commonly removed from their mothers and natal group at about 6 months of age (Honess et al. 2010). Removal from the natal group and manipulations in the early rearing environment are stressful and result in long-term alterations in the animals' immune system and its regulation (Coe et al. 1989). More specifically, weaning and removal of the mother are known to have immunosuppressive effects (Coe et al. 1987). Toxicologists testing new pharmaceuticals that are likely to alter immunological parameters in macaques should be aware of the potential confounding effects of differential rearing histories when selecting research subjects (Tasker 2012).

Table 2 Example life cycle of a cynomolgus macaque used in toxicology in the UK, with opportunities for refinement at all stages. PRT: positive reinforcement training

Potential negative welfare impact	Opportunity for refinement
Birth and rearing environment	
Prenatal stress (i.e., during gestation) can negatively impact stress responsivity of offspring (Clarke et al. 1994) potentially making the primate less "fit for purpose."	Appropriate breeding and rearing environment (see below)
Unnatural social group and inappropriate housing and husbandry, and/or overcrowding leading to stress and poor welfare (e.g., Buchanan-Smith et al. 2004).	Natural social group in appropriate housing, providing complexity, choice, and control, with visual barriers, increasing resilience and ability to cope with challenges associated with laboratory research and testing (Buchanan-Smith 2010a, b). Human socialization and PRT—visual and auditory cues well established and required in husbandry, research, and testing.
Weaning and transport	
Capture, handling, health screening, separation, and early weaning from groups (for adverse consequences of early weaning, see Prescott et al. 2012). In captive breeding, infants are often removed early from the mother and natal group, enforcing abrupt weaning (Honess et al. 2010).	Capture, handling, and health screening are facilitated by previous PRT and good human socialization (e.g., articles in Prescott and Buchanan-Smith 2003, 2007). Decisions on timing of separation from natal group should be based on numerous factors including age, but also behavioral and health considerations (see Prescott et al. 2012). Keeping weaned macaques with familiar compatible conspecifics provides social buffering and reduces stress (Gust 1996 for rhesus macaques).
Primates imported for research may have journeys up to 58 h with evidence of heightened levels of stress for over one month after arrival at the new establishment (Prescott 2001; Honess et al. 2004).	Efforts should be made to encourage social stability, before, during, and after transport, by housing animals in socially appropriate groups, allowing environmental conditions (light, heat, etc.) to vary in a natural daily rhythm, preventing boredom with suitable environmental enrichment and sufficient space. The total duration of transport should be minimized and conditions at the destination, should as a minimum, be at least as good as those at the source (Honess et al. 2004).
Where used, a holding facility may have physically smaller enclosures, limiting range of behavior, change in routines, diet, and changes in social grouping and hierarchy	Continuation of housing and husbandry provision from breeding facility (e.g., familiarity of diet). Continuation of signals for PRT, and any prestudy training as appropriate.
Designated environment for research	
Behavioral restriction, change in routines, diet, and changes in social grouping and hierarchy.	There are a number of considerations that should determine cage size (Buchanan-Smith et al. 2004) with possibilities for enrichment, exercise areas, and providing choice, complexity, and control (Buchanan-

(continued)

Table 2 (continued)

Potential negative welfare impact	Opportunity for refinement
	Smith 2010b). Clear temporal and signaled predictability to learn new routines (Bassett and Buchanan-Smith 2007), with possibilities for accelerated acclimatization. Positive staff interactions and socialization, using PRT (Tasker 2012; Ash and Buchanan-Smith 2016). Playback of affiliative vocalizations at a natural frequency improves welfare (Watson et al. 2014 for marmosets).
Study protocol	
Uncertainty, capture, restraint, sham dosing, dosing, effects of toxicology (see Table 1).	Use of reliable signals to inform primates of events, refined methods of capture and restraint, facilitated by socialization and PRT, removal of sham dosing that appears to sensitize primates (Tasker 2012). Quiet secluded area given to primates suffering from adverse effects of test substance.
Re-use raises particular welfare issues arising from inappropriate housing and husbandry, and their use in scientific procedures may prolong negative welfare states and impact on model suitability (see Morton et al. 2003).	Morton et al. (2003) include the following recommendations for re-use of primates in telemetry studies, over and above legal compliance: • All the ethical and welfare issues are fully addressed when making decisions about re-use, in addition to the scientific issues. • Ensure recovery and wash-out periods are adequate. • A system is set up where authority for re-use depends on veritable certification of health status that includes an assessment of behavioral, physical, psychological, and social well-being. • Consider all the potential welfare costs to each individual, including those associated with housing and husbandry, when making a decision about re-use or continued use.
Death or moribundity is used as the end points.	See Table 1

There are also potential problems with the rearing environments of common marmosets, the most frequently used New World primate. This species is characterized by twin births and the cooperation of all members of the family in rearing the young until independence and natural weaning, which occurs at approximately 8 weeks of age (Yamamoto 1993; Buchanan-Smith 2010a). In captivity, dams have higher weights than in the wild, which is associated with larger litters, birth complications, and increased infant mortality (Ash and Buchanan-Smith 2014). Supplemental feeding of litters of three or more, involving removal of the infant or infants from the natal group for hand-feeding, is often practiced during the first 2 months after birth to reduce infant mortality. Depending on how this is done, it has the potential to affect development and confound scientific output (Ash and Buchanan-Smith 2016).

Ash and Buchanan-Smith (2016) used a battery of tests to determine the impact of rearing environment in common marmoset infants in 3 conditions: family-reared marmoset twin pairs, family-reared marmosets from triplet litters where only 2 remain (2 stays), and supplementary-fed triplets. The supplementary-fed triplet infants were never isolated except for very short periods for weighing and had positive experiences with humans from an early age. Furthermore, they are naturally adapted to being passed between carriers (Ingram 1977). The infants were also returned to their family group as soon as possible after feeding, and so spent most of their time with their family group. This supplementary feeding rearing practice had no adverse effects on behavior/cognition, neophobia, nor affective state (Ash and Buchanan-Smith 2016). However, primate infants that are hand-raised entirely by humans have reduced reproductive success and often experience adverse welfare such as increased self-directed behaviors, abnormal behaviors, and inappropriate aggressive behavior (e.g., Porton and Niebruegge 2006). Dettling et al. (2002, 2007) found that early parental deprivation in common marmosets impacts endocrine responses (lower basal cortisol) and several behavioral responses (e.g., they are less mobile and make fewer contact calls than controls in response to social separation/exposure to novelty), as well as blood pressure, which is higher than in controls. These changes make them unfit models of normal healthy humans.

Ideally, the purpose for which primates are bred and subsequently used as research subjects should be known, so practices can be put in place to ensure the animals are "fit for purpose." This might, for example, involve human socialization and positive reinforcement training for certain husbandry practices and procedures, or exposure to a range of stimuli likely to be encountered, paired with rewards to desensitize the primates. However, many primates are bred in special centers that are often overseas and require transport to the laboratory (Prescott 2001) where they will be used in research (Ha and Sussman 2022). The laboratory of end use may not have direct control over, or the ability to monitor, social groups, weaning age, and conditions, although some countries have legislation to cover designated breeding centers. Ideally, research laboratories should have a coordinated approach with the breeding centers that supply their animals to promote welfare, as well as ensure the primates are "fit for purpose" and to minimize confounding factors that may introduce unwanted variability in scientific output.

5   Importance of Welfare for Quality of Scientific Output

To achieve high-quality and reliable science, several essential conditions must be satisfied. The experiment must yield unambiguous results by minimizing unwanted variation, there must be an absence of confounding factors (Poole 1997), and the study must be undertaken to a required standard (e.g., Organisation for Economic Cooperation and Development (OECD) 1997). In addition to aspects of quality pertaining to the study, quality in its fullest sense should also include the impact of research using animals (see Bateson et al. 2011)—that is, the application of knowledge resulting from research findings.

Poole's (1997) seminal paper "Happy animals make good science" argues that good laboratory animal science is based upon normal, healthy, and happy animals, unless illness or alleviation of stress is the subject of study. Although the effects of stress and disease are easy to identify, and their confounding effects are well known (e.g., Reinhardt et al. 1995; Festing and Altman 2002; Hall 2007), Poole (1997) argues that unhappiness is also a confounding variable because its effects on biological variables produce increased variation in the data output. He goes further, asserting that most scientists working with animals assume that they have normal physiology and behavior (e.g., heart rates, blood pressure, blood values, metabolism, hormones, and immunological competence). However, these parameters can be dramatically altered by the conditions in which the animals were bred, reared, kept, transported, and the way in which experimental procedures were conducted. Experimenters may assume these parameters to be normal because they commonly encounter them and have no reference for comparison (Tasker 2012). If animals are not well acclimatized, properly characterized, and stable, there is a major risk of confounding and under- or overestimating the treatment effect with no predictive validity to the clinic (Graham and Schuurman 2017). For example, restraint may lead to maximum heart rates, preventing the detection of arrhythmias (Tasker 2012) or significant changes in glycemic control, blunting the response to treatment (Graham and Schuurman 2017). The link between welfare and scientific output is covered in Schapiro and Hau (2022).

There are strong ethical, scientific, and economic arguments to suggest that "no data are better than poor data." If the data are not of good scientific quality and results are therefore potentially unreliable, inaccurate, or inconclusive, then the primates used in the research will have suffered in vain, violating utilitarian principles underlying animal use. Poor animal welfare and quality of science may also cause delays or lead the research down the wrong path, with more animals being used (going against the reduction principle) and more unnecessary suffering. This not only has ethical implications for the animals, but also wastes time and money. Indeed, Bains (2004) estimated that it takes an average of 12.5 years and $1 billion to take a new drug to market. Recognizing the dependence of reliability of scientific outcomes on animal welfare, it is logical to conclude that these costs are likely to be reduced with improved animal welfare.

6 Welfare Assessment

Given that scientific procedures directly impact negatively on welfare, it is critical for there to be ways of accurately assessing welfare. The list of factors in Table 1 illustrates that primates in laboratories often have reduced welfare, especially in the absence of refinement. However, assessment should include measuring positive welfare states, such as comfort and contentment, as well as negatives ones such as boredom, fear, pain, and/or suffering (Buchanan-Smith et al. 2005). We should focus not only on welfare as a snapshot of the animal in time, but view the animal, as much as is possible, from birth to death, 24/7, across the lifespan (Brando and Buchanan-Smith 2018). This includes day and nighttime assessments, and assessments over weekdays and weekends. For example, Lambeth et al. (1997) found higher wounding rates for chimpanzee during weekdays when care staff are present than at weekends, suggesting something about the weekday routines, such as elevated activities of caregiving, veterinary, research, and other personnel, were causing tension. As well as this welfare may vary across seasons, particularly if there is a mating season when male aggression rises. There are also individual requirements across the different life stages. Younger individuals require special provisioning to allow them to engage more in play, to explore, and to learn contingencies between behaviors and their outcomes. It is known that having control over aspects of the environment improves the welfare of younger individuals more than it does for older individuals (Badihi 2006). Waitt et al. (2010) provide a list of considerations for designing environments for aged primates that include accessibility issues, positioning, size, and type of furnishings, to avoid poor welfare related to age-related arthritis, deteriorating vision, difficulties in thermoregulation, etc. Furthermore, given individual differences in the propensity for welfare states due to personality differences (e.g., King and Landau 2003), we must consider individuals as being unique (Robinson and Weiss 2022).

Hawkins et al. (2011) provide an excellent review of assessment of welfare in laboratory-housed animals. Several practical issues were raised in this review, including how to set up and operate effective protocols for the welfare assessment. The need to tailor welfare assessment protocols to individual animals, as well as individual projects, is emphasized, too, in this review, together with the need to quantify objectively measures relating to the welfare state, and to intervene early to alleviate negative states and minimize them worsening. The problem is that even in our closest living relatives, the primates, it can be challenging to recognize internal states such as pain and suffering (e.g., Flecknell et al. 2011; Sneddon et al. 2014), and although the use of analgesics following potentially painful procedures is improving in primates, it is still not optimal (Coulter et al. 2009). Section 2 of this book provides a comprehensive review of the methods used to assess primate welfare.

7   The Role of Human Behavior Change in Refinement

From birth to death, the lives of primates in designated breeding and supplying establishments are under absolute human control. When seeking welfare improvements, a fully integrated approach is required to ensure refinements are implemented at every stage of the life cycle (e.g., Table 1). The stakeholders include the scientists, study directors, advising statisticians, ethical review panel staff, the veterinarians, the animal technicians, and care staff who are all responsible for the primates' day-to-day needs—they all have a stake in implementing positive welfare change.

It is often the case that primates spend a rather small amount of time directly engaged in scientific research. An exception is neuroscience, where the primates (usually macaques) may be food and fluid-controlled, restrained in chairs, and tested from 2 to 8 h at a time, 5–7 days a week depending on the requirements of the experiment, and use continues for a number of years (Prescott et al. 2005). But for most primates, the majority of time is spent living in enclosures. Given this, the social and physical complexity of the enclosures, and the control and choices that the primates can make directly impact their welfare. In all cases, the behavior of humans is critical for promoting the welfare of primates housed in laboratories (Rennie and Buchanan-Smith 2006a).

Human behavior change is a growing discipline. It refers to a process that translates knowledge into actions, so that targeted change is implemented. This process is underpinned by multidisciplinary scientific approaches and theoretical frameworks (Michie et al. 2014); it has considerable merit in improving human health (e.g., Ory 2002) and is gaining momentum as a practical way for advancing animal welfare (e.g., Van Dyne and Pierce 2004; van Dijk et al. 2013; Whay and Main 2015). However, traditional approaches to implementing refinements have not focused on stimulating changes in human behavior.

Broadly speaking, two types of intervention are employed to improve animal welfare. These are (1) enforcement of legislation, codes of practice, and supplementary, voluntary accreditation schemes or standards, and/or (2) encouragement, which includes promoting innovation that exceeds minimum standards, and regularly accessing and implementing new knowledge and scientific findings.

In the UK, the appointment of Named Animal Care and Welfare Officers and, in Europe, the appointment of the Institutional Care and Animal Welfare Responsible Person provide oversight and, together with mandatory staff training, ensure minimum standards are met. Pharmaceutical companies and Contract Research Organizations (CROs) are committed to improving animal welfare. They often undergo voluntary Assessment and Accreditation of Laboratory Animal Care accreditation to demonstrate that they meet the minimum standards required by law and are also going the extra step to improve animal care and use. Most large pharmaceutical companies and CROs are signatories of the Concordat on Openness in Animal Research, an agreement across the biomedical sciences sector to improve communication and increase transparency to the public about animal research. There is therefore considerable enthusiasm from the industry to improve animal care and use now and in future.

Good primate welfare is dependent upon creating a strong culture of care. To achieve this, we must overcome barriers, including a lack of knowledge/resources/skills, provide a robust scientific evidence base for recommendations, give ownership of improvements, and the recognition, support, and reward system for those who effectively engage with the refinements we are proposing. To tackle the lack of knowledge and skills, we see communication, training, and dissemination of findings as fundamental to moving the refinement agenda forward.

Given the range of standards worldwide, it is also critical that training is pitched at the right level for dissemination, and the effectiveness of training resources may be enhanced if created with an understanding of models of human behavior change. Considering that primate use is likely to continue until alternative technologies are developed (e.g., see Burm et al. 2014), we need to disseminate evidence-based practice and empower individuals to lead on refinement. The launches of websites, such as marmosetcare.com and the macaque website (nc3rs.org.uk/macaques/), together with other online resources (see Prescott 2016), are a good step toward this goal.

The second approach, encouragement, is where we see considerable opportunities for innovation and improvement, to promote sustainable human behavior changes that result in positive impacts on primate welfare. The cornerstone of encouraging behavior change is to transfer ownership of both the problem and solution to a person responsible for implementing change (Whay and Main 2015). There are two important components necessary for change to happen. The first is appreciating the relevance of the desired behavior change, which must be coupled with taking ownership of the process of change, rather than being told what to do, or even through demonstration. Therefore, creating opportunities for colleagues to explore issues and come up with their own solutions will be more effective than simply being instructed without individual motivation and responsibility (Cunningham et al. 2002). Indeed, there is no one-size-fits-all approach to refinement.

All key stakeholders need to be empowered toward improving welfare, while at the same time fully appreciating experimental aims and impact. The ultimate goal is to synergize better welfare and better science to elevate the quality of the research. Understanding and seeking out people who have powers of influence are important, as change requires targeting several levels. Behavior change may be encouraged using three broad approaches, namely social marketing (an extension of principles used in marketing and advertising to promote change among groups), participatory methods (such as those used in the community development sector), and the creation of action groups (self-help or discussion groups). Useful examples of how these approaches have been implemented and tested in the agricultural sector are discussed by Whay and Main (2015). In practice, both combining and coordinating approaches (e.g., Van Dyne and Pierce 2004; Pritchard et al. 2012; van Dijk et al. 2013) are required to improve welfare. We are keen to advance the evidence base showing that human behavior change techniques improve sustainable uptake of refinements for primates; hence, making primates better models for research.

8 Conclusions

There are specific welfare issues for primates used in laboratory research and testing, including painful procedures and a restricted environment. It is imperative that we apply refinement throughout laboratory primates' lives. Specifically, efforts must be made to integrate all stages of a primate's life to improve their welfare by providing opportunities for positive experiences and conditions that enable them cope with challenges. If primates are to be used as models of normal functioning humans, the promotion of welfare will also help ensure that they are "fit for purpose" and will avoid situations where a negative welfare state confounds biological data and leads to research with unreliable or faulty conclusions. By giving ownership of the resources to target audiences, and providing the evidence base underpinning benefits (welfare, scientific output, and financial), knowledge exchange within and across facilities should continue to improve, and with it animal welfare and quality of scientific output.

Acknowledgements We thank Lauren Robinson and Alex Weiss for the invitation to contribute this chapter, and their helpful editorial assistance, past and present members of the Behaviour and Evolution Research group at the University of Stirling for discussion on the topic of primate welfare, and Suzanne Rogers from Learning About Animals for useful comments on the manuscript.

References

Adams DL, Economides JR, Jocson CM, Horton JC (2007) A biocompatible titanium headpost for stabilizing behaving monkeys. J Neurophysiol 98(2):993–1001. https://doi.org/10.1152/jn.00102.2007

Ash H, Buchanan-Smith HM (2014) Long-term data on reproductive output and longevity in captive female common marmosets (*Callithrix jacchus*). Am J Primatol 76(11):1062–1073. https://doi.org/10.1002/ajp.22293

Ash H, Buchanan-Smith HM (2016) The long-term impact of infant rearing background on the affective state of adult common marmosets (*Callithrix jacchus*). Appl Anim Behav Sci 174: 128–136. https://doi.org/10.1016/j.applanim.2015.10.009

Ash H, Smith TE, Knight S, Buchanan-Smith HM (2018) Measuring physiological stress in the common marmoset (*Callithrix jacchus*): validation of a salivary cortisol collection and assay technique. Physiol Behav 185:14–22. https://doi.org/10.1016/j.physbeh.2017.12.018

Badihi I (2006) The effects of complexity, choice and control on the behaviour and the welfare of captive common marmosets (*Callithrix jacchus*). University of Stirling

Bains W (2004) Failure rates in drug discovery and development: will we ever get any better? Drug Discov World 9. https://www.ddw-online.com/failure-rates-in-drug-discovery-and-development-will-we-ever-get-any-better-1027-200410/

Bassett L, Buchanan-Smith HM (2007) Effects of predictability on the welfare of captive animals. Appl Anim Behav Sci 102(3–4):223–245. https://doi.org/10.1016/j.applanim.2006.05.029

Bateson P, Johansen-Berg H, Jones DK, et al (2011) Review of research using non-human primates: Report of a panel chaired by Professor Sir Patrick Bateson FRS. London

Bayne KAL (2003) Environmental enrichment of nonhuman primates, dogs and rabbits used in toxicology studies. Toxicol Pathol 31(Supplement):132–137. https://doi.org/10.1080/01926230390175020

Bayne K, Morris TH, France MP (2010) Legislation and oversight of the conduct of research using animals: a global overview. In: Hubrecht R, Kirkwood J (eds) The UFAW handbook on the care and management of laboratory and other research animals. Wiley-Blackwell, Oxford, pp 107–123

Bayne K, Hau J, Morris T (2022) The welfare impact of regulations, policies, guidelines and directives and nonhuman primate welfare. In: Robinson LM, Weiss A (eds) Nonhuman primate welfare: from history, science, and ethics to practice. Springer, Cham, pp 629–646

Bekoff M (2002) Awareness: Animal reflections. Nature 419(6904):255–255. https://doi.org/10.1038/419255a

Brando S, Buchanan-Smith HM (2018) The 24/7 approach to promoting optimal welfare for captive wild animals. Behav Process 156:83–95. https://doi.org/10.1016/j.beproc.2017.09.010

Brent L (2007) Life-long well being: applying animal welfare science to nonhuman primates in sanctuaries. J Appl Anim Welf Sci 10(1):55–61. https://doi.org/10.1080/10888700701277626

Broom DM (1986) Indicators of poor welfare. Br Vet J 142(6):524–526. https://doi.org/10.1016/0007-1935(86)90109-0

Broom DM (1999) Animal welfare: the concept and the issues. In: Dolins FL (ed) Attitudes to animals: views in animal welfare. Cambridge University Press, Cambridge, pp 129–143

Broom DM (2010) Cognitive ability and awareness in domestic animals and decisions about obligations to animals. Appl Anim Behav Sci 126(1–2):1–11. https://doi.org/10.1016/j.applanim.2010.05.001

Buchanan-Smith HM (2010a) Marmosets and tamarins. In: Hubrecht R, Kirkwood J (eds) The UFAW handbook on the care and management of laboratory and other research animals. Wiley-Blackwell, Oxford, pp 543–563

Buchanan-Smith HM (2010b) Environmental enrichment for primates in laboratories. Adv Sci Res 5(1):41–56. https://doi.org/10.5194/asr-5-41-2010

Buchanan-Smith HM, Badihi I (2012) The psychology of control: effects of control over supplementary light on welfare of marmosets. Appl Anim Behav Sci 137(3–4):166–174. https://doi.org/10.1016/j.applanim.2011.07.002

Buchanan-Smith HM, Prescott MJ, Cross NJ (2004) What factors should determine cage sizes for primates in the laboratory? Anim Welf 13(Supplement 1):S197–S201

Buchanan-Smith HM, Rennie AE, Vitale A et al (2005) Harmonising the definition of refinement. Anim Welf 14(4):379–384

Burm SM, Prins J-B, Langermans J, Bajramovic JJ (2014) Alternative methods for the use of non-human primates in biomedical research. ALTEX-Altern Anim Ex 31(4):520–529. https://doi.org/10.14573/altex.1406231

Capitanio JP, Mendoza SP, McChesney M (1996) Influences of blood sampling procedures on basal hypothalamic-pituitary-adrenal hormone levels and leukocyte values in rhesus macaques (*Macaca mulatta*). J Med Primatol 25(1):26–33. https://doi.org/10.1111/j.1600-0684.1996.tb00189.x

Carlsson H, Schapiro SJ, Farah I, Hau J (2004) Use of primates in research: a global overview. Am J Primatol 63(4):225–237. https://doi.org/10.1002/ajp.20054

Caron A, Lelong C, Pascual M-H, Benning V (2015) Miniaturized blood sampling techniques to benefit reduction in mice and refinement in nonhuman primates: applications to bioanalysis in toxicity studies with antibody-drug conjugates. J Am Assoc Lab Anim Sci 54(2):145–152

Chapman KL, Pullen N, Andrews L, Ragan I (2010) The future of non-human primate use in mAb development. Drug Discov Today 15(5–6):235–242. https://doi.org/10.1016/j.drudis.2010.01.002

Clarke AS, Wittwer DJ, Abbott DH, Schneider ML (1994) Long-term effects of prenatal stress on HPA axis activity in juvenile rhesus monkeys. Dev Psychobiol 27(5):257–269. https://doi.org/10.1002/dev.420270502

Coe CL, Rosenberg LT, Fischer M, Levine S (1987) Psychological factors capable of preventing the inhibition of antibody responses in separated infant monkeys. Child Dev 58(6):1420–1430. https://doi.org/10.1111/j.1467-8624.1987.tb03855.x

Coe CL, Lubach GR, Ershler WB, Klopp RG (1989) Influence of early rearing on lymphocyte proliferation responses in juvenile rhesus monkeys. Brain Behav Immun 3(1):47–60. https://doi.org/10.1016/0889-1591(89)90005-6

Coleman K, Timmel G, Prongay K, Baker KC (2022) Common husbandry, housing, and animal care practices. In: Robinson LM, Weiss A (eds) Nonhuman primate welfare: from history, science, and ethics to practice. Springer, Cham, pp 317–348

Coulter CA, Flecknell PA, Richardson CA (2009) Reported analgesic administration to rabbits, pigs, sheep, dogs and non-human primates undergoing experimental surgical procedures. Lab Anim 43(3):232–238. https://doi.org/10.1258/la.2008.008021

Crockett CM, Bowers CL, Shimoji M et al (1995) Behavioral responses of longtailed macaques to different cage sizes and common laboratory experiences. J Comp Psychol 109(4):368–383. https://doi.org/10.1037/0735-7036.109.4.368

Crockett CM, Shimoji M, Bowden DM (2000) Behavior, appetite, and urinary cortisol responses by adult female pigtailed macaques to cage size, cage level, room change, and ketamine sedation. Am J Primatol 52(2):63–80. https://doi.org/10.1002/1098-2345(200010)52:2<63::AID-AJP1>3.0.CO;2-K

Cunningham CE, Woodward CA, Shannon HS et al (2002) Readiness for organizational change: a longitudinal study of workplace, psychological and behavioural correlates. J Occup Organ Psychol 75(4):377–392. https://doi.org/10.1348/096317902321119637

Dawkins MS (2004) Using behaviour to assess animal welfare. Anim Welf 13(Supplement 1):S3–S7

de Boo MJ, Rennie AE, Buchanan-Smith HM, Hendriksen CFM (2005) The interplay between replacement, reduction and refinement: considerations where the Three Rs interact. Anim Welf 14(4):327–332

Dennis MB (2002) Welfare issues of genetically modified animals. ILAR J 43(2):100–109. https://doi.org/10.1093/ilar.43.2.100

Dettling AC, Feldon J, Pryce CR (2002) Early deprivation and behavioral and physiological responses to social separation/novelty in the marmoset. Pharmacol Biochem Behav 73(1): 259–269. https://doi.org/10.1016/S0091-3057(02)00785-2

Dettling AC, Schnell CR, Maier C et al (2007) Behavioral and physiological effects of an infant-neglect manipulation in a bi-parental, twinning primate: Impact is dependent on familial factors. Psychoneuroendocrinology 32(4):331–349. https://doi.org/10.1016/j.psyneuen.2007.01.005

Dunbar RI, Shultz S (2007) Understanding primate brain evolution. Philos T Roy Soc B 362(1480): 649–658. https://doi.org/10.1098/rstb.2006.2001

Epple G (1970) Quantitative studies on scent marking in the marmoset. (*Callithrix jacchus*). Folia Primatol 13(1):48–62. https://doi.org/10.1159/000155308

European Commission (2010) Directive 2010/63/EU of the European Parliament and of the Council of 22 September 2010 on the protection of animals used for scientific purposes OJL276/33

Festing MFW, Altman DG (2002) Guidelines for the design and statistical analysis of experiments using laboratory animals. ILAR J 43(4):244–258. https://doi.org/10.1093/ilar.43.4.244

Flecknell P, Leach M, Bateson M (2011) Affective state and quality of life in mice. Pain 152(5): 963–964. https://doi.org/10.1016/j.pain.2011.01.030

Flow BL, Jaques JT (1997) Effect of room arrangement and blood sample collection sequence on serum thyroid hormone and cortisol concentrations in cynomolgus macaques (*Macaca fascicularis*). Contemp Top Lab Anim Sci 36(1):65–68

Fraser D (2009) Assessing animal welfare: different philosophies, different scientific approaches. Zoo Biol 28(6):507–518. https://doi.org/10.1002/zoo.20253

Graham ML, Prescott MJ (2015) The multifactorial role of the 3Rs in shifting the harm-benefit analysis in animal models of disease. Eur J Pharmacol 759:19–29. https://doi.org/10.1016/j.ejphar.2015.03.040

Graham ML, Schuurman H-J (2017) Pancreatic islet xenotransplantation. Drug Discov Today Dis Model 23:43–50. https://doi.org/10.1016/j.ddmod.2017.11.004

Grimm D (2015) New rules may end U.S. chimpanzee research. Science 349(6250):777. https://doi.org/10.1126/science.349.6250.777

Gust D (1996) Effect of companions in modulating stress associated with new group formation in juvenile rhesus macaques. Physiol Behav 59(4–5):941–945. https://doi.org/10.1016/0031-9384(95)02164-7

Ha JC, Sussman AF (2022) Primate breeding colonies: colony management and welfare. In: Robinson LM, Weiss A (eds) Nonhuman primate welfare: from history, science, and ethics to practice. Springer, Cham, pp 303–316

Hall RL (2007) Clinical pathology of laboratory animals. In: Gad S (ed) Animal models in toxicology, 2nd edn. CRC Press, Boca Raton, FL, pp 826–828

Hall LE, Robinson S, Buchanan-Smith HM (2015) Refining dosing by oral gavage in the dog: a protocol to harmonise welfare. J Pharmacol Toxicol Methods 72:35–46. https://doi.org/10.1016/j.vascn.2014.12.007

Hartner M, Hall J, Penderghest J, Clark LP (2001) Group-housing subadult male cynomolgus macaques in a pharmaceutical environment. Lab Anim (NY) 30(8):53–57. https://doi.org/10.1038/5000167

Harvey PH, Martin RD, Clutton-Brock TH (1987) Life histories in comparative perspective. In: Smuts BB, Cheyney DL, Seyfarth RM et al (eds) Primate societies. University of Chicago Press, Chicago, pp 181–196

Hau J, Farah IO, Carlsson H-E, Hagelin J (2000) Opponents' statement: non-human primates must remain accessible for vital biomedical research. In: Balls M, van Zeller AM, Halder M (eds) Progress in the reduction, refinement and replacement of animal experimentation: Developments in animal and veterinary sciences, book 31b. Elsevier, Oxford, pp 1593–1601

Hawkins P, Morton DB, Burman O et al (2011) A guide to defining and implementing protocols for the welfare assessment of laboratory animals: eleventh report of the BVAAWF/FRAME/RSPCA/UFAW Joint Working Group on Refinement. Lab Anim 45(1):1–13. https://doi.org/10.1258/la.2010.010031

Hevesi R (2022) The welfare of primates kept as pets and entertainers. In: Robinson LM, Weiss A (eds) Nonhuman primate welfare: from history, science, and ethics to practice. Springer, Cham, pp 121–144

Home Office (1986) Animals (Scientific Procedures) Act 1986. Her Majesty's Stationary Office, London

Honess P, Wolfensohn S (2010) The extended welfare assessment grid: a matrix for the assessment of welfare and cumulative suffering in experimental animals. ATLA-Altern to Lab Anim 38(3): 205–212. https://doi.org/10.1177/026119291003800304

Honess PE, Johnson PJ, Wolfensohn SE (2004) A study of behavioural responses of non-human primates to air transport and re-housing. Lab Anim 38(3):119–132. https://doi.org/10.1258/002367704322968795

Honess P, Stanley-Griffiths MA, Narainapoulle S et al (2010) Selective breeding of primates for use in research: consequences and challenges. Anim Welf 19(Supplement 1):57–65

Hosey G (2000) Zoo animals and their human audiences: what is the visitor effect? Anim Welf 9(4): 343–357

Hosey G (2022) The history of primates in Zoos. In: Robinson LM, Weiss A (eds) Nonhuman primate welfare: from history, science, and ethics to practice. Springer, Cham, pp 3–30

Howard BR (2002) Control of variability. ILAR J 43(4):194–201. https://doi.org/10.1093/ilar.43.4.194

Hubrecht RC (2014) The welfare of animals used in research: practice and ethics. Wiley, Blackwell, Chichester, West Sussex

Ingram JC (1977) Interactions between parents and infants, and the development of independence in the common marmoset (*Callithrix jacchus*). Anim Behav 25(4):811–827. https://doi.org/10.1016/0003-3472(77)90035-5

Jennings M, Prescott MJ (eds) (2009) Refinements in husbandry, care and common procedures for non-human primates. Lab Anim 43(Supplement 1):1–47. https://doi.org/10.1258/la.2008.007143

Kaiser J (2013) NIH to phase out most chimp research. Science 341(6141):17–18. https://doi.org/10.1126/science.341.6141.17

King JE, Landau VI (2003) Can chimpanzee (*Pan troglodytes*) happiness be estimated by human raters? J Res Pers 37(1):1–15. https://doi.org/10.1016/S0092-6566(02)00527-5

Lambeth SP, Bloomsmith MA, Alford PL (1997) Effects of human activity on chimpanzee wounding. Zoo Biol 16(4):327–333. https://doi.org/10.1002/(SICI)1098-2361(1997)16:4<327::AID-ZOO4>3.0.CO;2-C

Leaman J, Latter J, Clemence M (2014) Attitudes to animal research in 2014. Ipsos Mori, 1–54

Lee PC (1996) The meanings of weaning: Growth, lactation, and life history. Evol Anthropol Issues News Rev 5(3):87–98. https://doi.org/10.1002/(SICI)1520-6505(1996)5:3<87::AID-EVAN4>3.0.CO;2-T

Lloyd MH, Foden BW, Wolfensohn SE (2008) Refinement: promoting the three Rs in practice. Lab Anim. 42(3):284–293.https://doi.org/10.1258/la.2007.007045

Lutz CK, Tiefenbacher S, Jorgensen MJ et al (2000) Techniques for collecting saliva from awake, unrestrained, adult monkeys for cortisol assay. Am J Primatol 52(2):93–99. https://doi.org/10.1002/1098-2345(200010)52:2<93::AID-AJP3>3.3.CO;2-2

Marini RP, Wachtman LM, Tardif SD et al (eds) (2019) The common marmoset in captivity and biomedical research. Academic, Cambridge, MA

McCann C, Buchanan-Smith HM, Jones-Engel L et al (2007) IPS International guidelines for the acquisition, care and breeding of nonhuman primates, 2nd edn. Primate Care Committee, International Primatological Society

McEwen BS (2006) Sleep deprivation as a neurobiologic and physiologic stressor: allostasis and allostatic load. Metabolism 55(Supplement 2):S20–S23. https://doi.org/10.1016/j.metabol.2006.07.008

McKinley J, Buchanan-Smith HM, Bassett L, Morris K (2003) Training common marmosets (*Callithrix jacchus*) to cooperate during routine laboratory procedures: ease of training and time investment. J Appl Anim Welf Sci 6(3):209–220. https://doi.org/10.1207/S15327604JAWS0603_06

Mendl M, Paul ES (2004) Consciousness, emotion and animal welfare: insights from cognitive science. Anim Welf 13(Supplement 1):17–25

Michie S, West R, Campbell R et al (2014) ABC of behaviour change theories: an essential resource for researchers, policy makers and practitioners. Silverback Publishing, London

Morton DB, Jennings M, Buckwell A et al (2001) Refining procedures for the administration of substances. Lab Anim 35(1):1–41. https://doi.org/10.1258/0023677011911345

Morton DB, Hawkins P, Bevan RM et al (2003) Refinements in telemetry procedures. Lab Anim 37:261–299

Organisation for Economic Cooperation and Development (1997) OECD series on principles of good laboratory practice and compliance monitoring. Number 1. OECD principles on good laboratory practice. Guideline

Organisation for Economic Cooperation and Development (2000) Guidance document on the recognition, assessment, and use of clinical signs as humane endpoints for experimental animals used in safety evaluation (ENV/JM/MONO(2000)7). ENV/JM/MONO(2000)7, Organisation for Economic Cooperation and Development, Paris

Ory MG (2002) The behavior change consortium: setting the stage for a new century of health behavior-change research. Health Educ Res 17(5):500–511. https://doi.org/10.1093/her/17.5.500

Parker KJ, Maestripieri D (2011) Identifying key features of early stressful experiences that produce stress vulnerability and resilience in primates. Neurosci Biobehav Rev 35(7):1466–1483. https://doi.org/10.1016/j.neubiorev.2010.09.003

Pickard J, Buchanan-Smith HM, Dennis M, et al (2013) Review of the assessment of cumulative severity and lifetime experience in non-human primates used in neuroscience research. Research report of the animal procedures committee's Primate Subcommittee Working Group. London

Poole T (1997) Happy animals make good science. Lab Anim 31(2):116–124. https://doi.org/10.1258/002367797780600198

Porton I, Niebruegge K (2006) The changing role of hand rearing in zoo-based primate breeding programs. In: Sackett GP, Ruppentahal GC, Elias K (eds) Nursery rearing of nonhuman primates in the 21st century. Developments in primatology: progress and prospects. Springer, Boston, pp 21–31

Prescott MJ (2001) Counting the cost: Welfare implications of the acquisition and transport of non-human primates for use in research and testing. Royal Society for the Prevention of Cruelty to Animals (RSPCA), Horsham, West Sussex

Prescott M (2006) Primate sensory capabilities and communication signals: implications for care and use in the laboratory. The National Centre for the Replacement, Refinement, and Reduction of Animals in Research, London

Prescott MJ (2016) Online resources for improving the care and use of non-human primates in research. Primate Biol 3(2):33–40. https://doi.org/10.5194/pb-3-33-2016

Prescott MJ, Buchanan-Smith HM (2003) Training nonhuman primates using positive reinforcement techniques: guest editors' introduction. J Appl Anim Welf Sci 6(3):157–161. https://doi.org/10.1207/S15327604JAWS0603_01

Prescott MJ, Buchanan-Smith HM (2007) Training laboratory-housed non-human primates, part I: A UK survey. Anim Welf 16(1):21–36

Prescott MJ, Buchanan-Smith HM, Smith AC (2005) Social interaction with non-averse groupmates modifies a learned food aversion in single- and mixed-species groups of tamarins (*Saguinus fuscicollis* and *S. labiatus*). Am J Primatol 65(4):313–326. https://doi.org/10.1002/ajp.20118

Prescott MJ, Brown VJ, Flecknell PA, Gaffan D, Garrod K, Lemon RN, Parker AJ, Ryder K, Schultz W, Scott L, Watson J, Whitfield L (2010) Refinement of the use of food and fluid control as motivational tools for macaques used in behavioural neuroscience research: report of a Working Group of the NC3Rs. J Neurosci Methods 193(2):167–188. https://doi.org/10.1016/j.jneumeth.2010.09.003

Prescott MJ, Nixon ME, Farningham DAH et al (2012) Laboratory macaques: when to wean? Appl Anim Behav Sci 137(3–4):194–207. https://doi.org/10.1016/j.applanim.2011.11.001

Pritchard JC, Van Dijk L, Ali M, Pradhan SK (2012) Non-economic incentives to improve animal welfare: positive competition as a driver for change among owners of draught and pack animals in India. Anim Welf. 21(Supplement 1):25–32 https://doi.org/10.7120/096272812X13345905673566

Reinhardt V, Liss C, Stevens C (1995) Restraint methods of laboratory non-human primates: a critical review. Anim Welf 4(3):221–238

Rennie AE, Buchanan-Smith HM (2006a) Refinement of the use of non-human primates in scientific research. Part I: the influence of humans. Anim Welf 15(3):203–213

Rensnie AE, Buchanan-Smith HM (2006b) Refinement of the use of non-human primates in scientific research. Part II: housing, husbandry and acquisition. Anim Welf 15(3):215–238

Rennie AE, Buchanan-Smith HM (2006c) Refinement of the use of non-human primates in scientific research. Part III: refinement of procedures. Anim Welf 15(3):239–261

Richter SH, Garner JP, Auer C et al (2010) Systematic variation improves reproducibility of animal experiments. Nat Methods 7(3):167–168. https://doi.org/10.1038/nmeth0310-167

Robinson LM, Weiss A (2022) Primate personality and welfare. In: Robinson LM, Weiss A (eds) Nonhuman primate welfare: from history, science, and ethics to practice. Springer, Cham, pp 387–402

Ross C (1992) Life history patterns and ecology of macaque species. Primates 33:207–215. https://doi.org/10.1007/BF02382750

Russell W, Burch R (1959) The principles of humane experimental technique. Methuen & Co, London

Sandøe P, Crisp R, Holtug N (1997) Ethics. In: Appleby MC, Hughes BO (eds) Animal welfare. CAB International, Wallingford, Oxford, pp 3–18

Schapiro SJ, Hau J (2022) Benefits of improving welfare in captive primates. In: Robinson LM, Weiss A (eds) Nonhuman primate welfare: from history, science, and ethics to practice. Springer, Cham, pp 433–450

Shively CA, Laber-Laird K, Anton RF (1997) Behavior and physiology of social stress and depression in female cynomolgus monkeys. Biol Psychiatry 41(8):871–882. https://doi.org/10.1016/S0006-3223(96)00185-0

Singer P (1975) Animal liberation. Random House, New York

Smith JA, Boyd KM (2002) The Boyd Group papers on: the use of non-human primates in research and testing. British Psychological Society Scientific Affairs Board Standing Advisory Committee on the Welfare of Animals in Psychology

Sneddon LU, Elwood RW, Adamo SA, Leach MC (2014) Defining and assessing animal pain. Anim Behav 97:201–212. https://doi.org/10.1016/j.anbehav.2014.09.007

Tasker L (2012) Linking welfare and quality of scientific output in cynomolgus macaques (*Macaca fascicularis*) used for regulatory toxicology. University of Stirling

Teixeira CP, de Azevedo CS, Mendl M et al (2007) Revisiting translocation and reintroduction programmes: the importance of considering stress. Anim Behav 73(1):1–13. https://doi.org/10.1016/j.anbehav.2006.06.002

Turner PV (2022) The history of chimpanzees in biomedical research. In: Robinson LM, Weiss A (eds) Nonhuman primate welfare: from history, science, and ethics to practice. Springer, Cham, pp 31–56

U.S. Department of Agriculture (2013) Animal Welfare Act and Animal Welfare Regulations, Section 3.81–Environmental enhancement to promote psychological well-being. Animal Welfare Act and Animal Welfare Regulations ("Blue Book"), USA

Universities Federation for Animal Welfare (2010) The UFAW handbook on the care and management of laboratory and other research animals, 8th edn. Wiley-Blackwell, Chichester, West Sussex

van Dijk L, Pradhan SK, Ali M (2013) Sustainable animal welfare: community-led action for improving care and livelihoods. Particip Learn Action 66:37–50

Van Dyne L, Pierce JL (2004) Psychological ownership and feelings of possession: three field studies predicting employee attitudes and organizational citizenship behavior. J Organ Behav 25(4):439–459. https://doi.org/10.1002/job.249

Vitale A, Manciocco A, Alleva E (2009) The 3R principle and the use of non-human primates in the study of neurodegenerative diseases: the case of Parkinson's disease. Neurosci Biobehav Rev 33(1):33–47. https://doi.org/10.1016/j.neubiorev.2008.08.006

Waitt CD, Bushmitz M, Honess PE (2010) Designing environments for aged primates. Lab Prim Newsl 49(3):5–9

Watson CFI, Buchanan-Smith HM, Caldwell CA (2014) Call playback artificially generates a temporary cultural style of high affiliation in marmosets. Anim Behav 93:163–171. https://doi.org/10.1016/j.anbehav.2014.04.027

Weatherall DJ, Goodfellow P, Harris J, et al (2006) The use of non-human primates in research: a working group report. London

Weber H, Berge E, Finch J et al (1999) Health monitoring of non-human primate colonies. Lab Anim 33(Supplement 1):S3–S18. https://doi.org/10.1258/002367799780640002

Whay HR, Main DCJ (2015) Improving animal welfare: practical approaches for achieving change. In: Grandin T (ed) Improving animal welfare: a practical approach, 2nd edn. CABI International, Wallingford, Oxford, pp 291–312

Wolfensohn S (2022) Humane end points and end of life in primates used in laboratories. In: Robinson LM, Weiss A (eds) Nonhuman primate welfare: from history, science, and ethics to practice. Springer, Cham, pp 369–386

Wolfensohn S, Sharpe S, Hall I et al (2015) Refinement of welfare through development of a quantitative system for assessment of lifetime experience. Anim Welf 24(2):139–149. https://doi.org/10.7120/09627286.24.2.139

Yamamoto ME (1993) From dependence to sexual maturity: the behavioural ontogeny of Callitrichidae. In: Rylands AB (ed) Marmosets and tamarins: systematics behaviour and ecology. Oxford University Press, Oxford, pp 235–254

Yang SH, Cheng PH, Banta H et al (2008) Towards a transgenic model of Huntington's disease in a non-human primate. Nature 453(7197):921–924. https://doi.org/10.1038/nature06975

The Welfare of Primates Kept as Pets and Entertainers

Rachel Hevesi

Abstract

The use of primates as pets and entertainers has a long history and spans many cultures; this perspective may confuse public understanding of primates as wild animals. Legislation concerning primate welfare varies wherever primates are kept in captivity. A growing concern about the psychological and physical impact of the pet and entertainment trades on primate welfare is reflected by calls from scientists and the public to end these practices. This chapter looks at the evidence from a variety of stakeholders and concludes that education and legislation are both necessary to prevent the suffering that is the inevitable outcome of primates being kept in a domestic setting or used as "actors."

Keywords

Primate welfare · primate pet trade · primate entertainers · animal actors

1 Introduction

Although the most controversial issues in animal welfare concern the way humans treat captive and domestic animals, the ways those animals respond are rooted in their evolutionary past and in how their wild ancestors responded to threats to their fitness. Behavioral ecologists thus have a major contribution to make to animal welfare science by connecting this evolutionary legacy to what now matters to the animals themselves. (Dawkins 2006)

R. Hevesi (✉)
Wild Futures, The Monkey Sanctuary, Cornwall, UK
e-mail: rachel_hevesi@wildfutures.org

The association between human and nonhuman primates (henceforth just primates) goes back millennia, with evidence mostly found in art and literature for the use of primates as pets and as entertainers (Agoramoorthy and Hsu 2005; Goudsmit and Brandon-Jones 1999; Osterberg and Nekaris 2015; Wich and Marshall 2016). In the twenty-first century, this tradition is still widespread, from countries with primate habitat (Ferreira et al. 2022) to countries where primates are imported or bred for the pet and/or entertainment trade. Assessing the welfare of primates kept as entertainers or as pets can be accomplished by examining the individual primate's ability to express normal behavior (Lutz and Baker 2022), its physical health, and its access to a suitable environment and nutrition, all within the context of the species in question. The personality structure of a species and individual differences in personality may also affect welfare outcomes (Robinson and Weiss 2022). Personality structure, for example, appears to reflect the physical and social environment in which species evolved (Gosling 2001; Weiss 2017) and may impact the ability for both a species and an individual to adapt positively to environments in which they are kept (Koene et al. 2016; Ferreira et al. 2016).

Primates kept as pets or used in entertainment are under the control and responsibility of humans. It is therefore imperative for their owners or keepers to be knowledgeable about and to understand individual primates and the characteristics of that species. This knowledge and understanding will enable owners or keepers to recognize normal and abnormal behaviors, the latter including rocking, pacing, and self-injury (Lutz and Baker 2022), and so will potentially benefit the welfare of individual primates. Increasingly, the suitability of primate species for use as pets and entertainers is being challenged by the scientific community and the general public (BVA 2014; IPS 2008; Schuppli and Fraser 2000; Soulsbury et al. 2009; Wild Futures and RSPCA 2012). This fact is reflected in the growing number of national legislatures that are limiting the trade and keeping of primates in the pet and entertainment industries, or banning it outright (Eurogroup for Animals 2013; Born Free US 2016). This chapter will review the welfare implications that have informed this understanding and legislative trend.

2   Definitions

For this chapter, the terms "pet," "companion animal," and "privately owned animal" are defined as animals in private ownership that are kept for leisure or hobby, and not as part of a recognized conservation program, licensed zoo, or scientific institution. The term "entertainers" is defined as primates used for noneducational display or exhibit, or as actors in film, television, or similar media, including advertising. Both terms exclude primates kept in rescue centers and sanctuaries. However, the definition of "education," "rescue," and "sanctuary" could be said to be open to interpretation. Some individuals and welfare organizations question the educational value of particular programs within zoos or of zoos in principle (Carr and Cohen 2011; Moss and Esson 2013). Many countries, including the UK, do not define rescue centers and sanctuaries in law. Therefore, motivation for private ownership and welfare standards by those that use the titles of rescue center and sanctuary vary enormously.

3 History

The trade in primates from native range states into Europe dates to at least the sixteenth century when it was fashionable for members of the Royal families and nobility to keep monkeys as pets (Hosey 2022). This was a time of extensive European exploration and colonization of Africa, Asia, and the New World. Catherine of Aragon, the Spanish wife of King Henry VIII of England, was often painted with one or more of her 14 pet monkeys. A portrait painted around 1525 by Lucas Horenbout appears to show Catherine holding a capuchin (*Cebus* spp.). The tradition of sailors and merchant seamen bringing back primates from their travels continued right up to the late twentieth century when primates remained popular as pets and in the entertainment industry, for example, in menageries, circuses, and then as photographers' props.

4 Domesticated or Wild

Of all 5416 extant or recently extinct mammals (Wilson and Reeder 2005), only a handful has been domesticated (Diamond 2002). Tame or habituated animals may be captive or wild-born but are genetically indistinct from their wild ancestors; there may be some individual behavioral adaptation to their environment, but their fundamental traits remain unchanged. Taming animals is necessary for domestication, but it is not domestication per se (Russell 2002). The predisposition of a species to domestication, and the degree to which they can be domesticated, is predicated around six criteria: social structure, food preferences, the ability to breed in captivity, aggression, growth rate and birth spacing, and temperament under stress (Clutton-Brock 1999; Zeder 2006).

Driscoll et al. (2009) summarized the favorable and unfavorable ecological and behavioral pre-adaptations to domestication. Table 1 uses these criteria to compare the suitability for domestication of species of the Cebus and Sapajus genera and of the family Callitrichidae.

Although primates and humans have had a close association for millennia, primates have never been domesticated. This is not surprising given that low aggression is probably the single most important pre-adaptive factor that makes animals candidates for domestication (Zeder 2006). High levels of aggression are frequently reported in wild populations of some commonly kept primates, including capuchins, callitrichids, and chimpanzees (Perry 1998; Lazaro-Perea 2001; Wilson 2007). Owners often attempt to mitigate the danger of primate aggression through veterinary interventions, such as teeth extraction and castration. The welfare implications of this are examined later in the chapter. It is generally accepted that domesticated animals such as cats and dogs can and do thrive in human

Table 1 Favorable and unfavorable ecological and behavioral pre-adaptations of Cebus, Sapajus, and callitrichids to domestication (adapted from Driscoll et al. 2009)

Favorable	*Cebus* and *Sapajus*	Callitrichids	Unfavorable	*Cebus* and *Sapajus*	Callitrichids
Social structure					
Dominance hierarchy	Y	Y	Territoriality	Y	Y
Large gregarious groups	N	N	Family groups or solitary	Y	Y
Male social group affiliation	Y	N	Males in separate groups	Y	Y
Persistent groups	N	Y	Open membership	Y	N
Food preferences					
Generalist herbivorous feeder or omnivore	Y	N	Dietary specialist or carnivore	N	Y
Captive breeding					
Polygamous/promiscuous mating	Y	N	Pair bonding prior to mating	Y	Y
Males dominant over females	N	N	Females dominant or males appease females	Y	Y
Males initiate	N	N	Females initiate	Y	Y
Movement or posture mating cues			Color or morphological mating cues		
Precocial young	N	N	Altricial young	Y	Y
Easy divestiture of young	N	N	Difficult divestiture of young	Y	Y
High meat yield per food/time	N	N	Low meat yield	Y	Y
Intra or interspecies aggressiveness					
Nonaggressive	N	N	Naturally aggressive	Y	Y
Tameable/readily habituated	Y	Y	Difficult to tame	N	N
Readily controlled	N	N	Difficult to control	Y	Y
Solicit attention	N	N	Avoids attention/independent	Y	Y
Captive temperament					
Low sensitivity to environmental change	N	N	High sensitivity to environmental change	Y	Y
Limited agility	N	N	Highly agile/difficult to contain	Y	Y
Small home range	N	N	Large home range	Y	Y
Wide environmental tolerance	N	N	Narrow environmental tolerance	Y	Y

(continued)

Table 1 (continued)

Favorable	Cebus and Sapajus	Callitrichids	Unfavorable	Cebus and Sapajus	Callitrichids
Nonshelter seeking	N	N	Shelter seeking	Y	Y
Implosive herd reaction to threat	N	N	Explosive herd reaction	Y	Y
Commensal initiative					
Exploits anthropic (human) environments	N	N	Avoids anthropic (human) environments	Y	Y

environments. The concern for primates kept in such environments is simply that they do not thrive.

5 Public Understanding of Primate Behavior

Primates have a wide repertoire of behaviors, varying between species and individuals. Most people, unless specialists working with or studying primates in captivity or in the field, have little to no knowledge of normal behavior in primates compared with their understanding of domesticated species, such as dogs with which even nonowners have probably had direct experience. There is no evidence that primates evolved to communicate with humans. However, dogs have a history of thousands of years of domestication and convergent evolution with humans (Hare et al. 2002). Consistent with this, a pilot study by Hanson (2014) suggested that there is a significant difference in humans' ability to recognize four common emotions in dogs and in primates. Private owners of primates are mostly nonspecialists who have not studied the behavior of the species in the wild. Owners often do not recognize disease or disorders, including abnormal behaviors (Hevesi 2005). They may not realize that stereotypic bouncing, pirouetting, or clapping are signs of stress, and find it comical rather than a cause for concern (Hevesi 2005). The entertainment industry capitalizes on the public's ignorance. For example, they portray chimpanzees and capuchins that are showing a fear grimace or bared teeth displays as smiling (Aldrich 2015). The owner of the primate "actor" may rebuff claims of poor welfare by claiming that the primate is trained to grimace on command and argue that the individual is not experiencing fear or stress. However, it is worth asking whether the repeated use of bared teeth displays encourages the viewer to believe that the actor's welfare is good and therefore enables further use of actors. "No animals were harmed" disclaimers in a film or television program's credits are provided by agencies such as American Humane Association. Aldrich (2015) questions whether these disclaimers are interpreted by the public as meaning that animals were not harmed by being actors or by being part of the entertainment industry. However, the disclaimer only relates to the time that the animal is on set. The viewer may be reassured and does not question how an animal has been habituated (by premature weaning, removal from social group, etc.) or how an animal is housed, transported, and trained offset.

The portrayal of animals in the media is known to affect the public's desire to acquire a species or breed (Schroepfer et al. 2011; Leighty et al. 2015; Lenzi et al. 2020). A rise in the number of licensed lemurs in the UK coincided with the popularity of the film *Madagascar*, featuring cartoon lemurs (Wild Futures unpublished). Other famous examples are the use of capuchins in the TV series *Friends* and more recently in films such as *Pirates of the Caribbean*, *Night at the Museum*, and *We Bought a Zoo*. One particular capuchin "actor," Crystal, has become famous and is photographed being feted at parties, wearing clothes, and waving from a red carpet. Research, including Borgi and Cirulli (2016) and Nittono et al. (2012), on human attraction to cute and infant-like features supports the hypothesis that a cartoon or puppet that captures the public's imagination has the same effect as a real animal in encouraging the desire to acquire a member of that species.

Entertainer primates, from the chimpanzees in the old tea adverts to orangutans in Southeast Asia, often star as human companions, friends, pets, and troublemakers, and are usually depicted as enjoying their status (Aldrich 2015; Agoramoorthy and Hsu 2005). The fascination with primates as a grotesque but cute version of humans is particularly prevalent on the Internet (Nekaris et al. 2013; Lenzi et al. 2020). Comments below videos are rarely concerned for the primates, whether they are eating junk food or displaying stereotypies. In a study that focused on a specific YouTube video, Nekaris et al. (2013) examined public knowledge and perceptions of conservation and welfare with respect to a video released in 2009, which featured a slow loris (*Nycticebus pygmaeus*) being tickled. The behavior of the slow loris was widely interpreted by viewers as positive; however, the interpretation by primate experts is that the slow loris was engaging in defensive behaviors. This lack of awareness, interest, or knowledge of primate welfare may be generalized across species. Aldrich (2015) suggests that false beliefs based on how primates are portrayed in human environments may help foster attitudes and behaviors detrimental to primate welfare. The portrayal of primates in a human environment, whether on screen or in print, has been shown to have a detrimental effect on the public's understanding of their endangered status in the wild (Schroepfer et al. 2011; Leighty et al. 2015). All four of these studies also conclude that portrayals of primates in a human environment make them appear as suitable pets.

6 Legislation, Regulation, and Numbers

Legislation and regulations pertinent to the welfare of primates traded and kept as pets or entertainers have various rationales. These include the desire to protect wild populations, ensuring and protecting the welfare of individual animals, and controlling animal–human health risks.

Legislation concerning the keeping of primates as pets varies throughout Europe and the U.S. report, "Wild Pets in the European Union" by Endcap (2012) states that, of the EU member states, ten prohibit the private keeping of primates and six have partial prohibition, i.e., whether it is prohibited often depends on the species or regional or federal laws in place. In 2019, in the UK, where there is no prohibition on privately owned primates, owners of some species are required to hold a license under the Dangerous Wild Animals Act 1976. Monitoring numbers in the UK is therefore hampered by the lack of a central registration scheme and high noncompliance of licensing. There is no monitoring system for those species that do not require any form of licensing. Wild Futures and the Royal Society for the Prevention of Cruelty to Animals (RSPCA) estimated there to be about 5000 privately owned primates in the UK (RSPCA 2016). In 2020, the Wild Futures submission to "The welfare of primates as pets in England: call for evidence" (DEFRA 2019) estimated that this figure may have fallen to under 4000 due to a fall in licensed primates. It is also difficult to ascertain the population of privately owned primates in Europe or in the USA. The Humane Society of the United States (HSUS) estimates that there are about 15,000 primates kept as pets in the USA (Humane Society International 2019).

In 2019, the UK government introduced a bill to ban the use of wild animals in circuses in England by 2020 (Ares 2019). This follows a similar ban in Scotland, and similar legislation is likely to be introduced in Wales.

The U.S. Animal Welfare Act 1966 (USAW) provides protection to primates used in research or kept by dealers and exhibitors (for review, see Bayne et al. 2022), but it does not extend to primates kept in private homes. There is no single database or central registration scheme for monitoring numbers or welfare of primates kept as pets in the USA. The use of wild animals in circuses in the USA is also being challenged. Protests by animal rights groups were partly responsible for the closure in 2017 of Ringling Brothers and Barnum & Bailey Circus. After performing for 146 years, the circus did not survive damaging revelations about its violations of the Animal Welfare Act (Wallenfeldt 2018). As noted later in this chapter, there is a big demand for wild animals as pets in the Middle East. In mitigation, in 2017 ownership of wild animals as pets became illegal in the United Arab Emirates. Most habitat countries limit or ban the capture, trade, and keeping of wild-born primates (for review, see Ferreira et al. 2022). This is primarily motivated by the need to conserve these species. However, the ability of many habitat countries to enforce protective legislation is compromised by economic and political situations. It is possible that the knowledge that it is legal to trade in primates in nonrange countries acts as incentive to illegal hunting and trade (Redmond 2005). The majority of legislatures around the world use a negative list to control the trading and keeping of species. In other words, unless specifically prohibited, any species may be kept, although some require a license. The opposite of this approach is the positive list, which has recently been adopted for mammals by Belgium and the Netherlands. This approach specifies that no species may be traded or kept unless that species has been approved (Vanautryve 2014; Eurogroup for Animals 2013).

7 Trade

In recent times, the pet trade in primate habitat range states appears to be associated with the bushmeat trade (United Nations 1992; Ceballos-Mago et al. 2010). Sales of wild animals as pets are an opportunistic by-product of hunting (Stiles et al. 2013). The welfare implications of this pet trade are not limited to individual animals taken for sale. If females become the preferred target for hunters because of the opportunities for selling young animals, the surviving population loses the individual female and the ability to maintain sustainable populations is reduced (Peres 2000). Medium- and large-bodied primate species are vulnerable to hunting because they produce more meat. This is problematic for conservation as these species tend to produce single offspring, have long interbirth intervals, prolonged infancies and juvenile periods, and are vulnerable to habitat and social disruption (Bodmer et al. 2002; United Nations Environment Programme 2010). The loss of key members of a group may also have detrimental effects on social cohesion, hierarchy, and vulnerability to disease and predation (United Nations Environment Programme 2010; Peres 1990).

There is evidence that the illegal trade in gorillas, chimpanzees, bonobos, and orangutans has shifted from being a by-product of deforestation, mining, and bushmeat hunting, to a more targeted trade dominated by organized illicit dealers who, in response to demand from an international market, trade in great apes. Due to demand for pets or zoo animals in Asia, Europe, and the Middle East, it is likely that as many as 22,218 wild great apes were lost to illegal trade between 2005 and 2011 (UN-Grasp 2016).

Transnational criminal networks can supply the tourist entertainment industry, disreputable zoos, and wealthy individuals, with primates. These illegally owned primates are used to attract tourists in the streets, entertainment facilities, parks, and circuses. They may be dressed in human clothes, trained for use by photographers or in boxing matches, or trained to roller skate (Stiles et al. 2013). The Apes Seizure Database, which includes records since 2005, was launched at the 17th CITES Conference of the Parties in 2016, in South Africa. It revealed that over 1800 great apes were seized from 23 nations, almost half of which were nonrange states from Asia, Europe, and the Middle East. Between 1999 and 2004, the number of orphaned chimpanzees in Pan African Alliance Sanctuaries rose significantly (Redmond 2005). The scale of this trade has implications for conservation of endangered species and the welfare of the animals caught in the trade. Furthermore, the situation appears more critical when mortality is used as a marker for welfare. Research estimates that for every wild-born primate captured and sold in the international trade, one to fifteen primates have died, and others may have been injured but not captured (United Nations Environment Programme 2010; Nijman 2009; Peres 1990).

The capture and sale of Barbary macaques (*Macaca sylvanus*) in Algeria and Morocco are a significant threat to their conservation. Consumption as food rarely appears to be the motivation for the capture and trade of this species (Lavieren 2008). In 2008, the Barbary macaque was classed as Endangered on the IUCN Red List of Threatened Species, with the international pet trade (mainly into Europe) cited as a significant reason for their decline. The IUCN reported that sanctuaries and rescue centers had become overstocked with these macaques from authorities or former owners. There are also reports of Barbary macaques being used for entertainment purposes, in circuses, and as fighting animals. In 2016, Barbary macaques were uplisted to CITES Appendix I (CITES 2016).

The trade in privately owned primates in nonrange states is most likely supplied by captive breeding. Primates are commonly sold as infants, from a few days to a few months old, frequently before weaning has occurred and almost always before social weaning has taken place (RSPCA 2016). Advertisements by vendors often cite hand rearing as an advantage as the juvenile primate is more acclimatized to handling and is therefore considered to be "tame." Most primates in the UK pet trade are believed to be captive-bred (Wild Futures 2014a; RSPCA 2016). This may also be true in mainland Europe, particularly for smaller species, such as the common marmoset (*Callithrix jacchus*). However, the porous borders of the European Union are a substantial problem for those who wish to control or monitor the trade of captive- or wild-born animals. Recently, a burgeoning number of vendors have been making sales via the Internet. This market is more accessible for private sales, and again, particularly sales of smaller species. Surveys by the International Fund for Animal Welfare (IFAW 2005), the RSPCA (2016), and the Blue Cross/PAAG (Blue Cross) all report that the Internet is a popular channel for advertising the sale of primates in the UK. In 2015, there were only three officially licensed breeders in the UK (Wild Futures unpublished data), but forums and Internet advertisements suggest that a large number of sales by nonlicensed breeders take place.

8 Housing

Primates evolved to survive in and engage with complex social and physical environments. Primates therefore require space and appropriate environments to enable natural behaviors, such as running, swinging, and jumping; resting and sleeping; foraging and exploring; and social interactions, such as grooming or playing. They need opportunities to escape or flee from aggressors or other threats, or to avoid contact with others. These requirements differ as a function of species-specific physiologies and behaviors. For example, in the wild, marmosets (*Callithrix* spp.) sleep in tree cavities or vine tangles and use vertical postures and space. On the other hand, wild Japanese macaques (*Macaca fuscata*) may sleep on the ground and adopt a predominantly horizontal posture. Species that evolved in environments with high predation rates, like patas monkeys (*Erythrocebus* spp.), may respond differently to threats than, for example, capuchin monkeys (*Sapajus* and *Cebus* spp.), which evolved in environments that were relatively free of predation threats. So, there may be intrinsic sensitivities to enclosure design, i.e., even if there is no predation threat, high perches and opportunities to scan the environment should be provided. Evidence suggests that the housing of primates kept as pets or entertainers is frequently, if not usually, inappropriate in size, structure, and environment. Cages are often very small and limited to the indoors. Cages for hamsters and parrots kept in hallways, living rooms, garages, and sheds are often used to house these primates

(Wild Futures 2014b; RSPCA 2016). This creates a problem for welfare as a lack of environmental stimulation and impoverished conditions in growing primates can result in poor brain development and impaired cognitive abilities (Sackett 1972; Davenport et al. 1973). Experiential deprivation during development may diminish several parameters in the primate brain (Kozorovitskiy et al. 2005).

Primate entertainers in traveling circuses or roadside attractions may be kept in travel boxes or cages for lengthy periods of time. These cages are often permanent enclosures. Primates can be housed in zoo-like conditions by agencies who keep private collections for the entertainment industry. However, these animals may also spend prolonged periods in travel boxes. For example, in 2009 two rhesus macaques (*Macaca mulatta*) hired for a London stage production spent three months commuting daily between Oxfordshire and London, a 257 km return journey (Wild Futures unpublished).

9 Temperature, Humidity, Light

Other constraints of a domestic setting, particularly in a temperate climate, are temperature and humidity. Temperature is a concern for primates living in nontropical countries, and especially if they are kept in sheds or aviaries. Fluctuating and low temperatures can result in the loss of digits and tail tips, as well as chilblains and other circulatory disorders. Many primates require humidity levels of up to 75% and are prone to respiratory problems if the atmosphere is too dry. Low humidity is associated with other problems, too. For example, in the case of common squirrel monkeys (*Saimiri sciureus*), humidity appears to be a regulatory factor in ovulation and seasonal cycles (Friedl and Holmes 1986). Given that many primate owners use central heating in their homes, the primates who share this space will inevitably be in a low humidity environment with potentially detrimental effects on their welfare.

Other issues include adequate light levels and cycles. Reproductive hormones, stress, and fecundity can be negatively impacted by low light intensity (Reinhardt 1997). Metabolic activity, hormone excretion, and endocrine function are factored by a 12-h day–night cycle, and disruption of circadian cycles can be detrimental to primates' health (Lewis et al. 2000). Access to sunlight to enable the production of vitamin D3 is vital, and supplementation by medication or artificial UV light is, for example, essential to prevent metabolic bone disease.

10 Noise

It is also important to understand primate sensitivity to the level and pitch of ambient sounds. The high-frequency hearing limit in humans is 17.6 kHz and that of primates can range from 28.5 kHz for chimpanzees to 65 kHz for bushbabies (galagos). Therefore, the high-pitched sounds emitted by computers and televisions that are in the same room as a cage may be disturbing to primates, and especially smaller ones given that they can hear very high-pitch sounds.

11   Inbreeding

If much of the European, UK, and North American trade depends on captive breeding, then the welfare implications of maintaining supply to the trade warrants investigation. It is known that the higher the genetic variation within a population, the less risk there is of inbreeding depression (Charpentier et al. 2007; Rails and Ballou 1982). Genetic diversity is essential to maintaining a healthy captive population and underpins zoos' breeding programs, such as the European Endangered Species Programs (EEPs), as well as in situ conservation policies, such as the Convention on Biological Diversity (United Nations 1992). So far as we are aware, there are no coordinated breeding programs in the primate pet trade. Individual animals are bought and sold with little or no known provenance or pedigree. The buyer may not know the full name and address of the vendor. They do not see the welfare of the breeding stock and may not be able to follow up any concerns after purchase. Buyers respond to advertisements, often using mobile phone numbers, and complete the exchange at a domestic dwelling, at a motorway service station, or some other neutral place. There is no requirement for paperwork, including proof of genetic lineage. As primates are bought, bred, sold, and exchanged, there is a high chance that closely related individuals will be mated to produce offspring. Thus, the nature of the unregulated trade of primates means that it is difficult to monitor the many adverse welfare outcomes that may come about due to inbreeding depression, and both infant mortality and pregnancy failures from these matings will remain hidden. Genetic disorders or diseases may also not be diagnosed as privately owned primates are seldom presented for veterinary care and rarely for postmortem examinations (Wild Futures and RSPCA 2012).

12   Diet and Nutrition

Diet and nutrition are vital to health and welfare. Species requirements vary from omnivores to folivores to generalists and specialists (Fleagle 1998). Time budgets in wild primates often differ from those of captive animals (Wolfensohn and Honess 2008). Primates in the wild often show great dietary flexibility and variability (Chapman and Chapman 1990). Nutrition in pet and entertainer primates is frequently compromised by a poor diet, which can contribute to health disorders and diseases, such as heart disease, marmoset wasting syndrome, metabolic bone disease, raised cholesterol, and diabetes (IFAW 2005; Wild Futures 2014a). In human children, malnourishment is linked to lower cognitive ability over a sustained period (Liu et al. 2003). If the same is true for nonhuman primates, this may also be a contributing factor for negative welfare. As the health of primates can be measured by individuals' ability to adapt to their environment (Broom 1991), the lack of dietary choice coupled with low nutrition could be another threat to pet and entertainer primate welfare.

13 Diabetes

Type 2 diabetes mellitus is a health issue commonly found in pet primates (Jungle Friends 2016; Kruzer 2016). In humans, type 2 diabetes is linked to a sedentary lifestyle, obesity, and a diet high in carbohydrates and, in particular, refined sugars. Many privately owned primates have poor diets and lack the opportunity to exercise. The long-term effects of diabetes can lead to progressive microvascular complications, including damage to the eyes (retinopathy) leading to blindness, the kidneys (nephropathy) leading to renal failure, and the nerves (neuropathy) leading to impotence and diabetic foot disorders, which include severe infections requiring amputation. People (and primates) with diabetes are also at increased risk of cardiovascular, peripheral vascular, and cerebrovascular disease (World Health Organization 2019). Acute fluctuations in blood sugar levels due to poor regulation by an underfunctioning pancreas can also impair cognitive function and cause changes in mood, such as irritability, anxiety, and aggression (Sommerfield et al. 2004).

14 Skeletal Disease

Metabolic bone disease is also a common health issue for pet primates. It is manifested in a wide range of disorders, including rickets (bowing of the long bones in young animals), osteomalacia (softening of the bone and decreased bone density due to decreased mineralization), and osteoporosis (thinning of the bone matrix that may be related to protein deficiency or lack of use of bones due to confinement in small cages). Symptoms can also include shortening of the jaw, curvature of the spine, tremors, and extreme lameness. The most common cause of metabolic bone disease is a lack of exposure to direct sunlight or ultraviolet light (UV-B) (Ludlage and Mansfield 2003; Kumar 2018). UV radiation enables the manufacture of vitamin D, which in turn enables the uptake and metabolism of calcium. Metabolic bone disease is particularly problematic for primates kept in temperate climates where UV radiation is not as high. Therefore, it is usually necessary to supplement captive primate diets with vitamin D and sometimes to provide artificial UV lighting. Vitamin D status may influence the risk of developing metabolic diseases such as type 2 diabetes in humans (Khan et al. 2013). Vitamin D deficiency is common in privately owned primates and may also contribute to the prevalence of hyperglycemia in primates rescued from the pet and entertainment trade.

15 Pregnancy and Weaning

There is evidence in human and animal studies to show that frequent and repeated (multiparous) pregnancies can be detrimental to maternal health (Dhawan et al. 2004; Bigiu 2018). If breeding primate females are subjected to the physical stress of frequent pregnancies that are attributable to the early weaning and removal of infants, it could be reasonable to assume that they could suffer similar health risks.

It is not unreasonable to think that nonhuman primates gain the same benefits from species-specific milk as do human babies, and other mammals, from their mothers. Early weaning of infant primates may affect short- and long-term health outcomes because the infant will not receive the antibodies, immune system stimuli, and species-specific nutrients contained in maternal milk (Admyre et al. 2019; Hanson and Söderström 1981). Short- and long-term resistance to disease is significantly increased in humans who are exclusively breastfed for the first four to six months (Renfrew et al. 2012; Ho et al. 2018). Analyses of the effect of breastfeeding versus not breastfeeding in human infants showed an association with a reduction in the risk of acute ear infections, nonspecific gastroenteritis, severe lower respiratory tract infections, atopic dermatitis, asthma (young children), obesity, type 1 and type 2 diabetes, childhood leukemia, sudden infant death syndrome, and necrotizing enterocolitis (Stuebe 2009; American Academy of Pediatrics 2012). If the causal relationship can be shown to be the same in hand-reared primates, this may explain similar health problems in some captive primates.

These morbidity issues are far from new. Examination of the preserved bodies of primates from ancient Egypt reveals that individuals had short lives and suffered from the same diseases and health problems that affect pet primates in modern times, including metabolic bone disease, malnourishment, and possibly tuberculosis *Mycobacterium tuberculosis* (Goudsmit and Brandon-Jones 1999).

16 Social Issues

The psychological impact of the repeated removal of infants from a mother may also be detrimental to welfare (Hevesi 2005). Socially reared primates have the opportunity to interact with conspecifics. The socially deprived primate does not have the opportunity to learn how and when to control their physical strength. This can have negative welfare outcomes. The pet primate who bites their owner may be housed more securely and lose the human contact that comprises their only social opportunity. Former pet primates often struggle to integrate into social groups because they lack the skills and understanding gained by socially reared primates (Soulsbury et al. 2009). Welfare issues persist even when pet and entertainer primates are kept with conspecifics in pairs or groups. Peer-raised primates present with more abnormal behaviors than those raised in a group comprising a range of ages, gender, and social relationships (Mason and Rushen 2008). Most primates in the wild live in groups of 5–20, or more, members. Primates have evolved to live in complex social hierarchies with varying levels of intimacy and types of relationships across gender and age ranges (Box 1999; Chapman and Rothman 2009). It is widely recognized that control and choice in the social, psychological, and physical environment are beneficial to captive primates (Bassett and Buchanan-smith 2007). The quality of social relationships has measurable fitness consequences (Silk 2007). Primates used in entertainment, who are usually young, are frequently removed and returned to their companions. This contrasts with the characteristic stability of infancy and juvenile periods in wild primates (McGrew 1981) and thus may have a detrimental effect on group stability (European Commission 2002).

17 'Undesirable' Behaviors and Interventions

In many species, scent marking, using specialized glands or urine, is normal (Heymann 2006; Roberts 2012), but scent marking is not often welcome in human domestic settings. The primate may be engaging in these behaviors to mark territory or assert affiliations and place in a perceived human hierarchy. The human response to this behavior is often to protect one's physical safety and property by restricting the primate access both to people and to places. Thus, a primate whose social affiliations are with people becomes further isolated and physically restricted.

Owners may attempt to modify or mitigate other unwanted behaviors. Methods may include punishing the primate, castration, or removal of teeth. The removal or cutting of teeth of pet monkeys in the USA is not uncommon (Jungle Friends Nonhuman Primates in the Private Sector 2020; Corning 2005). In the entertainment industry in Southeast Asia, slow lorises are routinely subjected to teeth cutting or pulling to protect humans from their toxic saliva (Nekaris et al. 2008). Consequently, slow lorises frequently suffer dental infections, which can become systemic and be fatal (Sanchez 2008). The removal of teeth has potential health implications other than injury or infection (Beerda et al. 1997; Nekaris et al. 2013). There may be a reduced ability to chew and therefore eat and digest a full range of foods and nutrients. This can lead to malnutrition, malabsorption, and periodontal disease, jaw muscle reduction, and bone loss (Chauncey et al. 1984; Silberg 2016). The effects of tooth extraction on bone health will be exacerbated by osteoporosis, a metabolic bone disease, which, as noted earlier, is common in privately kept primates (Ludlage and Mansfield 2003).

The American Veterinary Medical Association (AVMA) has a policy that addresses the immediate issues of welfare: "The AVMA is opposed to removal or reduction of healthy teeth in nonhuman primates ... (which) ... may also create oral pathologic conditions." (AVMA 2012). In this policy, the AVMA highlights their concern that removal or modification of teeth does not address the cause of behavioral issues and recommends assessment of husbandry and environment.

Castration is used to control behavior and unwanted aggression. It is often ineffective, perhaps in part due to the complex causal influences of aggressive behavior (Dixson 1980). Veterinarians who castrate primates may not account for the fact that, unlike most other mammals, primate copulatory ability is emancipated from hormonal control; this allows sexual behavior to be used in nonreproductive contexts (Miller 1931; Wallen 1990, 2001). Castration has little effect on aggression in many primate species (Wilson and Vessey 1968), and this appears true whether the castration is performed before or after puberty (Dixson 1980). Castration may create social problems for the individual, with sanctuaries reporting retarded social development in castrated individuals (Hevesi 2005). Castration is also linked to chronic changes in bone density that makes individuals susceptible to osteoporosis (Kessler et al. 2016).

A frequent solution to unwanted behavior is for the owner to sell the primate or move it to a new home, and so many primates have had multiple owners by the time they are received by rescue centers and sanctuaries. It can be surmised that these situations of social isolation and transfers are forms of social privation and may contribute to the abnormal behaviors frequently seen in these captive primates. This pattern of bonding between owner and primate followed by, from the human perspective, inappropriate and unwanted expressions of natural behaviors, leads to rejection and separation. This may also contribute to ex-pet primates having a conflicted relationship with people, demonstrating a desire for attention from people and, at the same time, apparent frustration and aggression toward people. In the case of great apes, the ability of a domestic dwelling to provide safe caging is more likely to become impossible and the individual may be sold to a roadside zoo or euthanized (Goodall 2016).

As a primate matures and becomes more aggressive, its relationships with other domestic pets can become agonistic, and this can endanger both the primate and the other pets. It is not uncommon for owners to report that their pet primates and other pets become jealous of attention shown by the owner to either the primates or other pets. Stories include cats afraid to enter the house after being attacked by a marmoset, or a capuchin having to be caged because of the aggressive behavior of a pet dog. Wild Futures reports that owners often recognize that primate pets may need more company than the owners can offer and try to resolve the situation inappropriately. For example, one capuchin that was rescued by Wild Futures some 16 years ago still had a heightened fear of dogs. The monkey had lived in a shed with a small aviary for 9 years. His owners thought that, when they were absent during the day, the monkey and their pet Staffordshire terrier could keep each other company, and so the dog was left the run of the garden. The dog appeared to have continuously circled the capuchin to the extent that it wore a path around the monkey's cage.

Abnormal behaviors are reported by sanctuaries as common in ex-pet and entertainer primates. For example, Wild Futures has recorded persistent abnormal behavior at the time of arrival in 100% of 33 individuals of two species (capuchins and Barbary macaques) of ex-pet monkeys (see Fig. 1). In comparison, in their survey of North American zoos, Bollen and Novak (2000) reported abnormal behavior in 497 individuals of 68 different species, giving an overall rate of 14%.

Psychological stress, such as that resulting from inappropriate housing, has a detrimental effect on the human immune system (Dantzer et al. 2018; Herbert and Cohen 1993; Segerstrom and Miller 2004). Ex-pet monkeys may similarly suffer chronic health problems as a result of a compromised immune system (Stuyven 2009, 2010) (Fig. 2).

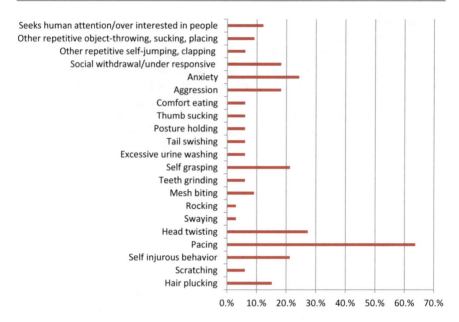

Fig. 1 Behavioral disorders in 33 ex-pet primates at Wild Futures Monkey Sanctuary. Percentage of 33 ex-pet primates with specific behavioral disorder. Total percent is greater than 100% as individuals often display more than one behavior

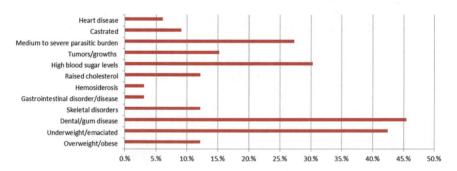

Fig. 2 Physical conditions or disorders in 33 ex-pet primates at Wild Futures Monkey Sanctuary. Percentage of 33 ex-pet primates with specific physical disorder or disease. Total percent is greater than 100% as individuals often display more than one disorder or disease

18   Transfer of Disease Between Humans and Primates

The physiological and genetic similarities between humans and anthropoid primates, and to a lesser extent simian primates, make the risk of disease transfer between humans and primates particularly high (Wolfe et al. 1998; Corning 2005). There is a disease risk where there is frequent physical contact between primates and humans, such as when primates are kept as pets or as entertainers. This risk may manifest itself as zoonoses (from primate to human) and as anthropozoonoses (from human to primate) with the potential of retransmission in the same or a mutated, and possibly more virulent, form (Corning 2005). Primates are more often threatened by microorganisms indigenous to humans than vice versa (Wolfe et al. 1998). In recognition of this risk people working with or near primates, in the field or in captivity, are usually required to be tested for tuberculosis (Lewis 2000). Risk for pet and entertainer primate welfare does not only come from humans, but from other organisms that share the human environment such as fungi, viruses, protozoa, helminths, and bacteria. Other local fauna can also lead to disease. In 2015, Wild Futures recorded a case of a pet capuchin monkey in the UK that was infected by tuberculosis, which was traced by the UK Government department, DEFRA, to the local badger population. It is likely that the capuchin, who had frequent contact with another capuchin, humans, domestic animals, and wildlife, harbored the disease for years. The monkey appeared healthy but was a potential vector for tuberculosis to any member of her species with whom she came into contact.

Access to veterinary care is also a recognized issue. Many owners are unwilling or unable to access qualified veterinarians. Veterinary experience with primates is often extremely limited or nonexistent. A survey by Hanson (2014) of 100 veterinary practices across five counties in the UK found that 87% reported being unable to treat a primate. Ninety-seven percent of the practices surveyed stated that they would not be able to offer advice on primate welfare and husbandry, as their knowledge and expertise only extended to common domesticated animals, such as pets and farm animals. The associations representing veterinarians in the UK, Europe, and the USA all have position statements expressing concern on keeping primates as pets or in private ownership (AVMA 2012; BVA 2014; Federation of European Vets 2013). Wild Futures has evidence of veterinarians misidentifying primate species and their sex as well as failing to diagnose serious diseases or disorders (Wild Futures 2014b), including metabolic bone disease, diabetes, and cleft lip and palate.

19 Rehabilitation Issues

Rescue centers and sanctuaries aim to improve and maintain quality of life for individual primates. The physical and psychological issues described here create a challenge to any such organization and the caregivers who work with the primates (Smith 2006). They must judge whether they can offer an acceptable quality of life. A physical disability may compromise an animal's ability to function safely in a social group. A chronic disease may require the primate to cooperate in accepting medication or frequent veterinary interventions. Insulin injections, for example, will test the relationship between caregiver and primate. A positive outcome relies on good training techniques but also the primate's ability to trust and adapt to the request for cooperation (Prescott et al. 2003; Westlund 2015). Caregivers must judge whether pain relief, for instance, to treat arthritis, is required and sufficiently effective. Rehabilitating and socializing hand-reared and/or socially deprived primates presents its own problems. Social rehabilitation is a potentially dangerous time for the individuals learning to live together. It may not be because the individual intends harm to new companions, but because it is ignorant of the appropriate and proportionate responses to social cues. Successful social rehabilitation depends in part on the history of the animal (Llorente et al. 2015; Lopresti-Goodman et al. 2012). Where the potential for good welfare is limited or absent, caregivers and veterinarians may consider euthanasia as the only appropriate option (Broom 2007; RSPCA 2016; Wolfensohn 2022).

20 Conclusion

Primates are intelligent, long-lived, often socially complex animals. Across species, they vary hugely in their morphology, ethology, and ecology. The difficulty of replicating their natural diets and nutrition, and of providing the psychological and physiological stimulation necessary to keep primates healthy, combined with the dangers of zoonoses and disease, makes it unlikely, or even impossible, for a person to keep a pet or entertainer primate in a welfare responsible or positive manner. Using parameters like the "five freedoms," there is considerable evidence that domesticated animals, like cats or dogs, often thrive in domestic environments, whereas there is no evidence that any primate species do so. Broom (1986) suggests that a wide range of welfare indicators should be used to measure welfare. The evidence suggests that whatever mitigation the owner offers the primate, poor welfare is inevitable.

There is still a wide gap between the expert opinion and a significant minority of the public who keep or promote the keeping of primates as pets or entertainers. The International Primatological Society, the World Organization for Animal Health, the American Zoological Association, the National Association of State Public Health Veterinarians, and the Council of State and Territorial Epidemiologists have position statements opposing private sector possession of nonhuman primates and/or their use as "actors." Over 360 primatologists, including Jane Goodall, Frans de Waal, Ian Redmond, Richard Wrangham, and Christophe Boesch, have signed a statement calling for an end to the keeping of primates as pets (Wild Futures and RSPCA 2012). Public petitions and polls also suggest that there is support for this view (RSPCA 2016). Despite this, the popularity of viral videos described earlier suggests that education is insufficient or inadequate. A large number of viewers will see a chimpanzee depicted as swaying and waving his arms as comical and not as distressing. It was once acceptable for people to pay to see the residents of the Bethlehem Hospital (Bedlam) as entertainment. This chapter started with a quote linking evolutionary legacy and welfare. We can end in a similar vein with McGrew (1981) who said "Primates in captivity that are socially or intellectually deprived are not realizing their evolutionary potential. Their behavior is abnormal in proportion to the degree to which such deprivation exists." Evidence of poor welfare outcomes has convinced many legislators that keeping primates as pets or as entertainers is unacceptable. For the individual primates caught in these industries, the trend to make this view the norm cannot come soon enough.

Acknowledgements The author thanks Sarah Hanson.

References

Admyre C, Johansson SM, Rahman K et al (2019) Exosomes with immune modulatory features are present in human breast milk. J Immunol 179(3):1969–1978. https://doi.org/10.4049/jimmunol.179.3.1969

Agoramoorthy G, Hsu MJ (2005) Use of nonhuman primates in entertainment in Southeast Asia. J Appl Anim Welf Sci 8(2):141–149. https://doi.org/10.1207/s15327604jaws0802_6

Aldrich BC (2015) Facial expressions in performing primates. Royal (Dick) School of Veterinary Studies. University of Edinburgh

American Academy of Pediatrics (2012) Breastfeeding and the use of human milk. Pediatrics 129 (3):e827–e841. https://doi.org/10.1542/peds.2011-3552

Ares BE (2019) Wild animals in circuses (No 2) Bill

AVMA (2012) Single policy on teeth reduction and removal adopted. Am Vet Med Assoc https://www.avma.org/News/JAVMANews/Pages/130115r.aspx. Accessed 5 Dec 2016

Bassett L, Buchanan-smith HM (2007) Effects of predictability on the welfare of captive animals §. Appl Anim Behav Sci 102(3–4):223–245. https://doi.org/10.1016/j.applanim.2006.05.029

Bayne K, Hau J, Morris T (2022) The welfare impact of regulations, policies, guidelines and directives and nonhuman primate welfare. In: Robinson LM, Weiss A (eds) Nonhuman primate welfare: from history, science, and ethics to practice. Springer, Cham, pp 629–646

Beerda B, Schilder MBH, van Hooff JARAM, de Vries HW (1997) Manifestations of chronic and acute stress in dogs. Appl Anim Behav Sci 52(3-4):307–319. https://doi.org/10.1016/S0168-1591(96)01131-8

Bigiu N et al (2018) Maternal depletion syndrome. Gineco.eu 11(2):98–103. https://doi.org/10.18643/gieu.2015.98

Blue Cross and The Born Free Foundation One Click Away, An investigation into the online sale of exotic animlas as pets. https://www.bornfree.org.uk/storage/media/content/files/Publications/ONE_CLICK_AWAY_REPORT_SPREADS.pdf

Bodmer RE, Eisenberg JF, Redford KH (2002) Hunting and the likelihood of extinction of Amazonian mammals. Conserv Biol 11(2):460–466. https://doi.org/10.1046/j.1523-1739.1997.96022.x

Bollen KS, Novak MA (2000) A survey of abnormal behavior in captive zoo primates. Am J Primatol 51(S1):47. https://doi.org/10.1002/(SICI)1098-2345(2000)51:1+%3C33::AID-AJP3%3E3.0.CO;2-F

Borgi M, Cirulli F (2016) Pet face: mechanisms underlying human-animal relationships. Front Psychol 7:298. https://doi.org/10.3389/fpsyg.2016.00298

Born Free USA Get The Facts: The Facts about Primates as 'Pets'. http://www.bornfreeusa.org/facts.php?p=2913&more=1. Accessed 30 Nov 2016

Box HO (1999) Temperament and socially mediated learning among primates. In: Box HO, Gibson KR (eds) Mammalian social learning: comparative and ecological perspectives. Cambridge University Press, Cambridge, pp 33–56

Broom DM (1986) Indicators of poor welfare. Br Vet J 142(6):524–526. https://doi.org/10.1016/0007-1935(86)90109-0

Broom DM (1991) Animal welfare: concepts and measurement. J Anim Sci 69(10):4167–4175. https://doi.org/10.2527/1991.69104167x

Broom D (2007) Quality of life means welfare: how is it related to other concepts and assessed? Anim Welf 16(9):45–53

BVA (2014) British Veterinary Association Policy statement. https://www.bva.co.uk/media/1160/primates-as-pets.pdf

Carr N, Cohen S (2011) The public face of zoos: images of entertainment education and conservation. Anthrozoös 24(2):175–189. https://doi.org/10.2752/175303711X12998632257620

Ceballos-Mago N, González C, Chivers D (2010) Impact of the pet trade on the Margarita capuchin monkey Cebus apella margaritae. Endanger Species Res 12:57–68. https://doi.org/10.3354/esr00289

Chapman CA, Chapman LJ (1990) Dietary variability in primate populations. Primates 31:121–128. https://doi.org/10.1007/BF02381035

Chapman CA, Rothman JM (2009) Within-species differences in primate social structure: evolution of plasticity and phylogenetic constraints. Primates:12–22. https://doi.org/10.1007/s10329-008-0123-0

Charpentier MJE, Widdig A, Alberts SC (2007) Inbreeding depression in non-human primates: a historical review of methods used and empirical data. Am J Primatol 69(12):1370–1386. https://doi.org/10.1002/ajp.20445

Chauncey HH, Muench ME, Kapur KKWA (1984) The effect of the loss of teeth on diet and nutrition. Int Dent J 34(2):98–104

CITES (2016) Notification to the Parties No. 2016/043. https://cites.org/sites/default/files/notif/E-Notif-2016-043.pdf

Clutton-Brock J (1999) A natural history of domesticated mammals. Cambridge University Press, Cambridge

European Commission (2002) The welfare of non-human primates used in research. Report of the Scientific Committee on Animal Health and Animal Welfare. Adopted on 17 December 2002

Corning S (2005) Public health and safety risks involved in the keeping and trade of primates as pets. Born to be wild primates are not pets international fund animal welfare

Dantzer R, Cohen S, Russo SJ, Dinan TG (2018) Resilience and immunity. Brain Behav Immun 74:28–42. https://doi.org/10.1016/j.bbi.2018.08.010

Davenport RK, Rogers CM, Rumbaugh DM (1973) Long-term cognitive deficits in chimpanzees associated with early impoverished rearing. Dev Psychol 9(3):343–347. https://doi.org/10.1037/h0034877

Dawkins MS (2006) A user's guide to animal welfare science. Trends Ecol Evol 21(2):77–82. https://doi.org/10.1016/J.TREE.2005.10.017

Defra (2019) The welfare of primates as pets in England: call for evidence. https://www.gov.uk/government/consultations/welfare-of-primates-as-pets-in-england-call-for-evidence/the-welfare-of-primates-as-pets-in-england-call-for-evidence. Accessed 17 Jan 2020

Dhawan V, Brookes ZLS, Kaufman S (2004) Long-term effects of repeated pregnancies (multiparity) on blood pressure regulation. Cardiovasc Res 64(1):179–186. https://doi.org/10.1016/j.cardiores.2004.06.018

Diamond J (2002) Evolution, consequences and future of plant and animal domestication. Nature 418(6898):700–707. https://doi.org/10.1038/nature01019

Dixson AF (1980) Androgens and aggressive in primates: a review. Aggress Behav 6(1):37–67. https://doi.org/10.1002/1098-2337(1980)6:1%3C37::AID-AB2480060106%3E3.0.CO;2-7

Driscoll CA, Macdonald DW, O'Brien SJ (2009) From wild animals to domestic pets, an evolutionary view of domestication. Proc Natl Acad Sci U S A 106(Suppl):9971–9978. https://doi.org/10.1073/pnas.0901586106

Endcap (2012) Wild pets in the European Union

Eurogroup for Animals (2013) Why Europe needs 'positive lists' to regulate the sale and keeping of exotic animals as pets. https://www.granalacantadvertiser.com/why-europe-needs-positive-lists. Accessed 18 May 2022

Ferreira RG, Mendl M, Wagner PGC et al (2016) Coping strategies in captive capuchin monkeys (*Sapajus* spp.). Appl Anim Behav Sci 176:120–127. https://doi.org/10.1016/j.applanim.2015.12.007

Ferreira RG, Ruiz-Miranda C, Sita S, Sánchez-López S, Pissinatti A, Corte S, Jerusalinsky L, Wagner PG, Maas C (2022) Primates under human care in developing countries: examples from

Latin America. In: Robinson LM, Weiss A (eds) Nonhuman primate welfare: from history, science, and ethics to practice. Springer, Cham, pp 145–170

Fleagle JG (1998) Primate adaptation and evolution, 2nd edn. Elsevier, Amsterdam

Friedl KE, Holmes WN (1986) The effect of relative humidity on osmoregulation in the squirrel monkey (*Saimiri sciureus*). Primates 27:465–470. https://doi.org/10.1007/BF02381891

Federation of European Vets (2013) Conference outcomes and recommendations. https://fve.org/cms/wp-content/uploads/CWA-Conference-outcomes-and-recommendations-June-2013-2.pdf. Accessed 20 Jun 2022

Goodall J. Chimpanzees in entertainment. http://www.janegoodall.org.uk/chimpanzees/chimpanzee-central/15-chimpanzees/chimpanzee-central/27-chimpanzees-in-entertainment. Accessed 30 Nov 2016

Gosling SD (2001) From mice to men: What can we learn about personality from animal research? Psychol Bull 127(1):45–86. https://doi.org/10.1037/0033-2909.127.1.45

Goudsmit J, Brandon-Jones D (1999) Mummies of olive baboons and Barbary macaques in the Baboon Catacomb of the Sacred Animal Necropolis at North Saqqara. J Egypt Archaeol: 45–53

Hanson S (2014) Primates as pets: investigating aspects of regulation throughout the European Union. MSc Primate Conservation, Oxford Brookes University, Oxford

Hanson LA, Söderström T (1981) Human milk: defense against infection. Prog Clin Biol Res 61:147–159

Hare B, Brown M, Williamson C, Tomasello M (2002) The domestication of social cognition in dogs. Science 298(5598):1634–1636. https://doi.org/10.1126/science.1072702

Herbert TB, Cohen S (1993) Stress and immunity in humans: a meta-analytic review. Psychosom Med 55(4):364–379. https://doi.org/10.1097/00006842-199307000-00004

Hevesi R (2005) Welfare and health implications for primates kept as pets. In: Born to be wild: Primates are not pets. International Fund for Animal Welfare (ed), London, pp 18–29

Heymann EW (2006) Special issue: the neglected sense — primate olfaction in primate behavior, ecology, and evolution. Am J Primatol 68(6):519–661. https://doi.org/10.1002/ajp.20249

Ho NT, Li F, Lee-Sarwar KA et al (2018) Effects of exclusive breastfeeding on infant gut microbiota: a meta-analysis across studies and populations. Nat Commun 9(1):4169. https://doi.org/10.1038/s41467-018-06473-x

Hosey G (2022) The history of primates in Zoos. In: Robinson LM, Weiss A (eds) Nonhuman primate welfare: from history, science, and ethics to practice. Springer, Cham, pp 3–30

Humane Society International (2019) No Title

IFAW (2005) In Born to be wild: primates are not pets. International Fund for Animal Welfare

International Primatological Society. http://18.117.158.79/policy-statements-and-guidelines/private-ownership-ofnonhuman-primates/, http://18.117.158.79/policy-statements-and-guidelines/nonhuman-primates-in-media/

Wallenfeldt J (2018) Ringling Bros. and Barnum & Bailey Circus. Encyclopedia Britannica. https://www.britannica.com/topic/Ringling-Bros-and-Barnum-and-Bailey-Combined-Shows

Jungle Friends (2016) Jungle Friends rescue centre. www.junglefriends.org. Accessed 12 Dec 2016

Jungle Friends Nonhuman Primates in the Private Sector. www.junglefriends.org/monkey-topics/understanding-the-issues/private-sector.html. Accessed 20 Oct 2020

Kessler MJ, Wang Q, Cerroni AM, Grynpas MD, Gonzalez Velez OD, Rawlins RG, Ethun KF, Wimsatt JH, Kensler TB, Pritzker KPH (2016) Long-term effects of castration on the skeleton of male rhesus monkeys (*Macaca mulatta*). Am J Primatol 78(1):152–166. https://doi.org/10.1002/ajp.22399

Khan H, Kunutsor S, Franco OH, Chowdhury R (2013) Vitamin D, type 2 diabetes and other metabolic outcomes: a systematic review and meta-analysis of prospective studies. Proc Nutr Soc 72(1):89–97. https://doi.org/10.1017/S0029665112002765

Koene P, de Mol RM, Ipema B (2016) Behavioral ecology of captive species: using bibliographic information to assess pet suitability of mammal species. Front Vet Sci 3:35. https://doi.org/10.3389/fvets.2016.00035

Kozorovitskiy Y, Gross CG, Kopil C et al (2005) Experience induces structural and biochemical changes in the adult primate brain. Proc Natl Acad Sci U S A 102(48):17478–17482. https://doi.org/10.1073/pnas.0508817102

Kruzer, AR (2016) Pet capuchin monkeys. http://exoticpets.com. Accessed 12 Dec 2016

Kumar R (2018) Metabolic bone diseases of captive mammal, reptile and birds. Approaches Poultry Dairy Vet Sci 3:1–5. https://doi.org/10.31031/apdv.2018.03.000563

Lazaro-Perea C (2001) Intergroup interactions in wild common marmosets, *Callithrix jacchus*: Territorial defence and assessment of neighbours. Anim Behav 62(1):11–21. https://doi.org/10.1006/anbe.2000.1726

Leighty KA, Valuska AJ, Grand AP et al (2015) Impact of visual context on public perceptions of non-human primate performers. PLoS One 10(2):e0118487. https://doi.org/10.1371/journal.pone.0118487

Lenzi C, Speiran SI, Grasso C (2020) "Let me take a selfie": reviewing the implications of social media for public perceptions of wild animals. Soc Anim. https://doi.org/10.1163/15685306-BJA10023

Lewis JCM (2000) Preventative health measures for primates and keeping staff in British and Irish zoological collections. A report to the British and Irish Primate Taxon Advisory Group (B&I PTAG), London

Lewis MH, Gluck JP, Petitto JM, Hensley LL, Ozer H (2000) Early social deprivation in nonhuman primates: long-term effects on survival and cell-mediated immunity. Biol Psychiatry 47(2):119–126. https://doi.org/10.1016/s0006-3223(99)00238-3

Liu J, Raine A, Venables PH et al (2003) Malnutrition at age 3 years and lower cognitive ability at age 11 years: independence from psychosocial adversity. Arch Pediatr Adolesc Med 157(6):593–600. https://doi.org/10.1001/archpedi.157.6.593

Llorente M, Riba D, Ballesta S et al (2015) Rehabilitation and socialization of chimpanzees (*Pan troglodytes*) used for entertainment and as pets: an 8-year study at Fundació Mona. Int J Primatol 36:605–624. https://doi.org/10.1007/s10764-015-9842-4

Lopresti-Goodman S, Kameka M, Dube A (2012) Stereotypical behaviors in chimpanzees rescued from the African bushmeat and pet trade. Behav Sci 3(1):1–20. https://doi.org/10.3390/bs3010001

Ludlage E, Mansfield K (2003) Overview clinical care and diseases of the common marmoset (*Callithrix jacchus*). Comp Med 53(4):369–382

Lutz CK, Baker KC (2022) Using behavior to assess primate welfare. In: Robinson LM, Weiss A (eds) Nonhuman primate welfare: from history, science, and ethics to practice. Springer, Cham, pp 171–206

Mason GJ, Rushen J (2008) Stereotypic animal behaviour: fundamentals and applications to welfare and beyond. CAB International, Cambridge, MA

McGrew WC (1981) Social and cognitive capabilities of nonhuman primates: lessons from the wild to captivity. Int J Study Anim Probl 2(3):138–149

Miller G (1931) The Primate basis of human sexual behavior. Q Rev Biol 6(4):379–410. https://doi.org/10.1086/394387

Moss A, Esson M (2013) The educational claims of zoos: where do we go from here? Zoo Biol 32(1):13–18. https://doi.org/10.1002/zoo.21025

Nekaris KAI, Sanchez KL, James S et al (2008) Javan slow loris *Nycticebus javanicus* É. Geoffroy, 1812 Indonesia. In: Mittermeier RA, Wallis J, Rylands AB, Ganzhorn JU, Oates JF, Williamson EA, Palacios E, Heymann EW, MCM K, Yongchen L, Supriatna J, Roos C, Walker S, Cortéz-Ortiz L, Schwitzer C (eds) 25 World's most endangered primates. IUCN/SSC Primate Specialist Group (PSG), International Primatological Society (IPS), and Conservation International (CI), Arlington, VA, pp 44–46

Nekaris BKAI, Campbell N, Coggins TG et al (2013) Tickled to death: analysing public perceptions of "cute" videos of threatened species (slow lorises - *Nycticebus* spp.) on web 2.0 sites. PLoS One 8(7):e69215. https://doi.org/10.1371/journal.pone.0069215

Nijman V (2009) An assessment of trade in gibbons and orang-utans in Sumatra, Indonesia. TRAFFIC Southeast Asia, Petaling Jaya, Selangor, Malaysia. https://www.traffic.org/site/assets/files/3986/sumatran-gibbons-orangutans.pdf

Nittono H, Fukushima M, Yano A, Moriya H (2012) The power of Kawaii: viewing cute images promotes a careful behavior and narrows attentional focus. PLoS One 7(9):e46362. https://doi.org/10.1371/journal.pone.0046362

Osterberg P, Nekaris KAI (2015) The use of animals as photo props to attract tourists in Thailand: a case study of the slow loris *Nycticebus* spp. TRAFFIC Bull 27(1):13–18

Peres CA (1990) Effects of hunting on western Amazonian primate communities. Biol Conserv 54:47–59. https://doi.org/10.1016/0006-3207(90)90041-M

Peres CA (2000) Effects of subsistence structure in hunting on vertebrate forests community. Conserv Biol 14(1):240–253. https://doi.org/10.1046/j.1523-1739.2000.98485.x

Perry S (1998) A case report of a male rank reversal in a group of wild white-faced capuchins (*Cebus capucinus*). Primates 39:51–70. https://doi.org/10.1007/BF02557743itle

Prescott MJ, Sussex W, Buchanan-smith HM (2003) Guest editors' introduction training nonhuman primates using positive reinforcement techniques. Anim Welf 6(3):157–161. https://doi.org/10.1207/S15327604JAWS0603

Rails K, Ballou J (1982) Effects of inbreeding on infant mortality in captive primates. Int J Primatol 3:491–505. https://doi.org/10.1007/BF02693747

Redmond I (2005) The primate pet trade and its impact on biodiversity conservation. In: Born to be wild: primates are not pets. International Fund for Animal Welfare (IFAW), London

Reinhardt V (1997) Lighting conditions for laboratory monkeys: are they accurate? AWIC Newsletter 8(2):3–6

Renfrew MJ et al (2012) Preventing disease and saving resources: the potential contribution of increasing breastfeeding rates in the UK. UNICEF UK. http://www.unicef.org.uk/Documents/Baby_Friendly/Research/Preventing_disease_saving_resources.pdf?epslanguage=en

Roberts SC (2012) On the relationship between scent-marking and territoriality in callitrichid primates. Int J Primatol 33:749–761. https://doi.org/10.1007/s10764-012-9604-5

Robinson LM, Weiss A (2022) Primate personality and welfare. In: Robinson LM, Weiss A (eds) Nonhuman primate welfare: from history, science, and ethics to practice. Springer, Cham, pp 387–402

RSPCA (2016) Do you give a monkey's? The need for a ban on pet primates

Russell N (2002) The wild side of domestication. Soc Anim 10(3):285–302. https://doi.org/10.1163/156853002320770083

Sackett GP (1972) Prospects for research on schizophrenia. 3. Neurophysiology. Isolation-rearing in primates. Neurosci Res Prog Bull 10(4):388–392

Sanchez KL (2008) Indonesia's slow lorises suffer in trade. International Primate Protection League News 35(2):10

Schroepfer KK, Rosati AG, Chartrand T, Hare B (2011) Use of "entertainment" chimpanzees in commercials distorts public perception regarding their conservation status. PLoS One 6(10): e26048. https://doi.org/10.1371/journal.pone.0026048

Schuppli CA, Fraser D (2000) A framework for assessing the suitability of different species as companion animals. Anim Welf 9:359–372

Segerstrom SC, Miller G (2004) Psychological stress and the human immune system: a meta-analytic study of 30 years of inquiry. Psychol Bull 130(4):601–630. https://doi.org/10.1037/0033-2909.130.4.601

Smith AS (2006) Optimalising the role of animal rescue centres by researching the source of the problem. In: Zgrabczynska E, Cwiertniz P, Ziomek J (eds) Animals, zoos and conservation. Zoological Garden in Poznan, Poznan, pp 111–117

Silberg MA (2016) The physical and psychological effects of tooth loss. http://www.pittsburghdentalimplants.com/

Silk JB (2007) Social components of fitness in primate groups. Science 317(5843):1347–1351. https://doi.org/10.1126/science.1140734

Sommerfield AJ, Deary IJ, Frier BM (2004) Acute hyperglycemia alters mood state and impairs cognitive performance in people with type 2 diabetes. Diabetes Care 27(10):2335–2340. https://doi.org/10.2337/diacare.27.10.2335

Soulsbury CD, Iossa G, Kennell S, Harris S (2009) The welfare and suitability of primates kept as pets. J Appl Anim Welf Sci 12(1):1–20. https://doi.org/10.1080/10888700802536483

Stiles D et al (2013) Stolen apes - the illicit trade in chimpanzees, gorillas, bonobos and orangutans. A Rapid Response Assessment. United Nations Environment Programme, GRID-Arendal. www.grida.no

Stuebe A (2009) The risks of not breastfeeding for mothers and infants. Rev Obstet Gynecol 2 (4):222–231. https://doi.org/10.3909/riog0093

Stuyven DE (2009) Faculteit diergeneeskunde academiejaar 2008–2009 oorzaken van een verzwakt immuunsysteem bij berberapen door ieteke verhoeven

Stuyven DE (2010) Faculteit diergeneeskunde academiejaar 2009–2010 het meten van immuniteitsproblemen bij berberapen die mogelijk gerelateerd zijn aan chronische stress door ieteke verhoeven

UN-Grasp (2016) The apes seizure database. https://database.un-grasp.org/

United Nations (1992) Convention on biological diversity. https://www.cbd.int/doc/legal/cbd-en.pdf

United Nations Environment Programme (2010) The last stand of the gorilla: environmental crime and conflict in the Congo basin. https://wedocs.unep.org/20.500.11822/7842

van Lavieren E (2008) The illegal trade in Barbary macaques from Morocco. TRAFFIC Bull 21 (3):123–130

Vanautryve E (2014) Safety FC Keeping of exotic animals in Belgium: the "positive list": 1–15

Wallen K (1990) Desire and ability: hormones and the regulation of female sexual-behavior. Neurosci Biobehav Rev 14(2):233–241. https://doi.org/10.1016/s0149-7634(05)80223-4

Wallen K (2001) Sex and context: hormones and primate sexual motivation. Horm Behav 40 (2):339–357. https://doi.org/10.1006/hbeh.2001.1696

Weiss A (2017) Personality traits: a view from the animal kingdom. J Pers 86(1):12–22. https://doi.org/10.1111/jopy.12310

Westlund K (2015) Training laboratory primates – benefits and techniques. Primate Biol 2 (1):119–132. https://doi.org/10.5194/pb-2-119-2015

Wich SA, Marshall AJ (2016) An introduction to primate conservation. Oxford University Press, Oxford

Wild Futures (2014a) It's time to end the primate pet trade

Wild Futures (2014b) Written Evidence for Environment, Food and Rural Affairs Committee. Eleventh report, primates as pets. Available at: http://data.parliament.uk/writtenevidence/WrittenEvidence.svc/EvidenceHtml/4964

Wild Futures and RSPCA (2012) Primates as pets: is there a case for regulation?

Wilson M (2007) Intergroup aggression in wild primates. Behaviour 144(12):1469–1471. https://doi.org/10.1163/156853907782512083

Wilson DE, Reeder DM eds. (2005) Mammal species of the world: a taxonomic and geographic reference (vol 1). Johns Hopkins University Press, Baltimore, MD

Wilson AP, Vessey SH (1968) Behavior of free-ranging castrated rhesus monkeys. Folia Primatol 9 (1):1–14. https://doi.org/10.1159/000155164

Wolfe ND, Escalante AA, Karesh WB et al (1998) Wild primate populations in emerging infectious disease research: The missing link? Emerg Infect Dis 4(2):149–158. https://doi.org/10.3201/eid0402.980202

Wolfensohn S (2022) Humane end points and end of life in primates used in laboratories. In: Robinson LM, Weiss A (eds) Nonhuman primate welfare: from history, science, and ethics to practice. Springer, Cham, pp 369–386

Wolfensohn S, Honess P (2008) Handbook of primate husbandry and welfare. Blackwell Publishing, Oxford

World Health Organization. Diabetes. https://www.who.int/diabetes/action_online/basics/en/index3.html. Accessed 3 Jun 2019

Zeder MA (2006) Central questions in the domestication of plants and animals. Evol Anthropol 15 (3):105–117. https://doi.org/10.1002/evan.20101

Primates Under Human Care in Developing Countries: Examples From Latin America

R. G. Ferreira, C. Ruiz-Miranda, S. Sita, S. Sánchez-López, A. Pissinatti, S. Corte, L. Jerusalinsky, P. G. Wagner, and C. Maas

Abstract

While primate populations in wild areas are facing intensive decline due to habitat loss and removal for illegal trade, the number of individuals kept in ex situ settings is sharply increasing. With limited facilities and relocation programs, rescue centers in Latin American countries are frequently overcrowded and rarely

R. G. Ferreira (✉) · S. Sita
Psychobiology Post-Graduate Program, Universidade Federal do Rio Grande do Norte, Natal, Rio Grande do Norte, Brazil
e-mail: renata.ferreira@ufrn.br

C. Ruiz-Miranda
Environmental Sciences Laboratory, Centro de Biociências e Biotecnologia, Universidade Estadual Norte Fluminense, Rio de Janeiro, Brazil

S. Sánchez-López
Instituto de Investigaciones Biológicas and Dirección del Área Biológica Agropecuaria, Universidad Veracruzana,, Xalapa, México

Psicología y Ciencias de la Educación, Universitat Oberta de Catalunya, Barcelona, Spain

A. Pissinatti
Centro de Primatologia do Rio Janeiro, Instituto Estadual do Ambiente, INEA, Rio de Janeiro, Brazil

S. Corte
Faculty of Sciences, Ethology, Biology Institute, Universidad de la República, Montevideo, Uruguay

L. Jerusalinsky
National Center for Research and Conservation of Brazilian Primates CPB, Rio de Janeiro, Brazil

P. G. Wagner
IBAMA-RS, Rio de Janeiro, Brazil

C. Maas
Brazilian Society for Aquarium and Zoos, Rio de Janeiro, Brazil

© Springer Nature Switzerland AG 2023
L. M. Robinson, A. Weiss (eds.), *Nonhuman Primate Welfare*,
https://doi.org/10.1007/978-3-030-82708-3_7

meet basic welfare conditions. In this chapter we provide an outlook of primate-human co-existence in Latin America and give an overview of the number of primates and how they are kept in zoos and rescue centers in Brazil, Mexico, and Uruguay. Impact on wild populations, recurrent diseases, and management strategies are presented as examples of the difficulties of caring for primates ex situ in developing countries. Legal protection and rules for trade in all Latin American countries are reviewed and suggestions for improvement are discussed. This chapter concludes with a call for more rehabilitation and releases of fauna in well-designed landscape restoration programs.

Keywords

Platyrrhine captive care · Latin America rescue centers · New World primates ex situ · Primate legal protection · Animal welfare legislation · Animal welfare Latin America

1 Introduction

Habitat loss, hunting, capture for the illegal trade, and conflicts with human populations are known and unsolved threats to wild nonhuman primate (primates, hereafter) populations (Chapman and Peres 2001; Estrada et al. 2017). At the same time, the number of primates kept in captivity under human care is growing. Here we offer a report on the status of primates, and in particular New World monkeys housed in captivity throughout Central and South America. Authors of this chapter represent three Latin American countries (Brazil, Mexico, and Uruguay) and offer perspectives from their work at universities, zoological societies, and both governmental and non-governmental organizations. The chapter concludes with a general discussion and possibilities for destinations of rescued primates.

2 History

There are presently around 165 primate species in the 26 Latin American and Caribbean countries. The proximity of primates and humans in Latin America dates back to about 10,000 years, with prehistoric paintings at Serra da Capivara (Caatinga biome-NE-Brazil) depicting capuchin monkeys alongside humans (Martin 1996). People in advanced Pre-Columbian cultures (e.g., Mayan, Aztec, Zapotecan, Teotihuacan cultures) venerated monkeys as evidenced by depictions of monkeys on household artifacts, religious calendars, funeral urns, decorative elements, reliefs, paintings, and engravings (González-Picazo et al. 2001). Among the Maya, monkeys were the patrons of dancers, musicians, and artists. Similarly, monkeys were associated with the god of pulque, an alcoholic beverage, and drunkenness for Mixtec people. For the Aztecs, monkeys were related to the god of games and were symbols of the joke and improvisation. The representation of monkeys in the

Fig. 1 Lid of pot. Image of monkey with a rope hanging cocoa pod. Museum Tonina, Chiapas. 600-900 A.C. (courtesy of Miller and Martin 2004)

Fig. 2 Composition of Frida Kahlo self-portraits. 1938, 1940, 1942 and 1943, respectively. Courtesy of http://www.fridakahlofans.com

mythology of Pre-Columbian cultures was also associated with food, such as cocoa (Fig. 1).

Historically, collections of animals were a form of entertainment for the aristocracy in Latin America (Hosey 2022; Babb Stanley 2002). For example, the House of Beasts of the emperor Moctezuma II Xocoyotzin held a variety of captive animals (Blanco et al. 2009). Later, throughout the colonial period and the period of independence, private reserves were created for the amusement of rulers (Babb Stanley 2002). Monkeys continue to enjoy a ubiquitous presence in contemporary Latin American cultures. One prominent example can be found in the self-portraits of the Mexican painter and poet Frida Kahlo (1907–1954), which feature the spider monkeys (*Ateles geoffroyi*) that she kept as pets (Fig. 2).

Monkeys are still widely kept as pets and hunted for food in Latin America. Ceballos-Mago et al. (2010) described how wedge-capped capuchin monkeys (*Cebus olivaceus*) can be illegally purchased for approximately US$298.00. The importation of this species via the illegal trade in the Margarita Islands constitutes a major threat via hybridization to the 300 exemplars of the subspecies *Cebus apella margaritae*. Indigenous people living in the Amazon region continue to hunt monkeys or keep them as pets (Ribeiro 1995; International Union for Conservation of Nature [IUCN] 2000). Amazon species such as howler monkeys (*Alouatta* spp.), spider monkeys (*Ateles* spp.), brown woolly monkeys (*Lagothrix lagotricha*), saki monkeys (*Chiropotes* spp.), and brown capuchin monkeys (*Sapajus apella*) are still hunted as a source of meat for local and indigenous populations (Nascimento 2009). A consequence of this close association between humans and primates is the sharing of pathogens. For example, malaria is currently present in 38 New World monkey species; however, Cormier (2010) argues that all three *Plasmodium* forms found in New World monkeys originated in post-Columbian periods due to the increase in human-primate contact. The recurrent yellow fever outbreaks throughout Latin America are evidence of this largely unstudied Anthropocene disease dynamic that threatens both New World monkeys and humans (Bicca-Marques et al. 2017).

3 Captive Nonhuman Primates in Numbers

Central and South America are the world's largest exporters of wild-born primates being responsible for 75.5% of all legal primate trade; Africa, in comparison, is responsible for 23.6% of these exports (Harrington 2015). Over half a million New World monkeys were exported from South America (notably Peru and Colombia) to the USA between 1964 and 1980 (Mack and Mittermeier 1984). A recent analysis indicates that 90,000 monkeys were legally exported from South American countries to 23 other countries between 1977 and 2013 (de Souza Fialho et al. 2016). According to the *Convention on International Trade in Endangered Species* of Wild Fauna and Flora (CITES) database, 2392 squirrel monkeys (*Saimiri* spp.), 1626 marmosets (*Callithrix* spp.), 898 capuchin monkeys (*Cebus* spp.), and 192 guenons (*Cercophithecus* spp.) were legally traded for personal and commercial purposes between 2006 and 2012. With the exception of marmosets, 94–99% of these primates came from wild populations (Harrington 2015). The trade of captive-born primates represents only 5% (about 5000 individuals in 26 years) of the legal primate trade (de Souza Fialho et al. 2016). This means that the legal trade is responsible for an annual take of thousands of New World monkeys from the wild.

The export of New World monkeys is believed to have begun in the 1940s, with the establishment of the first commercial flight between Quito, Ecuador; Peru, and Miami, USA. Scientific (especially biomedical) research has been the major driver for this trade. Approximately 30,000 Amazonian primates are exported annually for biomedical research (RENCTAS 2000; Torres et al. 2010). Although most of the exportation of New World monkeys is to developed countries, trade among Latin American countries also happens. For instance, over 4000 night monkeys (*Aotus* spp.) are traded per year from Brazil and Peru into Colombia to supply biomedical research (Maldonado et al. 2009; Nijman et al. 2011). Nearly all countries doubled the use of New World monkeys on biomedical research over the last decades (Torres et al. 2010). The family Cebidae, which includes the genera *Cebus*, *Sapajus*, and *Saimiri*, represents over 50% of the New World monkeys used in scientific research over the last 40 years worldwide. However, there is no published estimate of the total number of primates held in research centers in Latin American countries to date. In the USA, 71,000 primates were used in research in 2010 (Miller-Spiegel 2011).

Capture for the illegal pet trade has been a continuing pressure on wild primate populations (Mack and Mittermeier 1984; RENCTAS 2000; Shanee 2012). In Brazil, 179 monkeys were confiscated from the illegal pet trade or from householders in two years (RENCTAS 2000). In a survey of the illegal trade in Northeastern Peru, Shanee (2012) found 279 primates of 12 species kept by humans mainly as pets or for the tourism trade (e.g., tourist photographs with monkeys). In Brazil, the primates most commonly confiscated from the illegal trade are *Callithrix* spp. (48%) and *Cebus/Sapajus* spp. (37%), whereas in Peru the most commonly confiscated primates are *Saimiri* spp. (32%) and *Cebus* spp. (25%) (RENCTAS 2000; Shanee 2012). Trayford and Farmer (2013) estimated that across North, Central, and South America there are only 1776 primates kept in captivity in 48 sanctuaries. These numbers are probably not representative of the actual number illegally traded every year because only a small proportion of these primates are confiscated by governmental authorities.

We collected data on the number of primates, the species, and the conditions in which primates are kept in captivity in Brazil, Mexico, and Uruguay. Brazil and Mexico are the Latin American countries with the largest gross domestic product and populations, and so the countries with presumably higher impact on primates in the wild and in captivity. Uruguay represents small Latin American countries with no natural occurrence of primates. This is a sample of the reality of primate welfare in Latin America, and these are countries for which the authors have expertise.

3.1 Brazil

Brazil is home to the highest number of primate species in the world. Of the 139 species that live in Brazil, 6 are critically endangered, 15 are endangered, and 15 are vulnerable (ICMBio Portaria n°444 17/12/2014). Primates in captivity in Brazil are located in rescue or rehabilitation centers, zoos, university research facilities, primate centers, wildlife sanctuaries (private), and private collections. Most of the captive primates registered in the National System for Management of Wildlife are in zoos (43.0%) or in breeding colonies for scientific research (30.8%). The remaining portion are kept in wildlife sanctuaries (13.5%) or are bred for commercial purposes (4.2%), including the legal pet market, or are in private rescue centers (3.0%) (see also Csermak 2007; IBAMA 2016, Relatório Técnico CETAS 2002–2014, unpublished—based on 2016 data). As of 2016, Brazilian authorities have records of 2763 licensed captive-housed primates (Fig. 3), excluding the number of primates kept in government rescue centers. *Cebus/Sapajus* spp. (23.4%), *Macaca* spp. (22.0%), and *Callithrix* spp. (16.4%) are most frequently held in private facilities; the most common species being rhesus macaques *Macaca mulatta* (18.7%), brown capuchin monkeys (13.4%), and black-tufted marmosets *Callithrix penicillata* (7.2%).

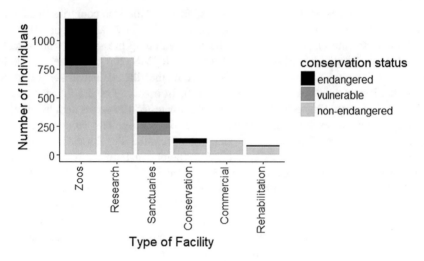

Fig. 3 Number of primates kept at private facilities. Conservation status: "endangered" groups species considered critically endangered or endangered are grouped; vulnerable; non-endangered groups least concern or data deficient, according to the Portaria n°44 (17/12/2014, Brazil). The number of primates is based on records of licenses provided by SISFAUNA (IBAMA 2016)

The Brazilian Ministry of the Environment has taken a proactive stance on conservation through the National Center for Research and Conservation of Brazilian Primates (CPB). Although the great focus of CPB is on conserving endangered species in situ, CPB is also concerned with the management of primates in captivity for two reasons: its role on biological invasions and its role on endangered species conservation. One way in which invasive populations have become established in Brazil has been the frequent disordered and unauthorized releases of captive specimens, which is a consequence of the illegal pet trade (Levacov et al. 2011; Oliveira and Grelle 2012). For example, areas, including protected areas, inhabited by two endangered species (the buffy-tufted marmoset *Callithrix aurita* and the golden-lion-tamarins *Leontopithecus rosalia*), have been invaded by introduced populations of common marmosets (*Callithrix jacchus*), black-tufted marmosets (*Callithrix penicillata*), and their hybrids (Pereira et al. 2008; Ruiz-Miranda et al. 2010). The ecological, behavioral, and genetic similarities between these allochthonous populations and resident species—especially for congeneric species—represent a serious threat through competition, introduction of pathogens, or hybridization, and may cause local extinctions (IUCN 2000). Hybridization with *Callithrix aurita* has already been recorded (Pereira et al. 2008).

The National Action Plans for the conservation of threatened primates in Brazil include, when appropriate, recommendations for captive management (CPB 2018). Adequate management of captive monkeys may contribute to the conservation of species. The Golden Lion Tamarin Conservation Program included captive management and reintroduction components (Beck et al. 1991; Stoinski et al. 2003; Kierulff et al. 2012). Currently, over 40% of the wild-living golden-lion-tamarins are descended from reintroduced animals, and this program contributed to the recovery of the species and the protection of over 3100 ha of forests on private land (Grativol et al. 2001; Procópio-de-Oliveira 2002; Ruiz-Miranda et al. 2010; Kierulff et al. 2012).

Wildlife Rescue and triage Centers (CETAS) and Wildlife Rehabilitation Centers constitute the official repositories of animals obtained from seizures by environmental law enforcement agencies and by the voluntary surrender of animals. The first governmental wildlife rescue and triage center dates to the 1970s. Brazil currently has 23 such centers spread across five regions. The region with the largest number of these centers is the Northeast (IBAMA 2016) where most wildlife confiscation occurs (RENCTAS 2000). CETAS receive on average 43,742 animals/year, totaling 568,645 between 2002 and 2014 (IBAMA 2016). Levacov et al. (2011) calculated that primates represent 38.2% of the mammals received by CETAS. *Callithrix jacchus* is the second most numerous mammal species rescued, and *Sapajus* spp. is the second most common primate rescued. Only in 2014, CETAS received 256 *Callithrix jacchus* and allocated 77% of them. Although efforts to release this species back to their habitat have been made, *Callithrix jacchus* are among the mammal species that accumulate and end up dying in CETAS facilities (Levacov et al. 2011).

Brazilian wildlife rescue center operations involve reception, screening, and destination. After a period in quarantine, animals are screened so as to identify their biome of origin, physical and behavioral characteristics (clinical status), and their epidemiological/health status, and to determine whether to reintroduce them to the wild (release), place them in captivity, or to euthanize them. The preferred options are to reintroduce the animal or to send it to a captive facility. Euthanasia is a last resort.

Primates are the main recipients of CETAS's resources in terms of infrastructure (especially for enclosures), feeding, and specialized personnel. Facilities at governmental rescue centers are meant to be transitory; enclosures are usually no greater than 5 m^2 and comprise concrete walls and floors. Due to the constant arrival of individuals and difficulty in performing adequate releases, primates accumulate in rescue centers. Most animals that arrive at CETAS are held for months or even years before they are transferred to other shelters or released. Clinical and behavioral conditions of animals in rescue centers are commonly compromised, and rehabilitation programs are still scarce (Ferreira et al. 2016). An estimate of 30% mortality 6 months after release (Sita 2016) and 20% confirmed survivorship 10 years after release is reported for *Sapajus libidinosus* (Ferreira, persn comm).

Primates are used in more than 33 research areas in Brazil, including cancerology, parasitology, hepatitis, virology, malaria, comparative physiology, dermatology, glomerulonephritis, ophthalmology, toxicology, pharmacology, microbiology, toxoplasmosis, arterial lesions, schistosomiasis, cell therapy, psychopharmacology, yellow fever, atherosclerosis, herpesviruses, lymphoblastoid cells, physiological stress, and cognition and the neurosciences (da Silva Barbosa et al. 2015). The two longest

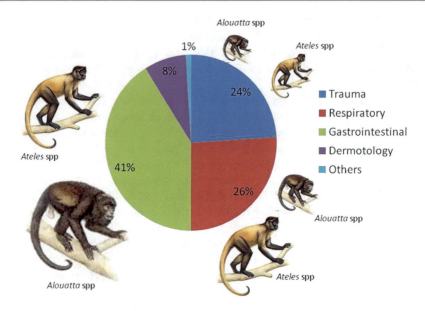

Fig. 4 Causes of death of the captive native monkeys. Sizes of primate pictures are proportional to causes of mortality affecting each species

established primate research centers in Brazil hold 774 primates currently. The first of these centers is Brazil's National Primate Center (CENP), which currently holds 560 primates from 22 New World monkey species (Torres et al. 2010). The second center is The Rio de Janeiro State Primatology Center holds 214 primates of 37 taxa (CPRJ 2016 annual report). Most of these monkeys are from the genus *Leontopithecus* (64 individuals) and 59% (126) of these primates comprise 8 critically endangered or endangered species, and 3 vulnerable species (Fig. 4).

Primates kept in captivity usually have higher cortisol levels, hematological changes, and nutritional deficiencies. Abnormal behavior and chronic stress can affect research validity, reliability, and replicability (Buchanan-Smith et al. 2022; Garner 2005). A study's internal and external validity can be affected when physiological and behavioral responses of stressed individuals differ from those of healthy individuals. These threats to research validity notwithstanding, an increasing number of captive primates, particularly New World monkeys given their smaller size and faster life history, are being used in research. Between 1966 and 2005, 10,814 studies using New World monkeys as models were published (Torres et al. 2010). From 2000 to 2005, the majority of those publications (54–73%) were related to neuroscience. Although many studies with New World monkeys have been conducted in South American countries, and the number has been increasing in recent decades, very few of these countries have specific regulations in place for the care and welfare of primates.

Since 2014, all research centers in Brazil that conduct experiments on animals are regulated by the National Council for Animal Experimentation Control (CONCEA).

Table 1 Spatial recommendations for laboratory primates

Animals	Maximum weight (kg)	Minimum floor area (m²)	Height (cm)
Cebuella spp., *Callithrix* spp., *Mico* spp., *Saguinus* spp., *Leontopithecus* spp., *Aotus* spp., *Callimico goeldii*	1.5	0.20	76.2
Cebus spp., *Sapajus* spp., *Saimiri* spp.	3	0.28	76.2
Alouatta spp., *Ateles* spp., *Chlorocebus aethiops*	10	0.40	76.2
Lagothrix spp., *Brachytheles* spp., *Macaca* spp., *Cercocebus* spp., *Erythrocebus* spp., *Papio anubis*	15	0.56	81.3
Group 5 Old World primates	20	0.74	91.4
Group 6 Old World primates	25	0.93	116.8
Group 7 Old World primates	30	1.40	116.8
Group 8 Old World primates	>30	≥2.32	152.4
Chimpanzees (Juveniles)	10	1.40	152.4
Chimpanzees (Adults)	>30	≥2.32	213.4

Source: CONCEA (2014). Groups 5–8 refers to non-specified Old World Primates whose weight fit into the correspondent category
ftp://ftp.saude.sp.gov.br/ftpsessp/bibliote/informe_eletronico/2015/iels.nov.15/Iels212/U_RN-MCTI-CONCEA-28_131115.pdf

A special guidebook provides regulations specific for research in primates, following mostly recommendations from the "Guide for the Care and Use for Laboratory Animals," which is published by the National Research Council, USA (2011). Recommendations include enclosures built in concrete to facilitate sanitation and suggestions that individuals should be housed in social groups. Nevertheless, primates under experimental protocols, especially for experimental studies of infectious and contagious diseases, tend to be housed in isolation and in environments with no access to sunlight or a view of the outside world (da Silva Barbosa et al. 2015). Table 1 summarizes the recommended minimal cage size for primates according to their maximum weight.

Zoos in Brazil keep a large number of Brazilian and exotic primate species. The Brazilian Society of Zoos and Aquariums (SZB) annually conducts a census to monitor the primate population. The diversity of species with specific physical and behavioral needs presents a husbandry challenge for zoos. SZB determined that the greatest problems in the maintenance of primates in Brazilian zoos are: (1) maintenance of unstable social groups and a lack of information about the geographical origin of individuals that compose the social groups; (2) little participation of ex situ populations in the conservation of their species; (3) absence of species-specific protocols relating to nutrition, preventive veterinary medicine, welfare programs, and studbooks; (4) little technical qualification of the teams (especially handlers) that maintain these animals, and the fact that this deficiency leads to the delay in the detection of problems and the implementation of solutions; and (5) the enclosures are small and there is little capacity to provide physical enrichment. The actions required to solve these five major challenges should be a priority for institutions in Brazil.

Brazilian legislation concerning recommendations for zoo enclosure area and height are based on the size of species and varies from 3 to 60 m^2 (e.g., 200 m^2 enclosures are recommended for gorillas) and 2.5–5 m of height. The regulation emphasizes the importance of the enclosure's ground being made of soft soil for all primate species and even recommends ground vegetation for a few species. Tree branches in abundance are recommended for all primate enclosures as well as a warm refuge for regions where temperatures can be low during winter (IBAMA IN 169/2008). Recommendations notwithstanding, many zoos across Brazil, including government owned zoos, house primates in concrete enclosures that are smaller than the recommended size, and provide little enrichment.

Despite the advances in knowledge about the management and pathologies of primates, a high percentage of the individuals in research and zoo colonies have respiratory problems, and this number is followed closely by the percentage of individuals that have nutritional and gastrointestinal problems, mainly due to parasites. Outbreaks of herpesvirus and toxoplasma are not uncommon either and are major sources of mortality in zoos and research colonies (Pissinatti et al. 2002; Vanstreels et al. 2011; Catão-Dias et al. 2013).

3.2 Mexico

In 1980, there were 26 public and two private zoos in Mexico. By 1990, their combined number nearly doubled to 51 (Babb Stanley 2002). As of 2003, there are 108 zoos in Mexico that are registered with the Directorate General of Wildlife Secretariat of Environment and Natural Resources. In addition, many native primate species (mantled howler monkeys *Alouatta palliata*, Guatemalan black howler monkeys *Alouatta pigra*, and black-handed spider monkeys *A. geoffroyi*) are kept across Mexico in breeding centers, circuses, animal facilities, private collections, and wildlife management units. The number of primates cannot be estimated because many individuals are kept in private shelters as pets (for review of primates in the pet trade see Hevesi 2022) and so are not accessible and/or available for census (Lascurain et al. 2009). Also, the Mexican authorities do not always have an updated inventory of confiscated animals (Aranda-Pérez et al. 2015).

The origin of the native monkeys kept in 18 surveyed zoos varies depending on the species. Just over half of the spider monkeys came from voluntary surrender (29%) or confiscation (22%), 32% were born in captivity, and 15% came from exchanges with other zoos. For black howler monkeys *A. pigra*, 62% were confiscated individuals and 38% were donated to zoos by people who had obtained the monkeys from the illegal pet trade (Gual-Sill and Rendón-Franco 2011). Of the three Mexican species, spider monkeys have been traditionally captured due to the

Table 2 Population of native nonhuman primates in captivity in Mexico

Place	A. geoffroyi				A. palliata				A. pigra			
	♂	♀	n.d.	N	♂	♀	n.d.	N	♂	♀	n.d.	N
Zoos	60	124	14	208	1	1	1	3	7	4	0	11
Breeding centers	19	29	1	49	2	9	17	28	0	2	0	2
CIVS	11	27	3	41	0	0	0	0	2	2	0	4
Circuses and tourist trade	81	120	21	222	0	0	0	0	0	0	1	1
Total	171	300	49	520	3	10	18	31	9	8	1	18
%	91.4				5.4				3.2			

Adapted from Gual-Sill and Rendón-Franco (2011). CIVS: wildlife research centers; ♂: male; ♀: female; n.d.: sex not determined

appeal of infants as pets, which has impacted native populations negatively (Kinzey 1997; Aranda-Pérez et al. 2015).

A recent study on the illegal trade in primates in Mexico emphasized that in the last 18 years, spider monkeys were the most affected. Of native species confiscated, 97% were spider monkeys versus only two were howler monkeys (one *A. palliatta* and one *A. pigra*) (Aranda-Pérez et al. 2015). Howler monkeys, especially those from the illegal pet trade, often die due to health problems related to early weaning, poor diet, social isolation, inappropriate facilities, and poor care. Those that survive do so in poor condition and are often donated to zoos. Consequently, the mortality rate within the first few weeks after they are donated is extremely high (Gual-Sill and Rendón-Franco 2011). Proper and quick management during this time is crucial for increasing survival.

Of the 122 facilities that house native Mexican monkeys comprising the data for Table 2, 119 keep *A. geoffroyi*, 6 keep *A. palliata*, and 11 keep *A. pigra* in all, 569 individuals. Because around 40% of captive spider monkeys are in circuses and tourist attractions, and these establishments do not adequately document their collections, it is not possible to ascertain the number of spider monkeys and their welfare status.

Differences in the number of primates kept in captivity may highlight species differences in how capable individuals are of coping with the captive environment and related stressors. For example, over one third (34.3%) of howler monkeys die during the first month of captivity (Gual-Sill and Rendón-Franco 2011). In general, for the three species of Mexican monkeys just discussed, gastrointestinal pathologies, followed by respiratory disorders and trauma, are the main causes of death (Fig. 4). A more detailed analysis indicates that the major causes of death for spider monkeys are respiratory (21%) and gastrointestinal disorders (21%), injuries (14%), and senility (10%). For howler monkeys, the leading causes of death are gastrointestinal (67.6%), respiratory or liver disorders (7.6%), and cardiovascular or nutritional disorders (2.9%) (Gual-Sill and Rendón-Franco 2011). More research on accurate gastrointestinal assessment and nutritional management during captivity would help to treat gastrointestinal or metabolic diseases.

3.3 Uruguay

Uruguay, like Chile, is a country in Latin America that is not home to populations of wild primates. Primates in Uruguay are either kept as pets in private homes (usually in precarious conditions) or housed in zoos. All of these individuals are monkeys; there are no more apes in zoos since the last two chimpanzees at Villa Dolores Zoo died from tuberculosis in the 1990s. Because keeping primates in homes is illegal in Uruguay, it is difficult to know how many or what species are being kept.

Corte et al. (2008) conducted a survey in seven zoos to analyze the characteristics of the primate facilities and the conditions of captivity. Each zoo's directors were interviewed and consulted about whether there were environmental enrichment programs. The enclosures for the primates were also measured and analyzed. The enclosure characteristics of interest included size in relation to number of individuals and species, design, whether there were sheltered areas, soil, vegetation, water availability, and the presence of enrichment items. The specimens and their enclosures were photographed and filmed. In the zoos surveyed, eight species of primates were registered, of which the most common and with the highest number of specimens were hamadryas baboons *Papio hamadryas* (17%) and *Cebus* (*Sapajus*) *apella* (22%). The few data provided by authorities on the origins of these monkeys indicated that most of the baboons were descendants of the same colony (Parque Lecocq) and that some specimens of *C. apella*, both specimens of *Chlorocebus aethiops*, and the original family of *Callithrix jacchus* were brought to the zoos by private individuals. However, 60% of specimens had an unknown origin, which indicates that they probably came from illegal traffic

All but two of the zoos that Corte et al. visited did not have an environmental enrichment program for the primates (or any other animals). The enclosures were generic and neither their size nor their complexity was related to the number or type of individuals housed within them. Some species' taxonomic classifications were wrong, and the sex of some individuals was also misclassified. The animals were fed, and basic sanitary measures were in place, but their behavioral needs were not attended to. More than 50% of the enclosures had no visual barriers between animals and visitors. The design (vertical versus horizontal) according to needs of the species (arboreal versus terrestrial) had also not been considered. There was a lack of records, management plans, and adequate information on the natural behavior of the species. Moreover, there was no systematic observation of behavior, although the existence of stereotyped behaviors, apathy, hyperactivity, and inadequate social groupings was recorded. There were also no biologists among staff. Overall, this suggests that the welfare needs of the primates, as well as other animals, in these zoos were not being met in these zoos. There were, however, two exceptions. The first was Reserva Tálice at Trinidad, Flores. At this zoo, enclosures were large and high, contained living trees, had earth flooring, and the zoos considered the animals' basic needs. Parque Lecocq Zoo at Montevideo, the only Uruguayan zoo listed in the Latin American Zoo and Aquarium Association was the other exception. This zoo included a new enclosure made specifically for their hamadryas baboon colony, which comprised 100–200 individuals. This installation was also much bigger than that found in most zoos in Uruguay and had indoor and outdoor facilities for colony management.

Uruguayans have been aware of the poor condition of captive animals in their country's zoos and many NGOs have pushed for the closure of these institutions. In response, some zoos have closed and recently 48 primates, 12 common brown lemurs *Eulemur fulvus*, 11 rhesus macaques, 3 black howler monkeys, 2 hamadryas baboons, and 20 brown capuchin monkeys, were relocated from one zoo to Reserva Talice. However, many of the primates held in these zoos would have no adequate place to go if released, and so there is major concern regarding the future of these individuals should the zoos close.

4 Endangered New World Monkeys in Captivity

The ex situ conservation of species in captivity refers to the maintenance of biological diversity outside the natural habitat to reduce the risk of extinction and, in some cases, restore wild populations (Hosey 2022; Lascurain et al. 2009). However, most zoos in Latin America have no record of individual lineages and consequently have difficulty in participating in collaborative breeding programs (Gual-Sill and Rendón-Franco 2011).

When the Rio de Janeiro State Primatology Center opened in November 1979, colonies of golden-headed lion tamarins *Leontopithecus chrysomelas*, golden lion tamarins *L. rosalia*, black lion tamarins *L. chrysopygus*, black-tufted marmosets *Callithrix penicillata*, common marmosets, and hybrids were brought from the Golden Lions Biological Bank, Tijuca, Rio de Janeiro. After that, groups of woolly spider monkeys (or muriquis) *Brachyteles* spp., bald uakari *Cacajao rubicundus*, black bearded saki monkeys *Chiropotes satanas*, pied tamarins *Saguinus bicolor*, Wied's marmoset *Callithrix kuhlii*, and Geoffroy's tamarin *Sanguinus geoffroyi* were formed. The main goals of establishing these colonies were: (a) to conduct biomedical studies, (b) to keep some colonies as a genetic backup to supply other captive colonies inside and outside Brazil, and (c) to prepare groups for in situ conservation. The recovery of the wild golden lion tamarin population is one of the best examples of how ex situ breeding coupled with the release of animals into the wild can bolster primate conservation efforts (Beck et al. 1991; Kleiman and Rylands 2002). However, it is estimated that a vertebrate ex situ breeding program requires between 250 and 500 animals to maintain genetic diversity for 100 years (Lascurain et al. 2009). This requires resources, specialist technical teams, and infrastructure that facilitate the recovery and survival of individuals or populations outside their habitat. Furthermore, maintaining wildlife in captivity could lead to problems, including infertility and hybridization.

There is no ex situ conservation program for any of the three threatened species of monkeys in Mexico (black-handed spider monkeys, mantled howler monkeys, and black howler monkeys), albeit there are individuals in captivity. The Mexican Primates' Conservation Assessment and Management Plan emphasizes that captivity is not considered an immediate tool for the long-term preservation of Mexican monkeys. Notwithstanding the critical and urgent situation with regard to wild populations, specialists decided it was more profitable to invest in the protection and development of in situ conservation programs (Rodríguez-Luna et al. 2009).

In Brazil, the yellow-breasted (*Sapajus xanthosternos*) and blond capuchin monkey (*S. flavius*) face a similar situation in that conservation efforts are focused on *in situ* habitat protection (Benoit Quintard, studbook keeper of *S. xanthosternos*, personal communication). Brazil has three national action plans in place to promote *in situ* conservation and ex situ management (CPB 2018). One of these plans is for muriqui, another is for pied tamarins, and the third is for five endangered primates found in northeast Brazil: red-handed howler monkeys (*Alouatta belzebul*), Barbara Brown's titi (*Callicebus barbarabrownae*), Coimbra Filho's titi (*Callicebus coimbrai*), blonde capuchin monkeys, and yellow-breasted capuchin monkeys.

5 Legal Protection for Primates in Latin American and Caribbean Countries

Legal protection is a cornerstone of wildlife conservation and the welfare of animals in captivity or under human management (for review of primate welfare legislation, see Bayne et al. 2022). The legal framework is a complex array of international treaties, national laws and regulations, and law enforcement. The legal panoply protects the physical and psychological well-being of primates through limits on research activities and by establishing husbandry and management practices. Laws and regulations can also guide and facilitate the development of captive facilities for research or conservation.

We reviewed the legal status of captive primates by surveying the laws and regulations that apply to wild primates and those in captivity (Appendix). We were interested in assessing which laws the protection of primates fall into, if the laws mention primates, or if the laws are specific about primates in captivity. We used Google to search for documents using keywords in Spanish, Portuguese, and English. Although the available information may not be exhaustive, we believe that it probably reflect the legal status of primates in captivity (Table 3). Some countries do not have native primates but were surveyed because they host primates in research facilities or zoos. We did not conduct in-depth content analysis of the laws and regulations and so interpretations of legal protection may differ if analyzed by an expert.

The only international treaties applicable are CITES and the Convention on Biological Diversity (CBD). There is no international treaty for animal welfare. All South American and Caribbean countries that we surveyed are signatories of CITES and CBD and have wildlife protection laws. All countries protect CITES-listed species. In most countries, the laws protect some wildlife species and allow for the exploitation of others. This is understandable because most of the countries have indigenous ethnic groups (native Americans) that hunt wild animals, including monkeys, for food or keep these animals as pets (Peres 1990; Ojasti 1996; Bodmer et al. 1997; Richard-Hansen et al. 2000; Drews 2002; de Araújo et al. 2008). Brazil

Table 3 Summary of legal protection for captive primates in Latin America

Country	N species	WP	AWP	RESL	Zoos	Captive	Circus	Primate facilities	IACUC	Min size	Social	Health
Brazil	75	Y	Y-W	Y	Y	Y	Y-Prohibits	Y	Y	Y	Y	Y
Colombia	36	Y	Y	AWP	AWP	Y	Y-Prohibits	N	Y			
Peru	32	Y	Y-W	AWP	AWP	Y	W-Prohibits	N	Y			
Bolivia	23	Y	Y	AWP	AWP	AWP	Y-Prohibits	N				
Ecuador	16	Y	Y-W	AWP	AWP	AWP	Y-No W	N	Y			
Venezuela	12	Y	Y	AWP	AWP	AWP	AWP	N	Y			
Guyana	8	Y-R	Y-R	AWP	AWP	N	N	N				
Suriname	8	Y	?		N	N	N	N				
French Guyana	7	N	N	AWP	AWP	N	N	N				
Panama	7	Y	Y	AWP	AWP	AWP	Y-No W	N	Y			
Costa Rica	6	Y	Y-W	Y	Y	Y	Y-Prohibits	Y	Y	Y		Y
Paraguay	5	Y	Y-R	AWP	AWP	AWP	AWP	N	Y		N	
Honduras	3	Y	Y-W	AWP	AWP	AWP	AWP-Prohibits	N	Y			
Nicaragua	3	Y	Y	AWP	AWP	Rescue	AWP	N	Y			
Argentina	2	Y	Y	Y	Y	AWP	AWP, St Proh	N	Y			
Belize	2	Y	Y-P	AWP	AWP	N	AWP	N				
El Salvador	2	N	Y	AWP	N	Zoocriador	W-Prohibits	N	Y			
Guatemala	2	Y	Y-R	AWP	AWP	AWP	AWP	N	Y			
Mexico	2	Y	Y	Y	AWP	Y	Y-No W	Y	Y	Y	Y	
Chile	0	Y	Y	AWP	AWP	Y	AWP	N	Y	N	N	N
Jamaica	0	Y	Y-W	AWP	N	N	NA	N	Y			

(continued)

Table 3 (continued)

Country	N species	WP	AWP	RESL	Zoos	Captive	Circus	Primate facilities	IACUC	Min size	Social	Health
Dominican Republic	0	Y general	Y-No W	AWP	N	N	N	N	Y	N	N	N
Uruguay	0	Y	Y	Y	AWP	Y	Y	N	Y			

The top line refers to the existence of laws specific to wildlife protection (WP), animal welfare (AWP), research laboratories (RESL), zoos, captive facilities (Captive), circuses, primate facilities; or the existence of IACUC committees, or if there is mention of international guidelines for primate husbandry or any guideline that specifies regulations on minimum cage size (min size), social environment, or health requirements for euthanasia

For WP we scored Y for yes and added if they are recent (< 3 years). For WP and AWP we scored Yes or No and if included wildlife (W), recent (R), partial (P) if it only included some elements of welfare or if it explicitly left wildlife out (No W)

For zoos we scored Yes for the presence of specific regulations, no for no legal regulation at all, and AWP for general animal welfare and protection law

For captive facilities, we scored Yes for a specific regulation, rescue for when it only was for rescue centers, zoocriador for when it referred only to that type of enterprise, No for no regulation and AWP for when the animal welfare law contemplated captive facilities

For circus we scored Y if there is a specific regulation, AWP for the animal welfare law, W for when the wildlife law prohibits animals in circuses, N for no measures

and Peru seem to be the only countries that differ in how they regulate the use of wildlife for indigenous and non-indigenous citizens. Brazil has a legal clause that states that all wildlife is under the protection of the state. And, as a consequence, the state is obligated to annually collect, receive, screen, treat, rehabilitate, and allocate hundreds of thousands of individual animals. Few countries have explicitly created or nominated an agency for the enforcement of wildlife laws. The lack of a clear enforcement mechanism is significant, because legal protection does not guarantee compliance (Peres and Terborgh 1995; Drews 2001, 2002; Rowcliffe et al. 2004). None of the wildlife protection laws cover animal welfare.

In nearly all countries, the wildlife laws require permits to keep native primates (or any wildlife) in captivity. Captivity can be for conservation, basic or biomedical research, or for keeping the animals as pets. Only Uruguay (Law 16.088 of 1989) clearly prohibits having primates as pets. Only Brazil, Peru, Mexico, Panama, Colombia, and Costa Rica have laws that regulate permits based on the CITES or IUCN status of the species. When law enforcement officers confiscate illegally kept primates, these primates are sent to triage or rescue centers, where they often stay for years. There does not seem to be any animal husbandry or welfare regulations, or any type of periodic inspections for those facilities.

The welfare of captive primates is subject to general animal welfare or animal protection laws present in all countries except for French Guiana and Guyana. Some of these laws, for example Ley Arouca (2008) of Brazil, Ley Orgánica de Bienestar Animal of Ecuador, and Peru's Animal Welfare Law (2017), are extensive. These animal welfare laws require that animals be provided proper housing, food, and treatment, and prohibit physical abuse and neglect. In seven countries, these laws explicitly mention wildlife, for the other countries, the laws deal exclusively with domestic animals. None have specific provisions for adequate social environments or specific guidelines for primates. None of the laws mention adherence to international guidelines for managing captive primates.

These general animal welfare laws apply to any form of captivity, including, but not limited to, being held in research facilities, zoos, or circuses. These laws, however, do not account for differences among the design and management of animals while in captivity that could affect primate well-being, such as enclosure size or social housing. One exception is the regulations for the use of laboratory animals, but only the regulations of Brazil and Mexico mention primates. Universities and research institutions have institutional animal care and use committees and it is assumed that these committees abide by international standards. Countries such as Brazil, Colombia, Costa Rica, Mexico, and Argentina have developed guidelines for keeping primates in captivity, albeit they have done so to differing degrees. Columbia and Brazil have the most specific guidelines (they are based on the Guide for the Care and Use for Laboratory Animals). Colombia has recommendations that cover enclosure size and details of the physical environment. When applying for a permit to keep primates, applicants must submit a management plan that includes details about the facilities, husbandry, and staff. The Brazilian guidelines for zoos specify criteria regarding the size of enclosures, density of animals, and the social system of the species, but it is not clear if these guidelines apply to research centers, breeding centers, or rehabilitation centers. Some countries, including Chile, Colombia, Peru, Ecuador, Mexico, and Uruguay, claim to follow international guidelines (National Research Council, Canadian Research Council, International Primatological Society, Wildlife Conservation Society) for keeping primates in captivity. Zoos are treated under the law together with other entertainment facilities, such as circuses, and, as such, are regulated by general animal protection and welfare laws. Brazil and Mexico do have specific separate laws for zoos and circuses. In eight countries the use of animals in circuses is prohibited.

In general, there is protection for CITES species living in the wild. The legal structure for obtaining permits varies among countries, and it is not clear for most countries, the exceptions being Brazil, Peru, Mexico, and Costa Rica, which species can be captured, by whom, and for what purpose. Captive primate husbandry is not regulated in most countries, falling under the umbrella of general welfare laws. The decisions regarding the enforcement of the laws or criteria for keeping monkeys in captivity seem to be left to expert opinion on a case-by-case basis. The lack of specific regulations or guidelines for primate husbandry makes it difficult to demand compliance with minimum standards for welfare. Also, the lack of clear and well-structured regulations may be one factor that hinders the development of captive facilities because it makes obtaining permits a highly bureaucratic process guided by individual interactions, context, and *ad hoc* expert opinion. The legal context also allows for the existence of informal and temporary captive settings that do not provide adequate welfare. Finally, the return of primates to their natural habitats is not facilitated by the legal-administrative processes. On the contrary, it is often strictly regulated to inhibit the formation of introduced populations out of the historic range of the species.

6   Human Care for Primates in Latin America, Where to Go from Here?

The constant removal of species of New World monkeys from the wild is a serious issue for conservation and animal welfare. Concurrently an increasing number of primates are kept ex situ, under human care, in rescue centers, zoos, research centers, and private collections. Although regulations of institutions that keep wildlife in captivity are usually present in Latin American countries' legislation, few are specific to or offer practical guidelines that ensure animal welfare. Moreover, in those countries, these laws are seldom enforced.

Unfortunately, the picture emerging from this chapter is that most captive primates in Latin America do not experience conditions that are likely to induce good welfare. Lack of funding to build new facilities or to improve existing facilities, to buy medicines or proper food, and need for specialized personnel are common complaints of zoos and commercial/research/rescue centers facility directors. Species-specific management protocols and little participation of ex situ populations in programs of conservation in situ are also difficulties faced by facility directors.

Rescue centers face unique challenges in developing countries because the number of animals that reach these centers is usually much larger and more constant than in centers within developed countries. Although some zoos and conservation centers have advocated for breeding programs of endangered species, there are too few rehabilitation centers for captive animals in Latin America and few examples of monitored reintroduction of endangered or non-endangered primate species. It has been argued that an alternative for these rescued populations could be the establishment of semi-free ranging colonies in forest fragments for educational and non-invasive research purposes (Kerwin 2010). Keeping primates in such fragments would decrease the need (and costs) for direct human care, improve the primates' welfare, and promote environmental education. For example, Tárano and López (2015) reported that wedge-capped capuchin monkeys (*Cebus olivaceus*) hosted in a semi-free ranging 15 ha exhibition showed behavioral patterns comparable to those of wild conspecifics, whereas groups living in enclosures (450 m^2 and 200 m^2) exhibited abnormal behavior patterns and increased aggression. Price et al. (1994) showed that visitors who observed semi-free ranging cotton-top tamarins (*Saguinus oedipus*) in a zoo made more comments about the animals behavior and their welfare when compared to visitors who observed a group in a cage.

From a conservation perspective, primates living in fragments face challenges such as the long-term population viability, more rapid spread of disease, and hybridization (see Marsh and Chapman 2013 for a review). Fragments, however, can offer a good opportunity for the scientific community to study these challenges and those areas might offer an opportunity to manage rescued animals *in situ*, which would promote welfare and conservation. By monitoring these fragments, scientists and institutions can develop essential and applied methods for increasing the viability of wild populations. Neutral or affiliative human interactions can also increase primate welfare in wild or semi-free ranging conditions. For example, wild silvery woolly monkeys (*Lagothrix poeppigii*) that are habituated to humans showed fewer threat displays in non-hunting than in hunting areas (Papworth et al. 2013). Mexican primatologists already acknowledge that moving animals between fragments or disturbed areas is a necessary strategy for ex situ and in situ conservation (Sánchez-López 2013; Rangel-Negrín et al. 2014).

The problems and challenges across Latin American countries in caring for primates ex situ are similar, and by sharing experiences and strategies that work we have the opportunity to solve welfare issues. Ensuring welfare of ex situ wildlife is a moving target that requires adaptive co-management between different actors. International organizations and scientific societies, both within and outside Latin America, with their organizational and networking capabilities are key players to promote these interactions and in finding solutions.

Acknowledgments This chapter is the result of a collaborative effort. Carlos Ruiz-Miranda conducted analyses on Legal Protection and revised the text. He was supported by a Senior Research Fellowship from the Brazilian higher Education Agency (CAPES 5575\15-0). Silvana Sita organized the quantitative data. Sonia Sanchez and Silvia Corte wrote reports on the status of monkeys in Mexico and Uruguay, respectively. Alcides Pissinatti offered the historical perspective, and raw data on numbers of primate in research centers. Leandro Jerusalinsky, Paulo Wagner, and Claudio Maas offered independent perspectives of legal agencies, care centers and zoos, respectively. Renata Ferreira conducted the general outline, organization, and writing of the chapter, revision the text (supported by Senior Research Fellowship CNPq Universal 443041-2014-8).

Appendix: Laws and Documents Consulted

Criterios Técnicos para la mantención y manejo de fauna silvestre en cautiverio (2013).

Criterios tecnicos para la mantencion y manejo de fauna silvestre en cautiverio (2014).

Ley 12.238. De aplicación para todos los parques zoológicos y establecimientos con animlaes vivos de la fauna silvestre en cautiverio o semicautiverio (1998).

Acuerdo 0062. Medidas para regular la actividad de circos, y prevenir el maltrato a los animales silvestres (2010).

Ley 17 (10 de septiembre de 2004). Ley Florestal y de conservación de areas naturales y vida silvestre (2004).

Lay 1638 del 27 de junio de 2013. Por medio de la cual se prohibe el uso de animales silvestees, ya sean nativos o exóticos, en circos fijos e itinerantes (2013).

X Taller de Bioética: Regulación del uso y cuidado de animales en investigación (2015).

NTON 05 023-03 Norma técnica para el establecimiento de centros de rescate y rehabilitacion de la fauna silvestre (2003).

Ley para la conservación y uso sustentable de la biodiversidad.

Reglamento para el establecimiento y manej o de zoocriaderos de especies de vida silvestre, decreto 57 de 2003 c.f.r. (2003).

Decreto 32633 (10 de marzo de 2005) (2005).

Decreto 156-2007. Ley florestal, áreas protegidas y vida silvestre (2007).

Norma oficial mexicana NOM-062-ZOO-1999, especificaciones técnicas para la producción, cuidado y uso de los animales de laboratório (1999).

Muñoz Moreno, D. M. (2011). Revisión de la normatividad para zoocriaderos en Colombia. Especialización en Ingenieria Ambiental Bachelor's Monograph Universidad Industrial de Santander, Bucaramanga.

Ley 30. de proteccion a los animales (1991).

Ley 24. Legislación de vida silvestre República de panmá y se dictan otras disposiciones (1995).

Ley de proteción a la fauna silvestre (1970).

Instrução normativa 7, 30 de abril de 2015. Instutui e normatiza as cateogrias de uso e manejo de fauna silvestre em cativeiro, e define, no ámbito do IBAMA, os procedimentos autorizativos para as categorias estabelecidas (2015).

Lei 11.794 de 8 de outubro de 2008. estabelecendo procedimentos para o uso científico de animais; revoga a Lei no 6.638, de 8 de maio de 1979; e dá outras providências. (2008).

Código Nacional de Recursos Naturales Renovables y Protección al Medio Ambiente, 2811 (18 de diciembre de 1974) C.F.R. (1974).
Decreto 1608 (31 de julio de 1978). Reglamento del Código Nacional de los Recursos Renovables y de Protección al Medio Ambiente, 1608 C.F.R. (1978).
Ley 27308 (15 de julio de 2000). Ley florestal y de fauna silvestre (2000).
Ley 96 de vida silvestre (1992).
Decreto Supremo 014-2001-AG (06 de abril de 2001). Reglamento de la ley florestal y de fauna silvestre (2001).
Ley 5.197 de 03 janeiro de 1967. Ley de proteção de Fauna (1967).
Ley 4840 de protección y bienestar animal (2013).
Decreto supremo (002 de 2015) que establece mnormas para el manejo y tenencia de fauna silvestre en cautiverio (2015).
Ley 7451. Bienestar de los animales.
Ley 7317 (30 de octubre de 1992). Ley de conservación de la vida silvestre (1992).
Guia 1. Guia técnica y operativa para el control del comercio y tráfico de fauna silvestre en Ecuador (2013).
Reglamento de la ley de protección a la fauna silvestre (1999).

References

Aranda-Pérez M, Sánchez-López S, González-Zamora. (2015) Illegal trade of primates in Mexico. MSC Dissertation. Fundació Universidad de Girona, España

Babb Stanley KA (2002) Los zoológicos en México: una visión del pasado y sus tareas actuales. In: Cabrales JA, Corona ME (eds) Relaciones hombre-fauna: Una zona interdisciplinaria de estudio. Instituto Nacional de Antropología e Historia, México, pp 51–62

Bayne K, Hau J, Morris T (2022) The welfare impact of regulations, policies, guidelines and directives and nonhuman primate welfare. In: Robinson LM, Weiss A (eds) Nonhuman primate welfare: from history, science, and ethics to practice. Springer, Cham, pp 629–646

Beck BB, Kleiman DG, Dietz JM et al (1991) Losses and reproduction in reintroduced golden lion tamarins, *Leontopithecus Rosalia*. Dodo 27:50–61

Bicca-Marques JC, Calegaro-Marques C, Rylands AB et al (2017) Yellow fever threatens Atlantic Forest primates. Sci Adv 3(1):e1600946/tab-e-letters

Blanco A, Pérez G, Rodríguez B et al (2009) El zoológico de Moctezuma: mito o realidad? AMMVEPE 20:29–39

Bodmer RE, Eisenberg JF, Redford KH (1997) Hunting and the likelihood of extinction of Amazonian mammals. Conserv Biol 11(2):460–466. https://doi.org/10.1046/j.1523-1739.1997.96022.x

Buchanan-Smith HM, Tasker L, Ash H, Graham ML (2022) Welfare of primates in laboratories: opportunities for improvement. In: Robinson LM, Weiss A (eds) Nonhuman primate welfare: from history, science, and ethics to practice. Springer, Cham, pp 97–120

Catão-Dias JL, Epiphanio S, Kierulff MCM (2013) Neotropical primates and their susceptibility to *Toxoplasma gondii*: new insights for an old problem. In: Brinkworth J, Pechnkina K (eds) Primates, pathogens, and evolution. Springer, New York, NY, pp 253–289

Ceballos-Mago N, González C, Chivers D (2010) Impact of the pet trade on the Margarita capuchin monkey *Cebus apella margaritae*. Endang Species Res 12:57–68. https://doi.org/10.3354/esr00289

Chapman CA, Peres CA (2001) Primate conservation in the new millennium: the role of scientists. Evol Anthropol 10(1):16–33. http://doi.org/10.1002/1520-6505(2001)10:1%3C16::AID-EVAN1010%3E3.0.CO;2-O

Cormier LA (2010) The historical ecology of human and wild primate malarias in the new world. Diversity 2(2):256–280. https://doi.org/10.3390/d2020256

Corte S, Minteguiaga M, Ventura V (2008) Etologia aplicada al manejo de especies salvajes en cautiverio: evaluación de la situacion de primates en Uruguay. Papel presente en I Congreso Latinoamericano de Etología Aplicada. Instituto de Investigaciones Biológicas Clemente Estable, Montevideo

CPB (2018) Planos de Ação Nacional do Centro de Primatas Brasileiros

CPRJ (2016) Relatório Anual. In: Machado T (ed) Instituto Estadual do Ambiente, INEA. Secretaria de Estado do Ambiente, Rio de Janeiro, p 292

Csermak AJ (2007) Fauna silvestres Brasileira em Cativeiro: Criação legalizada, distribuição geográfica e políticas públicas. Universidade Federal de Viçosa

da Silva Barbosa A, Pissinatti A, Dib LV et al (2015) *Balantidium coli* and other gastrointestinal parasites in captives non-human primates of the Rio de Janeiro, Brazil. J Med Primatol 44 (1):18–26. https://doi.org/10.1111/jmp.12140

de Araújo RM, de Souza MB, Ruiz-Miranda CR (2008) Density and population size of game mammals in two Conservation Units of the State of Rio de Janeiro, Brazil. Iheringia Série Zool 98(3):391–396. https://doi.org/10.1590/S0073-47212008000300014

de Souza Fialho M, Ludwig G, Valença-Montenegro MM (2016) Legal international trade in live Neotropical primates originating from South America. Primate Conserv 30:1–6

Drews C (2001) Wild animals and other pets kept in Costa Rican households: Incidence, species and numbers. Soc Anim 9(2):107–126. https://doi.org/10.1163/156853001753639233

Drews C (2002) Attitudes, knowledge and wild animals as pets in Costa Rica. Anthrozoös 15 (2):119–138. https://doi.org/10.2752/089279302786992630

Estrada A, Garber PA, Rylands AB et al (2017) Impending extinction crisis of the world's primates: Why primates matter. Sci Adv 3(1):e1600946. https://doi.org/10.1126/sciadv.1600946

Ferreira RG, Mendl M, Wagner PGC et al (2016) Coping strategies in captive capuchin monkeys (*Sapajus* spp.). Appl Anim Behav Sci 176:120–127. https://doi.org/10.1016/j.applanim.2015.12.007

Garner JP (2005) Stereotypies and other abnormal repetitive behaviors: potential impact on validity, reliability, and replicability of scientific outcomes. ILAR J 46(2):106–117. https://doi.org/10.1093/ilar.46.2.106

González-Picazo H, Estrada A, Coates-Estrada R, Ortíz-Martínez T (2001) Consistencias y variaciones en el uso de recursos alimentarios utilizados por una tropa de monos aulladores (*Alouatta palliata*) y deterioro del hábitat en Los Tuxtlas, Veracruz, México. Ecosistemas y Recur Agropecu 17(33):27–36. https://doi.org/10.19136/era.a17n33.194

Grativol AD, Ballou JD, Fleischer RC (2001) Microsatellite variation within and among recently fragmented populations of the golden lion tamarin (*Leontopithecus rosalia*). Conserv Genet 2:1–9. https://doi.org/10.1023/A:1011543401239

Gual-Sill F, Rendón-Franco E (2011) Primates mexicanos en cautiverio. In: Dias P, Negrín A, Espinosa D (eds) La conservación de los primates en México. Veracruz, p 199

Harrington LA (2015) International commercial trade in live carnivores and primates 2006-2012: response to Bush et al. 2014. Conserv Biol 29(1):293–296. https://doi.org/10.1111/cobi.12448

Hevesi R (2022) The welfare of primates kept as pets and entertainers. In: Robinson LM, Weiss A (eds) Nonhuman primate welfare: from history, science, and ethics to practice. Springer, Cham, pp 121–144

Hosey G (2022) The History of Primates in Zoos, Chapter 1. In: Robinson LM, Weiss A (eds) Nonhuman primate welfare: from history, science, and ethics to practice. Springer, Cham, pp 3–30

IBAMA IN 169/2008 IBAMA IN 169/2008. https://www.ibama.gov.br/component/legislacao/?view=legislacao&legislacao=113878

ICMBio Portaria n°444 17/12/2014 Portaria n°444, 17/12/2014
International Union for Conservation of Nature (2000) IUCN guidelines for the prevention of biodiversity loss caused by alien invasive species
Kerwin A (2010) Overcoming the barriers to the retirement of Old and New World monkeys from research facilities. J Appl Anim Welf Sci 9(4):337–347. https://doi.org/10.1207/s15327604jaws0904_9
Kierulff MCM, Ruiz-Miranda CR, Oliveira PP et al (2012) The Golden lion tamarin *Leontopithecus rosalia*: a conservation success story. Int Zoo Yearb 46(1):36–45. https://doi.org/10.1111/j.1748-1090.2012.00170.x
Kinzey WG (1997) Ateles. In: Kinzey KW (ed) New World primates: ecology, evolution, and behavior. Aldine de Gruyter, New York, pp 192–199
Kleiman D, Rylands AB (2002) Lion tamarin biology and conservation: a synthesis and challenges for the future. In: Kleiman D, Rylands AB (eds) Lion tamarins: biology and conservation. Smithsonian Institution Press, Washington, DC, pp 336–343
Lascurain M, List L, Barraza E et al (2009) Conservacion de especies ex situ. In: Sarukhan J (ed) Capital natural de México, Estado de conservación y tendencias de cambio, vol II. CONABIO, México, pp 517–544
Levacov D, Jerusalinsky L, de Souza Fialho M (2011) Levantamento dos primatas recebidos em Centros de Triagem e sua relação com o tráfico de animais silvestres no Brasil. In: Melo F, Mourthé I (eds) A primatologia do Brasil, vol 11. Sociedade Brasileira de Primatologia, Belo Horizonte, pp 281–305
Mack D, Mittermeier R (eds) (1984) The international primate trade. TRAFFIC USA. World Wildlife Fund US, IUCN/SSC Primate Specialist Group, Washington, DC
Maldonado A, Nijman V, Bearder S (2009) Trade in night monkeys *Aotus* spp. in the Brazil–Colombia–Peru tri-border area: international wildlife trade regulations are ineffectively enforced. Endanger Species Res 9:143–149. https://doi.org/10.3354/esr00209
Marsh LK, Chapman CA (2013) Primates in fragments: complexity and resilience. Springer, New York
Martin G (1996) Pré-história do Nordeste do Brasil. Editora Universitária UFPE
Miller-Spiegel C (2011) Animal research: the peaceful approach. Nature 471:449. https://doi.org/10.1038/471449b
Miller M, Martin S (2004) Courtly art of the ancient Maya. Thames & Hudson, London
Nascimento C (2009) Histórico oficial do comércio ilegal de fauna no estado do Amazonas. Universidade Federal do Amazonas
National Research Council (2011) Guide for the care and use of laboratory animals, 8th edn. The National Academies Press, Washington, DC
Nijman V, Nekaris KAI, Donati G et al (2011) Primate conservation: measuring and mitigating trade in primates. Endanger Species Res 13(2):159–161. http://doi.org/10.3354/esr00336
Ojasti J (1996) Wildlife utilization in Latin America: Current situation and prospects for sustainable management. Food & Agriculture Organization, Rome
Oliveira LC, Grelle CEV (2012) Introduced primate species of an Atlantic Forest region in Brazil: present and future implications for the native fauna. Trop Conserv Sci 5(1):112–120. https://doi.org/10.1177/194008291200500110
Papworth S, Milner-Gulland EJ, Slocombe K (2013) Hunted woolly monkeys (*Lagothrix poeppigii*) show threat-sensitive responses to human presence. PLoS One 8(4):e62000. https://doi.org/10.1371/journal.pone.0062000
Pereira DG, De Oliveira MEA, Ruiz-Miranda CR (2008) Interações Entre Calitriquídeos Exóticos E Nativos No Parque Nacional Da Serra Dos Órgãos - Rj. Espaço e Geogr 11:67–94
Peres CA (1990) Effects of hunting on Western Amazonian primate communities. Biol Conserv 54 (1):47–59. https://doi.org/10.1016/0006-3207(90)90041-M
Peres CA, Terborgh JW (1995) Amazonian nature reserves: an analysis of the defensibility status of existing conservation units and design criteria for the future. Conserv Biol 9(1):34–46. https://doi.org/10.1046/j.1523-1739.1995.09010034.x

Pissinatti A, Montali R, Simon F (2002) Diseases of lion tamarins. In: Kleiman D, Rylands AB (eds) Lion tamarins: biology and conservation. Smithsonian Institution Press, Washington, DC, pp 255–268

Price E, Ashmore L, Mcgivern AM (1994) Reaction of visitors to free-ranging monkeys. Zoo Biol 13(4):355–373. https://doi.org/10.1002/zoo.1430130409

Procópio-de-Oliveira P (2002) Ecologia Alimentar, Dieta e Área de uso de Micos-Leões-Dourados Translocados e sua Relação com a Distribuição Espacial e Temporal de Recursos Alimentares na Reserva Biológica União, RJ. Universidade Federal de Minas Gerais, Belo Horizonte

Rangel-Negrín A, Coyohua-Fuentes A, Chavira R et al (2014) Primates living outside protected habitats are more stressed: the case of Black howler monkeys in the yucatá n peninsula. PLoS One 9(11):e112329. https://doi.org/10.1371/journal.pone.0112329

RENCTAS (2000) No Rede Nacional de Combate ao Tráfico de Animais Silvestres. 1º relatório nacional sobre o tráfico de fauna silvestre, Brasília. http://www.renctas.org.br/wp-content/uploads/2014/02/REL_RENCTAS_pt_final.pdf

Ribeiro D (1995) O povo brasileiro: a formação e o sentido do Brasil, 2nd edn. Cia das Letras, São Paulo

Richard-Hansen C, Vié J-C, de Thoisy B (2000) Translocation of red howler monkeys (*Alouatta seniculus*) in French Guiana. Biol Conserv 93(2):247–253. https://doi.org/10.1016/S0006-3207(99)00136-6

Rodríguez-Luna E, Solórzano-García B, Shedden A, Rangel-Negrín A, Días PAD, Cristóbal-Azkárate J et al (2009) Taller de Conservación, Análisis y Manejo Planificado para los Primates Mexicanos. Universidad Veracruzana and Conservation Breeding Specialist Group (CBSG), Veracruz; Apple Valley, MN

Rowcliffe JM, de Merode E, Cowlishaw G (2004) Do wildlife laws work? Species protection and the application of a prey choice model to poaching decisions. Proc R Soc London Ser B Biol Sci 271(1557):2631–2636. https://doi.org/10.1098/rspb.2004.2915

Ruiz-Miranda C, Beck B, Kleiman D et al (2010) Re-introduction and translocation of golden lion tamarins, Atlantic Coastal Forest, Brazil: the creation of a metapopulation. IUCN/SSC Re-introduction Specialist Group Abu Dhabi, Abu Dhabi

Sánchez-López S (2013) Rehabilitación de primates para la reintroducción de espécies. In: Libro De Resúmenes Del II Congresso Latinoamericano Y XV Congresoo Brasileiro De Primatologia

Shanee N (2012) Trends in local wildlife hunting, trade and control in the Tropical Andes biodiversity hotspot, Northeastern Peru. Endanger Species Res 19:177–186. https://doi.org/10.3354/esr00469

Sita S (2016) Back to the wild: individual differences in capuchin monkey rehabilitation and relocation. Universidade Federal do Rio Grande do Norte

Stoinski TS, Beck BB, Bloomsmith MA, Maple TL (2003) A behavioral comparison of captive-born, reintroduced golden lion tamarins and their wild-born offspring. Behaviour 140(2):137–160. https://doi.org/10.1163/156853903321671479

Torres LB, Silva Araujo BH, Gomes de Castro PH et al (2010) The use of new world primates for biomedical research: An overview of the last four decades. Am J Primatol 72(12):1055–1061. https://doi.org/10.1002/ajp.20864

Tárano Z, López MC (2015) Behavioural repertoires and time budgets of semi-free-ranging and captive groups of wedge-capped capuchin monkeys, *Cebus olivaceus*, in zoo exhibits in Venezuela. Folia Primatol 86(3):203–222. https://doi.org/10.1159/000381397

Trayford HR, Farmer KH (2013) Putting the spotlight on internally displaced animals (IDAs): a survey of primate sanctuaries in Africa, Asia, and the Americas. Am J Primatol 75(2):116–134. https://doi.org/10.1002/ajp.22090

Vanstreels RET, Teixeira RHF, Camargo LC et al (2011) Revisão das causas de mortalidade de primatas neotropicais (Primates: Platyrrhini no Parque Zoológico Municipal Quinzinho de Barros, Sorocaba-SP). Clínica Veterinária 16(90):46–52

Part II
Assessing Nonhuman Primate Welfare

Using Behavior to Assess Primate Welfare

Corrine K. Lutz and Kate C. Baker

Abstract

Promoting welfare should be a goal of all facilities housing nonhuman primates. However, determining whether that goal has been met can be challenging. One means of measuring primate welfare is by assessing the animal's behavior. Herein, we review commonly used behavioral indices for measuring welfare. The first is abnormal behavior, which is defined as behavior that differs in kind or degree from natural behavior. Abnormal behavior can indicate past or present adverse experiences, but it is also impacted by intrinsic factors such as species, temperament, age, and sex. Although abnormal behavior may in some way help an animal to cope with its environment, the presence of abnormal behavior is of concern and interventions may be warranted. Low well-being can also be measured by the display of anxiety-related self-directed behaviors such as scratching and yawning, as well as fear-related facial expressions and vocalizations, freezing, and fleeing. The benefit of utilizing normal species-appropriate anxiety behaviors is that, unlike with abnormal behavior, they are ubiquitous and can function as a "warning system," which allows for earlier identification of environmental deficiencies and intervention. Species normative behaviors that are reflective of positive emotional states can be used to identify animals experiencing positive welfare, but determining appropriate levels of these behaviors in captivity can be challenging. Regardless of the behaviors being assessed, an understanding of the species' behavioral repertoire is critical when using behavior as a measure of welfare. When accurately assessed, an animal's

C. K. Lutz (✉)
Southwest National Primate Research Center, Texas Biomedical Research Institute, San Antonio, TX, USA

K. C. Baker
Tulane National Primate Research Center, Covington, LA, USA

behavior, whether normal or abnormal, can be utilized as an indicator of well-being in nonhuman primates.

Keywords

Abnormal behavior · Anxiety-related behavior · Species normative behavior · Psychological well-being · Positive welfare

1 Introduction

Since the term "psychological well-being" (herein used interchangeably with "welfare") was introduced in the 1985 amendment to the Animal Welfare Act, there has been considerable discussion regarding how to best define the term. Many definitions have been built around "the Five Freedoms" (Brambell Committee 1965): (1) freedom from thirst, hunger, and malnutrition, (2) freedom from discomfort, (3) freedom from pain, injury, and disease, (4) freedom to express natural behavior, and (5) freedom from fear and distress. Freedom from boredom (Ryder 1998), the principles of coping ability (Broom and Johnson 1993), and the ability to predict and control environments (Wiepkema and Koolhaas 1993) have also been raised as components of psychological well-being. Although the European Commission defines welfare using the Five Freedoms, several other definitions are widely used in zoos and laboratories. For example, the Association of Zoos and Aquariums defines welfare as "an animal's collective physical, mental, and emotional states over a period of time" and states that an animal possesses good welfare if it is "healthy, comfortable, well-nourished, safe, able to develop and express species-typical relationships, behaviors, and cognitive abilities, and not suffering from unpleasant states such as pain, fear, or distress" (Association of Zoos and Aquariums Welfare Committee 2016). The European Association of Zoos and Aquaria defines psychological well-being as "coping, both mentally and physically, at a particular point in time" (European Association of Zoos and Aquaria 2016). For laboratories that house primates, an animal is considered to have good welfare if it is (1) experiencing good physical health, (2) able to engage in a substantial range of beneficial species-typical activities and does not display high levels of abnormal behavior, (3) not in a chronic state of distress, and (4) able to cope effectively with day-to-day changes in its social and physical environment (with reference to meeting its own needs; Novak and Suomi 1988).

There are numerous means available for measuring welfare, many of which are addressed in this volume. Herein, we concentrate on commonly used behavioral indices for measuring welfare: (1) identifying the presence, absence, or quantity of behaviors that are abnormal (i.e., not species-appropriate); (2) assessing behaviors indicative of anxiety or fear, which can be useful for measuring distress; and (3) measuring indices of positive welfare, including normalized activity budgets and species-appropriate behaviors.

2 Abnormal Behavior

Abnormal behavior broadly defined includes any behavior that deviates from the norm for that species. Behaviors are considered abnormal if they are qualitatively different (i.e., differ in kind) or quantitatively different (i.e., differ in degree) from behaviors typically exhibited by that species (Erwin and Deni 1979). Abnormal behavior has been reported in animals across the primate order, including prosimians (Tarou et al. 2005), monkeys (Lutz et al. 2003; Vandeleest et al. 2011; Pomerantz et al. 2012; Conti et al. 2012; Camus et al. 2013; Crast et al. 2014), and apes (Pazol and Bloomsmith 1993; Nash et al. 1999; Hook et al. 2002; Birkett and Newton-Fisher 2011). In both zoo and laboratory conditions, abnormal behavior is often idiosyncratic, and there are behaviors that are considered to be abnormal both within and across primate populations (Marriner and Drickamer 1994; Bollen and Novak 2000; Lutz et al. 2003). Therefore, to better characterize abnormal behavior, we will subdivide it into stereotypies and pathological behavior (Novak et al. 2012).

2.1 Stereotypies

Stereotypies are a heterogeneous group of repetitive behavior patterns that have no clear goal or function (Mason 1991). They vary in form and frequency and can be further classified based on the physical characteristics of the behavior, such as whole-body and self-directed movements (Bayne and Novak 1998; Novak et al. 2012). Whole-body or motor stereotypies may include such behaviors as pacing, swinging, flipping, and rocking (Fritz et al. 1992; Pazol and Bloomsmith 1993; Lutz et al. 2003; Vandeleest et al. 2011). In contrast, abnormal self-directed behaviors or fine motor movements may include behaviors such as digit sucking, eye poking (saluting), self-clasping, or hair pulling (Davenport 1963; Lutz et al. 2003). Some types of stereotypies may be an exaggerated form of normal behavior, and normal behaviors can be stereotyped, making it difficult to determine at what point normal behavior ends and abnormal behavior begins (Mason 1993). For example, behaviors such as rocking in chimpanzees (*Pan troglodytes*) are difficult to distinguish from agonistic display behavior (Ross and Bloomsmith 2011). Moreover, stereotypies typically do not cause injury and may be adaptive in certain contexts (Mason and Latham 2004). Therefore, the presence of stereotypies may not be a reliable indicator of an animal's current well-being.

2.2 Pathological Behavior

Abnormal behaviors are considered pathological if they occupy a significant portion of the animal's time budget, interfere with normal behavior such as eating, or cause tissue damage (Bayne and Novak 1998). Although pathological behavior is here categorized separately from stereotypies, it can also be stereotyped in form. For example, self-biting in stump-tailed macaques (*Macaca arctoides*) was often directed to one particular side of the body (Anderson and Chamove 1985). Pathological behavior occurs in a small percentage of captive nonhuman primates and includes more severe forms of abnormal behavior that may cause injury. The term "self-injurious behavior" refers specifically to those behaviors that either cause, or have the potential to cause, injury. Behaviors that are considered to be self-injurious include hair pulling (Bayne and Novak 1998), self-biting (Fittinghoff et al. 1974; Lutz et al. 2003; Crast et al. 2014), head banging (Levison 1979), and self-slapping (Fittinghoff et al. 1974). However, although these behaviors have the potential for self-injury, actual injury is comparatively rare (Anderson and Chamove 1980; Lutz et al. 2003). Even though self-inflicted injury is relatively uncommon, unlike with stereotypies, pathological behavior has a clear potential negative impact on animal well-being.

2.3 The Prevalence of Abnormal Behavior

Abnormal behavior can vary in type and frequency across taxonomic groups of nonhuman primates. For example, coprophagy (consumption of feces) is commonly reported in chimpanzees (Nash et al. 1999; Birkett and Newton-Fisher 2011), while "wiggle digits" (fluttering fingers in front of the face) is typically observed in baboons (*Papio* spp.: Lutz et al. 2014). In contrast, pacing or other stereotypic locomotion has consistently been reported to be one of the most common abnormal behaviors across many species of captive nonhuman primates (prosimians: Tarou et al. 2005; macaques: Bellanca and Crockett 2002; Lutz et al. 2003; Vandeleest et al. 2011; Pomerantz et al. 2012; sooty mangabeys *Cercocebus atys*: Crast et al. 2014). The overall incidence of abnormal behavior has been reported at many facilities that house captive nonhuman primates, including zoos and laboratories. For example, in a cross-center survey of 108 zoos, Bollen and Novak (2000) reported that overall 14% of the nonhuman primates exhibited some form of abnormal behavior. In this survey, apes exhibited the highest percentage (40%), New World monkeys and prosimians exhibited the lowest (6–7%), and Old World monkeys exhibited an intermediate level (14%). In similar cross-center surveys focusing on specific genera or species at zoos, the numbers reported were higher for chimpanzees (64–100%; Birkett and Newton-Fisher 2011; Jacobson et al. 2016) and prosimians (14%; Tarou et al. 2005).

In contrast to surveys of zoo populations, surveys of laboratory-housed nonhuman primates focused on more restrictively housed animals, such as those that were singly housed, and therefore may not be representative of the facility's population. For example, the percentages of abnormal behavior in singly housed animals were often higher than what was observed in group-housed animals (Bayne et al. 1992b). Surveys of singly housed nonhuman primates have reported that as many as 89–100% of macaques, 83% of sooty mangabeys, and 26% of baboons exhibited some form of abnormal behavior (Bayne et al. 1992b; Lutz et al. 2003, 2014; Camus et al. 2013; Crast et al. 2014); however, the number was much lower for marmosets (*Callithrix jacchus*; ~1%, Berkson et al. 1966). Even in instances where a large number of animals exhibited abnormal behavior, it generally did not constitute a large portion of the animals' time budgets (Marriner and Drickamer 1994).

2.4 Risk Factors/Causes of Abnormal Behavior

Abnormal behavior can result from a combination of factors including environmental experiences, physiological characteristics, and genetic risk factors (Suomi 2006) as well as frustration, stress, or lack of stimulation (Mason 1991). For this review, potential causes are divided into extrinsic (environmental) and intrinsic (biological) categories.

2.4.1 Extrinsic Effects

Some instances of abnormal behavior may indicate that the environment is lacking in characteristics that fulfill a behavioral need for the animal; the captive environment may provide few opportunities to elicit normal species-typical behavioral patterns. Therefore, there is a lack of behavioral competition with stereotyped abnormal behavior, which can bring about the repetitiveness of stereotyped movements (Mason and Turner 1993).

2.4.2 Rearing

How an animal is reared can play a significant role in its subsequent behavior. The impact of rearing practices, ranging from nursery rearing to rearing in more natural groups with mothers and peers, has been assessed in several studies. Although infants reared in a nursery with their peers can develop many species-typical behaviors and relatively normal social responses (Harlow and Harlow 1965), they can still demonstrate higher levels of abnormal behavior than those seen in their mother-reared counterparts both in laboratories (Bellanca and Crockett 2002; Conti et al. 2012; Gottlieb et al. 2013a; Crast et al. 2014) and in zoos (Marriner and Drickamer 1994). In comparison with those that were mother-reared, nursery-reared monkeys and chimpanzees were more likely to exhibit behaviors such as clasping, digit sucking, rocking, repetitive movements, and self-biting (Dienske and Griffin 1978; Snyder et al. 1984; Maki et al. 1993; Lutz et al. 2007; Rommeck et al. 2009; Gottlieb et al. 2013a; Fig. 1). In one study, self-biting began as early as 32 days of age (Rommeck et al. 2009). However, rearing does not influence all abnormal behavior in the same way. For example, chimpanzees that were not mother-reared showed elevated levels of many abnormal behaviors, but not coprophagy, which was seen at higher levels in mother-reared chimpanzees and may be a socially learned behavior (Jacobson et al. 2016).

There are also differences in behavioral and physiological outcomes between nursery rearing conditions. For example, in rhesus monkeys, peer-only rearing (housed together 24 h per day) resulted in less self-biting and floating limb behavior than did surrogate-peer rearing (housed with an artificial surrogate, but given daily contact with other infants; Lutz et al. 2007; Rommeck et al. 2009), but it also resulted in greater amounts of partner clinging (Rommeck et al. 2009). Peer-only-reared

Fig. 1 Mother rearing and social housing help to prevent abnormal behavior. Photo courtesy of Texas Biomedical Research Institute

infants also showed heightened anxiety and physiological responses to a relocation stressor (Dettmer et al. 2012).

2.4.3 Later Environmental Restriction

Individual or restrictive housing later in life can also promote the expression of abnormal behavior such as motor stereotypies, self-directed behavior, and self-injurious behavior (Lutz et al. 2003, 2014; Gottlieb et al. 2013a; Crast et al. 2014). In a survey of 630 zoo primates, 40% of the animals housed alone, 5% of paired animals, and none of the animals housed in groups of four or more exhibited abnormal behavior (Trollope 1977). Similarly, singly housed rhesus monkeys exhibited a greater frequency of abnormal behavior than did socially housed animals (Bayne et al. 1992b) and they expressed more motor stereotypy than did pair-housed animals (Gottlieb et al. 2013a). Young adult rhesus monkeys separated from their family group and subsequently housed singly also exhibited higher levels of stereotypy and self-clasping than those that were separated but housed with familiar subjects (Suomi et al. 1975). The younger an animal is when first singly housed and the greater the proportion of the early years (i.e., the first 4 years) housed singly, the more likely it is to display abnormal behavior (Bellanca and Crockett 2002; Lutz et al. 2003, 2014). Restrictions such as indoor housing can also play a role in the development of abnormal behavior. For example, stereotyped behavior increased when chimpanzees were singly housed indoors (Brent et al. 1989). Similarly, in rhesus macaques, the greater the proportion of an individual's life spent indoors and/or singly housed, the greater the risk of motor stereotypy (Vandeleest et al. 2011), and stereotypies and self-biting were reduced for every year an animal spent outdoors (Gottlieb et al. 2013a).

2.4.4 Stress Due to Clinical Procedures, Human Activity, and Husbandry

In some situations, potentially stressful environmental factors such as routine husbandry, clinical procedures, or human activity may play a role in the development or maintenance of abnormal behavior (Coleman et al. 2022). For example, routine clinical procedures such as anesthesia events and blood draws were associated with an increase in stereotypies and self-injury (Lutz et al. 2003; Vandeleest et al. 2011). However, this effect was not detected in all animals (Bellanca and Crockett 2002; Crast et al. 2014). Human activity, such as visitors at a zoo, can result in increased levels of abnormal behavior (black-capped capuchins [*Cebus apella*], Sherwen et al. 2015; gorillas [*Gorilla gorilla*], Blaney and Wells 2004; Wells 2005; lion-tailed macaques [*Macaca silenus*], Mallapur et al. 2005; spider monkeys [*Ateles geoffroyi*], Davis et al. 2005; several species, Chamove et al. 1988). Even alterations in routine husbandry procedures, such as feeding, were associated with stereotypies. For example, stereotyped behavior in stump-tailed macaques was prolonged when feeding time was delayed (Waitt and Buchanan-Smith 2001), and rhesus macaques were less likely to perform stereotypies when feeding time was made more predictable (Gottlieb et al. 2013b). However, in chimpanzees, levels of abnormal behavior (primarily coprophagy) were highest under a predictable feeding schedule (Bloomsmith and Lambeth 1995), suggesting that not all animals or all abnormal behaviors respond equally to environmental influences.

Simple location changes can also impact behavior. For example, the number of cage relocations showed a significant positive relationship with self-injurious behavior in rhesus macaques (Rommeck et al. 2009) and motor stereotyped behavior increased by 3% for each relocation to a new room (Gottlieb et al. 2013a). Similarly, rhesus monkeys that experienced a higher number of pair separations exhibited higher rates of motor stereotypy and self-biting (Gottlieb et al. 2013a). Self-biting behavior in adult male rhesus macaques also increased from pre-move to post-move and remained elevated for up to one year later. The levels of salivary, serum, and hair cortisol (measures of stress) were also elevated post-move (Davenport et al. 2008). Even prior to moving, cage location can impact the animal's behavior. For example, motor stereotypy and self-biting were lower in cages further away from the entrance to the room and also lower in cages that were located on the top rack (Gottlieb et al. 2013a).

2.5 Intrinsic Effects

The environment does not impact behavior in isolation; it works in concert with other factors such as the animal's genetics and personality as well as age and sex. Although it is important to identify risk factors for abnormal behavior to determine appropriate management practices, comparing the prevalence of abnormal behavior across facilities may be confounded by intrinsic factors and their interactions with environmental experiences.

2.6 Species

Not all animals similarly reared or housed develop abnormal behavior; abnormal behavior can differ in frequency both across and within a species. For example, prosimians in the genera *Varecia* and *Microcebus* were more likely to exhibit stereotypies than were those of other genera (Tarou et al. 2005). Similarly, when comparing genera of Old World monkeys, macaques exhibited an overall higher level of abnormal behavior and different patterns of abnormal behavior than did baboons (Lutz 2018). Within the genus *Macaca*, rhesus infants reared in isolation had more than double the probability of abnormal behavior when compared to similarly reared pigtail infants (*Macaca nemestrina*; Sackett et al. 1976)

2.7 Temperament

An individual's temperament or personality can also play a role in the display of abnormal behavior. For example, rhesus monkeys scoring low on gentle temperament, high on activity during a human intruder challenge, and high on levels of contact with novel objects were more likely to exhibit motor stereotypy (Gottlieb et al. 2013a). In another study, scoring high on gentle and nervous temperament was a risk factor for motor stereotypy, but only in indoor-reared monkeys (Vandeleest et al. 2011). Such differences in abnormal behavior may have a genetic component (Tiefenbacher et al. 2005b; Chen et al. 2010). For a review of the links between personality and welfare, see Robinson and Weiss (2022).

2.8 Age

Age can play a role in the types and frequencies of abnormal behavior, in part due to the stage of the animal's physical development and the animal's current behavioral repertoire (Mason 1993). For example, in a survey of nonhuman primates housed in zoos, infants and juveniles were reported to have fewer behavioral abnormalities than were subadults and adults (Trollope 1977). In addition, younger animals tended to exhibit more active abnormal behaviors such as pacing, body flipping, and swinging, while older animals tended to exhibit more sedentary or self-directed abnormal behaviors such as eye poking, hair pulling, and self-injurious behavior (Lutz et al. 2003; Gottlieb et al. 2013a). In rhesus monkeys reared in partial isolation, self-sucking and self-grasping decreased with age, while chewing orality and self-biting increased with age (Cross and Harlow 1965; Suomi et al. 1971). Nursery-reared chimpanzees showed a decline in rocking with age, but the very youngest of the chimpanzees (less than 6 weeks of age) did not display this behavior, possibly due to motor immaturity (Pazol and Bloomsmith 1993). However, age did not have a reported effect on abnormal behavior in all studies (Hook et al. 2002; Tarou et al. 2005; Birkett and Newton-Fisher 2011).

2.9 Sex

The relationship between sex and levels of abnormal behavior is complex. Although one study of eight zoo primate species including prosimians, monkeys, and apes reported no overall sex difference in abnormal behavior (Marriner and Drickamer 1994), other studies reported sex differences that varied by species and behavior category. For example, in rhesus macaques, males were more likely to exhibit abnormal behavior than were females (Cross and Harlow 1965; Suomi et al. 1971; Gluck and Sackett 1974; Novak et al. 2002; Lutz et al. 2003; Rommeck et al. 2009; Vandeleest et al. 2011; Gottlieb et al. 2013a). However, some studies of macaques also reported no sex difference in either a few behaviors (e.g., salute, digit suck; Lutz et al. 2003) or all recorded behaviors (Hook et al. 2002). In baboons, males also exhibited higher overall levels of abnormal behavior than did females (Brent and Hughes 1997). When the behavioral categories were assessed separately, male baboons exhibited higher levels of abnormal appetitive behavior (e.g., coprophagy, hair eating, regurgitating), but there was no sex difference in other behavioral categories (Lutz et al. 2014). In contrast, chimpanzee and sooty mangabey females were more likely to exhibit higher overall levels of abnormal behavior than were males (Hook et al. 2002; Crast et al. 2014; Jacobson et al. 2016). However, in chimpanzees, the level also varied by behavior. For example, female chimpanzees showed a higher prevalence of coprophagy and self-clinging, while males showed a higher prevalence of rocking (Fritz et al. 1992; Nash et al. 1999; Jacobson et al. 2016). Although sex differences in abnormal behavior can occur across species, the reason for these differences remains unclear.

3 The Function of Abnormal Behavior

Abnormal behavior is typically defined as having no obvious purpose or function (Mason and Latham 2004). In some cases, abnormal behavior may be considered to be maladaptive, interfering with normal activities or biological functions or resulting in injury (Novak et al. 2012). In other cases, abnormal or stereotypic behavior may indicate a coping strategy utilized to reduce stress (Mason and Latham 2004; Novak et al. 2012). Alternatively, abnormal behavior may simply be acquired through social learning (Nash et al. 1999; Hook et al. 2002; Less et al. 2013; Hopper et al. 2016). Lastly, abnormal behavior may in some cases be unrelated to stress (Rushen 1993), affected by changes in brain chemistry or morphology (Kraemer and Clarke 1990; Tiefenbacher et al. 2005b), and/or may simply be an automatic behavior that persists long after the initial trigger has ended (Mason and Latham 2004).

3.1 Maladaptive

Abnormal behavior may have no function and may instead be a maladaptive response to poor physical and/or psychological well-being (Novak et al. 2012). The behavior could be due to a clinical condition or illness, with behaviors such as pacing and self-biting occurring as the result of pain or discomfort (Bourgeois et al. 2007). Such conditions can often be treated with medication or other methods of clinical care. Alternatively, the behavior could be associated with psychopathology. The "integrated developmental-neurochemical hypothesis" is one explanation for these behaviors. It proposes that adverse early experience and later stressors result in alterations in neurological systems. The subsequent dysregulation of these systems contributes to episodes of anxiety, leading to abnormal behavior (Tiefenbacher et al. 2005c). Abnormal behavior associated with psychopathology can be treated by environmental changes (e.g., socialization or enrichment) or pharmacotherapy, but is typically not eliminated (Novak 2003).

3.2 Coping Hypothesis

The presence of abnormal behavior may also be indicative of a coping strategy and performed in response to environmental deficits. Abnormal behavior may be used to either reduce or increase arousal (Mason 1991) and may be viewed as a U-shaped curve, increasing in both impoverished and overstimulating environments (Anderson and Chamove 1981). On one end of this curve, repetitive abnormal behavior may be a form of self-stimulation (Berkson 1983) or "do-it-yourself enrichments" (Mason and Latham 2004, p. S60). On the other end of this curve, repetitive abnormal behavior may be used to reduce arousal, as demonstrated physiologically in several studies (Soussignan and Koch 1985; Novak 2003; Major et al. 2009; Pomerantz et al. 2012). It could be argued that it is counterproductive to attempt to thwart abnormal behaviors that lack deleterious impacts on health. However, for behaviors that result in injury or other health problems, the fact that they may serve some anxiety-relieving function does not obviate vigorous intervention. If animals do not exhibit abnormal behavior, it is not clear as to whether they are not coping or simply not stressed (Rushen 1984). To avoid circular reasoning (e.g., the environment is assumed to be nonstimulating because the behavior occurs), the level of stimulation needs to be defined separately and not based on the amount of stereotyped behavior the individual performs (Baumeister and Rollings 1976).

4 Using Abnormal Behaviors to Measure Welfare

Most forms of abnormal behaviors are robust indicators of past or current poor welfare and therefore cannot necessarily be taken as a signal of poor welfare at the time of performance. As discussed above, some abnormal behaviors may function to reduce stress. It is when abnormal behavior *cannot* satisfy a motivation that it is more clearly linked to frustration and poor welfare. Also, some abnormal behaviors appear to be learned, and differ from other abnormal behaviors in risk factors and correlate with poor environments. However, abnormal behaviors are clearly deleterious when they compete with or prevent the expression of beneficial normal behavior (such as social interaction), are associated with signs of distress, or impose pain and health issues. Although abnormal behavior should not be indiscriminately associated with poor welfare in an animal, it is clear that abnormal behavior is an invaluable measure for refining behavioral management practices.

5 Interventions for Abnormal Behavior

5.1 Social Housing

The absence of companions, even in adulthood, has been shown to be a strong predictor for the development of aberrant behavior in macaques and chimpanzees (Bayne et al. 1992a; Nash et al. 1999; Bellanca and Crockett 2002; Lutz et al. 2003; Rommeck et al. 2009; Gottlieb et al. 2013a, 2015). Social housing is therefore one of the most effective interventions for abnormal behavior. Housing with a social partner can be unpredictable, stimulates the senses, and is less likely to produce habituation than are other types of enrichment (Novak and Suomi 1991). Simply moving individuals from single housing to pair housing reduced abnormal behavior in rhesus macaques (Weed et al. 2003; Baker et al. 2012a). However, this finding is not consistently observed among male rhesus macaques; a smaller magnitude of reduction was seen in one study (Baker et al. 2012a) and the reduction did not persist in another (Doyle et al. 2008). Furthermore, no change was observed in a third study (Baker et al. 2014). However, behavioral benefits in terms of reduced anxiety and increased activity were consistently seen among males in these studies. Unlike in macaques, pair housing does not appear sufficient to alter levels of abnormal behavior in common chimpanzees (Baker 1996); a larger social environment may be required if abnormal behavior is to be reduced in this species. Another interesting species difference has been seen between rhesus and long-tailed macaques (*Macaca fascicularis*). In caged rhesus macaques, higher levels of abnormal behavior were seen with partial social contact (i.e., interacting through partitions) in comparison with full contact in both sexes (Baker et al. 2012b, 2014); however, this partial social restriction was not associated with higher levels of abnormal behavior in female long-tailed macaques (Baker et al. 2012b).

5.2 Inanimate Enrichment

Enriching the environment by adding toys, structures, and foraging devices is one means of promoting species-typical behavior and reducing abnormal behavior in nonhuman primates. However, the effectiveness of enrichment can vary depending

Fig. 2 Enriching the environment is one means of promoting species-typical behavior. Photo courtesy of Texas Biomedical Research Institute

on the species, the individual's history, the type of abnormal behavior, and the enrichment type (Lutz and Novak 2005). The provisioning of manipulable items, such as PVC tubes and chew toys, has been shown to decrease abnormal behavior in numerous primate species including macaques, baboons, capuchins, and chimpanzees (Brent et al. 1989; Brent and Belik 1997; Kessel and Brent 1998; Boinski et al. 1999). However, reductions in abnormal behavior did not occur in all cases (Schapiro and Bloomsmith 1995), and Line et al. (1991) showed no relationship between toy use and abnormal stereotyped behavior in rhesus monkeys. One beneficial use of a chew toy was reported in a laboratory baboon with self-injurious behavior; the self-biting was redirected away from the body and onto the toy (Crockett and Gough 2002). However, other animals simply incorporated the toy into their pacing or biting rituals (Bayne 1989; Anderson and Stoppa 1991). Habituation to toys can also occur, though rotation of toys may help to maintain novelty (Paquette and Prescott 1988; Bloomsmith et al. 1990; Brent and Eichberg 1991; Line et al. 1991; Fig. 2).

The use of forage in devices or in a substrate has also been shown to reduce abnormal behavior. Providing animals with an opportunity to forage in wood chips, alfalfa, or straw (Chamove et al. 1982, 1984; Baker 1997; Boinski et al. 1999) or via foraging devices, such as fleece boards, puzzle feeders, or turf boards (Maki et al. 1989; Bayne et al. 1991, 1992a; Lam et al. 1991; Roberts et al. 1999), resulted in significant reductions in abnormal behavior in nonhuman primates. However, although abnormal behavior was shown to decrease as foraging behavior increased (Bayne et al. 1991, 1992a), foraging opportunities did not decrease abnormal behavior in all animals (Brent and Eichberg 1991; Byrne and Suomi 1991; Lutz and Farrow 1996; Fekete et al. 2000) and often the behavior was reduced only when the animals were working to extract the food (Novak et al. 1998), suggesting that there is not always a carryover effect (i.e., outside of the session).

One reason for the variable behavioral responses to enrichment may be individual differences in usage (Bloomstrand et al. 1986). For example, toys may be less effective with older animals (Line et al. 1991) or with older stereotypies, which can be more resistant to change (Mason 1991). Different types of abnormal behaviors can also respond to enrichment in different ways. Devices such as chew toys, mirrors, and foraging devices help reduce less severe forms of abnormal behavior such as motor stereotypies, but they are not as effective in reducing the more severe forms such as self-injurious behavior (Novak et al. 1998; Rommeck et al. 2009). Moreover, environmental enrichment may aid in the reduction in many types of abnormal behavior, but the behavior is typically not eliminated.

5.3   Increased Cage Size

Although there are legal requirements for minimum cage size for nonhuman primates (U.S. Department of Agriculture 2013), cages larger than the minimum standards are often provided. Studies assessing the effect of cage size on abnormal behavior have varying results, which may be due in part to the confounding effect of a change in cage location along with an increase in cage size. For example, animals may be moved from small indoor cages to larger outdoor cages. In this example, improvements in behavior may not be due exclusively to cage size, but also to the environment outside of the cage. The range of cage sizes evaluated also can vary across studies, making direct comparisons difficult. However, in general, it appears that small increases in cage size have a lesser impact on behavior than significant increases in cage size. Relatively small increases in cage size generally have been shown to have negligible effects on abnormal behavior (Line et al. 1990; Crockett and Bowden 1994; Crockett et al. 1995), and in one study, some abnormal behaviors actually increased in the slightly larger cage (Bayne and McCully 1989). However, motor stereotypies were shown to significantly decrease in common marmosets when they were placed in a cage with more vertical space (Kitchen and Martin 1996).

Larger increases in cage size appear to have a greater impact on abnormal behavior. For example, in rhesus monkeys, housing in a significantly larger cage (e.g., over 100 times larger) resulted in virtually no stereotypies (Draper and Bernstein 1963). Similarly, a sixfold increase in cage size resulted in decreased levels of stereotyped abnormal behaviors; however, other types of abnormal behavior, such as eye poke ("salute") or self-biting, were not affected by cage size (Paulk et al. 1977). In one study, singly housed rhesus monkeys were moved from pens into smaller cages located within the pens to avoid a location change confound. Although this alteration in cage size did not result in a change in abnormal behavior, levels of tension-related behavior (e.g., threat, yawn, scratch, cage shake, and aggression) were lower in the larger pen (Kaufman et al. 2004). Additional alterations to housing, such as the provisioning of a "play cage" or the addition of "porch" space to the home cage, have also been shown to have an impact on abnormal behavior. For example, temporary housing in a larger play cage resulted in a reduction in abnormal behavior in macaque monkeys (Tustin et al. 1996; Griffis et al. 2013), and providing additional space and visual opportunities via a "porch" resulted in decreased feces painting in rhesus monkeys (Gottlieb et al. 2014).

5.4 Outdoor Housing

As with cage size, comparisons between outdoor and indoor housing are also confounded. Animals housed outdoors tend to be in larger cages and housed in larger groups than those housed indoors. However, research shows that outdoor housing has a positive impact on behavior. Juvenile rhesus monkeys housed outside showed reduced levels of self-mouthing in comparison with those that remained indoors (O'Neill et al. 1991) and moving animals outdoors either singly or in groups significantly reduced self-biting and self-directed stereotypies (Fontenot et al. 2006). Being housed indoors is a risk factor for motor stereotypy (Vandeleest et al. 2011), and mother-reared rhesus monkeys raised outdoors were significantly less likely to exhibit self-injurious behavior than those similarly raised indoors (Rommeck et al. 2009; Fig. 3).

5.5 Positive Reinforcement Training

Positive reinforcement training (PRT), which consists of reinforcing animals with rewards for exhibiting desired behavioral responses, has been evaluated as an intervention for abnormal behavior. Much of the application of this technique has been approached as case studies (a drill, *Mandrillus leucophaeus*, Desmond et al.

Fig. 3 Outdoor housing has a positive impact on behavior. Photo courtesy of Kathy West Studios

1987; a chimpanzee, Morgan et al. 1993; a gorilla, *Gorilla gorilla*, Leeds et al. 2016; and an orangutan, *Pongo pygmaeus*, Raper et al. 2002) sometimes in conjunction with other attempted interventions such as drug therapy and environmental enrichment (chimpanzee; Bourgeois et al. 2007). Results have been mixed; carryover effects or persistent effects after treatment may not be seen in many species. However, a study of carryover effects in rhesus macaques detected benefits of PRT outside of training sessions, but only for individuals with high levels of abnormal behavior (Baker et al. 2009).

5.6 Drug Therapy

Because of the possible physiological and behavioral effects of psychoactive drugs, pharmacotherapy is generally a method of last resort for treating abnormal behavior (McKinney et al. 1980). Pharmacotherapy is therefore typically used to treat only the more severe forms of abnormal behavior, such as self-biting and self-injury. A wide variety of drug therapies have been utilized to treat abnormal behavior, ranging from L-tryptophan (a serotonin precursor), to fluoxetine (a selective serotonin reuptake inhibitor), to cyproterone acetate (a synthetic steroid), and to naltrexone (an opioid antagonist; Weld et al. 1998; Eaton et al. 1999; Macy et al. 2000; Fontenot et al. 2005; Crockett et al. 2007; McCoy et al. 2009; Kempf et al. 2012; Freeman et al. 2015). The number of drug therapies utilized to treat abnormal behavior is indicative of the number of routes through which abnormal behavior can develop and/or be maintained. For example, in a study testing diazepam as a treatment for self-injurious behavior, half of the subjects responded to the treatment while half did not, suggesting that there may be subtypes of animals that respond to treatments differentially (Tiefenbacher et al. 2005a). Given the different behavioral profiles and responses to treatment, therapeutic strategies will likely need to be tailored to individual animals rather than behavioral categories. In addition, the long-term effects of pharmacotherapy have not been thoroughly examined. Additional research is warranted to further investigate the efficacy of pharmacotherapy as a treatment for abnormal behavior.

6 Anxiety-Related Behavior

Well-being is of course conceptualized across more than the normal/abnormal axis. In practice, considerable attention is paid to preventing the development of or reducing levels of abnormal behavior. However, to address multiple facets of well-being, it is important to look beyond the expression of abnormal behaviors. An individual's ability to cope with stressors and the absence of distress both figure into many definitions of welfare. Animals may experience distress if their ability to cope with the environment exceeds their adaptive capacity due to the severity or frequency of a stressor. Distress has been proposed as the most salient factor in defining welfare in that it is less subjective than other concepts (Moberg 1985). Fortunately, there are many behaviors that can be used to measure distress in a more reliable way than some abnormal behaviors.

6.1 Behavioral Measures of Anxiety

There is a constellation of normal forms of self-directed behaviors that allow us to measure negative emotional states, such as anxiety, fear, conflict, and frustration, associated with aversive events (reviewed in Maestripieri et al. 1992). It has been established through behavioral and pharmacological studies that self-directed behavior (principally scratching, self-grooming, yawning, and body shaking) is related to anxiety (Fig. 4). For example, levels of anxiety-related behaviors increase with risk of attack, or following an attack (Troisi and Schino 1987; Aureli and van Schaik 1991; Aureli 1992, 1997; Baker and Aureli 1997; Castles and Whiten 1998). These indices of anxiety have been pharmacologically validated in a dose-dependent fashion using anxiolytic and anxiogenic medications (Ninan et al. 1982; Insel et al. 1984; Crawley 1985; Schino et al. 1991, 1996; Maestripieri et al. 1991). Anxiety-related behaviors are often (Watson et al. 1999; Ulyan et al. 2006), though not always (Peel et al. 2005), seemingly dissociated from activation of the hypothalamic–pituitary–adrenal (HPA) axis. This could be taken to undermine the validity of self-directed behaviors as accurate metrics of stress. However, there are a

Fig. 4 Self-grooming can be an expression of normal self-maintenance or an expression of anxiety. Photo courtesy of Texas Biomedical Research Institute

variety of potential explanations for this dissociation of anxiety-related behaviors and HPA measures. For example, there are the methodological difficulties of age effects, the diurnal effects on cortisol, and short-term cortisol increases in association with capture for sampling. It is also possible that self-directed behaviors may reduce anxiety in the same way as stereotypic abnormal behavior.

In addition to self-directed behaviors, other behaviors are used to infer anxiety, some perhaps better characterized as an expression of fear, i.e., a reaction to a threatening stimulus rather than a prolonged state of distress, including outside of the presence of the stimulus. Expressions of fear are salient measures of welfare and are often incorporated as measures of anxiety. These behaviors have not been validated in the same manner as self-directed behaviors, but are commonly observed in the wild in response to danger. Such behaviors are species-appropriate and are used as measures in the scientific literature. Examples include vocalizations and facial expressions (e.g., Barros and Tomaz 2002; Kalin 2004; Rogers et al. 2008; Coleman et al. 2011; Corcoran et al. 2012), freezing (Costall et al. 1992; Kalin et al. 1998), aggression directed toward humans (e.g., Coleman et al. 2003; Baker 2004; Kinnally et al. 2010; Corcoran et al. 2012), and fleeing or withdrawing to the maximal distance from the negative stimuli (Oler et al. 2009; Agustín-Pavón et al. 2012). Because these behaviors are so frequently employed as measures of anxiety, and because both anxiety and fear represent negative emotional states associated with aversive events, we will herein refer to both self-directed behaviors and fearful behaviors as "anxiety behaviors."

6.2 Testing Anxiety and Identifying the Anxious Phenotype

There are numerous strategies and tools for measuring anxiety (see Coleman and Pierre 2014 and Gonzales et al. 2016 for review). Several tools have been used to evoke anxiety by adding a negative stimulus to the environment. One commonly used tool is the human intruder paradigm (Kalin and Shelton 1989, 2003). This test is well represented in the applied behavioral management literature (for recent examples, see Capitanio et al. 2017; Hamel et al. 2017; Peterson et al. 2017) and is best conducted using an unfamiliar human "intruder" (Peterson et al. 2017). Other techniques for eliciting anxiety include social separation (e.g., mother–infant separations; McCormack et al. 2009), removal from familiar social groups (Miczek et al. 1995; McKenzie-Quirk and Miczek 2008; Kato et al. 2014), and the presentation of aversive stimuli (e.g., noise: Antoniadis et al. 2009; Willette et al. 2011) or neutral novel objects (Belzung and Le Pape 1994; McClintick and Grant 2016).

As with abnormal behavior, the nonhuman primate anxious phenotype appears to arise from adverse rearing experiences, such as nursery rearing (Kraemer et al. 1989; Higley et al. 1991; Suomi 1991; Dettmer et al. 2012) and mother-rearing paradigms that disturb the mother–infant relationship, whether induced (e.g., Coplan et al. 1998) or naturally occurring (e.g., variations in maternal care: McCormack et al. 2009). It also has clear genetic underpinnings, influenced by polymorphisms in the serotonin transporter (5HTTLPR) and monoamine oxidase A (MAOA-LPR) genes, as well as interactions between genetics and experience (Bennett et al. 2002; Spinelli et al. 2007; McCormack et al. 2009; Karere et al. 2009; Santangelo et al. 2016). This body of research is particularly useful for behavioral management in that it can aid in identifying the most vulnerable individuals.

6.3 Using Anxiety Behaviors to Measure Welfare

Although the presence of abnormal behavior is a strong indicator of impaired well-being, anxiety- and fear-related behaviors are also salient measures of welfare. For one, they provide a ubiquitous signal associated with well-being. Certain populations, such as those in large naturalistic and socially rich environments, may contain few individuals that express abnormal behavior, but all individuals therein will express species-appropriate anxiety- and fear-related behaviors. Second, using these behaviors to measure well-being may allow for early identification of individuals that need specialized management or intervention before species-inappropriate and sometimes physically damaging behaviors develop (Tiefenbacher et al. 2005c; Major et al. 2009; Vandeleest et al. 2011). Third, anxiety- and fear-related behaviors are indicators of current experience and may be more relevant to the present environment than is abnormal behavior, which may be an indicator of past welfare (Mason and Latham 2004).

Using anxious and fearful behaviors for measuring well-being is significantly different from using abnormal behavior. As species-appropriate behaviors, they are evolved responses and part of adaptation. It is also important to realize that particularly when measuring anxiety in the home environment and not in the context of experimentally applied stressors, self-directed behaviors may be performed as part of normal self-maintenance or social regulation (but see Baker and Aureli 1997). However, the behaviors are phenotypically identical regardless of whether they are performed as self-maintenance or as an expression of anxiety. Therefore, *changes* in levels, as opposed to *absolute* levels of these behaviors, may be more relevant for tracking welfare. Fear-related behaviors are likewise species-appropriate and cannot be expected to be absent, but dysregulated responses are maladaptive. In addition, measures that are social in nature and measured in person may be influenced by factors such as the identity of or the degree of habituation to the observer.

Self-directed behaviors may also be provoked by fleeting emotional states that have little impact on welfare. These behaviors may be provoked in circumstances that are intrinsic to activities that one could argue are stimulating challenges, rather than negative stressors. For example, self-directed behaviors were observed to correlate with errors or escalating difficulty during computerized tasks (Itakura 1993; Leavens et al. 2001, 2004; Yamanashi and Matsuzawa 2010; Wagner et al. 2016; but see Herrelko et al. 2012). Thus, computerized testing may incite transient stress associated with challenge and control, two features strongly associated with welfare. Such a situation may be a good example of "eustress," a term introduced by Selye (1976). In fact, the expression of self-directed behaviors during positive arousal (reviewed in Neal and Caine 2016) suggests that they may sometimes be associated with excitement rather than anxiety (Frankenhaeuser 1982). In addition, behaviors may not map onto emotional experience in the same manner in all primate species. For example, unlike scratching in macaques and chimpanzees, scratching in marmosets was found to be expressed at *lower* levels in anxiety-provoking contexts (Neal and Caine 2016).

6.4 Anxiety Behaviors and Welfare-Related Studies

Although there are complexities involved in relating anxiety behaviors to emotional states that contribute to poor well-being, this association is commonly used in the applied literature for evaluating environments and measuring welfare. While not utilized as a measure of welfare as pervasively as abnormal behavior, anxiety behavior has been employed in multiple zoological and laboratory studies. Most commonly, these studies rely on spontaneous, as opposed to provoked, behavior. Although an extensive literature review including negative findings exceeds the scope of this chapter, anxiety behaviors have provided insights into many aspects of behavioral management and captive care.

Anxiety behaviors in great apes have frequently been used to detect the negative impact of humans, including zoo visitors (reviewed in Hosey 2000; see also Wells 2005; Carder and Semple 2008; Clark 2011; Amrein et al. 2014; Bonnie et al. 2016; Hosey 2022; Baker and Farmer 2022). Reduced anxiety can be garnered from interaction with familiar caregivers or trainers in a zoo (Pomerantz and Terkel 2009; Chelluri et al. 2013). Aspects of enclosure design and social housing have been evaluated with respect to their effect on anxiety behaviors. For example, providing additional space and environmental complexity has been shown to reduce anxiety-related behaviors in singly housed rhesus macaques (Griffis et al. 2013) and zoo exhibit designs permitting fission–fusion grouping reduced stress in orangutans (Amrein et al. 2014). Levels of anxiety-related behaviors have also been shown to vary between single housing and social housing (Baker 1996; Doyle et al. 2008; Gilbert and Baker 2011; Baker et al. 2012a, 2014). A variety of inanimate enrichments ranging from simple toys, to foraging, to complex cognitive enrichment, have been found to influence anxiety behavior. For example, providing foraging boxes and additional toys to singly housed capuchins resulted in a reduction in anxiety-related behavior (Boinski et al. 1999), and providing opportunities to forage in bedding reduced anxiety in socially housed chimpanzees (Baker 1997). Many events that occur in the captive environment have been shown to provoke anxiety. Anxiety behaviors increase in response to relocation (Dettmer et al. 2012) and are elicited by common laboratory procedures, both intrinsically aversive (e.g., capture and research procedures: Clarke et al. 1988; Balcombe et al. 2004; Cagni et al. 2009) and nonaversive (e.g., feeding: Waitt and Buchanan-Smith 2001). Anxiety responses in singly housed macaques have also been shown in response to positive but unpredictable events such as feeding and enrichment distribution, but increased predictability ameliorates these responses (Gottlieb et al. 2013b). However, predictability of feeding was not found to influence anxiety behavior in group-housed chimpanzees (Bloomsmith and Lambeth 1995). Welfare impacts of behavioral or cognitive testing have also been evaluated in several species. Voluntary removal from social groups for testing was not found to result in increased anxiety upon return to the social group in either capuchins (Ruby and Buchanan-Smith 2015) or crested macaques (*Macaca nigra*; Whitehouse et al. 2013), and levels of anxiety behavior fell in group-housed Guinea baboons (*Papio papio*) in association with cognitive testing (Fagot et al. 2014).

7 Measures of Good Welfare

Much of the scientific welfare literature and regulatory approaches to psychological well-being stress the amelioration of behaviors indicative of poor welfare. In fact, among the "Five Freedoms," all but one (freedom to express natural behavior) relate to the absence of negative states. However, a pervasive view in behavioral management is that environments should allow nonhuman primates to express a wide range of their natural behavioral repertoire and to fill evolved behavioral needs (Carlstead 1996). Affective states relate to motivations that lead animals to perform essential behaviors and can be employed as benchmarks for welfare. Positive emotional states can be detected by correlating behavior with the physiological correlates in humans and/or reactions to stimuli that are intrinsically rewarding (reviewed for multiple taxa in Boissy et al. 2007). Examples of such natural behaviors include feeding and some social behaviors such as grooming (Boccia et al. 1989; Aureli et al. 1999; Aureli and Smucny 2000; Shutt et al. 2007) and play (see Ahloy-Dallaire et al. 2018 for a recent review). For example, it appears clear that social grooming is intrinsically rewarding and reduces stress (Boccia et al. 1989; Aureli et al. 1999; Aureli and Smucny 2000; Shutt et al. 2007). Similarly, play is actively sought by primates, suggesting a strong motivation, and therefore, the expression of play can be inferred to be a positive experience. This suggestion is supported by studies exploring pharmacological effects on the μ-opioid system (Schino and Troisi 1992; Loseth et al. 2014). However, levels of play are not consistently related to less ambiguous measures of welfare (e.g., increased grooming and decreased anxious or abnormal behavior; Baker 1997; Blois-Heulin et al. 2015). This may be due to how the function of play can shift over an individual's development. Across taxa, play occurs at relatively high levels in juveniles. It has long been posited that play allows immature animals to learn and practice social/locomotor skills and develop quick responses to changes in social or physical conditions (see Carpenter 1934; Jay 1965; Southwick et al. 1965 for early examples). However, in adults, play may function to reinforce social bonds and reduce tensions in stressful social situations (Nieuwenhuijsen and de Waal 1982; Spijkerman et al. 1994; Palagi et al. 2006; Videan and Fritz 2007; Ross et al. 2010; Norscia and Palagi 2011). Play therefore may be best viewed as a *mechanism* for improving welfare rather than as an unambiguous *signal* of good welfare. The appropriate use of play behavior as a metric of welfare may vary with the type of play, species, and age.

8 Species-Appropriate Behaviors as Benchmarks for Welfare

The view that environments supporting psychological well-being must meet evolved behavioral needs leads to the use of observed species-appropriate behaviors as an indicator of welfare (Markowitz and Spinelli 1986). This approach to measuring welfare relies upon thorough knowledge of an individual species' behavioral ecology and social organization. At its simplest, behavior may be evaluated on a presence-versus-absence basis. Is the animal able to perform the types of locomotion commonly observed in that species? Is it able to interact appropriately with conspecifics and employ evolved communication mechanisms (e.g., scent marking)? Does it have opportunities to pursue species-appropriate foraging strategies? In appropriate species, is it able to nest and pursue antipredator strategies such as hiding or retreating upward?

While a valid underlying concept in measuring well-being, using the expression of species normative behavior as indicators of welfare requires careful consideration from a number of perspectives. First, one cannot assume that all behaviors, just because they are natural, are associated with welfare as they may be performed in the wild under conditions that reduce welfare (Veasey et al. 1996). Second, some behaviors observed in the wild, such as regurgitation/reingestion and coprophagy in chimpanzees (Krief et al. 2004; Sakamaki 2010; Bertolani and Pruetz 2011), are typically classified as *qualitatively* abnormal but are in fact seen with some regularity in the wild. They are therefore better characterized as *quantitatively* different, and undesirable due to their potential negative health or social effects. Third, whether the inability to perform a behavior in captivity impacts welfare depends upon whether there is an underlying behavioral need (i.e., whether it is motivation-driven) as opposed to merely being triggered by a stimulus (Stolba and Wood-Gush 1984). As argued by Veasey et al. (1996), the absence of a particular behavior is not necessarily a signal of poor welfare.

Another complexity in employing the expression of natural behavior as a hallmark of welfare arises when, rather than simply attending to presence or absence, one *quantifies* behavior and attempts to replicate time budgets in the wild. However, using levels of behavior in the wild as a yardstick against which captive behavior can be measured is not straightforward given the within-species socioecological flexibility and variability, and the role of demographic variables and social dynamics (e.g., sex, age, and rank). There is no one species-appropriate level of any behavior. As an underlying assumption, this framework for measuring welfare is also challenging in that behavior in the wild is heavily influenced by unarguably negative conditions, many of which are largely absent in captivity, including food scarcity, intergroup fighting, and predation. The removal of these conditions may skew activity budgets in captivity, but it is difficult to argue that this is an adverse behavioral change.

A strict use of wild behavior as a template for activity budgets reflecting good welfare may limit options for effective behavioral management techniques. As one example, pair housing is a common form of social housing for research subjects. Pairs of male rhesus macaques have been found to affiliate at levels indistinguishable from pairs of females (Baker et al. 2012a). However, in several macaque species, adult males in the wild spend very little time engaged in affiliative behavior with one another. While this pattern appears to vary with group size and presence of provisioning, several long-term studies report no instances of male–male grooming (reviewed in Hill 1994; Van Hooff and Hill 1994). Despite the level of grooming observed in paired male macaques being "unnatural," when social housing in larger groups is not an option, it would be difficult to argue that pair housing is detrimental and that single housing is preferable. Pairing programs for adult male macaques are considered to promote psychological well-being, not only on the basis of affiliative behavior being desirable and in its fulfillment of social needs, but also through the use of other behavioral metrics.

9 Conclusions

In this chapter, we have outlined the utility of and challenges associated with employing several categories of behaviors as welfare metrics. Given the difficulties in defining welfare, using abnormal behaviors, anxiety behaviors, or species-appropriate natural behaviors to measure well-being is no simple proposition. Good welfare is supported by environments that provide an animal the opportunity to express a range of species normative behaviors. However, many questions arise from providing outlets for natural behavior which cannot be answered without the use of other metrics of well-being, such as abnormal, anxiety behaviors, and others that have been described in this volume. For example, social housing provides an opportunity for expressing desirable behaviors, but it may also be accompanied by the opportunity to express undesirable behaviors. Do these benefits outweigh the risks of wounding and food monopolization? The expression of abnormal and anxiety behaviors can help answer this question. Would thwarting the ability of animals to perform an abnormal behavior benefit their welfare? Evaluating anxiety could address this question. Is observed anxiety deriving from distress associated with persistent social strife or from ephemeral challenge? Observing levels of natural behavior can answer this question. How do we in behavioral management strike an optimal balance between stimulation, frustration, achievement, and risk? As has been suggested by many, the multidimensional aspect of welfare requires multimodal assessment including the behaviors discussed in this chapter and the measures discussed in chapters by Bethell and Pfefferle (2022), Capitanio et al. (2022), and Gartner (2022).

References

Agustín-Pavón C, Braesicke K, Shiba Y et al (2012) Lesions of ventrolateral prefrontal or anterior orbitofrontal cortex in primates heighten negative emotion. Biol Psychiatry 72(4):266–272. https://doi.org/10.1016/j.biopsych.2012.03.007

Ahloy-Dallaire J, Espinosa J, Mason G (2018) Play and optimal welfare: Does play indicate the presence of positive affective states? Behav Process 156:3–15. https://doi.org/10.1016/j.beproc.2017.11.011

Amrein M, Heistermann M, Weingrill T (2014) The effect of fission–fusion zoo housing on hormonal and behavioral indicators of stress in Bornean orangutans (*Pongo pygmaeus*). Int J Primatol 35:509–528. https://doi.org/10.1007/s10764-014-9765-5

Anderson JR, Chamove AS (1980) Self-aggression and social aggression in laboratory-reared macaques. J Abnorm Psychol 89(4):539–550. https://doi.org/10.1037/0021-843X.89.4.539

Anderson JR, Chamove AS (1981) Self-aggressive behaviour in monkeys. Curr Psychol Rev 1:139–158. https://doi.org/10.1007/BF02979261

Anderson JR, Chamove AS (1985) Early social experience and the development of self-aggression in monkeys. Biol Behav 10(2):147–157

Anderson JR, Stoppa F (1991) Incorporating objects into sequences of aggression and self-aggression by *Macaca arctoides*: an unusual form of tool use. Lab Primate Newsl 30(3):1–3

Antoniadis EA, Winslow JT, Davis M, Amaral DG (2009) The non human primate amygdala is necessary for the acquisition but not the retention of fear-potentiated startle. Biol Psychiatry 65 (3):241–248. https://doi.org/10.1016/j.biopsych.2008.07.007

Association of Zoos & Aquariums Welfare Committee (2016). https://www.aza.org/animal_welfare_committee. Retrieved 9/22/2016

Aureli F (1992) Post-conflict behaviour among wild long-tailed macaques (*Macaca fascicularis*). Behav Ecol Sociobiol 31:329–337. https://doi.org/10.1007/BF00177773

Aureli F (1997) Post-conflict anxiety in nonhuman primates: The mediating role of emotion in conflict resolution. Aggress Behav 23(5):315–328. https://doi.org/10.1002/(SICI)1098-2337 (1997)23:5<315::AID-AB2>3.0.CO;2-H

Aureli F, Smucny DA (2000) The role of emotion in conflict and conflict resolution. In: Aureli F, de Waal F (eds) Natural conflict resolution. University of California Press, Berkeley, CA, pp 199–224

Aureli F, van Schaik CP (1991) Post-conflict behavior in long-tailed macaques (*Macaca fascicularis*). II. Coping with the uncertainty. Ethology 89(2):101–114. https://doi.org/10.1111/j.1439-0310.1991.tb00297.x

Aureli F, Preston SD, de Waal FBM (1999) Heart rate responses to social interactions in free-moving rhesus macaques (*Macaca mulatta*): a pilot study. J Comp Psychol 113:(1)59–65. https://doi.org/10.1037/0735-7036.113.1.59

Baker KC (1996) Chimpanzees in single cages and small social groups: effects on behavior and well-being. Contemp Top Lab Anim Sci 35:61–64

Baker KC (1997) Straw and forage material ameliorate abnormal behaviors in adult chimpanzees. Zoo Biol 16(3):225–236. https://doi.org/10.1002/(SICI)1098-2361(1997)16:3<225::AID-ZOO3>3.3.CO;2-H

Baker KC (2004) Benefits of positive human interaction for socially housed chimpanzees. Anim Welf 13(2):239–245

Baker KC, Aureli F (1997) Behavioural indicators of anxiety: an empirical test in chimpanzees. Behaviour 134(13–14):1031–1050. https://doi.org/10.1163/156853997X00386

Baker KR, Farmer HL (2022) The welfare of primates in Zoo. In: Robinson LM, Weiss A (eds) Nonhuman primate welfare: from history, science, and ethics to practice. Springer, Cham, pp 79–96

Baker KC, Bloomsmith MA, Neu K et al (2009) Positive reinforcement training moderates only high levels of abnormal behavior in singly housed rhesus macaques. J Appl Anim Welf Sci 12 (3):236–252. https://doi.org/10.1080/10888700902956011

Baker KC, Bloomsmith MA, Oettinger BC et al (2012a) Benefits of pair housing are consistent across a diverse population of rhesus macaques. Appl Anim Behav Sci 137(3–4):148–156. https://doi.org/10.1016/j.applanim.2011.09.010

Baker KC, Crockett CM, Lee GH et al (2012b) Pair housing for caged female longtailed and rhesus macaques: Behavior in protected contact versus full contact. J Appl Anim Welf Sci 15 (2):126–143. https://doi.org/10.1080/10888705.2012.658330

Baker KC, Bloomsmith MA, Oettinger BC et al (2014) Comparing options for pair housing rhesus macaques using behavioral welfare measures. Am J Primatol 76(1):30–42. https://doi.org/10.1002/ajp.22190

Balcombe JP, Barnard ND, Sandusky C (2004) Laboratory routines cause animal stress. J Am Assoc Lab Anim Sci 43(6):42–51

Barros M, Tomaz C (2002) Non-human primate models for investigating fear and anxiety. Neurosci Biobehav Rev 26(2):187–201. https://doi.org/10.1016/S0149-7634(01)00064-1

Baumeister AA, Rollings JP (1976) Self-injurious behavior. In: Ellis NR (ed) International review of research in mental retardation, vol 8. Academic Press, New York, pp 1–34

Bayne K (1989) Nylon balls revisited. Lab Primate Newsl 28:5–6

Bayne K, McCully C (1989) The effect of cage size on the behavior of individually housed rhesus monkeys. Lab Anim (NY) 18(1):25–28

Bayne K, Novak MA (1998) Behavioral disorders. In: Bennett B, Abee C, Henrickson R (eds) Nonhuman primates in biomedical research. Academic Press, New York, pp 485–500

Bayne K, Mainzer H, Dexter S et al (1991) The reduction of abnormal behaviors in individually housed rhesus monkeys (*Macaca mulatta*) with a foraging/grooming board. Am J Primatol 23(1):23–35. https://doi.org/10.1002/ajp.1350230104

Bayne K, Dexter S, Mainzer H et al (1992a) The use of artificial turf as a foraging substrate for individually housed rhesus monkeys (*Macaca mulatta*). Anim Welf 1(1):39–53

Bayne K, Dexter S, Suomi SJ (1992b) A preliminary survey of the incidence of abnormal behavior in rhesus monkeys (*Macaca mulatta*) relative to housing condition. Lab Anim (NY) 21(5):38–46

Bellanca R, Crockett CM (2002) Factors predicting increased incidence of abnormal behavior in male pigtailed macaques. Am J Primatol 58(2):57–69. https://doi.org/10.1002/ajp.10052

Belzung C, Le Pape G (1994) Comparison of different behavioral test situations used in psychopharmacology for measurement of anxiety. Physiol Behav 56(3):623–628. https://doi.org/10.1016/0031-9384(94)90311-5

Bennett AJ, Lesch KP, Heils A et al (2002) Early experience and serotonin transporter gene variation interact to influence primate CNS function. Mol Psychiatry 7:118–122. https://doi.org/10.1038/sj.mp.4000949

Berkson G (1983) Repetitive stereotyped behaviors. Am J Ment Defic 88(3):239–246

Berkson G, Goodrich J, Kraft I (1966) Abnormal stereotyped movements of marmosets. Percept Mot Skills 23(2):491–498. https://psycnet.apa.org/doi/10.2466/pms.1966.23.2.491

Bertolani P, Pruetz JD (2011) Seed reingestion in Savannah chimpanzees (*Pan troglodytes verus*) at Fongoli, Senegal. Int J Primatol 32:1123–1132. https://doi.org/10.1007/s10764-011-9528-5

Bethell EJ, Pfefferle D (2022) Cognitive bias tasks: a new set of approaches to assess welfare in nonhuman primates. In: Robinson LM, Weiss A (eds) Nonhuman primate welfare: from history, science, and ethics to practice. Springer, Cham, pp 207–230

Birkett LP, Newton-Fisher NE (2011) How abnormal is the behaviour of captive, zoo-living chimpanzees? PLoS One 6(6):e20101. https://doi.org/10.1371/journal.pone.0020101

Blaney EC, Wells DL (2004) The influence of a camouflage net barrier on the behaviour, welfare, and public perceptions of zoo-housed gorillas. Anim Welf 13(2):111–118

Blois-Heulin C, Rochais C, Camus SMJ et al (2015) Animal welfare: Could adult play be a false friend? Anim Behav Cogn 2(2):156–185. https://doi.org/10.12966/abc.05.04.2015

Bloomsmith MA, Lambeth SP (1995) Effects of predictable versus unpredictable feeding schedules on chimpanzee behavior. Appl Anim Behav Sci 44(1):65–74. https://doi.org/10.1016/0168-1591(95)00570-I

Bloomsmith MA, Finlay TW, Merhalski JJ, Maple TL (1990) Rigid plastic balls as enrichment devices for captive chimpanzees. Lab Anim Sci 40(3):319–322

Bloomstrand M, Riddle K, Alford P, Maple TL (1986) Objective evaluation of a behavioral enrichment device for captive chimpanzees (*Pan troglodytes*). Zoo Biol 5(3):293–300. https://doi.org/10.1002/zoo.1430050307

Boccia ML, Reite M, Laudenslager M (1989) On the physiology of grooming in a pigtail macaque. Physiol Behav 45(3):667–670. https://doi.org/10.1016/0031-9384(89)90089-9

Boinski S, Swing SP, Gross TS, Davis JK (1999) Environmental enrichment of brown capuchins (*Cebus apella*): Behavioral and plasma and fecal cortisol measures of effectiveness. Am J Primatol 48(1):49–68. https://doi.org/10.1002/(SICI)1098-2345(1999)48:1%3C49::AID-AJP4%3E3.0.CO;2-6

Boissy A, Manteuffel G, Jensen MB et al (2007) Assessment of positive emotions in animals to improve their welfare. Physiol Behav 92(3):375–397. https://doi.org/10.1016/j.physbeh.2007.02.003

Bollen KS, Novak MA (2000) A survey of abnormal behavior in captive zoo primates. Am J Primatol 51(S1):47

Bonnie KE, Ang MYL, Ross SR (2016) Effects of crowd size on exhibit use by and behavior of chimpanzees (*Pan troglodytes*) and Western lowland gorillas (*Gorilla gorilla*) at a zoo. Appl Anim Behav Sci 178:102–119. https://doi.org/10.1016/j.applanim.2016.03.003

Bourgeois SR, Vazquez M, Brasky K (2007) Combination therapy reduces self-injurious behavior in a chimpanzee (*Pan troglodytes troglodytes*): a case report. J Appl Anim Welf Sci 10 (2):123–140. https://doi.org/10.1080/10888700701313454

Brambell Committee (1965) Report of the technical committee to enquire into the welfare of animals kept under intensive livestock husbandry systems. London

Brent L, Belik M (1997) The response of group-housed baboons to three enrichment toys. Lab Anim 31(1):81–85. https://doi.org/10.1258/002367797780600305

Brent L, Eichberg JW (1991) Primate puzzleboard: a simple environmental enrichment device for captive chimpanzees. Zoo Biol 10(4):353–360. https://doi.org/10.1002/zoo.1430100409

Brent L, Hughes A (1997) The occurrence of abnormal behavior in group-housed baboons. Am J Primatol 42(2):96–97

Brent L, Lee DR, Eichberg JW (1989) Evaluation of two environment enrichment devices for singly caged chimpanzees (*Pan troglodytes*). Am J Primatol 19(S1):65–70

Broom DM, Johnson KG (1993) Stress and animal welfare. Springer, Dordrecht

Byrne GD, Suomi SJ (1991) Effects of woodchips and buried food on behavior patterns and psychological well-being of captive rhesus monkeys. Am J Primatol 23(3):141–151. https://doi.org/10.1002/ajp.1350230302

Cagni P, Gonçalves I Jr, Ziller F et al (2009) Humans and natural predators induce different fear/anxiety reactions and response pattern to diazepam in marmoset monkeys. Pharmacol Biochem Behav 93(2):134–140. https://doi.org/10.1016/j.pbb.2009.04.020

Camus SMJ, Blois-Heulin C, Li Q et al (2013) Behavioural profiles in captive-bred cynomolgus macaques: Towards monkey models of mental disorders? PLoS One 8(4):e62141. https://doi.org/10.1371/journal.pone.0062141

Capitanio JP, Blozis SA, Snarr J et al (2017) Do "birds of a feather flock together" or do "opposites attract"? Behavioral responses and temperament predict success in pairings of rhesus monkeys in a laboratory setting. Am J Primatol 79(1):e22464. https://doi.org/10.1002/ajp.22464

Capitanio JP, Vandeleest J, Hannibal DL (2022) Physiological measures of welfare. In: Robinson LM, Weiss A (eds) Nonhuman primate welfare: from history, science, and ethics to practice. Springer, Cham, pp 231–254

Carder G, Semple S (2008) Visitor effects on anxiety in two captive groups of western lowland gorillas. Appl Anim Behav Sci 115(3–4):211–220. https://doi.org/10.1016/j.applanim.2008.06.001

Carlstead K (1996) Effects of captivity on the behavior of wild mammals. In: Kleiman D (ed) Wild mammals in captivity: principles and techniques. University of Chicago Press, Chicago, pp 317–333

Carpenter CR (1934) A field study of the behavior and social relationships of howling monkeys. J Mammal 15(4):324–336. https://doi.org/10.2307/1374520

Castles DL, Whiten A (1998) Post-conflict behaviour of wild olive baboons. II. Stress and self-directed behaviour. Ethology 104(2):148–160. https://doi.org/10.1111/j.1439-0310.1998.tb00058.x

Chamove AS, Anderson JR, Morgan-jones SC, Jones SP (1982) Deep woodchip litter: hygiene, feeding, and behavioral enhancement in eight primate species. Int J Study Anim Probl 3 (4):308–318

Chamove AS, Anderson JR, Nash VJ (1984) Social and environmental influences on self-aggression in monkeys. Primates 25:319–325. https://doi.org/10.1007/BF02382270

Chamove AS, Hosey GR, Schaetzel P (1988) Visitors excite primates in zoos. Zoo Biol 7 (4):359–369

Chelluri GI, Ross SR, Wagner KE (2013) Behavioral correlates and welfare implications of informal interactions between caretakers and zoo-housed chimpanzees and gorillas. Appl Anim Behav Sci 147(3–4):306–315. https://doi.org/10.1016/j.applanim.2012.06.008

Chen G-L, Novak MA, Meyer JS et al (2010) TPH2 5′- and 3′-regulatory polymorphisms are differentially associated with HPA axis function and self-injurious behavior in rhesus monkeys. Genes Brain Behav 9(3):335–347. https://doi.org/10.1111/j.1601-183X.2010.00564.x

Clark FE (2011) Space to choose: network analysis of social preferences in a captive chimpanzee community, and implications for management. Am J Primatol 73(8):748–757. https://doi.org/10.1002/ajp.20903

Clarke AS, Mason WA, Moberg GP (1988) Differentiall behavioral and adrenocortical responses to stress among three macaque species. Am J Primatol 14(1):37–52. https://doi.org/10.1002/ajp.1350140104

Coleman K, Pierre PJ (2014) Assessing anxiety in nonhuman primates. ILAR J 55(2):333–346. https://doi.org/10.1093/ilar/ilu019

Coleman K, Dahl RE, Ryan ND, Cameron JL (2003) Growth hormone response to growth hormone-releasing hormone and clonidine in young monkeys: correlation with behavioral characteristics. J Child Adolesc Psychopharmacol 13(3):227–241. https://doi.org/10.1089/104454603322572561

Coleman K, Robertson ND, Bethea CL (2011) Long-term ovariectomy alters social and anxious behaviors in semi-free ranging Japanese macaques. Behav Brain Res 225(1):317–327. https://doi.org/10.1016/j.bbr.2011.07.046

Coleman K, Timmel G, Prongay K, Baker KC (2022) Common husbandry, housing, and animal care practices. In: Robinson LM, Weiss A (eds) Nonhuman primate welfare: from history, science, and ethics to practice. Springer, Cham, pp 317–348

Conti G, Hansman C, Heckman JJ et al (2012) Primate evidence on the late health effects of early-life adversity. Proc Natl Acad Sci U S A 109(23):8866–8871. https://doi.org/10.1073/pnas.1205340109

Coplan JD, Trost RC, Owens MJ et al (1998) Cerebrospinal fluid concentrations of somatostatin and biogenic amines in grown primates reared by mothers exposed to manipulated foraging conditions. Arch Gen Psychiatry 55(5):473. https://doi.org/10.1001/archpsyc.55.5.473

Corcoran CA, Pierre PJ, Haddad T et al (2012) Long-term effects of differential early rearing in rhesus macaques: behavioral reactivity in adulthood. Dev Psychobiol 54(5):546–555. https://doi.org/10.1002/dev.20613

Costall B, Domeney AM, Farre AJ et al (1992) Profile of action of a novel 5-hydroxytryptamine (1A) receptor ligand E-4424 to inhibit aversive behavior in the mouse, rat and marmoset. J Pharmacol Exp Ther 262(1):90–98

Crast J, Bloomsmith MA, Perlman J et al (2014) Abnormal behaviour in captive sooty mangabeys. Anim Welf 23(2):167–177. https://doi.org/10.7120/09627286.23.2.167

Crawley JN (1985) Exploratory behavior models of anxiety in mice. Neurosci Biobehav Rev 9 (1):37–44. https://doi.org/10.1016/0149-7634(85)90030-2

Crockett CM, Bowden DM (1994) Challenging conventional wisdom for housing monkeys. Lab Anim (NY):29–33

Crockett CM, Gough GM (2002) Onset of aggressive toy biting by a laboratory baboon coincides with cessation of self-injurious behavior. Am J Primatol 57(S1):39

Crockett CM, Bowers CL, Shimoji M et al (1995) Behavioral responses of longtailed macaques to different cage sizes and common laboratory experiences. J Comp Psychol 109(4):368–383. https://doi.org/10.1037/0735-7036.109.4.368

Crockett CM, Sackett GP, Sandman CA et al (2007) Beta-endorphin levels in longtailed and pigtailed macaques vary by abnormal behavior rating and sex. Peptides 28(10):1987–1997. https://doi.org/10.1016/j.peptides.2007.07.014

Cross HA, Harlow HF (1965) Prolonged and progressive effects of partial isolation on the behavior of macaque monkeys. J Exp Res Personal 1(1):39–49

Davenport RK (1963) Stereotyped behavior of the infant chimpanzee. Arch Gen Psychiatry 8(1):99. https://doi.org/10.1001/archpsyc.1963.01720070101013

Davenport MD, Lutz CK, Tiefenbacher S et al (2008) A rhesus monkey model of self-injury: effects of relocation stress on behavior and neuroendocrine function. Biol Psychiatry 63(10):990–996. https://doi.org/10.1016/j.biopsych.2007.10.025

Davis N, Schaffner CM, Smith TE (2005) Evidence that zoo visitors influence HPA activity in spider monkeys (*Ateles geoffroyii rufiventris*). Appl Anim Behav Sci 90(2):131–141. https://doi.org/10.1016/j.applanim.2004.08.020

Desmond T, Laule G, McNary J (1987) Training to enhance socialization and reproduction in drills. In: American Zoo and Aquarium Association (AZA) regional conference proceedings. Wheeling, WV, pp 352–358

Dettmer AM, Novak MA, Suomi SJ, Meyer JS (2012) Physiological and behavioral adaptation to relocation stress in differentially reared rhesus monkeys: hair cortisol as a biomarker for anxiety-related responses. Psychoneuroendocrinology 37(2):191–199. https://doi.org/10.1016/j.psyneuen.2011.06.003

Dienske H, Griffin R (1978) Abnormal behaviour patterns developing in chimpanzee infants during nursery care: A note. J Child Psychol Psychiatry 19(4):387–391. https://doi.org/10.1111/j.1469-7610.1978.tb00485.x

Doyle LA, Baker KC, Cox LD (2008) Physiological and behavioral effects of social introduction on adult male rhesus macaques. Am J Primatol 70(6):542–550. https://doi.org/10.1002/ajp.20526

Draper WA, Bernstein IS (1963) Stereotyped behavior and cage size. Percept Mot Skills 16(1):231–234. https://doi.org/10.2466/pms.1963.16.1.231

Eaton GG, Worlein JM, Kelley ST et al (1999) Self-injurious behavior Is decreased by cyproterone acetate in adult male rhesus (*Macaca mulatta*). Horm Behav 35(2):195–203. https://doi.org/10.1006/hbeh.1999.1513

Erwin J, Deni R (1979) Strangers in a strange land: abnormal behaviors or abnormal environment? In: Erwin J, Maple TL, Mitchell G (eds) Captivity and behavior: primates in breeding colonies, laboratories, and zoos. Van Nostrand Reinhold, New York, pp 1–28

European Association of Zoos and Aquaria (2016) Animal welfare. http://www.eaza.net/about–us/animal–welfare/. Retrieved 9/22/2016

Fagot J, Gullstrand J, Kemp C et al (2014) Effects of freely accessible computerized test systems on the spontaneous behaviors and stress level of Guinea baboons (*Papio papio*). Am J Primatol 76(1):56–64. https://doi.org/10.1002/ajp.22193

Fekete JM, Norcross JL, Newman JD (2000) Artificial turf foraging boards as environmental enrichment for pair-housed female squirrel monkeys. Contemp Top Lab Anim Sci 39(2):22–26

Fittinghoff NA, Lindburg DG, Gomber J, Mitchell G (1974) Consistency and variability in the behavior of mature, isolation-reared, male rhesus macaques. Primates 15:111–139. https://doi.org/10.1007/BF01742276

Fontenot MB, Padgett EE, Dupuy AM et al (2005) The effects of fluoxetine and buspirone on self-injurious and stereotypic behavior in adult male rhesus macaques. Comp Med 55:67–74

Fontenot MB, Wilkes MN, Lynch CS (2006) Effects of outdoor housing on self-injurious and stereotypic behavior in adult male rhesus macaques (*Macaca mulatta*). J Am Assoc Lab Anim Sci 45(5):35–43

Frankenhaeuser M (1982) Challenge-control interaction as reflected in sympathetic-adrenal and pituitary-adrenal activity: comparison between the sexes. Scand J Psychol 23(S1):158–164. https://doi.org/10.1111/j.1467-9450.1982.tb00466.x

Freeman ZT, Rice KA, Soto PL et al (2015) Neurocognitive dysfunction and pharmacological intervention using guanfacine in a rhesus macaque model of self-injurious behavior. Transl Psychiatry 5:e567. https://doi.org/10.1038/tp.2015.61

Fritz J, Nash LT, Alford PL, Bowen JA (1992) Abnormal behaviors, with a special focus on rocking, and reproductive competence in a large sample of captive chimpanzees (*Pan troglodytes*). Am J Primatol 27(3):161–176. https://doi.org/10.1002/ajp.1350270302

Gartner MC (2022) Questionnaires and their use in primate welfare. In: Robinson LM, Weiss A (eds) Nonhuman primate welfare: from history, science, and ethics to practice. Springer, Cham, pp 255–266

Gilbert MH, Baker KC (2011) Social buffering in adult male rhesus macaques (*Macaca mulatta*): Effects of stressful events in single vs. pair housing. J Med Primatol 40(2):71–78. https://doi.org/10.1111/j.1600-0684.2010.00447.x

Gluck JP, Sackett GP (1974) Frustration and self-aggression in social isolate rhesus monkeys. J Abnorm Psychol 83(3):331–334. https://doi.org/10.1037/h0036584

Gonzales HK, O'Reilly M, Lang R et al (2016) Research involving anxiety in non-human primates has potential implications for the assessment and treatment of anxiety in autism spectrum disorder: a translational literature review. Dev Neurorehabil 19(3):175–192. https://doi.org/10.3109/17518423.2014.941117

Gottlieb DH, Capitanio JP, McCowan BJ (2013a) Risk factors for stereotypic behavior and self-biting in rhesus macaques (*Macaca mulatta*): animal's history, current environment, and personality. Am J Primatol 75(10):995–1008. https://doi.org/10.1002/ajp.22161

Gottlieb DH, Coleman K, McCowan BJ (2013b) The effects of predictability in daily husbandry routines on captive rhesus macaques (*Macaca mulatta*). Appl Anim Behav Sci 143(2–4):117–127. https://doi.org/10.1016/j.applanim.2012.10.010

Gottlieb DH, O'Connor JR, Coleman K (2014) Using porches to decrease feces painting in rhesus macaques (*Macaca mulatta*). J Am Assoc Lab Anim Sci 53(6):653–656

Gottlieb DH, Maier A, Coleman K (2015) Evaluation of environmental and intrinsic factors that contribute to stereotypic behavior in captive rhesus macaques (*Macaca mulatta*). Appl Anim Behav Sci 171:184–191. https://doi.org/10.1016/j.applanim.2015.08.005

Griffis CM, Martin AL, Perlman JE, Bloomsmith MA (2013) Play caging benefits the behavior of singly housed laboratory rhesus macaques (*Macaca mulatta*). J Am Assoc Lab Anim Sci 52(5):534–540

Hamel AF, Lutz CK, Coleman K et al (2017) Responses to the Human Intruder Test are related to hair cortisol phenotype and sex in rhesus macaques (*Macaca mulatta*). Am J Primatol 79(1): e22526. https://doi.org/10.1002/ajp.22526

Harlow HF, Harlow MK (1965) The effect of rearing conditions on behavior. Int J Psychiatry 1:43–51

Herrelko ES, Vick S-J, Buchanan-Smith HM (2012) Cognitive research in zoo-housed chimpanzees: influence of personality and impact on welfare. Am J Primatol 74(9):828–840. https://doi.org/10.1002/ajp.22036

Higley JD, Hasert MF, Suomi SJ, Linnoila MV (1991) Nonhuman primate model of alcohol abuse: effects of early experience, personality, and stress on alcohol consumption. Proc Natl Acad Sci U S A 88(16):7261–7265. https://doi.org/10.1073/pnas.88.16.7261

Hill DA (1994) Affiliative behaviour between adult males of the genus *Macaca*. Behaviour 130 (3–4):293–308. https://doi.org/10.1163/156853994X00578

Hook M, Lambeth SP, Perlman JE et al (2002) Inter-group variation in abnormal behavior in chimpanzees (*Pan troglodytes*) and rhesus macaques (*Macaca mulatta*). Appl Anim Behav Sci 76(2):165–176. https://doi.org/10.1016/S0168-1591(02)00005-9

Hopper LM, Freeman HD, Ross SR (2016) Reconsidering coprophagy as an indicator of negative welfare for captive chimpanzees. Appl Anim Behav Sci 176:112–119. https://doi.org/10.1016/j. applanim.2016.01.002

Hosey GR (2000) Zoo animals and their human audiences: What is the visitor effect? Anim Welf 9 (4):343–357

Hosey G (2022) The history of primates in zoos. In: Robinson LM, Weiss A (eds) Nonhuman primate welfare: from history, science, and ethics to practice. Springer, Cham, pp 3–30

Insel TR, Ninan PT, Aloi J et al (1984) A benzodiazepine receptor-mediated model of anxiety: studies in nonhuman primates and clinical implications. Arch Gen Psychiatry 41(8):741–750. https://doi.org/10.1001/archpsyc.1984.01790190015002

Itakura S (1993) Emotional behavior during the learning of a contingency task in a chimpanzee. Percept Mot Skills 76(2):563–566. https://doi.org/10.2466/pms.1993.76.2.563

Jacobson SL, Ross SR, Bloomsmith MA (2016) Characterizing abnormal behavior in a large population of zoo-housed chimpanzees: prevalence and potential influencing factors. PeerJ 4: e2225. https://doi.org/10.7717/peerj.2225

Jay LC (1965) The common langur of north India. In: Devore I (ed) Primate behavior: field studies of monkey and apes. Holt, Rinehart, and Winston, New York, pp 197–249

Kalin NH (2004) The role of the central nucleus of the amygdala in mediating fear and anxiety in the primate. J Neurosci 24(24):5506–5515. https://doi.org/10.1523/JNEUROSCI.0292-04.2004

Kalin N, Shelton S (1989) Defensive behaviors in infant rhesus monkeys: Environmental cues and neurochemical regulation. Science 243(4899):1718–1721. https://doi.org/10.1126/science. 2564702

Kalin NH, Shelton SE (2003) Nonhuman primate models to study anxiety, emotion regulation, and psychopathology. Ann N Y Acad Sci 1008(1):189–200. https://doi.org/10.1196/annals.1301. 021

Kalin NH, Shelton SE, Rickman M, Davidson RJ (1998) Individual differences in freezing and cortisol in infant and mother rhesus monkeys. Behav Neurosci 112(1):251–254. https://doi.org/ 10.1037/0735-7044.112.1.251

Karere GM, Kinnally EL, Sanchez JN et al (2009) What is an "adverse" environment? Interactions of rearing experiences and MAOA genotype in rhesus monkeys. Biol Psychiatry 65 (9):770–777. https://doi.org/10.1016/j.biopsych.2008.11.004

Kato Y, Gokan H, Oh-Nishi A et al (2014) Vocalizations associated with anxiety and fear in the common marmoset (*Callithrix jacchus*). Behav Brain Res 275:43–52. https://doi.org/10.1016/j. bbr.2014.08.047

Kaufman BM, Pouliot AL, Tiefenbacher S, Novak MA (2004) Short and long-term effects of a substantial change in cage size on individually housed, adult male rhesus monkeys (*Macaca mulatta*). Appl Anim Behav Sci 88(3–4):319–330. https://doi.org/10.1016/j.applanim.2004.03. 012

Kempf DJ, Baker KC, Gilbert MH et al (2012) Effects of extended-release injectable naltrexone on self-injurious behavior in rhesus macaques (*Macaca mulatta*). Comp Med 62(3):1–9

Kessel AL, Brent L (1998) Cage toys reduce abnormal behavior in individually housed pigtail macaques. J Appl Anim Welf Sci 1(3):227–234. https://doi.org/10.1207/s15327604jaws0103_3

Kinnally EL, Karere GM, Lyons LA et al (2010) Serotonin pathway gene–gene and gene–environment interactions influence behavioral stress response in infant rhesus macaques. Dev Psychopathol 22(1):35–44. https://doi.org/10.1017/S0954579409990241

Kitchen AM, Martin AA (1996) The effects of cage size and complexity on the behaviour of captive common marmosets, *Callithrix jacchus jacchus*. Lab Anim 30(4):317–326. https://doi.org/10. 1258/002367796780739853

Kraemer GW, Clarke AS (1990) The behavioral neurobiology of self-injurious behavior in rhesus monkeys. Prog Neuro-Psychopharmacol Biol Psychiatry 14(S1):S141–S168. https://doi.org/10. 1016/0278-5846(90)90092-U

Kraemer GW, Ebert MH, Schmidt DE, McKinney WT (1989) A longitudinal study of the effect of different social rearing conditions on cerebrospinal fluid norepinephrine and biogenic amine metabolites in rhesus monkeys. Neuropsychopharmacology 2(3):175–189. https://doi.org/10.1016/0893-133X(89)90021-3

Krief S, Jamart A, Hladik C-M (2004) On the possible adaptive value of coprophagy in free-ranging chimpanzees. Primates 45:141–145. https://doi.org/10.1007/s10329-003-0074-4

Lam K, Rupniak NM, Iversen SD (1991) Use of a grooming and foraging substrate to reduce cage stereotypies in macaques. J Med Primatol 20(3):104–109

Leavens DA, Aureli F, Hopkins WD, Hyatt CW (2001) Effects of cognitive challenge on self-directed behaviors by chimpanzees (*Pan troglodytes*). Am J Primatol 55(1):1–14. https://doi.org/10.1002/ajp.1034

Leavens DA, Hopkins WD, Aureli F (2004) Behavioral evidence for the cutaneous expression of emotion in a chimpanzee (*Pan Troglodytes*). Behaviour 141(8):979–997. https://doi.org/10.1163/1568539042360189

Leeds A, Elsner R, Lukas K (2016) The effect of positive reinforcement training on an adult female western lowland gorilla's (*Gorilla gorilla gorilla*) rate of abnormal and aggressive behavior. Anim Behav Cogn 3(2):78–87. https://doi.org/10.12966/abc.02.05.2016

Less E, Kuhar C, Lukas K (2013) Assessing the prevalence and characteristics of hair-plucking behaviour in captive western lowland gorillas (*Gorilla gorilla gorilla*). Anim Welf 22 (2):175–183. https://doi.org/10.7120/09627286.22.2.175

Levison CA (1979) The development of head banging in a young rhesus monkey. In: Keeh JD (ed) Origins of madness. Pergamon Press, Oxford, pp 343–349

Line SW, Morgan KN, Markowitz H, Strong S (1990) Increased cage size does not alter heart rate or behavior in female rhesus monkeys. Am J Primatol 20(2):107–113. https://doi.org/10.1002/ajp.1350200205

Line SW, Morgan KN, Markowitz H (1991) Simple toys do not alter the behavior of aged rhesus monkeys. Zoo Biol 10(6):473–484. https://doi.org/10.1002/zoo.1430100606

Loseth GE, Ellingsen D-M, Leknes S (2014) State-dependent μ-opioid modulation of social motivation. Front Behav Neurosci 8:430. https://doi.org/10.3389/fnbeh.2014.00430

Lutz CK (2018) A cross-species comparison of abnormal behavior in three species of singly-housed old world monkeys. Appl Anim Behav Sci 199:52–58. https://doi.org/10.1016/j.applanim.2017.10.010

Lutz CK, Farrow RA (1996) Foraging device for singly housed longtailed macaques does not reduce stereotypies. Contemp Top Lab Anim Sci 35(3):75–78

Lutz CK, Novak MA (2005) Environmental enrichment for nonhuman primates: theory and application. ILAR J 46(2):178–191. https://doi.org/10.1093/ilar.46.2.178

Lutz CK, Well A, Novak MA (2003) Stereotypic and self-injurious behavior in rhesus macaques: a survey and retrospective analysis of environment and early experience. Am J Primatol 60 (1):1–15. https://doi.org/10.1002/ajp.10075

Lutz CK, Davis EB, Ruggiero AM, Suomi SJ (2007) Early predictors of self-biting in socially-housed rhesus macaques (*Macaca mulatta*). Am J Primatol 69(5):584–590. https://doi.org/10.1002/ajp.20370

Lutz CK, Williams PC, Sharp RM (2014) Abnormal behavior and associated risk factors in captive baboons (*Papio hamadryas spp.*). Am J Primatol 76(4):355–361. https://doi.org/10.1002/ajp.22239

Macy J, Beattie TA, Morgenstern SE, Arnsten AFT (2000) Use of guanfacine to control self-injurious behavior in two rhesus macaques (*Macaca mulatta*) and one baboon (*Papio anubis*). Comp Med 50(4):419–425

Maestripieri D, Martel FL, Nevison CM et al (1991) Anxiety in rhesus monkey infants in relation to interactions with their mother and other social companions. Dev Psychobiol 24(8):571–581. https://doi.org/10.1002/dev.420240805

Maestripieri D, Schino G, Aureli F, Troisi A (1992) A modest proposal: displacement activities as an indicator of emotions in primates. Anim Behav 44(5):967–979. https://doi.org/10.1016/S0003-3472(05)80592-5

Major CA, Kelly BJ, Novak MA et al (2009) The anxiogenic drug FG7142 increases self-injurious behavior in male rhesus monkeys (*Macaca mulatta*). Life Sci 85(21–22):753–758. https://doi.org/10.1016/j.lfs.2009.10.003

Maki S, Alford P, Bloomsmith MA, Franklin J (1989) Food puzzle device simulating termite fishing for captive chimpanzees (*Pan troglodytes*). Am J Primatol 19(S1):71–78

Maki S, Fritz J, England N (1993) An assessment of early differential rearing conditions on later behavioral development in captive chimpanzees. Infant Behav Dev 16(3):373–381. https://doi.org/10.1016/0163-6383(93)80042-7

Mallapur A, Sinha A, Waran N (2005) Influence of visitor presence on the behaviour of captive lion-railed macaques (*Macaca silenus*) housed in Indian zoos. Appl Anim Behav Sci 94 (3–4):341–352. https://doi.org/10.1016/j.applanim.2005.02.012

Markowitz H, Spinelli JS (1986) Environmental engineering for primates. In: Benirschke K (ed) Primates: the road to self-sustaining populations. Springer, New York, pp 489–498

Marriner LM, Drickamer LC (1994) Factors influencing stereotyped behavior of primates in a zoo. Zoo Biol 13(3):267–275. https://doi.org/10.1002/zoo.1430130308

Mason GJ (1991) Stereotypies: a critical review. Anim Behav 41(6):1015–1037. https://doi.org/10.1016/S0003-3472(05)80640-2

Mason GJ (1993) Forms of stereotypic behaviour. In: Lawrence AB, Rushen J (eds) Stereotypic animal behaviour: Fundamentals and applications to welfare. CAB International, Wallingford, pp 7–40

Mason GJ, Latham NR (2004) Can't stop, won't stop: Is stereotypy a reliable animal welfare indicator? Anim Welf 13(Supplement):S57–S69

Mason GJ, Turner MA (1993) Mechanisms involved in the development and control of stereotypies. In: Bateson PPG et al (eds) Perspectives in ethology, Behavior and evolution, vol 10. Plenum, New York, pp 53–85

McClintick MN, Grant KA (2016) Aggressive temperament predicts ethanol self-administration in late adolescent male and female rhesus macaques. Psychopharmacology 233:3965–3976. https://doi.org/10.1007/s00213-016-4427-2

McCormack K, Newman TK, Higley JD et al (2009) Serotonin transporter gene variation, infant abuse, and responsiveness to stress in rhesus macaque mothers and infants. Horm Behav 55 (4):538–547. https://doi.org/10.1016/j.yhbeh.2009.01.009

McCoy JG, Fontenot MB, Hanbury DB et al (2009) L-tryptophan and correlates of self- injurious behavior in small-eared bushbabies (*Otolemur garnettii*). J Am Assoc Lab Anim Sci 48(2):185–191

McKenzie-Quirk SD, Miczek KA (2008) Social rank and social separation as determinants of alcohol drinking in squirrel monkeys. Psychopharmacology 201:137–145. https://doi.org/10.1007/s00213-008-1256-y

McKinney WT, Moran EC, Kraemer GW, Prange AJ (1980) Long-term chlorpromazine in rhesus monkeys: production of dyskinesias and changes in social behavior. Psychopharmacology 72:35–39. https://doi.org/10.1007/BF00433805

Miczek KA, Weerts EM, Vivian JA, Barros HM (1995) Aggression, anxiety and vocalizations in animals: GABAA and 5-HT anxiolytics. Psychopharmacology 121:38–56. https://doi.org/10.1007/BF02245590

Moberg GP (1985) Biological response to stress: key to assessment of animal well-being? In: Moberg GP (ed) Animal stress. American Physiological Society, Bethesda, MD, pp 27–49

Morgan L, Howell SM, Fritz J (1993) Regurgitation and reingestion in a captive chimpanzee (*Pan troglodytes*). Lab Anim (NY) 22:42–45

Nash LT, Fritz J, Alford PA, Brent L (1999) Variables influencing the origins of diverse abnormal behaviors in a large sample of captive chimpanzees (*Pan troglodytes*). Am J Primatol 48 (1):15–29. https://doi.org/10.1002/(SICI)1098-2345(1999)48:1<15::AID-AJP2>3.0.CO;2-R

Neal SJ, Caine NG (2016) Scratching under positive and negative arousal in common marmosets (*Callithrix jacchus*). Am J Primatol 78(2):216–226. https://doi.org/10.1002/ajp.22498

Nieuwenhuijsen K, de Waal FBM (1982) Effects of spatial crowding on social behavior in a chimpanzee colony. Zoo Biol 1(1):5–28. https://doi.org/10.1002/zoo.1430010103

Ninan P, Insel T, Cohen R et al (1982) Benzodiazepine receptor-mediated experimental "anxiety" in primates. Science 218(4579):1332–1334. https://doi.org/10.1126/science.6293059

Norscia I, Palagi E (2011) When play is a family business: adult play, hierarchy, and possible stress reduction in common marmosets. Primates 52:101–104. https://doi.org/10.1007/s10329-010-0228-0

Novak MA (2003) Self-injurious behavior in rhesus monkeys: new insights into its etiology, physiology, and treatment. Am J Primatol 59(1):3–19. https://doi.org/10.1002/ajp.10063

Novak MA, Suomi SJ (1988) Psychological well-being of primates in captivity. Am Psychol 43 (10):765–773. https://doi.org/10.1037/0003-066X.43.10.765

Novak MA, Suomi SJ (1991) Social interaction in nonhuman primates: an underlying theme for primate research. Lab Anim Sci 41(4):308–314

Novak MA, Kinsey JH, Jorgensen MJ, Hazen TJ (1998) Effects of puzzle feeders on pathological behavior in individually housed rhesus monkeys. Am J Primatol 46(3):213–227. https://doi.org/10.1002/(SICI)1098-2345(1998)46:3<213::AID-AJP3>3.0.CO;2-L

Novak MA, Crockett CM, Sackett GP (2002) Self-injurious behavior in captive macaque monkeys. In: Schroeder S, Oster-Granite M, Thomson T (eds) Self-injurious behavior gene-brain-behavior relationships. American Psychological Association, Washington, DC, pp 151–161

Novak MA, Kelly BJ, Bayne K, Meyer JS (2012) Behavioral disorders of nonhuman primates. In: Abee C, Mansfield K, Tardif S, Morris T (eds) Nonhuman primates in biomedical research: biology and management, vol 1. Elsevier, Waltham, MA, pp 177–196

O'Neill PL, Novak MA, Suomi SJ (1991) Normalizing laboratory-reared rhesus macaque (*Macaca mulatta*) behavior with exposure to complex outdoor enclosures. Zoo Biol 10(3):237–245. https://doi.org/10.1002/zoo.1430100307

Oler JA, Fox AS, Shelton SE et al (2009) Serotonin transporter availability in the amygdala and bed nucleus of the stria terminalis predicts anxious temperament and brain glucose metabolic activity. J Neurosci 29(32):9961–9966. https://doi.org/10.1523/JNEUROSCI.0795-09.2009

Palagi E, Paoli T, Tarli SB (2006) Short-term benefits of play behavior and conflict prevention in *Pan paniscus*. Int J Primatol 27:1257–1270. https://doi.org/10.1007/s10764-006-9071-y

Paquette D, Prescott J (1988) Use of novel objects to enhance environments of captive chimpanzees. Zoo Biol 7(1):15–23. https://doi.org/10.1002/zoo.1430070103

Paulk HH, Dienske H, Ribbens LG (1977) Abnormal behavior in relation to cage size in rhesus monkeys. J Abnorm Psychol 86(1):87–92. https://doi.org/10.1037/0021-843X.86.1.87

Pazol KA, Bloomsmith MA (1993) The development of stereotyped body rocking in chimpanzees (*Pan troglodytes*) reared in a variety of nursery settings. Anim Welf 2(2):113–129

Peel AJ, Vogelnest L, Finnigan M et al (2005) Non-invasive fecal hormone analysis and behavioral observations for monitoring stress responses in captive western lowland gorillas (*Gorilla gorilla gorilla*). Zoo Biol 24(5):431–445. https://doi.org/10.1002/zoo.20055

Peterson EJ, Worlein JM, Lee GH et al (2017) Rhesus macaques (*Macaca mulatta*) with self-injurious behavior show less behavioral anxiety during the human intruder test. Am J Primatol 79(1):e22569. https://doi.org/10.1002/ajp.22569

Pomerantz O, Terkel J (2009) Effects of positive reinforcement training techniques on the psychological welfare of zoo-housed chimpanzees (*Pan troglodytes*). Am J Primatol 71(8):687–695. https://doi.org/10.1002/ajp.20703

Pomerantz O, Paukner A, Terkel J (2012) Some stereotypic behaviors in rhesus macaques (*Macaca mulatta*) are correlated with both perseveration and the ability to cope with acute stressors. Behav Brain Res 230(1):274–280. https://doi.org/10.1016/j.bbr.2012.02.019

Raper JR, Bloomsmith MA, Stone A, Mayo L (2002) Use of positive reinforcement training to decrease stereotypic behaviors in a pair of orangutans (*Pongo pygmaeus*). Am J Primatol 57 (S1):70–71. https://doi.org/10.1002/ajp.1091

Roberts RL, Roytburd LA, Newman JD (1999) Puzzle feeders and gum feeders as environmental enrichment for common marmosets. Contemp Top Lab Anim Sci 38(5):27–31

Robinson LM, Weiss A (2022) Primate personality and welfare. In: Robinson LM, Weiss A (eds) Nonhuman primate welfare: from history, science, and ethics to practice. Springer, Cham, pp 387–402

Rogers J, Shelton SE, Shelledy W et al (2008) Genetic influences on behavioral inhibition and anxiety in juvenile rhesus macaques. Genes Brain Behav 7(4):463–469. https://doi.org/10.1111/j.1601-183X.2007.00381.x

Rommeck I, Gottlieb DH, Strand SC, McCowan B (2009) The effects of four nursery rearing strategies on infant behavioral development in rhesus macaques (*Macaca mulatta*). J Am Assoc Lab Anim Sci 48(4):395–401

Ross SR, Bloomsmith MA (2011) A comment on Birkett & Newton-Fisher. PLoS One 6:e20101

Ross SR, Wagner KE, Schapiro SJ, Hau J (2010) Ape behavior in two alternating environments: comparing exhibit and short-term holding areas. Am J Primatol 72(11):951–959. https://doi.org/10.1002/ajp.20857

Ruby S, Buchanan-Smith HM (2015) The effects of individual cubicle research on the social interactions and individual behavior of brown capuchin monkeys (*Sapajus apella*). Am J Primatol 77(10):1097–1108. https://doi.org/10.1002/ajp.22444

Rushen J (1984) Stereotyped behaviour, adjunctive drinking and the feeding periods of tethered sows. Anim Behav 32(4):1059–1067. https://doi.org/10.1016/S0003-3472(84)80222-5

Rushen J (1993) The "coping" hypothesis of stereotypic behaviour. Anim Behav 45(3):613–615. https://doi.org/10.1006/anbe.1993.1071

Ryder RD (1998) Measuring animal welfare. J Appl Anim Welf Sci 1(1):75–80. https://doi.org/10.1207/s15327604jaws0101_7

Sackett GP, Holm RA, Ruppenthal GC (1976) Social isolation rearing: species differences in behavior of macaque monkeys. Dev Psychol 12(4):283–288. https://doi.org/10.1037/0012-1649.12.4.283

Sakamaki T (2010) Coprophagy in wild bonobos (*Pan paniscus*) at Wamba in the Democratic Republic of the Congo: a possibly adaptive strategy? Primates 51:87–90. https://doi.org/10.1007/s10329-009-0167-9

Santangelo AM, Ito M, Shiba Y et al (2016) Novel primate model of serotonin transporter genetic polymorphisms associated with gene expression, anxiety, and sensitivity to antidepressants. Neuropsychopharmacology 41:2366–2376. https://doi.org/10.1038/npp.2016.41

Schapiro SJ, Bloomsmith MA (1995) Behavioral effects of enrichment on singly-housed, yearling rhesus monkeys: an analysis including three enrichment conditions and a control group. Am J Primatol 35(2):89–101. https://doi.org/10.1002/ajp.1350350202

Schino G, Troisi A (1992) Opiate receptor blockade in juvenile macaques: effect on affiliative interactions with their mothers and group companions. Brain Res 576(1):125–130. https://doi.org/10.1016/0006-8993(92)90617-I

Schino G, Troisi A, Perretta G, Monaco V (1991) Measuring anxiety in nonhuman primates: effect of lorazepam on macaque scratching. Pharmacol Biochem Behav 38(4):889–891. https://doi.org/10.1016/0091-3057(91)90258-4

Schino G, Perretta G, Taglioni AM et al (1996) Primate displacement activities as an ethopharmacological model of anxiety. Anxiety 2(4):186–191. https://doi.org/10.1002/(SICI)1522-7154(1996)2:4<186::AID-ANXI5>3.0.CO;2-M

Selye H (1976) Stress without distress. In: Serban G (ed) Psychopathology of human adaptation. Springer, Boston, MA, pp 137–146

Sherwen SL, Harvey TJ, Magrath MJ, Butler KL, Fanson KV, Hemsworth PH (2015) Effects of visual contact with zoo visitors on black-capped capuchin welfare. Appl Anim Behav Sci 167:65–73. https://doi.org/10.1016/j.applanim.2015.03.004

Shutt K, MacLarnon A, Heistermann M, Semple S (2007) Grooming in Barbary macaques: better to give than to receive? Biol Lett 3(3):231–233. https://doi.org/10.1098/rsbl.2007.0052

Snyder DS, Graham CE, Bowen JA, Reite M (1984) Peer separation in infant chimpanzees, a pilot study. Primates 25:78–88. https://doi.org/10.1007/BF02382297

Soussignan R, Koch P (1985) Rhythmical stereotypies (leg-swinging) associated with reductions in heart-rate in normal school children. Biol Psychol 21(3):161–167. https://doi.org/10.1016/0301-0511(85)90027-4

Southwick C, Beg M, Siddiqi M (1965) Rhesus monkeys in north India. In: DeVore I (ed) Primate behavior: field studies of monkey and apes. Holt, Rinehart and Winston, New York, pp 111–159

Spijkerman R, Dienske H, Van Hooff JARAM, Jens W (1994) Causes of body rocking in chimpanzees (*Pan troglodytes*). Anim Welf 3(3):193–211

Spinelli S, Schwandt ML, Lindell SG et al (2007) Association between the recombinant human serotonin transporter linked promoter region polymorphism and behavior in rhesus macaques during a separation paradigm. Dev Psychopathol 19(4):977–987. https://doi.org/10.1017/S095457940700048X

Stolba A, Wood-Gush DG (1984) The identification of behavioural key features and their incorporation into a housing design for pigs. Ann Rech Vet 15(2):287–302

Suomi SJ (1991) Early stress and adult emotional reactivity in rhesus monkeys. In: Ciba Foundation (ed) The childhood environment and adult disease. Wiley, Guildford, pp 171–188

Suomi SJ (2006) Risk, resilience, and gene x environment interactions in rhesus monkeys. Ann N Y Acad Sci 1094(1):52–62. https://doi.org/10.1196/annals.1376.006

Suomi SJ, Harlow HF, Kimball SD (1971) Behavioral effects of prolonged partial social isolation in the rhesus monkey. Psychol Rep 29(3_suppl):1171–1177. https://doi.org/10.2466/pr0.1971.29.3f.1171

Suomi SJ, Eisele CD, Grady SA, Harlow HF (1975) Depressive behavior in adult monkeys following separation from family environment. J Abnorm Psychol 84(5):576–578. https://doi.org/10.1037/h0077066

Tarou LR, Bloomsmith MA, Maple TL (2005) Survey of stereotypic behavior in prosimians. Am J Primatol 65(2):181–196. https://doi.org/10.1002/ajp.20107

Tiefenbacher S, Fahey MA, Rowlett JK et al (2005a) The efficacy of diazepam treatment for the management of acute wounding episodes in captive rhesus macaques. Comp Med 55(4):387–392

Tiefenbacher S, Newman TK, Davenport MD et al (2005b) The role of two serotonin pathway gene polymorphisms in self-injurious behavior in singly housed *Macaca mulatta*. Am J Primatol 66 (S1):91. https://doi.org/10.1002/ajp.20150

Tiefenbacher S, Novak MA, Lutz CK, Meyer JS (2005c) The physiology and neurochemistry of self-injurious behavior: a nonhuman primate model. Front Biosci 10(1):1–11. https://doi.org/10.2741/1500

Troisi A, Schino G (1987) Environmental and social influences on autogrooming behaviour in a captive group of Java monkeys. Behaviour 100(1–4):292–302. https://doi.org/10.1163/156853987X00161

Trollope J (1977) A preliminary survey of behavioural stereotypes in captive primates. Lab Anim 11(3):195–196. https://doi.org/10.1258/002367777780936666

Tustin GW, Williams LE, Brady AG (1996) Rotational use of a recreational cage for the environmental enrichment of Japanese macaques (*Macaca fuscata*). Lab Primate Newsl 35(1):5–7

U.S. Department of Agriculture (2013) Animal Welfare Act and Animal Welfare Regulations, Section 3.81–Environmental enhancement to promote psychological well-being. Animal Welfare Act and Animal Welfare Regulations ("Blue Book")

Ulyan MJ, Burrows AE, Buzzell CA et al (2006) The effects of predictable and unpredictable feeding schedules on the behavior and physiology of captive brown capuchins (*Cebus apella*). Appl Anim Behav Sci 101(1–2):154–160. https://doi.org/10.1016/j.applanim.2006.01.010

Van Hooff JARAM, Hill DA (1994) Affiliative relationships between males in groups of nonhuman primates: a summary. Behaviour 130(3–4):143–149. https://doi.org/10.1163/156853994X00497

Vandeleest JJ, McCowan BJ, Capitanio JP (2011) Early rearing interacts with temperament and housing to influence the risk for motor stereotypy in rhesus monkeys (*Macaca mulatta*). Appl Anim Behav Sci 132(1–2):81–89. https://doi.org/10.1016/j.applanim.2011.02.010

Veasey JS, Waran NK, Young RJ (1996) On comparing the behaviour of zoo housed animals with wild conspecifics as a welfare indicator. Anim Welf 5:13–24

Videan EN, Fritz J (2007) Effects of short- and long-term changes in spatial density on the social behavior of captive chimpanzees (*Pan troglodytes*). Appl Anim Behav Sci 102(1–2):95–105. https://doi.org/10.1016/j.applanim.2006.03.011

Wagner KE, Hopper LM, Ross SR (2016) Asymmetries in the production of self-directed behavior by chimpanzees and gorillas during a computerized cognitive test. Anim Cogn 19:343–350. https://doi.org/10.1007/s10071-015-0937-2

Waitt C, Buchanan-Smith HM (2001) What time is feeding? How delays and anticipation of feeding schedules affect stump-tailed macaque behavior. Appl Anim Behav Sci 75(1):75–85. https://doi.org/10.1016/S0168-1591(01)00174-5

Watson SL, Ward JP, Davis KB, Stavisky RC (1999) Scent-marking and cortisol response in the small-eared bushbaby (*Otolemur garnettii*). Physiol Behav 66(4):695–699. https://doi.org/10.1016/S0031-9384(99)00005-0

Weed JL, Wagner PO, Byrum R et al (2003) Treatment of persistent self-injurious behavior in rhesus monkeys through socialization: a preliminary report. Contemp Top Lab Anim Sci 42(5):21–23

Weld KP, Mench JA, Woodward RA et al (1998) Effect of tryptophan treatment on self-biting and central nervous system serotonin metabolism in rhesus monkeys (*Macaca mulatta*). Neuropsychopharmacology 19:314–321. https://doi.org/10.1016/S0893-133X(98)00026-8

Wells DL (2005) A note on the influence of visitors on the behaviour and welfare of zoo-housed gorillas. Appl Anim Behav Sci 93(1–2):13–17. https://doi.org/10.1016/j.applanim.2005.06.019

Whitehouse J, Micheletta J, Powell LE et al (2013) The impact of cognitive testing on the welfare of group housed primates. PLoS One 8(11):e78308. https://doi.org/10.1371/journal.pone.0078308

Wiepkema PR, Koolhaas JM (1993) Stress and animal welfare. Anim Welf 2(3):195–218

Willette AA, Lubach GR, Knickmeyer RC et al (2011) Brain enlargement and increased behavioral and cytokine reactivity in infant monkeys following acute prenatal endotoxemia. Behav Brain Res 219(1):108–115. https://doi.org/10.1016/j.bbr.2010.12.023

Yamanashi Y, Matsuzawa T (2010) Emotional consequences when chimpanzees (*Pan troglodytes*) face challenges: Individual differences in self-directed behaviours during cognitive tasks. Anim Welf 19:25–30

Cognitive Bias Tasks: A New Set of Approaches to Assess Welfare in Nonhuman Primates

Emily J. Bethell and Dana Pfefferle

Abstract

At the start of the new millennium, the "cognitive bias" paradigm emerged as a new approach to assessing animal emotion. In the animal welfare literature, cognitive bias describes how emotions such as anxiety and depression are associated with changes in the way the brain processes information. For example, studies with humans have long demonstrated that anxious people are more vigilant for negative cues and depressed people interpret the proverbial glass of water as "half empty" rather than "half full." In this chapter, we review how methods developed to study cognitive bias in humans have been adapted to measure the interaction between emotion and cognition in nonhuman primates. We focus on judgment bias and attention bias tasks and discuss study design, controls, confounds, and advantages and limitations of each. We also indicate future research directions. This chapter is intended to introduce readers with little or no experience of cognitive bias tasks to theory and practical considerations around designing these tasks.

E. J. Bethell (✉)
Research Centre in Brain and Behaviour, School of Psychology, Liverpool John Moores University, Liverpool, UK

Research Centre for Biological Anthropology, School of Biological and Environmental Sciences, Liverpool John Moores University, Liverpool, UK
e-mail: E.J.Bethell@ljmu.ac.uk

D. Pfefferle
Welfare and Cognition Group, Cognitive Neuroscience Laboratory, German Primate Center, Göttingen, Germany

Leibniz-ScienceCampus Primate Cognition, German Primate Center, Göttingen, Germany
e-mail: dpfefferle@dpz.eu

Keywords

Cognitive bias · Judgment bias · Attention bias · Nonhuman primate · Monkey · Welfare assessment

1 The Potential of Cognitive Bias Tasks for Assessing Nonhuman Primate Welfare

Evaluating the psychological component of welfare ("psychological well-being") in animals who cannot tell us directly about how they feel is a key challenge in animal welfare research (Russell and Burch 1959; Dawkins 2017). Welfare researchers commonly acknowledge that it is justifiable on ethical grounds to discuss the potential for subjective experience in other species (e.g., Goodall 1986; de Waal 2006; Broom 2010; Panksepp 2011). However, scientific debate about the capacity of animals to suffer has been limited by lack of quantifiable measures. Recent developments in animal welfare science, borrowing theory and methods from the field of human cognitive psychology, have resulted in the development of novel measures of underlying emotional and cognitive processes, specifically, cognitive biases (Harding et al. 2004). In humans, cognitive biases can be measured objectively using experimental tasks in the laboratory. Studies have shown that cognitive biases may occur outside of awareness and are reliably associated with people's self-reported (subjectively experienced) feelings (Mendl and Paul 2004; Mathews and MacLeod 2005; Evans 2008). If these cognitive bias measures in humans are associated with subjective experience, can cognitive bias methods provide a window into the subjective worlds of other species?

Cognitive bias, in its broadest formulation, describes a tendency for the brain to process information in a way that deviates from a presumed norm (Tversky and Kahneman 1974). Applied to human and animal welfare, cognitive bias describes the way in which changes in mood are associated with changes in cognitive processes, i.e., what individuals look at, what they think about it, and how they remember it (MacLeod et al. 1986; Mathews 1990; Williams et al. 1996; Bradley et al. 1998; Clark 1999; Harding et al. 2004; Mendl et al. 2009; Bethell et al. 2012a). This rapidly expanding field of research is grounded in evolutionary theory, draws on earlier work in classical conditioning, and incorporates more recent methodological developments from the biological and cognitive sciences (Mendl and Paul 2004). Current evidence shows that basic emotions (e.g., fear, anger, lust) are common mechanisms that are essential for survival and reproduction across animal taxa (LeDoux 1996; Panksepp 1998; Lang et al. 2000). Emotions are now considered central to a scientific understanding of human and animal cognition and behavior; cognition cannot be studied without considering emotion, and vice versa (LeDoux 1996; Crump et al. 2020; Lang et al. 2000; Seligman and Csikszentmihalyi 2000).

When terms are adopted from one literature to another, their definitions can change. Used in the human literature, the term cognitive bias covers a broad range of contexts that do not necessarily imply a role of emotion. Tversky and Kahneman (1974), for example, discussed cognitive biases in terms of how human brains in general processes information, separate from (and without mention of) emotions. In the animal welfare literature, the term is used to mean "emotion-mediated cognitive bias." From herein, we use the term cognitive bias broadly to cover a number of experimental paradigms that assess the relationship between emotion and cognition.

Cognitive bias methods present a valuable approach to assessing animal welfare because we already know so much about how cognitive biases manifest in humans and their relationship with subjective well-being. In clinical studies, people suffering with anxiety or depression exhibit different thought patterns to people who are not anxious or depressed (Tversky and Kahneman 1974; Williams et al. 1996; Bar-Haim et al. 2007; Yiend 2010). Anxious people are vigilant for negative information (e.g., they are faster at detecting an angry face in an array of smiling faces and look at such faces for longer) compared with nonanxious people (Bar-Haim et al. 2007). Depressed people tend to interpret ambiguous information more negatively and to recall more negative past events than do nondepressed people (Mathews and MacLeod 2005). Importantly, people are often not aware of emotional responses to stimuli shown outside of awareness (e.g., when shown stimuli for such a short amount of time that they cannot say what was shown), although they may demonstrate a shift in their cognitive bias when tested on automated tasks and report a general shift in their feeling state (Evans 2008). The relationship between emotions and cognitive processes is therefore bidirectional. In essence, having negative thoughts induces negative emotions that, in turn, induce negative thoughts, and this becomes a feedback loop colloquially referred to as the "downward spiral" of anxiety or depression.

The potential of cognitive bias tasks for assessing nonhuman primate welfare is promising, but in these early days of the application of these methods, it is important to ensure studies are well designed, definitions are clear, and that we take heed of lessons from the human literature. Here, we present a summary of the current methods used to study cognitive bias in nonhuman primates with a focus on study design. We discuss limitations and indicate future research directions.

2 Cognitive Bias Tasks Explained: Design and Application

The adaptation of cognitive bias methods developed with humans for use with other species has been rapid (Harding et al. 2004; Mendl et al. 2009; Panksepp 2014; Bethell 2015; Dawkins 2017), and meta-analytical studies are now becoming possible (e.g., Neville et al. 2020). In the animal welfare literature, the terms "cognitive bias" and "judgment bias" have been used interchangeably as a consequence of the use of the term cognitive bias in the first study presenting what has now been termed the judgment bias task (Harding et al. 2004). In this chapter, the experimental paradigms we discuss under the umbrella term cognitive bias are judgment bias tasks and attention bias tasks.

3 Judgment Bias

3.1 What Is It?

Judgment bias describes the way in which changes in emotional state, or mood, are associated with changes in judgments individuals make about ambiguous information (Mathews and MacLeod 2005). Consider a day when you are in a good mood and compare it to a day when you are in a bad mood. On your "good mood" day, you are more likely to make positive judgments about otherwise ambiguous events (an odd comment from a work colleague or email from a stranger) than you would on a "bad mood" day. The judgment bias paradigm tests for similar changes in judgments about ambiguous events in animals. In a recent review (Bethell 2015), 64 studies of judgment bias across taxa were identified, overall highlighting the suitability of the paradigm for use across species and contexts.

3.2 Method: The Basics

For the judgment bias task, simple operant conditioning is used to train the subjects that one cue (e.g., a particular color such as black) signals a reward if they make response "A," while a different cue (e.g., color white) signals low-reward, nonreward or punishment if, or unless, the subject performs response "B." We therefore have two responses, "A" and "B," which reflect whether our subject expects reward or not. Typically, response "A" is a "Go" response (e.g., approach to touch the black cue) and response "B" is a "No Go" (e.g., do not approach or touch the white cue). Cues intermediate in form to the learned cues (e.g., intermediate shades of gray) are then presented. To assess whether animals have a more positive or negative judgment bias, we calculate the proportion of "A" responses (e.g., "Go") made to the intermediate cues. A greater proportion of "A" responses indicates a more positive judgment bias.

3.3 Examples of Application for Measuring Primate Welfare

Since Harding et al.'s (2004) seminal paper that described the development of the judgment bias task with rats, the judgment bias task has been adapted to assess welfare in a range of nonhuman primates (marmosets: Gordon and Rogers 2015; capuchins: Pomerantz et al. 2012; Schino et al. 2016; macaques: Bethell et al. 2012a; McGuire et al. 2017 and chimpanzees, Bateson and Nettle 2015). We describe them here as examples of the different experimental designs and applications.

Currently, two designs of the judgment bias task have been used with nonhuman primates, i.e., the "Go-No Go" discrimination task and the "active choice" task. The "Go-No Go" discrimination task is the classic judgment bias paradigm in which one cue is presented on each trial. By contrast, the active choice task requires participants to choose one of two cues presented at the same time. Cues may vary in characteristics such as size, color, or location. Suitable cue types should be determined by pilot work as to which cues a given species can discriminate with good accuracy (we recommend >80%).

(a) The Classic Task: "Go-No Go"—Size Cues
The first study to apply the judgment bias paradigm to nonhuman primates was by Bethell et al. (2012a). In this study, we created a visual analog of the original auditory judgment bias "Go-No Go" task by Harding et al. (2004; Fig. 1). We first trained 12 adult male rhesus macaques (*Macaca mulatta*), who were naïve to cognitive testing, to touch a line presented on a touch screen to gain a reward. This required shaping monkeys' responses from initially reaching toward the screen to touching the exact location of a stimulus shown on the screen. First, we habituated monkeys to the apparatus by covering an old (sacrificial) computer monitor with honey and leaves and allowing monkeys to feed freely on these food items. Touches anywhere on the screen were rewarded with a secondary reinforcing tone and delivery of a food pellet into a tray in front of the monkey, both of which were manually triggered by the experimenter sitting in an adjacent room and watching progress via a live video link. Once a monkey was reliably retrieving leaves from the monitor, we connected the monitor to a laptop and introduced a white square that appeared at random locations on the screen. When the monkey now touched the screen to retrieve a leaf around the location of the white square, the reinforcing tone and pellet were delivered. Over successive days, depending on each monkey's rate of progress, we reduced the amount of honey and leaves until none was required for a monkey to touch the screen, and only touches at the exact location of the white square were rewarded. Likewise, we reduced the duration the white square remained on the screen from several minutes to several seconds. Touches were reinforced on 100% of touches with the tone and on 30–50% of touches with a food pellet. Once the monkey was touching the white square on the monitor for pellets, we introduced the automated touch screen and, following a period of habituation to the new screen, the monkey was ready for cognitive bias testing.

Once monkeys had learned to work with the touch screen, we began training on the judgment bias discrimination task. During training, monkeys were shown a single yellow line on the screen for 2 s on each trial, for 62 trials per day. Two training stimuli were used: a long line and a short line (Fig. 1a, b). For three monkeys, the rewarded stimulus was the long line and the unrewarded stimulus was the short line. For four monkeys, this was reversed. Touching the rewarded line within 2 s resulted in receiving the tone (on all trials) and two small food pellets (on 40% of trials). Touching the unrewarded line resulted in a mildly punishing 16 s delay to the next trial. Not touching either the rewarded or the unrewarded line did not result in any feedback, and after 5–6 s, the next trial started automatically. Seven out of 12 monkeys reached our training criterion (70% correct responses for "Go" and "No Go" trials) taking between 19 and 43 daily training sessions to learn the judgment bias task. All monkeys who

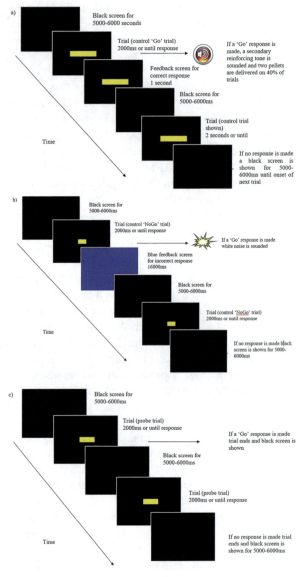

Fig. 1 A visual analog of the classic "Go-No Go" judgment bias task, adapted for use with nonhuman primates. (**a**) The experimental procedure for S+ ("Go") trials, showing the long line as S+ and outcomes following a response (correct), and no response (incorrect). (**b**) The experimental procedure for "No Go" control trials, showing the short line as the S- and outcomes following a response (incorrect), and no response (correct). (**c**) The experimental procedure for ambiguous probe trials, showing a line of intermediate length to the S+ and S, and outcomes following a response and no response (neither response rewarded nor punished). Figure adapted with permission from Bethell (2012a)

learned the task were under 7.5 years of age (range 3.6–7.4 years). The monkeys who failed to learn the "Go-No Go" task discrimination were mostly older individuals (range 9.9–25 years).

Our aim was to test the validity of the judgment bias task as a measure of emotional stress, and so we tested monkeys on presumably low-stress and high-stress days. We opted to run our high-stress testing sessions in the week during which monkeys underwent their quarterly veterinary health checks. Low-stress sessions were run during a quiet week without the presence of the veterinarian or any other disruptions to the usual daily routine, and with some additional enrichment.

During testing sessions, monkeys were shown the same lines as in training (Fig. 1a), interspersed with trials on which ambiguous probes—lines of intermediate length—were shown (Fig. 1b). We predicted that during "low-stress" weeks monkeys would make more positive judgments about the ambiguous probes, and this would result in more touches to the probes. During high-stress weeks, we predicted monkeys would make more negative judgments, leading to fewer touches to the probes. Responses were as predicted. We interpreted this as monkeys becoming more pessimistic about ambiguous cues when they had recently been visited by the veterinarian.

(b) The Classic Task: "Go-No Go"—Color Cues

A subsequent study using the "Go-No Go" discrimination task shows how the task can be applied using cues of different colors, as well as assessing personality effects on judgment bias. Gordon and Rogers (2015) trained common marmosets (*Callithrix jacchus*) to lift lids on bowls to access food rewards. During training, one cue was a bowl with a black lid and the other a bowl with a white lid. Marmosets learned that the bowl with one color lid contained food, while the bowl with the other color lid was always empty. Each trial lasted 2 min, and marmosets completed 20 trials per day. It took 10–22 days for 12 marmosets to reach the 85% accuracy level for testing. During testing, marmosets were presented with a bowl with a gray lid and the proportion of trials on which the lid was lifted, and time taken, were recorded. Monkeys who were left-handed were less likely to remove the gray lid to inspect the ambiguous bowl than monkeys who were right-handed, irrespective of rate of learning, sex, or age. Retrospective assessment of colony records revealed that the left-handed monkeys received more aggression than did right-handed individuals, likely altering their emotional state, and possibly reflecting hemispheric differences in processing and behavior between left- and right-handed marmosets.

(c) Active Choice Task: "Go-No Go"—Various Cue Types

Active choice tasks require selecting one of two or more cues presented simultaneously. McGuire et al. (2017) published an "ambiguous cue task," which was designed to maintain the ambiguity of the probes over repeated presentations. The authors attempted to train three lowland gorillas (*Gorilla gorilla*) to discriminate between the conditions under which three stimuli had unique reward associations. Stimuli were colored geometric shapes (Fig. 2). For each gorilla, one stimulus was always rewarded when selected on a touch screen monitor ("P"), one stimulus was never rewarded when touched ("N"), and a third stimulus was rewarded when paired with "N" but not rewarded when paired with "P," and was therefore an ambiguous cue ("A") with a context-dependent reward value. Interpretation of ambiguity could therefore be measured by recording the proportion of touches to "A" when it was paired with novel shapes that had not previously been seen. An optimistic interpretation of "A" would be

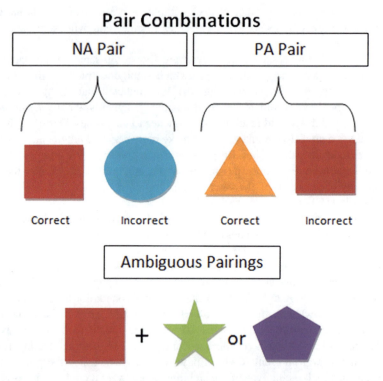

Fig. 2 Example of stimulus pair combinations used in the ambiguous cue paradigm. In this example, the red square is the rewarded stimulus ("Go") when paired with the blue circle (NA pair), but it becomes the nonrewarded ("No Go") stimulus when paired with the orange triangle (PA pair). On test trials, the red square is shown paired with previously unseen (i.e., ambiguous) colored shapes. Figure reproduced with permission from McGuire et al. (2017)

evident in a greater proportion of touches to "A," while a pessimistic interpretation of "A" would lead to gorillas selecting the novel shape. Test trials were always rewarded regardless of the selection made, presumably to maintain motivation on the task. Overall, the gorillas failed to adequately learn the initial ambiguous cue task. It was therefore not possible for the authors to test for changes in interpretation of the ambiguous cue following presumed manipulation of affective state. The test condition would have been a phase of enhanced forage enrichments vs. a phase with a standard forage baseline. This study highlights the possibility for innovative experimental designs that can address some of the limitations of existing tasks (e.g., habituation and learning of the intermediate probes), and may reflect variation between species in training success on complex tasks (see also Allritz et al. 2016 and Cronin et al. 2018).

(d) Active Choice Task: "Go-Go"—Size Cues

An alternative to the "Go-No Go" paradigm is the "Go-Go" paradigm in which individuals choose between two cues signaling higher and lower reward values (Fig. 3). Pomerantz et al. (2012) presented rectangular blocks of different

Fig. 3 Example of a "Go-Go" version of the judgment bias task. (**a**) The long rectangular block indicates a high-value reward (S++) under the dark lid, shown here on the left. (**b**) The short rectangular block indicates a lower value reward (S+) under the light lid, shown here on the right. (**c**) The block of intermediate length represents an ambiguous cue.
Figure reproduced with permission from Pomerantz et al. (2012)

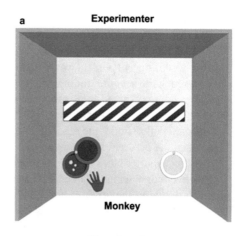
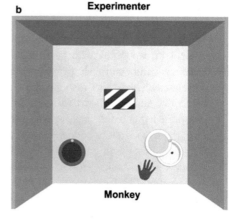
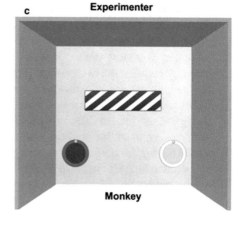

lengths to train tufted capuchins (*Sapajus apella*) on an active choice "Go-Go" discrimination task in which monkeys could gain preferred or less preferred (but edible) foods. Rectangular blocks were presented on a plastic board in front of two wells, one of which was covered with a black sliding lid and the other covered with a white sliding lid. In the "Go-Go" task, a line of one length indicated a presumed high-value reward (marshmallow) under one color of lid (Fig. 3a) and a line of another length indicated a reward of lower value (banana-flavored pellet) under the other color of lid (Fig. 3b). Twelve out of 16 monkeys learned the task and went on to the testing stage. Monkeys were then presented with rectangular blocks of intermediate length and indicated their more-or-less positive judgment about likely reward value by reaching toward either the high-reward lid or the low-reward lid (Fig. 3c). The researchers also recorded stereotypical head twirling and pacing. There was a significant negative correlation between rates of stereotypical head twirling and the probability of choosing the lid associated with the preferred reward when the ambiguous intermediate length line was presented. Monkeys who performed more head twirls were therefore considered to have a more negative judgment bias. No such bias was evident in capuchins who performed more stereotypical pacing, even though head twirling and pacing were themselves significantly and positively correlated. There was also no relationship between any of the stereotypical behaviors and likelihood to learn the initial discrimination task. The authors interpreted this result as suggesting that some forms of stereotypical behavior may more accurately reflect current negative mood than others and that the judgment bias paradigm may provide a means of distinguishing between these.

(e) Active Choice Task: "Go-Go"—Location

Schino et al. (2016) applied a spatial version of the "Go-Go" active choice task with 5 male and 8 female tufted capuchins. In this variant of the task, the presence of a preferred or nonpreferred reward in one of two differently colored cups placed next to each other (28 cm apart) was indicated by placing a rectangular block next to the rewarded cup. Once monkeys had learned the "Go-Go" discrimination task for the greater or lesser reward, they were presented with test trials. On test trials, the rectangular block was placed centrally between the two cups. An optimistic choice was indicated by reaching for the cup that had contained the preferred reward during training, and a less optimistic choice was indicated by reaching for the cup that had contained the less preferred reward. The researchers also recorded grooming behavior and dominance rank of the monkeys to test the hypothesis that a system of emotional 'bookkeeping' underlies the capacity of group-living animals to reciprocate cooperative interactions. Monkeys who received grooming prior to testing did not, as predicted, show a more optimistic judgment bias for the ambiguously placed rectangle. However, there was evidence for a more optimistic response to the central rectangle in dominant monkeys and those who received more grooming overall.

3.4 Advantages and Limitations

(a) Advantages

The handful of judgment bias studies performed with nonhuman primates to date suggest that the judgment bias task is adaptable for use with a range of primate species, across a number of contexts, and is sensitive to both short-term emotion and more stable trait characteristics. Enough studies have now been published across taxonomic groups ($n = 64$ in 2015, Bethell 2015) that meta-analyses are becoming viable. Neville et al. (2020) conducted the first meta-analysis of judgment bias studies, focusing only on studies in which a pharmacological manipulation had been used to alter affective state. While none of the 20 articles included in the analysis involved nonhuman primates, trends in task outcomes were identified across taxonomic groups commonly used in research that can be informative for designing and interpreting future primate studies. Further large-scale analyses will speed progression of research applying judgment bias tasks.

(b) Limitations

There are several caveats around design and application of the judgment bias task. Firstly, operant training usually leads to attrition of subjects; some individuals will fail to learn the task (as occurred in all the studies described above), subordinates may not gain access to the apparatus, and some individuals will lose interest once trained. Active choice tests may be at greater risk of attrition since the distinction between two levels of positive reward is more subtle than the distinction between reward and nonreward. In either case, this results in a self-selected sample in which we may be missing those individuals with poorest welfare. We are aware of several unpublished studies, from our own and others' groups, where researchers have attempted to measure judgment bias in nonhuman primates with inconclusive or variable results (as in McGuire et al. 2017, discussed above), either because of limitations on the time for initial discrimination training or because of attrition of subjects during the initial training stage (Katie Cronin, personal communication). Are animals failing to learn because they are already in a negative emotional state (Mendl 1999)? The well-documented publishing bias for significant results (Rosenthal 1979) means there may be much unpublished data showing no evidence for a bias, hampering efforts to understand why or under which conditions biases are not seen.

Secondly, judgment bias tests may be prone to habituation and learning effects. For example, Perdue (2017) trained rhesus macaques and tufted capuchin monkeys on a "Go-Go" judgment bias task. There was no manipulation of emotional state, but over time, rhesus macaques and capuchin monkeys became less likely to respond to the intermediate probes. It is not surprising that over repeated presentations, the ambiguity of probes is lost and learning then influences rate of responding. While the monkeys completed thousands of trials, more than run in most published studies, this paper highlights that the effects of repeated testing certainly need more consideration in judgment bias studies. It is therefore preferable to limit the number of times each probe is presented, for example, by increasing the number of different probes used, or to reduce learning with designs that maintain ambiguity using context-dependent contingencies (as in McGuire et al. 2017).

Thirdly, we have yet to fully test the extent to which the judgment bias paradigm does indeed measure biases in judgment about ambiguity rather than other processes (that are not necessarily mutually exclusive to judgment bias), such as biased attention, risk-taking behavior, and/or arousal. The "Go-No Go" tasks may be at greatest risk of these confounds since they rely on two types of response. A nonresponse is interpreted as reflecting a judgment of nonreward when it might instead reflect reduced arousal, distraction, confusion, or lost motivation. In the active choice task, nonresponses due to confounds should occur more equally for all trial types and therefore be less likely to skew interpretation in terms of judgments. Potential confounds are sometimes not controlled for in judgment bias studies. It is essential to check for, and report on, any changes in responses to the conditioned stimuli (S+ and S-, or S++ and S+) to identify the likelihood of confounds. For example, Bethell et al. (2012a) found no change in responses on S+ ("Go") and S- ("No Go") trials during testing sessions, adding support to the interpretation of change in responding to the probes in terms of judgments about their ambiguity. By contrast, Harding et al. (2004) found a reduction in "Go" responses to the S+ during their stress condition, leading to the possibility that the reduction in responding to the probes could have been partially driven by factors not related to ambiguity such as reduced motivation or arousal.

Fourthly, results are reported in categorical terms such as "optimism" and "pessimism," yet in the absence of a baseline, we are truly testing relative shifts in tendency to respond. There have not yet been enough studies with nonhuman primates to identify whether there are species-typical ranges of response, for example, or thresholds beyond which we can say an animal's welfare is compromised. Nor have there been enough to fully understand the sensitivity of the paradigm to variation in transient emotions compared with longer-term moods, within or between individuals. While the "Go-No Go" tasks may measure shifts in expectation of both negative and positive outcomes (e.g., Bethell and Koyama 2015 found shifts in response at the two probes closest to the conditioned stimuli but not the central probe), active choice paradigms may only be sensitive to shifts in expectation of reward, lacking sensitivity altogether to expectation of negative events (discussed in Bethell 2015).

In summary, judgment bias tasks are useful tools for testing hypotheses about the influence of emotion on judgments about ambiguous information, the likelihood of positive outcomes following responses, and identifying differences in optimistic/pessimistic interpretation between individual primates. They may not, however, be the most efficient approach for measuring emotions in real-world settings due to the need for initial training. Nevertheless, in the context of findings from the broader animal literature, the few studies with small numbers of nonhuman primates indicate that the judgment bias task does measure something that maps onto the presumed underlying emotional state. And for now, that is a good start.

4 Attention Bias

4.1 What Is It?

"Attention bias" describes how emotional state or mood can influence which cues in the environment individuals attend to (Mogg and Bradley 1999). For example, as noted earlier, in laboratory tests, anxious people are faster at detecting threatening faces in a crowd of neutral faces than are nonanxious people (Gilboa-Schechtman et al. 1999; Mogg and Bradley 1999). This bias for detecting threat is evident even when stimuli are presented subliminally, i.e., outside of awareness (Mogg and Bradley 1999). A characteristic of phobias in humans is an inability to stop attending to the target of the phobia (Gilboa-Schechtman et al. 1999), and this manifests in various coping strategies to avoid phobia-related cues, such as total avoidance (Chen and Clarke 2017). A basis of many meditation and therapeutic practices is the deliberate shifting of attention away from "negative" and toward "positive" thoughts, and the field of positive psychology is based on the premise that what individuals attend to can influence their sense of psychological well-being (Seligman and Csikszentmihalyi 2000).

Our research groups have been exploring the utility of attention bias tasks for assessing emotion in nonhuman primates in free-ranging, zoo, and laboratory settings. Attention bias tasks can require less training than judgment bias tasks, and the use of biologically relevant stimuli should reduce the likelihood of habituation and learning effects. In a recent review, Crump et al. (2018) identified 12 studies that used attention bias paradigms to measure emotion in animals, highlighting that this field is at an early stage of development.

4.2 Method: The Basics

Attention bias tasks, for the purpose of welfare assessment, can be split into two broad categories: those that do not require prior operant training and those that do. The first category comprises (a) the attention bias preferential-looking task (Calvo and Avero 2005; Garner et al. 2006; Bethell et al. 2012b). This is a simple paradigm in which the subject is shown two pictures that vary in valence (e.g., a negative vs. a neutral picture) side by side, and the direction and duration of gaze toward the pictures is filmed or tracked with an eye tracker. Attention bias is assessed as the difference in looking time toward the emotional compared to the neutral picture. For a recent review of the use of preferential-looking attention bias in animal welfare assessment, see Crump et al. (2018). The second category includes tasks in which subjects are required to make a manual response, the speed of which is used to assess biases in attention. These are (b) dot-probe tasks (MacLeod et al. 1986) and (c) visual search tasks (Öhman et al. 2001). In both paradigms, participants are instructed (in the case of humans) or rewarded (e.g., in nonhuman primates) for responding accurately and as quickly as possible to a cue when one or more emotional distractor cues have been shown. For a good review of dot-probe studies for assessing animal welfare, we recommend van Rooijen et al. (2017).

4.3 Examples of Application for Measuring Primate Welfare

(a) *Preferential looking*

Preferential-looking tasks have been used to identify cues of interest to nonhuman primates (e.g., Gerald et al. 2009; Parr 2011; Pfefferle et al. 2014; Parr et al. 2016). The first study to apply a preferential-looking task to explicitly test emotion-mediated attention bias in a nonhuman primate was conducted with adult male rhesus macaques (Bethell et al. 2012b). Seven monkeys were shown picture pairs of conspecific faces (one "threat" face and one "neutral" face) on two adjacent computer monitors and the direction and duration of looks toward the two face pictures was filmed. As with our test of judgment bias, each monkey was tested at baseline and after the statutory veterinary examination to provide a within-subjects comparison of presumed emotional state.

Monkeys were faster to direct initial gaze toward threat than neutral faces suggesting a general vigilance for threat irrespective of underlying emotional state. This finding is in line with earlier studies demonstrating a general enhanced attention to threatening or agonistic images and scenes in nonhuman primates (Parr and Hopkins 2000; Kano and Tomonaga 2010; Watson et al. 2012). Subsequent looking responses, however, were dependent on condition. During the baseline condition, monkeys continued to look toward the threat face, demonstrating sustained vigilance for threat. However, after the health check those same monkeys rapidly disengaged their gaze from threat faces and spent less time looking toward threat faces overall. We interpreted this as an avoidant attention bias in macaques when stressed. In a later study of juvenile macaques, conducted using the same stimulus pairs, Mandalaywala et al. (2014) found that offspring of protective and dominant mothers had an enhanced attention bias toward threat faces compared to offspring of less protective and more subordinate mothers. The preferential-looking paradigm shows great promise as a measure of attention bias as it requires limited training, is relatively simple to execute, and has been widely validated with nonhuman primates to assess the types of stimuli that capture and hold attention (e.g., Waitt et al. 2003; Waitt and Buchanan-Smith 2006; Watson et al. 2012).

(b) *Dot probe*

The dot-probe task provides another means of testing attention bias but requires prior operant training. Two stimuli (one "emotional" and the other "neutral") are presented simultaneously, usually on a touch screen. After a predefined period (which can range between 14 ms and 2000 ms depending on what point in the attention time course is of interest), the stimuli are removed from the screen and a probe (e.g., a circle, arrow, or other neutral cues) appears in the location of one of two previously shown stimuli. Each subject's task is to respond to the dot probe typically either by pressing a key on a keyboard (in the case of humans) or by touching the probe on the screen (in the case of nonhuman primates). Reaction time for detecting the dot probe will be faster at the attended location than at the unattended location. By comparing response time on trials in which the probe was at the location of the emotional stimulus (congruent trials) with reaction time at the neutral stimulus (incongruent trials), biases in attention toward or away from emotional stimuli can be detected. By manipulating the stimulus duration and the positive or negative value of the emotional stimulus, researchers can distinguish between early attention capture, dwell, and later avoidance of positive and negative information. Studies with humans have revealed that anxious people are generally faster to detect probes that appear at the location of faces with negative or threatening expression compared with probes that appear at the location of faces with neutral expression (MacLeod et al. 2002; Chen et al. 2002; Mogg et al. 2004; Bar-Haim et al. 2007; Beevers et al. 2007; Carlson et al. 2012; Hommer et al. 2014).

Several published studies have tested for attention bias to emotional stimuli in nonhuman primates using dot-probe tasks (King et al. 2012; Kret et al. 2016; Wilson and Tomonaga 2018). However, at the time of writing, no published studies have tested an effect of emotional state on attention bias using this paradigm. Here, we review the dot-probe studies published to date.

Kret et al. (2016) adapted the dot-probe task for use with four bonobos (*Pan paniscus*). Stimuli were pictures of bonobo social scenes or faces rated by four experts as emotional or neutral. In order to begin each trial, bonobos were required to touch a start cue that appeared at the center of a touch screen. A stimulus pair was then shown for 300 ms, one image on the left and one on the right of the screen, followed by the dot probe at the location of one of the two stimuli for 2 s. The dot probe was congruent to the location of the emotional stimulus on 50% of trials and occurred at the location of the neutral stimulus in the other 50% of trials. All four bonobos were significantly faster to touch the probe when it followed the location of the emotional stimulus than the neutral stimulus, with fastest responses at the location of high-intensity emotional stimuli. When the authors attempted to categorize the emotional stimuli in terms of valence, the results were less clear; bonobos showed greatest attention bias toward social stimuli containing grooming, sex, and yawning, but not aggression or distress. This study demonstrates that the dot-probe task can be adapted for use with bonobos, with further work on categorization of stimuli in terms of emotionality needed.

Wilson and Tomonaga (2018) trained eight chimpanzees (*Pan troglodytes*) on a dot-probe task. Stimuli were grayscale pictures of chimpanzee threat faces of high intensity (scream face) and lower intensity (bared teeth), paired either with chimpanzee faces with neutral expressions, or face images that had been scrambled so that they were no longer recognizable as faces but contained the same color information as the original pictures. Stimuli were shown on a touch screen monitor. To start a trial, the chimpanzee had to touch a start cue presented just below the center of the screen. Two stimuli were then shown for 150 ms, one stimulus at each of the left and right screen locations. Stimulus pairs were a threat face paired with a neutral face, or a threat face paired with a scrambled image. After 150 ms, the two stimuli disappeared from the screen and the dot probe appeared, congruent to the emotional stimulus on 50% of trials. Attention bias was determined as faster responses to the probe when it appeared at the location where the emotional face had previously been shown. No attention bias was evident for the threat-neutral face pairs. Chimpanzees were as fast to touch the probe when it appeared at the location of the neutral face as when it appeared at the location of the threat face. However, an attention bias was evident when threat faces were paired with scrambled images. Chimpanzees were faster to touch probes appearing at the location of the threat face than the scrambled face, indicating that they were looking toward the location of the threat faces when the probe appeared.

(c) Visual search

The visual search task has been widely applied to measure attention bias to threat in humans (Gilboa-Schechtman et al. 1999; Yiend 2010). Participants are asked to detect, as quickly as possible, the location of one "target" picture (e.g., a frowning face) in an array of other pictures (e.g., neutral faces). The speed with which the target picture is detected reveals how quickly and strongly the target pictured captured the viewer's attention. Anxious people are faster to detect a negative face in an array of neutral faces than are nonanxious controls (Gilboa-Schechtman et al. 1999; Yiend 2010).

The visual search task has been adapted for use with nonhuman primates (Öhman and Mineka 2003; Kawai and Koda 2016; Kawai et al. 2016), although it has never been applied to assess emotional states. These studies reveal that monkeys are faster to detect snakes in arrays of flowers, herbivorous mammals, and spiders than vice versa (Shibasaki and Kawai 2009; Kawai and Koda 2016), and to detect conspecific threat faces among an array of neutral faces (Kawai et al. 2016).

Shibasaki and Kawai (2009) trained three Japanese macaques (*Macaca fuscata*) on a visual search task using a touch screen. Monkeys were first trained with color patches in which they learned to touch the "odd one out" in order to receive a small food reward. Stimulus arrays were either 2×2 or 3×3. Once the basic task was learned, pictures of snakes and flowers were introduced to test for differential ability to locate a fear-relevant stimulus in an array of nonfear stimuli and vice versa. Monkeys were required to touch a rectangular white start cue at the center of the screen to start each trial. After 500 ms, an array of four or nine pictures appeared and stayed on the screen until the monkey touched one of the pictures. A correct touch to the odd one out resulted in a small food reward. Overall monkeys were faster to find snakes in flower arrays than vice versa, both for pictures in color and grayscale. This effect was seen only when snakes were the fear-relevant stimulus and not when pictures of spiders were used (Kawai and Koda 2016). The same research group later tested another three Japanese macaques on a visual search task using conspecific threat and neutral faces (Kawai et al. 2016). Macaques were faster to find the threat face in the neutral array than vice versa. These results suggest that fear-relevant stimuli, like snakes and aggressive faces, capture attention in nonhuman primates. The extent to which this is mediated by emotional state, as it does in humans, remains to be tested.

4.4 Advantages and Limitations

(a) *Advantages*

Advantages of attention bias tasks are simpler training protocols and fewer trials overall compared to the judgment bias task, due to the use of biologically relevant stimuli. The preferential-looking task requires no training beyond that required to habituate animals to the researcher and apparatus. For example, Mandalaywala et al. (2013) recorded attention bias for threat-neutral face pictures in free-ranging monkeys at the Cayo Santiago field station simply by approaching animals that were resting and presenting the pictures while filming which pictures the macaque looked at. An advantage of the dot-probe task is that the actual dot probe has nothing to do with the preceding emotional stimuli that are used to induce attention bias. Further, it can provide information about the time course of attention. Adjustment of duration for which stimuli are shown on the screen can be used to assess attention bias at earlier stages of attention (e.g., to subliminally presented stimuli) separate from later stages of attention (e.g., maintenance or disengagement of attention). Attention bias tasks can be less at risk of attrition of subjects, learning effects, confounding factors, such as arousal, or interpretation in terms of categories of emotion (since responses are relative time looking at either stimulus in the preferential-looking task), compared with judgment bias tasks.

(b) *Limitations*

There have been a limited number of published studies of attention bias, and the sample sizes are small. For the preferential-looking paradigm, our experience is that monkeys show the most pronounced looking on the first trial, and steadily lose interest in subsequent trials. Therefore, running only one trial at a time may work best (e.g., Mandalaywala et al. 2014). By contrast, the dot-probe task, due to the need to make responses to receive rewards, is more engaging and therefore suitable for repeated trials within a session. Coding trials from video can be time-consuming, while eye trackers are expensive, often difficult to calibrate, and are only available in a few primate facilities. Dot-probe tasks require automated systems, which are available in a limited number of facilities being that they are relatively expensive to set up and maintain.

There are well-documented visual field (hemispheric laterality) effects for processing emotional information in both humans (Sato and Aoki 2006) and nonhuman primates (Lindell 2013). We recommend including location as a control variable when analyzing data so potential laterality effects can be accounted for. For the dot-probe task, it is essential to counterbalance the side on which the emotional face is shown, as well as congruency of the subsequent probe location.

A challenge is to accurately categorize the emotional content of stimuli and acknowledge that the perception of emotional content may vary between individuals. In a recent study (Bethell et al. 2019b), we tested the influence of fearful temperament on reaction times to touch different categories of stimuli in adult male rhesus macaques. Stimuli were categorized as those we predicted to be negative in meaning (conspecific male faces with direct stare, pictures of husbandry items including a veterinary glove, brush and net, and pictures of a human wearing a mask). Monkeys with the most fearful temperament showed the slowest responses to the conspecific faces. There was some limited slowing of response to pictures of husbandry items and no slowing of response to the masked human. It is important to understand how individual monkeys perceive the stimuli that we classify as threatening.

Creating adequate stimulus sets can be time-consuming and challenging. Obtaining clear face pictures of the same individual nonhuman primate displaying different facial expressions is difficult. Stimuli must be equated for properties such as color, luminance, and contrast energy (Waitt and Buchanan-Smith 2006). This is more easily done with pictures in grayscale than color, but color information is then lost, which may reduce salience. Open-access online resources like the Macaque Stimulus Set (Witham and Bethell 2019) and the Facial Stimuli—Macaques Sets (Pfefferle 2020) provide a platform for researchers across fields to share stimuli, as well as to download stimuli used in published studies. For example, the stimuli used in Bethell et al. (2012b, 2016, 2019a), Mandalaywala et al. (2013), and Pfefferle et al. (2014) can be downloaded from these online resources. Shared resources enhance quality of science by allowing for more concise replication of studies across contexts.

Adequate controls are essential for interpreting results. The dot-probe study by Wilson and Tomonaga (2018) is a good example of how adequate controls are needed to interpret results. In that study, Wilson and Tomonaga (2018) found no evidence for an attention bias to chimpanzee threat faces that were paired with neutral chimpanzee faces but did find an attention bias for threat faces paired with a scrambled stimulus. Without a neutral-scrambled pair comparison, it is not clear from the results whether the attention bias seen in threat-scrambled trials was driven by the threat value of the face or the presence of social information. In control trials that took place during the same study, the chimpanzees were also faster to detect probes following a picture of a chair when paired with a scrambled stimulus. The latter is an example of a good control insofar as it provides information about how to interpret data. In this case, chimpanzees showed an attention bias for threat faces and chairs, relative to scrambled images. We cannot interpret the bias for threat faces solely in terms of attention bias for threat without also considering the chimpanzees found chairs threatening; it is as likely that pictures of faces and objects are simply more interesting to look at than abstract scrambled images. The lack of attention bias for threat faces when paired with neutral faces could indicate the absence of an attention bias for threat faces in those chimpanzees, or the use of stimuli not perceived by the chimpanzees to vary significantly in their threatening or neutral value to elicit a bias in looking. Without adequate controls, we are limited in what we can conclude from any data set.

5 Take Home and Considerations for Study Design

We encourage facilities housing nonhuman primates to consider how they can incorporate cognitive bias testing into their welfare assessment protocols. There is not a one-size-fits-all design for measuring cognitive bias, but all designs must be fit for purpose with adequate controls and consideration around interpretation of results.

1. Can all or most animals be trained on an operant task? If not, simpler attention bias tasks may be best.
2. What discrimination tasks can they learn? Ability to discriminate between two stimuli is required for judgments bias tasks ("Go-No Go" and "Go-Go" tasks). If not, simpler attention bias tasks require learning to touch one stimulus (as in dot-probe tests) and "the odd one out" in visual search.
3. What stimuli to use? Stimuli need to be species-appropriate and, where they are intended to be valenced, to be perceived as such by subjects.
4. What controls are needed? Do you have a baseline against which to compare responses? In the judgment bias task, check for any change in "Go" or "No Go" responses to the conditioned stimuli. In attention bias tasks, control for visual field and basic stimulus characteristics such as color, luminance, and contrast energy that might bias looking.
5. How many trials to run? For judgment bias, tasks limit presentations of the probes to maintain ambiguity (e.g., by having multiple probe forms presented at low frequency among presentations of the conditioned stimuli). In the preferential-looking task, macaques have been shown to lose interest in face images after the first trial. In dot-probe and visual search tasks, nonhuman primates will perform many more trials due to rewards, but initial training is required.
6. What reward contingency to use? In the judgment bias task, using a variable reward ratio (e.g., 40% in Bethell et al. 2012a) may reduce the likelihood, or speed, of learning that probe trials are not rewarded. Consider that using punishers in "Go-No Go" designs may confound the effects of any positive mood manipulations. In this case, a "Go-Go" task may be more suitable. For attention bias tasks, use any reward contingency that keeps subjects working.
7. How to interpret data? With caution and in consideration with other measures of welfare. Currently, there is no categorical measure of emotional valence (e.g., optimist/pessimist, stressed/unstressed) that can be extracted from cognitive bias data. However, building up pictures of individual profiles can help identify changes in animals over time, or distinguish between more vs. less optimistic individuals in a group. As with many health measures, changes may be more informative than absolute values.

Acknowledgments We thank the funders of our research: NC3Rs (grant#: NC/L000539/1), European Cooperation in Science and Technology (COST) primTRAIN program (http://www.cost.eu/COST_Actions/ca/CA15131,COSTSTSMCA15131-36153), Leibniz-Science Campus Primate Cognition (https://www.primate-cognition.eu/en/funding-measures.html), German Research Foundation (https://www.dfg.de/) Research Unit 2591, the Primate Society of Great Britain, and the Universities Federation for Animal Welfare. We thank all of our collaborators, the Caribbean Primate Research Centre, and MRC Harwell Centre for Macaques for supporting our work. We also thank the editors and reviewers for their thoughtful comments on an earlier draft of this chapter. This chapter is dedicated to the memory of Jaak Panksepp and Corri Waitt.

References

Allritz M, Call J, Borkenau P (2016) How chimpanzees (*Pan troglodytes*) perform in a modified emotional Stroop task. Anim Cogn 19:435–449. https://doi.org/10.1007/s10071-015-0944-3

Bar-Haim Y, Lamy D, Pergamin L et al (2007) Threat-related attentional bias in anxious and nonanxious individuals: a meta-analytic study. Psychol Bull 133(1):1–24. https://doi.org/10.1037/0033-2909.133.1.1

Bateson M, Nettle D (2015) Development of a cognitive bias methodology for measuring low mood in chimpanzees. PeerJ 3:e998. https://doi.org/10.7717/peerj.998

Beevers CG, Gibb BE, McGeary JE, Miller IW (2007) Serotonin transporter genetic variation and biased attention for emotional word stimuli among psychiatric inpatients. J Abnorm Psychol 116(1):208–212. https://doi.org/10.1037/0021-843X.116.1.208

Bethell EJ (2015) A "how-to" guide for designing judgment bias studies to assess captive animal welfare. J Appl Anim Welf Sci 18(S1):S18–S42. https://doi.org/10.1080/10888705.2015.1075833

Bethell EJ, Koyama NF (2015) Happy hamsters? Enrichment induces positive judgement bias for mildly (but not truly) ambiguous cues to reward and punishment in *Mesocricetus auratus*. R Soc Open Sci 2(7):140399. https://doi.org/10.1098/rsos.140399

Bethell EJ, Holmes A, MacLarnon A, Semple S (2012a) Cognitive bias in a non-human primate: husbandry procedures influence cognitive indicators of psychological well-being in captive rhesus macaques. Anim Welf 21(2):185–195. https://doi.org/10.7120/09627286.21.2.185

Bethell EJ, Holmes A, MacLarnon A, Semple S (2012b) Evidence that emotion mediates social attention in rhesus macaques. PLoS One 7(8):e44387. https://doi.org/10.1371/journal.pone.0044387

Bethell EJ, Holmes A, MacLarnon A, Semple S (2016) Emotion evaluation and response slowing in a non-human primate: new directions for cognitive bias measures of animal emotion? Behav Sci 6(1):2. https://doi.org/10.3390/bs6010002

Bethell EJ, Kemp C, Thatcher H et al (2019a) Heritability and maternal effects on social attention during an attention bias task in a non-human primate, *Macaca mulatta*. EcoEvoRxiv Prepr. https://doi.org/10.32942/osf.io/5nzd4

Bethell EJ, Cassidy LC, Brockhausen RR, Pfefferle D (2019b) Toward a standardized test of fearful temperament in primates: a sensitive alternative to the human intruder task for laboratory-housed rhesus macaques (*Macaca mulatta*). Front Psychol 10:1051. https://doi.org/10.3389/fpsyg.2019.01051

Bradley BP, Mogg K, Falla SJ, Hamilton LR (1998) Attentional bias for threatening facial expressions in anxiety: manipulation of stimulus duration. Cogn Emot 12(6):737–753. https://doi.org/10.1080/026999398379411

Broom DM (2010) Cognitive ability and awareness in domestic animals and decisions about obligations to animals. Appl Anim Behav Sci 126(1–2):1–11. https://doi.org/10.1016/j.applanim.2010.05.001

Calvo MG, Avero P (2005) Time course of attentional bias to emotional scenes in anxiety: Gaze direction and duration. Cogn Emot 19(3):433–451. https://doi.org/10.1080/02699930441000157

Carlson JM, Mujica-Parodi LR, Harmon-Jones E, Hajcak G (2012) The orienting of spatial attention to backward masked fearful faces is associated with variation in the serotonin transporter gene. Emotion 12(2):203–207. https://doi.org/10.1037/a0025170

Chen NTM, Clarke PJF (2017) Gaze-based assessments of vigilance and avoidance in social anxiety: a review. Curr Psychiatry Rep 19:59. https://doi.org/10.1007/s11920-017-0808-4

Chen Y, Ehlers A, Clark D, Mansell W (2002) Patients with generalized social phobia direct their attention away from faces. Behav Res Ther 40(6):677–687. https://doi.org/10.1016/S0005-7967(01)00086-9

Clark DM (1999) Anxiety disorders: why they persist and how to treat them. Behav Res Ther 37(S1):S5–S27. https://doi.org/10.1016/S0005-7967(99)00048-0

Cronin KA, Bethell EJ, Jacobson SL, Egelkamp C, Hopper LM, Ross SR (2018) Evaluating mood changes in response to anthropogenic noise with a response-slowing task in three species of zoo-housed primates. Anim Behav Cogn 5(2):209–221. https://doi.org/10.26451/abc.05.02.03.2018

Crump A, Arnott G, Bethell E (2018) Affect-driven attention biases as animal welfare indicators: review and methods. Animals 8(8):136. https://doi.org/10.3390/ani8080136

Crump A, Bethell EJ, Earley R, Lee VE, Mendl M, Oldham L, Turner SP, Arnott G (2020) Emotion in animal contests. Proc R Soc B Biol Sci 287(1939):20201715. https://doi.org/10.1098/rspb.2020.1715

Dawkins MS (2017) Animal welfare with and without consciousness. J Zool 301(1):1–10. https://doi.org/10.1111/jzo.12434

de Waal FBM (2006) Primates and philosophers: how morality evolved. Princeton, Princeton Science Library

Evans JSBT (2008) Dual-processing accounts of reasoning, judgment, and social cognition. Annu Rev Psychol 59:255–278. https://doi.org/10.1146/annurev.psych.59.103006.093629

Garner M, Mogg K, Bradley BP (2006) Orienting and maintenance of gaze to facial expressions in social anxiety. J Abnorm Psychol 115(4):760–770. https://doi.org/10.1037/0021-843X.115.4.760

Gerald MS, Waitt C, Little AC (2009) Pregnancy coloration in macaques may act as a warning signal to reduce antagonism by conspecifics. Behav Process 80(1):7–11. https://doi.org/10.1016/j.beproc.2008.08.001

Gilboa-Schechtman E, Foa EB, Amir N (1999) Attentional biases for facial expressions in social phobia: the face-in-the-crowd paradigm. Cogn Emot 13(3):305–318. https://doi.org/10.1080/026999399379294

Goodall J (1986) The chimpanzees of Gombe: patterns of behaviour. Harvard University Press, Cambridge

Gordon DJ, Rogers LJ (2015) Cognitive bias, hand preference and welfare of common marmosets. Behav Brain Res 287:100–108. https://doi.org/10.1016/j.bbr.2015.03.037

Harding EJ, Paul ES, Mendl M (2004) Cognitive bias and affective state. Nature 427:312. https://doi.org/10.1038/427312a

Hommer RE, Meyer A, Stoddard J et al (2014) Attention bias to threat faces in severe mood dysregulation. Depress Anxiety 31(7):559–565. https://doi.org/10.1002/da.22145

Kano F, Tomonaga M (2010) Attention to emotional scenes including whole-body expressions in chimpanzees (*Pan troglodytes*). J Comp Psychol 124(3):287–294. https://doi.org/10.1037/a0019146

Kawai N, Koda H (2016) Japanese monkeys (*Macaca fuscata*) quickly detect snakes but not spiders: evolutionary origins of fear-relevant animals. J Comp Psychol 130(3):299–303. https://doi.org/10.1037/com0000032

Kawai N, Kubo K, Masataka N, Hayakawa S (2016) Conserved evolutionary history for quick detection of threatening faces. Anim Cogn 19:655–660. https://doi.org/10.1007/s10071-015-0949-y

King HM, Kurdziel LB, Meyer JS, Lacreuse A (2012) Effects of testosterone on attention and memory for emotional stimuli in male rhesus monkeys. Psychoneuroendocrinology 37 (3):396–409. https://doi.org/10.1016/j.psyneuen.2011.07.010

Kret ME, Jaasma L, Bionda T, Wijnen JG (2016) Bonobos (*Pan paniscus*) show an attentional bias toward conspecifics' emotions. Proc Natl Acad Sci U S A 113(14):3761–3766. https://doi.org/10.1073/pnas.1522060113

Lang PJ, Davis M, Öhman A (2000) Fear and anxiety: animal models and human cognitive psychophysiology. J Affect Disord 61(3):137–159. https://doi.org/10.1016/S0165-0327(00)00343-8

LeDoux J (1996) The emotional brain: the mysterious underpinnings of emotional life. Simon & Schuster, New York

Lindell AK (2013) Continuities in emotion lateralization in human and non-human primates. Front Hum Neurosci 7:464. https://doi.org/10.3389/fnhum.2013.00464

MacLeod C, Mathews A, Tata P (1986) Attentional bias in emotional disorders. J Abnorm Psychol 95(1):15–20. https://doi.org/10.1037/0021-843X.95.1.15

MacLeod C, Rutherford E, Campbell L et al (2002) Selective attention and emotional vulnerability: assessing the causal basis of their association through the experimental manipulation of attentional bias. J Abnorm Psychol 111(1):107–123. https://doi.org/10.1037/0021-843X.111.1.107

Mandalaywala TM, Bethell EJ, Parker KJ, Maestripieri D (2013) Negativity bias in free-ranging infant rhesus macaques (*Macaca mulatta*) on Cayo Santiago, Puerto Rico. Am J Primatol 75 (S1):43. https://doi.org/10.1002/ajp.22188

Mandalaywala TM, Parker KJ, Maestripieri D (2014) Early experience affects the strength of vigilance for threat in rhesus monkey infants. Psychol Sci 25(10):1893–1902. https://doi.org/10.1177/0956797614544175

Mathews A (1990) Why worry? The cognitive function of anxiety. Behav Res Ther 28(6):455–468. https://doi.org/10.1016/0005-7967(90)90132-3

Mathews A, MacLeod C (2005) Cognitive vulnerability to emotional disorders. Annu Rev Clin Psychol 1:167–195. https://doi.org/10.1146/annurev.clinpsy.1.102803.143916

McGuire MC, Vonk J, Fuller G, Allard S (2017) Using an ambiguous cue paradigm to assess cognitive bias in gorillas (*Gorilla gorilla gorilla*) during a forage manipulation. Anim Behav Cogn 4(1):91–104. https://doi.org/10.12966/abc.06.02.2017

Mendl M (1999) Performing under pressure: stress and cognitive function. Appl Anim Behav Sci 65(3):221–244. https://doi.org/10.1016/S0168-1591(99)00088-X

Mendl M, Paul ES (2004) Consciousness, emotion and animal welfare: insights from cognitive science. Anim Welf 13(Suppl):S17–S25

Mendl MT, Burman OHP, Parker RMA, Paul ES (2009) Cognitive bias as an indicator of animal emotion and welfare: emerging evidence and underlying mechanisms. Appl Anim Behav Sci 118(3–4):161–181. https://doi.org/10.1016/j.applanim.2009.02.023

Mogg K, Bradley BP (1999) Orienting of attention to threatening facial expressions presented under conditions of restricted awareness. Cogn Emot 13(6):713–740. https://doi.org/10.1080/026999399379050

Mogg K, Bradley BP, Miles F, Dixon R (2004) Time course of attentional bias for threat scenes: Testing the vigilance-avoidance hypothesis. Cogn Emot 18(5):689–700. https://doi.org/10.1080/02699930341000158

Neville V, Nakagawa S, Zidar J et al (2020) Pharmacological manipulations of judgement bias: a systematic review and meta-analysis. Neurosci Biobehav Rev 108:269–286. https://doi.org/10.1016/j.neubiorev.2019.11.008

Öhman A, Mineka S (2003) The malicious serpent: snakes as a prototypical stimulus for an evolved module of fear. Curr Dir Psychol Sci 12(1):5–9. https://doi.org/10.1111/1467-8721.01211

Öhman A, Flykt A, Esteves F (2001) Emotion drives attention: detecting the snake in the grass. J Exp Psychol Gen 130(3):466–478. https://doi.org/10.1037/0096-3445.130.3.466

Panksepp J (1998) Affective neuroscience: the foundations of human and animal emotions. Oxford University Press, New York

Panksepp J (2011) The basic emotional circuits of mammalian brains: do animals have affective lives? Neurosci Biobehav Rev 35(9):1791–1804. https://doi.org/10.1016/j.neubiorev.2011.08.003

Panksepp J (2014) Integrating bottom-up internalist views of emotional feelings with top-down externalist views: might brain affective changes constitute reward and punishment effects within animal brains? Cortex 59:208–213. https://doi.org/10.1016/j.cortex.2014.04.015

Parr LA (2011) The evolution of face processing in primates. Philos Trans R Soc B Biol Sci 366 (1571):1764–1777. https://doi.org/10.1098/rstb.2010.0358

Parr LA, Hopkins W (2000) Brain temperature asymmetries and emotional perception in chimpanzees, *Pan troglodytes*. Physiol Behav 71(3–4):363–371. https://doi.org/10.1016/S0031-9384(00)00349-8

Parr LA, Murphy L, Feczko E et al (2016) Experience-dependent changes in the development of face preferences in infant rhesus monkeys. Dev Psychobiol 58(8):1002–1018. https://doi.org/10.1002/dev.21434

Perdue B (2017) Mechanisms underlying cognitive bias in nonhuman primates. Anim Behav Cogn 4(1):105–118. https://doi.org/10.12966/abc.08.02.2017

Pfefferle D (2020) Facial Stimuli - Macaques. figshare. Dataset. https://doi.org/10.6084/m9.figshare.13227779

Pfefferle D, Kazem AJN, Brockhausen RR et al (2014) Monkeys spontaneously discriminate their unfamiliar paternal kin under natural conditions using facial cues. Curr Biol 24(15):1806–1810. https://doi.org/10.1016/j.cub.2014.06.058

Pomerantz O, Terkel J, Suomi SJ, Paukner A (2012) Stereotypic head twirls, but not pacing, are related to a 'pessimistic'-like judgment bias among captive tufted capuchins (*Cebus apella*). Anim Cogn 15:689–698. https://doi.org/10.1007/s10071-012-0497-7

Rosenthal R (1979) The file drawer problem and tolerance for null results. Psychol Bull 86 (3):638–641. https://doi.org/10.1037/0033-2909.86.3.638

Russell W, Burch R (1959) The principles of humane experimental technique. Methuen & Co, London

Sato W, Aoki S (2006) Right hemispheric dominance in processing of unconscious negative emotion. Brain Cogn 62(3):261–266. https://doi.org/10.1016/j.bandc.2006.06.006

Schino G, Massimei R, Pinzaglia M, Addessi E (2016) Grooming, social rank and 'optimism' in tufted capuchin monkeys: a study of judgement bias. Anim Behav 119:11–16. https://doi.org/10.1016/j.anbehav.2016.06.017

Seligman MEP, Csikszentmihalyi M (2000) Positive psychology: an introduction. Am Psychol 55 (1):5–14. https://doi.org/10.1037/0003-066X.55.1.5

Shibasaki M, Kawai N (2009) Rapid detection of snakes by Japanese monkeys (*Macaca fuscata*): an evolutionarily predisposed visual system. J Comp Psychol 123(2):131–135. https://doi.org/10.1037/a0015095

Tversky A, Kahneman D (1974) Judgment under uncertainty: heuristics and biases. Science 185 (4157):1124–1131. https://doi.org/10.1126/science.185.4157.1124

van Rooijen R, Ploeger A, Kret ME (2017) The dot-probe task to measure emotional attention: a suitable measure in comparative studies? Psychon Bull Rev 24:1686–1717. https://doi.org/10.3758/s13423-016-1224-1

Waitt C, Buchanan-Smith HM (2006) Perceptual considerations in the use of colored photographic and video stimuli to study nonhuman primate behavior. Am J Primatol 68(11):1054–1067. https://doi.org/10.1002/ajp.20303

Waitt C, Little AC, Wolfensohn SE et al (2003) Evidence from rhesus macaques suggests that male coloration plays a role in female primate mate choice. Proc R Soc London B Biol Sci 270 (Suppl_2):S144–S146. https://doi.org/10.1098/rsbl.2003.0065

Watson KK, Ghodasra JH, Furlong MA, Platt ML (2012) Visual preferences for sex and status in female rhesus macaques. Anim Cogn 15:401–407. https://doi.org/10.1007/s10071-011-0467-5

Williams JMG, Mathews A, MacLeod C (1996) The emotional Stroop task and psychopathology. Psychol Bull 120(1):3–24. https://doi.org/10.1037/0033-2909.120.1.3

Wilson DA, Tomonaga M (2018) Exploring attentional bias towards threatening faces in chimpanzees using the dot probe task. PLoS One 13(11):e0207378. https://doi.org/10.1371/journal.pone.0207378

Witham C, Bethell EJ (2019) Macaque faces. figshare. Dataset. https://doi.org/10.6084/m9.figshare.9862586

Yiend J (2010) The effects of emotion on attention: a review of attentional processing of emotional information. Cogn Emot 24(1):3–47. https://doi.org/10.1080/02699930903205698

Physiological Measures of Welfare

John P. Capitanio, Jessica Vandeleest, and Darcy L. Hannibal

Abstract

Animal well-being can be assessed in a variety of ways; in this chapter, we focus on physiological measures of welfare and review advantages and disadvantages of measures assessing hypothalamic–pituitary–adrenal, sympathetic–adrenal–medullary, cardiovascular, and immune function. For each physiological system, measurement issues are discussed. Throughout the discussion, a distinction is made between assessing *levels* of a measure versus assessing *regulation* of the system. We argue that, because "well-being" reflects a persisting state or condition that one is in, measurement of an indicator of that state at one point in time (i.e., measurement of levels) may be of limited value. We propose that assessment of regulatory aspects of biological systems may be more congruent with the concept of well-being.

Keywords

Welfare · HPA axis · Immune system · Sympathetic nervous system · Cardiovascular system

1 Introduction

It is generally agreed that promotion of the welfare of an individual involves attention to a variety of factors. Environmental, nutritional, social, health, cognitive, and structural (i.e., the physical space) considerations are among the important issues that must be addressed. A critically important aspect of welfare involves

J. P. Capitanio (✉) · J. Vandeleest · D. L. Hannibal
California National Primate Research Center, University of California, Davis, Davis, CA, USA
e-mail: jpcapitanio@ucdavis.edu

psychological health; one could argue, in fact, that the psychological well-being of the animal is the successful outcome of proper attention being paid to the factors just listed.

There are many ways to assess the well-being of an animal. One might think, given that "well-being" is a psychological state, that behavioral measures might be the most valuable (Lutz and Baker 2022). There is considerable merit to this argument; behavior is easily observable, and there is a sense that it is a direct link to mental processes. However, behavior is a tricky phenomenon. It is the principal means by which an animal interacts with its environment, but any given behavior pattern can be displayed for multiple reasons—that is, behavior shows equifinality: Different internal, psychological conditions can lead to the same behavioral outcome. Moreover, which behavior the animal chooses to display will be determined by the animal's needs, which could be very different from the needs of others, including our own. For example, we would love to know when an animal is beginning to feel poorly, so that we could quickly intervene. However, it is likely that selection favored a more stoic approach to feeling ill—animals that too quickly showed they were ill might have gotten left behind, lost rank, been selected as prey, etc. (Weary et al. 2006). Those who "covered" by adopting a more stoic approach may have been better at survival and reproduction, and so that strategy could get passed on. So, while the display of abnormal/rare behaviors might indicate poor well-being, the lack of such behaviors does not necessarily indicate positive well-being.

Considerations like these have led some to focus less on behavioral indicators of welfare and more on physiological indicators. We believe this is a very useful approach, and it is not uncommon these days to see welfare-associated papers that utilize measures such as cortisol concentrations. Like behavior, however, this set of measures has its own set of complexities. Consider an example (Capitanio et al. 1998). In an experimental study of the role of social factors in the progression of immunodeficiency virus disease in rhesus monkeys, one set of animals got 100 min of daily experience in social groups in which the same animals met with each other every day (stable groups). A second set of animals met daily for an equal amount of time in social groups whose members changed every day (unstable groups). Baseline cortisol concentrations were measured on an approximately monthly basis, not while the animals were in their social groups, but while they were individually housed. We found that one group had significantly lower cortisol concentrations than the other group. One might expect that the animals in the stable social groups had the lower basal cortisol concentrations, but that was not the case—it was the animals in the unstable social groups that had the lower cortisol levels. Pharmacological probing of the hypothalamic–pituitary–adrenal (HPA) system, which is responsible for cortisol release (see below), revealed why: the social manipulation had altered the regulation of the HPA axis, one consequence of which was reduced basal levels of cortisol. This result was identical with results that have been found in humans that have post-traumatic stress disorder (e.g., Yehuda et al. 1995). Thus, this example indicates that physiological measures may not be easy or simple indicators of welfare.

This example illustrates a critical point that we will return to, namely the distinction between *levels* of an analyte and *regulation* of that analyte. Because well-being reflects a persisting state or condition that one is in, measurement of an indicator of that state at one point in time, can be of limited value and can lead to simplistic statements like "the animal is fine because its cortisol is low." As the example in the last paragraph shows, this may not be the case. It is our strong opinion that challenging a system (whether a behavioral or physiological system) to reveal how the system is regulated can reveal much more about the functioning of that system. Furthermore, we propose that statements about regulation (rather than level) better capture the idea of a psychological state such as well-being. A focus on regulation allows for transient deviations in level; for example, if construction is going on in a housing room next door, animals may well show elevated cortisol concentrations for that day; the animals may be disturbed, but is their well-being compromised? As long as that situation does not persist, the *regulation* of the HPA system will not change. We would argue that when one sees alterations in regulation of a physiological system, then that may be the best physiological indicator of poor well-being.

In this chapter, we will describe some physiological measures that have been, or might be, employed to index psychological well-being. For each system that we discuss, we will also examine methodological issues pertaining to sample collection that can affect the utility of these measures. Where possible, we will try to highlight the value of taking a *regulation* rather than a *levels* approach.

2   Hypothalamic–Pituitary–Adrenal Axis

One of the most commonly used physiological indicators of welfare is the end product of the hypothalamic–pituitary–adrenal (HPA) axis, namely concentrations of cortisol. The reason for this is because of the HPA axis' involvement in the physiological stress response (Novak et al. 2013). Stress is generally considered inimical to well-being, although we must recognize a distinction between eustress ("good" stress, such as running and jumping during play or arousal during mating) and distress ("bad" stress, such as the threat of impending danger; this is usually what people are referring to by their use of the word "stress") (Selye 1979). There are several issues, however, that must be considered when using cortisol to measure welfare. Below, we briefly describe the HPA axis and its functions, after which we discuss the different methods for measuring the activity and regulation of the HPA axis as well as important factors to consider when doing so.

The HPA axis is a neuroendocrine system that is important for both the daily regulation of metabolism and the physiological response to stress (Fig. 1). Activation of the axis begins when the hypothalamus releases corticotropin-releasing hormone (CRH), which stimulates the anterior pituitary to release adrenocorticotropic hormone (ACTH) into the circulatory system. Upon reaching the cortex of the adrenal gland, ACTH stimulates the release of glucocorticoids (cortisol is the principal glucocorticoid in primates). Cortisol release is regulated through a negative feedback

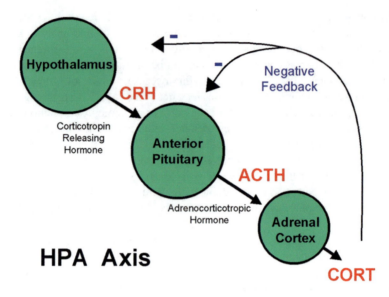

Fig. 1 Schematic diagram of the hypothalamic–pituitary–adrenal axis. Original artwork by Brian M. Sweis, reproduced under Creative Commons license

loop involving tissues at multiple levels (e.g., hippocampus, hypothalamus, pituitary: Dallman et al. 1994; Herman et al. 1996, 2003): Elevated cortisol concentrations feed back onto the system to dampen subsequent cortisol release. On a day-to-day basis, cortisol plays a critical role in regulating metabolism. Cortisol secretion follows a daily cycle in diurnal animals, with high levels during the morning and the lowest levels in the late afternoon. Once secreted, cortisol attaches to two types of receptors that differ both in their affinity for cortisol and in their functions (see Sapolsky et al. 2000 for more information). The daily variation in cortisol plays an important role in metabolism due to cortisol's function in mobilizing sugars into the circulatory system and the fact that cortisol levels are altered after consuming food (Sapolsky et al. 2000; Lemmens et al. 2011).

While primarily a metabolic hormone, cortisol is also critical in the physiological stress response. The HPA axis supports the physiological response to short-term, acute stressors. During an acute stressor (e.g., encountering a bear in the woods), the initial stress response begins within seconds and is primarily driven through neural pathways, namely through the sympathetic–adrenal–medullary system (SAM; see below). This response elevates heart rate, increases blood flow and glucose delivery to skeletal muscle, stimulates immune function, increases cerebral blood flow, and decreases appetite (Sapolsky et al. 2000). A few minutes later, cortisol also is released into the circulatory system and serves to regulate the SAM effects on the body (e.g., by suppressing aspects of the immune system or by increasing appetite), and to facilitate responses to and recovery from the stressor (e.g., by increasing blood glucose). Although the effects of the initial stress response occur within seconds, the effects of cortisol on the body often do not appear until an hour or more after the onset of the stressor (Sapolsky et al. 2000). This is because, as a steroid hormone, cortisol's principal mechanism of action is to regulate gene transcription and protein production—processes that take time to accomplish. Although the HPA axis evolved to be adaptive in response to acute stressors, its effects can lead to many negative outcomes when activation of the axis is chronic or frequent. Effects of chronic stress include suppression of reproductive behavior and physiology, and of immune function.

Because of cortisol's involvement in the stress response, it is a commonly used measure in welfare research (e.g., Clarke et al. 1995; Ruys et al. 2004; Baker et al. 2012; Novak et al. 2013). However, there are multiple ways to assess the activity of the HPA axis, each with its own advantages and disadvantages; broadly speaking, the two main approaches, as described above, are those examining cortisol *levels* and those examining the *regulation* of the HPA axis. Levels are commonly assessed through single biological samples and are the most commonly reported type of cortisol measure used in studies of welfare. In these studies, there is often the assumption that higher cortisol levels reflect greater stress experienced by the individual. While this can be true, it is important to note that higher cortisol levels can be caused by a number of other factors (discussed in more detail below) and that chronic stress can lead to *lower* cortisol levels as well, as described above (Capitanio et al. 1998; Yehuda et al. 2004; Yehuda 2006). The second method of assessing the HPA axis involves examining the regulation of the HPA axis. This approach requires multiple biological samples and can require the administration of pharmaceuticals to probe the system (e.g., ACTH or dexamethasone). Examination of the regulation of the HPA axis is important when understanding welfare because it provides more sensitive information about the physiological mechanisms that lead to higher or lower cortisol levels. For example, in humans, both major depression and PTSD have been associated with elevated CRH; however, depression is often associated with higher cortisol levels, while PTSD is associated with lower levels (Baker et al. 1999; Holsboer 2000; Yehuda et al. 2004). Only by examining the regulation of the HPA axis, we can understand how the HPA axis responds differently in these two stress-related disorders. Moreover, as discussed earlier, the concept of well-being does not necessarily imply an organism never experiences stressful conditions; rather, it implies that the stress that it does experience is not sustained. How much stress is "too much"? The answer to that question is unknown; we would argue, however, that a change to the regulation of a physiological system may indicate that some critical threshold has been reached and that this may indicate that the well-being of the animal is compromised.

2.1 Measuring Cortisol Levels

Cortisol can be sampled from multiple biological fluids and excreta including plasma, saliva, urine, and feces. It can also be sampled from hair. Each of these matrices differs in the timescale of HPA axis activation that they capture, ranging from minutes to months.

2.1.1 Short Time Scale (i.e., Point Samples): Blood and Saliva

Upon activation of the HPA axis, cortisol is released into the circulation within minutes. Cortisol sampled from blood or saliva reflects the amount of cortisol circulating at the time when the sample was collected. This type of sampling is often referred to as "point sampling" because it reflects the state of an individual at a particular point in time. In the bloodstream, cortisol can either be floating free or bound to corticosteroid-binding globulin. Generally, cortisol assays of plasma or serum measure free and bound cortisol, while cortisol measured in saliva only reflects the free cortisol that is able to passively diffuse into saliva (Teruhisa et al. 1981). Despite this difference, studies have shown salivary and plasma cortisol levels to be highly correlated (Gozansky et al. 2005).

Although point samples are often used, collection of these samples must be carefully controlled because they are also the most susceptible to environmental or contextual effects. For example, cortisol levels in blood and saliva are affected by eating, changes in ambient temperature, time of day (due to the diurnal cycle), position in a housing room, the amount of time it takes to obtain a sample, anesthesia, and some eustressful experiences (Capitanio et al. 1998; Ange-van Heugten et al. 2009; Lemmens et al. 2011; Menicucci et al. 2013; Vandeleest et al. 2013). Collection of blood samples is invasive due to the necessity of venipuncture for sample collection, although training animals to cooperate can ameliorate much of the stress of sampling (Reinhardt et al. 1990; Coleman et al. 2008). Once collected, serum or plasma needs to be separated from the other components of blood and stored at $-20\ ^{\circ}\mathrm{C}$ or less for long-term storage (Stroud et al. 2007; Sheriff et al. 2011). Obtaining salivary samples is less invasive than blood samples but necessitates the cooperation of the animal (Lutz et al. 2000). In addition, for salivary samples the presence of any bleeding injury to the gums or mouth can lead to increased levels of cortisol due to the presence of corticosteroids in blood (Novak et al. 2013). Saliva samples have been shown to be stable for up to 3 months at $5\ ^{\circ}\mathrm{C}$, but should be stored at least at $-20\ ^{\circ}\mathrm{C}$ if longer time periods are needed (Garde and Hansen 2005). Appropriate use of point samples (blood or saliva) can provide detailed information on how animals respond to specific conditions (e.g., housing changes), provided efforts to reduce the influence of confounding variables (described above) and the stress of sampling are thoroughly considered. For example, blood and saliva samples have been useful in demonstrating that cortisol levels respond to feeding demands, rearing history, social separation, and permanent relocations (Laudenslager et al. 1995; Champoux et al. 2001; Capitanio et al. 2005; Davenport et al. 2008).

2.1.2 Medium Time Scale: Urine and Feces

Cortisol also accumulates in urine and feces with cortisol levels reflecting average cortisol secretion over a period of hours to days, depending on the length of time the excreta were produced. Urinary or fecal sample collection to assess cortisol is often employed because these substances can be collected noninvasively and relatively easily (particularly for feces), and they are somewhat insensitive to sampling-associated confounding compared to point samples. Collection of urinary samples can pose challenges. The collection of urine, particularly from Old World monkeys, often requires single housing and appropriate collection pans that reduce cross-contamination from feces and from the urine of other animals (Setchell et al. 1977; Tiefenbacher et al. 2004). A second option is often employed with New World monkeys and chimpanzees that have been trained to voluntarily give urine samples, allowing for collection under group living conditions (Ziegler et al. 1995; Smith and French 1997; Muller and Wrangham 2004; Anestis 2005; Jarcho et al. 2012). There are additional concerns regarding the assay of urinary cortisol: Care must be taken to measure creatinine to control for individual or species differences in urine output or hydration. Finally, much of the cortisol in urine is in a conjugated form and researchers may or may not need to remove the conjugates prior to assay depending on their research questions and the species that they study (Bahr et al. 2000; Novak et al. 2013).

Fecal cortisol is most often measured in the field as it is the least invasive and most easily accessible excreta. Care must be taken, however, in how samples are collected and stored, particularly under the constraints of field conditions (as reviewed in Touma and Palme 2005; Sheriff et al. 2011). Additionally, the choice of assay for fecal cortisol levels requires caution due to the fact that cortisol, corticosterone (a similar molecule, which in primates is present in small concentrations), and their metabolites are present in feces. Notably, cortisol metabolism differs between species, which means that selecting the best hormone or metabolite to measure depends on your species of interest (Bahr et al. 2000; Heistermann et al. 2006; Sheriff et al. 2011). Fecal cortisol levels have been shown to be influenced by habitat (conserved or fragmented forests and captivity), season, food availability, and tourist exposure (Rangel-Negrín et al. 2009; Behie et al. 2010).

2.1.3 Long Time Scale: Hair

Previously, the measurement of long-term, chronic cortisol secretion was difficult and often was done by averaging multiple short or medium time scale samples. This changed, however, when researchers found a way to identify and measure physiological concentrations of cortisol in hair (Koren et al. 2002; Raul et al. 2004). The concentrations of cortisol in hair are thought to reflect the average activity of the HPA axis over a period of months. Hair is increasingly being used because it is easy and noninvasive to collect, can be stored at room temperature, and requires only one sample to provide a measure of long-term activation of the system. Although not yet well understood, there are multiple pathways by which cortisol is thought to be incorporated in hair. At the level of the hair follicle, it is thought that free cortisol can passively diffuse from the bloodstream to be incorporated into actively growing hair, similar to how cortisol gets into saliva. Cortisol is also contained in sweat and sebaceous gland secretions, although it is unclear if this cortisol is only deposited externally on the hair or if it is incorporated into the shaft (Meyer and Novak 2012). Finally, there is evidence that the skin and follicle contain a peripheral mini-HPA axis that can synthesize and secrete cortisol (Ito et al. 2005). Much of the validation for hair cortisol has been accomplished through correlations with multiple measurements of saliva (Davenport et al. 2006; Bennett and Hayssen 2010; D'Anna-Hernandez et al. 2011), or with urinary or fecal cortisol levels (Sauvé et al. 2007; Accorsi et al. 2008). As the use of hair to assess cortisol levels is a relatively new technique, factors that influence cortisol concentrations are still being explored. For example, hair color has been shown to be associated with hair cortisol levels in dogs but not humans (Sauvé et al. 2007; Kirschbaum et al. 2009), and location of hair collection has been shown to impact hair cortisol concentrations in grizzly bears and Canada lynx (Macbeth et al. 2010; Terwissen et al. 2013) but not humans (Sauvé et al. 2007). To address these potential issues, hair should be collected from a standardized location (often the posterior vertex of the head). Hair cortisol levels are altered by rearing history, relocation, and population density (Davenport et al. 2008; Dettmer et al. 2012, 2014).

2.2 Measuring Regulation of the HPA Axis

Regulation of the HPA axis can be assessed in two ways, (1) examining the diurnal cycle of endogenous cortisol secretion and (2) administration of exogenous hormones to test aspects of HPA axis regulation.

Measuring the diurnal cycle of cortisol secretion is most frequently accomplished using blood or saliva samples collected in the morning and afternoon. Since these are point samples that are susceptible to confounding events (see above), it is recommended that samples be collected on more than 1 day to reduce the impact of transient events on the measurements. The combination of both the level of cortisol and the presence or absence of a decline over the day can provide important information regarding alterations to the regulation of the HPA axis that can be caused by stress. For example, both PTSD and some types of depression in humans are characterized by flattened diurnal rhythms. These flattened rhythms are likely due to different mechanisms, however. Individuals with depression can exhibit elevated afternoon levels (hypercortisolemia). In contrast, individuals with PTSD often have lower cortisol levels in the morning resulting in overall lower levels of cortisol (hypocortisolemia) (Yehuda et al. 2004). While flattened rhythms are a symptom of altered regulation of the HPA axis, it is not possible to determine what aspect of the HPA axis is dysregulated from diurnal rhythms alone.

Pharmacological manipulation of the HPA axis provides the most detailed information regarding how chronic stress impacts the HPA axis. This approach introduces exogenous HPA hormones or analogs to test how particular tissues respond. The administration of dexamethasone, a synthetic glucocorticoid, is the most commonly used and provides information regarding the effectiveness of HPA axis negative feedback (e.g., Capitanio et al. 1998). Generally, a low dose of dexamethasone is administered in the afternoon or evening and then cortisol levels are sampled the next morning. If negative feedback is operating, then the dexamethasone should inhibit somewhat the secretion of endogenous cortisol resulting in lower than normal cortisol levels the next morning. As stated above, flattened diurnal rhythms can be a symptom of alterations to the regulation of the HPA axis. Evidence from the dexamethasone test suggests that both the flattened diurnal rhythms seen in individual with PTSD and depression can be explained by alterations to the negative feedback of the HPA axis. In depression, hypercortisolemia is due to a failure of negative feedback (i.e., dexamethasone does not suppress cortisol secretion (Yehuda et al. 2004). Conversely, in the case of PTSD, there is enhanced cortisol suppression after administration of dexamethasone, an indicator of particularly strong negative feedback (Yehuda et al. 2004).

A second pharmacological approach can involve the administration of exogenous ACTH to assess how strongly the adrenal gland responds. A strong cortisol response after administration of ACTH could indicate a history of frequent or chronic activation of the HPA axis. Studies in rats have shown that with chronic or repeated exposure to stressors, the adrenal gland becomes more responsive to ACTH and may even become enlarged (i.e., adrenal hypertrophy and hypoplasia: Ulrich-Lai et al. 2006).

3   Sympathetic–Adrenal–Medullary System

The second major stress response system in the body is the sympathetic–adrenal–medullary (SAM) system. The sympathetic nervous system (SNS) is one branch of the autonomic nervous system; the other branch is the parasympathetic nervous system (PNS). The autonomic nervous system innervates and helps regulate internal organs—heart, blood vessels, kidneys, gastrointestinal (GI) tract, sexual organs, etc.—and it generally accomplishes this regulation via parallel innervation of these target organs by fibers of both the SNS and PNS.

The PNS is concerned with so-called vegetative responses—those physiological processes that are typically active when an organism is in a low-activity state. These processes involve digestion of food in the GI tract and storage of nutrients such as glucose (in the form of glycogen), urination, defecation, and sexual arousal. The PNS fibers that synapse onto effector organs employ acetylcholine as the neurotransmitter. The SNS is more concerned with physiological processes that occur when the body is in an activated state, whether due to eustress or distress—the response when the SNS is activated is often referred to as the "fight or flight" response. The SNS will speed up heart rate and dilate blood vessels to deliver blood to skeletal muscles, suppress actions in the GI tract, initiate sweating, etc.—all functions that quickly prepare the body for action. The principal neurotransmitter used by the SNS on effector organs is norepinephrine (NE). Importantly, the PNS and SNS do not operate in an all-or-none fashion; rather, they act in a complementary fashion—we can say that, at any particular point in time, one branch of the autonomic nervous system "predominates." While reading this chapter, it is likely that your PNS is predominant, but should someone in the office next door shout "Fire!" your SNS would come to the fore to prepare you for action.

One organ that the SNS innervates is the adrenal medulla. Recall from above that the adrenal cortex is where glucocorticoids are synthesized and secreted. The adrenal medulla is the inner part of the adrenal gland, and when stimulated by the SNS, secretes into the circulation epinephrine (E; also called adrenalin) in large quantities, and NE (also called noradrenaline) in smaller quantities. E and NE, secreted as hormones by the adrenal medulla, can enhance the activity of the target organs that were being stimulated directly by fibers of the SNS, which release NE, hence the name of this stress-responsive axis as the sympathetic–adrenal–medullary system. Once epinephrine and norepinephrine are released, they are metabolized by an enzyme, catechol-O-methyltransferase (COMT), which is located in the extracellular space. The metabolites are metanephrine and normetanephrine, respectively, which are excreted in urine and can be measured relatively easily.

In contrast to the HPA system, which releases its effector molecule, cortisol, over a period of minutes, the SAM system releases its effectors on the order of seconds. This can provide some challenges to using blood-borne E and NE as indicators of welfare—one runs the risk, especially with untrained animals, that the process of collecting a blood sample could by itself cause levels of these hormones to become elevated. Training the animals (e.g., Bloomsmith et al. 2022) to extend their arms for phlebotomy, however, may prevent this rise. In one of our studies (Capitanio and Cole 2015), we drew blood on well-trained animals in the mid-afternoon to assess E and NE levels in control monkeys, in contrast to those that had been receiving methamphetamine, a drug that can mimic (and stimulate) SAM activity. Our results (Capitanio and Cole 2015, Fig. 5) showed that methamphetamine did indeed result in elevated E and NE levels; more importantly, however, the levels for the control animals were well within the range of basal concentrations reported in the classic reference in this area (Mason et al. 1961). Training for phlebotomy, then, may result in plasma concentrations of E and NE being useful point measures of activation of the SAM system.

A better measure, however, may be urinary levels of metabolites (which, as with urinary cortisol measurement, must be corrected for differences in urinary output between animals, typically by assessing creatinine and expressing concentrations relative to this measure). In a different study reported in a paper referenced above (Capitanio and Cole 2015), adult male rhesus monkeys were placed into stable social groups for a period of several weeks. Animals were then switched to socially stressful conditions for a five-week period, after which they were switched back to stable social conditions. Animals experienced either stable or unstable (stressful) social conditions for 100 min/day; the remainder of the time, they were housed individually. Urine was collected by placing a pan under their cage at approximately 1600 h, then collecting the urine at 0700 h the next morning. Our results (Capitanio and Cole 2015, Fig. 2) demonstrated significantly elevated concentrations of E and NE metabolites in urine after only 2 weeks of daily social stress; these levels returned to baseline when stable conditions were re-imposed. It is important to note that urine collection occurred several hours after the animals had experienced their social conditions, and almost certainly, they had urinated at least once between the end of their social session and the placement of the pan under their cage, suggesting that our urine collection was not capturing the acute effects of the social manipulations. Thus, while the dynamics of urinary concentration of the metabolites of substances like E and NE are not well known, it is possible that urinary measures, collected in this fashion, may indicate altered regulation of the SAM system resulting from the experience of chronically stressful conditions. As such, measures of urinary metabolites of SAM activity may be useful as indicators of well-being in captive animals.

Finally, a measure of SAM activity that is of growing interest in the field of human psychosomatic medicine is alpha-amylase concentrations in saliva. Fibers of both the sympathetic and parasympathetic branches of the autonomic nervous system (ANS) innervate salivary glands, and sympathetic activation increases secretion of salivary proteins, one of which is alpha-amylase. Evidence suggests that plasma NE levels and salivary alpha-amylase levels are correlated and that both can be elevated by acute stress in humans (Rohleder et al. 2004; Nater and Rohleder 2009). To date, the only study in nonhuman primates that we are aware of that examined this measure was by Petrullo et al. (2016). They found that alpha-amylase concentration was associated with exposure to stressful situations, as well as to variation in early life adversity in rhesus monkeys on Cayo Santiago. These investigators also measured salivary cortisol concentrations and found an interesting dissociation between cortisol and alpha-amylase levels in animals that had experienced early life stress (specifically, maternal abuse). It is possible that, in a welfare context, examination of both HPA and SNS measures together in saliva could provide a window into regulation of stress–response systems.

4 Cardiovascular System

As just described, the two branches of the ANS, namely the sympathetic and the parasympathetic, regulate the internal organs (Jänig 2008). Stress directly and immediately impacts these functions, which support an animal's response to the stressor. When in distress, these responses prepare an animal for defense, flight, or to freeze until the threat either passes or requires action (Cannon 1953). Baseline characteristics of ANS activity are predictive of stress responses, how successfully animals cope with stress, and whether they develop stress-related pathologies (Porges 1995). Measuring animal welfare using the effects of stress on functions governed by the ANS has been limited due to the largely transient nature of ANS responses, as described in the previous section (Moberg 2000). Once the acute period subsides, animals generally return to their homeostatically regulated baseline. However, evidence suggests that in animals (including humans) with anxiety-related psychopathologies, profiles of key ANS measures (primarily heart rate and heart rate variability) are distinct from each other and vary separately in both stressful and nonstressful conditions (examples discussed below) (Friedman and Thayer 1998; Bachmann et al. 2003). Thus, both the concurrent and poststressor basal conditions of the ANS are becoming more widely used to assess the acute and prolonged stress responses of animals, including nonhuman primates, to the conditions and procedures experienced in captivity (e.g., von Borell et al. 2007; Grandi and Ishida 2015).

Heart rate patterns and changes have been used as physiological measures of stress responses and welfare in a handful of primate studies (Rasmussen and Suomi 1989; Boccia et al. 1989; Line et al. 1990; Aureli et al. 1999; Novak 2003; Doyle et al. 2008). Indoor-housed monkeys that engage in self-injurious behavior (SIB) (see Lutz and Baker 2022) tend to have experienced stressful procedures and management practices early in life and repeatedly (Novak 2003; Gottlieb et al. 2013). Rhesus monkeys with SIB exhibit increased HR shortly before self-biting, a further increase in HR during biting, and then decrease back to baseline HR after biting themselves (Novak 2003). Indoor-housed rhesus monkeys have heightened HR during pair introductions, but then HR returns to baseline after introductions; after a few months of pairing, HR can fall below baseline levels (Doyle et al. 2008). Heart rate increases, however, are also expected during eustress or minor distress (Hainsworth 1995). An animal with increased heart rate due to physical exertion while playing or mating could not be said to be suffering reduced welfare, for example. This means that it is difficult to interpret whether increased heart rate associated with a normal but stressful event, such as an approach by a dominant individual (Aureli et al. 1999), is actually distressing, or whether the lack of a change in heart rate with a change in environment indicates it has no impact on welfare (e.g., increased cage size examined in Line et al. 1990).

Heart rate variability (HRV), which is a measure of the changes in time elapsed between two beats (referred to as R-R intervals), has been shown to distinguish between eustress and distress in a wide variety of mammals and has been used as a measure of welfare in captivity (e.g., Sgoifo et al. 2001; Mohr et al. 2002; Geverink et al. 2002; Bachmann et al. 2003; Rietmann et al. 2004; von Borell et al. 2007), and HRV is determined by both vagal (parasympathetic) activity and sympathetic activity and increased or decreased activity in either of these nerves changes the balance that affects HRV; as such, HRV has been investigated as a useful measure of autonomic regulation of the cardiovascular system. Healthy individuals exhibit highly variable R-R intervals in response to the environment and internal state of the animal. It appears that HRV may be used as a predictive measure of how well an animal will cope with stress and an outcome for how distressed they are by their environment. Distressing experiences suppress vagal tone, increase HR, and reduce HRV; when such experiences are chronic, these cardiovascular effects can develop into pathological disease states, such as diarrhea and cardiovascular disease (von Borell et al. 2007). High vagal tone is associated with better behavioral regulation and responsiveness to stress and changes in the environment. Infants and children who are behaviorally inhibited tend to have high HR, low HRV, and vagal tone, and are more likely to develop anxiety-related psychopathology later in life (Friedman and Thayer 1998).

Measuring HR and HRV in primates usually requires that subjects are either immobilized in some manner (e.g., chair restraint: Bliss-Moreau et al. 2013; Grandi and Ishida 2015) or have a surgically implanted telemetry device, which allows data collection while the animal engages in normal activities (e.g., Boccia et al. 1989; Doyle et al. 2008). Chair restraint allows for external monitoring and thus does not require surgery, but eliminates normal activity patterns and interaction with the environment and can only be used to measure ANS responses to controlled stimuli (e.g., video playback, human–primate grooming). Implantable devices prevent subjects and conspecifics from pulling at or removing the electrodes and wires measuring cardiovascular function and allow the animals to engage in regular activity during data collection but require a surgical procedure and recovery time. Implanting a measurement device is of course more expensive in both time and money, but it provides a more portable device that can be used over a longer period of time. Data recorded by the devices can be transmitted to and stored on a computer. Details of the output variables to create a HRV measure are provided in von Borell et al. (2007) and Grandi and Ishida (2015).

5   Immune System

The notion that psychological factors can impact health and disease guides the field of psychosomatic medicine. This idea also drives the field of psychoneuroimmunology, although in this case, the focus is more on the mechanisms by which psychological factors affect neural (including neuroendocrine) and immune function. Given that the goal of captive management of nonhuman primates is to keep the animals both psychologically and physically healthy, one might expect that measures of immunity might be usefully employed to assess well-being. In fact, there are many studies that have been done with nonhuman primates that suggest measures that might be useful. As with the previous sections, we will first provide a brief overview of the immune system and then discuss measures that could be used in a welfare context. While we will mention a variety of measures, our emphasis will be on easily obtainable ones; as the reader might expect, much of immunology involves tissue culture that requires considerable technical skill and specialized (and expensive) equipment. We will describe these measures only briefly and emphasize simpler, less expensive measures.

Given that the major function of the immune system is to protect the organism from attack by potentially harmful bacteria, viruses, toxins, etc., and that these substances could enter the body anywhere over the surface of the individual, it should come as no surprise that tissues associated with the immune system are widely distributed throughout the body. Primary lymphoid tissue—bone marrow and the thymus—is where immune cells, particularly lymphocytes, are generated from progenitor cells and mature. Secondary lymphoid tissue—lymph nodes, spleen, and mucosa-associated lymphoid tissue [MALT], such as in the tonsils and lining of the gastrointestinal tract—is much more distributed and is generally "where the action" is: at these sites, immune cells encounter antigens (defined as a toxin or foreign substance, such as a protein on a virus, that generates an immune response) and initiate immune responses. Most widely distributed are the immune cells themselves (usually referred to as leukocytes or white blood cells), which circulate through the blood and the lymphatic system, serving an essential surveillance role, and traveling to sites of injury or to secondary lymphoid tissue to generate responses.

What kinds of responses are generated by the immune system? In general, immune responses are classified into two types, innate and adaptive, and different leukocytes are associated with each. We note, however, that this is a somewhat artificial distinction, inasmuch as cells of these two systems usually work together to generate immune responses. Nevertheless, there are some important differences. Innate immune responses are usually the first line of defense against a pathogen, allowing the organism to combat the microbe while more targeted, specific (i.e., adaptive), immune responses develop. For example, within hours of becoming infected with a virus, a cell will produce high levels of proteins called interferons, which inhibit replication of the virus and generally induce an antiviral state in the affected cell. Interferons also activate natural killer (NK) cells, which are leukocytes and which kill virally infected cells. These processes are "innate" in the sense that they occur relatively automatically, and with consistent "strength" to virtually every viral infection. This is in contrast to adaptive immunity, the key cell type of which is the lymphocyte. When a pathogen is first encountered, only a few lymphocytes are present in the organism that can detect that pathogen, and these cells become activated and clonally expand. Over the course of the next couple of weeks, a highly specific response develops to that specific microbe (or more accurately, to a protein portion of the microbe—the antigen). Once the microbe has been eliminated, specific immune responses linger—lymphocytes demonstrate "memory." Should that same microbe be encountered a second time, months, or years later, memory lymphocytes will be able to quickly upregulate their responses, and so a very intense and targeted immune response will occur much more rapidly. This is the rationale behind vaccination—present the organism with an attenuated or killed version of a pathogen in order to develop a specific immune response; later, should the organism encounter the wild-type pathogen, the resources of specific immunity can be brought to bear virtually immediately.

One final topic that should be addressed in our overview of the immune system is a description of the soluble mediators of immunity. These are proteins that can be found in blood that are the effector molecules of immune cells or that permit communication between immune cells. We will consider two types of soluble mediators. Cytokines (which includes interleukins, chemokines, and interferons) are proteins that are produced by a variety of cell types, including immune and nonimmune cells, and they serve communicative functions, generally promoting or dampening immune responses. For example, interleukin-6 (IL-6) is a cytokine that generally promotes inflammation (i.e., it is pro-inflammatory), while IL-10 is often considered an anti-inflammatory cytokine. Inflammation is a process that has been linked to a variety of diseases, such as heart disease, atherosclerosis, and asthma. The second type of soluble mediator to discuss is antibody (also referred to as immunoglobulin or Ig). Antibody is produced and secreted by B cells, which are a subset of lymphocytes, and its main function is to identify and neutralize pathogens. Antibodies are an important component of the vaccination response described in the previous paragraph. They are also specific to antigens. For example, if you are

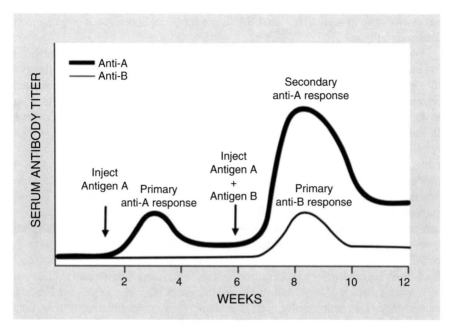

Fig. 2 Idealized illustration of generation of antibody responses to immunizations. Injection of antigen A results in a modest antibody response that is specific to that antigen (i.e., antibody to antigen B is unaffected) that wanes over time. Re-exposure to that antigen at the second time point, however, results in a faster and larger antibody response that is more persistent. Artwork by O. Flores, adapted from Abbas et al. (1991)

immunized/vaccinated against tetanus, you will generate antibodies that will help identify the toxin that is produced by the bacterium that causes tetanus. But that immunization will not help you against pertussis, or diphtheria, or other diseases. Often, once one is exposed to an antigen, antibodies will develop and will remain in circulation for your lifetime (see Fig. 2). In other cases, immunity needs to be renewed via "booster" immunizations.

This description is very simplistic, but it does provide a background to discuss measures that might be useful in a welfare context. One point that must be emphasized, however, is that the immune system does not operate in isolation, although up until the last few decades, most physiologists believed that it did. We now know that cells of the immune system contain receptors for a variety of nonimmune substances. For example, nearly every cell type in your body contains receptors for glucocorticoids such as cortisol. In immune cells, glucocorticoids, which as steroids function to influence transcription of genes, generally have an anti-inflammatory effect—that is, when glucocorticoid levels are high, immune cells are more likely to decrease transcription and production of pro-inflammatory cytokines. Similarly, immune cells also have adrenergic receptors, which are particularly sensitive to norepinephrine (Lorton and Bellinger 2015). Thus, the two major stress–response systems, HPA and SAM, are intimately involved in regulation of immune responses.

With this as a background, what types of immune-related measures might be valuable in assessing psychological well-being? As described at the beginning of this section, there is a variety of measures that one could use. Many such measures require advanced technical skills and/or specialized equipment. Examples of these more complex measures that have been shown to be sensitive to chronic stress include assessing the density of sympathetic innervation in lymph nodes (Sloan et al. 2008); in vitro stimulation of cells to determine their responsiveness to an antigen (e.g., Chun et al. 2013); or examination, in vitro, of how effectively cytokine production is regulated by glucocorticoids (e.g., Hennessy et al. 2016). More practical measures can be obtained via blood samples, although we note that all such measures can be affected by many factors, such as the time of day the sample is taken, illness, pregnancy, and eustressful experiences such as exercise; sampling protocols must attend to these factors. Perhaps the simplest measures might be counting the number of leukocytes (and subsets of leukocytes) in peripheral blood. It has been known for decades that stress can result in changes in leukocyte numbers, increasing numbers of some cell types, and decreasing numbers of others. This was demonstrated experimentally, for example, by administering exogenous glucocorticoids (Fauci and Dale 1974), or exogenous epinephrine (Crary et al. 1983). These cells, which are constantly circulating and performing a surveillance function, are not being newly created or destroyed by the presence of these hormones; rather, the change in numbers results from cells that move into and out of the circulation.

The use of cell counts as indicators of welfare has been proposed before (e.g., Coe and Scheffler 1989) and a literature review described studies in which immune measures were used to examine effects of relocations and group formations (Capitanio 1998). It is important to realize, however, that the very fact that immune cell numbers are sensitive to stress-related hormones such as cortisol or epinephrine implies that careful attention must be paid to sampling conditions. For example, in our laboratory, whenever we draw blood from animals, especially when we are drawing blood from multiple animals that are living in the same room, we always collect timing information and make a distinction between disturbance time and draw time. Consider a situation where there are nine animals housed individually in a single room (Capitanio et al. 1996). We start a stopwatch as soon as we open the door to the room (door-opening time). Then, as soon as we step in front of the first cage, we record the time (cage-front time), and when the needle is withdrawn from the animal's arm (as with nearly all of our studies, we train animals to extend their arms for blood sampling), we record the time again (needle time). Then, we move on to the second animal, recording the cage-front and needle times, and so forth until we withdraw the needle from the ninth animal. The draw time is calculated as the amount of time we interact with each specific animal: the needle time minus the cage-front time. The disturbance time reflects how long we have been "disturbing" the room, and is the difference between the needle time and the door-opening time. Draw time is usually on the order of about 2 min for each animal, while disturbance time cumulates—the disturbance time for the ninth animal will be substantially greater than the disturbance time for the fourth animal, although the draw times for each may be the same. When we analyzed the effects of draw and disturbance times on the measures of interest to us, we found an important difference for cortisol concentrations versus cell counts: As others have found, cortisol concentrations are most sensitive to the amount of time it takes to obtain blood from the specific animal (i.e., the draw time); samples obtained within a few minutes of first contacting the animal typically do not result in cortisol elevations. In other words, the cortisol concentrations for the ninth animal and for the first animal were equivalent (except of course for individual variation); there was no effect of disturbance time on cortisol. In contrast, significant effects of disturbance time were seen for counts of leukocytes (Capitanio et al. 1996, Fig. 1): Neutrophil numbers were significantly elevated after 12 min of disturbance time, while total lymphocyte numbers (and especially the CD8+ lymphocyte subset) were elevated after 6 min of disturbance (i.e., by the second or third of the nine animals). Because cortisol concentrations did not change based on disturbance time, the elevated numbers of neutrophils and CD8+ lymphocytes probably were related to elevations in concentrations of epinephrine (which we did not measure in this study). These results indicate that careful attention must be paid to the timing associated with sample collection when using cell counts.

A second set of measures one could easily obtain from blood sampling includes concentrations of cytokines. Although levels of some cytokines in blood may be too low for detection in many individuals, studies have found that levels of some cytokines are indeed stress-responsive (though, again, as with cell counts, they are also responsive to a host of other processes: Zhou et al. 2010). For example, in a meta-analysis of human studies, Steptoe et al. (2007) reported that the pro-inflammatory cytokines IL-6 and IL-1-beta show robust increases in response to acute stress. It is unclear to us, however, what information cytokine concentrations might provide that is better than what one might obtain from examining cell counts, or even cortisol concentrations.

A final blood-borne measure that merits strong consideration, in our opinion, is the antibody response to vaccination (Fig. 2). As described earlier, antibody production involves the coordination of sets of immune cells—the antigen is "processed" by circulating antigen-presenting cells like monocytes, which then relocate to secondary lymphoid tissue where interactions with B cells (which produce antibody) occur; usually, T cells provide help (see Siegrist 2008 for an excellent discussion). An antibody response to an administered vaccine, then, provides information on how well the various components of immunity are working; as such, antibody responses to vaccines get at aspects of regulation. Moreover, because the time frame for developing an antibody response is on the order of many days to a few weeks, antibody responses can reflect relatively long-term psychological conditions that the animal may be experiencing, such as poor welfare; in this way, antibody responses are similar to glucocorticoid measures one can obtain from hair samples, as described above. Assessment of antibody levels to common vaccines like tetanus toxoid can be easily accomplished with commercially available assay kits.

So far, the measures we have been describing have all been obtained from blood samples. Are there potentially useful measures found in other biological fluids? A growing interest in the field of primate disease ecology has resulted in examination of measures of immunity in urine or feces, substances that are more easily obtained from animals in the field than is blood. Two recent papers (Higham et al. 2015; Heistermann and Higham 2015) have examined measures of immunity in urine, feces, and blood, and one measure in particular shows strong correlations between blood and urine: neopterin. Neopterin is synthesized by monocytes upon stimulation with interferon-gamma, a cytokine that is involved in inflammation, and neopterin concentrations are typically characterized as reflecting general activation of the immune system. Neopterin concentrations have been measured in urine for decades in humans as evidence of disease (e.g., HIV disease: Fuchs et al. 1990), but have not been used extensively as indicators of stress (though see Dunbar et al. 1993). While the use of urinary and/or fecal measures of immunity remains rare in nonhuman primate studies, we believe they hold promise for assessment of welfare in the future, as we learn more about the dynamics of immune measures in these excreta.

6   Conclusion

As indicated in our introduction, we believe that physiological measures—particularly if they are assessing regulatory aspects of biological systems, rather than simply levels of analytes—may be especially useful measures of welfare, either alone or combined with other measures, because they provide information on the internal state of the animal that might not be apparent in its behavior. Moreover, the ability to assess many of these measures in substances other than blood provides for measurement in longer time frames, which may parallel more closely what we believe is an important aspect of "well-being"—a persistent psychological state that is maintained despite occasional, acutely stressful experiences. Interestingly, progress in developing new measures that can be assessed noninvasively may come from the growing interest in primate disease ecology—a discipline in which options for sample collection are limited. That primate disease ecology and captive animal management share similar goals and problems highlights, in our opinion, the highly integrative nature of primatology, and suggests closer communication between these two areas might be mutually beneficial.

Acknowledgments Preparation of this chapter was supported by grants OD010962 (JPC) and OD011107 (JPC, JV, DLH).

References

Abbas AK, Lichtman AH, Pober JS (1991) Cellular and molecular immunology. W.B. Saunders Company, Philadelphia, PA

Accorsi PA, Carloni E, Valsecchi P et al (2008) Cortisol determination in hair and faeces from domestic cats and dogs. Gen Comp Endocrinol 155(2):398–402. https://doi.org/10.1016/j.ygcen.2007.07.002

Anestis SF (2005) Behavioral style, dominance rank, and urinary cortisol in young chimpanzees (*Pan troglodytes*). Behaviour 142(9–10):1245–1268. https://doi.org/10.1163/156853905774539418

Ange-van Heugten KD, van Heugten E, Timmer S, et al (2009) Fecal and salivary cortisol concentrations in woolly (*Lagothrix* ssp.) and spider monkeys (*Ateles* spp.). Int J Zool 127852. https://doi.org/10.1155/2009/127852

Aureli F, Preston SD, de Waal FBM (1999) Heart rate responses to social interactions in free-moving rhesus macaques (*Macaca mulatta*): a pilot study. J Comp Psychol 113(1):59–65. https://doi.org/10.1037/0735-7036.113.1.59

Bachmann I, Bernasconi P, Herrmann R et al (2003) Behavioural and physiological responses to an acute stressor in crib-biting and control horses. Appl Anim Behav Sci 82(4):297–311. https://doi.org/10.1016/S0168-1591(03)00086-8

Bahr NI, Palme R, Möhle U et al (2000) Comparative aspects of the metabolism and excretion of cortisol in three individual nonhuman primates. Gen Comp Endocrinol 117(3):427–438. https://doi.org/10.1006/gcen.1999.7431

Baker DG, West SA, Nicholson WE et al (1999) Serial CSF corticotropin-releasing hormone levels and adrenocortical activity in combat veterans with posttraumatic stress disorder. Am J Psychiatry 156(4):585–588. https://doi.org/10.1176/ajp.156.4.585

Baker KC, Bloomsmith MA, Oettinger BC et al (2012) Benefits of pair housing are consistent across a diverse population of rhesus macaques. Appl Anim Behav Sci 137(3–4):148–156. https://doi.org/10.1016/j.applanim.2011.09.010

Behie AM, Pavelka MSM, Chapman CA (2010) Sources of variation in fecal cortisol levels in howler monkeys in Belize. Am J Primatol 72(7):600–606. https://doi.org/10.1002/ajp.20813

Bennett A, Hayssen V (2010) Measuring cortisol in hair and saliva from dogs: coat color and pigment differences. Domest Anim Endocrinol 39(3):171–180. https://doi.org/10.1016/j.domaniend.2010.04.003

Bliss-Moreau E, Machado CJ, Amaral DG (2013) Macaque cardiac physiology is sensitive to the valence of passively viewed sensory stimuli. PLoS One 8(8):e71170. https://doi.org/10.1371/journal.pone.0071170

Bloomsmith M, Perlman J, Franklin A, Martin AL (2022) Training research primates. In: Robinson LM, Weiss A (eds) Nonhuman primate welfare: from history, science, and ethics to practice. Springer, Cham, pp 517–546

Boccia ML, Reite M, Laudenslager M (1989) On the physiology of grooming in a pigtail macaque. Physiol Behav 45(3):667–670. https://doi.org/10.1016/0031-9384(89)90089-9

Cannon WB (1953) Bodily changes in pain, hunger, fear and rage: an account of recent researches into the function of emotional excitement, 2nd edn. Branford, Boston

Capitanio JP (1998) Social experience and immune system measures in laboratory-housed macaques: implications for management and research. ILAR J 39(1):12–20. https://doi.org/10.1093/ilar.39.1.12

Capitanio JP, Cole SW (2015) Social instability and immunity in rhesus monkeys: the role of the sympathetic nervous system. Philos Trans R Soc B Biol Sci 370(1669):20140104. https://doi.org/10.1098/rstb.2014.0104

Capitanio JP, Mendoza SP, McChesney M (1996) Influences of blood sampling procedures on basal hypothalamic-pituitary-adrenal hormone levels and leukocyte values in rhesus macaques (*Macaca mulatta*). J Med Primatol 25(1):26–33. https://doi.org/10.1111/j.1600-0684.1996.tb00189.x

Capitanio JP, Mendoza SP, Lerche NW, Mason WA (1998) Social stress results in altered glucocorticoid regulation and shorter survival in simian acquired immune deficiency syndrome. Proc Natl Acad Sci U S A 95(8):4714–4719. https://doi.org/10.1073/pnas.95.8.4714

Capitanio JP, Mendoza SP, Mason WA, Maninger N (2005) Rearing environment and hypothalamic-pituitary-adrenal regulation in young rhesus monkeys (*Macaca mulatta*). Dev Psychobiol 46(4):318–330. https://doi.org/10.1002/dev.20067

Champoux M, Hwang L, Lang O, Levine S (2001) Feeding demand conditions and plasma cortisol in socially-housed squirrel monkey mother–infant dyads. Psychoneuroendocrinology 26 (5):461–477. https://doi.org/10.1016/S0306-4530(01)00006-3

Chun K, Miller LA, Schelegle ES et al (2013) Behavioral inhibition in rhesus monkeys (*Macaca mulatta*) is related to the airways response, but not immune measures, commonly associated with asthma. PLoS One 8(8):e71575. https://doi.org/10.1371/journal.pone.0071575

Clarke AS, Czekala NM, Lindburg DG (1995) Behavioral and adrenocortical responses of male cynomolgus and lion-tailed macaques to social stimulation and group formation. Primates 36:41–56. https://doi.org/10.1007/bf02381914

Coe CL, Scheffler J (1989) Utility of immune measures for evaluating psychological well-being in nonhuman primates. Zoo Biol 8(S1):89–99. https://doi.org/10.1002/zoo.1430080510

Coleman K, Pranger L, Maier A et al (2008) Training rhesus macaques for venipuncture using positive reinforcement techniques: a comparison with chimpanzees. J Am Assoc Lab Anim Sci. 47(1):37–41

Crary B, Hauser SL, Borysenko M et al (1983) Epinephrine-induced changes in the distribution of lymphocyte subsets in peripheral blood of humans. J Immunol 131(3):1178–1181

D'Anna-Hernandez KL, Ross RG, Natvig CL, Laudenslager ML (2011) Hair cortisol levels as a retrospective marker of hypothalamic-pituitary axis activity throughout pregnancy: comparison to salivary cortisol. Physiol Behav 104(2):348–353. https://doi.org/10.1016/j.physbeh.2011.02.041

Dallman MF, Akana SF, Levin N et al (1994) Corticosteroids and the control of function in the hypothalamo-pituitary-adrenal (HPA) Axis. Ann N Y Acad Sci 746(1):22–31. https://doi.org/10.1111/j.1749-6632.1994.tb39206.x

Davenport MD, Tiefenbacher S, Lutz CK et al (2006) Analysis of endogenous cortisol concentrations in the hair of rhesus macaques. Gen Comp Endocrinol 147(3):255–261. https://doi.org/10.1016/j.ygcen.2006.01.005

Davenport MD, Lutz CK, Tiefenbacher S et al (2008) A rhesus monkey model of self-injury: effects of relocation stress on behavior and neuroendocrine function. Biol Psychiatry 63(10):990–996. https://doi.org/10.1016/j.biopsych.2007.10.025

Dettmer AM, Novak MA, Suomi SJ, Meyer JS (2012) Physiological and behavioral adaptation to relocation stress in differentially reared rhesus monkeys: hair cortisol as a biomarker for anxiety-related responses. Psychoneuroendocrinology 37(2):191–199. https://doi.org/10.1016/j.psyneuen.2011.06.003

Dettmer AM, Novak MA, Meyer JS, Suomi SJ (2014) Population density-dependent hair cortisol concentrations in rhesus monkeys (*Macaca mulatta*). Psychoneuroendocrinology 42:59–67. https://doi.org/10.1016/j.psyneuen.2014.01.002

Doyle LA, Baker KC, Cox LD (2008) Physiological and behavioral effects of social introduction on adult male rhesus macaques. Am J Primatol 70(6):542–550. https://doi.org/10.1002/ajp.20526

Dunbar PR, Hill J, Neale TJ (1993) Urinary neopterin quantification indicates altered cell-mediated immunity in healthy subjects under psychological stress. Aust New Zeal J Psychiatry 27 (3):495–501. https://doi.org/10.3109/00048679309075808

Fauci AS, Dale DC (1974) The effect of in vivo hydrocortisone on subpopulations of human lymphocytes. J Clin Invest 53(1):240–246. https://doi.org/10.1172/JCI107544

Friedman BH, Thayer JF (1998) Autonomic balance revisited: panic anxiety and heart rate variability. J Psychosom Res 44(1):133–151. https://doi.org/10.1016/S0022-3999(97)00202-X

Fuchs D, Artner-Dworzak E, Hausen A et al (1990) Urinary excretion of porphyrins is increased in patients with HIV-1 infection. AIDS 4(4):341–344. https://doi.org/10.1097/00002030-199004000-00009

Garde AH, Hansen ÅM (2005) Long-term stability of salivary cortisol. Scand J Clin Lab Invest 65 (5):433–436. https://doi.org/10.1080/00365510510025773

Geverink N, Schouten WG, Gort G, Wiegant V (2002) Individual differences in behavioral and physiological responses to restraint stress in pigs. Physiol Behav 77(2–3):451–457. https://doi.org/10.1016/S0031-9384(02)00877-6

Gottlieb DH, Capitanio JP, McCowan BJ (2013) Risk factors for stereotypic behavior and self-biting in rhesus macaques (*Macaca mulatta*): animal's history, current environment, and personality. Am J Primatol 75(10):995–1008. https://doi.org/10.1002/ajp.22161

Gozansky WS, Lynn JS, Laudenslager ML, Kohrt WM (2005) Salivary cortisol determined by enzyme immunoassay is preferable to serum total cortisol for assessment of dynamic hypothalamic-pituitary-adrenal axis activity. Clin Endocrinol 63(3):336–341. https://doi.org/10.1111/j.1365-2265.2005.02349.x

Grandi LC, Ishida H (2015) The physiological effect of human grooming on the heart rate and the heart rate variability of laboratory non-human primates: a pilot study in male rhesus monkeys. Front Vet Sci 2:50. https://doi.org/10.3389/fvets.2015.00050

Hainsworth R (1995) The control and physiological importance of heart rate. In: Malik M, Camm A (eds) Heart rate variability. Futura Publishing Company Inc, Armonk, pp 3–19

Heistermann M, Higham JP (2015) Urinary neopterin, a non-invasive marker of mammalian cellular immune activation, is highly stable under field conditions. Sci Rep 5:16308. https://doi.org/10.1038/srep16308

Heistermann M, Palme R, Ganswindt A (2006) Comparison of different enzyme-immunoassays for assessment of adrenocortical activity in primates based on fecal analysis. Am J Primatol 68(3):257–273. https://doi.org/10.1002/ajp.20222

Hennessy MB, Chun K, Capitanio JP (2016) Depressive-like behavior, its sensitization, social buffering, and altered cytokine responses in rhesus macaques moved from outdoor social groups to indoor housing. Soc Neurosci 12(1):65–75. https://doi.org/10.1080/17470919.2016.1145595

Herman JP, Prewitt CMF, Cullinan WE (1996) Neuronal circuit regulation of the hypothalamo-pituitary-adrenocortical stress axis. Crit Rev Neurobiol 10(3–4):371–394. https://doi.org/10.1615/CritRevNeurobiol.v10.i3-4.50

Herman JP, Figueiredo H, Mueller NK et al (2003) Central mechanisms of stress integration: hierarchical circuitry controlling hypothalamo–pituitary–adrenocortical responsiveness. Front Neuroendocrinol 24(3):151–180. https://doi.org/10.1016/j.yfrne.2003.07.001

Higham JP, Kraus C, Stahl-Hennig C et al (2015) Evaluating noninvasive markers of nonhuman primate immune activation and inflammation. Am J Phys Anthropol 158(4):673–684. https://doi.org/10.1002/ajpa.22821

Holsboer F (2000) The corticosteroid receptor hypothesis of depression. Neuropsychopharmacology 23:477–501. https://doi.org/10.1016/S0893-133X(00)00159-7

Ito N, Ito T, Kromminga A et al (2005) Human hair follicles display a functional equivalent of the hypothalamic-pituitary-adrenal axis and synthesize cortisol. FASEB J 19(10):1332–1334. https://doi.org/10.1096/fj.04-1968fje

Jänig W (2008) The integrative action of the autonomic nervous system: neurobiology of homeostasis. Cambridge University Press, New York

Jarcho MR, Mendoza SP, Bales KL (2012) Hormonal and experiential predictors of infant survivorship and maternal behavior in a monogamous primate (*Callicebus cupreus*). Am J Primatol 74(5):462–470. https://doi.org/10.1002/ajp.22003

Kirschbaum C, Tietze A, Skoluda N, Dettenborn L (2009) Hair as a retrospective calendar of cortisol production-increased cortisol incorporation into hair in the third trimester of pregnancy. Psychoneuroendocrinology 34(1):32–37. https://doi.org/10.1016/j.psyneuen.2008.08.024

Koren L, Mokady O, Karaskov T et al (2002) A novel method using hair for determining hormonal levels in wildlife. Anim Behav 63:403–440. https://doi.org/10.1006/anbe.2001.1907

Laudenslager ML, Boccia ML, Berger CL et al (1995) Total cortisol, free cortisol, and growth hormone associated with brief social separation experiences in young macaques. Dev Psychobiol 28(4):199–211. https://doi.org/10.1002/dev.420280402

Lemmens SG, Born JM, Martens EA et al (2011) Influence of consumption of a high-protein- vs. high-carbohydrate meal on the physiological cortisol and psychological mood response in men and women. PLoS One 6(2):e16826. https://doi.org/10.1371/journal.pone.0016826

Line SW, Morgan KN, Markowitz H, Strong S (1990) Increased cage size does not alter heart rate or behavior in female rhesus monkeys. Am J Primatol 20(2):107–113. https://doi.org/10.1002/ajp.1350200205

Lorton D, Bellinger DL (2015) Molecular mechanisms underlying β-adrenergic receptor-mediated cross-talk between sympathetic neurons and immune cells. Int J Mol Sci 16(3):5635–5665. https://doi.org/10.3390/ijms16035635

Lutz CK, Baker KC (2022) Using behavior to assess primate welfare. In: Robinson LM, Weiss A (eds) Nonhuman primate welfare: from history, science, and ethics to practice. Springer, Cham, pp 171–206

Lutz CK, Tiefenbacher S, Jorgensen MJ et al (2000) Techniques for collecting saliva from awake, unrestrained, adult monkeys for cortisol assay. Am J Primatol 52(2):93–99. https://doi.org/10.1002/1098-2345(200010)52:2<93::AID-AJP3>3.3.CO;2-2

Macbeth BJ, Cattet MRL, Stenhouse GB et al (2010) Hair cortisol concentration as a noninvasive measure of long-term stress in free-ranging grizzly bears (*Ursus arctos*): considerations with implications for other wildlife. Can J Zool 88(10):935–949. https://doi.org/10.1139/Z10-057

Mason JW, Mangan G, Brady JV et al (1961) Concurrent plasma epinephrine, norepinephrine and 17-hydroxycorticosteroid levels during conditioned emotional disturbances in monkeys. Psychosom Med 23(4):344–353. https://doi.org/10.1097/00006842-196107000-00011

Menicucci D, Piarulli A, Mastorci F et al (2013) Interactions between immune, stress-related hormonal and cardiovascular systems following strenuous physical exercise. Arch Ital Biol 151(3):126–136. https://doi.org/10.4449/aib.v151i3.1523

Meyer JS, Novak MA (2012) Minireview: hair cortisol: a novel biomarker of hypothalamic-pituitary- adrenocortical activity. Endocrinology 153(9):4120–4127. https://doi.org/10.1210/en.2012-1226

Moberg GP (2000) Biological response to stress: implications for animal welfare. In: Moberg GP, Mench JA (eds) The biology of animal stress: basic principles and implications for animal welfare. CABI, New York, pp 1–21

Mohr E, Langbein J, Nürnberg G (2002) Heart rate variability: a noninvasive approach to measure stress in calves and cows. Physiol Behav 75(1–2):251–259. https://doi.org/10.1016/S0031-9384(01)00651-5

Muller MN, Wrangham RW (2004) Dominance, cortisol and stress in wild chimpanzees (*Pan troglodytes schweinfurthii*). Behav Ecol Sociobiol 55:332–340. https://doi.org/10.1007/s00265-003-0713-1

Nater UM, Rohleder N (2009) Salivary alpha-amylase as a non-invasive biomarker for the sympathetic nervous system: current state of research. Psychoneuroendocrinology 34(4):486–496. https://doi.org/10.1016/j.psyneuen.2009.01.014

Novak MA (2003) Self-injurious behavior in rhesus monkeys: new insights into its etiology, physiology, and treatment. Am J Primatol 59(1):3–19. https://doi.org/10.1002/ajp.10063

Novak MA, Hamel AF, Kelly BJ et al (2013) Stress, the HPA axis, and nonhuman primate well-being: a review. Appl Anim Behav Sci 143(2–4):135–149. https://doi.org/10.1016/j.applanim.2012.10.012

Petrullo LA, Mandalaywala TM, Parker KJ et al (2016) Effects of early life adversity on cortisol/salivary alpha-amylase symmetry in free-ranging juvenile rhesus macaques. Horm Behav 86:78–84. https://doi.org/10.1016/j.yhbeh.2016.05.004

Porges SW (1995) Cardiac vagal tone: a physiological index of stress. Neurosci Biobehav Rev 19(2):225–233. https://doi.org/10.1016/0149-7634(94)00066-A

Rangel-Negrín A, Alfaro JL, Valdez RA et al (2009) Stress in Yucatan spider monkeys: effects of environmental conditions on fecal cortisol levels in wild and captive populations. Anim Conserv 12(5):496–502. https://doi.org/10.1111/j.1469-1795.2009.00280.x

Rasmussen KL, Suomi SJ (1989) Heart rate and endocrine responses to stress in adolescent male rhesus monkeys on Cayo Santiago. P R Health Sci J 8(1):65–71

Raul J-S, Cirimele V, Ludes B, Kintz P (2004) Detection of physiological concentrations of cortisol and cortisone in human hair. Clin Biochem 37(12):1105–1111. https://doi.org/10.1016/j.clinbiochem.2004.02.010

Reinhardt V, Cowley D, Scheffler J et al (1990) Cortisol response of female rhesus monkeys to venipuncture in homecage versus venipuncture in restraint apparatus. J Med Primatol 19(6):601–606

Rietmann TR, Stuart AEA, Bernasconi P et al (2004) Assessment of mental stress in warmblood horses: heart rate variability in comparison to heart rate and selected behavioural parameters. Appl Anim Behav Sci 88(1–2):121–136. https://doi.org/10.1016/j.applanim.2004.02.016

Rohleder N, Nater UM, Wolf JM et al (2004) Psychosocial stress-induced activation of salivary alpha-amylase: an indicator of sympathetic activity? Ann N Y Acad Sci 1032(1):258–263. https://doi.org/10.1196/annals.1314.033

Ruys J, Mendoza S, Capitanio J, Mason W (2004) Behavioral and physiological adaptation to repeated chair restraint in rhesus macaques. Physiol Behav 82(2–3):205–213. https://doi.org/10.1016/j.physbeh.2004.02.031

Sapolsky RM, Romero LM, Munck AU (2000) How do glucocorticoids influence stress responses? Integrating permissive, suppressive, stimulatory, and preparative actions. Endocr Rev 21(1):55–89. https://doi.org/10.1210/edrv.21.1.0389

Sauvé B, Koren G, Walsh G et al (2007) Measurement of cortisol in human hair as a biomarker of systemic exposure. Clin Investig Med 30(5):183. https://doi.org/10.25011/cim.v30i5.2894

Selye H (1979) The stress concept and some of its implications. In: Hamilton V, Varburton D (eds) Human stress and cognition. Wiley, New York, pp 11–30

Setchell KDR, Chua KS, Himsworth RL (1977) Urinary steroid excretion by the squirrel monkey (*Saimuri sciureus*). J Endocrinol 73(2):365–375. https://doi.org/10.1677/joe.0.0730365

Sgoifo A, Pozzato C, Costoli T et al (2001) Cardiac autonomic responses to intermittent social conflict in rats. Physiol Behav 73(3):343–349. https://doi.org/10.1016/S0031-9384(01)00455-3

Sheriff MJ, Dantzer B, Delehanty B et al (2011) Measuring stress in wildlife: techniques for quantifying glucocorticoids. Oecologia 166:869–887. https://doi.org/10.1007/s00442-011-1943-y

Siegrist CA (2008) Vaccine immunology. In: Plotkin S, Orenstein W, PA Offit (eds) Vaccines, 5th. Elsevier, New York, pp. 17–36

Sloan EK, Capitanio JP, Cole SW (2008) Stress-induced remodeling of lymphoid innervation. Brain Behav Immun 22(1):15–21. https://doi.org/10.1016/j.bbi.2007.06.011

Smith T, French J (1997) Psychosocial stress and urinary cortisol excretion in marmoset monkeys (*Callithrix kuhli*). Physiol Behav 62(2):225–232. https://doi.org/10.1016/S0031-9384(97)00103-0

Steptoe A, Hamer M, Chida Y (2007) The effects of acute psychological stress on circulating inflammatory factors in humans: a review and meta-analysis. Brain Behav Immun 21(7):901–912. https://doi.org/10.1016/j.bbi.2007.03.011

Stroud LR, Solomon C, Shenassa E et al (2007) Long-term stability of maternal prenatal steroid hormones from the National Collaborative Perinatal Project: still valid after all these years. Psychoneuroendocrinology 32(2):140–150. https://doi.org/10.1016/j.psyneuen.2006.11.008

Teruhisa U, Ryoji H, Taisuke I et al (1981) Use of saliva for monitoring unbound free cortisol levels in serum. Clin Chim Acta 110(2–3):245–253. https://doi.org/10.1016/0009-8981(81)90353-3

Terwissen CV, Mastromonaco GF, Murray DL (2013) Influence of adrenocorticotrophin hormone challenge and external factors (age, sex, and body region) on hair cortisol concentration in Canada lynx (*Lynx canadensis*). Gen Comp Endocrinol 194:162–167. https://doi.org/10.1016/j.ygcen.2013.09.010

Tiefenbacher S, Novak MA, Marinus LM et al (2004) Altered hypothalamic–pituitary–adrenocortical function in rhesus monkeys (*Macaca mulatta*) with self-injurious behavior. Psychoneuroendocrinology 29(4):501–515. https://doi.org/10.1016/S0306-4530(03)00068-4

Touma C, Palme R (2005) Measuring fecal glucocorticoid metabolites in mammals and birds: the importance of validation. Ann N Y Acad Sci 1046(1):54–74. https://doi.org/10.1196/annals.1343.006

Ulrich-Lai YM, Figueiredo HF, Ostrander MM et al (2006) Chronic stress induces adrenal hyperplasia and hypertrophy in a subregion-specific manner. Am J Physiol Metab 291(5):E965–E973. https://doi.org/10.1152/ajpendo.00070.2006

Vandeleest JJ, Blozis SA, Mendoza SP, Capitanio JP (2013) The effects of birth timing and ambient temperature on the hypothalamic–pituitary–adrenal axis in 3–4 month old rhesus monkeys. Psychoneuroendocrinology 38(11):2705–2712. https://doi.org/10.1016/j.psyneuen.2013.06.029

von Borell E, Langbein J, Després G et al (2007) Heart rate variability as a measure of autonomic regulation of cardiac activity for assessing stress and welfare in farm animals—a review. Physiol Behav 92(3):293–316. https://doi.org/10.1016/j.physbeh.2007.01.007

Weary DM, Niel L, Flower FC, Fraser D (2006) Identifying and preventing pain in animals. Appl Anim Behav Sci 100(1–2):64–76. https://doi.org/10.1016/j.applanim.2006.04.013

Yehuda R (2006) Advances in understanding neuroendocrine alterations in PTSD and their therapeutic implications. Ann N Y Acad Sci 1071(1):137–166. https://doi.org/10.1196/annals.1364.012

Yehuda R, Kahana B, Binder-Brynes K et al (1995) Low urinary cortisol excretion in holocaust survivors with posttraumatic stress disorder. Am J Psychiatry 152(7):982–986. https://doi.org/10.1176/ajp.152.7.982

Yehuda R, Halligan SL, Golier JA et al (2004) Effects of trauma exposure on the cortisol response to dexamethasone administration in PTSD and major depressive disorder. Psychoneuroendocrinology 29(3):389–404. https://doi.org/10.1016/S0306-4530(03)00052-0

Zhou X, Fragala MS, McElhaney JE, Kuchel GA (2010) Conceptual and methodological issues relevant to cytokine and inflammatory marker measurements in clinical research. Curr Opin Clin Nutr Metab Care 13(5):541–547. https://doi.org/10.1097/MCO.0b013e32833cf3bc

Ziegler TE, Scheffler G, Snowdon CT (1995) The relationship of cortisol levels to social environment and reproductive functioning in female cotton-top tamarins, *Saguinus oedipus*. Horm Behav 29(3):407–424. https://doi.org/10.1006/hbeh.1995.1028

Questionnaires and Their Use in Primate Welfare

Marieke Cassia Gartner

Abstract

A variety of methods can be used to measure primate welfare, including, but not limited to, behavioral observations, cognitive bias tests, and questionnaires. The latter allows scientists to explore the individual primate's attitudes and behavior, as well as trends across groups. Questionnaires have been used to obtain an estimate of animals' personality, behavior, health, well-being, and more. Quantitative questionnaires allow for comparisons between individuals, but also across species. For primate welfare, they are a good way to access caretaker knowledge—this is a vital measure, as caregivers are the best placed to assess welfare. Like any methodology, questionnaires are not perfect, but with good design they may be a useful tool among other measures to assess welfare. Questionnaires have not been used extensively in primate welfare research, but there is evidence to support their expanded use as they offer many benefits, including a broad assessment of a state of being, economic efficiency, noninvasiveness, reliability, and validity. A variety of questionnaire types are described, as well as pitfalls of design and ways to address them.

Keywords

Questionnaires · Surveys · Primate · Welfare

M. C. Gartner (✉)
Zoo Atlanta, Atlanta, GA, USA
e-mail: mgartner@zooatlanta.org

1 Introduction

Welfare is notoriously hard to measure (Mason and Mendl 1993). It is comprised of several facets, including physical, emotional, and psychological health although some scientists recently have argued that welfare is purely comprised of affect (McMillan 2000; Whitham and Wielebnowski 2009). Physical health is self-explanatory, but the difference between emotional and psychological health is often confused. Emotional health is about expressing emotions appropriately; in contrast, psychological health is cognitive—it is about focus, processing, and understanding information. While it has been studied in a variety of forms in humans (not often called welfare, but well-being), for years, scientists would not attempt to understand emotional or psychological health in nonhuman animals, arguing that it was impossible to measure (Vanderwolf 1998; Fraser 1999). However, following the work of Jane Goodall and others in individualizing animals, this attitude has changed dramatically in the last 50 years. While some researchers still resist this type of science, many are now considering how to measure emotional and psychological health in nonhuman animals (e.g., Mendl et al. 2009).

Like in humans, physical health is addressed by medical professionals. When it comes to emotional and psychological health, we must discover ways to "ask" animals what their state of mind is and to get reliable answers. New ways of doing this are being tested, including studying personality (see Robinson and Weiss 2022), which has been shown to be related to well-being in a number of species, including chimpanzees (*Pan troglodytes* King and Landau 2003), orangutans (*Pongo* spp., Weiss et al. 2006), and rhesus macaques (*Macaca mulatta* Weiss et al. 2011b); measuring anticipatory behavior, which is goal-directed behavior that occurs during the appetitive phase, and which is tied to both the dopaminergic and opioid systems and therefore has implications for welfare (Watters 2014); and cognitive bias (see chapter by Bethell and Pfefferle 2022), which can reveal a positive or negative state of mind, and has been shown in a variety of species from bees (Bateson et al. 2011) to dogs (Mendl et al. 2010) to chimpanzees (Bateson and Nettle 2015). These newer ideas can be assessed with measures that have been used historically, such as behavioral observations and questionnaires.

Behavioral observations are an important aspect of understanding welfare (Lutz and Baker 2022). However, this type of methodology is subjective (although it is usually seen as objective; Vazire et al. 2007) in that interpretation of behavior is individual and limited to the number of observations made. Complex decisions must be made in the creation of a behavioral coding system and in its use, for example, in the definition and coding of ambiguous behaviors (Vazire et al. 2007). In addition, if the behavioral observations are being used in an experiment with treatments, blinded studies are recommended to ensure cueing is not occurring. Because it is often assumed to be objective, behavior coding methodology is rarely reported and is not always tested for reliability (Vazire et al. 2007). Observation times are rarely often enough to counter the problem of measuring over context and time. To counter this, some researchers limit observations to one observer, use basic behaviors to avoid disagreement, and test reliability among observers when there is more than one. Behavioral coding requires expertise in a specific species' behavior and therefore does not necessarily create a whole picture of welfare when used alone (Meagher 2009). Questionnaires, however, allow caretakers to use their accumulated knowledge of behavior and physical and mental states over different contexts and time in order to give a multifaceted picture of welfare. They are mostly tested for reliability, and methodology is usually outlined in detail. While the methodology of questionnaires has been brought into question anecdotally in regard to subjectivity, research has shown that they are not, in fact, subjective (Vazire et al. 2007) and are not subject to anthropomorphism (Kwan et al. 2008). More research demonstrating these ideas would be useful, to build a robust argument for their use. Questionnaires can therefore be a useful tool for studying nonhuman animals, allowing for a very focused (e.g., what is this animal's personality?) or a multidimensional approach (e.g., what is this animal's state of welfare?). When it comes to measuring welfare, however, it is often advised to use more than one measure, to ensure that the interpretation of the measures all coincides. Behavioral observations may offer information that questionnaires do not, and vice versa. Because welfare is a relatively new science, and no one measure has been proven to measure welfare perfectly for every individual without a doubt, pairing different methodologies may allow for a fuller picture of welfare.

2 Design and Analysis

Questionnaires have been commonly used in psychology since at least the early 1800s (Gault 1907) through to the present day for a variety of purposes, from census statistics to the measure of pain intensity to personality assessment. While the use of questionnaires can be widely defined, here, I mean information that is gathered by at least two observers about a single phenomenon (de Vaus 2014). This allows an evaluation of characteristics of the subject at hand while addressing reliability of the data. This type of methodology allows scientists to understand individual attitudes, differences, behavior, and trends across groups comprised of these individuals. Questionnaires may be conducted as one-on-one interviews or can be completed anonymously, allowing for collection of both qualitative and quantitative data. The former allows for more information, and possibly more complex responses, while the latter allows for more privacy and is easier to administer (Marks 2004). In addition, questionnaires allow for a broad collection of data that may include more than one type of psychological construct. One does not have to just measure anxiety, for example, but can measure welfare as a whole. Finally, questionnaires are always noninvasive, which is of utmost importance when measuring welfare. First, noninvasive measures, by definition, do not decrease welfare. Next, questionnaires, unlike invasive methodologies, do not interfere with the measurement of welfare itself. That is, the welfare of an animal suffering physically or mentally from an invasive treatment may not bare data that are reflective of the animal's emotional state or normal behaviors.

While generally considered a reliable and valid method of assessment (Donsbach 1997), questionnaires are not without their problems—for example, response effects (de Vaus 2014) or responder bias (Meagher 2009) may affect data and therefore analysis and results. This requires questionnaires to be created very carefully to avoid response effects due to word choice, closed or open format questions, question order, length of questionnaire, and which subjects complete it (Anastasi and Urbina 1997). For example, a closed format question, which may require a yes/no response, may not allow for enough information to assess the phenomenon at hand, as well as encouraging a response whether one is known or not (e.g., Waterman et al. 2001), while an open format question allows the responder to give not only more information, but also more nuanced information, although it must be worded carefully to elicit the needed data (Schuman and Scott 1987; Schwarz 1999). The benefits of closed format questions—ease of use and analysis—are not enough to counter the problems they pose—especially in the form of leading or confusing questions, or not giving enough information to get the answer needed. Generally speaking, then, open format questions are preferred, if designed correctly. Without attention to these aspects of questionnaire design, bias may be introduced, and answers may not be accurate. In addition, response biases may prove problematic, especially when asking questions about welfare. Caretakers may feel pressured to answer more positively if they are worried that a low welfare score may reflect badly on them. This can be addressed by having multiple raters—usually biases are not consistent across raters (Block 1961). Other ways to avoid responder bias are using careful wording, blinding respondents to the study goals, and, if possible, using disinterested observers (Meagher 2009), and by omitting identifying information so that respondents may be kept anonymous.

Clarity and conciseness help to ensure comprehension and that you are asking exactly what you want to know. Length is important so that subjects will fill the questionnaire out and not get mental fatigue, but long enough to get at what you want. This is true, of course, of any methodology—it must be well structured in order to ensure the resulting data are reflective of the question being asked. There are many articles and books that focus on designing questionnaires and carrying out analyses (e.g., de Vaus 2014), so I will not focus on that here. Instead, I will focus on the use of questionnaires for assessing primate welfare.

Like any methodology, choosing the right analytic tools is essential to using questionnaires. The same questions may be asked of quantitative questionnaire data that are asked of any data—are the data normally distributed? Should I use parametric or nonparametric tests? Should I reduce the data to a smaller number of variables or try to define it as is? As with any data, reliability and validity should always be tested.

3 The Need for Reliable Methodology in Studying Welfare

Primates are used extensively in laboratories (61,950 were reported in use during fiscal year 2015 in the USA alone; U.S. Department of Agriculture 2016), for both behavioral and medical science (for review, see Prescott 2022; Buchanan-Smith et al. 2022). In addition, they are widely kept in zoos (Baker and Farmer 2022). Because humans have put them in these situations, we therefore have the responsibility of addressing their welfare so that they are living the best lives they can. Assessing primate welfare poses interesting challenges. Because nonhuman primates are so closely related to us, welfare indicators, such as huddling in a corner or play behavior with social partners, may be easier to identify; however, the risk of anthropomorphism is high (anecdotally, zoo visitors often assume primate facial expressions are similar to ours) and must be addressed to ensure that welfare is being assessed correctly. This is why it is so important that people filing out the questionnaires have extensive experience with the individual animals being assessed.

There is evidence that human caregivers (e.g., parents of a sick child) are better than clinicians at assessing psychological states (as opposed to physical ones) such as anxiety or depression (Bryan et al. 2005). Similarly, animal caretakers spend most days with their animals, and know them intimately, making them the best possible assessors of an animal's state across context and time (Whitham and Wielebnowski 2009; Meagher 2009). For example, inter-rater agreement on rhinoceros behavior across 19 facilities was significant a majority of the time (Carlstead et al. 1999). While the study of welfare focused on problematic or abnormal behaviors for years, from reducing pain and suffering to addressing basic needs (Mench 1998), more recently, welfare scientists are trying to measure welfare as a whole by including positive and negative states. This has led to the study of well-being and quality of life in addition to stereotypies and other negative behavior. Questionnaires are useful for this all-around approach. They allow for a deeper assessment beyond common psychological testing of anxiety and fear, which, by exposing animals to a stressor, defeats the purpose of trying to improve their welfare, as well as missing an opportunity to assess overall welfare (Boissy et al. 2007). The flexibility of a questionnaire—the ability to ask any number of questions on any number of topics—lends itself to a well-rounded assessment of welfare that takes into account both negative and positive experiences. Of course, questionnaires are not the only tools used to measure positive welfare—as the focus on positive welfare has increased, so has the experimentation with tools to measure it, including cognitive bias testing, anticipatory behavioral testing, hormonal testing (e.g., IgA, alpha-amylase, and oxytocin), and sensory testing.

4 Different Types of Questionnaires for Studying Primate Welfare

Questionnaires have been used with animals to measure physical traits for decades and began to be used for behavioral traits first on chimpanzees (Crawford 1938) and then more widely in the 1970s, when Buirski et al. (1973) measured emotional and social behavior in baboons, and Buirski et al. (1978) and Stevenson-Hinde and Zunz (1978) measured chimpanzee and rhesus macaque personality, respectively. Since then, it has become the method of choice of many personality scientists. However, questionnaires have not been used extensively to assess primate welfare (Robinson et al. 2016), but in the cases where they have, they have shown to be effective. The study of subjective well-being offers a good example. Subjective well-being in humans is comprised of emotional responses, domain satisfactions (such as work, family, leisure, health, finances, self, and group), and global judgments of life satisfaction (Diener et al. 1999) and can generally be defined as happiness. King and Landau (2003) showed that well-being can be reliably assessed in chimpanzees by creating a questionnaire based on several aspects of well-being studied in humans. Caretakers were asked to rate chimpanzees on four items on a seven-point scale. The first, overall mood, asks raters to assess the balance of positive vs negative moods in the individual. The second, the effect of social interactions, asks raters how satisfying social interactions are for the individual. The third measures personal control by asking how successful the individual is at achieving its personal goals, and the fourth asks how happy the rater would be if they were the individual for a week. They found that all four components were represented by a single variable that described chimpanzee well-being.

Weiss et al. (2006) measured well-being reliably in orangutans using a similar methodology. Subsequently, Weiss et al. (2011a) showed that, as in humans, well-being affects other aspects of orangutan life: Happier orangutans (as measured by the subjective well-being questionnaire mentioned above) live longer lives. Similarly, rhesus macaques can also be reliably rated on well-being, which is stable across time in this species (Weiss et al. 2011b). These data can be useful in not only assessing primate welfare, but also addressing individuals who may be at greater risk for decreased welfare, as well-being is heritable not only in humans (Bartels 2015), but also in nonhuman primates (Weiss et al. 2002; Adams et al. 2012). That is, genetic factors explain about 35% of the variance in well-being in humans (Bartels 2015) and 40% in chimpanzees (Weiss et al. 2002). Thus, some animals may start out at greater risk due to their heredity—if this is known, attempts may be made to address the problem or possible problems before they start.

Another use of questionnaires to assess welfare was developed by the Chicago Zoological Society. The WelfareTrak® system was designed to track the welfare status of individuals over time (Whitham and Wielebnowski 2009). Caretakers complete species-specific questionnaires weekly, which include 10–15 indicators of both positive and negative physical and emotional/psychological well-being. The system flags any scores that are especially positive or negative, allowing the zoo to track individual preferences and states over time. This allows for quicker responses to potential problem behaviors, but also assessment of interventions that may be having a positive impact. Anecdotal evidence suggests that the system may work to increase welfare by proactively identifying welfare issues, evaluating successes, promoting discussion, and gaining insight into individuals.

To try to assess overall welfare, Robinson et al. (2016) developed a 12-item questionnaire specifically designed to assess welfare and tested it on brown capuchin monkeys (*Sapajus apella*). In addition to questions on physical health, quality-of-life measures (McMillan 2005), comprised of social relationships, mental stimulation, health, stress, and control over both the social and physical environment, were also included. These data were then compared with subjective well-being ratings as described earlier, locomotor stereotypy, and personality traits. They found high inter-rater reliability and that happiness correlated with welfare to form a single component. They also found that capuchins exhibiting stereotypies were rated lower in welfare. These results are useful in that a short questionnaire can be administered to assess welfare and therefore give a tool to caretakers to address any problems found.

A similar study was conducted with chimpanzees (Robinson et al. 2017). Using welfare, well-being, and personality questionnaires, the authors found that welfare and well-being were related and that a lower rating on the combined welfare/well-being rated was associated with regurgitation, coprophagy, urophagy, and decreased proximity to nearest neighbor. In addition, higher extraversion and lower neuroticism were related to higher welfare/well-being.

A slightly different methodology for assessing welfare that also employs a questionnaire is qualitative behavior analysis (Wemelsfelder et al. 2001), which was developed for farm animals and has not yet been used in studies of primates. Using free choice profiling, observers come up with their own words to describe an animal after observing them for a certain period, and then, they rate the animals using their own words. Independent observers consistently show agreement in their descriptions and ratings. While the method uses personality terminology, it extends its use to assess a current state of welfare through behavior. For example, if a sheep is alert and active, is it also calm, or is it fearful (Wemelsfelder 2007)? How that animal is handled would be dependent on the latter assessment, and without knowledge of both the species and the individual, that animal's welfare could be impacted.

This method has shown high interobserver reliability, repeatability, and observer detection of individual differences (Wemelsfelder et al. 2001) and has been validated with correlations to physical parameters including core body temperature, heart rate, plasma glucose, and the neutrophil:lymphocyte ratio in cattle. In sheep, it has been validated with heart rate, heart rate variability, and core body temperature. It also is related to experience vs. naiveté of road transport in cattle (Stockman et al. 2011) and sheep (Wickham et al. 2012). Similarly, an association was found between traits defined with this method and the before-and-after results of a drug known to decrease aggression and stress in pigs (Rutherford et al. 2012). Qualitative behavior analysis could clearly be used with primates in any captive environment since its requirements could easily be handled by caretakers. Its advantages include quick assessment (most studies had subjects observe animals for only 10 min) and a methodology that has been repeatedly validated and has proved reliable for assessing an animal's state of welfare.

Most recently, scientists have been attempting to assess what they are calling "quality of life" (QoL). This is mostly an equivalent to welfare or well-being (McMillan 2000), and its measurement is usually based on human scales, mostly in questionnaire form (McMillan 2000). These questionnaires tend to be similar to well-being and welfare measures and attempt to assess an overall state of an animal. Historically, quality of life was used to assess suffering in animals, as a timeline for euthanasia. As the focus on positive welfare has grown across fields, quality-of-life measures have expanded to include affective state and physical state, bringing the assessments closer to those of well-being and welfare. Just as in well-being and welfare, quality of life can be referred to as happiness (Belshaw 2018). There is no consistent methodology for quality of life, as it has not been used as formally as welfare and well-being have. It is important, as this tool develops, to distinguish it from those two measures, in order to allow comparison and to ensure that we are discussing the same thing when comparing methodologies.

Clearly, there is more work to be done on assessing primate welfare. Keeping in mind good design, and the possible pitfalls of a poorly created questionnaire, questionnaires can be a useful tool to aid in this goal, incorporating caretaker knowledge in a quick, economical way to reliably assess welfare. Because welfare is still a new science, it is nevertheless recommended to combine methodologies, from physiological measures to behavioral observations, to ensure that measurement and interpretation are as accurate as possible.

References

Adams MJ, King JE, Weiss A (2012) The majority of genetic variation in orangutan personality and subjective well-being is nonadditive. Behav Genet 42:675–686. https://doi.org/10.1007/s10519-012-9537-y

Anastasi A, Urbina S (1997) Psychological testing, 7th edn. Pearson, New York

Baker KR, Farmer HL (2022) The welfare of primates in Zoo. In: Robinson LM, Weiss A (eds) Nonhuman primate welfare: from history, science, and ethics to practice. Springer, Cham, pp 79–96

Bartels M (2015) Genetics of wellbeing and its components satisfaction with life, happiness, and quality of life: a review and meta-analysis of heritability studies. Behav Genet 45:137–156. https://doi.org/10.1007/s10519-015-9713-y

Bateson M, Nettle D (2015) Development of a cognitive bias methodology for measuring low mood in chimpanzees. PeerJ 3:e998. https://doi.org/10.7717/peerj.998

Bateson M, Desire S, Gartside SE, Wright GA (2011) Agitated honeybees exhibit pessimistic cognitive biases. Curr Biol 21(12):1070–1073. https://doi.org/10.1016/j.cub.2011.05.017

Belshaw Z (2018) Quality of life assessment in companion animals: what, why, who, when and how. Companion Anim 23(5):264–268. https://doi.org/10.12968/coan.2018.23.5.264

Bethell EJ, Pfefferle D (2022) Cognitive bias tasks: a new set of approaches to assess welfare in nonhuman primates. In: Robinson LM, Weiss A (eds) Nonhuman primate welfare: from history, science, and ethics to practice. Springer, Cham, pp 207–230

Block J (1961) The Q-Sort method in personality assessment and psychiatric research. Charles C Thomas Publisher, Springfield, IL

Boissy A, Manteuffel G, Jensen MB et al (2007) Assessment of positive emotions in animals to improve their welfare. Physiol Behav 92(3):375–397. https://doi.org/10.1016/j.physbeh.2007.02.003

Bryan S, Hardyman W, Bentham P et al (2005) Proxy completion of EQ-5D in patients with dementia. Qual Life Res 14:107–118. https://doi.org/10.1007/s11136-004-1920-6

Buchanan-Smith HM, Tasker L, Ash H, Graham ML (2022) Welfare of primates in laboratories: opportunities for improvement. In: Robinson LM, Weiss A (eds) Nonhuman primate welfare: from history, science, and ethics to practice. Springer, Cham, pp 97–120

Buirski P, Kellerman H, Plutchik R et al (1973) A field study of emotions, dominance, and social behavior in a group of baboons (*Papio anubis*). Primates 14:67–78. https://doi.org/10.1007/BF01730516

Buirski P, Plutchik R, Kellerman H (1978) Sex differences, dominance, and personality in the chimpanzee. Anim Behav 26(Part 1):123–129. https://doi.org/10.1016/0003-3472(78)90011-8

Carlstead K, Mellen J, Kleiman DG (1999) Black rhinoceros (*Diceros bicornis*) in US zoos: I. Individual behavior profiles and their relationship to breeding success. Zoo Biol 18(1):17–34. https://doi.org/10.1002/(SICI)1098-2361(1999)18:1%3C17::AID-ZOO4%3E3.0.CO;2-K

Crawford MP (1938) A behavior rating scale for young chimpanzees. J Comp Psychol 26(1):79–92. https://doi.org/10.1037/h0054503

de Vaus D (2014) Surveys in social research, 6th edn. Routledge, Oxon

Diener E, Suh EM, Lucas RE, Smith HL (1999) Subjective well-being: three decades of progress. Psychol Bull 125(2):276–302. https://doi.org/10.1037/0033-2909.125.2.276

Donsbach W (1997) Survey research at the end of the twentieth century: theses and antitheses. Int J Public Opin Res 9(1):17–28. https://doi.org/10.1093/ijpor/9.1.17

Fraser D (1999) Animal ethics and animal welfare science: bridging the two cultures. Appl Anim Behav Sci 65(3):171–189. https://doi.org/10.1016/S0168-1591(99)00090-8

Gault RH (1907) A history of the questionnaire method of research in psychology. Pedagog Semin 14(3):366–383. https://doi.org/10.1080/08919402.1907.10532551

King JE, Landau VI (2003) Can chimpanzee (*Pan troglodytes*) happiness be estimated by human raters? J Res Pers 37(1):1–15. https://doi.org/10.1016/S0092-6566(02)00527-5

Kwan VSY, Gosling SD, John OP (2008) Anthropomorphism as a special case of social perception: a cross-species social relations model analysis of humans and dogs. Soc Cogn 26(2):129–142. https://doi.org/10.1521/soco.2008.26.2.129

Lutz CK, Baker KC (2022) Using behavior to assess primate welfare. In: Robinson LM, Weiss A (eds) Nonhuman primate welfare: from history, science, and ethics to practice. Springer, Cham, pp 171–206

Marks DF (2004) Questionnaires and surveys. In: Marks DF, Yardley L (eds) Research methods for clinical and health psychology. Routledge, London, pp 122–144

Mason G, Mendl M (1993) Why is there no simple way of measuring animal welfare? Anim Welf 2(4):301–319

McMillan FD (2000) Quality of life in animals. J Am Vet Med Assoc 216(12):1904–1910. https://doi.org/10.2460/javma.2000.216.1904

McMillan FD (2005) Mental wellness: the concept of quality of life in animals. In: McMillan FD (ed) Mental health and well-being in animals. Blackwell Publishing, Ames

Meagher RK (2009) Observer ratings: validity and value as a tool for animal welfare research. Appl Anim Behav Sci 119(1–2):1–14. https://doi.org/10.1016/j.applanim.2009.02.026

Mench JA (1998) Thirty years after Brambell: whither animal welfare science? J Appl Anim Welf Sci 1(2):91–102. https://doi.org/10.1207/s15327604jaws0102_1

Mendl M, Burman OHP, Parker RMA, Paul ES (2009) Cognitive bias as an indicator of animal emotion and welfare: emerging evidence and underlying mechanisms. Appl Anim Behav Sci 118(3–4):161–181. https://doi.org/10.1016/j.applanim.2009.02.023

Mendl M, Brooks J, Basse C et al (2010) Dogs showing separation-related behaviour exhibit a 'pessimistic' cognitive bias. Curr Biol 20(19):R839–R840. https://doi.org/10.1016/j.cub.2010.08.030

Prescott MJ (2022) Using primates in captivity: research, conservation, and education. In: Robinson LM, Weiss A (eds) Nonhuman primate welfare: from history, science, and ethics to practice. Springer, Cham, pp 57–78

Robinson LM, Weiss A (2022) Primate personality and welfare. In: Robinson LM, Weiss A (eds) Nonhuman primate welfare: from history, science, and ethics to practice. Springer, Cham, pp 387–402

Robinson LM, Waran NK, Leach MC et al (2016) Happiness is positive welfare in brown capuchins (*Sapajus apella*). Appl Anim Behav Sci 181:145–151. https://doi.org/10.1016/j.applanim.2016.05.029

Robinson LM, Altschul DM, Wallace EK et al (2017) Chimpanzees with positive welfare are happier, extraverted, and emotionally stable. Appl Anim Behav Sci 191:90–97. https://doi.org/10.1016/j.applanim.2017.02.008

Rutherford KMD, Donald RD, Lawrence AB, Wemelsfelder F (2012) Qualitative behavioural assessment of emotionality in pigs. Appl Anim Behav Sci 139(3–4):218–224. https://doi.org/10.1016/j.applanim.2012.04.004

Schuman H, Scott J (1987) Problems in the use of survey questions to measure public opinion. Science 236(4804):957–959. https://doi.org/10.1126/science.236.4804.957

Schwarz N (1999) Self-reports: how the questions shape the answers. Am Psychol 54(2):93–105. https://doi.org/10.1037/0003-066X.54.2.93

Stevenson-Hinde J, Zunz M (1978) Subjective assessment of individual rhesus monkeys. Primates 19:473–482. https://doi.org/10.1007/BF02373309

Stockman CA, Collins T, Barnes AL et al (2011) Qualitative behavioural assessment and quantitative physiological measurement of cattle naïve and habituated to road transport. Anim Prod Sci 51(3):240. https://doi.org/10.1071/AN10122

United States Department of Agriculture (2016) Retrieved from https://www.aphis.usda.gov/animal_welfare/downloads/7023/Annual-Reports-FY2015.pdf. Accessed 14 October 2016

Vanderwolf C (1998) Brain, behavior, and mind: what do we know and what can we know? Neurosci Biobehav Rev 22(2):125–142. https://doi.org/10.1016/S0149-7634(97)00009-2

Vazire S, Gosling SD, Dickey AS, Schapiro SJ (2007) Measuring personality in nonhuman animals. In: Robins R, Fraley RC, Krueger RF (eds) Handbook of research methods in personality psychology. The Guildford Press, New York, pp 190–207

Waterman AH, Blades M, Spencer C (2001) Interviewing children and adults: the effect of question format on the tendency to speculate. Appl Cogn Psychol 15(5):521–531. https://doi.org/10.1002/acp.741

Watters JV (2014) Searching for behavioral indicators of welfare in zoos: uncovering anticipatory behavior. Zoo Biol 33(4):251–256. https://doi.org/10.1002/zoo.21144

Weiss A, King JE, Enns RM (2002) Subjective well-being is heritable and genetically correlated with dominance in chimpanzees (*Pan troglodytes*). J Pers Soc Psychol 83(5):1141–1149. https://doi.org/10.1037/0022-3514.83.5.1141

Weiss A, King JE, Perkins L (2006) Personality and subjective well-being in orangutans (*Pongo pygmaeus* and *Pongo abelii*). J Pers Soc Psychol 90(3):501–511. https://doi.org/10.1037/0022-3514.90.3.501

Weiss A, Adams MJ, King JE (2011a) Happy orang-utans live longer lives. Biol Lett 7(6):872–874. https://doi.org/10.1098/rsbl.2011.0543

Weiss A, Adams MJ, Widdig A, Gerald MS (2011b) Rhesus macaques (*Macaca mulatta*) as living fossils of hominoid personality and subjective well-being. J Comp Psychol 125(1):72–83. https://doi.org/10.1037/a0021187

Wemelsfelder F (2007) Qualitative behaviour assessment: application to welfare assessment in extensively managed sheep. In: Proceedings of welfare goals from the perspective of extensively managed sheep. The Macaulay Institute, Aberdeen, UK, pp 57–62

Wemelsfelder F, Hunter TEA, Mendl MT, Lawrence AB (2001) Assessing the 'whole animal': a free choice profiling approach. Anim Behav 62(2):209–220. https://doi.org/10.1006/anbe.2001.1741

Whitham JC, Wielebnowski N (2009) Animal-based welfare monitoring: using keeper ratings as an assessment tool. Zoo Biol 28(6):545–560. https://doi.org/10.1002/zoo.20281

Wickham SL, Collins T, Barnes AL et al (2012) Qualitative behavioral assessment of transport-naïve and transport-habituated sheep. J Anim Sci 90(12):4523–4535. https://doi.org/10.2527/jas.2010-3451

Part III
Nonhuman Primate Housing and Husbandry

Meeting Cognitive, Behavioral, and Social Needs of Primates in Captivity

Catherine F. Talbot, Lisa A. Reamer, Susan P. Lambeth, Steven J. Schapiro, and Sarah F. Brosnan

Abstract

Addressing the welfare needs of nonhuman primates in captivity is a significant challenge due to the differences among different species, sexes, ages, dominance groups, and individuals. Interventions that increase species typical behaviors and/or reduce atypical behaviors or stress for one species, group, individual, or context may cause the opposite in others. One area that has recently gained substantial attention is species' cognitive needs, which are particularly important for highly encephalized species such as nonhuman primates. In addition, behavioral and social needs are critical for species that have extended life spans and live in complex social groups. In this chapter, we summarize the cognitive, behavioral, and social needs of primates, focusing on the ways in which they vary among species. As we cannot cover each of the hundreds of primate species, we focus on issues that are likely to be important for most of us involved in research

C. F. Talbot (✉)
Department of Psychology, Florida Institute of Technology, Melbourne, FL, USA
e-mail: ctalbot@fit.edu

L. A. Reamer · S. P. Lambeth
Department of Comparative Medicine, UT MD Anderson Cancer Center, Michale E. Keeling Center for Comparative Medicine and Research, Bastrop, TX, USA

S. J. Schapiro
Department of Comparative Medicine, UT MD Anderson Cancer Center, Michale E. Keeling Center for Comparative Medicine and Research, Bastrop, TX, USA

Department of Experimental Medicine, University of Copenhagen, Copenhagen, Denmark

S. F. Brosnan
Department of Comparative Medicine, UT MD Anderson Cancer Center, Michale E. Keeling Center for Comparative Medicine and Research, Bastrop, TX, USA

Department of Psychology, Neuroscience Institute, and Center for Behavioral Neuroscience, Georgia State University, Atlanta, GA, USA

and captive management and draw examples from taxa commonly held in captivity, including apes, macaques, callitrichids, and capuchin monkeys. In each section, we outline the ecological or evolutionary basis of primates' needs and discuss behavioral management strategies designed to meet these needs in captive settings. We hope that this summary of the cognitive, behavioral, and social welfare needs of captive nonhuman primates is useful in informing which enrichment strategies will be the most effective for which species and context.

Keywords

Delay of gratification · Planning · Memory · Tool use · Theory of mind · Social learning · Social interaction · Welfare

1 Introduction

Nonhuman primates (hereafter "primates") are a diverse group of species whose evolutionary proximity to humans (who are primates) and large brain size led to the hypothesis that they have similarly complex needs as humans that must be understood and addressed. Indeed, in their natural habitat, primates engage in numerous behaviors that indicate advanced cognitive ability, such as self-control, causal understanding, tool use, planning, and problem-solving (Tomasello and Call 1997). This is particularly true in the social domain as primates engage in social learning, triadic awareness, and even deception (Byrne and Whiten 1988; Whiten and Byrne 1997). The lack of these challenges in captivity may therefore lead to boredom and frustration (Meehan and Mench 2007). Thus, a recent focus in primate welfare has been on addressing cognitive needs by providing captive primates with ecologically appropriate challenges and social complexity (Clark 2011; Schapiro et al. 2014). However, while both are clearly important, they also vary substantially in the Primate order due to differences in ecology and life history, meaning that there is no one-size-fits-all solution to these problems. The goal of our chapter is to outline cognitive abilities and social and behavioral needs in primates, which we see as a foundation on which decisions about enrichment can be made. While we do make some recommendations for enrichment and welfare, we refer readers to other chapters within this volume for additional details and evidenced-based practices.

The goal of enrichment is to provide a living situation that functionally simulates natural environments and provides individuals with opportunities to perform species-typical behaviors at rates similar to their wild-living counterparts (Schapiro et al. 2014). However, the species-specific nature of primates' cognitive and behavioral needs is a challenge to addressing this. Indeed, interventions that benefit one species may increase stress in another. For example, novel objects are standard enrichment for most primates; however, owl monkeys are susceptible to neophobia and caution must be used when introducing novel items to their environment (Ehrlich 1970; Tardif et al. 2006). Moreover, primates exhibit individual differences in cognition, behavior, and sociability, both across demographic categories and as idiosyncratic differences. Thus, aside from considering species-level differences, maintaining well-being should incorporate individuals' preferences.

In this chapter, we summarize how each of the abovementioned needs is related to the environment in which the species evolved. We choose a subset of interventions and discuss the ways in which they have (or have not) helped and provide some general guidelines on how to determine whether an intervention will benefit the animals. Although most studies focus on a select few, commonly studied groups (great apes, particularly chimpanzees; macaques, particularly rhesus macaques; and select New World species, particularly callitrichids and capuchin monkeys), here we provide extensive tables that, for a broad array of species, outline basic social and ecological factors that affect their natural behavior in the wild with the hope that they will provide a starting point for those interested in improving welfare and enrichment in their species. The overall goal of this chapter is to allow individuals who are interested in beginning or improving a cognitive enrichment program to better understand the context-specific issues involved. We hope this will promote educated decisions on how to best incorporate species' natural socioecology and evolutionary history.

2   The Influence of Ecology and Evolution on the Needs of Captive Primates

2.1   Cognitive Needs

A unifying feature of primates is their impressive cognitive abilities. Cognition is broadly defined as how sensory input is acquired, processed, stored, and acted upon, and involves processes such as learning, memory, causal reasoning, and problem-solving (Shettleworth 2010). Primates' aptitude for complex cognition makes them particularly interesting as subjects for studies of the evolution of intelligence, but it also means that they are predisposed to get bored and require substantial cognitive enrichment. "Occupational enrichment," a term that emerged in the 1980s, emphasizes providing animals with opportunities to voluntarily "work" and occupy their time with species-relevant problems (Schapiro et al. 2014). Over the past decade, scientists have seen the emergence of "cognitive enrichment," a subset of enrichment that specifically aims to stimulate the cognitive skills of animals.

Clark (2011) defined cognitive enrichment as "tasks whose use (1) engages evolved cognitive skills by providing opportunities to solve problems or control some aspect of the environment, and (2) is correlated to one or more validated measures of well-being." To evaluate cognitive enrichment, we must first know and understand the cognitive skills that we are trying to stimulate (Tables 1, 2, 3 and 4). Therefore, in the following section we outline primary cognitive needs of primates and comment on ways in which we can best meet those needs in captivity.

Table 1 Relevant information on foraging behavior and environment for the behavioral management of captive nonhuman primates

	Foraging behavior		Environment	
Captive recommendation	Provide enrichment that simulates natural feeding behavior		Ensure enclosures have both adequate floor space and climbing/resting structures off the ground	
Taxa	Foraging	Citations	Habitat	Citations
Chimpanzees	Extractive foragers, primarily frugivores and omnivores	Goodall (1986)	Terrestrial/arboreal	Wild: Doran (1996)
				Captive: Pruetz and McGrew (2001); Reamer et al. (2017)
Gorillas	Folivores	Robbins (2011)	Terrestrial/arboreal (some)	Wild: Robbins (2011)
				Captive: Ross et al. (2011)
Orangutans	Primarily frugivores/folivores	Knot and Kahlenberg (2011)	Arboreal	Wild: Knot and Kahlenberg (2011)
Gibbons	Primarily frugivores	Bartlett (2011)	Arboreal	Wild: Bartlett (2011)
Colobus monkeys	Folivores	Fashing (2011)	Arboreal	Wild: Fashing (2011)
Patas monkeys/vervets/guenons	Primarily frugivores/folivores; omnivores	Jaffe and Isbell (2011)	Terrestrial/arboreal (mostly)	Wild: Jaffe and Isbell (2011)
Mangabeys	Omnivores	Swedell (2011)	Terrestrial/arboreal	Wild: Swedell (2011)
Mandrills/drills	Omnivores	Swedell (2011)	Terrestrial/arboreal	Wild: Swedell (2011)
Baboons	Extractive foragers, omnivores	Dunbar (1992); Swedell (2011)	Mostly terrestrial	Wild: Swedell (2011)
				Captive: Lutz and Neville (2017)
Macaques	Primarily frugivores/folivores	Thierry (2011)	Terrestrial/arboreal	Wild: Thierry (2011)
				Captive: See Winnicker et al. (2013)
Spider monkeys	Frugivores	DiFiore et al. (2011)	Arboreal	Wild: Di Fiore et al. (2011)
Howler monkeys	Folivores	DiFiore et al. (2011)	Arboreal	Wild: Di Fiore et al. (2011)
Squirrel monkeys	Primarily insectivores and frugivores	Jack (2011)	Arboreal	Wild: Jack (2011)
				Captive: Williams et al. (1988)
Capuchins	Extractive/destructive foragers, omnivores	Dunbar (1992); Jack (2011); McKinney (2011)	Arboreal	Wild: Jack (2011)

(continued)

Table 1 (continued)

	Foraging behavior		Environment	
Captive recommendation	Provide enrichment that simulates natural feeding behavior		Ensure enclosures have both adequate floor space and climbing/resting structures off the ground	
Taxa	Foraging	Citations	Habitat	Citations
Titi monkeys	Primarily frugivores	Norconk (2011)	Arboreal	Wild: Norconk (2011)
Owl monkeys	Primarily frugivores	Fernandez-Duque (2011)	Arboreal	Wild: Fernandez-Duque (2011)
				Captive: Baer et al. (2012)
Marmosets/ tamarins	Marmosets: Extractive foragers, exudativores/ insectivores; Tamarins: Omnivores	Dunbar (1992); Digby et al. (2011)	Arboreal	Wild: Digby et al. (2011)
Lemurs	Primarily frugivores/ omnivores	Gould et al. (2011)	Arboreal	Wild: Gould et al. (2011)

2.1.1 Physical Cognition

Primates in general have complex cognitive abilities, which they use to manipulate their environment. Physical cognition includes the skills that organisms use to interact with their physical (as opposed to social) environments. Much of this takes place in the context of food acquisition, no doubt in large part because in the wild, food acquisition dominates most species' time budget (Table 5). This becomes a problem in captivity, where humans provision food, giving primates little opportunity to "exercise" these cognitive skills. Below, we briefly describe primates' capacities in several areas of physical cognition and indicate the type of enrichment that may be useful.

2.1.1.1 Delay of Gratification and Self-Control

Delay of gratification, sometimes called self-control, is the ability to forego an immediate reward to obtain a better but delayed reward (Rachlin and Green 1972), and is considered a prerequisite for complex goal-directed behavior, such as cooperation and planning (Mischel 1974). Delay of gratification plays a role in foraging in some species and is likely to be of importance in all species in social contexts. Several paradigms have been designed to examine self-control in captive primates, including "smaller sooner—larger later" tasks (Bramlett et al. 2012), delayed exchange tasks (Judge and Essler 2013), and accumulation tasks (Evans and Beran

Table 2 Relevant information on hunting and tool use for the behavioral management of captive nonhuman primates

	Hunting		Tool use	
Captive recommendation	Offer cognitive studies that simulate this behavior		Provide enrichment that requires tool use or offer cognitive studies that utilize tools	
Taxa	Trait present	Citations	Trait present	Citations
Chimpanzees	Yes	Wild: Goodall (1986); Pruetz and Bertolani (2007)	Yes	Wild: Goodall (1986); Pruetz and Bertolani (2007)
		Captive: Johnson et al. (2013); Reamer et al. (2017)		Captive: Pruetz and McGrew (2001); Reamer et al. (2017)
Gorillas	No	Wild: Robbins (2011)	Yes	Wild: Breuer (2005)
				Captive: Lonsdorf et al. (2009); Mulcahy et al. (2005); Fontaine et al. (1995)
Orangutans	Yes, "slow loris capture and eat"	Wild: Russon et al. (2009)	Yes	Wild: Galdikas (1989); van Schaik et al. (1999); van Schaik and Knott (2001); Fox and Bin'Muhammad (2002)
				Captive: Call and Tomasello (1994)
Gibbons	No	–	Yes	Wild: Baldwin and Teleki (1976); van Schaik et al. (1999)
				Captive: Cunningham et al. (2006); Hill et al. (2011)
Colobus monkeys	No	–	Yes	Wild: Starin (1990); Struthsaker (1975)
Patas monkeys/vervets/guenons	No	–	Yes	Wild: Pollack (1998); van Schaik et al. (1999); Worch (2001)
				Captive: Santos et al. (2006)
Mangabeys	No	–	Yes	Wild: van Schaik et al. (1999)
				Captive: Kyes (1988)
Mandrills/drills	No	–	Yes	Wild: van Schaik et al. (1999)
Baboons	Yes	Wild: Harding (1975); Butynski (1982)	Yes	Wild: Hamilton and Tilson (1985); van Schaik et al. (1999)
				Captive: Benhar and Samuel (1978); Beck (1973a, b)

(continued)

Table 2 (continued)

	Hunting		Tool use	
Captive recommendation	Offer cognitive studies that simulate this behavior		Provide enrichment that requires tool use or offer cognitive studies that utilize tools	
Taxa	Trait present	Citations	Trait present	Citations
Macaques	Some species	Wild: Sushma and Singh (2008); Young et al. (2012)	Yes	Wild: Chiang (1967); van Schaik et al. (1999)
				Captive: Anderson (1985); Iriki et al. (1996); Zuberbühler et al. (1996); Beck (1976)
Spider monkeys	No	–	Yes	Wild: Rodrigues and Lindshield (2007); Lindshield and Rodrigues (2009)
				Captive: (No) Gibson (1990)
Howler monkeys	No	–	Yes	Wild: Richard-Hansen (1998); van Schaik et al. (1999)
Squirrel monkeys	No	–	Yes	Wild: van Schaik et al. (1999)
				Captive: (Yes) Buckmaster et al. (2015); (No) Westergaard and Fragaszy (1987); Gibson (1990)
Capuchins	Yes	Wild: Rose (1997)	Yes	Wild: Boinski (1988); van Schaik et al. (1999); Canale et al. (2009)
				Captive: Anderson and Henneman (1994); Westergaard and Fragaszy (1987)
Titi monkeys	No	–	No	Wild: van Schaik et al. (1999)
Owl monkeys	No	–	No	Wild: van Schaik et al. (1999)
Marmosets/ tamarins	No	–	Yes	Wild: Stoinski and Beck (2001)
				Yamazaki et al. (2011); Hauser et al. (2002); Santos et al. (2006)
Lemurs	No	–	No	Wild: van Schaik et al. (1999)
				Yes: Santos et al. (2005)

2007). The first two require subjects to choose between a smaller/less valuable reward (or token, for exchange tasks) sooner and a larger/more valuable one later, while the latter requires subjects to avoid reaching for accumulating rewards to obtain additional ones.

A species' ability to delay gratification in these paradigms is hypothesized to relate to the selective pressures a species encounters in its natural environment, and researchers have demonstrated relationships between delay of gratification and

Table 3 Relevant information on mirror self-recognition and cooperation for the behavioral management of captive nonhuman primates

Captive recommendation	Mirror self-recognition		Cooperation	
	Provide a mirror for enrichment within or near enclosure, allowing both self-inspection and visual access to otherwise inaccessible areas		Provide enrichment that requires cooperation or offer cognitive studies that require cooperation with a social partner	
Taxa	Trait present	Citations	Trait present	Citations
Chimpanzees	Yes	Gallup (1970); Suárez and Gallup (1981); Inoue-Nakamura (1997); Lethmate and Ducker (1973); Tomasello and Call (1997); Schapiro and Lambeth (2010); Menzel et al. (1985); Lambeth and Bloomsmith (1992)	Yes	Wild: Boesch (2002); Jaeggi et al. (2013)
				Captive: Brosnan et al. (2011); Proctor et al. (2013); Suchak et al. (2014)
Gorillas	Yes	Patterson (1984); Inoue-Nakamura (1997); debated: Suárez and Gallup (1981); Tomasello and Call (1997); Zaragoza et al. (2011)	Yes	Wild: Harcourt and Stuart (2007)
Orangutans	Yes	Lethmate and Ducker (1973); Suárez and Gallup (1981); Inoue-Nakamura (1997); Tomasello and Call (1997)	Yes	Captive: Chalmeau et al. (1997); Call and Tomasello (1994)
Gibbons	No	Lethmate and Ducker (1973); Inoue-Nakamura (1997); Tomasello and Call (1997)	–	–
Colobus monkeys	No	Shaffer and Renner (2000)	–	–
Patas monkeys/vervets/guenons	No	Hall (1962); Harris and Edwards (2004)	–	–
Mangabeys	No	Escuin, unpublished video data (2014)	–	–
Mandrills/drills	No	Lethmate and Ducker (1973); Tomasello and Call (1997)	–	–
Baboons	No	Lethmate and Ducker (1973)	Yes	
				Captive: Beck (1973a, b)
Macaques	No	Lethmate and Ducker (1973); debated: Anderson and Gallup (2011); Inoue-Nakamura (1997); Tomasello and Call (1997); Coleman et al. (2012); Itakura (1987); Anderson (1986)	Yes	Wild: Berghänel et al. (2011) (coalitions)
Spider monkeys	No	Lethmate and Ducker (1973); Tomasello and Call (1997)	–	–

(continued)

Table 3 (continued)

	Mirror self-recognition		Cooperation	
Captive recommendation	Provide a mirror for enrichment within or near enclosure, allowing both self-inspection and visual access to otherwise inaccessible areas		Provide enrichment that requires cooperation or offer cognitive studies that require cooperation with a social partner	
Taxa	Trait present	Citations	Trait present	Citations
Howler monkeys	–	–	–	Wild: Pope (2000)
Squirrel monkeys	No	Inoue-Nakamura (1997)	Yes	Wild: Boinski (1987); Mitchell (1994)
				Captive: Talbot et al. (2014) (limited)
Capuchins	No	Lethmate and Ducker (1973); Anderson and Roeder (1989); Anderson et al. (2009); Inoue-Nakamura (1997); Tomasello and Call (1997); Anderson and Roeder (1989); Marchal and Anderson (1993)	Yes	Rose (1997)
				Takimoto and Fujita (2011); Mendres and de Waal (2000);
Titi monkeys	No	Fisher-Phelps et al. (2015)	–	–
Owl monkeys	No	–	–	–
Marmosets/tamarins	No	Inoue-Nakamura (1997); debated: Anderson and Gallup (1997)	Yes	Wild: Garber (1997)
				Captive: Werdenich and Huber (2002)
Lemurs	No	Inoue-Nakamura (1997)	–	–

species' allometric variables (e.g., body size, metabolic rates, and life span), social dynamics, and ecology. If the hypothesis is correct, higher metabolic rates, due to either smaller bodies or shorter life spans, should correlate with choosing fewer delayed rewards, which indeed has been found (Stevens 2014). Similarly, subjects living in more complex social groups are hypothesized to have better delay of gratification, and indeed, primates living in fission–fusion groups show greater behavioral inhibition than primates that live in more cohesive groups (Amici et al. 2008). Finally, ecology is hypothesized to play a role; for example, marmosets, who feed on plant exudates, show significantly longer delay skills than tamarins, who specialize on insects, the hunting of which may require greater impulsivity (Stevens et al. 2005). Thus, when choosing enrichment devices, it is important to consider allometric variables and choose devices that are sufficiently, but not overly, challenging (e.g., artificial gum feeders for captive callitrichids) (Roberts et al. 1999).

Table 4 Relevant information on problem-solving behavior and other species-specific behavior for the behavioral management of captive nonhuman primates

	Problem-solving		Other species-specific behavior	
Captive recommendation	Provide occupational enrichment and offer cognitive studies that stimulate this behavior		See specific recommendations below	
Taxa	Trait present	Citations	Trait present	Captive analog/ citations
Chimpanzees	Yes	Wild: Boesch (1991)	Leaf-dipping	Provide enrichment that simulates this behavior (e.g., juice trough/ 'wadger' with butcher paper or pond in enclosure)
		Captive: Bloomsmith and Else (2005); Hopper et al. (2014)		Wild: Sousa et al. (2009); Captive: Schapiro and Lambeth (2010); Reamer et al. (2017)
Gorillas	Yes	Wild: Byrne (1996)	Extensive food processing (e.g., nettles)	Offer a variety of difficult to process foods as well as puzzle boxes/artificial fruit
		Captive: Mulcahy et al. (2005)		Wild: Byrne (1996); Captive: Mulcahy et al. (2005)
Orangutans	Yes	Wild: van Schaik and Knott (2001)	–	–
		Captive: Chalmeau et al. (1997); Call and Tomasello (1994)		
Gibbons	Yes	Wild: Cheyne et al. (2012)	Duets/ Songs	Offer playback studies and provide music for auditory enrichment
		Captive: Beck (1967); Cunningham et al. (2006); Hill et al. (2011)		Wild: Mitani (1985); Captive: Shepherdson et al. (1989)
Colobus monkeys	Yes	Captive: Dickie (1998)	–	–
Patas monkeys/ vervets/guenons	Yes	Captive: Santos et al. (2006); van de Waal et al. (2013)	–	–
Mangabeys	Yes	Captive: Crast et al. (2016)	–	–
Mandrills/drills	Yes	Captive: Clark (2017)	–	–

(continued)

Table 4 (continued)

	Problem-solving		Other species-specific behavior	
Captive recommendation	Provide occupational enrichment and offer cognitive studies that stimulate this behavior		See specific recommendations below	
Taxa	Trait present	Citations	Trait present	Captive analog/citations
Baboons	Yes	Captive: Beck (1973a, b)	–	–
Macaques	Yes		Food washing; Stone handling	Provide enrichment that stimulates these behaviors (e.g., pools, tubs of water, stones)
		Captive: Beck (1976); Novak et al. (1998); Reinhardt (1994)		Wild: Nakamichi et al. (1998); Leca et al. (2007); Captive: Visalberghi and Fragaszy (1990); Leca et al. (2008)
Spider monkeys	Yes	Captive: Hill et al. (2011)	–	–
Howler monkeys	–	–	–	–
Squirrel monkeys	Yes	Captive: Visalberghi and Mason (1983)	–	–
Capuchins	Yes		Hammer and anvil nut cracking	Provide nuts/coconuts for cracking
		Captive: Anderson and Henneman (1994)		Wild: Visalberghi et al. (2007); Captive: Fragaszy and Visalberghi (1989)
Titi monkeys	Yes	Captive: Visalberghi and Mason (1983)	–	–
Owl monkeys	Yes	–	Nocturnal	Ensure light cycle is adequate and conducive for uninterrupted sleep patterns
				Wild: Fernandez-Duque (2011); Captive: Erkert (1989)
Marmosets/tamarins	Yes	Captive: Hauser et al. (2002); Santos et al. (2006); Cameron and Rogers (1999); Roberts et al. (1999)	–	–

(continued)

Table 4 (continued)

	Problem-solving		Other species-specific behavior	
Captive recommendation	Provide occupational enrichment and offer cognitive studies that stimulate this behavior		See specific recommendations below	
Taxa	Trait present	Citations	Trait present	Captive analog/ citations
Lemurs	Yes		Sunbathing	Provide natural sunlight
		Captive: Santos et al. (2005); Larson (2011)		Wild: Morland (1993); Captive: Hedge (2005)

2.1.1.2 Memory and Planning

Primates have a keen ability to store, recall, and integrate different types of information about their environment and apply this knowledge flexibly. This includes activities such as mental mapping, in which primates remember the details of their environment and plan their daily foraging path accordingly to maximize their intake of the best foods (Janson 1998; Zuberbühler and Janmaat 2010). For example, frugivores integrate "when" information with "what" information when foraging. In the rainforest, many trees produce fruits concurrently, with fruit production peaking at some point during the year. Frugivores use their knowledge of the synchronicity of fruit emergence as an indicator for the presence of fruit in other trees of the same species (i.e., "when" information) and then employ their spatial memory to relocate fruit-bearing trees (i.e., "what" information; Menzel 1991). Furthermore, some species may rely on long-term spatial memory to monitor food sources and previous feeding experiences at these locations across fruiting seasons (Janmaat et al. 2013). Primates can also plan for the future. For instance, male chimpanzees proactively go on border patrols or hunts, convening and leaving in atypically quiet groups prior to any evidence (that researchers can see) that they have specific knowledge of a stranger in their territory (i.e., they have not heard calls; Bates and Byrne 2009) and chacma baboons take detours in order to avoid locations where they have encountered neighboring social groups (Noser and Byrne 2007).

In captivity, we see experimental evidence for these skills. Considering working memory, primates can recall serial lists of images that include as many as nine items (Inoue and Matsuzawa 2007). They also show evidence of skills related to mental mapping; chimpanzees who have seen miniaturized models of their enclosure can use what they see in the model to find food hidden in their real-life environments (Kuhlmeier et al. 1999). Researchers have also taken advantage of computerized testing procedures to use virtual reality to show that primates can navigate mazes (Dolins et al. 2014). Finally, primates plan in several domains. They will choose a tool that they will need in a future food acquisition task (Mulcahy and Call 2006), and a chimpanzee at a zoo in Sweden demonstrated that chimpanzees may plan for entertainment purposes as he saved up rocks that he later used to throw at zoo visitors (Osvath 2009). Many common enrichment activities, such as adding physical complexity to housing or offering opportunities for choice and control, tap into

Table 5 Average daily activity budget information for wild primates to guide behavioral management of nonhuman primates in captivity

Taxa	Rest	Forage	Travel	Social/other	Reference(s)
Chimpanzees	13–37%	30–56%	8–20%	6–14%	Pruetz and McGrew (2001)
Gorillas	21–34%	55–67%	7–12%	1–4%	Watts (1988), Masi et al. (2009)
Orangutans	20–50%	33–60%	9–19%	5–30%	Morrogh-Bernard et al. (2009)
Gibbons	31–51%	29–33%	14–25%	1–17%	Daoying (1989), Bartlett (2003), Fan et al. (2013)
Colobus monkeys	32–61%	19–45%	2–25%	1–8%	Fashing (2011)
Guenons	25–40%	19–32%	26–39%	5–12%	Tashiro (2006), Baldellou and Adan (1997)
Mangabeys	8–38%	25–74%	2–31%	5–24%	Swedell (2011)
Mandrills/drills	22–42%[a]	4–64%[a]	8–33%[a]	4–45%[a]	[a]Norris (1988), [a]Walters (1989), [a]Chang et al. (1999), [a]Zhang et al. (2003)
Baboons	2–32%	21–80%	7–58%	2–21%	Swedell (2011)
Macaques	17–40%	19–55%	15–37%	4–44%	Maruhashi (1981), Kurup and Kumar (1993), Menard and Vallet (1997), O'Brien and Kinnaird (1997), Agetsuma and Nakagawa (1998), Machairas et al. (2003), Hanya (2004), Hambali et al. (2012)
Spider monkeys	45–61%	17–51%	10–36%	0.1–8%	Di Fiore et al. (2011)
Howler monkeys	56–80%	8–25%	2–27%	1–17%	Di Fiore et al. (2011)
Squirrel monkeys	3–22%	49–63%	23–29%	4–6%	Stone (2007), Pinheiro et al. (2013), Terborgh (2014)
Capuchins	5–33%	21–66%	19–41%	6–33%	Sabbatini et al. (2008), Matthews (2009), McKinney (2011), Terborgh (2014)
Titi monkeys	30–56%	17–35%	13–24%	8–14%	Caselli and Setz (2011), Terborgh (2014), Kulp and Heymann (2015)
Owl monkeys	22–38%	27–53%	21–32%	1–3%[a]	Wright (1978), Huck (2014), [a]Milozzi et al. (2012), [a]Case (2013)
Marmosets/tamarins	7–44%	21–55%	20–32%	3–20%	Garber (1988), Passamani (1998), Correa et al. (2000), Raboy and Dietz (2004), Terborgh (2014)

(continued)

Table 5 (continued)

Taxa	Rest	Forage	Travel	Social/other	Reference(s)
Lemurs	36–57%	22–46%	7–25%	5–7%	Britt (2000), Vasey (2005), Millette et al. (2009), Eppley et al. (2011), Thoren et al. (2011)

[a]Indicates captive data

these capacities. There may also be easy tweaks to existing enrichment activities that provide additional planning opportunities, such as placing the tool in a different location than where it is used to acquire food (e.g., indoors vs outdoors).

2.1.1.3 Tool Use and Causal Understanding

Many primates manufacture, modify, and use tools flexibly, demonstrating an understanding of the physical properties of objects and the causal relations between them (i.e., causal understanding). Chimpanzees cracking nuts will modify branches to use as tools and bring appropriate tools with them if needed; for instance, when cracking a hard species of nut, chimpanzees always get a stone, because only a hard, heavy object will open the nut (Boesch and Boesch 1982). Similarly, capuchin monkeys select stones of specific weight and material, transport them, and use them to crack nuts effectively (Visalberghi et al. 2009). These skills seem to be learned throughout the course of development as individuals become more selective of their tools and materials with age, improving their effectiveness (Pouydebat et al. 2006). Of course, these skills may be learned through both personal interaction and social observation (i.e., social learning). Once learned, however, the lack of trial and error behavior observed in these situations suggests that primates are mentally representing their needs before they go about choosing or making their tools.

Primates with more restrictive diets (e.g., Prosimians or leaf-eating monkeys; Tables 1, 2, 3 and 4) tend to manipulate objects less and, after controlling for body size, have smaller brains than extractive foragers that process embedded foods (e.g., roots, insects, hard shell nuts, and fruits; DeCasien et al. 2017). Of course, a species' failure to exhibit tool use does not necessarily mean that they are not capable of understanding causal relations or using tools but may mean that they lack the motivation to do so. Accordingly, a number of primates that have only been observed using tools on rare occasions in the wild regularly use tools in captivity (Tables 1, 2, 3 and 4). Guinea baboons, for instance, learn to use sticks to rake in out-of-reach food in captivity (Beck 1973a, b). This exemplifies that many behaviors are stimulus-driven. Although behavioral comparisons between captive animals and their wild counterparts provide useful indicators of behavioral repertoires, taken alone, they are not always a reliable assessment of animals' welfare (Veasey et al. 1996). Therefore, regardless of whether a behavior frequently occurs in the wild, providing primates in captivity with the opportunity to use tools may provide substantial cognitive benefits.

Understanding a species' feeding ecology can better inform what type of foraging devices may help meet the species-specific cognitive needs of primates in captivity. Extractive foragers (e.g., chimpanzees; Tables 1, 2, 3 and 4) may receive the most benefit from foraging devices that require treats to be extracted. To stimulate the ant-dipping behavior of chimpanzees (Whiten et al. 1999), for example, "probe feeders" are stuffed with substances, such as mashed bananas, barbeque sauce, or mustard and attached to enclosures (Maki et al. 1989). Bamboo shoots or other sticks are often distributed in conjunction with probe feeders so that apes can modify and use them to extract the food from the device. In contrast, destructive foragers (e.g., capuchin monkeys) may benefit most from destructible enrichment, or devices that require one to break open a contraption to obtain treats (e.g., coconuts and forage boxes).

Food puzzles may provide a higher level of cognitive challenge, as they require subjects to use their fingers or tools to poke/lift/rotate mechanisms to move visible or hidden food items through a multistep contraption to gain access to food. Fairly complex food puzzles have been successfully used with apes and macaques to evaluate aptitude for problem-solving skills (Heath et al. 1992; Hopper et al. 2014) and reduce abnormal behavior incompatible with foraging (Novak et al. 1998). Of course, food puzzles are most likely to positively influence welfare if the level and type of cognitive challenge are appropriate for the skills possessed by a given species. Simpler puzzles can also be enriching, especially when complex ones cause frustration (de Rosa et al. 2003). Because dominant individuals tend to monopolize access to devices like puzzle boxes (Bloomstrand et al. 1986), it is best to provide multiple devices (perhaps of varying difficulty) to large social groups.

2.1.2 Social Cognition

Primates' social organizations range from the monogamous pairings of owl monkeys and gibbons to the large, complex societies of baboons and macaques to the fission–fusion societies of spider monkeys and chimpanzees (Smuts et al. 1987). One thing that all primates have in common is the need to recognize group members, learn from previous social interactions, and predict what others may do in a variety of social contexts. Of course, primates show many abilities in the context of social cognition that are beyond the scope of this chapter. Below, we highlight a select few social cognitive skills that we think are particularly consistent with high welfare.

2.1.2.1 Social Interaction

Social interaction and communication provide the foundation of relationships (e.g., kinship, dominance, coalitions, alliances, and friendships) and are therefore integral components of any primate's life. Primates communicate with one another across a variety of modalities (olfactory, visual, auditory, even tactile); interact through positive interactions, such as grooming and play, and negative ones, such as fighting and aggression; and show extended periods of mother–offspring care. Moreover, primates show complex social interactions, including cooperation, food sharing, reconciliation, and deception. We do not have space to go into all of these but refer interested readers to one of several good reviews of primate social cognition

(Tomasello and Call 1997; Shettleworth 2010; Vonk and Shackelford 2012; see also Tables 1, 2, 3 and 4). Below, we summarize a few basic aspects of social interactions that are particularly relevant to welfare and enrichment.

As is well recognized, providing opportunities for species-appropriate social interactions is probably the single most important aspect of establishing, maintaining, and improving welfare for captive primates (Lutz and Novak 2005). Indeed, mere visual exposure to a conspecific is inherently rewarding to primates, more so than nonsocial stimuli and even food (see Anderson 1998, for a review, but see Bloomsmith and Lambeth 2000, for an exception). When social groups are not an option, however, video is a relatively simple and cost-effective means of visual enrichment that has been associated with a reduction in stereotypic behavior (Platt and Novak 1997; but see Washburn et al. 1997). One difficulty associated with implementing visual enrichment is that animals intrinsically habituate to the stimuli. Some research suggests that the novelty of the content may be the most important factor in the attractiveness of visual stimuli, and primates prefer viewing novel stimuli in a variety of contexts (Platt and Novak 1997). For instance, the mean duration socially housed chimpanzees watched a videotape dropped 40% from the first presentation to the fourth (Bloomsmith et al. 1990). Subjects can also benefit from auditory exposure (i.e., music), although the benefits of this are much less certain (Schapiro et al. 2014; Wallace et al. 2017 but see Novak and Drewsen 1989; Howell et al. 2003).

Simply having things to watch or hear is very basic, and not very interactive. Thus, if inactivity is a problem, video may not be a great enrichment option. If possible, primates should also be given access to complex social interactions that simulate the cooperation and competition seen in the wild. For instance, neighboring groups can be given visual access to one another, or play "tug of war" between enclosures. Of course, this should only be done if appropriate; highly territorial primates, like owl monkeys, may show acute stress responses if they can hear or see neighboring groups (Fernandez-Duque et al. 2012). Nonetheless, giving animals the opportunity to at least try such activities may increase their long-term well-being, even if it increases transient stress responses in the short term.

2.1.2.2 Social Learning

Learning from others allows individuals to avoid time-consuming and potentially dangerous trial-and-error learning and allows for the nongenetic transmission of information between individuals and across generations. Many of these behaviors may become traditions, or culture, in which the behavior of a single individual shifts the behavioral patterns seen in a substantial percentage of a social group. In one of the more compelling examples, an 18-month-old female Japanese macaque, named Imo, began washing provisioned sweet potatoes in water, following which the behavior eventually spread to most of the rest of the group (Itani and Nishimura 1973). Similar reports of tool use and potential socially learned behavioral traditions among wild populations are widespread across primates, both in the wild and in captivity (Tomasello and Call 1997).

Social learning can occur via several different methods that vary in their cognitive complexity. "Simpler" forms of social learning include contagion and enhancement, both of which are based on attraction to conspecifics' actions. In contagion, animals perform actions in synchrony with another that they observe, such as fleeing when others flee (Thorpe 1963). In enhancement, an individual's attention is drawn to a location (i.e., local enhancement) or an object (i.e., stimulus enhancement) because of the proximity or actions of another individual (Whiten and Ham 1992). "Higher-level" social learning processes include those that result in behavior matching; emulation occurs when an observer achieves the end goal of a demonstrated action through different means (Wood 1989), whereas in imitation an individual copies both the actions and the goals of a model (Tomasello and Call 1997). Typically, imitation is considered to be the most complex form of social learning as it may require perspective taking, though there is still debate surrounding the underlying cognitive mechanisms of these two processes (see Hopper and Whiten 2012 for further discussion; Hopper et al. 2016). With regard to welfare, it is important to know that social learning is widespread and can be very useful, both to teach animals new skills (i.e., to work with veterinarians for noninvasive medical examinations) and as enrichment in and of itself. Still, it is important to tap into the appropriate mechanisms when designing apparatuses. For example, a complex imitation task may be beyond the ability of some monkeys who nonetheless benefit from opportunities to learn through stimulus enhancement.

2.1.2.3 Mirror Self-Recognition and Theory of Mind

Mirror self-recognition has long been considered a hallmark of intelligence and self-awareness. Human babies begin to recognize themselves in a mirror between 18 and 24 months of age (Amsterdam 1972), and mirrors have been used for many years as a tool for testing self-recognition and self-awareness in nonhuman species on the assumption that it relates to other complex cognitive abilities, such as theory of mind (Premack and Woodruff 1978) and empathy (de Waal 1996). In a seminal paper, Gallup (1970) presented chimpanzees with a mirror and recorded their behavioral responses. Not only did chimpanzees explore the contingencies between their movements and the mirror, but also they used the mirror to explore areas of their bodies that they could not see. In an experimental condition, when the mirror was present, chimpanzees were more likely to investigate a mark that could not be seen without the mirror. Most apes do recognize themselves in the mirror, although evidence is mixed for gorillas, but monkeys do not, unless they have extensive training (Tomasello and Call 1997).

Whether primates recognize themselves in mirrors or not, mirrors can be highly entertaining to captive primates. Even in species that probably lack mirror self-recognition, the presence of mirrors may elicit species-typical behaviors. For instance, when stump-tailed macaques were presented with mirrors, the frequency and duration of their social behaviors increased, including submissive and aggressive responses (Straumann and Anderson 1991). Often times, the physical and structural environment of a facility may limit captive primates' opportunity to view other conspecifics. In this case, mirrors can also be used to view other animals in a room that they may otherwise not be able to see, which may be particularly important in rooms with poor sight lines. In chimpanzees, mirrors increase facial expressions, and sexual and agonistic behavior (Lambeth and Bloomsmith 1992). As with other enrichment, monkeys may become habituated to mirrors (Clarke et al. 1995), although interest can be restored simply by placing the mirror in a new location on the cage (Suárez and Gallup 1981). Note, too, that responses directed toward mirrors are higher when the mirrors are attached to cage mesh than when they are placed outside of the cage (Anderson and Roeder 1989).

2.1.2.4 Behavioral Testing as Cognitive Enrichment

Finally, behavioral and cognitive research itself can be greatly enriching to captive primates (see Ross 2010; Clark 2017; Hopkins and Latzman 2017; Schapiro et al. 2017 for reviews). For instance, Washburn and Rumbaugh (1992) found in macaques that removal of a computerized system that offered a variety of cognitive tests led to increases in stereotypic behaviors. Importantly, when given a choice between receiving pellets for free and completing trials for the pellets, the macaques chose to work for their reward (Markowitz 1982), a phenomenon known as contrafreeloading (Jensen 1963). Furthermore, rather than always selecting the easiest task that produced the most pellets per minute, the macaques chose to work on tasks with various levels of difficulty (although this could also be due to the matching law, or the relative rates of reinforcement).

Computerized testing has also been shown to promote species-typical behavior. Yamanashi and Hayashi (2011) assessed the effects of cognitive experiments by directly comparing the activity budgets of three populations of chimpanzees: wild chimpanzees in Bossou, captive chimpanzees that participated in computer-controlled cognitive experiments, and captive chimpanzees that did not participate in cognitive experiments. The feeding and resting times of the chimpanzees participating in cognitive experiments were almost the same as their wild counterparts (Table 5), whereas the chimpanzees that did not participate in experiments exhibited significantly shorter feeding times and significantly longer resting times. Other work shows that socially housed chimpanzees exhibited higher activity levels during behavioral testing whether or not they participated in the task (Hopper et al. 2016). This benefit extends beyond chimpanzees; socially housed crested macaques given access to cognitive testing showed increased affiliation and decreased aggression, as well as increased association between dyads on testing days, possibly reflecting reunion behavior (Whitehouse et al. 2013).

One challenge to this is that implementing computerized test systems is both expensive and time-consuming. Most systems either require individual computers or automated access systems (e.g., using passive integrated transponder, or PIT, tags), as well as automatic feeders, all of which are relatively costly. In addition, subjects must be trained on the systems in order to benefit from them. Of course, the training itself is enriching, so there is likely little down side to the primates, and several well-validated training programs exist (Washburn et al. 1992; Evans et al. 2008; Cronin et al. 2017). However, it requires a substantial investment of time by staff or researchers that may not be possible in times of tight budgets. It is our hope that as technology becomes cheaper (i.e., by using programmable Raspberry Pi systems instead of more expensive personal computers) and the benefits become more widely recognized and accepted, computerized testing may become a viable option for promoting species-typical behavior in many captive settings.

3 Behavioral Needs

Captive environments are by their nature artificial and different from natural environments. In natural environments, primates have access to much larger home ranges than are possible in captivity, and captive environments may lack key elements, such as the opportunity to explore or get away from group mates. Still, many features from natural environments can be replicated, or analogues that functionally simulate species-typical behavior can be provided. Here, we focus on two important topics when considering the behavioral needs of captive primates: space and activity levels. These categories are not mutually exclusive and often overlap with other strategies for improving the welfare of captive primates.

3.1 Space

Although the focus is often on amount of space, we must also consider the quality of that space. For many primates, this includes three-dimensional volume with opportunities to climb (i.e., primarily arboreal New World monkeys and apes; Table 6). Species' patterns of space use in captivity mimic preferences exhibited in the wild. For instance, chimpanzees, which can spend up to 68% of their time in trees (Doran and Hunt 1994), spend more time in the highest tier of their exhibit as compared to gorillas, the least arboreal species of great ape (Ross and Lukas 2006). Thus, if there is a trade-off between square feet and height, arboreal species (Tables 1, 2, 3 and 4) are likely better off with fewer square feet, but higher ceilings and complex climbing structures. Supporting this, numerous studies have found no meaningful behavioral differences between large enclosures and smaller enclosures, if the environments are comparably complex (Reamer et al. 2015; Ross et al. 2011).

Individuals also need to be able to hide from, or be out of view of, conspecifics. This is presumably most critical for fission–fusion species, which are accustomed to fluid social arrangements and likely benefit cognitively from changing social dynamics, but the ability to escape, particularly from social tension, is presumably important for virtually all species. One easy way to accomplish this is visual barriers (Boere 2001), but nesting boxes, barrels, boxes to hide in, or enclosures that are structured with multiple rooms or spaces may also provide temporary seclusion. Moreover, for species that rarely encounter other groups in the wild (Table 6), such as owl monkeys, large visual barriers and waterfalls can be used to prevent family units from seeing and hearing other family units (Abee 2011).

Table 6 Relevant social information of wild primates to guide behavioral management of NHPs in captivity

	Social structure	Group size	Dispersal	Dominance	Home range	Fission–fusion	Reference(s)
Chimpanzees	Multimale–multifemale	30–100+	Females	Male and female hierarchies	400–500 ha	Yes	Reamer et al. (2017), Stumpf (2011)
Gorillas	*G. gorilla*: Unimale–multifemale; *G. beringei*: multimale–multifemale	8–10	Males and females	Male is dominant	300–3200 ha	No	Robbins (2011)
Orangutans	Solitary/temporary feeding aggregates/consortships	1–2	Males	Male hierarchy within territory overlap	200–850 ha	Yes	Knott and Kahlenberg (2011), Fox (2002), Amici et al. (2008)
Gibbons	Pair-bonded adults with offspring	2–5	Males and females	None	15–129 ha	No	Bartlett (2011), Gittins and Raemaekers (1980)
Colobus monkeys	Multimale–multifemale/unimale–multifemale	2–80	Males and females—varies based on species	Male and female hierarchies	5–2440 ha	Depends on habitat size/quality	Fashing (2011), Struhsaker (2010)
Guenons	Unimale–multifemale, with exceptions in some species	3–64	Males	None, except *C. aethiops*	4–335 ha; *E. patas*: 440–5200 ha	No	Jaffe and Isbell (2011)
Mangabeys	Multimale–multifemale	3–60	Males	Male and female hierarchies	13–492 ha	Yes	Swedell (2011)
Mandrills/drills	Solitary males outside of breeding season/multimale–multifemale during breeding season	5–845	Males	Male and female hierarchies	Data not available	No	Swedell (2011), Hoshino et al. (1984)

	Social organization	Group size	Dispersing sex	Dominance	Home range	Fission-fusion	References
Baboons	Multimale–multifemale/unimale–multifemale	4–262	Males	Male and female hierarchies	54–4375 ha	Yes	Swedell (2011)
Macaques	Multimale–multifemale	15–50	Males, with exceptions in some species	Male and female hierarchies	3–2200 ha	In some species	Thierry (2011), Amici et al. (2008)
Spider monkeys	Multimale–multifemale	16–55	Females	Male and female hierarchies	80–963 ha	Yes	Di Fiore et al. (2011), Amici et al. (2008)
Howler monkeys	Multimale–multifemale/unimale–multifemale	16–21	Males and females	Male and female hierarchies	2–182 ha	No	Di Fiore et al. (2011), Milton (1980)
Squirrel monkeys	Multimale–multifemale	15–50	Males and females—varies based on species	Male and female hierarchies	17–250 ha	No	Jack (2011), Rowe (1996), Boinski and Mitchell (1997)
Capuchins	Multimale–multifemale	16–75	Males	Male and female hierarchies	25–300 ha	No	Jack (2011), Rowe (1996), Amici et al. (2008)
Titi monkeys	Pair-bonded adults with offspring	2–7	Males and females	No	1–29 ha	No	Norconk (2011), Kinzey (1981)
Owl monkeys	Monogamous pairs	2–4	Males and females	Male slightly dominant over female	4–17 ha	No	Fernandez-Duque (2011), Moynihan (1964)
Marmosets/tamarins	Multimale–multifemale/unimale–multifemale/monogamous pairs	2–20	Males and females	None or weak	1–394 ha	No	Digby et al. (2011)
Lemurs	Multimale–multifemale/unimale–multifemale/monogamous pairs	1–31	Males and females—varies based on species	None or weak	<1–400 ha	Some species	Gould et al. (2011)

3.2 Activity

Animals in captivity are generally found to spend a dramatically smaller percentage of their day active (and a correspondingly larger component inactive or at rest) than wild living animals. This leads to a tableau of negative health outcomes similar to those seen in humans with sedentary lifestyles (Ely et al. 2013). Although it is impossible to replicate many components of life in the wild that lead to high levels of activity (such as foraging over large areas or evading predators), it is possible to simulate some of them, thereby increasing activity levels and reducing concomitant health problems. In the wild, primates spend considerable time and effort foraging for food (Table 5). Therefore, one way to increase activity is to increase the time and effort it takes for captive primates to obtain their food. For example, food can be distributed across the enclosure or spread through wood chips/straw on the floor (Byrne and Suomi 1991), rather than provided in a single location. Although providing chopped fruits and vegetables may increase search time and food availability, food is unlikely to be presented in bite-sized pieces in the wild and many primates exhibit specific food-handling behaviors. Indeed, some primate species spend significantly more time engaged in feeding behaviors when presented with whole foods as opposed to chopped (Smith et al. 1989). To increase climbing and sustained activity (e.g., hanging), food can be placed up high, scattered on platforms, or suspended, such as from a chain-link ceiling, which also requires additional manipulation to access it (Britt 1998).

Recently, some facilities have explored more innovative approaches to curtail obesity, which often leads to morbidity and mortality in captive primates (Ely et al. 2013). For instance, simple modifications requiring chimpanzees to climb and support their own weight while working harder to access high-calorie primate biscuits led to a significant increase in foraging times (and possibly reduced feeding), and ultimately to weight loss for overweight individuals (yet no weight changes for healthy weight chimpanzees) (Bridges et al. 2013). This may also help alleviate the common issue of higher-ranking individuals dominating access to food. In another example, several zoos tried providing their gorillas with only leafy greens (their natural food source) in hopes of controlling weight gain. In an astonishing test of this hypothesis, the gorillas ate more calories, but lost weight and became healthier. Aside from weight loss benefits, the removal of biscuits from their diet also led to a reduction in regurgitation and reingestion (Less et al. 2014), a stereotypic behavior often associated with social deficits in early development, diet, and/or boredom (Gould and Bres 1986; Lukas 1999).

Primates also frequently explore and manipulate different objects in their natural environment. Japanese macaques gather, roll, rub, and carry stones (Huffman and Quiatt 1986). Chimpanzees in the Mahale mountains in Tanzania "leaf-pile pull"; as they descend a mountain, they sometimes stop and walk backward, raking handfuls of leaves with both arms and then either walk or somersault through the pile of leaves that was created (Nishida and Wallauer 2003). Therefore, providing objects (e.g., rubber dog toys, reflective balls, and browse) may increase activity by providing primates with the opportunity to explore and manipulate objects. Again, habituation occurs, so rotating the objects presented in the enclosure or providing a wide variety of toys increases object use and play for most species (Schapiro and Bloomsmith 1995; Sanz et al. 1999). Additionally, not all primates in the wild show these behaviors, and different individuals will have different preferences. It is important to provide a variety of options and monitor to see who is benefitting from the provided enrichment.

3.3 Social Needs

As has been reiterated throughout this chapter, sociality is central to the behavior and management of primates, and providing primates with social stimulation in the form of social housing is widely regarded as the most effective form of enrichment (Lutz and Novak 2005). Overall, the literature suggests that increasing social stimulation (moving from single housing to pair housing) improves welfare, whereas the reduction in social stimulation diminishes welfare. Because it is difficult to evaluate positive evidence of welfare, this is typically evaluated by the development of abnormal behavior when social stimulation is limited or lacking entirely (Harlow and Harlow 1962; Vandeleest et al. 2011). More recent studies have assessed how suboptimal social housing, or changes in the social environment, impacts physiological responses (e.g., stress and immune function), which is highly informative for the welfare of primates and the participation of primates in biological research as models (Schapiro et al. 1993, 2000, 2012; Williams et al. 2010; Capitanio and Cole 2015).

The natural socioecology of a species can greatly inform captive group size and composition. For instance, a pair-bonded species (e.g., callitrichids) should be housed as a bonded male and female and their offspring, whereas gregarious species, such as rhesus macaques or squirrel monkeys, which may live with scores or even hundreds of others in the wild, will do very well when housed in large groups (Table 6). Decisions about which animals are moved and when should be based whenever possible on which sex emigrates from the natal group, and at what age they do so (Table 6). This will help to mitigate social tension and aggression following relocations. Additionally, when an adult male or male coalition takes over a group, the other males generally either leave or are killed. Therefore, when putting new males into a group of females, it may be necessary to remove the existing adult males first (Cooper et al. 2001). If this is not possible, the group should be carefully watched after for unexpected takeovers by existing group males.

Group composition is also an important factor to consider when moving individuals between groups. Sometimes moves are critical, because of fighting between animals, and sometimes they are desirable, when they offer the chance to combine two smaller groups or to separate a particular age–sex class (e.g., a bachelor group or an "elderly" group), or for the species survival plan (Price and Stoinski 2007). Regardless of whether the animals are relocated within a facility or to a new facility, they will potentially be separated from familiar companions, introduced to new companions, and may have to deal with changes in housing or routine. The few studies that have been conducted conclude that relocation negatively impacts primates' welfare (Watson et al. 2005; Kim et al. 2005; Koban et al. 2010; Williams et al. 2010; Schapiro et al. 2012; Gottlieb et al. 2013; Nehete et al. 2017).

When moves are essential, one way to ameliorate their impact is by allowing animals sufficient time to acclimate to their new environment (e.g., Schel et al. 2013), particularly prior to beginning any research studies. Unfortunately, very few studies have attempted to quantify the amount of time primates need to acclimate to their new environment (Capitanio et al. 2006; Obernier and Baldwin 2006). Although there is no clear-cut answer, a period of up to 3 months has been recommended for laboratory species (Capitanio et al. 2006; Schapiro et al. 2012). Additionally, providing a regular care schedule with little exposure to novel or uncertain situations may reduce stress and aid acclimation. Of course, there will be individual differences in responses based on the animal's genotype, temperament, early rearing history, and/or other factors, so it is important to collect individual baseline measures before and after relocation (Boccia et al. 1995; Bennett et al. 2002). Captive management would benefit from future studies that evaluate the best management practices for acclimation once animals have been relocated.

4 Conclusions and Future Directions

Compared to the complex natural environment in which primates evolved, captive conditions may include fewer cognitive challenges, restricted space, and limited social opportunities. Therefore, behaviors exhibited by wild primates cannot always be replicated in captivity. However, we can functionally simulate natural conditions to provide captive primates with opportunities that will stimulate the performance of species-typical behaviors. It is now widely recognized that mental stimulation and social stimulation are critical needs. As a result, an emerging goal is to provide more extensive cognitive and behavioral enrichment, as well as other sorts of physical and social enrichment. This is a challenge for two reasons. First, cognitive and behavioral needs are tightly tied to species' socioecology, making general recommendations for the whole order Primates impossible, or even counterproductive. Second, as this is an emerging area, we lack consensus on objective measures to assess the effectiveness of cognitive enrichment (Clark 2011; Hopkins and Latzman 2017; Schapiro et al. 2017). There is clearly a need for additional research to clarify which interventions stimulate species-typical behavior and cognition and reduce the effects of stress and boredom. In particular, we hope that researchers continue to seek opportunities and collaborations to further examine the impact of behavioral and cognitive testing on the welfare of individual species of captive primates. Together, we can advance the field while bettering the lives of captive primates.

Acknowledgments We would like to thank the researchers and caregivers that have devoted their careers to improving the welfare of captive nonhuman primates. CFT was supported by NIH R01 HD087048, and SFB was supported by NSF SES 1425216 and NSF SES 1123897.

Bibliography

Abee CR (2011) Creating better primate housing: dialog, design, and funding. Am J Primatol 73(S1):33
Agetsuma N, Nakagawa N (1998) Effects of habitat differences on feeding behaviors of Japanese monkeys: comparison between Yakushima and Kinkazan. Primates 39:275–289. https://doi.org/10.1007/BF02573077
Amici F, Aureli F, Call J (2008) Fission-fusion dynamics, behavioral flexibility, and inhibitory control in primates. Curr Biol 18(18):1415–1419. https://doi.org/10.1016/J.CUB.2008.08.020
Amsterdam B (1972) Mirror self-image reactions before age two. Dev Psychobiol 5(4):297–305. https://doi.org/10.1002/dev.420050403
Anderson JR (1985) Development of tool-use to obtain food in a captive group of *Macaca tonkeana*. J Hum Evol 14(7):637–645. https://doi.org/10.1016/S0047-2484(85)80072-5
Anderson JR (1986) Mirror-mediated finding of hidden food by monkeys (*Macaca tonkeana* and *M. fascicularis*). J Comp Psychol 100(3):237–242. https://doi.org/10.1037/0735-7036.100.3.237
Anderson JR (1998) Social stimuli and social rewards in primate learning and cognition. Behav Process 42(2–3):159–175. https://doi.org/10.1016/S0376-6357(97)00074-0
Anderson JR, Gallup GG (1997) Self-recognition in *Saguinus*? A critical essay. Anim Behav 54(6):1563–1567. https://doi.org/10.1006/anbe.1997.0548
Anderson JR, Gallup GG (2011) Do rhesus monkeys recognize themselves in mirrors? Am J Primatol 73(7):603–606. https://doi.org/10.1002/ajp.20950
Anderson JR, Henneman MC (1994) Solutions to a tool-use problem in a pair of *Cebus apella*. Mammalia 58(3):351–362. https://doi.org/10.1515/mamm.1994.58.3.351
Anderson JR, Roeder J-J (1989) Responses of capuchin monkeys (*Cebus apella*) to different conditions of mirror-image stimulation. Primates 30:581–587. https://doi.org/10.1007/BF02380884
Anderson JR, Kuroshima H, Paukner A, Fujita K (2009) Capuchin monkeys (*Cebus apella*) respond to video images of themselves. Anim Cogn 12:55–62. https://doi.org/10.1007/s10071-008-0170-3
Baer JF, Weller RE, Kakoma, I. (Eds.). (2012) *Aotus*: the owl monkey. Academic Press, New York
Baldellou M, Adan ANA (1997) Time, gender, and seasonality in vervet activity: a chronobiological approach. Primates 38:31–43. https://doi.org/10.1007/BF02385920
Baldwin LA, Teleki G (1976) Patterns of gibbon behavior on Hall's Island, Bermuda: a preliminary ethogram for *Hylobates lar*. In: Rumbaugh DM (ed) Gibbon and siamang, vol 4. Karger, Basel, pp 21–105
Bartlett TQ (2003) Intragroup and intergroup social interactions in white-handed gibbons. Int J Primatol 24:239–259. https://doi.org/10.1023/A:1023088814263
Bartlett TQ (2011) The Hylobatidae: small apes of Asia. In: Campbell CJ, Fuentes A, MacKinnon KC, Bearder SK, Stumpf SM (eds) Primates in perspective. Oxford University Press, Oxford, pp 301–312
Bates LA, Byrne RW (2009) Sex differences in the movement patterns of free-ranging chimpanzees (*Pan troglodytes schweinfurthii*): foraging and border checking. Behav Ecol Sociobiol 64:247–255. https://doi.org/10.1007/s00265-009-0841-3
Beck BB (1967) A study of problem solving by gibbons. Behaviour 28(1–2):95–109. https://doi.org/10.1163/156853967X00190
Beck BB (1973a) Cooperative tool use by captive hamadryas baboons. Science 182(4112):594–597. https://doi.org/10.1126/science.182.4112.594
Beck BB (1973b) Observation learning of tool use by captive Guinea baboons (*Papio papio*). Am J Phys Anthropol 38(2):579–582. https://doi.org/10.1002/ajpa.1330380270
Beck BB (1976) Tool use by captive pigtailed macaques. Primates 17(3):301–310. https://doi.org/10.1007/BF02382787

Benhar EE, Samuel D (1978) A case of tool use in captive olive baboons (*Papio anubis*). Primates 19:385–389. https://doi.org/10.1007/BF02382807

Bennett AJ, Lesch KP, Heils A et al (2002) Early experience and serotonin transporter gene variation interact to influence primate CNS function. Mol Psychiatry 7:118–122. https://doi.org/10.1038/sj.mp.4000949

Berghänel A, Ostner J, Schröder U, Schülke O (2011) Social bonds predict future cooperation in male Barbary macaques, *Macaca sylvanus*. Anim Behav 81(6):1109–1116. https://doi.org/10.1016/j.anbehav.2011.02.009

Bernstein IS (1962) Response to nesting materials of wild born and captive born chimpanzees. Anim Behav 10(1–2):1–6. https://doi.org/10.1016/0003-3472(62)90123-9

Bloomsmith MA, Else JG (2005) Behavioral management of chimpanzees in biomedical research facilities: the state of the science. ILAR J 46(2):192–201. https://doi.org/10.1093/ilar.46.2.192

Bloomsmith MA, Lambeth SP (2000) Videotapes as enrichment for captive chimpanzees (*Pan troglodytes*). Zoo Biol 19(6):541–551. https://doi.org/10.1002/1098-2361(2000)19:6<541::AID-ZOO6>3.0.CO;2-3

Bloomsmith MA, Keeling ME, Lambeth SP (1990) Videotapes: environmental enrichment for singly housed chimpanzees. Lab Anim (NY) 19(1):42–46

Bloomstrand M, Riddle K, Alford P, Maple TL (1986) Objective evaluation of a behavioral enrichment device for captive chimpanzees (*Pan troglodytes*). Zoo Biol 5(3):293–300. https://doi.org/10.1002/zoo.1430050307

Boccia ML, Laudenslager ML, Reite ML (1995) Individual differences in macaques' responses to stressors based on social and physiological factors: implications for primate welfare and research outcomes. Lab Anim 29(3):250–257. https://doi.org/10.1258/002367795781088315

Boere V (2001) Environmental enrichment for neotropical primates in captivity. Ciência Rural 31(3):543–551. https://doi.org/10.1590/S0103-84782001000300031

Boesch C (1991) Teaching among wild chimpanzees. Anim Behav 41(3):530–532. https://doi.org/10.1016/S0003-3472(05)80857-7

Boesch C (2002) Cooperative hunting roles among Taï chimpanzees. Hum Nat 13(1):27–46. https://doi.org/10.1007/s12110-002-1013-6

Boesch C, Boesch H (1982) Optimisation of nut-cracking with natural hammers by wild chimpanzees. Behaviour 83(3–4):265–286. https://doi.org/10.1163/156853983X00192

Boinski S (1987) Mating patterns in squirrel monkeys (Saimiri oerstedi). Behav Ecol Sociobiol 21(1):13–21. https://doi.org/10.1007/BF00324430

Boinski S (1988) Use of a club by a wild white-faced capuchin (*Cebus capucinus*) to attack a venomous snake (*Bothrops asper*). Am J Primatol 14(2):177–179. https://doi.org/10.1002/ajp.1350140208

Boinski S, Mitchell CL (1997) Chuck vocalizations of wild female squirrel monkeys (*Saimiri sciureus*) contain information on caller identity and foraging activity. Int J Primatol 18:975–993. https://doi.org/10.1023/A:1026300314739

Bramlett JL, Perdue BM, Evans TA, Beran MJ (2012) Capuchin monkeys (*Cebus apella*) let lesser rewards pass them by to get better rewards. Anim Cogn 15:963–969. https://doi.org/10.1007/s10071-012-0522-x

Breuer T, Ndoundou-Hockemba, & M., Fishlock, V (2005) First observation of tool use in wild gorillas. PLoS Biol 3(11):e380. https://doi.org/10.1371/journal.pbio.0030380

Bridges JP, Mocarski EC, Reamer LA, Lambeth SP, Schapiro SJ (2013) Weight management in captive chimpanzees (*Pan troglodytes*) using a modified feeding device. Am J Primatol 75(S1):51

Britt A (1998) Encouraging natural feeding behavior in captive-bred black and white ruffed lemurs (*Varecia variegata variegata*). Zoo Biol 17(5):379–392. https://doi.org/10.1002/(SICI)1098-2361(1998)17:5<379::AID-ZOO3>3.0.CO;2-X

Britt A (2000) Diet and feeding behaviour of the black-and-white ruffed lemur (*Varecia variegata variegata*) in the Betampona Reserve, eastern Madagascar. Folia Primatol 71(3):133–141. https://doi.org/10.1159/000021741

Brosnan S, Parrish A, Beran M, Flemming T, Heimbauer L, Talbot C, Wilson, B (2011) Responses to the assurance game in monkeys, apes, and humans using equivalent procedures. Proc Natl Acad Sci U S A 108:3442–3447. https://doi.org/10.1073/pnas.1016269108

Buckmaster CL, Parker KJ, Lyons DM (2015) Preliminary observations of social interaction during spontaneous cup tool use by captive-born adult female squirrel monkeys (*Saimiri sciureus*). Am J Primatol 77(S1):38

Butynski TM (1982) Vertebrate predation by primates: a review of hunting patterns and prey. J Hum Evol 11(5):421–430. https://doi.org/10.1016/S0047-2484(82)80095-X

Byrne RW (1996) The misunderstood ape: cognitive skills of the gorilla. In: Russon AE, Bard KA, Parker ST (eds) Reaching into thought: the minds of the great apes. Cambridge University Press, Cambridge, pp 111–130

Byrne GD, Suomi SJ (1991) Effects of woodchips and buried food on behavior patterns and psychological well-being of captive rhesus monkeys. Am J Primatol 23(3):141–151. https://doi.org/10.1002/ajp.1350230302

Byrne RW, Whiten A (eds) (1988) Machiavellian intelligence: social expertise and the evolution of intellect in monkeys, apes, and humans. Oxford University Press, Oxford

Call J, Tomasello M (1994) The social learning of tool use by orangutans (*Pongo pygmaeus*). Hum Evol 9:297–313. https://doi.org/10.1007/BF02435516

Cameron R, Rogers LJ (1999) Hand preference of the common marmoset (*Callithrix jacchus*): problem solving and responses in a novel setting. J Comp Psychol 113(2):149–157. https://doi.org/10.1037/0735-7036.113.2.149

Canale GR, Guidorizzi CE, Kierulff MCM, Gatto CAFR (2009) First record of tool use by wild populations of the yellow-breasted capuchin monkey (*Cebus xanthosternos*) and new records for the bearded capuchin (*Cebus libidinosus*). Am J Primatol 71(5):366–372. https://doi.org/10.1002/ajp.20648

Capitanio JP, Cole SW (2015) Social instability and immunity in rhesus monkeys: the role of the sympathetic nervous system. Philos Trans R Soc B Biol Sci 370(1669):20140104. https://doi.org/10.1098/rstb.2014.0104

Capitanio JP, Kyes RC, Fairbanks LA (2006) Considerations in the selection and conditioning of Old World monkeys for laboratory research: animals from domestic sources. ILAR J 47(4):294–306. https://doi.org/10.1093/ilar.47.4.294

Case L (2013) Understanding behavior and nest box usage in three species of owl monkeys: Azara's owl monkey (*Aotus azarai*), Spix's owl monkey (*A. vociferans*) and Nancy Ma's owl monkey (*A. nancymaae*). Doctoral dissertation, Texas State University-San Marcos

Caselli CB, Setz EZ (2011) Feeding ecology and activity pattern of black-fronted titi monkeys (*Callicebus nigrifrons*) in a semideciduous tropical forest of southern Brazil. Primates 52(4):351–359. https://doi.org/10.1007/s10329-011-0266-2

Chalmeau R, Lardeux K, Brandibas P, Gallo A (1997) Cooperative problem solving by orangutans (*Pongo pygmaeus*). Int J Primatol 18(1):23–32. https://doi.org/10.1023/A:1026337006136

Chang TR, Forthman DL, Maple TL (1999) Comparison of confined mandrill (*Mandrillus sphinx*) behavior in traditional and "ecologically representative" exhibits. Zoo Biol 18(3):163–176. https://doi.org/10.1002/(SICI)1098-2361(1999)18:3<163::AID-ZOO1>3.0.CO;2-T

Cheyne SM, Höing A, Rinear J, Sheeran LK (2012) Sleeping site selection by agile gibbons: the influence of tree stability, fruit availability and predation risk. Folia Primatol 83(3–6):299–311. https://doi.org/10.1159/000342145

Chiang M (1967) Use of tools by wild macaque monkeys in Singapore. Nature 214:1258–1259

Clark FE (2011) Great ape cognition and captive care: can cognitive challenges enhance well-being? Appl Anim Behav Sci 135(1–2):1–12. https://doi.org/10.1016/J.APPLANIM.2011.10.010

Clark FE (2017) Cognitive enrichment and welfare: current approaches and future directions. Anim Behav Cogn 4(1):52–71. https://doi.org/10.12966/abc.05.02.2017

Clarke AS, Czekala NM, Lindburg DG (1995) Behavioral and adrenocortical responses of male cynomolgus and lion-tailed macaques to social stimulation and group formation. Primates 36(1):41–56. https://doi.org/10.1007/bf02381914

Coleman K, Bloomsmith MA, Crockett CM, Weed JL, Schapiro SJ (2012) Behavioral management, enrichment, and psychological well-being of laboratory nonhuman primates. In: Abee CR, Keith M, Suzette T, Timothy M (eds) Nonhuman primates in biomedical research, vol 1, 2nd edn. Academic Press, Boston, pp 149–176

Cooper MA, Bernstein IS, Fragaszy DM, de Waal FBM (2001) Integration of new males into four social groups of tufted capuchins (*Cebus apella*). Int J Primatol 22(4):663–683. https://doi.org/10.1023/A:1010745803740

Corrêa HK, Coutinho PE, Ferrari SF (2000) Between-year differences in the feeding ecology of highland marmosets (*Callithrix aurita* and *Callithrix flaviceps*) in south-eastern Brazil. J Zool 252(4):421–427

Crast J, Bloomsmith MA, Jonesteller TJ (2016) Behavioral effects of an enhanced enrichment program for group-housed sooty mangabeys (*Cercocebus atys*). J Am Assoc Lab Anim Sci 55(6):756–764

Cronin KA, Jacobson SL, Bonnie KE, Hopper LM (2017) Studying primate cognition in a social setting to improve validity and welfare: a literature review highlighting successful approaches. PeerJ 5:e3649. https://doi.org/10.7717/peerj.3649

Cunningham CL, Anderson JR, Mootnick AR (2006) Object manipulation to obtain a food reward in hoolock gibbons, *Bunopithecus hoolock*. Anim Behav 71(3):621–629

Custance D, Whiten A, Fredman T (1999) Social learning of an artificial fruit task in capuchin monkeys (*Cebus apella*). J Comp Psychol 113(1):13–23. https://doi.org/10.1037/0735-7036.113.1.13

Daoying L (1989) Preliminary study on the group composition behavior and ecology of the black gibbons (*Hylobates concolor*) in southwest Yunnan. Zool Res 10(zk):119–126

de Rosa C, Vitale A, Puopolo M (2003) The puzzle-feeder as feeding enrichment for common marmosets (*Callithrix jacchus*): a pilot study. Lab Anim 37(2):100–107. https://doi.org/10.1258/00236770360563732

de Waal FBM (1996) Good natured: the origins of right and wrong in humans and other animals. Harvard University Press, Cambridge, MA

DeCasien AR, Williams SA, Higham JP (2017) Primate brain size is predicted by diet but not sociality. Nat Ecol Evol 1:0112. https://doi.org/10.1038/s41559-017-0112

Di Fiore A, Link A, Campbell CJ (2011) The Atelines: behavioral and socioecological diversity in a New World Monkey radiation. In: Campbell CJ, Fuentes A, MacKinnon KC, Bearder SK, Stumpf SM (eds) Primates in perspective. Oxford University Press, Oxford, pp 155–188

Dickie L (1998) Environmental enrichment for Old World primates with reference to the primate collection at Edinburgh Zoo. Int Zoo Yearb 36(1):131–139. http://doi.org/10.1111/j.1748-1090.1998.tb02895.x

Digby LJ, Ferrari SF, Saltzman W (2011) Callitrichines: the role of competition in cooperatively breeding species. In: Campbell CJ, Fuentes A, MacKinnon KC, Bearder SK, Stumpf SM (eds) Primates in perspective. Oxford University Press, Oxford, pp 91–107

Dolins FL, Klimowicz C, Kelley J, Menzel CR (2014) Using virtual reality to investigate comparative spatial cognitive abilities in chimpanzees and humans. Am J Primatol 76(5):496–513. https://doi.org/10.1002/ajp.22252

Doran DM (1996) Comparative positional behavior of the African apes. In: McGrew WC, Marchant LF, Nishida T (eds) Great Ape societies. Cambridge University Press, Cambridge, pp 213–224

Doran DM, Hunt KD (1994) Comparative locomotor behavior of chimpanzees and bonobos. In: Wrangham RW, McGrew WC, de Waal FBM, Heltne PG (eds) Chimpanzee cultures. Chicago Academy of Sciences, Chicago, pp 93–108

Dunbar RIM (1992) Neocortex size as a constraint on group size in primates. J Hum Evol 22(6):469–493. https://doi.org/10.1016/0047-2484(92)90081-J

Ehrlich A (1970) Response to novel objects in three lower primates: greater galago, slow loris, and owl monkey. Behaviour 37(1–2):55–63. https://doi.org/10.1163/156853970X00231

Ely JJ, Zavaskis T, Lammey ML (2013) Hypertension increases with aging and obesity in chimpanzees (*Pan troglodytes*). Zoo Biol 32(1):79–87. https://doi.org/10.1002/zoo.21044

Eppley TM, Verjans E, Donati G (2011) Coping with low-quality diets: a first account of the feeding ecology of the southern gentle lemur, *Hapalemur meridionalis*, in the Mandena littoral forest, southeast Madagascar. Primates 52(1):7–13. https://doi.org/10.1007/s10329-010-0225-3

Erkert HG (1989) Lighting requirements of nocturnal primates in captivity: a chronobiological approach. Zoo Biol 8(2):179–191. https://doi.org/10.1002/zoo.1430080209

Escuin A Unpublished video data. Published on July 23, 2014. Research on personality in white-naped mangabey (*Cercocebus atys lunulatus*) by Anna Escuin, Barbara Sansone e Iván García Nisa. June–July 2014 - Barcelona Zoo (Spain). https://www.youtube.com/watch?v=6YWQAeC6-SQ

Evans TA, Beran MJ (2007) Delay of gratification and delay maintenance by rhesus macaques (*Macaca Mulatta*). J Gen Psychol 134(2):199–216. https://doi.org/10.3200/GENP.134.2.199-216

Evans TA, Beran MJ, Chan B et al (2008) An efficient computerized testing method for the capuchin monkey (*Cebus apella*): adaptation of the LRC-CTS to a socially housed nonhuman primate species. Behav Res Methods 40(2):590–596. https://doi.org/10.3758/BRM.40.2.590

Fan PF, Ai HS, Fei HL, Zhang D, Yuan SD (2013) Seasonal variation of diet and time budget of Eastern hoolock gibbons (*Hoolock leuconedys*) living in a northern montane forest. Primates 54(2):137–146. https://doi.org/10.1007/s10329-012-0336-0

Fashing PJ (2011) The African colobines: their behavior, ecology, and conservation. In: Campbell CJ, Fuentes A, MacKinnon KC, Bearder SK, Stumpf SM (eds) Primates in perspective. Oxford University Press, Oxford, pp 203–229

Fernandez-Duque E (2011) Aotinae: social monogamy in the only nocturnal anthropoid. In: Campbell CJ, Fuentes A, MacKinnon KC, Bearder SK, Stumpf SM (eds) Primates in perspective. Oxford University Press, Oxford, pp 140–155

Fernandez-Duque E, Di Fiore A, Huck M (2012) The behavior, ecology, and social evolution of New World monkeys. In: Call J, Kappeler PM, Palombit RA et al (eds) Primate societies. University of Chicago Press, Chicago, pp 43–55

Fisher-Phelps ML, Mendoza SP, Serna S, Griffin LL, Schaefer TJ, Jarcho MR et al (2015) Laboratory simulations of mate-guarding as a component of the pair-bond in male titi monkeys, *Callicebus cupreus*. Am J Primatol 78(5):573–582. https://doi.org/10.1002/ajp.22483

Fontaine B, Moisson PY, Wickings EJ (1995) Observations of spontaneous tool making and tool use in a captive group of western lowland gorillas (*Gorilla gorilla gorilla*). Folia Primatol 65(4):219–223. https://doi.org/10.1159/000156892

Fox EA (2002) Female tactics to reduce sexual harassment in the Sumatran orangutan (*Pongo pygmaeus abelii*). Behav Ecol Sociobiol 52(2):93–101. https://doi.org/10.1007/s00265-002-0495-x

Fox EA, Bin'Muhammad I (2002) New tool use by wild Sumatran orangutans (*Pongo pygmaeus abelii*). Am J Phys Anthropol 119(2):186–188. https://doi.org/10.1002/ajpa.10105

Fragaszy D, Visalberghi E (1989) Social influences on the acquisition and use of tools in tufted capuchin monkeys (*Cebus apella*). J Comp Psychol 103(2):159–170. https://doi.org/10.1037/0735-7036.103.2.159

Galdikas BMF (1989) Orangutan tool use. Science 243(4888):152. https://doi.org/10.1126/science.2911726

Gallup GG (1970) Chimpanzees: self-recognition. Science 167(3914):86–87. https://doi.org/10.1126/SCIENCE.167.3914.86

Garber PA (1988) Diet, foraging patterns, and resource defense in a mixed species troop of *Saguinus mystax* and *Saguinus fuscicollis* in Amazonian Peru. Behaviour 105(1–2):18–34. https://doi.org/10.1163/156853988X00421

Garber PA (1997) One for all and breeding for one: cooperation and competition as a tamarin reproductive strategy. Evol Anthropol Issues News Rev 5(6):187–199. https://doi.org/10.1002/(SICI)1520-6505(1997)5:6<187::AID-EVAN1>3.0.CO;2-A

Gibson KR (1990) Tool use, imitation, and deception in a captive cebus monkey. In: Parker ST, Gibson KR (eds) 'Language' and intelligence in monkeys and Apes: comparative developmental perspectives. Cambridge University Press, New York, pp 205–218

Gittins SP, Raemaekers JJ (1980) Siamang, lar, and agile gibbons. In: Chivers DJ (ed) Malayan forest primates: ten years' study in tropical rain forest. Plenum Press, New York, pp 63–105

Goodall J (1986) The chimpanzees of Gombe: patterns of behavior. Harvard University Press, Cambridge

Gottlieb DH, Capitanio JP, McCowan BJ (2013) Risk factors for stereotypic behavior and self-biting in rhesus macaques (*Macaca mulatta*): animal's history, current environment, and personality. Am J Primatol 75(10):995–1008. https://doi.org/10.1002/ajp.22161

Gould E, Bres M (1986) Regurgitation and reingestion in captive gorillas: description and intervention. Zoo Biol 5(3):241–250. https://doi.org/10.1002/zoo.1430050302

Gould L, Sauther M, Cameron A (2011) Lemuriformes. In: Campbell CJ, Fuentes A, MacKinnon KC, Bearder SK, Stumpf SM (eds) Primates in perspective. Oxford University Press, Oxford, pp 55–79

Hall KRL (1962) Behaviour of monkeys towards mirror-images. Nature 196:1258–1261. https://doi.org/10.1038/1961258a0

Hambali K, Ismail A, Md-Zain BM (2012) Daily activity budget of long-tailed macaques (*Macaca fascicularis*) in Kuala Selangor Nature Park. Int J Basic Appl Sci 12(4):47–52

Hamilton WJ, Tilson RL (1985) Fishing baboons at desert waterholes. Am J Primatol 8(3):255–257. https://doi.org/10.1002/ajp.1350080308

Hanya G (2004) Seasonal variations in the activity budget of Japanese macaques in the coniferous forest of Yakushima: effects of food and temperature. Am J Primatol 63(3):165–177. https://doi.org/10.1002/ajp.20049

Harcourt AH, Stewart KJ (2007) Gorilla society: conflict, compromise, and cooperation between the sexes. University of Chicago Press, Chicago

Harding RS (1975) Meat eating and hunting in baboons. In: Tuttle RH (ed) Socioecology and psychology of primates. De Gruyter Mouton, The Hague, pp 245–257

Harlow HF, Harlow MK (1962) Social deprivation in monkeys. Sci Am 207:136–146. https://doi.org/10.1038/scientificamerican1162-136

Harris HG, Edwards AJ (2004) Mirrors as environmental enrichment for African green monkeys. Am J Primatol 64(4):459–467. https://doi.org/10.1002/ajp.20092

Hauser M, Pearson H, Seelig D (2002) Ontogeny of tool use in cottontop tamarins, *Saguinus oedipus*: innate recognition of functionally relevant features. Anim Behav 64(2):299–311. https://doi.org/10.1006/anbe.2002.3068

Heath S, Shimoji M, Tumanguil J, Crockett C (1992) Peanut puzzle solvers quickly demonstrate aptitude. Lab Prim Newsl 31:12–13

Hedge Z (2005) Furniture usage and activity budgets of captive black and white ruffed lemurs (*Varecia variegata variegata*) and ring-tailed lemurs (*Lemur catta*) at Bramble Park Zoo, Watertown, South Dakota. J Undergraduate Res 3:69–80

Hill A, Collier-Baker E, Suddendorf T (2011) Inferential reasoning by exclusion in great apes, lesser apes, and spider monkeys. J Comp Psychol 125(1):91–103. https://doi.org/10.1037/a0020867

Hopkins WD, Latzman RD (2017) Future research with captive chimpanzees in the United States: integrating scientific programs with behavioral management. In: Schapiro SJ (ed) Handbook of primate behavioral management. CRC Press, Boca Raton, FL, pp 141–155

Hopper LM, Whiten A (2012) The evolutionary and comparative psychology of social learning and culture. In: Vonk J, Shackelford TK (eds) The Oxford handbook of comparative evolutionary psychology. Oxford University Press, New York, pp 451–473

Hopper LM, Price SA, Freeman HD, Lambeth SP, Schapiro SJ, Kendal RL (2014) Influence of personality, age, sex, and estrous state on chimpanzee problem-solving success. Anim Cogn 17 (4):835–847. https://doi.org/10.1007/s10071-013-0715-y

Hopper LM, Shender MA, Ross SR (2016) Behavioral research as physical enrichment for captive chimpanzees. Zoo Biol 35(4):293–297. https://doi.org/10.1002/zoo.21297

Hoshino J, Mori A, Kudo H, Kawai M (1984) Preliminary report on the grouping of mandrills (*Mandrillus sphinx*) in Cameroon. Primates 25(3):295–307

Howell S, Schwandt M, Fritz J et al (2003) A stereo music system as environmental enrichment for captive chimpanzees. Lab Anim (NY) 32(10):31–36. https://doi.org/10.1038/laban1103-31

Huck M, Van Lunenburg M, Dávalos V, Rotundo M, Di Fiore A, Fernandez-Duque E (2014) Double effort: parental behavior of wild Azara's owl monkeys in the face of twins. Am J Primatol 76(7):629–639. https://doi.org/10.1002/ajp.22256

Huffman MA, Quiatt D (1986) Stone handling by Japanese macaques (*Macaca fuscata*): implications for tool use of stone. Primates 27(4):413–423. https://doi.org/10.1007/BF02381887

Inoue S, Matsuzawa T (2007) Working memory of numerals in chimpanzees. Curr Biol 17(23): R1004–R1005. https://doi.org/10.1016/J.CUB.2007.10.027

Inoue-Nakamura N (1997) Mirror self-recognition in nonhuman primates: a phylogenetic approach. Jpn Psychol Res 39(3):266–275. https://doi.org/10.1111/1468-5884.00059

Iriki A, Tanaka M, Iwamura Y (1996) Coding of modified body schema during tool use by macaque postcentral neurons. Neuroreport 7(14):2325–2330. https://doi.org/10.1097/00001756-199610020-00010

Itakura S (1987) Mirror guided behavior in Japanese monkeys (*Macaca fuscata fuscata*). Primates 28(2):149–161. https://doi.org/10.1007/BF02382568

Itani J, Nishimura A (1973) The study of infra-human culture in Japan. In: Menzel EWJ (ed) Precultural primate behaviour. Karger, Basel, Switzerland, pp 26–50

Jack KM (2011) The Cebines: toward an explanation of variable social structure. In: Campbell CJ, Fuentes A, MacKinnon KC, Bearder SK, Stumpf SM (eds) Primates in perspective. Oxford University Press, Oxford, pp 108–122

Jaeggi AV, Dunkel LP, Van Noordwijk MA, Wich SA, Sura AA, Van Schaik CP (2010) Social learning of diet and foraging skills by wild immature Bornean orangutans: implications for culture. Am J Primatol 72:62–71

Jaeggi A, De Groot E, Stevens J, Van Schaik C (2013) Mechanisms of reciprocity in primates: testing for short-term contingency of grooming and food sharing in bonobos and chimpanzees. Evol Hum Behav 34:69–77

Jaffe KE, Isbell LA (2011) The Guenons: polyspecific associations in socioecological perspective. In: Campbell CJ, Fuentes A, MacKinnon KC, Bearder SK, Stumpf SM (eds) Primates in perspective. Oxford University Press, Oxford, pp 277–300

Janmaat KRL, Ban SD, Boesch C (2013) Chimpanzees use long-term spatial memory to monitor large fruit trees and remember feeding experiences across seasons. Anim Behav 86(6):1183–1205. https://doi.org/10.1016/J.ANBEHAV.2013.09.021

Janson CH (1998) Experimental evidence for spatial memory in foraging wild capuchin monkeys, *Cebus apella*. Anim Behav 55(5):1229–1243. https://doi.org/10.1006/ANBE.1997.0688

Jensen GD (1963) Preference for bar pressing over "freeloading" as a function of number of rewarded presses. J Exp Psychol 65(5):451–454. https://doi.org/10.1037/h0049174

Johnson ET, Lynch PA, Schapiro SJ (2013) Hunting, cooperation and sharing in captive chimpanzees. In: Proceedings of the 59th annual convention of the Southwestern Psychological Association, Ft Worth, TX

Judge PG, Essler JL (2013) Capuchin monkeys exercise self-control by choosing token exchange over an immediate reward. Int J Comp Psychol 26:256–266

Kim C-Y, Han JS, Suzuki T, Han S-S (2005) Indirect indicator of transport stress in hematological values in newly acquired cynomolgus monkeys. J Med Primatol 34(4):188–192. https://doi.org/10.1111/j.1600-0684.2005.00116.x

Kinzey WG (1981) The titi monkeys, genus *Callicebus*: I. Description of the species. In: Coimbra-Filho AF, Mittermeier RA (eds) Ecology and behavior of neotropical primates, vol 1. Academia Brasileira de Ciências, Rio de Janeiro, pp 241–276

Knott CD, Kahlenberg SM (2011) Orangutans: understanding forced copulations. In: Campbell CJ, Fuentes A, MacKinnon KC, Bearder SK, Stumpf SM (eds) Primates in perspective. Oxford University Press, Oxford, pp 313–338

Koban TL, Schapiro SJ, Kusznir T et al (2010) Effects of international transit and relocation on cortisol values in cynomolgus macaques (*Macaca fascicularis*). Am J Primatol 72(S1):51

Kuhlmeier VA, Boysen ST, Mukobi KL (1999) Scale-model comprehension by chimpanzees (*Pan troglodytes*). J Comp Psychol 113(4):396–402. https://doi.org/10.1037/0735-7036.113.4.396

Kulp J, Heymann EW (2015) Ranging, activity budget, and diet composition of red titi monkeys (*Callicebus cupreus*). Primates 56(3):273–278. https://doi.org/10.1007/s10329-015-0471-5

Kurup GU, Kumar A (1993) Time budget and activity patterns of the lion-tailed macaque (*Macaca silenus*). Int J Primatol 14(1):27–39. https://doi.org/10.1007/BF02196501

Kyes RC (1988) Grooming with a stone in sooty mangabeys (*Cercocebus atys*). Am J Primatol 16(2):171–175. https://doi.org/10.1002/ajp.1350160208

Lambeth SP, Bloomsmith MA (1992) Mirrors as enrichment for captive chimpanzees (*Pan troglodytes*). Lab Anim Sci 42(3):261–266

Larson TM (2011) A cross-species study of cognitive ability within the order primates. Western Illinois University, Macomb, IL

Leca JB, Gunst N, Huffman MA (2007) Japanese macaque cultures: inter-and intra-troop behavioural variability of stone handling patterns across 10 troops. Behaviour 144(3):251–281. https://doi.org/10.1163/156853907780425712

Leca JB, Gunst N, Huffman MA (2008) Of stones and monkeys: testing ecological constraints on stone handling, a behavioral tradition in Japanese macaques. Am J Phys Anthropol 135(2):233–244. https://doi.org/10.1002/ajpa.20726

Less EH, Bergl R, Ball R et al (2014) Implementing a low-starch biscuit-free diet in zoo gorillas: the impact on behavior. Zoo Biol 33(1):63–73. https://doi.org/10.1002/zoo.21116

Lethmate J, Ducker G (1973) Experiments on self-recognition in a mirror in orangutans, chimpanzees, gibbons and several monkey species. Z Tierpsychol 33(3–4):248–269. https://doi.org/10.1111/j.1439-0310.1973.tb02094.x

Lindshield SM, Rodrigues MA (2009) Tool use in wild spider monkeys (*Ateles geoffroyi*). Primates 50(3):269–272. https://doi.org/10.1007/s10329-009-0144-3

Lonsdorf EV, Ross SR, Linick SA, Milstein MS, Melber TN (2009) An experimental, comparative investigation of tool use in chimpanzees and gorillas. Anim Behav 77(5):1119–1126. https://doi.org/10.1016/j.anbehav.2009.01.020

Lukas KE (1999) A review of nutritional and motivational factors contributing to the performance of regurgitation and reingestion in captive lowland gorillas (*Gorilla gorilla gorilla*). Appl Anim Behav Sci 63(3):237–249. https://doi.org/10.1016/S0168-1591(98)00239-1

Lutz CK, Nevil CH (2017) Behavioral management of *Papio* spp. In: Schapiro SJ (ed) Handbook of primate behavioral management. Taylor & Francis, Boca Raton, FL

Lutz CK, Novak MA (2005) Environmental enrichment for nonhuman primates: theory and application. ILAR J 46(2):178–191. https://doi.org/10.1093/ilar.46.2.178

Lutz CK, Davis EB, Ruggiero AM, Suomi SJ (2007) Early predictors of self-biting in socially housed rhesus macaques (*Macaca mulatta*). Am J Primatol 69(5):584–590. https://doi.org/10.1002/ajp.20370

Machairas I, Camperio Ciani A, Sgardelis S (2003) Interpopulation differences in activity patterns of *Macaca sylvanus* in the Moroccan Middle Atlas. Hum Evol 18(3–4):185–202

Maki S, Alford P, Bloomsmith MA, Franklin J (1989) Food puzzle device simulating termite fishing for captive chimpanzees (*Pan troglodytes*). Am J Primatol 19(S1):71–78

Marchal P, Anderson JR (1993) Mirror-image responses in capuchin monkeys (*Cebus capucinus*): social responses and use of reflected environmental information. Folia Primatol 61(3):165–173. https://doi.org/10.1159/000156745

Markowitz H (1982) Behavioral enrichment in the zoo. Van Nostrand Reinhold, New York

Maruhashi T (1981) Activity patterns of a troop of Japanese monkeys (*Macaca fuscata yakui*) on Yakushima Island, Japan. Primates 22(1):1–4. https://doi.org/10.1007/BF02382552

Masi S, Cipolletta C, Robbins MM (2009) Western lowland gorillas (*Gorilla gorilla gorilla*) change their activity patterns in response to frugivory. Am J Primatol 71(2):91–100. https://doi.org/10.1002/ajp.20629

Matthews LJ (2009) Activity patterns, home range size, and intergroup encounters in *Cebus albifrons* support existing models of capuchin socioecology. Int J Primatol 30(5):709–728. https://doi.org/10.1007/s10764-009-9370-1

McKinney T (2011) The effects of provisioning and crop-raiding on the diet and foraging activities of human-commensal white-faced capuchins (*Cebus capucinus*). Am J Primatol 73(5):439–448. https://doi.org/10.1002/ajp.20919

Meehan CL, Mench JA (2007) The challenge of challenge: can problem solving opportunities enhance animal welfare? Appl Anim Behav Sci 102(3–4):246–261. https://doi.org/10.1016/J.APPLANIM.2006.05.031

Mehlman PT, Doran DM (2002) Influencing western gorilla nest construction at Mondika Research Center. Int J Primatol 23(6):1257–1285. https://doi.org/10.1023/A:1021126920753

Ménard N, Vallet D (1997) Behavioral responses of Barbary macaques (*Macaca sylvanus*) to variations in environmental conditions in Algeria. Am J Primatol 43(4):285–304. https://doi.org/10.1002/(SICI)1098-2345(1997)43:4<285::AID-AJP1>3.0.CO;2-T

Mendres KA, de Waal FBM (2000) Capuchins do cooperate: the advantage of an intuitive task. Anim Behav 60(4):523–529. https://doi.org/10.1006/anbe.2000.1512

Menzel CR (1991) Cognitive aspects of foraging in Japanese monkeys. Anim Behav 41(3):397–402. https://doi.org/10.1016/S0003-3472(05)80840-1

Menzel EW, Savage-Rumbaugh ES, Lawson J (1985) Chimpanzee (*Pan troglodytes*) spatial problem solving with the use of mirrors and televised equivalents of mirrors. J Comp Psychol 99(2):211–217. https://doi.org/10.1037/0735-7036.99.2.211

Millette JB, Sauther ML, Cuozzo FP (2009) Behavioral responses to tooth loss in wild ring-tailed lemurs (*Lemur catta*) at the Beza Mahafaly Special Reserve, Madagascar. Am J Phys Anthropol 140(1):120–134. https://doi.org/10.1002/ajpa.21045

Milozzi C, Steinberg ER, Mudry MD (2012) Use of genetic and behavioral analysis in the captive management of night monkeys *Aotus azarae* (*Platyrrhini*: *Cebidae*). InVet 14(1):9–18

Milton K (1980) The foraging strategy of howler monkeys: a study in primate economics. Columbia University Press, New York

Mischel W (1974) Processes in delay of gratification. Adv Exp Soc Psychol 7:249–292. https://doi.org/10.1016/S0065-2601(08)60039-8

Mitani JC (1985) Gibbon song duets and intergroup spacing. Behaviour 92(1–2):59–96. https://doi.org/10.1163/156853985X00389

Mitchell CL (1994) Migration alliances and coalitions among adult male South American squirrel monkeys (*Saimiri sciureus*). Behaviour 130(3-4):169–190. https://doi.org/10.1163/156853994X00514

Morland HS (1993) Seasonal behavioral variation and its relationship to thermoregulation in ruffed lemurs (*Varecia variegata variegata*). In: Kappeler PM, Ganzhorn JU (eds) Lemur social systems and their ecological basis. Plenum Press, New York, pp 193–203

Morrogh-Bernard HC, Husson SJ, Knott CD, Wich SA, van Schaik CP, van Noordwijk MA, Lackman-Ancrenaz I, Marshall AJ, Kanamori T, Kuze N, bin Sakong R (2009) Orangutan activity budgets and diet. In: Wich SA (ed) Orangutans: geographic variation in behavioral ecology and conservation. Oxford University Press, Oxford, pp 119–133

Moynihan M (1964) Some behavior patterns of platyrrhine monkeys I. The night monkeys (*Aotus trivirgatus*). Smithson Misc Collect 146(5):1–84

Mulcahy NJ, Call J (2006) Apes save tools for future use. Science 312(5776):1038–1040. https://doi.org/10.1126/science.1125456

Mulcahy NJ, Call J, Dunbar RIM (2005) Gorillas (*Gorilla gorilla*) and orangutans (*Pongo pygmaeus*) encode relevant problem features in a tool-using task. J Comp Psychol 119(1):23–32. https://doi.org/10.1037/0735-7036.119.1.23

Nakamichi M, Kato E, Kojima Y, Itoigawa N (1998) Carrying and washing of grass roots by free-ranging Japanese macaques at Katsuyama. Folia Primatol 69(1):35–40. https://doi.org/10.1159/000021561

Nehete PN, Shelton KA, Nehete BP et al (2017) Effects of transportation, relocation, and acclimation on phenotypes and functional characteristics of peripheral blood lymphocytes in rhesus monkeys (*Macaca mulatta*). PLoS One 12(12):e0188694. https://doi.org/10.1371/journal.pone.0188694

Nishida T, Wallauer W (2003) Leaf-pile pulling: an unusual play pattern in wild chimpanzees. Am J Primatol 60(4):167–173. https://doi.org/10.1002/ajp.10099

Norconk M (2011) Sakis, Uakaris, and Titi Monkeys: behavioral diversity in a radiation of primate seed predators. In: Campbell CJ, Fuentes A, MacKinnon KC, Bearder SK, Stumpf SM (eds) *Primates in perspective*. Oxford University Press, Oxford, pp 122–140

Norris J (1988) Diet and feeding behavior of semi-free ranging mandrills in an enclosed Gabonais forest. Primates 29(4):449–463. https://doi.org/10.1007/BF02381133

Noser R, Byrne RW (2007) Mental maps in chacma baboons (*Papio ursinus*): using inter-group encounters as a natural experiment. Anim Cogn 10:(3)331–340. https://doi.org/10.1007/s10071-006-0068-x

Novak MA, Drewsen KH (1989) Enriching the lives of captive primates: issues and problems. In: Segal EF (ed) Housing, care and psychological wellbeing of captive and laboratory primates. Noyes Publications, Park Ridge, NJ, pp 161–182

Novak MA, Kinsey JH, Jorgensen MJ, Hazen TJ (1998) Effects of puzzle feeders on pathological behavior in individually housed rhesus monkeys. Am J Primatol 46(3):213–227. https://doi.org/10.1002/(SICI)1098-2345(1998)46:3<213::AID-AJP3>3.0.CO;2-L

O'Brien TG, Kinnaird MF (1997) Behavior, diet, and movements of the Sulawesi crested black macaque (*Macaca nigra*). Int J Primatol 18(3):321–351. https://doi.org/10.1023/A:1026330332061

Obernier JA, Baldwin RL (2006) Establishing an appropriate period of acclimatization following transportation of laboratory animals. ILAR J 47(4):364–369. https://doi.org/10.1093/ilar.47.4.364

Onderdonk DA, Chapman CA (2000) Coping with forest fragmentation: the primates of Kibale National Park, Uganda. Int J Primatol 21(4):587–611. https://doi.org/10.1023/A:1005509119693

Osvath M (2009) Spontaneous planning for future stone throwing by a male chimpanzee. Curr Biol 19(5):R190–R191. https://doi.org/10.1016/j.cub.2009.01.010

Passamani M (1998) Activity budget of Geoffroy's marmoset (*Callithrix geoffroyi*) in an Atlantic forest in southeastern Brazil. Am J Primatol 46(4):333–340. https://doi.org/10.1002/(SICI)1098-2345(1998)46:4<333::AID-AJP5>3.0.CO;2-7

Patterson F (1984) Self-recognition by *Gorilla gorilla gorilla*. Gorilla 7:2–3

Pinheiro T, Ferrari SF, Lopes MA (2013) Activity budget, diet, and use of space by two groups of squirrel monkeys (*Saimiri sciureus*) in eastern Amazonia. Primates 54(3):301–308. https://doi.org/10.1007/s10329-013-0351-9

Platt DM, Novak MA (1997) Video stimulation as enrichment for captive rhesus monkeys (*Macaca mulatta*). Appl Anim Behav Sci 52(1–2):139–155. https://doi.org/10.1016/S0168-1591(96)01093-3

Pollack D (1998) Spontaneous tool use in a vervet monkey (*Cercopithecus aethiops sabaeus*). Am J Primatol 45(2):201

Pope TR (2000) Reproductive success increases with degree of kinship in cooperative coalitions of female red howler monkeys (*Alouatta seniculus*). Behav Ecol Sociobiol 48(4):253–267. https://doi.org/10.1007/s002650000236

Pouydebat E, Gorce P, Bels V, Coppens Y (2006) Substrate optimization in nut cracking by capuchin monkeys (*Cebus apella*). Am J Primatol 68(10):1017–1024. https://doi.org/10.1002/ajp.20291

Premack D, Woodruff G (1978) Does the chimpanzee have a theory of mind? Behav Brain Sci 1(4): 515–526. https://doi.org/10.1017/S0140525X00076512

Price EE, Stoinski TS (2007) Group size: determinants in the wild and implications for the captive housing of wild mammals in zoos. Appl Anim Behav Sci 103(3–4):255–264. https://doi.org/10.1016/J.APPLANIM.2006.05.021

Proctor D, Williamson R, de Waal F, Brosnan S (2013) Chimpanzees play the ultimatum game. Proc Natl Acad Sci U S A 110:2070–2075. https://doi.org/10.1073/pnas.1220806110

Pruetz JD, Bertolani P (2007) Savanna chimpanzees, *Pan troglodytes verus*, hunt with tools. Curr Biol 17(5):412–417. https://doi.org/10.1016/j.cub.2006.12.042

Pruetz J, McGrew W (2001) What does a chimpanzee need? Using natural behavior to guide the care and management of captive populations. In: Brent L (ed) The care and management of captive chimpanzees. The American Society of Primatologists, San Antonio, pp 17–37

Raboy BE, Dietz JM (2004) Diet, foraging, and use of space in wild golden-headed lion tamarins. Am J Primatol 63(1):1–5. https://doi.org/10.1002/ajp.20032

Rachlin H, Green L (1972) Commitment, choice and self-control. J Exp Anal Behav 17(1):15–22. https://doi.org/10.1901/jeab.1972.17-15

Reamer LA, Talbot CF, Hopper LM, Mareno MC, Hall K, Brosnan SF, Lambeth SP, Schapiro SJ (2015) Assessing quantity of space for captive chimpanzee welfare. Am J Primatol 77(S1):84–85

Reamer LA, Haller RL, Lambeth SP, Schapiro SJ (2017) Behavioral management of Pan spp. In: Schapiro SJ (ed) Handbook of primate behavioral management. CRC Press, Boca Raton, FL, pp 385–408

Reinhardt V (1994) Caged rhesus macaques voluntarily work for ordinary food. Primates 35(1):95–98. https://doi.org/10.1007/BF02381490

Richard-Hansen C, Bello N, Vié JC (1998) Tool use by a red howler monkey (*Alouatta seniculus*) towards a two-toed sloth (*Choloepus didactylus*). Primates 39(4):545–548. https://doi.org/10.1007/BF02557575

Robbins MM (2011) Gorillas: diversity in ecology and behavior. In: Campbell CJ, Fuentes A, MacKinnon KC, Bearder SK, Stumpf SM (eds) Primates in perspective. Oxford University Press, Oxford, pp 326–339

Roberts RL, Roytburd LA, Newman JD (1999) Puzzle feeders and gum feeders as environmental enrichment for common marmosets. Contemp Top Lab Anim Sci 38(5):27–31

Rodrigues MR, Lindshield SL (2007) Scratching the surface: observations of tool use in wild spider monkeys. Am J Phys Anthropol S4:201–202

Rose LM (1997) Vertebrate predation and food-sharing in *Cebus* and *Pan*. Int J Primatol 18(5): 727–765. https://doi.org/10.1023/A:1026343812980

Ross SR (2010) How cognitive studies help shape our obligation for ethical care of chimpanzees. In: Lonsdorf EV, Ross SR, Matsuzawa T (eds) The mind of the chimpanzees: ecological and experimental perspectives. The University of Chicago Press, Chicago, pp 309–319

Ross SR, Lukas KE (2006) Use of space in a non-naturalistic environment by chimpanzees (*Pan troglodytes*) and lowland gorillas (*Gorilla gorilla gorilla*). Appl Anim Behav Sci 96(1–2):143–152. https://doi.org/10.1016/J.APPLANIM.2005.06.005

Ross SR, Calcutt S, Schapiro SJ, Hau J (2011) Space use selectivity by chimpanzees and gorillas in an indoor-outdoor enclosure. Am J Primatol 73(2):197–208. https://doi.org/10.1002/ajp.20891

Rowe N (1996) Pictorial guide to the living primates. Pogonias Press, New York, pp 93–110

Russon AE, van Schaik CP, Kuncoro P, Ferisa A, Handayani DP, Van Noordwijk MA (2009) Innovation and intelligence in orangutans. In: Wich SA (ed) Orangutans: geographic variation in behavioral ecology and conservation. Oxford University Press, Oxford, pp 279–298

Sabbatini G, Stammati M, Tavares MC, Visalberghi E (2008) Behavioral flexibility of a group of bearded capuchin monkeys (*Cebus libidinosus*) in the National Park of Brasília (Brazil): consequences of cohabitation with visitors. Braz J Biol 68(4):685–693. https://doi.org/10.1590/s1519-69842008000400002

Santos LR, Mahajan N, Barnes JL (2005) How prosimian primates represent tools: experiments with two lemur species (*Eulemur fulvus* and *Lemur catta*). J Comp Psychol 119(4):394–403. https://doi.org/10.1037/0735-7036.119.4.394

Santos LR, Pearson HM, Spaepen GM, Tsao F, Hauser MD (2006) Probing the limits of tool competence: experiments with two non-tool-using species (*Cercopithecus aethiops* and *Saguinus oedipus*). Anim Cogn 9(2):94–109. https://doi.org/10.1007/s10071-005-0001-8

Sanz C, Blicher A, Dalke K et al (1999) Use of temporary and semipermanent enrichment objects by five chimpanzees. J Appl Anim Welf Sci 2(1):1–11. https://doi.org/10.1207/s15327604jaws0201_1

Schapiro SJ, Bloomsmith MA (1995) Behavioral effects of enrichment on singly-housed, yearling rhesus monkeys: an analysis including three enrichment conditions and a control group. Am J Primatol 35(2):89–101. https://doi.org/10.1002/ajp.1350350202

Schapiro SJ, Lambeth SP (2010) Chimpanzees. In: Hubrecht R, Kirkwood J (eds) The UFAW handbook on the care and management of laboratory animals, 8th edn. Oxford, Wiley-Blackwell, pp 618–634

Schapiro SJ, Bloomsmith MA, Kessel AL, Shively CA (1993) Effects of enrichment and housing on cortisol response in juvenile rhesus monkeys. Appl Anim Behav Sci 37(3):251–263. https://doi.org/10.1016/0168-1591(93)90115-6

Schapiro SJ, Nehete PN, Perlman JE, Sastry KJ (2000) A comparison of cell-mediated immune responses in rhesus macaques housed singly, in pairs, or in groups. Appl Anim Behav Sci 68(1):67–84. https://doi.org/10.1016/s0168-1591(00)00090-3

Schapiro SJ, Lambeth SP, Jacobsen KR et al (2012) Physiological and welfare consequences of transport, relocation, and acclimatization of chimpanzees (*Pan troglodytes*). Appl Anim Behav Sci 137(3–4):183–193. https://doi.org/10.1016/J.APPLANIM.2011.11.004

Schapiro SJ, Coleman K, Akinyi M et al (2014) Nonhuman primate welfare in the research environment. In: Bayne K, Turner PV (eds) Laboratory animal welfare. Academic Press, San Diego, pp 197–212

Schapiro SJ, Brosnan SF, Hopkins WD et al (2017) Collaborative research and behavioral management. In: Schapiro SJ (ed) Handbook of primate behavioral management. CRC Press, Boca Raton, FL, pp 243–254

Schel AM, Rawlings B, Claidière N et al (2013) Network analysis of social changes in a captive chimpanzee community following the successful integration of two adult groups. Am J Primatol 75(3):254–266. https://doi.org/10.1002/ajp.22101

Shaffer VA, Renner MJ (2000) Black-and-white colobus monkeys (*Colobus guereza*) do not show mirror self-recognition. Int J Comp Psychol 13(3):154–160

Shepherdson D, Bemment N, Carman M, Reynolds S (1989) Auditory enrichment for Lar gibbons *Hylobates lar* at London Zoo. Int Zoo Yearb 28(1):256–260. https://doi.org/10.1111/j.1748-1090.1989.tb03294.x

Shettleworth SJ (2010) Cognition, evolution, and behavior. Oxford University Press, Oxford

Smith A, Lindburg DG, Vehrencamp S (1989) Effect of food preparation on feeding behavior of lion-tailed macaques. Zoo Biol 8(1):57–65. https://doi.org/10.1002/zoo.1430080108

Smuts BB, Cheney D, Seyfarth R et al (1987) Primate societies. University of Chicago Press, Chicago

Sousa C, Biro D, Matsuzawa T (2009) Leaf-tool use for drinking water by wild chimpanzees (*Pan troglodytes*): acquisition patterns and handedness. Anim Cogn 12(Suppl 1):115–125. https://doi.org/10.1007/s10071-009-0278-0

Starin ED (1990) Object manipulation by wild red colobus monkeys living in the Abuko Nature Reserve, The Gambia. Primates 31(3):385–391. https://doi.org/10.1007/BF02381109

Stevens JR (2014) Evolutionary pressures on primate intertemporal choice. Proc R Soc B Biol Sci 281(1786):20140499. https://doi.org/10.1098/rspb.2014.0499

Stevens JR, Hallinan EV, Hauser MD (2005) The ecology and evolution of patience in two New World monkeys. Biol Lett 1(2):223–226

Stoinski TS, Beck BB (2001) Spontaneous tool use in captive, free-ranging golden lion tamarins (*Leontopithecus rosalia rosalia*). Primates 42(4):319–326. https://doi.org/10.1007/BF02629623

Stoinski TS, Whiten A (2003) Social learning by orangutans (*Pongo abelii* and *Pongo pygmaeus*) in a simulated food-processing task. J Comp Psychol 117(3):272–282. https://doi.org/10.1037/0735-7036.117.3.272

Stoinski TS, Wrate JL, Ure N, Whiten A (2001) Imitative learning by captive western lowland gorillas (*Gorilla gorilla gorilla*) in a simulated food-processing task. J Comp Psychol 115(3):272–281. https://doi.org/10.1037/0735-7036.115.3.272

Stone AI (2007) Responses of squirrel monkeys to seasonal changes in food availability in an eastern Amazonian forest. Am J Primatol 69(2):142–157. https://doi.org/10.1002/ajp.20335

Straumann C, Anderson JR (1991) Mirror-induced social facilitation in stumptailed macaques (*Macaca arctoides*). Am J Primatol 25(2):125–132. https://doi.org/10.1002/ajp.1350250206

Struhsaker TT (1975) The red colobus monkey. University of Chicago Press, Chicago

Struhsaker TT (2010) The red colobus monkeys: variation in demography, behavior, and ecology of endangered species. Oxford University Press, Oxford

Stumpf RM (2011) Chimpanzees and Bonobos: inter- and intraspecies diversity. In: Campbell CJ, Fuentes A, MacKinnon KC, Bearder SK, Stumpf SM (eds) Primates in perspective. Oxford University Press, Oxford, pp 340–356

Suárez SD, Gallup GG (1981) Self-recognition in chimpanzees and orangutans, but not gorillas. J Hum Evol 10(2):175–188. https://doi.org/10.1016/S0047-2484(81)80016-4

Suchak M, Eppley TM, Campbell MW, de Waal FB (2014) Ape duos and trios: spontaneous cooperation with free partner choice in chimpanzees. PeerJ 2:e417

Sushma HS, Singh M (2008) Hunting of Indian giant squirrel (*Ratufa indica*) by the lion-tailed macaque (*Macaca silenus*) in the Western Ghats, India. Curr Sci 95(11):1535–1536

Swedell L (2011) African papionins: diversity of social organization and ecological flexibility. In: Campbell CJ, Fuentes A, MacKinnon KC, Bearder SK, Stumpf SM (eds) Primates in perspective. Oxford University Press, Oxford, pp 241–277

Takimoto A, Fujita K (2011) I acknowledge your help: Capuchin monkeys' sensitivity to others' labor. Anim Cogn 14(5):715–725. https://doi.org/10.1007/s10071-011-0406-5

Talbot CF, Hall K, Williams LE, Brosnan SF (2014) Squirrel monkeys coordinate on a cooperative bar pull task. Am J Primatol 76(S1):82

Tardif S, Bales K, Williams L et al (2006) Preparing New World monkeys for laboratory research. ILAR J 47(4):307–315. https://doi.org/10.1093/ilar.47.4.307

Tashiro Y (2006) Frequent insectivory by two guenons (*Cercopithecus lhoesti* and *Cercopithecus mitis*) in the Kalinzu Forest, Uganda. Primates 47(2):170–173. https://doi.org/10.1007/s10329-005-0160-x

Terborgh J (2014) Five New World primates: a study in comparative ecology. Princeton University Press, Princeton, NJ

Thierry BT (2011) The macaques: a double-layered social organization. In: Campbell CJ, Fuentes A, MacKinnon KC, Bearder SK, Stumpf SM (eds) Primates in perspective. Oxford University Press, Oxford, pp 229–241

Thorén S, Quietzsch F, Schwochow D, Sehen L, Meusel C, Meares K, Radespiel U (2011) Seasonal changes in feeding ecology and activity patterns of two sympatric mouse lemur species, the gray mouse lemur (*Microcebus murinus*) and the golden-brown mouse lemur (*M. ravelobensis*), in northwestern Madagascar. Int J Primatol 32(3):566–586. https://doi.org/10.1007/s10764-010-9488-1

Thorpe WH (1963) Learning and instinct in animals. Methuen, London

Tomasello M, Call J (1997) Primate cognition. Oxford University Press, New York

Tutin CE, Fernandez M (1992) Insect-eating by sympatric Lowland gorillas (*Gorilla g. gorilla*) and chimpanzees (*Pan t. troglodytes*) in the Lopé Reserve, Gabon. Am J Primatol 28(1):29–40. https://doi.org/10.1002/ajp.1350280103

van de Waal E, Claidière N, Whiten A (2013) Social learning and spread of alternative means of opening an artificial fruit in four groups of vervet monkeys. Anim Behav 85(1):71–76. https://doi.org/10.1016/j.anbehav.2012.10.008

Van Schaik CP, Knott CD (2001) Geographic variation in tool use on Neesia fruits in orangutans. Am J Phys Anthropol 114(4):331–342. https://doi.org/10.1002/ajpa.1045

Van Schaik CP, Deaner RO, Merrill MY (1999) The conditions for tool use in primates: implications for the evolution of material culture. J Hum Evol 36(6):719–741. https://doi.org/10.1006/jhev.1999.0304

Vandeleest JJ, McCowan BJ, Capitanio JP (2011) Early rearing interacts with temperament and housing to influence the risk for motor stereotypy in rhesus monkeys (*Macaca mulatta*). Appl Anim Behav Sci 132(1–2):81–89. https://doi.org/10.1016/j.applanim.2011.02.010

Vasey N (2005) Activity budgets and activity rhythms in red ruffed lemurs (*Varecia rubra*) on the Masoala Peninsula, Madagascar: seasonality and reproductive energetics. Am J Primatol 66(1): 23–44. https://doi.org/10.1002/ajp.20126

Veasey JS, Waran NK, Young RJ (1996) On comparing the behaviour of zoo housed animals with wild conspecifics as a welfare indicator. Anim Welf 5:13–24

Visalberghi E, Fragaszy DM (1990) Food-washing behaviour in tufted capuchin monkeys, *Cebus apella*, and crabeating macaques, *Macaca fascicularis*. Anim Behav 40(5):829–836. https://doi.org/10.1016/S0003-3472(05)80983-2

Visalberghi E, Mason WA (1983) Determinants of problem-solving success in *Saimiri* and *Callicebus*. Primates 24(3):385–396. https://doi.org/10.1007/BF02381983

Visalberghi E, Fragaszy D, Ottoni E, Izar P, de Oliveira MG, Andrade FRD (2007) Characteristics of hammer stones and anvils used by wild bearded capuchin monkeys (*Cebus libidinosus*) to crack open palm nuts. Am J Phys Anthropol 132(3):426–444. https://doi.org/10.1002/ajpa.20546

Visalberghi E, Addessi E, Truppa V et al (2009) Selection of effective stone tools by wild bearded capuchin monkeys. Curr Biol 19(3):213–217. https://doi.org/10.1016/J.CUB.2008.11.064

Vonk J, Shackelford TK (2012) The Oxford handbook of comparative evolutionary psychology. Oxford University Press, New York

Wallace EK, Altschul D, Körfer K et al (2017) Is music enriching for group-housed captive chimpanzees (*Pan troglodytes*)? PLoS One 12(3):e0172672. https://doi.org/10.1371/journal.pone.0172672

Walters BS (1989) Social interactions in a mixed group of mandrills (*Mandrillus sphinx*) and mangabeys (*Cercocebus atys*) in a captive setting. Dissertation from University of Illinois

Washburn DA, Rumbaugh DM (1992) Investigations of rhesus monkey video-task performance: evidence for enrichment. Contemp Top Lab Anim Sci 31(5):6–11

Washburn DA, Rumbaugh DM, Richardson WK (1992) The Language Research Center's Computerized Test System for environmental enrichment and psychological assessment. Contemp Top Lab Anim Sci 31(6):11–15

Washburn DA, Gulledge JP, Rumbaugh DM (1997) The heuristic and motivational value of video reinforcement. Learn Motiv 28(4):510–520. https://doi.org/10.1006/LMOT.1997.0981

Watson SL, McCoy JG, Stavisky RC et al (2005) Cortisol response to relocation stress in Garnett's bushbaby (*Otolemur garnettii*). Contemp Top Lab Anim Sci 44(3):22–24

Watts DP (1988) Environmental influences on mountain gorilla time budgets. Am J Primatol 15(3): 195–211

Werdenich D, Huber L (2002) Social factors determine cooperation in marmosets. Anim Behav 64 (5):771–781. https://doi.org/10.1006/anbe.2002.9001

Westergaard GC, Fragaszy DM (1987) The manufacture and use of tools by capuchin monkeys (*Cebus apella*). J Comp Psychol 101(2):159–168. https://doi.org/10.1037/0735-7036.101.2.159

Whitehouse J, Micheletta J, Powell LE et al (2013) The impact of cognitive testing on the welfare of group housed primates. PLoS One 8(11):e78308. https://doi.org/10.1371/journal.pone.0078308

Whiten A, Byrne RW (eds) (1997) Machiavellian intelligence II: extensions and evaluations. Cambridge University Press, Cambridge

Whiten A, Ham R (1992) On the nature and evolution of imitation in the animal kingdom: reappraisal of a century of research. In: Slater P, Beer C, Milinski M, Rosenblatt J (eds) Advances in the study of behavior. Academic Press, San Diego, CA, pp 239–283

Whiten A, Goodall J, McGrew WC, Nishida T, Reynolds V, Sugiyama Y et al (1999) Cultures in chimpanzees. Nature 399:682–685. https://doi.org/10.1038/21415

Whiten A, Spiteri A, Horner V, Bonnie K, Lambeth S, Schapiro S, de Waal F (2007) Transmission of multiple traditions within and between chimpanzee groups. Curr Biol 17(12):1038–1043. https://doi.org/10.1016/j.cub.2007.05.031

Williams LE, Abee CR, Barnes SR, Ricker RB (1988) Cage design and configuration for an arboreal species of primate. Lab Anim Sci 38(3):289–291

Williams LE, Nehete PN, Schapiro SJ, Lambeth SP (2010) Effects of relocation on immunological measures in two captive nonhuman primate species: squirrel monkeys and owl monkeys. Am J Primatol 72(S1):28

Winnicker C, Honess P, Schapiro SJ, Bloomsmith MA, Lee DR, McCowan B et al (2013) A guide to the behavior and enrichment of laboratory macaques. Charles River Laboratories International Publishing, Wilmington, MA

Wood D (1989) Social interaction as tutoring. In: Bornstein MH, Burner JS (eds) Interaction in human development. Lawrence Erlbaum, Hillside, NJ, pp 59–80

Worch EA (2001) Simple tool use by a red-tailed monkey (*Cercopithecus ascanius*) in Kibale Forest, Uganda. Folia Primatol 72(5):304–306. https://doi.org/10.1159/000049953

Wright PC (1978) Home range, activity pattern, and agonistic encounters of a group of night monkeys (*Aotus trivirgatus*) in Peru. Folia Primatol 29(1):43–55. https://doi.org/10.1159/000155825

Yamanashi Y, Hayashi M (2011) Assessing the effects of cognitive experiments on the welfare of captive chimpanzees (*Pan troglodytes*) by direct comparison of activity budget between wild and captive chimpanzees. Am J Primatol 73(12):1231–1238. https://doi.org/10.1002/ajp.20995

Yamazaki Y, Echigo C, Saiki M, Inada M, Watanabe S, Iriki A (2011) Tool-use learning by common marmosets (*Callithrix jacchus*). Exp Brain Res 213(1):63–71. https://doi.org/10.1007/s00221-011-2778-9

Young C, Schülke O, Ostner J, Majolo B (2012) Consumption of unusual prey items in the Barbary macaque (*Macaca sylvanus*). Afr Primates 7(2):224–229

Zaragoza F, Ibáñez M, Mas B, Laiglesia S, Anzola B (2011) Influence of environmental enrichment in captive chimpanzees (*Pan troglodytes* spp.) and gorillas (*Gorilla gorilla gorilla*): behavior and faecal cortisol levels. Revista Científica 21(5):447–456

Zhang J, Hu JC, Zhong SL, Fei LS, Wang Q, Chen HW, Deng JB (2003) Observation on Behavior of Mandrill (*Mandrillus sphinx*) in captive. Sichuan J Zool 2:69–72

Zuberbühler K, Janmaat KR (2010) Foraging cognition in non human primates. In: Platt ML, Ghazanfar AA (eds) Primate neuroethology. Oxford University Press, Oxford, pp 64–83

Zuberbühler K, Gygax L, Harley N, Kummer H (1996) Stimulus enhancement and spread of a spontaneous tool use in a colony of long-tailed macaques. Primates 37(1):1–12. https://doi.org/10.1007/BF02382915

Primate Breeding Colonies: Colony Management and Welfare

James C. Ha and Adrienne F. Sussman

Abstract

We describe the basic and specialized requirements specifically for primate breeding colonies, including veterinary care, housing, welfare and enrichment, and management of social behavior. We emphasize that all four areas are of critical importance to the ethical and healthy productions of these animals. We explore in detail the needs for record keeping and pedigree maintenance when breeding primates, including genetic and demographic management.

Keywords

Nonhuman primate · Colony management · Welfare · Enrichment · Housing · Pedigree management · Genetic management · Infant rearing · Nursery rearing

1 Introduction

Nonhuman primates (hereafter referred to as "primates") are used as study systems for modern neuroscience, biomedicine, and behavioral research, and, as such, are found in research facilities around the USA, Europe, and Canada. In recent decades, the primates used for such research are increasingly being sourced from captive breeding colonies, rather than from the wild. In turn, this has necessitated the development of better tools and standards of care for colony-housed primates, especially in terms of physical and mental outcomes for colony-housed primates. Just as our coauthors in this book have carefully considered and documented good colony management principles for primates in other captive settings, here, we aim to do the same for those housed in breeding groups. There is an excellent literature on

---

J. C. Ha (✉) · A. F. Sussman
Department of Psychology, University of Washington, Seattle, WA, USA
e-mail: jcha@uw.edu

this topic (e.g., Abee et al. 2012); our goal here is to summarize the basic requirements for colony management and then to discuss aspects of colony management and welfare that are perhaps unique to the more-demanding social and physiological conditions of a breeding colony.

Primates can be housed in groups in several contexts, such as zoos or laboratories, as described in other chapters of this book. For this section, we will focus on breeding colonies, which are a unique type of housing necessitating specialized care and management. For the purposes of this discussion, we will define breeding colonies as facilities housing multiple social groups of the same species with the goal of increasing the population so that individuals can be removed to use for research. Most breeding colonies have the primary goal of increasing the population of research primates. Breeding primates domestically has several benefits over importing wild-caught primates for research, including reducing the stress of transportation and allowing researchers to select for specific-pathogen-free (SPF) populations without transmissible zoonotic diseases such as simian immunodeficiency virus or herpes B. In this chapter, we will focus primarily on macaque breeding colonies, as macaques are the most commonly used primate model group in the USA and the species that we have the most extensive experience with at the University of Washington National Primate Research Center.

In this chapter, we will explore the basic and specialized requirements for caring for primate breeding colonies, including veterinary care, housing, welfare, and enrichment, and management of social behavior. We will also explore in detail the needs for record keeping and pedigree maintenance when breeding primates, including genetic and demographic management.

2   Basic Requirements of Colony Management

In our experience, high-quality primate colony management requires the seamless integration of expertise in several fields. The best primate colony management teams require effort in team communication and training, and especially ongoing continuing education.

The five main components of a colony management program are as follows: (1) veterinary care, (2) housing, (3) behavior monitoring and environmental enrichment, (4) transport, and (5) record keeping.

The fundamental principles of most of these components in a colony management program are described in the following chapter, primarily for the most common primate housing situation, nonbreeding research housing, or "residential" housing. Most primate housing does not involve breeding, and the goal is simply to house primates for research purposes, or future research purposes. This is residential housing, as opposed to housing for breeding purposes, which has some different requirements and issues. Our goal in the remainder of this chapter is to document the aspects of each of these components that are specialized or especially problematic in the less common primate breeding circumstances. The transport facet of a colony management program is highly individual because of species-specific requirements, legal requirements, and the distance/duration of the transportation, and we will not comment on that topic.

3 Specialized Aspects of a Primate Breeding Colony Management Program

3.1 Veterinary Care

Veterinary care for a breeding colony is concerned with health, physical and mental, of individuals, and the fundamentals of this care are described elsewhere in detail (Coleman et al. 2022). However, veterinary care in a breeding colony facility includes a few additional issues, which fall into two basic topics: prenatal care and neonatal care. While zoos and laboratories may occasionally care for pregnant females or infant primates, these individuals are essential components of a breeding colony population, by definition.

Prenatal care involves any care of the breeding female prior to the birth event and may include specialized health issues even prior to conception. For instance, nutrition plays a significant role in healthy estrus and fertilization; females with severe protein or caloric deficiencies may not exhibit normal reproductive cycles and may spontaneously abort early embryos (Kohrs et al. 1976; Hendrickx and Binkerd 1980). Specialized nutritional issues continue to occur during pregnancies. Gestational diabetes is also of great concern, often resulting in postmature fetuses and postnatal physical and psychological deficiencies (Schwartz and Susa 1980). Adequate maternal nutrition during pregnancy is also crucial to normal neural development for macaques, which, unlike humans, undergo their major brain growth spurt prenatally (Morgane et al. 1993).

Prenatal veterinary care in the primate breeding colony should also include prenatal examinations, which are predominantly carried out by palpation, although the use of ultrasound is increasing. Ultrasound examination is in some cases a trained, conscious procedure for the subject but most often performed under sedation. Ultrasound provides information on growth rates, as well as the position and general health of the fetus, and is therefore much more informative than palpation alone (e.g., Conrad et al. 1995). Physical measurements obtained via ultrasound can be used to refine pregnancy end dates as well. Physical position information via palpation or ultrasound can be used to monitor for pregnancy issues such as breech delivery and other in utero issues, which may dictate a surgical intervention such as Cesarean section. Deliveries via C-section are frequently required in research settings either due to maternal complications from stress, nutritional issues, such as gestational diabetes, or research requirements (e.g., prevention of infant infection with maternal pathogens during delivery; Stockinger et al. 2011). Specialized post-C-section housing is required for monitoring of recent surgery subjects; this will be similar to housing for recipients of other invasive surgeries and may last several weeks.

Following modern ethical standards, every effort is made to retain the newborn infant with its dam. A dam-raised infant is in almost all cases healthier. However, in some clinical and research situations, this is contraindicated, perhaps due to neonatal critical condition, dam condition or behavior, or research requirements as approved by an ethics review board (e.g., animal care and use committee). While every effort is made to minimize these situations, in some cases the neonate must be separated and raised in a nursery setting.

Neonatal care largely resembles the more familiar juvenile or adult veterinary care but may require adaptations in treatments due to infants' immature immune systems, smaller and weaker bones, and other limitations (e.g., medication may need to be delivered as liquids rather than pills) (see Ruppenthal 1979 for a description of some aspects of neonatal care). Curious infant monkeys with smaller anatomies and developing bones may need specialized housing to prevent injuries. Young animals may fall more easily or be injured while exploring their enclosures or mastering new movements. While young animals may recover quickly from such injuries, caging should be designed to prevent tiny limbs and fingers from catching in cracks.

Fig. 1 *Left*, Young pigtail macaque experiencing social contact and play at the University of Washington's Infant Primate Research Laboratory; *right*, subadult pigtail macaques playing in an enclosure structure in a breeding colony at the Washington National Primate Research Center (Photograph by Dennis Raines)

Nutrition becomes an issue again for the neonate, especially when there is a medical or research need to separate the infant from its mother (e.g., when breeding for an SPF population or when a mother suffers a medical issue during pregnancy). Specialized diets and formula (milk) may be required for non-mother-reared animals.

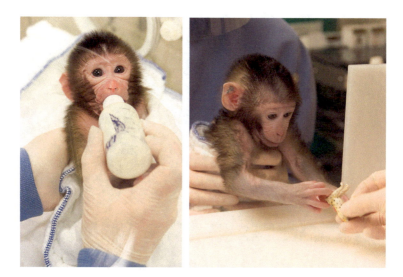

Fig. 2 *Left*, Infant pigtail macaque being bottle-fed at the University of Washington's Infant Primate Research Laboratory; *right*, infant pigtail macaque undergoing object permanence cognitive assessment at the University of Washington's Infant Primate Research Laboratory

3.2 Housing

For animals that are as social, reactive, and visually oriented as primates, housing and its layout, design, and spacing are critical for well-being. Like veterinary care, most housing requirements and concerns for a breeding colony are like those for a residential colony (Coleman et al. 2022), although there are a few vital differences.

Fig. 3 Breeding pair of pigtail macaques in a residential compound at the Washington National Primate Research Center (Photograph by Dennis Raines)

Typically, breeding groups are based on more natural social group sizes for the species, rather than the 1–2 animals per cage standard for residential systems. This can range from small family groups in squirrel monkeys and marmosets to harem-based social groups of a single male and 3–6 females, plus their offspring, to 30 or more animals in a single socially housed group. One exception is highly controlled research-based mating systems for carefully cycled females for projects in which it is desired to know the exact gestation age. In all cases, the size of the group's enclosure depends on the species and number of individuals in the group. The USDA has welfare legislation (though limited research) on minimum space requirements for Primates in groups of various sizes (U.S. Department of Agriculture 2017).

Some older research indicates the relative importance of not only the total enclosure space, but also of visibility (Sackett et al. 1975; Erwin and Erwin 1976; Oswald and Erwin 1976; Erwin and Sackett 1990). This research suggests that female–female aggression in harem-housed macaques is better predicted by lack of visibility of the male in the enclosure. Based on this research, spaces for harem-housed groups should avoid visual barriers that may prevent the male group member from observing and policing the females in the enclosure. However, Bushmitz (personal communication) and others consider that visual barriers in relatively high-density housing of macaques can reduce the effects of crowding on stress. It seems likely that unseen group members no longer contribute to an individual's social stress load.

Housing is a tremendous investment on the part of primate colonies, and major issues compete for attention in the design of any such facility. Safety for both humans and animals, as well as staff work efficiency, ease of cleaning, and access to animals are all major factors. Still, the conflicting results in the literature suggest that the design of the housing may vary greatly by the species and type of breeding group involved; a much greater investment in research is needed to understand how housing design and implementation can contribute to behavioral, health, and breeding outcomes for Primates.

3.3 Social Behavior

Behavior issues, especially social behavior, become far more complex in breeding situations. Again, the basics of behavior assessment and management, primarily for residential pair-housed (at most) animals, are covered elsewhere in this volume (Lutz and Baker 2022; Farmer et al. 2022). Many principles remain the same in group-housed and breeding-colony-housed animals. But additional complexities arise when the number of group members increases.

The number of possible interactive pairs increases exponentially in social groups greater than two, and therefore, conflicts are far more likely to occur in groups than in pairs. Requirements for forming these groups, of choosing individuals to form breeding groups, are increasingly complex due to the need to account for space available, disease transmission and SPF colony mechanics, and the population genetics of such a colony (see further comments below), as well as social compatibility. Again, many of these factors are addressed in other chapters (Coleman et al. 2022; Beisner et al. 2022).

In practice, forming successful social groups often requires trial and error; after the genetic, demographic, and medical considerations are considered, colony managers often assign groups somewhat arbitrarily and hope they "all get along." However, this approach often leads to heightened levels of aggression within the group, in turn reducing the fecundity in the breeding programs (Ha et al. 2000). While managers may wish to avoid this outcome, the tools to make better decisions about social aspects of housing are lacking. Little or no research has investigated optimal methods of (social) group formation in macaques, beyond early reports like that of Erwin and Sackett (1990), who found that aggressive interactions were more frequent in larger groups, but most importantly, that aggressive encounters were significantly more likely when females could escape the controlling influence of the breeding male, i.e., aggression occurred among females when out of sight of the male.

The latest, and, as yet, not entirely satisfying, approaches seek to identify combinations of individual differences in temperament and personality which, when merged, form compatible social groups. John Capitanio at the California National Primate Research Center has explored this avenue (Capitanio 1999, 2004), as has a recent doctoral dissertation at the University of Washington by Adrienne Sussman (2014). In the former study, the personalities of group members were found to be moderately predictive of aggressive behaviors in a small social group of similarly aged conspecifics. However, in the latter study, personalities were found to predict social behaviors in pairs of macaques, but not in large social groups.

It seems that as groups grow larger, their social dynamics appear to grow more and more complex, and are no longer significantly predicted by the personalities of the group members.

Interestingly, individual temperaments or personalities strongly predict group behaviors in some other model systems, such as water striders, guppies, and mountain gorillas (Sih and Watters 2005; Croft et al. 2009; Racevska, and Hill 2017). The answer may lie in the complexity of primate social behavior. Most studies show a larger number of personality dimensions for primates compared to other species (Gosling and John 1999; Gosling 2001), leading to a greater number of possible personality combinations and interactions in a primate group. Moreover, personality changes predictably with development, and the typical personality of an infant primate is not necessarily the same as that of an adult (Sussman et al. 2014; Smith and Weiss 2017). Previous approaches to predicting group behavior based on personality may have been overly simplistic, and more sophisticated computational modeling techniques may be needed to account for the many factors that can influence social behavior.

The already complex behavioral dynamics of a healthy social group are made even more complicated by the presence of cycling and pregnant females. The reproductive status of females in the group affects aggression and stress within the group (Ha et al. 2000), with greater proportions of pregnant females reducing aggression. These effects warrant frequent and careful monitoring of the social interactions within a group in an active breeding program, as the social structures may change suddenly as females enter estrus or become pregnant. Colony managers may be surprised to see that compatible individuals develop new antagonistic or aggressive relationships when in estrous season.

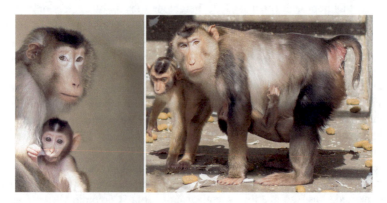

Fig. 4 *Left*, A pigtail macaque breeding colony female nursing her infant at the Washington National Primate Research Center; *right*, a pigtail macaque breeding colony female and her infant at the Washington National Primate Research Center (Photographs by Dennis Raines)

Finally, social group behavior will be influenced in manners beyond typical single or pair housing by the adult, juvenile, and infant behavioral dynamics. Infants, for instance, can become a source of conflict within a group. Unrelated female group members may engage in infant handling, or even kidnapping; either kind of interaction can lead to conflict between the adult females, and, occasionally, injury to the infant (Silk 1980, 1999). Males of some macaque species, including Barbary macaques, also utilize infants to develop alliances with other males or females, a behavior sometimes called "agonistic buffering" (Taub 1980). Overall, the presence of infants can trigger new types of interactions among the adults in a group, and colony managers must be vigilant to detect any new potential aggressive dynamics.

Juveniles spend a significant amount of time engaged in play behavior, including play aggression; typically, the playful intention of these encounters is signaled behaviorally during such encounters (Pellis and Pellis 1996). However, if these signals are missed or misinterpreted, a playful approach can be met with aggression by a conspecific or cause unwarranted concern from caretaker staff, given that play aggression rarely results in injury.

Social aspects of behavior are exaggerated and become potential issues in a breeding colony environment as opposed to single- or pair-housed residential populations.

3.4 Welfare and Environment Enrichment

Significant efforts to provide environments designed to optimize psychological well-being are mandated by law in the USA and elsewhere, although the specific requirements are sometimes vague. There is good science to support such efforts (Schapiro and Hau 2022): Stress, isolated development, and abnormal behaviors all will be reflected in animals, which can no longer be successful representatives of their species for use in research. A great deal of research has been performed in assessing the need for, and the effectiveness of, environmental enrichment in providing optimal psychological well-being for laboratory-housed Primates, and this information is well-summarized in chapters by Kemp (2022) and Morimura et al. (2022).

Enrichment is usually provided by introducing stimulating objects, such as food puzzles, foraging boards, swings, and ropes, into the housing environment. Competition for access to these enrichment objects can occur even in paired housing situations, with a dominant individual monopolizing access. In larger social groups, the potential for such competitions rises substantially, sometimes leading to aggression and injury. Caretakers must monitor groups when introducing new enrichment items to make sure contests over exciting new objects do not lead to contact aggression. They also must plan for mechanisms of controlling such aggressive situations, by separating individuals using gates or other control systems if necessary. In more elaborate housing situations, it is often advisable to (temporarily) divide larger social groups into smaller and/or more compatible groups for enrichment periods. Again, social behavior grows more complex and unpredictable in larger groups, and it can be safer to keep groups small while interacting with a novel enrichment item.

4 Record Keeping and Pedigree Management

Record keeping tends to be the ignored component to management of any colony of animals, including primates. We have addressed the principles of colony records management elsewhere (Ha and Davis 2006; Ha 2012) and direct the reader there for further information. However, since record management for primate colonies is not covered elsewhere in this book, we will provide some additional information about record keeping for pedigree management, including the applications and usefulness of demographic and pedigree monitoring.

4.1 Population Genetics and Demography

One important objective of a captive breeding program is to maintain demographically stable populations of sufficient size to preserve a high level of genetic diversity (traditionally 90%) over a long period (>100 years), while keeping animal breeding rates sufficiently high to make animals available for research assignment. Unlike other captive breeding programs, such as programs for the conservation of endangered species, colonies that breed animals for use in research must factor in the need to periodically remove animals for research, while still maintaining a stable breeding population.

There are three phases to the lifespan of a breeding colony: founding, growth, and maintenance. The founding and early growth of the University of Washington's National Primate Research Center's breeding colony has been described in Ha et al. (2000). Population maintenance size must be large enough to support the desired genetic variability and the target harvest numbers. This required population size should be determined by computer modeling, as described below. Modeling in general should be used in the growth and maintenance phases for three purposes: population growth measurement and projection, management decision making for zero population growth, and determining how to optimally breed animals to maintain desired levels of genetic diversity.

4.2 Demographic Management

Two types of monitoring and modeling are required in any captive colony: population demographics and genetics, both of which should eventually be merged into second-generation supermodels. Given the close link between demographics and genetics, the development of supermodels, which combine the two effects on a population's viability in a practical way, is sorely needed. Demographic modeling has three objectives: measurement of population growth rates, determination of how to modify reproduction and survival rates to meet colony objectives (e.g., the desired levels of animal removal for research), and evaluation of vital rates and age structure for potential problems. To meet these objectives, age-specific fertility and mortality rates need to be determined at regular intervals. This information can then be applied to computer models of population change.

An addition to this approach that was proposed by Ballou (1997) is the ability to incorporate variability, confidence interval estimation, and probability of achieving goal estimation into current modeling software. This computational approach greatly improved the accuracy of modeling, and by extension, breeding colony productivity. This is now more readily available through modeling software add-ins like Crystal Ball™ (Akkoç and Williams 2005) or iterative techniques like agent-based modeling.

Another factor that very few colonies incorporate into their management plans, especially on a formal basis, is behavioral considerations. Behavioral (social) factors play an important role in the demographics of any population, and an even more important role in the closed, high-density environment of a captive breeding facility. In studies of our own breeding colony, several social factors and several factors related to social behavior (the control role of the male, the effect of repeated disturbance of social relationships) have been demonstrated to have significant effects (Ha et al. 1999), and yet, these factors are often not considered in colony management decisions. It would be valuable to monitor these factors, to continue to research their effects (under outdoor housing conditions, for instance), and to incorporate this information into colony husbandry decision making (see Morimura et al. 2022.

4.3   Genetic Management

The objective for a genetic management plan in a captive research colony is to maintain sufficiently high genetic diversity and avoid inbreeding or any kind of inadvertent directional selection. Mismanagement of a colony's genetics could lead to animals who are behaviorally atypical, unhealthy, or otherwise unrepresentative of their species and no longer suitable for research. Genetic management has direct implications for animal welfare: Inbreeding depression can increase the rates of genetic deformities, and inadvertent directional selection can increase nonoptimal behaviors, such as aggression and parenting behavior.

To maintain genetic diversity, a colony manager must specify not only which animals can breed, but also how often, and with whom, they can breed. These decisions are based on the genetic importance of each individual, which, in turn, is based on the pedigree structure of the population. Maintenance of the colony pedigree is crucial in determining the genetic importance of breeding animals. This is very similar to the Species Survival Plan program in Association of Zoos and Aquariums-accredited public facilities (Association of Zoos and Aquariums 2018), and in fact, much of what we know about the demographic and genetic management of captive exotic populations comes from zoo and aquarium research.

The genetic importance of an individual is determined in two ways: mean kinship and genome uniqueness. Mean kinship is the average kinship (measured as Wright's coefficient of relatedness, or the proportion of genes shared, on average, by two individuals) of the focal animals with every other animal. Animals with the lowest mean kinship (the fewest near relatives) are the most valuable; they share fewer genes with other individuals and hence are less likely to produce offspring exhibiting inbreeding disorders. A breeding program that maintains the minimum mean kinship will have the maximum genetic diversity (Ballou and Lacy 1995).

Genome uniqueness is the proportion of an individual's genes, which are unique in the population. This approach prevents the loss of genes that are at a high risk for being lost in the colony. Determining genome uniqueness is computationally intensive and involves a statistical technique called "gene-drop pedigree analysis." (MacCluer et al. 1986). Software to conduct these analyses is readily available in the public domain but increased computational power may be needed.

While mean kinship and genome uniqueness are established techniques for genetic management, a third technique has been introduced more recently. Management of founder representation (Willoughby et al. 2015) is an approach that attempts to create a uniform distribution of the representation of the founding genes in a population and is probably a proxy measure of genomic uniqueness (although the precise relationship between the two measures remains to be investigated). A full pedigree is required to determine founder representation, but if not available, computer simulations that are underway indicate that maintaining a uniform founder representation may maximize genetic variability while minimizing the loss of desired recessive genes. Further work is needed to assess the efficiency of this approach, especially under different husbandry schemes.

4.4 Integrating Demographics and Genetics into Breeding Plans

Eventually, demographic and genetic information and animal breeding decisions must be integrated into a single coherent breeding plan. In the past, this integration has not been approached with a high level of sophistication, often due to limited resources or options. Few captive research primate colonies perform this kind of analysis, and those emphasize the understanding of sustainable animal removal levels (for research assignment) while checking for genetic problems using modeling. A new approach, agent-based or individual-based modeling, is available now and is proposed for use in modeling the genetic outcomes for breeding groups. This approach involves simulations in which individual "virtual animals" interact based on simple rules of reproduction (age-specific fertility) and mortality (age-specific survival) while maintaining their genetic identities. Through iterative simulation techniques, answers (compromises) that maximize genetic diversity and maximize colony productivity can be obtained. This is a promising approach for balancing the genetic and demographic needs of a colony in a single computational model and could provide a new tool for colony husbandry, and a new methodological resource for investigators.

5 Closing

Our goal in this chapter was to provide an overview of the welfare issues involved in managing a strong primate breeding colony, defined as one that has a strong breeding record of physically and mentally healthy animals, characterized in such a way as to be useful to the research community. Several chapters in this volume provide excellent materials which focused primarily on the generic, or residential, primate population. For each of the topics discussed: veterinary care, housing, behavior monitoring and environmental enrichment, and record keeping for demographics and genetics, we have attempted to address issues above and beyond those that are the same for residential primate colonies, and well-addressed by our coauthors.

Acknowledgements We would like to acknowledge Dr. Tom Burbacher for his assistance in preparing this chapter, especially in acquisition of the photos. We appreciate the strong editing help by the volume's co-editors. Finally, photos in this chapter were provided by the Washington National Primate Research Center (WaNPRC). They are acknowledged as following: The WaNPRC is supported by the National Institutes of Health (NIH) Office of Research Infrastructure Programs (ORIP) under award number P51OD010425. The WaNPRC SPF M. nemestrina colony is supported by grant U42OD011123 from the NIH Office of Research Infrastructure Programs.

References

Abee CR, Mansfield K, Tardif S, Morris T (2012) Primates in biomedical research, 2nd edn. Academic Press, London

Akkoç CC, Williams LE (2005) Population modeling for a captive squirrel monkey colony. Am J Primatol 65:239–254. https://doi.org/10.1002/ajp.20112

Association of Zoos and Aquariums (2018) Species Survival Plan® (SSP) Program handbook

Ballou JD (1997) Genetic and demographic modeling for animal colony and population management. ILAR J 38:69–75. https://doi.org/10.1093/ilar.38.2.69

Ballou JD, Lacy RC (1995) Identifying genetically important individuals for management of genetic variation in pedigreed populations. In: Ballou JD, Lacy RC (eds) Population management for survival and recovery: analytical methods and strategies in small population conservation. Columbia University Press, New York, pp 76–111

Beisner BA, Hannibal DL, Vandeleest JJ, McCowan B (2022) Sociality, health, and welfare in nonhuman primates. In: Robinson LM, Weiss A (eds) Nonhuman primate welfare. Springer, Heidelberg, pp 403–432

Capitanio JP (1999) Personality dimensions in adult male rhesus macaques: prediction of behaviors across time and situation. Am J Primatol 47:299–320. https://doi.org/10.1002/(SICI)1098-2345(1999)47:4<299::AID-AJP3>3.0.CO;2-P

Capitanio JP (2004) 2 personality factors between and within species. In: Thierry B, Singh M, Kaumanns W (eds) Macaque societies: a model for the study of social organization. Cambridge University Press, Cambridge, pp 13–37

Coleman K, Timmel G, Prongay K, Baker KC (2022) Common husbandry, housing, and animal care practices. In: Robinson LM, Weiss A (eds) Nonhuman primate welfare: from history, science, and ethics to practice. Springer, Cham, pp 317–348

Conrad S, Ha JC, Lohr C, Sackett GP (1995) Ultrasound measurement of fetal growth in *Macaca nemestrina*. Am J Primatol 36:15–35

Croft DP, Krause J, Darden SK et al (2009) Behavioural trait assortment in a social network: patterns and implications. Behav Ecol Sociobiol 63:1495–1503. https://doi.org/10.1007/s00265-009-0802-x

Erwin N, Erwin J (1976) Social density and aggression in captive groups of pigtail monkeys (*Macaca nemestrina*). Appl Anim Ethol 2:265–269. https://doi.org/10.1016/0304-3762(76)90059-6

Erwin J, Sackett GP (1990) Effects of management methods, social organization, and physical space on primate behavior and health. Am J Primatol 20:23–30. https://doi.org/10.1002/ajp.1350200104

Farmer HL, Baker KR, Cabana F (2022) Housing and husbandry for primates in zoos. In: Robinson LM, Weiss A (eds) Nonhuman primate welfare: from history, science, and ethics to practice. Springer, Cham, pp 349–368

Gosling SD (2001) From mice to men: what can we learn about personality from animal research? Psychol Bull 127:45–86. https://doi.org/10.1037/0033-2909.127.1.45

Gosling SD, John OP (1999) Personality dimensions in nonhuman animals: a cross-species review. Curr Dir Psychol Sci 8:69–75. https://doi.org/10.1111/1467-8721.00017

Ha JC (2012) Animal identification and record keeping: current practice and use. In: Abee CR, Mansfield K, Tardif SD, Morris K (eds) Primates in biomedical research: biology and management, vol 1, 2nd edn. Elsevier, London, pp 287–292

Ha JC, Davis AE (2006) Data management for the primate nursery. In: Sackett GP, Ruppenthal GC, Elias K (eds) Nursery rearing of primates in the 21st century. Springer, New York, pp 49–64

Ha JC, Robinette RL, Sackett GP (1999) Social housing and pregnancy outcome in captive pigtailed macaques. Am J Primatol 47:153–163. https://doi.org/10.1002/(SICI)1098-2345(1999)47:2<153::AID-AJP5>3.0.CO;2-D

Ha JC, Robinette RL, Sackett GP (2000) Demographic analysis of the Washington Regional Primate Research Center pigtailed macaque colony, 1967-1996. Am J Primatol 52:187–198. https://doi.org/10.1002/1098-2345(200012)52:4<187::AID-AJP3>3.0.CO;2-C

Hendrickx AG, Binkerd PE (1980) Fetal deaths in primates. In: Porter IH, Hook EB (eds) Human embryonic and fetal death. Academic Press, New York, pp 45–69

Kemp C (2022) Enrichment. In: Robinson LM, Weiss A (eds) Nonhuman primate welfare: from history, science, and ethics to practice. Springer, Cham, pp 451–488

Kohrs MB, Harper AE, Kerr GR (1976) Effects of a low-protein diet during pregnancy of the rhesus monkey I. Reproductive efficiency. Am J Clin Nutr 29:136–145. https://doi.org/10.1093/ajcn/29.2.136

Lutz CK, Baker KC (2022) Using behavior to assess primate welfare. In: Robinson LM, Weiss A (eds) Nonhuman primate welfare: from history, science, and ethics to practice. Springer, Cham, pp 171–206

MacCluer JW, VandeBerg JL, Read B, Ryder OA (1986) Pedigree analysis by computer simulation. Zoo Biol 5:147–160. https://doi.org/10.1002/zoo.1430050209

Morgane PJ, Austin-LaFrance R, Bronzino J et al (1993) Prenatal malnutrition and development of the brain. Neurosci Biobehav Rev 17:91–128. https://doi.org/10.1016/S0149-7634(05)80234-9

Morimura N, Hirata S, Matsuzawa T (2022) Challenging cognitive enrichment: examples from caring for the chimpanzees in the Kumamoto Sanctuary, Japan and Bossou, Guinea. In: Robinson LM, Weiss A (eds) Nonhuman primate welfare: from history, science, and ethics to practice. Springer, Cham, pp 489–516

Oswald M, Erwin J (1976) Control of intragroup aggression by male pigtail monkeys (*Macaca nemestrina*). Nature 262:686–688. https://doi.org/10.1038/262686a0

Pellis SM, Pellis VC (1996) On knowing it's only play: the role of play signals in play fighting. Aggress Violent Behav 1:249–268. https://doi.org/10.1016/1359-1789(95)00016-X

Racevska E, Hill CM (2017) Personality and social dynamics of zoo-housed western lowland gorillas (*Gorilla gorilla* gorilla). J Zoo Aquarium Res 5(3):116–122. https://doi.org/10.19227/jzar.v5i3.275

Ruppenthal GC (1979) Survey of protocols for nursery-rearing infant macaques. In: Ruppenthal GC (ed) Nursery care of primates. Springer, Boston, MA, pp 49–64

Sackett DP, Oswald M, Erwin J (1975) Aggression among captive female pigtail monkeys in all-female and harem groups. J Biol Psychol 17:17–20

Schapiro SJ, Hau J (2022) Benefits of improving welfare in captive primates. In: Robinson LM, Weiss A (eds) Nonhuman primate welfare: from history, science, and ethics to practice. Springer, Cham, pp 433–450

Schwartz R, Susa J (1980) Fetal macrosomia—animal models. Diabetes Care 3:430–432. https://doi.org/10.2337/diacare.3.3.430

Sih A, Watters JV (2005) The mix matters: behavioural types and group dynamics in water striders. Behaviour 142:1417–1431. https://doi.org/10.1163/156853905774539454

Silk JB (1980) Kidnapping and female competition among captive bonnet macaques. Primates 21:100–110. https://doi.org/10.1007/BF02383827

Silk JB (1999) Why are infants so attractive to others? The form and function of infant handling in bonnet macaques. Anim Behav 57:1021–1032. https://doi.org/10.1006/anbe.1998.1065

Smith CG, Weiss A (2017) Evolutionary aspects of personality development. In: Specht J (ed) Personality development across the lifespan. Academic Press, London, pp 139–156

Stockinger DE, Torrence AE, Hukkanen RR et al (2011) Risk factors for dystocia in pigtailed macaques (*Macaca nemestrina*). Comp Med 61:170–175

Sussman AF (2014) Macaque personality: structure, development, and relationship to social behavior. University of Washington, Seattle, WA

Sussman AF, Mates EA, Ha JC et al (2014) Tenure in current captive setting and age predict personality changes in adult pigtailed macaques. Anim Behav 89:23–30. https://doi.org/10.1016/j.anbehav.2013.12.009

Taub DM (1980) Testing the 'agonistic buffering' hypothesis. Behav Ecol Sociobiol 6:187–197

U.S. Department of Agriculture (2017) The "Blue Book": Animal Welfare Act and Animal Welfare Regulations. https://www.aphis.usda.gov/animal_welfare/downloads/AC_BlueBook_AWA_FINAL_2017_508comp.pdf

Willoughby JR, Fernandez NB, Lamb MC et al (2015) The impacts of inbreeding, drift and selection on genetic diversity in captive breeding populations. Mol Ecol 24:98–110. https://doi.org/10.1111/mec.13020

Common Husbandry, Housing, and Animal Care Practices

Kristine Coleman, Gregory Timmel, Kamm Prongay, and Kate C. Baker

Abstract

Animal care in facilities housing nonhuman primates has undergone a transformation in the past two decades, as the scientific community has learned more about the effects of husbandry practices on behavior and physiology of captive nonhuman primates. Today, husbandry consists of more than simply feeding animals and removing waste. Husbandry practices cover all aspects of animal care, from ensuring the animals' nutritional needs to providing adequate shelter, monitoring the health of the animals, and refining procedures so that they reduce stress and distress. As such, husbandry is integral to providing optimal animal welfare. This chapter will summarize four main areas of husbandry (health monitoring, nutrition, shelter, and humane handling) and discuss how each can influence aspects of the welfare of captive nonhuman primates living in a research environment.

Keywords

Health monitoring · Nutrition · Caging · Macroenvironment · Humane handling

---

K. Coleman (✉) · G. Timmel · K. Prongay
Oregon National Primate Research Center, Beaverton, OR, USA
e-mail: colemank@ohsu.edu

K. C. Baker
Tulane National Primate Research Center, Covington, LA, USA

© Springer Nature Switzerland AG 2023
L. M. Robinson, A. Weiss (eds.), *Nonhuman Primate Welfare*,
https://doi.org/10.1007/978-3-030-82708-3_14

1 Introduction

Animal husbandry is often thought of as the care and breeding of domestic animals. Traditionally, this has included feeding, watering, and cleaning. Today, animal husbandry is considered more broadly, and includes all practices critical to promoting and maintaining animal health, from ensuring the animals' nutritional needs to providing adequate shelter, monitoring health, and providing preventative medicine. The separation between husbandry—the treatment that an animal receives—and welfare—the physical and mental needs of the animal—is somewhat artificial. The link between animal husbandry and welfare can be illustrated by examining specific elements that contribute to the animal's quality of life, including those known as the "Five Freedoms" (Brambrell 1965). For example, health monitoring protects animals from pain, injury, and disease, and providing a proper diet and access to water provides freedom from hunger, thirst, and malnutrition. Appropriate shelter addresses both freedom from physical and thermal discomfort and freedom to express normal patterns of behavior such as locomotion, climbing, and play. Lastly, humane handling protects animals from fear and distress, and can promote positive welfare by creating an environment in which the animals are calm and relaxed. In this chapter, we will introduce four main areas of husbandry (health monitoring, nutrition, shelter, and humane handling) and discuss how each can influence aspects of the welfare of captive nonhuman primates living in a research environment.

2 Animal Husbandry

2.1 Husbandry Components in a Research Setting

Animal husbandry is a general term, used to describe many aspects of animal care. The four basic husbandry components are similar across environments in which captive nonhuman primates are housed, including research environments, zoos, and sanctuaries, although specifics about how they are achieved may differ somewhat depending on species and type of housing (Morimura et al. 2022). For example, monkeys can be maintained indoors in cages or pens, or they can be housed in large outdoor corrals consisting of several hundred individuals. Gaining access to an individual in a cage is often easier than accessing an individual in a two-acre corral. However, animals in both environments still participate in the health monitoring program of the facility. These husbandry components are influenced by local regulations and the behavioral biology of the species.

2.2 Regulations

Animal care practices for captive nonhuman primates, hereafter referred to as primates, in research facilities are overseen by various regulatory and/or accreditation bodies (for review, see Bayne et al. 2022). These regulations cover topics such as cage size, airflow, temperature, and humidity. In addition, facilities receiving funding from U.S. governmental agencies (e.g., the National Institutes of Health or National Science Foundation), and/or those that are accredited by AAALAC, International, must be compliant with the *Guide for the Care and Use of Laboratory Animals* (the *"Guide"*; National Research Council 2011), which provides guidance for husbandry practices.

Regulations guiding animal care are typically based on principles known as the "3Rs": *replacement, reduction,* and *refinement* (Russell and Burch 1959). *Replacement* and *reduction* involve using only the minimum number of animals necessary in research when other methods, such as computer simulations, are insufficient. *Refinement* is of particular importance to animal care, emphasizing ways to minimize the stress and distress experienced by the animals while enhancing their well-being (Jennings et al. 2009). As we discuss below, many current husbandry practices are refinements to animal care.

2.3 Understanding the Behavioral Biology

A key component for caring for captive animals is familiarity with the physiological, nutritional, and behavioral needs of the species. New World monkeys and Old World monkeys differ with respect to family structure, feeding patterns, and illness susceptibility, all of which affect their care. Even within taxonomic groups, there can be important differences between closely related species. For example, Hamadryas baboons (*Papio hamadryas*) have a social system which differs from that of other baboons and Old World monkeys, and consequently, they should be housed in harems instead of large multimale, multifemale groups. It is critical, therefore, that staff receive training in the natural history and behavioral biology of the species with which they work.

3 Health Monitoring/Disease Prevention

A robust health monitoring and disease prevention program promotes the welfare of primates by ensuring that they are free from pain and disease. Daily health checks, routine health screening, and preventative care practices are core components of a wellness program.

3.1 Daily Health Checks

Captive primates are typically observed at least daily by trained personnel, including clinical, husbandry, and/or investigative staff. Many New World monkey species (e.g., squirrel monkeys, *Saimiri* spp., and marmosets, *Callithrix* spp.) can experience rapid declines in health, and thus should be monitored at least twice daily (Williams et al. 2010). During daily health observations, personnel carefully examine primates for signs of overt injury or illness, such as the presence of blood, wounds, limb guarding, abnormal postures or movement, vomitus or diarrhea, and decreased appetite. Treats can be used to lure animals to the front of the cage to allow a closer look. Because most primate species are stoic, often hiding injuries or pain, it is important to also look for subtle indicators of pain and/or distress, such as the individual's attitude or general behavior. Recently, studies have correlated specific facial expressions with pain, opening up new opportunities for assessing pain (Descovich et al. 2017).

Daily health checks should include assessments of behavioral well-being (Baker 2016). The attitude of healthy animals can typically be described as bright, alert, and responsive (BAR). Animals that are quiet (i.e., withdrawn) or obtunded (e.g., hunched over, reluctant to move) may be ill or injured. Additional behavioral indicators of pain or distress include changes in activity (e.g., excessive activity or lethargy), changes in vocalizations, and changes in temperament (e.g., animals may become more aggressive or less interactive); one hallmark sign of pain in animals is the absence of normal behavior (e.g., a normally calm animal acts aggressively, or a normally active animal remains passive in the back of the cage). Because behaviors normal to one animal can be abnormal to another, daily observations are most effective when performed by those who know the behavioral characteristics of each individual animal. Health concerns noted during daily observations should be reported to clinical staff.

Performing health observations on animals living in large groups can be challenging, particularly if the observer cannot distinguish among individuals. Mildly to moderately injured animals often behave "normally" when stressed, which can make accurate assessment difficult. However, after several minutes, animals typically acclimate to the presence of the observer, at which time injuries may become more apparent. Removing an animal from an established group for clinical purposes can disrupt the group's social dynamics, particularly if the subject to be removed is high ranking (e.g., Ehardt and Bernstein 1986; Flack et al. 2005). For this reason, the threshold for intervention (i.e., when to treat) for group-housed primates might be different than that for caged primates. The potential consequences of removing an individual should be weighed against the potential harm to that animal and remaining group members if the individual is untreated. If a subject presents with a relatively minor wound, but removal from the group may cause significant social unrest, leaving the animal in the group and monitoring carefully may be appropriate. "Satellite cages" that permit visual access to the group can be useful for treating certain animals. When animals must be removed from their group for treatment, it should be for the shortest amount of time possible and in a way that minimizes group stress. For example, training monkeys to enter transfer tunnels or chutes for removal is generally less stressful than capturing them with a net (Luttrell et al. 1994).

Decisions about whether to remove an animal and/or whether to return the animal to the group after extended treatment are often best accomplished by taking a holistic, team approach (e.g., Hawkins et al. 2011; Gottlieb et al. 2017). This kind of approach promotes communication among invested parties and ensures that all aspects of the individual's welfare (e.g., clinical, social, behavioral) are considered. The Oregon National Primate Research Center (ONPRC) has relied on such team-based decisions regarding the care of group-housed rhesus macaques that comprise their breeding colony for several years (Gottlieb et al. 2017). A team consisting of clinical, behavioral, husbandry, and animal resource staff meet at least weekly to discuss monkeys requiring veterinary care and determine whether or not to return them to their social group. Individuals who are not suited for breeding groups, for either clinical or behavioral reasons, are reassigned to smaller groups or appropriate protocols. Additionally, the team discusses social group dynamics and reviews morbidity data from the previous weeks in an effort to formulate appropriate animal management plans. Creating an environment in which everyone who works with the animals is able to openly discuss concerns and contribute to solutions that are agreed upon leads to improved animal welfare.

3.2 Routine Health Monitoring

Regular physical examinations are another essential part of routine health monitoring. Examinations should be performed at least annually, but quarterly or bi-annually is preferred. Physical examinations often coincide with tuberculosis testing (see below). In most situations, hands-on physical examinations are performed on heavily sedated or anesthetized animals by veterinarians or trained veterinary technicians. A systematic approach, in which body condition (Clingerman and Summers 2005), weight, body temperature, coat condition (e.g., Steinmetz et al. 2006; Honess et al. 2005; Baker et al. 2017), and general appearance are assessed, followed by a comprehensive review of major body systems, ensures completeness and consistency (Kramer et al. 2012). Females should be palpated for pregnancy if appropriate. Any abnormalities found during the physical examination should be documented and treated by clinical staff.

3.3 Disease Prevention

Pathogens can be spread to primates by numerous organisms, including conspecifics, other primate species, animal care staff (or other humans), and other animals (e.g., birds, rodents, insects). Further, disease can also be spread from primates to humans. There are a wide range of zoonotic agents with varying degrees of morbidity. One of the most potentially devastating diseases for monkeys is tuberculosis, caused by *Mycobacterium tuberculosis*. Tuberculosis (TB) is highly contagious and often fatal and can be transmitted to monkeys from humans or other monkeys. Other zoonotic agents of concern include measles, viral or bacterial pneumonia, and enteric pathogens, such as *Campylobacter* spp., *Shigella*, and *Yersinia* (Kramer et al. 2012). Enteric diseases may cause severe diarrhea and gastrointestinal distress, and are a significant cause of morbidity and mortality in many primate colonies, particularly for young animals living outdoors (Prongay et al. 2013). Preventing the transmission of disease is necessary to the health and welfare of all individuals, whether human or nonhuman primate. There are several components of a successful disease prevention program.

3.3.1 Personal Protective Equipment

An important part of preventing transmission of zoonotic agents between humans and nonhuman primates is personal protective equipment (PPE). Requirements for the type of PPE vary depending on the species with which one is working, but generally include a dedicated uniform and/or outer clothing (e.g., laboratory coat or tyvek), latex gloves, dedicated close-toed shoes, a face mask, and eye protection (eye shield or safety glasses). In some situations, such as quarantine or when working with infected subjects, additional respiratory protection (e.g., a powered air-purifying respirator (PAPR) or N95 mask) may be required. Standards for PPE are usually established by the institutional Occupational Health and Safety group. PPE should not be worn outside the facility and should be changed in between animals of differing viral status, e.g., specific pathogen-free (SPF) to non-SPF. PPE should also be changed before working with species susceptible to zoonotic agents from other species of monkeys. For example, squirrel monkeys are carriers of *Herpesvirus saimiri 1*, which can be fatal to owl monkeys (*Aotus* spp.), and thus, PPE used while working with the former should not be worn in rooms containing the latter. Similarly, it is often desirable to have dedicated instruments and equipment (e.g., clinical, surgical, behavioral) for animal groups based on their viral status and/or species.

3.3.2 Disease Surveillance

Pathogen surveillance is essential to preventing disease and ensuring the health of the colony. While it may not be feasible to screen the colony for every possible zoonotic agent, most facilities have some sort of surveillance program. Due to its virulence, all facilities screen for TB at the very least. Primates undergo TB testing at least annually, although many facilities screen bi-annually or even quarterly. Most facilities utilize the tuberculin skin test, an injection of mammalian old tuberculin, typically administered intradermally at the edge of the upper eyelid. This location is favored due to ease of observation without restraint (Kramer et al. 2012). Trained technicians read the tuberculin test 24, 48, and 72 h after administration, looking for swelling and erythema. The reactions are given scores from 1 to 5, with 1–2 considered negative, 3 considered indeterminate, and 4–5 considered positive (Kramer et al. 2012). Tests resulting in indeterminate results should be repeated in the other eyelid. In most cases, animals with confirmed positive TB results (after confirmation by other sources, see Kramer et al. 2012 for details) are humanely euthanized to prevent disease transmission to the rest of the colony. In addition to the animals, personnel are also screened regularly for TB as part of the institution's Occupational Health and Safety program.

Other pathogens that undergo routine surveillance, at least for macaque species, include *Macacine herpesvirus 1* (commonly called "Herpes B virus"). This virus is endemic in wild macaque populations, although many research facilities have eliminated it from their colonies (e.g., "specific pathogen-free" colonies). While infection with this virus does not typically cause serious illness for the macaque, Herpes B can be fatal to humans and other primate species. In addition to this virus, colonies housing SPF macaques commonly screen for simian T-lymphotropic virus (STLV), simian immunodeficiency virus (SIV), and simian retrovirus type D (SRV-D).

3.3.3 Vaccination

Vaccination is another part of many preventative medicine programs. Both personnel working with primates and the primates themselves can be vaccinated against pathogens, such as the agents that cause measles and tetanus. Some facilities vaccinate primates against rabies as well, particularly if they house animals in outdoor enclosures.

3.4　Quarantine

The quarantine process typically begins well before animals arrive at the receiving facility. Pre-shipment physical examinations and testing for TB and other specified disease agents should be accomplished by the facility sending the animals. Medical records should be reviewed by the receiving facility, which may request that the animals are vaccinated against measles or other diseases. The transfer of information pertaining to rearing and housing history, as well as behavioral profiles, can support continuity of behavioral care to meet the needs of individual animals (National Research Council 2011; Baker et al. 2017). Both parties should confirm completion of required importation permits. Appropriate transportation following regulatory requirements should be agreed upon prior to shipment. This has become increasingly challenging as most commercial airlines will no longer transport primates due to pressure from animal activist groups (Wadman 2012). This pressure has led to animals often having to be shipped long distances by truck, resulting in increased travel time and stress.

The quarantine facility itself should be separated from other animal holding areas and ideally should have a separate HVAC system and equipment (X-ray, etc.) to minimize the risk of cross-contamination. Preferably, well-trained, dedicated staff will care for quarantined animals; alternatively, staff should care for quarantined primates last and wear PPE dedicated to the quarantine area.

Upon arrival, primates should be visually assessed and treated if indicated. A 2- to 3-day acclimation period is appropriate prior to a comprehensive arrival examination. The quarantine period should be a minimum of 31 days and may be longer if there is any indication of illness (Roberts and Andrews 2008). The arrival evaluation includes physical examination, TB testing (including with thoracic radiographs), and typically blood collection for complete blood count and serum chemistry analysis. Internal and external parasites should be checked for and parasiticides administered as appropriate or as a part of the quarantine program. A fecal culture for enteric pathogens such as *Shigella* spp. should be considered. TB testing is typically performed three times at 2-week intervals prior to the quarantined group being released. In the USA, international quarantines are regulated by the Centers for Disease Control and require more stringent facilities and practices than domestic quarantines (Roberts and Andrews 2008).

3.5 Sedation/Anesthesia

While physical examinations and zoonotic screening are vital parts of animal care programs, they need to be done thoughtfully, so as not to negatively impact welfare. In particular, the sedation typically associated with routine health screening can impact welfare. Ketamine hydrochloride, a dissociative agent, is commonly used for immobilization (Murphy et al. 2012). However, while ketamine is routinely administered, it is not necessarily benign and can have long-lasting effects on the welfare of the individual. Several studies have shown that ketamine decreases appetite the day following administration in macaques and African green monkeys (Crockett et al. 2000; Springer and Baker 2007). While few studies have examined long-term effects of ketamine, there is evidence that it is associated with alopecia in rhesus macaques (Lutz et al. 2016). Anesthetics and sedatives may be particularly harmful for younger animals; ketamine given early in life may produce long-term cognitive deficits in rhesus macaques (Paule et al. 2011). Reducing the use of sedatives, either through the strategic scheduling of procedures or through the use of positive reinforcement training (see Bloomsmith et al. 2022), can improve welfare for captive primates.

3.6 Identification

Being able to identify individuals is an essential requirement for husbandry practices. Primates are typically given unique identifiers, often in the form of an alpha-numeric code, that remain with them for their whole lives. These codes may be tattooed on the sedated animal, either on the inner thigh or across the chest. Animals may also have microchips implanted that enable people to more easily monitor food intake or access their electronic health records. While useful, these microchips do not provide visual identification. New World Monkeys are often given collars with attached identification tags. Unique dye marks may be used for group-housed macaques, which can greatly aid in physical health and behavior observations.

While not universally accepted, many facilities promote naming of individual monkeys. In some facilities, naming monkeys is discouraged out of concern for preferential treatment causing experimental confounds (Erard 2015). However, this attitude has changed as the scientific community learns more about the effects of husbandry practices on the behavior and physiology of the monkeys (e.g., Coleman and Heagerty 2019). Names help staff identify animals and strengthen the animal–caretaker bond (Rennie and Buchanan-Smith 2006), which likely results in improved animal care.

4 Nutrition

Meeting the nutritional needs of captive animals and ensuring they are free from hunger, thirst, and malnutrition are essential to promoting their welfare. Primate species have diverse feeding ecologies and digestive strategies. Most of the species used in research facilities are omnivorous, consuming a variety of food including fruits, leaves, insects, and gum. Dietary requirements for each species are based on several factors, including gastrointestinal tract structure, natural feeding behavior, and nutrients found in natural food sources. A detailed list of nutrient requirements for various primate species can be found in *Nutrient Requirements of Nonhuman Primates* (National Research Council 2003).

4.1 Standard Diet

All captive nonhuman primates should receive nutritionally balanced and appropriate food, which is most often commercially produced in pellets or chow. Standard chow should be provided at least twice a day (Röder and Timmermans 2002), although many species benefit by being fed small meals throughout the day (e.g., Schapiro and Lambeth 2010). This standard diet should be supplemented with items such as fresh fruit, vegetables, leaves, insects, and/or gum, depending on the species. Because these supplements tend to be more palatable than chow, they should be provided in moderation; animals presented with an abundance of palatable items may gain weight and/or not eat their standard diet. Supplements may be used for underweight or stressed animals to ensure adequate caloric intake.

Fresh water should be available ad libitum. Water can be provided in bottles or through automatic watering systems. The latter are easier to maintain than bottles, although lines must be checked daily and should be flushed at regular intervals to prevent the growth of biofilm. Water quality should be assessed regularly. Animals that do not have access to water typically decrease their food consumption; thus, if an otherwise healthy individual reduces food intake, the water supply should be checked.

4.2 Food Presentation

Foraging is a major part of the behavioral repertoire of most species. In the wild, nonhuman primates spend considerable time searching for and processing food (e.g., removing skin from fruit or obtaining meat from nuts). This time is often greatly reduced in the laboratory setting, where food is readily available. Increasing the foraging time for animals can provide them with a "job" and can help reduce boredom and the development of stereotypic behaviors (e.g., Lutz and Novak 2005; Lutz and Baker 2022; Kemp 2022). Even simple practices such as providing fruits, such as bananas and oranges with their outer skins intact, can increase the time it takes the animals to process those items. Standard monkey chow can be presented in a way that promotes foraging as well. Puzzle feeders or other devices used to provide enrichment can be modified for use with monkey chow (Bennett et al. 2016).

When possible, food should be presented in a way that approximates natural conditions. For example, New World monkeys are arboreal, rarely coming to the ground to forage. Therefore, their food should be provided toward the top of the cage and not scattered on the enclosure floor. Whole fruits can be speared and hung from the cage, promoting naturalistic foraging behavior (Buchanan-Smith 2010). Exudate from trees, or gum, is a staple of the callitrichid diet. Marmosets may therefore benefit from artificial "gum trees" (Röder and Timmermans 2002) in which commercially available gum Arabic is placed on branches in the enclosure (Buchanan-Smith 2010).

Regardless of the species, standard monkey pellets and all supplements should be provided in a clean part of the enclosure. When more than one animal is living in a group or cage, food should be provided in multiple places to avoid monopolization by one or more individuals (usually the dominant, or alpha, individual).

4.3 Developmental Needs

Energy requirements vary and depend not only on species, but also on the sex and age of the individual (e.g., Wolfensohn 2010). Growing infants have a higher energy requirement (per body weight) than adult conspecifics (National Research Council 2003). Further, reproductive status can also affect energy requirements. Lactating females, for example, have an energy expenditure significantly higher than that of nonpregnant/lactating females (National Research Council 2003). The amount of food offered should take these developmental needs into account.

Proper nutrition is especially important to infant primates. Infants typically get most of their caloric intake through nursing. When infants have to be hand-reared, for either clinical (e.g., abandonment or death of the mother) or research purposes, they are often fed human infant formula. It is important to make sure that the formula utilized has appropriate supplements, as protein or vitamin D levels in human formula may be too low for some species (National Research Council 2003). Formula should also include omega 3-fatty acids, which have been shown to play a role in enhanced vision (Neuringer et al. 1984; Champoux et al. 2002), motor skills (Champoux et al. 2002), and brain connectivity (Grayson et al. 2014) in rhesus macaque infants. Probiotics may also be an important additive. Rhesus macaques reared on infant formula fortified with probiotics had better immune function and recovered more quickly from diarrhea than those fed control formula (Kelleher et al. 2002).

There is increasing evidence that the mother's diet can affect the physiology and behavior of the offspring. For example, infant Japanese macaques born to mothers given a high-fat diet (32% of calories from fat) had perturbations in the serotonergic system and increased anxiety compared to those on a standard diet (12% of calories from fat; Sullivan et al. 2010). Thus, foods high in fat, as may be the case with certain enrichment items, should be provided in moderation to pregnant primates.

5 Shelter

One key component to husbandry practices is the provision of adequate housing. Shelter includes both the microenvironment (the cage or pen) and the macroenvironment (the room or outdoor location). Appropriate shelter should provide animals with physical and thermal comfort and the ability to express species-typical behaviors.

There are many housing options for captive primates. Caged primates are typically housed singly, or in pairs (for review see Buchanan-Smith et al. 2022), and group-housed primates are maintained in larger enclosures such as pens, corncribs, runs, or corrals. Cages are typically maintained indoors, although some facilities may utilize protected outdoor areas for caged animals. Group enclosures can range in size from relatively small indoor runs or pens to multi-acre outdoor corrals. The kind of housing provided can depend on factors such as location of the facility and reason for housing (e.g., breeding vs experimental purposes). While runs, pens, and corrals are typically associated with breeding, they may be appropriate for housing research subjects as well, depending on the degree and frequency of access required to individual animals. If such enclosures are used for experiments in which individual animals will need to be removed from and returned to the social group on a frequent basis, the potential for distress experienced by other group members and the risk of social destabilization must be weighed against the benefits to the individuals. Positive reinforcement training (see Bloomsmith et al. 2022) can help reduce this stress.

5.1 Microenvironment (Enclosure)

5.1.1 Caging

Much attention has been paid recently to the size and configuration of primate cages (for a review of the legislation, see Bayne et al. 2022). Standards for cage size both in the U.S. and Europe mandate that regulations state that cages should allow individuals to engage in normal posturing, including standing and stretching. European standards for cage sizes are substantially larger than those in the USA (European Parliament 2010) and emphasize that caging dimensions should be based on the specific needs and characteristics of each species.

The effect of cage floor space on welfare has been studied in macaques. Short-term confinement to smaller cages was associated with stereotypical behavior in rhesus macaques (Draper and Bernstein 1963), and short-term increases in cage size decreased these behaviors (Paulk et al. 1977). However, no behavioral effects were observed when tested over a matter of weeks or more (Line et al. 1990). Cage size did not alter urinary cortisol measures in either pigtailed (Crockett et al. 2000) or long-tailed macaques (Crockett et al. 1993), and small increases in available floor space (with nearly identical height) produced no behavioral changes in either species (Crockett et al. 1995; Crockett et al. 2000). Male rhesus macaques housed in either a large cage or pen (with a sixfold increase in floor space over the cage) showed less anxiety behavior in the larger enclosure (Kaufman et al. 2004). In these studies, however, animals were single-housed, and the level of environmental complexity was not clear. It is possible that cage size plays a larger role in welfare when the environment, including the social environment, is complex. More work is needed to examine this hypothesis.

In addition to floor space, vertical space is an important component of overall cage size. The use of vertical space has been studied in rhesus macaques and callitrichids, both of which have been shown to prefer elevated areas (marmosets: Buchanan-Smith et al. 2002; macaques: Clarence et al. 2006; Griffis et al. 2013; MacLean et al. 2009). Vertical space confers benefits by at least two routes, providing for species-appropriate locomotion and exercise, and moderation of stressful stimuli by permitting vertical flight. Vertical space could be assumed to be particularly impactful for species that are highly arboreal and subject to high levels of predation in the wild, although findings in these species are inconsistent. For marmosets, taller cage dimensions increased locomotion, foraging, grooming, and affiliative contact, and decreased abnormal behavior, aggression, and startle responses (Kitchen and Martin 1996). A study of cotton-top tamarins found no significant differences in behavior with higher cage ceilings, but when both the height and floor space were increased, the tamarins displayed higher levels of activity (Box and Rohrhuber 1993). Rhesus macaques moved to cages that were significantly smaller in both height and floor space exhibited no changes in abnormal behavior but tension-related behaviors decreased when they were returned to larger enclosures (Kaufman et al. 2004). The negative or inconsistent finding concerning the welfare impacts of cage dimensions may relate to the relatively small magnitude of change.

In addition to cage ceiling height, the configuration of cages in an animal room, i.e., how high cages themselves are mounted relative to the floor, can potentially play a role in welfare. In the USA and elsewhere caging is often configured in a "two-tiered" manner consisting of two horizontal rows of cages within an animal room, such that some animals are housed directly above others. Alternatively, some facilities use a "one-tiered" configuration, which affords the ability to provide at least twice the vertical space to individual animals. Several caging companies manufacture cages that have a removable floor between the upper and lower cages to allow vertical access. There have been relatively few studies examining the effect of cage configuration on welfare. Tamarins housed in lower-row cages displayed less activity and physical contact than those housed in upper rows (Box and Rohrhuber 1993). In macaques, results are mixed. While some studies detected no effects of cage location in macaques (Crockett et al. 1993, 2000; Schapiro et al. 2000), another detected lower levels of stereotypic behavior among animal housed in the upper row (Gottlieb et al. 2013a). This finding was also seen in vervet monkeys (Seier et al. 2011). There are several potential reasons for these behavioral differences. Primates housed in research facilities often lack opportunities to withdraw from stressors, especially humans and their activities. This problem may be exacerbated for those living on the bottom rows. Further, animals on the bottom tier may receive less positive attention from humans because the lower rack is somewhat more difficult for people to attend to. Humans may inadvertently "loom over" animals in these cages. A potential confound in evaluating mechanisms by which cage levels influence well-being is the fact that cages in lower rows are darker (see Sect. 5.2.2 for discussion of the role of light).

Despite the benefits of providing animals with vertical height, one-tiered caging is often not adopted due to logistical and financial challenges (Bloomsmith 2017). Still, steps can be taken to periodically provide animals with greater ceiling height. One method is the provision of exercise cages or enclosures (see below). Additionally, animals can be rotated between the top and bottom rows, although this may not be ideal in all situations. Relocations are a risk factor for the development of abnormal behaviors (Gottlieb et al. 2013a; Novak 2003; Rommeck et al. 2009). Awareness of the potential link between row height and welfare is best approached in terms of individual animal needs. For instance, moving animals to the upper tier may help mitigate abnormal or fearful behavior.

5.1.2 Other Enclosure Types

Primates may be housed in enclosure such as pens, runs, corncribs, and corrals, which differ from cages in several important respects. They are larger and afford opportunities for complex structural enrichment and a variety of types of ground cover. With respect to enclosure sizes, few regulations stipulate minimum size requirements per se. The U.S. Animal Welfare Act stipulates the minimum space requirements for individual animals rather than absolute enclosure volume. Typically, in enclosures such as runs and pens, the space per animal vastly exceeds regulatory minima for each individual should it be caged alone, so the issue of social density is more salient than absolute enclosure size. There are also few published guidelines for density. For example, the *Guide* (National Research Council 2011) states that the enclosure size for a social group may not need to represent sum of the space required for each individual animal. Instead, the *Guide* emphasizes functional standards, such that the enclosure meets the needs of individual animals. In an effort to socially house as many individuals as possible, the number of animals in groups often exceeds the number found in natural populations. Space must be sufficient to hinder monopolization of resources, to allow animals to flee from other group members, and to permit the inclusion of features such as visual barriers and other structural complexities necessary to regulate aggression. While captive primates have conflict avoidance and a tension reduction strategies for moderating aggression when living in socially dense housing (e.g., Aureli and de Waal 1997; Judge and de Waal 1993, 1997; van Wolkenten et al. 2006), high density can nonetheless be stressful (Dettmer et al. 2014; Judge and de Waal 1997; Lee et al. 2018; Pearson et al. 2015), necessitating features (e.g., visual barriers) that can moderate social stress.

Providing access to the outside environment can improve welfare for a variety of species. Outdoor enclosures typically involve social housing and have more space for exercise and sensory stimulation than indoor housing. Outdoor housing is associated with reduced physiological signs of stress (Schapiro et al. 1993) and lower prevalence and expression of abnormal behavior (Fontenot et al. 2006, Gottlieb et al. 2013a; Schapiro et al. 1995; Vandeleest et al. 2011). However, outdoor housing is not without risks. For example, there is the risk of potential disease transmission from rodents or birds. Environmental factors, such as extreme temperatures, may also be a concern (see Sect. 5.2.1).

5.1.3 Cage Complexity

While cage size is often the focus of animal welfare concerns, cage complexity is arguably as important in promoting welfare (Line et al. 1991). The primary enclosure should allow for the expression of species-specific behavior. Resting areas, such as perches, hammocks, or nest boxes, should be provided. Cage perches should be mounted at different levels within cages, to promote climbing and to avoid monopolization by individuals. Perches are often part of the cage (e.g., metal), but wood branches and/or other materials (e.g., fire hoses) can be used to make resting areas. Regardless of the material, resting surfaces should be adapted for the physiology and behavior of the species; for example, macaques sit on perches, but squirrel monkeys perch with their hands and feet, and thus should be provided with round rather than flat perching options (Williams et al. 1988). Nest boxes, for species that utilize them (e.g., *Aotus* spp), are also important features of the primary enclosure. Callitrichids live in small family units (a male and female and their offspring), which sleep huddled together at night to avoid predation. These species should be provided with sleeping boxes large enough for the whole family to utilize (Röder and

Fig. 1 Cynomolgus macaques (*Macaca fascicularis*) with a tunnel connecting upper- and lower-level cages

Timmermans 2002). Multilevel structures and benches often supplement or replace perches in larger enclosures.

Primates should have the ability to move normally and climb in their primary enclosure. Obviously, large enclosures provide more opportunity for walking and running than small cages. However, cage furniture such as perches or "tunnels" that connect bottom and upper cages (Fig. 1) can promote climbing in cages. Swings and moveable objects such as exercise wheels are often found in larger enclosures. These sorts of items are readily used by most primates and are relatively resistant to habituation (Coleman et al. 2012). Periodically rearranging these items can help keep them novel and can foster spatial learning (Honess and Marin 2006). It is important to ensure that these items are sturdy enough to withstand use, and they should be replaced when necessary.

Another relatively simple way to increase the complexity of a standard cages is by including a porch, or cage extender (Fig. 2; Gottlieb et al. 2014). Macaques typically spend a great deal of time in porches when they are present (personal observation), presumably because they provide a relatively wide view of the room. Further, porches have been found to reduce the occurrence of some abnormal behaviors in

Fig. 2 Caged rhesus macaque (*Macaca mulatta*) with a "porch". Photo reprinted from Gottlieb et al. (2014) (Fig. 1, p. 654) with permission from the *Journal of the American Association of Laboratory Animal Science*

macaques, including feces painting (Gottlieb et al. 2014) and stereotypy (personal observation). Play or exercise cages (e.g., Griffis et al. 2013), which can attach directly to the primary enclosure, can also provide additional complexity for the animals (Fig. 3). Last, the caging itself can be constructed such that foraging surfaces are integral to the caging and minimize the workload associated with enrichment devices, which may require repair or reinstallation when detaching from caging. Tabs and slots can be constructed for similar purposes.

The primary enclosure should also contain ways for animals to avoid visual contact with others. Visual barriers, including walls or cylinders, have been found to reduce aggression in pair (Reinhardt and Reinhardt 1991) and group-housed (Crast et al. 2015; Maninger et al. 1998) primates. They may be particularly beneficial during group formation, though see Erwin et al. (1976). Some animals, particularly timid or stress-sensitive individuals, living in cages may also benefit from having a "privacy panel," which allows them to avoid visual contact with other individuals in their room. Privacy panels can be made of plastic or other materials and hung on the cage (Fig. 4). Interconnecting cages and racks can be designed such that when slides are open between animals, part of the sidewall remains solid.

5.1.4 Flooring and Substrates

Along with the variety of enclosure types housing laboratory primates, there are a number of different flooring types as well, ranging from sealed artificial flooring

Common Husbandry, Housing, and Animal Care Practices

Fig. 3 Exercise pen for squirrel monkeys (*Saimiri boliviensis*), attached to their home cages

Fig. 4 Caged rhesus macaque (*Macaca mulatta*) utilizing a privacy panel

(concrete, epoxy, and tile) to naturalistic ground cover. Sealed flooring has some advantages, including ease of water disposal via drain, the ability to provide heat, and the prompt detection of clinical problems such as diarrhea. However, sealed flooring does not provide opportunities for foraging. Adding substrates to concrete or other smooth flooring has been associated with changes in behavioral indices of welfare. A number of types of substrates have been studied, including woodchips, dried ground corncob, wood wool, garden peat, and straw. Benefits of substrate include higher levels of foraging, activity, and prosocial behavior, as well as lower levels of abnormal behavior and aggression (Baker 1997; Blois-Heulin and Jubin 2004; Boccia 1989; Boccia and Hijazi 1998; Byrne and Suomi 1991; Chamove et al. 1982; Doane et al. 2013; Ludes and Anderson 1996). While most of these studies included both substrate and edible foraging material, even those that focused on the substrate alone found benefits (Blois-Heulin and Jubin 2004; Boccia and Hijazi 1998; Chamove et al. 1982).

Common challenges associated with substrates include plumbing, sanitation, and workload concerns. However, in the few studies that included a thorough assessment of these issues, successful drain covers were designed, and microbial analysis demonstrated that enclosures can be adequately sanitized (Bennett et al. 2010; Chamove et al. 1982). Furthermore, incorporating substrate reduces water and chemical use (Bennett et al. 2010; Doane et al. 2013), and has been determined to either save husbandry time (Bennett et al. 2010) or at least not increase it (Doane et al. 2013). These are important findings given widespread beliefs that providing substrates increases the labor associated with husbandry. Another concern regarding the substrate is that it will be consumed, leading to clinical problems, but to the authors' knowledge this has not been demonstrated.

Outdoor-housed primates may be housed directly on the ground as well. The nature of the ground cover can have differential effects on welfare. Macaques living in enclosures with natural ground cover foraged more but affiliated less than those housed with gravel (Beisner and Isbell 2008). Both aggression and hair loss attributed to excessive grooming are seen at lower levels in the former (Beisner and Isbell 2011).

5.2 Macroenvironment

5.2.1 Temperature

Failure to maintain an environment consistent with the natural biology of a species will negatively impact animal health and welfare. For individuals living indoors, temperature is typically controlled automatically. Appropriate temperature ranges vary by taxa; for example, recommended temperatures for Old World monkeys are generally in the 18–24 °C range (Wolfensohn 2010), while those for New World monkeys are higher (e.g., squirrel monkeys: 24–27 °C, Williams et al. 2010; marmosets: 23–28 °C, Buchanan-Smith 2010). The range of appropriate humidity can also vary between species.

It is clearly more challenging to control temperature for animals living in outdoor environments. All animals housed outdoors should have access to an environmentally controlled indoor area when temperatures reach a critical point. Heat support should be provided during periods of low temperatures. Provisions should also be in place for instances in which temperatures get too warm (e.g., misters, pools, and ice blocks). Such features that moderate weather extremes should be plentiful enough to allow use by all animals in a social group and spacious enough to be sufficiently ventilated and cleaned (National Research Council 2011).

5.2.2 Lighting

Lighting is often automatically controlled for animals living indoors. Light cycles vary, but typically include a 12:12 cycle, in which lights are on for 12 h a day. This automatic lighting differs from outdoor lighting, which, in most climates, changes with time of year. Further, unlike outdoors, where the sun slowly sets and rises, indoor lights typically turn on and off abruptly, which can result in increased alarm calls (personal observation). Artificial lights can be configured to simulate dusk and dawn (e.g., Steinmetz et al. 2006), although there is a paucity of information regarding the benefit of doing so.

There have been relatively few studies examining the relationship between light levels and welfare, and many of these are confounded by issues of cage size and location (i.e., top or bottom tier). Cages in the top tier are typically brighter than those in the bottom tier. Whether these differences in lighting affect behavior is unclear; studies have shown no behavioral differences between macaques in top versus bottom row cages (Crockett et al. 2000; Schapiro and Bloomsmith 2001), although other studies have found that animals living in cages in the top row are less likely than those in the bottom to have behavioral issues (Gottlieb et al. 2013a). However, this finding may be due to the location rather than lighting per se; cynomolgus macaques prefer to spend time in upper-row cages compared to lower-row cages regardless of available light (MacLean et al. 2009).

Equally important to welfare is darkness. Constant lightness, as may occur with broken lighting, faulty timers, or with constant outdoor lights, can interfere with sleep, causing stress and increased aggression (Morgan and Tromborg 2007). Thus, it is important to ensure that animals have access to darkness as well as light.

5.2.3 Noise

Noise in research facilities may consist of ambient sound pressure or sudden noises, which may arise from human activity or inanimate sources such as laboratory equipment, machinery, or even construction. Unpredictable and startling noise is a form of environmental perturbation that can have lasting negative effects on welfare. Offspring of mothers exposed to unpredictable noise during gestation showed behavioral differences compared to those born to mothers without such noise, including increased abnormal social behavior and decreased normal social behavior (Clarke and Schneider 1993). Rhesus macaques exposed to unpredictable loud noises, including construction, have higher cortisol levels compared to those without noise or those with control over the noise (Hanson et al. 1976; Westlund et al. 2012)

Noise engendered by routine husbandry activity can also have negative effects. Caged rhesus macaques have higher heart rates during times of day in which noises associated with husbandry activities are pervasive compared to quieter times (Doyle et al. 2008).

Noise abatement steps can be taken during facility design or as part of modification to existing enclosures. Machinery that is noisy or is the focus of noisy human activity (e.g., cage washers) should not be located close to animal holding rooms. Noise-generating areas should be separated by multiple doors in an effort to reduce both overall noise and sudden increases in noise (Bohm et al. 2009). Information on sound control for behavioral laboratories (Howard and Foucher 2009) may include strategies that can be adopted for primary housing. Soundproofing rooms, installing acoustic door seals, or using nonmetal material for caging components may pose a challenge in light of the need for pathogen control and sanitization, but relatively straightforward steps such as avoiding doors that slam shut are compatible with these needs.

Noise can also be reduced by training staff to talk quietly and avoid banging cages and equipment as much as possible, or by innovations that help make equipment run more quietly. Further, some sounds are intrinsically more stress invoking than others. The highest audible frequency for humans is 20 kHz, but it is much higher (approximately 42 kHz) for many nonhuman primate species, including squirrel monkeys and macaques. These species can hear ultrasonic sounds such as fluorescent lighting, ultrasonic cleaners, and AC adapters, and these sounds can cause stress and anxiety (Morgan and Tromborg 2007). Thus, monitoring sound levels and frequencies is an important factor in improving the welfare of captive primates.

5.3 Cleaning

The primary enclosure for captive primates must be clean and sanitized. Cages should be cleaned daily and disinfected on a regular basis, such as every 14 days. Chemicals and/or hot water can be used to disinfect the primary enclosure (National Research Council 2011). Daily cleanings should include removal of soiled bedding, feces, and uneaten food.

It is important to keep in mind natural behavior of the species when cleaning. Some New World Monkey species, such as common marmosets (*Callithrix jacchus*), rely heavily on scent marking. Frequent cage cleaning can eliminate olfactory signals, the absence of which may then disrupt social behavior and dominance hierarchies in these species (National Research Council 1998). Therefore, enclosures and objects on which these species scent mark, such as branches or porches, are often cleaned on alternate schedules, to ensure that familiar scents are not completely eliminated during cleaning (e.g., Buchanan-Smith 2010).

6 Humane Handling

Many of the husbandry events discussed thus far can be somewhat stressful for captive animals. In a broad sense, humane handling refers to husbandry practices designed to reduce fear and distress for the animals. Below, we describe some of these practices, which we have loosely organized into those that focus on animal management and those that focus on human–animal interactions.

6.1 Animal Management

6.1.1 Appropriate Rearing

Appropriate early rearing is one of the most important ways to promote psychological well-being and decrease distress for primates. Indeed, one of the greatest risk factors for the development of behavioral pathologies such as self-injurious behavior and stereotypy is being reared in a nursery (e.g., Novak 2003). Nursery rearing is also associated with physiological problems including diarrhea (Elmore et al. 1992). For these reasons, nursery rearing should be avoided when possible. However, there are situations in which infants are abandoned, neglected, or orphaned by their mothers, which necessitate intervention. One option in these situations is to find a lactating female, such as one who recently lost her infant to illness, to serve as "foster mother." While many such lactating females are willing, and often eager, to accept a new infant, the number of such females is often quite low. Another option is to train non-lactating dams to serve as foster mothers. The ONPRC uses operant conditioning to train non-lactating females to allow infants to come to the front of the cage to nurse from a bottle (Gottlieb et al. 2017). To date, they raised over 23 infants in this manner, none of whom have developed behavioral problems, such as stereotypies or self-injurious behavior, suggesting that this type of rearing might be useful for abandoned or orphaned infants.

Infants reared with their mothers in cages are usually removed (e.g., artificially weaned) at some point (e.g., 6–12 months of age) for management purposes. Infant macaques should be allowed to stay with their mothers for at least 10–14 months (McCann et al. 2007), and longer if the animals are living in a group. This time with the mother affords infants the opportunity to become behaviorally independent (Prescott et al. 2012), and to learn appropriate monkey behavior. If young animals (e.g., weanlings) are grouped together, including an adult female or male can help reduce aggression and teach them proper adult behavior (Champoux et al. 1989).

6.1.2 Reducing Relocations

In many research facilities, monkeys get relocated on a regular basis. A recent study (Lutz et al. 2016) found that animals in some facilities may move as often as four times a year. Such changes in location can include movement of animals from social groups to cages, from cages to social groups, from one room to another room, or within a room. These relocations can result from scientific needs, facility needs (e.g., repairs), or social needs. Regardless of the reason, relocations often result in animals being moved into groups or rooms with unfamiliar individuals, which can be stressful (e.g., Bethea et al. 2005). Frequent moves have been correlated with increased hair cortisol levels (Lutz et al. 2016) and increased risk of stereotypical behavior (Gottlieb et al. 2013a). Further, they can affect immune function. Capitanio et al. (1998) found that rhesus macaques subjected to frequent housing relocation and social separations around the time of infection with simian immunodeficiency virus had shorter survival times compared to those not exposed to these common stressors. Reducing the number of times individuals must move may help to reduce stress and improve well-being. Further, when animals do have to be moved, they should be given time to acclimate to their new environment before taking part in research projects.

6.1.3 Make Handling Predictable and Consistent

It is widely established that having a sense of control is important to the psychological well-being of primates and other animals (e.g., Mineka et al. 1986). Lack of control, on the other hand, can lead to stress and learned helplessness (e.g., Maier and Seligman 1976). There are many husbandry events over which research primates have no control, including cage washing, feeding, restraint for injection, or other procedures. These events are often preceded by a signal; for example, the sound of a food cart in the hallway typically indicates that food is about to be delivered. However, many signals are somewhat ambiguous. A husbandry technician entering the room can precede a positive event (e.g., feeding), a negative event (e.g., cage washing), or a neutral event (e.g., daily health observations; although these may be negative for some individuals). This ambiguity can lead to stress as the animals cannot predict what will happen to them. Making these events predictable to the animals can help reduce associated anticipatory stress with them, thus improving welfare.

One way to increase the predictability of husbandry events is by performing them at the same time each day (i.e., temporal predictability). Making husbandry events, such as cage cleaning or feeding, temporally predictable has been found to reduce stress and improve well-being in capuchins (Ulyan et al. 2006) and macaques (Gottlieb et al. 2013b; Waitt and Buchanan-Smith 2001). These findings are not universal; group-housed primates have been found to benefit from an unpredictable feeding schedule (e.g., Bloomsmith and Lambeth 1995). Importantly, when temporally predictable schedules are changed (e.g., food that is usually delivered at a specific time is late), it can increase stress (Waitt and Buchanan-Smith 2001).

It can be challenging to keep husbandry events temporally predictable in a research setting. Unexpected factors, such as staff call-outs or power failures, inevitably result in changes in the timing of husbandry events. Another way to reduce the stress of these events is by providing a signal before the event is to occur (i.e., signaled predictability). In a study of zoo-housed capuchins, care staff knocked on the door to the animal's room prior to husbandry events. This simple signal reduced anxiety associated with these husbandry events (Rimpley and Buchanan-Smith 2013). Signals can be auditory (e.g., buzzer or doorbell) or visual (e.g., colored sign). The benefits of adding a predictable signal to aversive events have largely been explained by the "safety signal hypothesis," which states that when an aversive event is reliably preceded by a signal, the absence of the signal communicates to the animal that it is a "safe" period, and the aversive event is not about to occur (Seligman 1968). Thus, if animals receive a reliable signal on days in which a negative event is to occur (e.g., physical exams), they will know on all other days that they are "safe," reducing anticipatory stress toward that event.

There are simple ways to increase signal reliability and thus make events more predictable to the animals. For example, at the ONPRC, technicians may wear clothing specific to different types of events, i.e., a scrub jacket for observations and/or giving out treats, and a standard waterproof gown for health examinations. By wearing the scrub jackets, the technicians are effectively communicating to the monkeys that they are not going to be doing anything aversive. The technicians noted that within a few weeks of this change, the monkeys were seemingly less anxious and reactive while they were in the room doing rounds (personal communication from A. Kvitky, ONPRC).

6.2 Human–Animal Interactions

6.2.1 Positive Relationship with Care Staff

Perhaps the most important way to improve humane handling is by fostering a positive relationship between the primates and the people taking care of them. Once discouraged and considered a potential threat to scientific objectivity (Wolfle 2002), such positive relationships are now encouraged by many facilities (Coleman 2011). Unstructured human interactions such as providing treats (Bayne et al. 1993) or playing with and/or talking to the animals (Manciocco et al. 2009; Baker 2004) can foster good relationships between care staff and animals. These unstructured interactions have been shown to reduce abnormal behavior, increase species-appropriate behaviors such as grooming, and improve well-being for a variety of primates, including marmosets, macaques, and chimpanzees (Reinhardt 1997; Manciocco et al. 2009; Waitt et al. 2002; Baker 2004; Bayne et al. 1993). Importantly, these relationships can also promote coping skills (Rennie and Buchanan-Smith 2006) and help mitigate stress reactivity toward novel objects or situations. For example, Miller et al. (1986) found that chimpanzees were less anxious when confronted with novelty in the presence of their trusted caretaker than when this caretaker was absent. The amount of positive interactions does not have to be great to have an effect; simply handing out treats for a few minutes several times a week reduced indicators of stress in cynomolgus macaques (Tasker 2012), and sitting quietly in the room observing behavior and occasionally handing out treats reduced abnormal behavior in rhesus macaques (Markowitz and Line 1989).

There are indirect benefits of promoting a positive relationship between care staff and primates as well. Primates are more likely to sit calmly in the front of their cage when they trust their caretakers than when they do not (Lehman 1992). This relaxed response to humans can facilitate daily observations and health checks as well

Fig. 5 Group-housed male Hamadryas baboon (*Papio hamadryas*) trained to come to the front of the enclosure and allow his nails to be trimmed

research procedures. Further, these relationships can also benefit the caretaker (Davis 2002; Rennie and Buchanan-Smith 2006). The opportunity to engage in positive interactions with the animals leads to increased morale and job satisfaction, which, in turn, lead to better care and improved animal well-being (Waitt et al. 2002; Coleman and Heagerty 2019).

6.2.2 Positive Reinforcement Training

Positive reinforcement training (PRT) is another example of a human–animal interaction that can foster humane handling. PRT techniques are a form of operant conditioning. In PRT, the subject is presented with a stimulus (e.g., a verbal command), responds by performing a specific behavior (e.g., move to the front of the cage and remain stationary), and is provided with reinforcement (e.g., food treat) when it has completed the specific behavior (see Bloomsmith et al. 2022).

There are many benefits associated with PRT. It desensitizes animals to potentially stressful stimuli, such as injections (Schapiro et al. 2005), thereby reducing associated stress. Further, it can reduce the need for physical or chemical restraint, which can negatively affect welfare. PRT can also allow group-housed animals to stay in their home environments (e.g., Fig. 5). For these reasons (and others, as detailed in Bloomsmith et al. 2022), PRT should be incorporated into husbandry practices whenever possible.

7 Conclusions

Animal care for captive nonhuman primates has undergone major changes in the past few decades, as the scientific community has learned more about behavioral biology and the effects of husbandry practices on behavior and physiology of captive nonhuman primates. Today, husbandry practices cover all aspects of animal care, from ensuring the animals' nutritional needs to providing adequate shelter, monitoring the health of the animals, and refining procedures so that they reduce stress and distress.

Husbandry is not performed in a vacuum; it takes a team to effectively care for captive primates. Most facilities have highly trained animal care technicians, veterinarians, and behaviorists working together to address the welfare of the animals. Having these disparate people, each with unique expertise, working together can be challenging in itself. However, an increasing number of facilities are using a team approach toward animal care (e.g., Hawkins et al. 2011; Gottlieb et al. 2017). Such an approach encourages people with different expertise, such as care staff, veterinary staff, behavioral staff, and research staff, to have a voice in decisions. It also provides an opportunity for clear communication among all parties. One example of how this kind of approach can be useful is the Quality of Life team established to assist in the decision-making process regarding euthanasia of primates at the University of Texas MD Anderson Cancer Center (Lambeth et al. 2013). This team, composed of veterinarians, veterinary technicians, and the trainer, colony manager, and care staff who work closely with the animal, relies on both quantifiable behavioral observations and clinical information to determine the quality of life for individual animals, which can aid in decisions regarding humane euthanasia. The use of a team approach has empowered care staff and has provided the veterinarian access to information he or she might not otherwise have from people who are most familiar with the animals (Lambeth et al. 2013).

Providing for the needs of captive primates is a dynamic process that must be continuously evaluated and refined. There has been a dramatic increase in the number of empirical scientific papers addressing different aspects of husbandry in the past two decades. This information should be incorporated into animal care programs to the greatest extent possible, to ensure that we are providing the highest quality of care for the animals.

Acknowledgements We thank Alona Kvitky, Daniel Gottlieb, and Allison Heagerty for useful discussions, and the dedicated animal care technicians at the Oregon and Tulane National Primate Research Centers, who provide the best possible environment for the monkeys. We would also like to thank our colleague, Dr. Kirk Andrews, who was always a strong advocate for the welfare of the animals. Support is acknowledged from the Oregon National Primate Research Center, 8P51OD011092, and the Tulane National Primate Research Center, 2P51 OD011104.

References

Aureli F, de Waal FB (1997) Inhibition of social behavior in chimpanzees under high-density conditions. Am J Primatol 41:213–228. https://doi.org/10.1002/(sici)1098-2345(1997)41:3%3C213::aid-ajp4%3E3.0.co;2-#

Baker KC (1997) Straw and forage material ameliorate abnormal behaviors in adult chimpanzees. Zoo Biol 16:225–236. https://doi.org/10.1002/(SICI)1098-2361(1997)16:3%3C225::AID-ZOO3%3E3.0.CO;2-C

Baker KC (2004) Benefits of positive human interaction for socially housed chimpanzees. Anim Welf 13:239–245

Baker KC (2016) Survey of 2014 behavioral management programs for laboratory primates in the United States. Am J Primatol 78(7):780–796. https://doi.org/10.1002/ajp.22543

Baker KC, Bloomsmith MA, Coleman K, Crockett CM, Lutz CK, McCowan B, Pierre PJ, Weed JL, Worlein JM (2017) The behavioral management consortium: a partnership for promoting consensus and best practices. In: Schapiro SJ (ed) The handbook of primate behavioral management. CRC Press, Boca Raton, FL, pp 9–23

Bayne K, Dexter SL, Strange GM (1993) The effects of food treat provisioning and human interaction on the behavioral well-being of rhesus monkeys. Contem Top Lab Anim Sci 32(2):6–9

Bayne K, Hau J, Morris T (2022) The welfare impact of regulations, policies, guidelines and directives and nonhuman primate welfare. In: Robinson LM, Weiss A (eds) Nonhuman primate welfare: from history, science, and ethics to practice. Springer, Cham, pp 629–646

Beisner BA, Isbell LA (2008) Ground substrate affects activity budgets and hair loss in outdoor captive groups of rhesus macaques (*Macaca mulatta*). Am J Primatol 70(12):1160–1168. https://doi.org/10.1002/ajp.20615

Beisner BA, Isbell LA (2011) Factors affecting aggression among females in captive groups of rhesus macaques (*Macaca mulatta*). Am J Primatol 73(11):1152–1159. https://doi.org/10.1002/ajp.20982

Bennett AJ, Corcoran CA, Hardy VA, Miller LR, Pierre PJ (2010) Multidimensional cost-benefit analysis to guide evidence-based environmental enrichment: providing bedding and foraging substrate to pen-housed monkeys. J Am Assoc Lab Anim Sci 49(5):571–577

Bennett AJ, Perkins CM, Tenpas PD, Reinebach AL, Pierre PJ (2016) Moving evidence into practice: cost analysis and assessment of macaques' sustained behavioral engagement with videogames and foraging devices. Am J Primatol 78(12):1250–1264. https://doi.org/10.1002/ajp.22579

Bethea CL, Pau FK, Fox S, Hess DL, Berga SL, Cameron JL (2005) Sensitivity to stress-induced reproductive dysfunction linked to activity of the serotonin system. Fertil Steril 83(1):148–155. https://doi.org/10.1016/j.fertnstert.2004.06.051

Blois-Heulin C, Jubin R (2004) Influence of the presence of seeds and litter on the behaviour of captive red-capped mangabeys *Cercocebus torquatus torquatus*. Appl Anim Behav Sci 85(3–4):349–362. https://doi.org/10.1016/j.applanim.2003.10.005

Bloomsmith MA (2017) Behavioral management of laboratory primates: principles and projections. In: Schapiro SJ (ed) The handbook of primate behavioral management. CRC Press, Boca Raton, FL, pp 497–513

Bloomsmith MA, Lambeth S (1995) Effects of predictable versus unpredictable feeding schedules on chimpanzee behavior. Appl Anim Behav Sci 44(1):65–74. https://doi.org/10.1016/0168-1591(95)00570-I

Bloomsmith M, Perlman J, Franklin A, Martin AL (2022) Training research primates. In: Robinson LM, Weiss A (eds) Nonhuman primate welfare: from history, science, and ethics to practice. Springer, Cham, pp 517–546

Boccia ML (1989) Preliminary report on the use of a natural foraging task to reduce aggression and stereotypies in socially housed pigtail macaques. Lab Prim Newsl 28(1):3–4

Boccia ML, Hijazi AS (1998) A foraging task reduces agonistic and stereotypic behaviors in pigtail macaque social groups. Lab Prim Newsl 37:1–4

Bohm RP, Kreitlein ES, Jack RH, Noel DML (2009) Facilities for non-human primates. In: Hessler JR, Lehner NDM (eds) Planning and designing research animal facilities. Academic Press, New York, pp 289–312

Box HO, Rohrhuber B (1993) Differences in behaviour among adult male, female pairs of cotton-top tamarins (*Saguinus oedipus*) in different conditions of housing. Anim Techn 44:19–30

Brambrell FWR (1965) Report of the technical committee to enquire into the welfare of animals kept under intensive livestock husbandry systems. Her Majesty's Stationery Office, London

Buchanan-Smith HM (2010) Marmosets and tamarins. In: Hubrecht R, Kirkwood J (eds) The UFAW handbook on the care and management of laboratory and other research animals, 8th edn. Wiley-Blackwell, Oxford, UK, pp 543–563

Buchanan-Smith HM, Shand C, Morris K (2002) Cage use and feeding height preferences of captive common marmosets (*Callithrix j. jacchus*) in two-tier cages. J Appl Welf Anim Sci 5(2):39–149. https://doi.org/10.1207/s15327604jaws0502_04

Buchanan-Smith HM, Tasker L, Ash H, Graham ML (2022) Welfare of primates in laboratories: opportunities for improvement. In: Robinson LM, Weiss A (eds) Nonhuman primate welfare: from history, science, and ethics to practice. Springer, Cham, pp 97–120

Byrne GD, Suomi SJ (1991) Effects of woodchips and buried food on behavior patterns and psychological well-being of captive rhesus monkeys. Am J Primatol 23(3):141–151. https://doi.org/10.1002/ajp.1350230302

Capitanio JP, Mendoza SP, Lerche NW, Mason WA (1998) Social stress results in altered glucocorticoid regulation and shorter survival in simian acquired immune deficiency syndrome. Proc Natl Acad Sci U S A 95(8):4714–4719. https://doi.org/10.1073/pnas.95.8.4714

Chamove AS, Anderson JR, Morgan-Jones SC, Jones SP (1982) Deep woodchip litter: hygiene, feeding and behavioral enhancement in eight primate species. Int J Stud Anim Prob 3:308–318

Champoux M, Metz B, Suomi SJ (1989) Rehousing nonreproductive rhesus macaques with weanlings: I. Behavior of adults toward weanlings. Lab Primate Newsl 28(4):4

Champoux M, Hibbeln JR, Shannon C, Majchrzak S, Suomi SJ, Salem N Jr, Higley JD (2002) Fatty acid formula supplementation and neuromotor development in rhesus monkey neonates. Pediatr Res 51(3):273–281. https://doi.org/10.1203/00006450-200203000-00003

Clarence WM, Scott JP, Dorris MC, Paré M (2006) Use of enclosures with functional vertical space by captive rhesus monkeys (*Macaca mulatta*) involved in biomedical research. J Am Assoc Lab Anim Sci 45(5):31–34

Clarke AS, Schneider ML (1993) Prenatal stress has long-term effects on behavioral responses to stress in juvenile rhesus monkeys. Dev Psychobiol 26(5):293–304. https://doi.org/10.1002/dev.420260506

Clingerman KJ, Summers L (2005) Development of a body condition scoring system for nonhuman primates using *Macaca mulatta* as a model. Lab Anim (NY) 34(5):31–36. https://doi.org/10.1038/laban0505-31

Coleman K (2011) Caring for nonhuman primates in biomedical research facilities: scientific, moral and emotional considerations. Am J Primatol 73(3):220–225. https://doi.org/10.1002/ajp.20855

Coleman K, Heagerty A (2019) Human-animal interactions in the research environment. In: Hosey G, Melfi V (eds) Anthrozoology. Oxford University Press, Oxford, pp 59–80

Coleman K, Bloomsmith MA, Crockett CM, Weed JL, Schapiro SJ (2012) Behavioral management, enrichment and psychological well-being of laboratory nonhuman primates. In: Abee CR, Mansfield K, Tardif S, Morris T (eds) Nonhuman primates in biomedical research. Academic Press, London, pp 149–176

Crast J, Bloomsmith MA, Jonesteller T (2015) Effects of changing housing conditions on mangabey behavior (*Cercocebus atys*): Spatial density, housing quality, and novelty effects. Am J Primatol 77(9):1001–1014. https://doi.org/10.1002/ajp.22430

Crockett CM, Bowers CL, Sackett GP, Bowden DM (1993) Urinary cortisol responses of longtailed macaques to five cage sizes, tethering, sedation, and room change. Am J Primatol 30(1):55–74. https://doi.org/10.1002/ajp.1350300105

Crockett CM, Bowers CL, Shimoji M, Leu M, Bowden DM, Sackett GP (1995) Behavioral responses of longtailed macaques to different cage sizes and common laboratory experiences. J Comp Psychol 109(4):368–383. https://doi.org/10.1037/0735-7036.109.4.368

Crockett CM, Shimoji M, Bowden DM (2000) Behavior, appetite, and urinary cortisol responses by adult female pigtailed macaques to cage size, cage level, room change, and ketamine sedation. Am J Primatol 52(2):63–80. https://doi.org/10.1002/1098-2345(200010)52:2%3C63::aid-ajp1%3E3.0.co;2-k

Davis H (2002) Prediction and preparation: Pavlovian implications of research animals discriminating among humans. ILAR J 43(1):19–26. https://doi.org/10.1093/ilar.43.1.19

Descovich KA, Wathan J, Leach MC, Buchanan-Smith HM, Flecknell P, Farningham D, Vick SJ (2017) Facial expression: an under-utilised tool for the assessment of welfare in mammals. ALTEX 34(3):409–429. https://doi.org/10.14573/altex.1607161

Dettmer AM, Novak MA, Meyer JS, Suomi SJ (2014) Population density-dependent hair cortisol concentrations in rhesus monkeys (*Macaca mulatta*). Psychoneuroendocrinology 42:59–67. https://doi.org/10.1016/j.psyneuen.2014.01.002

Doane CJ, Andrews K, Schaefer LJ, Morelli N, McAllister S, Coleman K (2013) Dry bedding provides cost-effective enrichment for group-housed rhesus macaques (*Macaca mulatta*). J Am Assoc Lab Anim Sci 52(3):247–252

Doyle LA, Baker KC, Cox LD (2008) Physiological and behavioral effects of social introduction on adult male rhesus macaques. Am J Primatol 70(6):542–550. https://doi.org/10.1002/ajp.20526

Draper WA, Bernstein IS (1963) Stereotyped behavior and cage size. Percept Motor Skills 16 (1):231–234. https://doi.org/10.2466/pms.1963.16.1.231

Ehardt CL, Bernstein IS (1986) Matrilineal overthrows in rhesus monkey groups. Int J Primatol 7:157–181. https://doi.org/10.1007/BF02692316

Elmore DB, Anderson JH, Hird DW, Sanders KD, Lerche NW (1992) Diarrhea rates and risk factors for developing chronic diarrhea in infant and juvenile rhesus monkeys. Lab Anim Sci 42 (4):356–359

Erard M (2015) What's in a name? Science 347(6225):941–943. https://doi.org/10.1126/science.347.6225.941

Erwin J, Anderson N, Erwin L, Lewis L, Flynn D (1976) Aggression in captive pigtail monkeys groups: effects of provision of cover. Percept Motor Skills 42(1):319–324. https://doi.org/10.2466/pms.1976.42.1.319

European Parliament (2010) Directive 2010/63/EU of the European Parliament and of the Council of 22 September 2010 on the protection of animals used for scientific purposes

Flack JC, Krakauer DC, de Waal FBM (2005) Robustness mechanisms in primate societies: a perturbation study. Proc Biol Sci 272(1568):1091–1099. https://doi.org/10.1098/rspb.2004.3019

Fontenot BM, Wilkes MN, Lynch CS (2006) Effects of outdoor housing on self-injurious and stereotypic behavior in adult male rhesus macaques (*Macaca mulatta*). J Am Assoc Lab Anim 45(5):35–43

Gottlieb DH, Capitanio JP, McCowan B (2013a) Risk factors for stereotypic behavior and self-biting in rhesus macaques (*Macaca mulatta*): animal's history, current environment, and personality. Am J Primatol 75(10):995–1008. https://doi.org/10.1002/ajp.22161

Gottlieb DH, Coleman K, McCowan B (2013b) The effects of predictability in daily husbandry routines on captive rhesus macaques (*Macaca mulatta*). Appl Anim Behav Sci 143 (2–4):117–127. https://doi.org/10.1016/j.applanim.2012.10.010

Gottlieb DH, O'Connor JR, Coleman K (2014) Using porches to decrease feces painting in rhesus macaques (*Macaca mulatta*). J Am Assoc Lab Anim Sci 53(6):653–656

Gottlieb DH, Coleman K, Prongay K (2017) Behavioral management of macaques. In: Schapiro SJ (ed) The handbook of primate behavioral management. CRC Press, Boca Raton, FL, pp 279–303

Grayson DS, Kroenke CD, Neuringer M, Fair DA (2014) Dietary omega-3 fatty acids modulate large-scale systems organization in the rhesus macaque brain. J Neurosci 34(6):2065–2074. https://doi.org/10.1523/JNEUROSCI.3038-13.2014

Griffis C, Martin AL, Perlman JE, Bloomsmith MA (2013) Play caging benefits the behavior of singly housed laboratory rhesus macaques (*Macaca mulatta*). J Am Assoc Lab Anim Sci 52 (5):534–540

Hanson JD, Larson ME, Snowdon CT (1976) The effects of control over high intensity noise on plasma cortisol levels in rhesus monkeys. Behav Biol 16(3):333–340. https://doi.org/10.1016/S0091-6773(76)91460-7

Hawkins P, Morton DB, Burman O, Dennison N, Honess P, Jennings M, Lane S, Middleton V, Roughan JV, Wells S, Westwood K, BVAAWF/FRAME/RSPCA/UFAW UKJWGoR (2011) A guide to defining and implementing protocols for the welfare assessment of laboratory animals: eleventh report of the BVAAWF/FRAME/RSPCA/UFAW Joint Working Group on Refinement. Lab Anim 45(1):1–13. https://doi.org/10.1258/la.2010.010031

Honess PE, Marin CM (2006) Behavioural and physiological aspects of stress and aggression in nonhuman primates. Neurosci Biobehav Rev 30(3):390–412. https://doi.org/10.1016/j.neubiorev.2005.04.003

Honess P, Gimpel J, Wolfensohn S, Mason G (2005) Alopecia scoring: the quantitative assessment of hair loss in captive macaques. Altern Lab Anim 33(3):193–206. https://doi.org/10.1177/026119290503300308

Howard H, Foucher YK (2009) Animal-use space. In: Hessler JR, Lehner NDM (eds) Planning and designing research animal facilities. Elsevier, San Diego, CA, pp 203–261

Jennings M, Prescott MJ, Members of the Joint Working Group on Refinement (Primates) (2009) Refinements in husbandry, care and common procedures for non-human primates: ninth report of the BVAAWF/FRAME/RSPCA/UFAW Joint Working Group on Refinement. Lab Anim 43 (Suppl 1):1–47. https://doi.org/10.1258/la.2008.007143

Judge PG, de Waal FBM (1993) Conflict avoidance among rhesus monkeys: coping with short-term crowding. Anim Behav 46(2):221–232. https://doi.org/10.1006/anbe.1993.1184

Judge PG, de Waal FBM (1997) Rhesus monkey behaviour under diverse population densities: coping with long-term crowding. Anim Behav 54(3):643–662. https://doi.org/10.1006/anbe.1997.0469

Kaufman BM, Pouliot AL, Tiefenbacher S, Novak MA (2004) Short and long-term effects of a substantial change in cage size on individually housed, adult male rhesus monkeys (*Macaca mulatta*). Appl Anim Behav Sci 88(3–4):319–330. https://doi.org/10.1016/j.applanim.2004.03.012

Kelleher SL, Casas I, Carbajal N, Lönnerdal B (2002) Supplementation of infant formula with the probiotic *Lactobacillus reuteri* and zinc: impact on enteric infection and nutrition in infant rhesus monkeys. J Ped Gastroenterol Nutr 35(2):162–168. https://doi.org/10.1097/01.MPG.0000019660.99636.68

Kemp C (2022) Enrichment. In: Robinson LM, Weiss A (eds) Nonhuman primate welfare: From history, science, and ethics to practice. Springer, Cham, pp 451–488

Kitchen AM, Martin AA (1996) The effects of cage size and complexity on the behaviour of captive common marmosets, *Callithrix jacchus jacchus*. Lab Anim 30(4):317–326. https://doi.org/10.1258/002367796780739853

Kramer JA, Ford EW, Capuano S (2012) Preventative medicine in nonhuman primates. In: Abee CR, Mansfield K, Tardif S, Morris T (eds) Nonhuman primates in biomedical research: biology and management, vol 1. Elsevier, London

Lambeth S, Schapiro S, Bernacky B, Wilkerson G (2013) Establishing 'quality of life' parameters using behavioural guidelines for humane euthanasia of captive non-human primates. Anim Welf 22(4):429–435. https://doi.org/10.7120/09627286.22.4.429

Lee YA, Obora T, Bondonny L, Toniolo A, Mivielle J, Yamaguchi Y, Goto Y (2018) The effects of housing density on social interactions and their correlations with serotonin in rodents and primates. Sci Rep 8(1):3497. https://doi.org/10.1038/s41598-018-21353-6

Lehman H (1992) Scientist-animal bonding: some philosophical reflections. In: Davis H, Balfour D (eds) The inevitable bond: examining scientist-animal interactions. Cambridge University Press, Cambridge, pp 383–396

Line SW, Morgan KN, Markowitz H, Strong S (1990) Increased cage size does not alter heart rate or behavior in female rhesus monkeys. Am J Primatol 20(2):107–113. https://doi.org/10.1002/ajp.1350200205

Line SW, Markowitz H, Morgan KN, Strong S (1991) Effects of cage size and environmental enrichment on behavioral and physiological responses of rhesus macaques to the stress of daily events. In: Novak MA, Petto AJ (eds) Through the looking glass: Issues of psychological well-being in captive nonhuman primates. American Psychological Association, Washington, DC, pp 160–179

Ludes E, Anderson JR (1996) Comparison of the behaviour of captive white-faced capuchin monkeys (*Cebus capucinus*) in the presence of four kinds of deep litter. Appl Anim Behav Sci 49(3):293–303. https://doi.org/10.1016/0168-1591(96)01056-8

Luttrell L, Acker L, Urben M, Reinhardt V (1994) Training a large troop of rhesus macaques to cooperate during catching: analysis of the time investment. Anim Welf 3(2):135–140

Lutz CK, Baker KC (2022) Using behavior to assess primate welfare. In: Robinson LM, Weiss A (eds) Nonhuman primate welfare: from history, science, and ethics to practice. Springer, Cham, pp 171–206

Lutz CK, Novak MA (2005) Environmental enrichment for nonhuman primates: Theory and application. ILAR J 46(2):178–191. https://doi.org/10.1093/ilar.46.2.178

Lutz CK, Coleman K, Worlein JM, Kroeker R, Menard MT, Rosenberg K, Meyer JS, Novak MA (2016) Factors influencing alopecia and hair cortisol in rhesus macaques (*Macaca mulatta*). J Med Primatol 45(4):180–188. https://doi.org/10.1111/jmp.12220

MacLean EL, Prior SR, Platt ML, Brannon EM (2009) Primate location preference in a double-tier cage: the effects of illumination and cage height. J Appl Anim Welf Sci 12(1):73–81. https://doi.org/10.1080/10888700802536822

Maier SF, Seligman ME (1976) Learned helplessness: theory and evidence. J Expt Psychol Gen 105 (1):3–46. https://doi.org/10.1037/0096-3445.105.1.3

Manciocco A, Chiarotti F, Vitale A (2009) Effects of positive interaction with caretakers on the behaviour of socially housed common marmosets (*Callithrix jacchus*). Appl Anim Behav Sci 120(1–2):100–107. https://doi.org/10.1016/j.applanim.2009.05.007

Maninger N, Kim JH, Ruppenthal GC (1998) The presence of visual barriers decreases agonism in group housed pigtail macaques (*Macaca nemestrina*). Am J Primatol 45(2):193–194. https://doi.org/10.1002/(SICI)1098-2345(1998)45:2%3C193::AID-AJP6%3E3.0.CO;2-R

Markowitz H, Line S (1989) Primate research models and environmental enrichment. In: Segal EF (ed) Housing, care and psychological wellbeing of captive and laboratory primates. Noyes Publication, Park Ridge, NJ, pp 203–212

McCann C, Buchanan-Smith H, Jones-Engel L, Farmer K, Prescott M, Fitch-Snyder H, Taylor S (2007) IPS international guidelines for the acquisition, care and breeding of nonhuman primates. International Primatological Society

Miller CL, Bard KA, Juno CJ, Nadler RD (1986) Behavioral responsiveness of young chimpanzees to a novel environment. Folia Primatol (Basel) 47(2–3):128–142. https://doi.org/10.1159/000156270

Mineka S, Gunnar M, Champoux M (1986) Control and early socioemotional development in infant rhesus macaques reared in controllable vs uncontrollable environments. Child Dev 57 (5):1241–1256. https://doi.org/10.2307/1130447

Morgan KN, Tromborg CT (2007) Sources of stress in captivity. Appl Anim Behav Sci 102 (3–4):262–302. https://doi.org/10.1016/j.applanim.2006.05.032

Morimura N, Hirata S, Matsuzawa T (2022) Challenging cognitive enrichment: examples from caring for the chimpanzees in the Kumamoto Sanctuary, Japan and Bossou, Guinea. In: Robinson LM, Weiss A (eds) Nonhuman primate welfare: from history, science, and ethics to practice. Springer, Cham, pp 489–516

Murphy KL, Baxter MG, Flecknell PA (2012) Anesthesia and analgesia in nonhuman primates. In: Abee CR, Mansfield K, Tardif S, Morris K (eds) Nonhuman primates in biomedical research: biology and management. Academic Press, London, pp 403–435

National Research Council (1998) The psychological well-being of nonhuman primates. National Academy Press, Washington, DC

National Research Council (2003) Nutrient requirements of nonhuman primates. National Academy Press, Washington, DC

National Research Council (2011) Guide for the care and use of laboratory animals. National Academic Press, Washington, DC

Neuringer M, Connor WE, Van Petten C, Barstad L (1984) Dietary omega-3 fatty acid deficiency and visual loss in infant rhesus monkeys. J Clin Invest 73(1):272–276. https://doi.org/10.1172/JCI111202

Novak MA (2003) Self-injurious behavior in rhesus monkeys: new insights into its etiology, physiology, and treatment. Am J Primatol 59(1):3–19. https://doi.org/10.1002/ajp.10063

Paule MG, Li M, Allen RR, Liu F, Zou X, Hotchkiss C, Hanig JP, Patterson TA, Slikker W Jr, Wang C (2011) Ketamine anesthesia during the first week of life can cause long-lasting cognitive deficits in rhesus monkeys. Neurotoxicol Teratol 33(2):220–230. https://doi.org/10.1016/j.ntt.2011.01.001

Paulk HH, Dienske H, Ribbens LG (1977) Abnormal behavior in relation to cage size in rhesus monkeys. J Abnorm Psychol 86(1):87–92. https://doi.org/10.1037/0021-843X.86.1.87

Pearson BL, Reeder DM, Judge PG (2015) Crowding increases salivary cortisol but not self-directed behavior in captive baboons. Am J Primatol 77(4):462–467. https://doi.org/10.1002/ajp.22363

Prescott MJ, Nixon ME, Farningham DAH, Naiken S, Griffiths M-A (2012) Laboratory macaques: when to wean? Appl Anim Behav Sci 137(3–4):194–207. https://doi.org/10.1016/j.applanim.2011.11.001

Prongay K, Park B, Murphy SJ (2013) Risk factor analysis may provide clues to diarrhea prevention in outdoor-housed rhesus macaques (*Macaca mulatta*). Am J Primatol 75(8):872–882. https://doi.org/10.1002/ajp.22150

Reinhardt V (1997) Refining the traditional housing and handling of laboratory rhesus macaques improves scientific methodology. Primate Rep 49:93–112

Reinhardt V, Reinhardt A (1991) Impact of a privacy panel on the behavior of caged female rhesus monkeys living in pairs. J Exp Anim Sci 34(2):55–58

Rennie A, Buchanan-Smith H (2006) Refinement of the use of non-human primates in scientific research. Part 1: the influence of humans. Anim Welf 15(3):203–213

Rimpley K, Buchanan-Smith HM (2013) Reliably signalling a startling husbandry event improves welfare of zoo-housed capuchins (*Sapajus apella*). Appl Anim Behav Sci 147(1–2):205–213. https://doi.org/10.1016/j.applanim.2013.04.017

Roberts JA, Andrews K (2008) Nonhuman primate quarantine: its evolution and practice. ILAR J 49(2):145–156. https://doi.org/10.1093/ilar.49.2.145

Röder EL, Timmermans PJA (2002) Housing and care of monkeys and apes in laboratories: adaptations allowing essential species-specific behaviour. Lab Anim 36(3):221–242. https://doi.org/10.1258/002367702320162360

Rommeck I, Anderson K, Heagerty A, Cameron A, McCowan B (2009) Risk factors and remediation of self-injurious and self-abuse behavior in rhesus macaques. J Appl Welf Anim Sci 12(1):61–72. https://doi.org/10.1080/10888700802536798

Russell WMS, Burch RL (1959) The principles of humane experimental technique. Methuen, London

Schapiro SJ, Bloomsmith M (2001) Lower-row caging in a two-tiered housing system does not affect the behaviour of young, singly housed rhesus macaques. Anim Welf 10(4):387–394

Schapiro SJ, Lambeth SP (2010) Chimpanzees. In: Hubrecht R, Kirkwood J (eds) The UFAW handbook on the care and management of laboratory and other research animals, 8th edn. Wiley-Blackwell, Oxford, UK, pp 618–633

Schapiro SJ, Bloomsmith MA, Kessel AL, Shively CA (1993) Effects of enrichment and housing on cortisol response in juvenile rhesus monkeys. Appl Anim Behav Sci 37(3):251–263. https://doi.org/10.1016/0168-1591(93)90115-6

Schapiro SJ, Porter LM, Suarez SA, Bloomsmith MA (1995) The behaviour of singly-caged, yearling rhesus monkeys is affected by the environment outside of the cage. Appl Anim Behav Sci 45:151–163. https://doi.org/10.1016/0168-1591(95)00597-L

Schapiro SJ, Stavisky R, Hook M (2000) The lower-row cage may be dark, but behaviour does not appear to be affected. Lab Prim Newsl 39(1):4–6

Schapiro SJ, Perlman JE, Thiele E, Lambeth S (2005) Training nonhuman primates to perform behaviors useful in biomedical research. Lab Anim (NY) 34(5):37–42. https://doi.org/10.1038/laban0505-37

Seier J, De Villiers C, Van Heerden J, Laubscher R (2011) The effect of housing and environmental enrichment on stereotyped behavior of adult vervet monkeys (*Chlorocebus aethiops*). Lab Anim (NY) 40(7):218–224. https://doi.org/10.1038/laban0711-218

Seligman ME (1968) Chronic fear produced by unpredictable electric shock. J Comp Physiol Psychol 66(2):402–411. https://doi.org/10.1037/h0026355

Springer DA, Baker KC (2007) Effect of ketamine anesthesia on daily food intake in *Macaca mulatta* and *Cercopithecus aethiops*. Am J Primatol 69(10):1080–1092. https://doi.org/10.1002/ajp.20421

Steinmetz HW, Kaumanns W, Dix I, Heistermann M, Fox M, Kaup FJ (2006) Coat condition, housing condition and measurement of faecal cortisol metabolites-a non-invasive study about alopecia in captive rhesus macaques (*Macaca mulatta*). J Med Primatol 35(1):3–11. https://doi.org/10.1111/j.1600-0684.2005.00141.x

Sullivan EL, Grayson B, Takahashi D, Robertson N, Maier A, Bethea CL, Smith MS, Coleman K, Grove KL (2010) Chronic consumption of a high-fat diet during pregnancy causes perturbations in the serotonergic system and increased anxiety-like behavior in nonhuman primate offspring. J Neurosci 30(10):3826–3830. https://doi.org/10.1523/JNEUROSCI.5560-09.2010

Tasker L (2012) Linking welfare and quality of scientific output in cynomolgus macaques (Macaca fascicularis) used for regulatory toxicology, Doctoral dissertation. University of Stirling

Ulyan MJ, Burrows AE, Buzzell CA, Raghanti MA, Marcinkiewicz JL, Phillips KA (2006) The effects of predictable and unpredictable feeding schedules on the behavior and physiology of captive brown capuchins (*Cebus apella*). Appl Anim Behav Sci 101(1–2):154–160. https://doi.org/10.1016/j.applanim.2006.01.010

van Wolkenten ML, Davis JM, Gong ML, de Waal FBM (2006) Coping with acute crowding by *Cebus apella*. Int J Primatol 27(5):1241–1256. https://doi.org/10.1007/s10764-006-9070-z

Vandeleest JJ, McCowan B, Capitanio JP (2011) Early rearing interacts with temperament and housing to influence the risk for motor stereotypy in rhesus monkeys (*Macaca mulatta*). Appl Anim Behav Sci 132(1–2):81–89. https://doi.org/10.1016/j.applanim.2011.02.010

Wadman M (2012) Activists ground primate flights. Nature 483:381–382. https://doi.org/10.1038/483381a

Waitt C, Buchanan-Smith HM (2001) What time is feeding? How delays and anticipation of feeding schedules affect stump-tailed macaque behavior. Appl Anim Behav Sci 75(1):75–85. https://doi.org/10.1016/S0168-1591(01)00174-5

Waitt C, Buchanan-Smith HM, Morris K (2002) The effects of caretaker-primate relationships on primates in the laboratory. J Appl Anim Welf Sci 5(4):309–319. https://doi.org/10.1207/s15327604jaws0504_05

Westlund K, Fernström AL, Wergård EM, Fredlund H, Hau J, Spångberg M (2012) Physiological and behavioural stress responses in cynomolgus macaques (*Macaca fascicularis*) to noise associated with construction work. Lab Anim 46(1):51–58. https://doi.org/10.1258/la.2011.011040

Williams LE, Abee CR, Barnes SR, Ricker RB (1988) Cage design and configuration for an arboreal species of primate. Lab Anim Sci 38(3):289–291

Williams LE, Brady AG, Abee CR (2010) Squirrel monkeys. In: Hubrecht R, Kirkwood J (eds) The UFAW handbook on the care and management of laboratory and other research animals, 8th edn. Wiley-Blackwell, Oxford, UK, pp 564–578

Wolfensohn S (2010) Old World monkeys. In: Hubrecht R, Kirkwood J (eds) The UFAW handbook on the care and management of laboratory and other research animals, 8th edn. Wiley-Blackwell, Oxford, UK, pp 592–617

Wolfle TL (2002) Implications of human-animal interactions and bonds in the laboratory. ILAR J 43(1):1–3. https://doi.org/10.1093/ilar.43.1.1

Housing and Husbandry for Primates in Zoos

H. L. Farmer, K. R. Baker, and F. Cabana

Abstract

Where once it was common to house primates singly in concrete enclosures, it has since been recognized that nonhuman primates have complex behavioral needs. The purpose of this chapter is to provide an overview of the main housing and husbandry considerations for zoo-housed primates. We first address the need to balance zoo goals, ensuring good animal welfare, conservation, education, research, and entertainment within enclosure design. We then discuss some specific design considerations such as space and complexity. We end on a brief discussion regarding diet and nutrition as this is perhaps an area which requires further investigation when considering the welfare of primates in zoos.

Keywords

Behavior · Enclosure design · Enclosure complexity · Diet and nutrition

1 Overview and Introduction

Finding a single solution to meet housing and husbandry considerations for primate species in zoos is a difficult if not impossible task. The sheer number of species held poses a problem; as of the end of 2019, the European Association of Zoos and Aquaria (EAZA) lists 81 primate species, including representatives from all families of nonhuman primates, for which breeding programs are established. Moreover,

---

H. L. Farmer (✉) · K. R. Baker
Wild Planet Trust, Newquay Zoo, Newquay, Cornwall, UK
e-mail: holly.farmer@wildplanettrust.org.uk

F. Cabana
Wildlife Reserves Singapore, Singapore, Singapore

there are many primate species for which established conservation breeding programs have not been established. Given morphological differences and differences in life histories, we need to devote considerable thought to how to care for individual species. For example, because of differences in body size, the absolute exhibit space required for ring-tailed lemurs (*Lemur catta*) will be less than that required for chimpanzees (*Pan troglodytes*). However, both exhibits would need to house a large social group and ensure space to allow appropriate social interactions and separation of individuals should that be required. Compare these to the requirements of orangutans, a primarily solitary, large, great ape, which may require many large interlinked enclosures to allow for appropriate social encounters while making it possible to separate individuals when that is required.

There is thus no single approach to primate housing and husbandry, but since zoos first opened there has been a move from menagerie-type enclosures to naturalistic and immersive exhibits, which focus on meeting good animal welfare standards (Coe 2003, Hosey 2022). With this progression, there has also come an emphasis on producing evidence-based housing and husbandry guidelines for individual primate species/taxa. These guidelines are often produced through behavioral research and multizoo questionnaires, or through consultation with industry experts.

2 Regulations, Guidelines, and Accreditation

Zoos operate within varying legislative frameworks, internationally, regionally, and nationally (Bayne et al. 2022). The legal side of zoo welfare provides a structured and simplified assessment of animal welfare in zoos. Therefore, species-specific best-practice guidelines are written for animals managed as part of captive breeding programs. These guidelines cover aspects of husbandry, such as nutritional requirements, enclosure facilities, disease intervention, and breeding behaviors. However, because the zoo industry tries to be inclusive with regard to what individual zoos can achieve, these guidelines are often produced by surveying current practice, and so provide only a minimum standard for welfare, and one that is not backed by scientific evidence (Melfi 2005). A recent paper by Veasey (2017) reviewed 19 Association of Zoos and Aquariums (AZA) care manuals and found that within these documents, animal training accounted for 15% of content, veterinary care for 26%, and only 16% was dedicated to the physical and social environment of the animal. Veasey argued that this imbalance is cause for concern as many veterinary practices can negatively affect animals' welfare. The paper concludes by recommending that zoos prioritize what is meaningful to an animal over what welfare indices zoos can measure.

3 Balancing Housing and Husbandry Requirements with Zoo Mission

As mentioned in our previous chapter (Baker and Farmer 2022), modern zoological collections have five primary goals, namely animal welfare, conservation, education, research, and entertainment. In terms of enclosure design, these goals could be thought of as distinct. For example, zoo visitors want to be able to see animals, but to achieve good animal welfare, zoos must provide areas where animals can be "off-show." These goals should not, however, be considered mutually exclusive for with careful design and planning, all five goals can be met. For example, Sulawesi crested macaques (*Macaca nigra*) are classed as critically endangered by the International Union for the Conservation of Nature (Lee et al. 2020). The Wild Planet Trust currently runs the EAZA Ex situ Programme (EEP) for the species, houses two breeding groups of *M. nigra* in their zoos (Paignton Zoo and Newquay Zoo, both in the UK), and initiated the Selamatkan Yaki in situ conservation project in Sulawesi. By contributing to in situ conservation efforts and ex situ breeding programs, the Trust is achieving its conservation aim. The two captive groups of Sulawesi macaques held in these zoos participate in research projects that cover topics including behavior and nutrition, and so the Trust also meets their welfare and research goals. Moreover, the Trust meets educational goals by means of educational interpretation boards at the enclosures, which convey information about the captive animals and conservation efforts in the wild, and by keeper talks, which are very popular. Both enclosures also have separate indoor and outdoor areas, which are structurally complex and allow individuals to be "off-show" if they choose, thus meeting animal welfare needs.

In articles that evaluate an extensive exhibit renovation for gorillas and chimpanzees at Lincoln Park Zoo, Chicago, Illinois, USA, the authors document how a change from a traditional to a naturalistic exhibit design affects animal behavior and welfare (Ross et al. 2009, 2011b), visitor behavior (Ross and Lukas 2005; Ross et al. 2012), and visitor knowledge and attitudes (Lukas and Ross 2014). Briefly, the Lester E. Fisher Great Ape House underwent significant renovation in 2002 and reopened in 2004 as the Regenstein Center for African Apes (RCAA). While the Great Ape House met the requirements for modern ape enclosures and was considered an innovative facility, the renovations increased both the available exhibit space and the complexity of furnishings. These improvements led to a decrease in abnormal behaviors and visual monitoring of the public and keepers (attention behaviors) in chimpanzees, and a decrease in agonistic, inactive, and attention behaviors in gorillas. In addition, visitors spent twice as long at the naturalistic exhibit than they did at the traditional exhibit. They also spent longer in absolute terms watching the apes and reading interpretation boards. On the other hand, the proportion of time visitors spent watching apes was less in the new exhibit. The authors suggest that the increased proportion of time that visitors spent engaged in "other" behaviors was due to people interacting with members of their group and demonstrated a comfort with their surroundings, both of which can be important for learning (Lukas and Ross 2014). There was also a decline in disruptive behavior by the visitors, such as banging on the exhibit's glass windows. When visitors' knowledge about and attitudes toward apes were surveyed, visitors demonstrated knowledge gains after their visit, but this gain was no higher than that for the traditional exhibit (Lukas and Ross 2014). However, visitors to the RCAA demonstrated more ecoscientistic attitudes, that is, they were more likely to agree with statements such as "I would enjoy learning about the ecosystem or population dynamics of wild gorillas/chimpanzees," toward chimpanzees, suggesting that naturalistic exhibits may be more effective in raising concerns for nature and conservation (Lukas and Ross 2014). These examples demonstrate how a shift to more functionally and aesthetically naturalistic enclosures (Coe 2003) may benefit both animal welfare and visitor experience alike, thus achieving the goals of modern zoological collections.

3.1 Behavioral Needs of Zoo-Housed Primates

As previously outlined, the behavioral needs of primates in zoos will differ within and between species, i.e., ape species ultimately need more space than callitrichid species, whereas differences in locomotor behavior may influence the furnishings required to meet behavioral needs; for example, arboreal primates will require tall climbing structures.

Rather than basing decisions on phylogenetic relatedness by, for example, assuming that all prosimian species will have similar requirements, enclosures must be designed (or adapted) for species based on evidence obtained from research on wild and captive members of that species. An example of this kind of research is a comparative study by Pomerantz et al. (2013) who identified factors associated with stereotypic hair pulling and pacing in zoo-housed primate species. They found that similarity in the degree of stereotypy that individuals exhibited was not affected by phylogenetic relatedness but by natural group size. This finding suggests that management decisions should be guided by a species' biology/ecology and not on what works for closely related species and what does not work for distantly related species.

We will next cover some specific issues in zoo primate housing and husbandry. Other chapters in this volume (e.g., Talbot et al. 2022) address species-specific cognitive, behavioral, social, and other needs, and each of these must also be considered in the housing and husbandry of zoo-housed primates.

3.2 Accommodating Zoos' Exhibition Needs

One of the largest and more unique challenges for zoos is coping with the daily influx of visitors and meeting their expectations. Visitors want to see animals, but in most zoos, primate species are behind barriers, such as mesh fencing or glass. Some smaller-bodied primate species; however, may have free-ranging access to all or specific parts of an enclosure. In the Secretary of State's Standards for Modern Zoo

Practice, UK (DEFRA 2015), zoo animals are categorized into risk levels "on the basis of the animal's likely ferocity and ability to cause harm to people, and the scale of harm if it should do so." For example, great ape species are classed as category 1 (greater risk), where "contact between the public and animals is likely to cause serious injury or be a serious threat to life." They must therefore "be separated from the public by a barrier of suitable design." Callitrichids are in category 2 (less risk), where "contact between the public and animals may result in injury or illness" and animals "would normally be separated from the public by a barrier" but "given adequate space and refuge."

In response to the 2005 International Symposium on Zoo Design, Melfi et al. (2005) suggested that many modern zoo enclosures do not meet the needs of animals and they do those of visitors and staff. The authors suggested that the visitor experience has become the priority for enclosure design considerations. This may be because visitors are the primary source of income for most zoos and, as such, visitor perceptions of enclosures are considered high priority, but it may also be because what an animal requires from an enclosure is mostly based on anecdote and not evidence.

Visitor perceptions of animal welfare and behavior are particularly important because the public will often evaluate the welfare of animals through visual interpretation of an enclosure. A survey conducted at Paignton Zoo asked 42 zoo visitors to evaluate photographs of different tiger and primate enclosures. The photographs showed enclosures with varying levels of vegetation, barriers, and naturalistic materials. The survey revealed that respondents viewed more naturalistic enclosures as being better; all respondents rated the greenest enclosure highly and thought its inhabitants would have the best welfare (Melfi et al. 2004). In response, it is suggested that enclosures that have more naturalistic planting and fewer physical barriers may be perceived by the public as more beneficial to animals; however, such enclosures, which include free-ranging exhibits, involve a potentially higher level of interaction between the animals and visitors, which poses welfare problems.

Zoo managers and marketing staff often favor housing primates in free-ranging exhibits. Their view seems to be that proximity to the animals will enhance the visitor experience. Research on three free-ranging exhibits at the Singapore Zoo which housed white-faced sakis (*Pithecia pithecia*), cotton-top tamarins (*Saguinus oedipus*), and orangutans (*Pongo pygmaeus*) found high rates of visitor–animal interactions, including negative interactions, at the cotton-top tamarin enclosure where visitors could get in closest contact with the animals. However, for all species, visitor responses were positive in terms of enjoyment, learning, and perceptions of good welfare (Mun et al. 2013). Along the same lines, Bryan et al. (2017) noted that, in addition to the benefits for visitors, free-ranging exhibits provide primates with a more naturalistic setting and may therefore promote species-typical behaviors and animal welfare.

3.3 Accommodating Zoo's Research Needs

Primates are popular research subjects and most zoos either carry out research or allow visiting researchers to use their collections. There have been recent collaborative efforts between zoos and academic institutions that demonstrate ways in which enclosure design can incorporate research needs and possibly improve welfare. The Living Links to Human Evolution Research Centre at the Edinburgh Zoo is managed in collaboration between the Royal Zoological Society of Scotland and the University of St Andrews, and the Scottish Primate Research Group. The center operates as a research center and educational exhibit, and it has resulted in high-quality research on mixed-species groups (Buchanan-Smith et al. 2013), social learning and traditions (e.g., Claidière et al. 2013), and personality (e.g., Morton et al. 2013). Researchers at Birmingham University are working alongside Twycross Zoo and the British and Irish Association of Zoos and Aquariums (BIAZA) to devise an Enclosure Design Tool for chimpanzees, to ensure that captive animal behavior mimics that of wild chimpanzees (University of Birmingham, 2020). These examples demonstrate the importance of primate research within zoos. Enclosures can be built or modified to enhance research, but high-quality studies can also be conducted in existing enclosures.

4 Specific Enclosure Design and Husbandry Considerations

There are a range of resources that need to be included within enclosures to enhance the management of zoo primates. Best-practice guidelines for primate enclosure design are achieved through communication between zoos, conference presentations, stakeholder meetings, and personal communication. Empirical evidence regarding the success or failure of specific resources should also feed into best-practice guidelines. We outline some of the main considerations below.

4.1 Space

While important resources, such as sleeping and feeding sites, are contained within enclosures, space itself is a valuable resource. Concerns over too little space in captive environments are often focused on taxa, such as carnivores, with large ranging patterns (Clubb and Mason 2003). The relationship between home range size and travel patterns, and optimal enclosure size, may also be applicable to captive primates (Pomerantz et al. 2013). While it is generally accepted that there must be minimum guidelines for space requirements per animal, what these requirements should be is difficult to agree on. This is true even for species such as chimpanzees, whose home range sizes and travel patterns in the wild are well-known. The American Association of Zoos and Aquariums (AZA) Ape Taxon Advisory Group recommends outdoor space of at least 185 m^2 for groups of five or fewer chimpanzees and an additional 93 m^2 for every additional individual. The National

Institutes of Health recommend that chimpanzees be provided with 23 m^2 per animal and the US Department of Agriculture states that spaces must exceed 2.32 m^2 per animal (Ross and Shender 2016).

While absolute space is important, preventing a perceived reduction in space, that is, when animals are forced into proximity with each other, is just as important a consideration in enclosure design. The impact of restricted space on primate species is complex because of their sophisticated social behavior; primates have numerous coping strategies that they use in times of social stress (de Waal 1989). Two such strategies are "tension reduction," which involves increasing positive social behaviors, and "conflict avoidance," which involves avoiding conspecifics in the same enclosure and increasing the frequency of nonsocial behaviors (Judge et al. 2006).

Individual primates in zoos will, at some point, experience changes in their social situation. This may happen naturally through births and deaths in the social group or come about because of the actions of human caregivers, such as zoo transfers. Ultimately, these changes require not only space for animals to retreat but also complexity in enclosure design that meets the needs of all individuals. A study examined the behaviors that took place during the introduction of 11 chimpanzees from Beekse Bergen Zoo in the Netherlands into Edinburgh Zoo's Budongo Trail enclosure, which held 11 chimpanzees. Budongo Trail is a purpose built enclosure incorporating many enclosure areas to allow for fission–fusion social dynamics. The researchers found that the number of accessible enclosure areas rather than the absolute amount of space had a greater effect in reducing the frequency of stress-related behaviors, such as scratching, when the introductions took place (Herrelko et al. 2015). This suggests that having an increased choice of locations ameliorates the impact of increased social density (Herrelko et al. 2015). Incorporating different areas in new enclosures like Budongo Trail is an important consideration but even in existing enclosures where development is limited due to physical or financial constraints, increasing the number of perceived enclosure locations through visual barriers such as vegetation screens may be a key management tool during social changes.

Zoos are often required to house animals temporarily within off-show holding areas for short periods as part of general husbandry procedures, or for longer periods, such as prior to or after transportation. These areas are often smaller in size than on-show facilities. During such times, depending on the composition of the group, there may be a period of stress and aggression. The behavior of gorillas and chimpanzees housed in Lincoln Park Zoo was compared when the animals were housed in their indoor on-show enclosure (109 and 124 m^2, respectively) to their comparatively small off-show holding exhibits, comprised of six 10.8 m^2 rooms. The gorillas displayed more activity and more affiliative behavior when housed in the smaller, off-show exhibit. The chimpanzees, on the other hand, displayed more aggressive behavior and engaged in more self-directed behavior (Ross et al. 2010) suggesting that zoo off-show facilities must cater for the behavioral needs of all animals.

It is widely recognized that, although the absolute space available must meet minimum requirements for species' health and well-being, the quality of the space is also critical (Hosey 2005; Ross et al. 2011a). Choice and control provided through exhibit design may be as simple as providing indoor and outdoor areas, to incorporating technologies that allow animals to control aspects of their environment such as temperature, humidity, or even whether music is playing. In the following sections, we discuss broad issues related to increasing the environmental complexity and the functionality of primate enclosures.

4.2 Retreat Areas

As reported by Baker and Farmer (2022) and above, visitor presence affects primates' welfare in positive, neutral, and negative ways. The most commonly reported visitor effect is negative (Hosey 2005), and so enclosures need to be designed to mitigate this effect. Kuhar (2008) reported that when they were exposed to large crowds, the gorillas (*Gorilla gorilla gorilla*) at Disney's Animal Kingdom hid behind denser vegetation in the enclosure. A simple way to reduce visitor effects is to provide a visual barrier between the visitors and animals; for example, camouflage netting was a successful addition to the gorilla enclosure at Belfast Zoo. After the barrier was implemented, there was a significant decrease in conspecific-directed aggression and stereotypies (Blaney and Wells 2004). This, and other examples (e.g., Sherwen et al. 2015), suggests that providing visual refuge may benefit animals that respond in a negative way to visitors, including animals housed in enclosures that promote animal–human proximity.

4.3 Indoor Versus Outdoor Facilities

As a rule, most zoos provide both indoor and outdoor areas for their primates. However, small neotropical primates and/or nocturnal primates can be kept solely in indoor enclosures, which come with their own challenges (see below).

For the moment, we will consider enclosures that provide indoor and outdoor space. The provision of indoor and outdoor areas is important for several reasons. Firstly, allowing animals choice to be inside or outside adds a level of complexity to the daily lives of the animals. Secondly, it provides areas where animals can escape to should they need or wish to remove themselves from stressful social situations. Thirdly, time spent outside exposes individuals to natural sunlight, which is important for health and physiological function (Coleman et al. 2022). The resources provided in indoor and outdoor areas may differ greatly. Indoor areas often have hard floors that may be covered in substrates, such as bark, or hay, incorporate dens/

sleeping sites, have artificial furnishings, and are close to management/service areas. Outdoor enclosures tend to be larger and have more natural substrates, such as grass/soil, and furnishings may be more natural.

Evaluating how zoo primates utilize indoor and outdoor areas is key to ensuring welfare and for improving enclosure designs in the future. For example, a multizoo study of Sulawesi macaques found that the animals spent an average of 44% of their time indoors regardless of season, yet indoor areas represented only 5% of the total enclosure area (Melfi and Feistner 2002). In other species or situations, different preferences may be seen. For example, common marmosets (*Callithrix jacchus*) at a university research facility showed a strong preference for outdoor areas, choosing to spend almost 70% of their time outdoors (Pines et al. 2007).

A recent in-depth study by Ross et al. (2011b) evaluated how two groups of chimpanzees ($n = 7$ and 12) and two groups of gorillas ($n = 7$ and 4) used enclosures at the Lincoln Park Zoo. All four groups had naturalistic enclosures consisting of an indoor dayroom (101.4 m^2) and an outdoor yard (585.5 m^2). The day rooms had a deep mulch substrate, 8-m-high meshed ceilings, elevated platforms, artificial climbing structures, ropes, and vines. The yards had grass substrate, varied topography, and both natural and artificial climbing apparatus. Because the apes were provided access to outdoor areas only when the temperature exceeded 5 °C, the authors were able to compare enclosure use under two conditions: (i) when access to the yard was restricted because it was too cold outside and (ii) when the apes had access to the yard. Over the four-year study period (2004–2008), chimpanzees were seen in 56% of available quadrants and spent over half of their time in just 3% of enclosure areas. Gorillas were seen in only 28% of quadrants and spent half their time in just 2% of the enclosure area. Moreover, when they were given access to outdoor areas, the chimpanzees spent about a third of their day in outdoor areas while the gorillas spent only 7% of their day outdoors. A later study compared the behavior of the same chimpanzees and gorillas at Lincoln Park Zoo when individuals could utilize outdoor areas and when they could not (Kurtycz et al. 2014). During times when individuals could access the outdoor areas, the chimpanzees exhibited more prosocial behavior, less inactivity, and more self-directed behavior during the same time, the gorillas manipulated objects, fed less often, and were less active.

These findings by Ross et al. (2011b) and Kurtycz et al. (2014) suggest that captive nonhuman primates may prefer indoor enclosures, which means that, to meet welfare needs, new enclosure designs must allocate more space to indoor areas, for example, by making use of vertical space (see below and Morimura et al. 2022). Moreover, the results of Ross et al. (2011b) and Kurtycz et al. (2014) suggest that, although the effects may differ between species, for example, between chimpanzees and gorillas, the ability to access outdoor areas may benefit primate welfare as it provides them with a choice of environment.

4.4 Indoor Enclosures

Providing access to outdoor enclosures can benefit the health of captive nonhuman primates. Exposure to ultraviolet (UV) light has many health benefits as it encourages the metabolism of vitamin D3. Little research has examined the effect of UV light on mammals. For many nonhuman primate species, including Japanese macaques (*Macaca fuscata* Hanya et al. 2007) and ruffed lemurs (*Varecia variegata* Morland 1993), there is evidence that basking is important for thermoregulation.

In the past, because of vitamin D deficiency, captive primates have suffered from defective bone growth and demineralization (Power et al. 1995), known commonly as rickets. For example, three juvenile colobus monkeys (*Colobus guereza kikuyuensi*) reared solely indoors with no access to UV light were reported to have developed rickets (Morrisey et al. 1994), with similar findings in chimpanzees (Junge et al. 2000). In captive primates, vitamin D supplementation reportedly eradicates and prevents the development of rickets (Power et al. 1995), and so vitamin D supplementation in captive primates is a common practice. However, there are risks associated with vitamin D supplementation, including hypervitaminosis D, hypercalcemia, and hyperphosphatemia, all of which lead to the mineralization of muscles and other soft tissues, which can be fatal (see review by Dittmer and Thompson 2011).

Many zoos that house species that are known to get exposure to sunlight in the wild are located in countries where UV levels are low for many months of the year. In response, many of these zoos have begun to investigate the benefits of providing artificial UV lighting. For example, a study at Jersey Zoo found that pied tamarins (*Saguinus bicolor*) that were exposed to UV without dietary supplementation had higher vitamin D3 serum levels than pied tamarins that were given dietary supplements (López et al. 2001). Findings such as this suggest that it may be preferable to provide animals with artificial UV lighting than to supplement their diet. However, there are drawbacks to this approach, including that of chronic exposure to unsafe levels of UV radiation, which has been shown to cause the development of corneal clouding, retinal lesions, and lens cataracts in rhesus macaques (*Macaca mulatta*; Zuclich 1984). Therefore, further research into the effects of UV provision is required to provide evidence-based guidelines for its application in captive settings.

Lighting is also a strong consideration for the exhibition of nocturnal primates such as potto (*Perodicticus potto*), aye-aye (*Daubentonia madagascariensis*), bush babies (Galago spp.), and species of loris (*Loris* and *Nycticebus* spp.). Nocturnal species are most commonly housed indoors under a reverse lighting system with a range of colors (red, blue, and/or white) provided in the dark phase. Research on aye-ayes at two US zoos found that in nocturnal exhibits, time spent active was significantly lower when animals were housed under blue compared to red light (Fuller et al. 2016). This study also found that salivary melatonin, which functions as a "time-keeping" hormone that maintains individuals' daily rhythms (Mirick and Davis 2008), was lower in the blue than in the red light condition (Fuller et al. 2016). A review of husbandry practices in North American zoos reported that loris species were exposed to equal amounts of red and blue lighting (Fuller et al. 2013), whereas the current husbandry manual recommends that captive lorises are exposed to a full light spectrum during the daytime and full spectrum or red light at night (Fitch-Snyder and Schultze 2001). These findings suggest that a wider review of the effect of current lighting practices on nocturnal primate behavior is necessity to promote animal welfare and the implementation of appropriate husbandry.

4.5 Three-Dimensional Structures

Many primate species are primarily arboreal. Most primate species also have a vertical flight response in that they climb up when alarmed (Milller and Treves 2006; Caws et al. 2008). Providing three-dimensional space for primate species is therefore as important as providing large areas in terms of square footage. For example, gibbon brachiation is an adaptation for life in an arboreal environment and zoo environments must enable this type of locomotion. Modifications of a habitat for two siamang gibbons (*Symphalangus syndactylus*) and six white cheeked gibbons (*Nomascus leucogenys*), at the Smithsonian's National Zoo, Washington, USA, involved the addition of a log bridge, hammocks, and a pulley system (to enable feeding at higher levels) within the enclosure. After these modifications to the enclosure, the gibbons spent more time at higher levels in the enclosure and more time active (Anderson 2014).

Zoo designers at the Fort Wayne Children's Zoo, Indiana, USA, came up with a unique orangutan enclosure in that the floor of the enclosure was flooded to replicate the orangutans' natural habitat and to encourage arboreal movement. Researchers studied the behavior of three adolescents (two females and one male) over a 10-month period. For their analysis, they split the enclosure into four vertical levels: top, upper canopy, lower canopy, and floor. All the orangutans used the upper canopy the most, followed by the lower canopy, and the top and bottom levels, respectively. There were also behavioral differences associated with the different locations. For example, orangutans were most solitary and inactive in the upper canopy and most social and active in the lower canopy (Hebert and Bard 2000).

Vertical structures may also meet an important social need as they provide escape routes for primates which are trying to avoid aggression from conspecifics. For example, during Sulawesi macaque introductions at Newquay Zoo, UK, keepers provided branches that would support the weight of adult females and juveniles but not the weight of the large dominant male (Baker, *personal observation*). After adding 50 vertical poles connected by ropes and nets to a chimpanzee exhibit at Chester Zoo, UK, the proportion of severe aggressive encounters, that is, those that included biting and physical fighting, initiated indoors, significantly declined because the recipient of aggression used these structures to escape (Caws et al. 2008).

4.6 Enclosure Materials and Substrates

The shift to more naturalistic enclosures in zoos is thought to convey a better message to the public and to better meet animal welfare needs. However, if enclosures are designed to be more naturalistic purely for aesthetic reasons, they may offer little to the animals (Sheperdson and Mellen 1999; Melfi 2005). To illustrate this, Fàbregas et al. (2012) evaluated 1381 enclosures in 63 Spanish zoos and rated them in terms of how naturalistic they were and whether they were suitable for the species. The degree of suitability in this instance was based on whether the enclosures provided environmental resources that were relevant to the species' biology. Of the 153 enclosures that were considered naturalistic, 78% were deemed suitable for the species, and of the 1228 enclosures that were considered non-naturalistic, 40% were deemed as suitable for the species. This suggests that naturalistic enclosures are better at meeting species needs, although it is worrying that the complex design process behind naturalistic enclosures is not meeting welfare needs in 22% of the cases. Moreover, the fact that 40% of non-naturalistic enclosures were classed as suitable indicates that, if they are designed with the species' biology in mind, non-naturalistic enclosures can meet welfare needs, too. Seventy-seven enclosures were for primate species and the aspect that was most commonly unfulfilled was the provision of suitable space and structures to allow species-specific locomotor behaviors. This highlights the need to consider elements such as three-dimensional space and complexity that we have discussed here.

Depending on a zoo's design ethos, materials used for structures, shelters, and other resources within an enclosure may be artificial (e.g., hammocks made from recycled firehose), natural (e.g., vegetation used as climbing structures), or artificial materials that are made to look natural (e.g., fake rockwork). Regardless of material, it is essential that these resources serve a biological purpose. For example, vegetation, rope, wood, and firehose can all be used for climbing but what is biologically relevant to the primates is that they can climb regardless of material.

That said, the choice of material should be guided by the species' biology. For example, a study of arboreal substrate use by grey mouse lemurs (*Microcebus murinus*) showed that their strategy for grasping small, mobile food items, such as insects, was affected by substrate diameter (Toussaint et al. 2013). Compared to a narrow substrate (one which individuals could wrap their hands and feet around), when foraging for prey on a wide substrate (one which individuals could not wrap their hands and feet around completely), there was greater variability in the grasping technique (Toussaint et al. 2013). While this study was conducted in an experimental facility of the Museum National d'Histoire Naturelle, France, for at least this species, there are implications for the design of zoo exhibits, namely that a range of substrate diameters should be provided to encourage a range of foraging strategies and locomotory behaviors.

Flooring substrate is another area of enclosure design that has received attention. Chamove et al. (1982) provided eight species of primate (moustached guenons *Cercopithecus cephus*, vervets *C. aethiops*, ring-tailed lemurs *Lemur catta*, stump-tailed macaques *Macaca arctoides*, squirrel monkeys *Saimiri sciureus*, black-capped capuchins *Cebus apella*, red-bellied tamarins *Saguinus labiatus*, and common marmosets *Callithric jacchus*) with deep litter woodchip substrate across three conditions: woodchip, woodchip and grain, and woodchip and mealworms. Compared to bare floor, all species spent more time on the ground when it was covered with woodchip. When grain or mealworms were added, the animals spent even more time on the ground. In addition, compared to the bare floor condition, the researchers observed more positive social behavior in the woodchip condition and fewer negative social interactions in all woodchip and woodchip plus food conditions.

Ensuring that enclosure structures and furnishings are appropriate for each species is a key task for those involved in enclosure design. While broad requirements may be similar for some species, consultation of a wide range of resources is needed to ensure that species-specific and individual needs are met.

4.7 Mixed-Species Exhibits

The housing of multiple primate species within the same exhibit is common practice for small-bodied species, such as callitrichids and lemurs. It is less common for larger species, such as apes. The decision on whether to house different species together is based on many factors. These include whether the species are found in the same habitat in the wild, the temperament of the species and individuals, and the ability for the zoo to reduce competition for resources, which may not be as much of a problem if the different species have different diets in the wild. If a decision is made to establish this type of exhibit, the housing design is essential to ensure success of the mix and that welfare needs are met. For example, all species must be able to use a large proportion of the exhibit area and have the choice to remove themselves from interspecies interactions if they choose to do so.

Successful mixed-species exhibits often have multiple exhibit areas. For example, the shared pygmy marmoset and Goeldi's monkey (*Callimico goeldii*) exhibit at Edinburgh Zoo, consists of five sets of enclosures with spacious indoor and outdoor areas joined by overhead ducting, and a wide variety of furnishings at different orientations (Dalton and Buchanan-Smith 2005). Even larger primate species can be successfully housed in mixed-species exhibits with careful management and planning. In the Adelaide Zoo, Australia, a female orangutan (of a pair) frequently engages in affiliative behaviors with a pair of siamang gibbons in the mixed exhibit (Pearson et al. 2010).

Wojciechowski (2004) details what happened when five red-capped mangabeys were introduced into a tropical house exhibit that housed three African primate species (black and white colobus monkeys *Colobus guereza*, mandrills *Mandrillus sphinx*, and sooty mangabeys *Cercocebus atys*). The introduction did not change the behavior of the mandrills and colobus monkeys, but the behavior of and use of the enclosure by the sooty mangabeys was significantly affected. An important note in this study is that all species were locked in species-specific holding areas at night and fed in these dens to reduce feeding competition. These findings indicate that conflict may arise when a new species is introduced into an exhibit that houses a closely related species. This may occur because both species have similar ecological niches and so they will tend to compete with one another.

4.8 Environmental Enrichment

Environmental enrichment is the addition of stimuli or the provision of choice and complexity within an enclosure. The types of enrichment that can be provided and the importance of a dynamic enrichment program are covered in detail elsewhere in this volume (Kemp 2022). One important role of enrichment in zoos is to stimulate natural behaviors, which may increase the likelihood that these animals will survive when they are introduced into the wild (Reading et al. 2013). Due to their complex cognitive and behavioral needs, many primate taxa receive extensive environmental enrichment compared to other mammals (Hoy et al. 2009).

In terms of zoo exhibit design, daily physical enrichment can be achieved by designing exhibits that are complex, dynamic environments. For an intervention to be classed as enrichment, it cannot be a constant fixture. Physical enrichment can therefore be achieved if there is an ability to change features of the exhibit or by the addition/removal of objects. For example, the use of dynamic branching in golden lion tamarin (*Leontopithecus rosalia*) enclosures enabled individuals to develop appropriate locomotory behaviors, which, alongside pre- and post-release training, led to the successful reintroduction of this species (Stoinski and Beck 2004). Similarly, if an enclosure includes natural plants, then seasonal changes in foliage will provide complexity.

4.9 Diet and Nutrition

Our knowledge of animal nutrition has evolved over the years, starting from focusing on how to feed agricultural production animals, to pets and laboratory animals, and finally to zoo animals. The sheer number of species housed in zoos coupled with the lack of published wild diet information or nutrient requirements meant that new methods have been developed to learn how to feed these animals. Relying on the strategies used for domestic or production animals is of little use due to a lack of reference materials. Physiological model species (usually domestic animals) have been identified but rather than base a model on their nutritional requirements or feeding ecology, they are often assigned to an exotic species based on phylogeny or physical appearance. Limited knowledge in the dietary requirements of wild primates has led to many captive species commonly being fed very similar diets (Plowman and Cabana 2019), even if they are completely different species with different feeding ecologies and feeding niches. Up until very recently, inadequate diets contributed significantly to chronic health problems, including heart disease, obesity, diabetes, gastrointestinal disorders, dental disease, and irritable bowel syndrome (Edwards and Ullrey 1999; Kuhar et al. 2013; Plowman 2013; Clayton et al. 2016). There is evidence that diet also has a strong effect on behavior and a change in diet can sometimes significantly reduce abnormal behavior patterns (Less et al. 2014). Fortunately, within the past 20 years, there has been a boom in research on the natural ecology of primates and on behavior of captive primates.

Feeding captive primates requires in-depth knowledge about their natural feeding ecology, digestive morphology, and alternative food items that would supply similar nutrients in appropriate concentrations. In the latter case, for example, the nutrients found in fruits eaten by wild primates are different from those available to primates in zoos (Oftedal and Allen 1997). Care must also be taken to not fall into feeding niche archetypes, such as frugivorous or insectivorous; well-known primates, such as gorillas, were assumed to be frugivores, which led to the belief that they should be fed fruits (Plowman 2013). With a few notable exceptions (such as tarsiers who are known to be exclusively carnivorous), primates are generalists, and so do not subsist on one kind of food. Most primates classified as frugivores, for example, will eat many other food items when fruit is unavailable (Russon et al. 2009).

The nutrients found within wild fruits are best replaced by providing vegetables (Schwitzer and Kaumanns 2003). Switching from a high fruit diet that is higher in sugar and lower in fiber to a diet that includes more vegetables and browse (tree leaves, branches, and other roughage), and thus is lower in sugar and higher in fiber, has marked health and behavior benefits for captive primates. A diet that includes more vegetables and browse also more closely matches the nutrient concentration ingested by wild primates.

Browse is also a great addition to the diet of any primate and is an essential part of the diet of folivores, such as gorillas (Doran-Sheehy et al. 2009) and colobines (Edwards and Ullrey 1999). Indeed, the nutrients found in browse are difficult to replace adequately, so if an institution cannot provide browse throughout the year to a folivorous primate, the institution should consider keeping a different species. Insects are another very popular, palatable, and important food for small primates. They are often dusted with calcium powder to even-out their mineral compositions, but proper gut loading of insects should also be employed. Gut loading occurs through feeding a food, which is highly concentrated in the nutrients that insects themselves lack (e.g., vitamins, calcium, omega-3 fatty acids) (Finke 2003). The most common food item in captive diets has been pellets (monkey chow). While pellets are not a bad food to include as part of an adequate diet, their inclusion should not comprise the majority of the diet. It is possible to have a healthy and even superior diet without including pellets. Using pellets decreases the variety of food types in the diet, which has been linked to decreased gut microbe diversity (Clayton et al. 2016). If adequate whole foods cannot be sourced, however, then pellets are a popular alternative, some of which may provide adequate nutrient concentration but rarely support natural feeding behaviors.

Some animals have adapted so that they can process particular foods. For example, slow lorises (*Nycticebus* spp.) and callitrichids (Callitrichinae spp.) have physiological adaptations to be able to harvest and digest tree gum (Cabana et al. 2018a, b) and capuchins have developed tool use behavior to break apart tough nuts (Wright et al. 2019). Feeding a food that a species has a physiological or behavioral adaptation for is essential for meeting welfare needs as they promote the expression of a natural feeding behavior and/or provide essential nutrients that are difficult to obtain from other foods.

Feeding naturalistic diets has many benefits, such as preventing diabetes and reducing the frequency of abnormal behaviors (Cabana et al. 2018c), the likelihood of wasting syndrome (Cabana et al. 2018e), and parasite load (Cabana et al. 2018d), support an ideal body condition (Cabana et al. 2018d), and decrease occurrence of dental disease and aggression (Plowman 2013; Britt et al. 2015). The latest evidence also shows that a more naturalistic diet selects for a commensal gut microbe community, and this has a host of benefits relating to health, reproduction, and behavior (Clayton et al. 2018; Cabana et al. 2019).

5 Summary

The variety of primate species housed in zoos worldwide makes generalizing about their housing and husbandry difficult. With more evidence-based husbandry recommendations regarding features, such as space and climbing structures, it should be easier to design good enclosures. Husbandry considerations such as moving animals, enrichment, and diet also benefit from increased evidence-based practice and communication between institutions with similar housing and husbandry requirements.

References

Anderson MR (2014) Reaching new heights: the effect of an environmentally enhanced outdoor enclosure on gibbons in a zoo setting. J Appl Anim Welf Sci 17:216–227. https://doi.org/10.1080/10888705.2014.916172

Baker KR, Farmer HL (2022) The welfare of primates in Zoo. In: Robinson LM, Weiss A (eds) Nonhuman primate welfare: from history, science, and ethics to practice. Springer, Cham, pp 79–96

Bayne K, Hau J, Morris T (2022) The welfare impact of regulations, policies, guidelines and directives and nonhuman primate welfare. In: Robinson LM, Weiss A (eds) Nonhuman primate welfare: from history, science, and ethics to practice. Springer, Cham, pp 629–646

Blaney EC, Wells DL (2004) The influence of a camouflage net barrier on the behaviour, welfare and public perceptions of zoo-housed gorillas. Anim Welf 13(2):111–118

Britt S, Cowlard K, Baker K, Plowman A (2015) Aggression and self-directed behaviour of captive lemurs (*Lemur catta, Varecia variegata, V. rubra and Eulemur coronatus*) is reduced by feeding fruit-free diets. J Zoo Aquarium Res 3(2):52–58. https://doi.org/10.19227/jzar.v3i2.119

Bryan K, Bremner-Harrison S, Price E, Wormell D (2017) The impact of exhibit type on behaviour of caged and free-ranging tamarins. Appl Anim Behav Sci 193:77–86. https://doi.org/10.1016/j.applanim.2017.03.013

Buchanan-Smith HM, Griciute J, Daoudi S et al (2013) Interspecific interactions and welfare implications in mixed species communities of capuchin (*Sapajus apella*) and squirrel monkeys (*Saimiri sciureus*) over 3 years. Appl Anim Behav Sci 147(3–4):324–333. https://doi.org/10.1016/j.applanim.2013.04.004

Cabana F, Dierenfeld E, Wirdateti W et al (2018a) Trialling nutrient recommendations for slow lorises (*Nycticebus* spp.) based on wild feeding ecology. J Anim Physiol Anim Nutr 102(1):e1–e10. https://doi.org/10.1111/jpn.12694

Cabana F, Dierenfeld ES, Wirdateti ES et al (2018b) Exploiting a readily available but hard to digest resource: a review of exudativorous mammals identified thus far and how they cope in captivity. Integr Zool 13(1):94–111. https://doi.org/10.1111/1749-4877.12264

Cabana F, Jasmi R, Maguire R (2018c) Great ape nutrition: low-sugar and high-fibre diets can lead to increased natural behaviours, decreased regurgitation and reingestion, and reversal of prediabetes. Int Zoo Yearb 52(1):48–61. https://doi.org/10.1111/izy.12172

Cabana F, Jayarajah P, Oh PY, Hsu C-D (2018d) Dietary management of a hamadryas baboon (*Papio hamadryas*) troop to improve body and coat condition and reduce parasite burden. J Zoo Aquarium Res 6(1):16–21. https://doi.org/10.19227/jzar.v6i1.306

Cabana F, Maguire R, Da Hsu C, Plowman A (2018e) Identification of possible nutritional and stress risk factors in the development of marmoset wasting syndrome. Zoo Biol 37(2):98–106. https://doi.org/10.1002/zoo.21398

Cabana F, Clayton JB, Nekaris KAI et al (2019) Nutrient-based diet modifications impact on the gut microbiome of the Javan slow loris (*Nycticebus javanicus*). Sci Rep 9:4078. https://doi.org/10.1038/s41598-019-40911-0

Caws CE, Wehnelt S, Aureli F (2008) The effect of a new vertical structure in mitigating aggressive behaviour in a large group of chimpanzees (*Pan troglodytes*). Anim Welf 17(2):149–154

Chamove AS, Anderson JR, Morgan-jones SC et al (1982) Deep woodchip litter: hygiene, feeding, and behavioral enhancement in eight primate species. Int J Study Anim Prob 3(4):308–318

Claidière N, Messer EJE, Hoppitt W, Whiten A (2013) Diffusion dynamics of socially learned foraging techniques in squirrel monkeys. Curr Biol 23(13):1251–1255. https://doi.org/10.1016/j.cub.2013.05.036

Clayton JB, Vangay P, Huang H et al (2016) Captivity humanizes the primate microbiome. Proc Natl Acad Sci U S A 113(37):10376–10381. https://doi.org/10.1073/pnas.1521835113

Clayton JB, Gomez A, Amato K et al (2018) The gut microbiome of nonhuman primates: lessons in ecology and evolution. Am J Primatol 80(6):e22867. https://doi.org/10.1002/ajp.22867

Clubb R, Mason G (2003) Captivity effects on wide-ranging carnivores. Nature 425:473–474. https://doi.org/10.1038/425473a

Coe JC (2003) Steering the ark toward Eden: design for animal well-being. J Am Vet Med Assoc 223(7):977–980. https://doi.org/10.2460/javma.2003.223.977

Coleman K, Timmel G, Prongay K, Baker KC (2022) Common husbandry, housing, and animal care practices. In: Robinson LM, Weiss A (eds) Nonhuman primate welfare: from history, science, and ethics to practice. Springer, Cham, pp 317–348

Dalton R, Buchanan-Smith HM (2005) A mixed-species exhibit for Goeldi's monkeys and pygmy marmosets *Callimico goeldii* and *Callithrix pygmaea* at Edinburgh Zoo. Int Zoo Yearb 39(1):176–184. https://doi.org/10.1111/j.1748-1090.2005.tb00017.x

de Waal FBM (1989) The myth of a simple relation between space and aggression in captive primates. Zoo Biol 8(S1):141–148. https://doi.org/10.1002/zoo.1430080514

DEFRA (2015) Secretary of State standards of modern zoo practice. https://assets.publishing.service.gov.uk/government/uploads/system/uploads/attachment_data/file/69596/standards-of-zoo-practice.pdf. Accessed 28 Oct 2020

Dittmer KE, Thompson KG (2011) Vitamin D metabolism and rickets in domestic animals: a review. Vet Pathol 48(2):389–407. https://doi.org/10.1177/0300985810375240

Doran-Sheehy D, Mongo P, Lodwick J, Conklin-Brittain NL (2009) Male and female western gorilla diet: preferred foods, use of fallback resources, and implications for ape versus old world monkey foraging strategies. Am J Phys Anthropol 140(4):727–738. https://doi.org/10.1002/ajpa.21118

Edwards MS, Ullrey DE (1999) Effect of dietary fiber concentration on apparent digestibility and digesta passage in non-human primates. II. Hindgut- and foregut-fermenting folivores. Zoo Biol 18(6):537–549. https://doi.org/10.1002/(SICI)1098-2361(1999)18:6<537::AID-ZOO8>3.0.CO;2-F

Fàbregas MC, Guillén-Salazar F, Garcés-Narro C (2012) Do naturalistic enclosures provide suitable environments for zoo animals? Zoo Biol 31(3):362–373. https://doi.org/10.1002/zoo.20404

Finke MD (2003) Gut loading to enhance the nutrient content of insects as food for reptiles: a mathematical approach. Zoo Biol 22(2):147–162. https://doi.org/10.1002/zoo.10082

Fitch-Snyder H, Schultze H (2001) Management of lorises in captivity. A husbandry manual for Asian lorisines. http://www.loris-conservation.org/database/captive_care/manual/PDF/1_intro_contents.pdf

Fuller G, Kuhar CW, Dennis PM, Lukas KE (2013) A survey of husbandry practices for lorisid primates in north American zoos and related facilities. Zoo Biol 32(1):88–100. https://doi.org/10.1002/zoo.21049

Fuller G, Raghanti MA, Dennis PM et al (2016) A comparison of nocturnal primate behavior in exhibits illuminated with red and blue light. Appl Anim Behav Sci 184:126–134. https://doi.org/10.1016/j.applanim.2016.08.011

Hanya G, Kiyono M, Hayaishi S (2007) Behavioral thermoregulation of wild Japanese macaques: comparisons between two subpopulations. Am J Primatol 69(7):802–815. https://doi.org/10.1002/ajp.20397

Hebert PL, Bard K (2000) Orangutan use of vertical space in an innovative habitat. Zoo Biol 19(4):239–251. https://doi.org/10.1002/1098-2361(2000)19:4<239::AID-ZOO2>3.0.CO;2-7

Herrelko ES, Buchanan-Smith HM, Vick SJ (2015) Perception of available space during chimpanzee introductions: number of accessible areas is more important than enclosure size. Zoo Biol 34(5):397–405. https://doi.org/10.1002/zoo.21234

Hosey GR (2005) How does the zoo environment affect the behaviour of captive primates? Appl Anim Behav Sci 90(2):107–129. https://doi.org/10.1016/j.applanim.2004.08.015

Hosey G (2022) The history of primates in zoos. In: Robinson LM, Weiss A (eds) Nonhuman primate welfare: from history, science, and ethics to practice. Springer, Cham, pp 3–30

Hoy JM, Murray PJ, Tribe A (2009) Thirty years later: enrichment practices for captive mammals. Zoo Biol 29(3):303–316. https://doi.org/10.1002/zoo.20254

Judge PG, Griffaton NS, Fincke AM (2006) Conflict management by hamadryas baboons (*Papio hamadryas hamadryas*) during crowding: a tension-reduction strategy. Am J Primatol 68(10):993–1006. https://doi.org/10.1002/ajp.20290

Junge RE, Gannon FH, Porton I, McAlister WHWM (2000) Management and prevention of vitamin D deficiency rickets in captive-born juvenile chimpanzees (*Pan troglodytes*). J Zoo Wildl Med 31(3):361–369. https://doi.org/10.1638/1042-7260(2000)031[0361:MAPOVD]2.0.CO;2

Kemp C (2022) Enrichment. In: Robinson LM, Weiss A (eds) Nonhuman primate welfare: from history, science, and ethics to practice. Springer, Cham, pp 451–488

Kuhar CW (2008) Group differences in captive gorillas' reaction to large crowds. Appl Anim Behav Sci 110(3–4):377–385. https://doi.org/10.1016/j.applanim.2007.04.011

Kuhar CW, Fuller GA, Dennis PM (2013) A survey of diabetes prevalence in zoo-housed primates. Zoo Biol 32(1):63–69. https://doi.org/10.1002/zoo.21038

Kurtycz LM, Wagner KE, Ross SR (2014) The choice to access outdoor areas affects the behavior of great apes. J Appl Anim Welf Sci 17(3):185–197. https://doi.org/10.1080/10888705.2014.896213

Lee R, Riley E, Sangermano F, Cannon C, Shekelle M (2020) *Macaca nigra*. The IUCN Red List of Threatened Species 2020:e.T12556A17950422. https://doi.org/10.2305/IUCN.UK.2020-3.RLTS.T12556A17950422.en

Less E, Bergl R, Ball R et al (2014) Implementing a low-starch biscuit-free diet in zoo gorillas: the impact on behavior. Zoo Biol 33(1):63–73. https://doi.org/10.1002/zoo.21116

López J, Wormell D, Rodríguez A (2001) Preliminary evaluation of the efficacy and safety of a UVB lamp used to prevent metabolic bone disease in pied tamarins *Saguinus bicolor* at Jersey Zoo. Dodo 37:41–49

Lukas KE, Ross SR (2014) Naturalistic exhibits may be more effective than traditional exhibits at improving zoo-visitor attitudes toward African apes. Anthrozoös 27(3):435–455. https://doi.org/10.2752/175303714X14023922797904

Melfi V (2005) The appliance of science to zoo-housed primates. Appl Anim Behav Sci 90(2):97–106. https://doi.org/10.1016/j.applanim.2004.08.017

Melfi VA, Feistner ATC (2002) A comparison of the activity budgets of wild and captive Sulawesi crested black macaques (*Macaca nigra*). Anim Welf 11(2):213–222

Melfi VA, McCormick W, Gibbs A (2004) A preliminary assessment of how zoo visitors evaluate animal welfare according to enclosure style and the expression of behavior. Anthrozoös 17(2):98–108. https://doi.org/10.2752/089279304786991792

Melfi V, Bowkett A, Plowman A, Pullen K (2005) Do zoo designers know enough about animals? Innovation or replication. In: Proceedings of the 6th International Symposium on Zoo Design, pp 119–127. https://doi.org/10.1017/CBO9781107415324.004

Milller LE, Treves A (2006) Predation in primates. In: Primates in perspective. Oxford University Press, New York, pp 525–543

Mirick DK, Davis S (2008) Melatonin as a biomarker of circadian dysregulation. Cancer Epidemiol Biomarkers Prev 17(12):3306–3313. https://doi.org/10.1158/1055-9965.epi-08-0605

Morimura N, Hirata S, Matsuzawa T (2022) Challenging cognitive enrichment: examples from caring for the chimpanzees in the Kumamoto Sanctuary, Japan and Bossou, Guinea. In: Robinson LM, Weiss A (eds) Nonhuman primate welfare: from history, science, and ethics to practice. Springer, Cham, pp 489–516

Morland HS (1993) Seasonal behavioral variation and its relationship to thermoregulation in ruffed lemurs (*Varecia Variegata Variegata*). In: Kappeler PM, Ganzhorn JU (eds) Lemur social systems and their ecological basis. Springer, Boston, MA. https://doi.org/10.1007/978-1-4899-2412-4_14

Morrisey J, Reichard T, Janssen D, Lloyd M (1994) Vitamin D deficiency rickets in Colobinae monkeys. In: AAZV Collected Annual Conference Proceedings

Morton FB, Lee PC, Buchanan-Smith HM (2013) Taking personality selection bias seriously in animal cognition research: a case study in capuchin monkeys (*Sapajus apella*). Anim Cogn 16(6):677–684. https://doi.org/10.1007/s10071-013-0603-5

Mun JSC, Kabilan B, Alagappasamy S, Guha B (2013) Benefits of naturalistic free-ranging primate displays and implications for increased human-primate interactions. Anthrozoös 26(1):13–26. https://doi.org/10.2752/175303713X13534238631353

Oftedal OT, Allen M (1997) The feeding and nutrition of omnivores with emphasis on primates. In: Kleiman DG, Allen ME, Thompson K (eds) Wild mammals in captivity: principles and techniques. University of Chicago Press, Chicago, pp 148–157

Pearson EL, Davis JM, Litchfield CA (2010) A case study of orangutan and siamang behavior within a mixed-species zoo exhibit. J Appl Anim Welf Sci 13(4):330–346. https://doi.org/10.1080/10888705.2010.507125

Pines MK, Kaplan G, Rogers LJ (2007) A note on indoor and outdoor housing preferences of common marmosets (*Callithrix jacchus*). Appl Anim Behav Sci 108(3):348–353. https://doi.org/10.1016/j.applanim.2006.12.001

Plowman A (2013) Diet review and change for monkeys at Paignton Zoo Environmental Park. J Zoo Aquarium Res 1(2):73–77. https://doi.org/10.19227/jzar.v1i2.35

Plowman A, Cabana F (2019) Transforming the nutrition of zoo primates (or How we became known as Loris Man and That Evil Banana Woman). In: Kaufman A, Bashaw M, Maple T (eds) Scientific foundations of zoos and aquariums: their role in conservation and research. Cambridge University Press, Cambridge, pp 274–303

Pomerantz O, Meiri S, Terkel J (2013) Socio-ecological factors correlate with levels of stereotypic behavior in zoo-housed primates. Behav Process 98:85–91. https://doi.org/10.1016/j.beproc.2013.05.005

Power ML, Oftedal O, Tardif SD, Allen ME (1995) Vitamin D and primates: recurring problems on a familiar theme. In: Proceedings of the First Annual Conference of the Nutrition Advisory Group

Reading RP, Miller B, Shepherdson D (2013) The value of enrichment to reintroduction success. Zoo Biol 32(3):322–341. https://doi.org/10.1002/zoo.21054

Ross SR, Lukas KE (2005) Zoo visitor behavior at an African ape exhibit. Visitor Stud Tod 8(1):4–12

Ross SR, Shender MA (2016) Daily travel distances of zoo-housed chimpanzees and gorillas: implications for welfare assessments and space requirements. Primates 57(3):395–401. https://doi.org/10.1007/s10329-016-0530-6

Ross SR, Schapiro SJ, Hau J, Lukas KE (2009) Space use as an indicator of enclosure appropriateness: a novel measure of captive animal welfare. Appl Anim Behav Sci 121(1):42–50. https://doi.org/10.1016/j.applanim.2009.08.007

Ross SR, Wagner KE, Schapiro SJ, Hau J (2010) Ape behavior in two alternating environments: comparing exhibit and short-term holding areas. Am J Primatol 72(11):951–959. https://doi.org/10.1002/ajp.20857

Ross SR, Calcutt S, Schapiro SJ, Hau J (2011a) Space use selectivity by chimpanzees and gorillas in an indoor-outdoor enclosure. Am J Primatol 73(2):197–208. https://doi.org/10.1002/ajp.20891

Ross SR, Wagner KE, Schapiro SJ et al (2011b) Transfer and acclimatization effects on the behavior of two species of African great ape (*Pan troglodytes* and *Gorilla gorilla gorilla*) moved to a novel and naturalistic zoo environment. Int J Primatol 32(1):99–117

Ross SR, Melber LM, Gillespie KL, Lukas KE (2012) The impact of a modern, naturalistic exhibit design on visitor behavior: a cross-facility comparison. Visitor Stud 15(1):3–15. https://doi.org/10.1080/10645578.2012.660838

Russon, A E; Wich, S A; Ancrenaz, M; Kanamori, T; Knott, C D; Kuze, N; Morrogh-Bernard, H C; Pratje, P; Ramlee, H; Rodman, P; Sawang, A; Sidiyasa, K; Singleton, I; van Schaik, C P (2009). Geographic variation in orangutan diets. In: Wich, S A; Utami Atmoko, S S; Mitra Setia, T; van Schaik, C P. Orangutans: geographic variation in behavioral ecology and conservation. New York: Oxford University Press, 135–156

Schwitzer C, Kaumanns W (2003) Foraging patterns of free-ranging and captive primatesimplications for captive feeding regimes. In: Fidgett A, Clauss M, Gansloßer U, Hatt J-M, Nijboer J (eds) Zoo Animal Nutrition, vol II. Filander Verlag Fürth, Fürth, pp 247–265

Sheperdson DJ, Mellen J (1999) Second nature: environmental enrichment for captive animals. Smithsonian Institution Press, Washington, DC

Sherwen SL, Harvey TJ, Magrath MJL et al (2015) Effects of visual contact with zoo visitors on black-capped capuchin welfare. Appl Anim Behav Sci 167:65–73. https://doi.org/10.1016/j.applanim.2015.03.004

Stoinski TS, Beck BB (2004) Changes in locomotor and foraging skills in captive-born reintroduced golden lion tamarins (*Leontopithecus rosalia rosalia*). Am J Primatol 62(1):1–13. https://doi.org/10.1002/ajp.20002

Talbot CF, Reamer LA, Lambeth SP, Schapiro SJ, Brosnan SF (2022) Meeting cognitive, behavioral, and social needs of primates in captivity. In: Robinson LM, Weiss A (eds) Nonhuman primate welfare: from history, science, and ethics to practice. Springer, Cham, pp 267–302

Toussaint S, Reghem E, Chotard H et al (2013) Food acquisition on arboreal substrates by the grey mouse lemur: implication for primate grasping evolution. J Zool 291(4):235–242. https://doi.org/10.1111/jzo.12073

Veasey JS (2017) In pursuit of peak animal welfare; the need to prioritize the meaningful over the measurable. Zoo Biol 36(6):413–425. https://doi.org/10.1002/zoo.21390

Wojciechowski S (2004) Introducing a fourth primate species to an established mixed-species exhibit of African monkeys. Zoo Biol 23(2):95–108. https://doi.org/10.1002/zoo.10128

Wright BW, Wright KA, Strait DS et al (2019) Taking a big bite: working together to better understand the evolution of feeding in primates. Am J Primatol 81(5):e2298. https://doi.org/10.1002/ajp.22981

Zuclich JA (1984) Ultraviolet induced damage in the primate cornea and retina. Curr Eye Res 3(1):27–34. https://doi.org/10.3109/02713688408997185

Humane Endpoints and End of Life in Primates Used in Laboratories

Sarah Wolfensohn

Abstract

Deciding when and how to end an animal's life are critical when managing its welfare. This chapter gives an overview of the use of nonhuman primates in science, and of the European regulations and the ethical perspective of justifying such use with a harm:benefit assessment. It reviews the definitions of a humane endpoint and considers how to set that endpoint to limit the harms by objectively assessing the animal's welfare. The setting of humane endpoints should be incorporated into management systems with other considerations around the end of life of nonhuman primates, and applied to their use in other contexts, such as in zoos and sanctuaries.

Keywords

Humane endpoint · Harm:benefit · Refinement · Euthanasia · Welfare assessment

1 Introduction

The decisions about when and how to end an animal's life are perhaps the two most important ones in terms of managing its welfare, whether it is in a laboratory setting, captive in a zoo or wild and free ranging (Wolfensohn et al. 2018). For nonhuman primates, particularly those with higher cognitive abilities, this is considered by many to become more critical, yet human emotion may delay us from intervening and preventing, or ending, the animal's suffering (Wolfensohn 2020). Any animal that is suffering severe pain, which cannot be alleviated, must be humanely killed immediately and not simply left to die. Decisions about end of life for primates used

S. Wolfensohn (✉)
University of Surrey School of Veterinary Medicine, Guildford, Surrey, UK
e-mail: s.wolfensohn@surrey.ac.uk

in research can be particularly complex as there is a need to balance the killing of the animal with the scientist's desire to acquire more data, so decisions will rest on the outcome of a harm:benefit ratio for the work. Objective measures of harms and benefits are therefore required in order to make the decision about precisely when to euthanize a primate. Euthanasia is defined as the bringing about of a gentle and easy death (Oxford English Dictionary); so should not, itself, cause pain or suffering. Killing an animal is never a pleasant task, but it does not have to be unpleasant for the animal, provided it is carried out competently and humanely. Having to kill a primate may be a more emotionally challenging experience for staff than having to kill some other species. The method of euthanasia must also be considered in the context of the type of experiment being carried out, the possible requirement to collect tissue post mortem and how to do so with the minimum suffering for the animal. However, the humane endpoint is not necessarily only the point at which the animal is humanely killed; it could be an intervention to alleviate the stress or pain of an experimental procedure, such as by stopping the procedural intervention earlier or providing better analgesics for longer. Humane endpoints are not necessarily based on clinical signs; they can be defined from preclinical signs or from physiological or molecular biomarkers that are associated with pain or distress that occurs later in the disease process. Humane endpoints must also be balanced against the scientific endpoints to be met since pain and distress might be intrinsic to an experimental model (such as evaluating the efficacy of analgesics), but they should never be beyond the scientific endpoint or beyond the level of moral justification.

2   An Overview of the Use of Nonhuman Primates in Science

In 2017, the total number of animals used for scientific purposes in the European Union were 9,388,162 of which 8235 (0.09%) were nonhuman primates. A summary of laboratory animal usage within the 27 states of the European Union can be found at https://eur-lex.europa.eu/legal-content/EN/TXT/?qid=1581689520921&uri=CELEX:52020SC0010 and the change in use of nonhuman primates between 2008 and 2017 is summarized in Table 1. Examination of the data relating to the uses of nonhuman primates shows that in 2017, 65% were to satisfy legislative requirements for medicinal products for human use (of these 64% are on studies

Table 1 A summary of the use of nonhuman primates comparing 2008, 2011, and 2017 within the European Union. Ten of the 27 states recorded use of nonhuman primates. No apes were used (from https://eur-lex.europa.eu/legal-content/EN/TXT/?qid=1581689520921&uri=CELEX:52020SC0010)

	2008	2011	2017	% change 2008–2017
Prosimians	1261	83	98	−92
New World monkeys (Ceboidea)	904	700	476	−47
Old World monkeys (Cercopithecoidea)	7404	5312	7661	+3
Total number of primates	8308	6095	8235	−0.9

for repeated dose toxicity and 19% for kinetics). In the areas of basic and applied research, nonhuman primates are mainly used for studying human infectious disorders (7% of all nonhuman primate uses), nervous system (3%), and nonregulatory toxicology and ecotoxicology (3%). Routine production, of mostly blood based products, represents 10% of nonhuman primate use. Some nonhuman primates were reported to have been used for the purpose of education and training, but since this is prohibited under the Directive, this may have taken place within the transitional provisions under Article 64 of the Directive allowed until 31.12.2017. In 2017, 54% of uses of nonhuman primates were reported as being of mild severity and 1.6% of uses were assessed as severe. For comparison, in the United States 74,498 nonhuman primates were used in 2017 out of a total of 777,960 animals (9%) https://www.aphis.usda.gov/aphis/ourfocus/animalwelfare/sa_obtain_research_facility_annual_report/ct_research_facility_annual_summary_reports. The regulations, norms, practices, and standards in animal research are not currently harmonized (Chatfield and Morton 2018), so comparable data from other countries are not available.

The source of the primates varies depending on the species. In 2017, the three main sources of nonhuman primates were Africa, Asia, and EU-registered breeders representing more than 97% of nonhuman primates used for scientific purposes. Of the 574 New World monkeys (Ceboidea) used in 2017 in the European Union, 483 (84%) came from within the European Union, and 91 originated from outside the European Union. About 31% of procedures with New World monkeys involved reuse. However, of the 7661 Old World monkeys (Cercopithecoidea) used in the European Union in 2017, only 574 (7%) came from within the European Union and 7087 (92%) originated from elsewhere. About 0.28% of procedures with Old World monkeys involved reuse. In 2017, cynomolgus monkeys represented 88% of the nonhuman primates used for the first time. These were sourced almost entirely from outside of the EU, but the majority of these were sourced either from self-sustaining colonies (30%) or as second or higher generation purpose-bred (53%). No nonhuman primates were sourced from the wild in 2017.

However, the critical question is not just how many animals are being used and why, but what is the beneficial outcome of that use and how much do the primates suffer in order to generate that knowledge. Is the use of these primates justified? And moving forward, how can we reduce any suffering caused and generate better, more useful data in order to improve the justification, or, even better, find replacements with less sentient animals lower down the phylogenetic scale or by using human clinical trials or computer modeling?

3 European Regulations and the Ethical Perspective

The use of animals in scientific procedures in the European Union is controlled by the Directive 2010/63/EU of the European Parliament and of the Council of 22 September 2010 on the Protection of Animals used for Scientific Purposes.

Each member state transposed the Directive into its own legislation (see review of legislation, see Bayne et al. 2022).

In 2007, the European Parliament adopted a written declaration urging the European Union to end the use of nonhuman primates in scientific experiments signed by 433 MEPs (out of 626). The European Commission's response was that, currently, the use of a limited number of nonhuman primates remained unavoidable for vital research into immune-mediated diseases such as multiple sclerosis, neurodegenerative disorders such as Parkinson's disease and Alzheimer's disease, and infectious diseases such as Human Immunodeficiency Virus, malaria, and tuberculosis (see t'Hart et al. 2022, for argument in favor of using primates to study autoimmune disorders). Furthermore, the European Commission stated that setting a deadline to phase out use of nonhuman primates was not possible because technological and scientific developments had not yet reached the stage that would make such a program realistic (see Bailey 2022, for argument in favor of alternatives). The revision of the Directive in 2010 was an opportunity to incorporate strong incentives, combined with a specific review clause, to provide an appropriate and effective mechanism to move toward the goal of phasing out the use of nonhuman primates in experiments. The 2010/63 Directive was reviewed in 2017 and no amendments to the Directive were proposed at this stage.

In the current Directive, there are several sections that are specifically relevant to experiments in nonhuman primates. Article 4, which covers the principle of replacement, reduction, and refinement, requires that member states shall ensure refinement of breeding, accommodation, and care; and refinement of methods used in procedures in order to eliminate, or reduce to the minimum, any possible pain, suffering, distress or lasting harm to the animals. Article 13, on the choice of methods, requires that death as the endpoint of a procedure shall be avoided as far as possible and replaced by early and humane endpoints. Where death as the endpoint is unavoidable, the procedure shall be designed so as to: (a) result in the deaths of as few animals as possible; and (b) reduce the duration and intensity of suffering to the animal to the minimum possible and, as far as possible, ensure a painless death. In Annex VIII to the Directive, it requires that the type of species and genotype, the maturity, age, and gender of the animal, the training experience of the animal with respect to the procedure, the actual severity of the previous procedures (if the animal is to be reused), the methods used to reduce or eliminate pain, suffering and distress, including refinement of housing, husbandry and care conditions, and application of humane endpoints will all be taken into account to determine the final severity classification of the procedure. Article 55 is a safeguard clause which allows a member state to adopt a provisional measure in order to use a procedure involving severe pain, suffering, or distress that is likely to be long-lasting and cannot be ameliorated on a nonhuman primate for exceptional and scientifically justifiable reasons. In Annex VIII, there are some examples of procedures assigned to the severe severity category including toxicity testing where death is the endpoint, or fatalities are to be expected and severe pathophysiological states are induced;

vaccine potency testing characterized by persistent impairment of the animal's condition, progressive disease leading to death, associated with long-lasting moderate pain, distress or suffering; organ transplantation where organ rejection is likely to lead to severe distress or impairment of the general condition of the animals (e.g., xenotransplantation); and complete isolation for prolonged periods of social species which includes nonhuman primates.

The regulations and legislation therefore clearly define the use of humane endpoints and the requirement to maximize welfare as much as possible and to reduce suffering. The use of animals in science requires a balance between the desire for scientific progress, the impact on animal welfare and the public acceptability of such work, much of which is funded by the public purse. It is also important to recognize that any data collected must be of adequate quality to be usable since, without that, the animal would have been used without leading to any scientific output and any suffering caused would have been pointless and unethical. The moral case for controlled experiments using animals can be defended if the harms that these experiments cause to the animals are outweighed by the benefits that these experiments bring to society, humans, and/or other animals (Davies et al. 2017). However, the individual animal's experience of pain and suffering is the same regardless of the context or "reason" for its life whether, for example, it is a pet animal, a farm animal, a wild animal, or an animal being used for research (Wolfensohn and Honess 2007). A laboratory animal will not have the benefit of knowing that its harm is "for a good cause" nor can it give informed consent, so the value of knowledge resulting from experiments cannot be the only determinant of whether they can be justified. Potential benefits of scientific experiments are easy to overestimate whereas potential harms may be easier to predict but can be difficult to measure. Limiting harms effectively are directly dependent on local standards of animal care put in place by management and the competence of staff to adhere to those standards. A judgment of the quality of life of animals used in experiments against the benefits to other animals, including humans, is the basis of the harm–benefit analysis and this must be carried out prospectively, and regularly reassessed through the duration of the study to ensure the continued justification for the use of the primate (Home Office 2015; Brønstad et al. 2016; Animal Science Committee 2017).

Well-designed experiments should detect early signs of any deterioration in the animal's welfare, allowing the definition of a point at which adequate data have been obtained, but ensuring that animal suffering is minimized (Morton 2000). The precise time at which to kill the animal must be based on appropriate and accurate clinical judgment, assessing the degree of suffering against the loss of potential data. This balance will have been assessed as part of the harm:benefit analysis required to justify the experiment in the first place. To go beyond what is required for a scientific outcome may cause unnecessary suffering and is therefore inhumane, unjustified, and unethical (Gilbert and Wolfensohn 2011). Ideally the maximum achievable information should be obtained from each animal while suffering and distress are kept to the minimum. Setting humane endpoints will refine the procedure and reduce the harms to the animals, improving the justification and the harm-benefit equation for the experiment.

For both ethical and experimental reasons, it is therefore important to keep adverse effects to a minimum. In order to demonstrate this, objective methods of assessing pain and distress in animals are necessary and these are also valuable as they allow potential refinements and new research techniques to be evaluated critically; they allow sound judgments to be made about the need for, and efficacy of, analgesics; and they enable the implementation of humane endpoints. This ensures a defined set of criteria at which the animal is taken off study, usually by euthanasia, when its welfare has become compromised beyond a certain point. This ensures that unnecessary suffering does not occur. It is unusual for laboratory animals to die in a laboratory as an expected consequence of a procedure without the intervention of euthanasia, and if they do so this is considered, under the UK legislation, to bring the protocol within the highest level of severity classification. Examples of biomedical research areas which can have a relatively high level of pain and distress, depending on the experimental design and implementation of refinements, are invasive models of neuroscience and vaccine challenge studies. It is therefore particularly important for these types of work to define clear humane endpoints with specific details of euthanasia implementation in order to remain within the legislative controls and to maintain welfare as much as possible.

4   What Is a Humane Endpoint?

There are three possible endpoints to every experiment: experimental, error, or humane. The experimental endpoint is when the experiment has run its course, and adequate experimental data have been collected. However, even when designed into a study, endpoints should not be slavishly followed if unpredicted events occur during the experiment. Judgment and experience must be used to halt studies before the anticipated endpoint if the objectives of the study have already been met, or if the study will clearly not meet its objectives. The error endpoint occurs when mistakes happen (such as the administration of the incorrect dose of a drug), which they do from time to time. Although there may not necessarily be a welfare impact, the errors may invalidate the experiment and therefore the experiment has to be ended early, there is no point in continuing the experiment since the data generated will be erroneous. The definitions of a humane endpoint are shown in Table 2 below with the references from which they are taken:

All of these definitions have in common the idea of defining a set intervention point at which to kill the animal. This allows the collection of quality scientific data but limits the amount of suffering, either contingent or direct (Russell and Burch 1959), to which the animal may be subjected. By definition, procedures carried out under the UK's Animals (Scientific Procedures) Act will have the potential to cause pain, suffering, distress or lasting harm, so it is necessary to address how these adverse effects might be controlled or, even better, prevented. However, the humane endpoint is not necessarily the point at which the animal is humanely killed; it could

Table 2 Definition of a humane endpoint

	References
The earliest indicator in an animal experiment of (potential) pain and/or distress that, within the context of moral justification and scientific endpoints to be met, can be used to avoid or limit pain and/or distress by taking actions such as humane killing or terminating or alleviating the pain and distress	Hendriksen et al. (2011)
Clear, predictable, and irreversible criteria that allow early termination of a procedure before an animal experiences harm that is not authorized or scientifically justified	Home Office (2014)
The earliest indicator in an animal of pain, distress, suffering, or impending death on the basis of which an animal is killed	Organisation for Economic Cooperation and Development (2000)
The point at which an experimental animal's pain and/or distress is terminated, minimized or reduced, by taking actions such as killing the animal humanely, terminating a painful procedure, or giving treatment to relieve pain and/or distress	Canadian Council on Animal Care (1998)
The limits placed on the amount of pain and distress any laboratory animal will be allowed to experience within the context of the scientific endpoints to be met	Wallace (2000)

be an intervention to alleviate the stress or pain of an experimental procedure, such as by moving the animal from the experimental cage to a well enriched environment, or by providing better analgesics for longer. Another type of endpoint is a simple temporal one; for example, the animal will be killed x hours/days/weeks after a technique is carried out at which point the adverse effects are not expected to cause a welfare problem but the time limit is set nonetheless, thus reducing the potential for contingent suffering. Alternatively there may be a defined measurable point relating to a physiological parameter, for example, when the blood glucose level reaches x mmol/l, which indicates that the animal is in the required physiological state to gather the data, but not yet experiencing deterioration in clinical condition (Wolfensohn and Lloyd 2013). Or, in some cases, the defined intervention point uses the alterations to the animal's usual state such as behavioral changes, changes in body temperature, body condition or weight loss. It may be based on a collection of scores cumulatively measured, but sometimes with additional scorings for more challenging clinical conditions, to ensure that potential suffering is limited to the minimum consistent with obtaining satisfactory data.

By defining humane endpoints potential suffering can be reduced or avoided. The application of a defined humane endpoint will thus utilize a lower severity level than would be needed if the procedure was allowed to run its course. Retrospective assessment of welfare to determine the level of severity should be carried out at regular intervals throughout the life of a project. This can be used to demonstrate how the ongoing implementation of the 3Rs is leading to a lower level severity classification and improved quality of life for the animals (Lloyd et al. 2008). Indeed regular assessment of welfare should be carried out throughout the lifetime of any primate to ensure maintenance of quality of life.

Scientists are often concerned that killing an animal too early when on procedure might miss the opportunity to observe that a sick animal will indeed respond to and improve with a new treatment. It is crucial in such cases to identify signs that point to *irreversible* decline in the level of welfare. These signs will vary depending on the species, its mental health, the environment in which it is kept, and the procedures being carried out. The use of the Animal Welfare Assessment Grid (Wolfensohn et al. 2015) will assist with identifying the decline in welfare. Some protocols will specify conditions under which pre-emptive euthanasia will be performed and may state that animals will be euthanized when they become "moribund." However, this term is poorly defined and arbitrary, and different people may have varying interpretations of its implications. An animal in a moribund state may be beyond suffering, i.e., they may be comatose, but there is an ongoing dispute over consciousness and awareness in human medicine (Boly et al. 2013). In people, as in nonhuman primates, lack of behavioral response to painful stimuli does not prove a lack of consciousness. A state of unresponsiveness can mimic neuromuscular blockade, in which awareness remains but motor responses cannot be generated (Sebel et al. 2004). Clinicians refer to this condition as "locked-in" syndrome (Smith and Delargy 2005). In addition to causing an unacceptable reduction in welfare, data collected under such conditions could prove variable or impossible to interpret, as the moribund state and imminent death might modify important physiological variables. Before the animal gets to the point of being moribund, judgments based on accurate observations of the animal could have set an earlier endpoint to reduce distress.

5   Setting the Humane Endpoint: Objective Assessments of Welfare

The application of the humane endpoint is not simply about recognizing pain. The harms that the animal experiences, relate to all five domains of animal welfare, i.e., nutrition, environment, health, behavior, and mental state (Mellor 2012). A key element of ensuring good welfare will be to provide an environment that allows the animals to express their natural repertoire of behaviors as far as possible (Honess and Wolfensohn 2010b). Welfare assessment should therefore include consideration of the environment and the behavior of the animals, not simply their physical condition. To assess the animal's welfare, it is necessary to have in place objective measures of suffering or of well-being, such as those described in Main et al. (2003) and Honess et al. (2005). An objective, quantifiable assessment is necessary to be able to judge whether welfare has been improved, and to ascertain whether the degree of pain and distress is within the severity classification of the protocols. The data collected on welfare must be robust so as to enable consistency over time and across groups and to remove any interobserver variability in making the assessment. This will ensure that decisions on euthanasia are consistently applied. The data may involve the use of scoring sheets and checklists, and should account for species differences, age or developmental differences, differences in sex and reproductive status, and differences in social status.

Welfare measurement can incorporate behavioral and endocrine data, together with analysis of production records. Schemes for objective assessment of adverse events have been published (see, e.g., Morton and Griffiths 1985; Carstens and Moberg 2000). Pain-scoring system for various species using facial expressions has been described (Langford et al. 2010; Sotocina et al. 2011; Gleerup et al. 2015). These and similar schemes facilitate monitoring of welfare, allowing an evaluation of responses to treatment, and logical decision making, on interventions including euthanasia. Most schemes, however, only record the detailed welfare state of the animal within a relatively short time frame (Wolfensohn et al. 2018); at one particular moment, or over the duration of a particular treatment, which is generally very short in proportion to its lifetime. These schemes rarely reflect cumulative suffering and the lifetime experience of the animal. When setting humane endpoints, the cumulative experience must be evaluated, in addition to the welfare at individual moments in time. This is particularly important since the European Directive 2010/63 refers to "...taking into account the life-time experience of the individual animal..." (para.25), "...to enhance the life-time experience of the animals..." (para.31), and to "...reduce the duration and intensity of suffering to the minimum possible..." (article 13.3b). Experiments are assigned a severity category, which takes into account the nature and intensity of pain, suffering, distress, and lasting harm; and the duration, frequency, and multiplicity of techniques and cumulative suffering within a procedure (Annex VIII). Combining a range of assessment parameters into one usable tool has been identified as an important goal in providing practical, objective, and robust assessments of lifetime experience and welfare in order to be able to use this to determine when it is appropriate to end an animal's life.

The use of the Animal Welfare Assessment Grid (Wolfensohn and Honess 2007; Honess and Wolfensohn 2010a; Wolfensohn et al. 2015) offers a schematic of parameters reflecting the five domains of animal welfare (Mellor 2012), including nutrition, environment, health, behavior, and mental state, by assessing factors to measure the physical condition of the animal, its psychological condition, the effect of procedures being carried out, and the quality of the environment. It records changes in the state of an animal over time allowing for predictive, retrospective, or scheduled event monitoring and illustrates the duration of the components of suffering. The lifetime experience graph can be investigated at any time point to produce a grid showing the effect of the specific components of the domains of welfare, allowing the effect of specific refinements to be clearly evaluated (see Figs. 1 and 2). The grid includes contingent as well as direct suffering to reflect cumulative harms and allows an evaluation of the animal's quality of life. It can demonstrate the welfare implications of research and the effect of refinements, both at the planning stage and when reviewing finished work to ensure that the harm: benefit assessment remains in order to justify the work. This assessment method is a valuable tool for those tasked with ensuring ethical oversight as well as for those planning the use of, or monitoring of, animals in research. It is particularly applicable

Fig. 1 Cumulative Welfare Assessment Score graph

to animals used in long-term studies. It has also been used to assess the welfare of primates held in zoological collection (Justice et al. 2017). The use of a welfare assessment grid can be helpful in balancing some of the difficult decisions about the future of an animal, whether in research or another context.

6 The Outcome of Welfare Assessment

The interpretation of what is meant by mild, moderate, or severe under the European Directive 2010/63 classification of procedures can be very variable and subject to interpretation. Objective assessments of welfare are needed but some researchers continue to use subjective assessments or indeed no method of assessment at all in some jurisdictions (Chatfield and Morton 2018). Pain and distress caused by scientific procedures are predictable and should be avoided or relieved, but even if there is unexpected distress, objectively measuring pain and distress in animals are not impossible (e.g., behaviorally; see Lutz and Baker 2022). The challenge for animal caretakers and researchers is to accurately interpret the state of the animal from the information available. If any doubt exists about how to interpret its behavior and clinical signs, then the maintenance of good welfare of the animal must come first, and the responsibility for this lies with the researcher. It is important to remember that the assessment of welfare alone is not sufficient. If welfare assessment is simply an academic exercise, it does nothing for the animal. Proper welfare assessment allows feedback on changes affecting the animal's welfare and then decisions must be made about what action should be taken. The outcome could be to do nothing, but

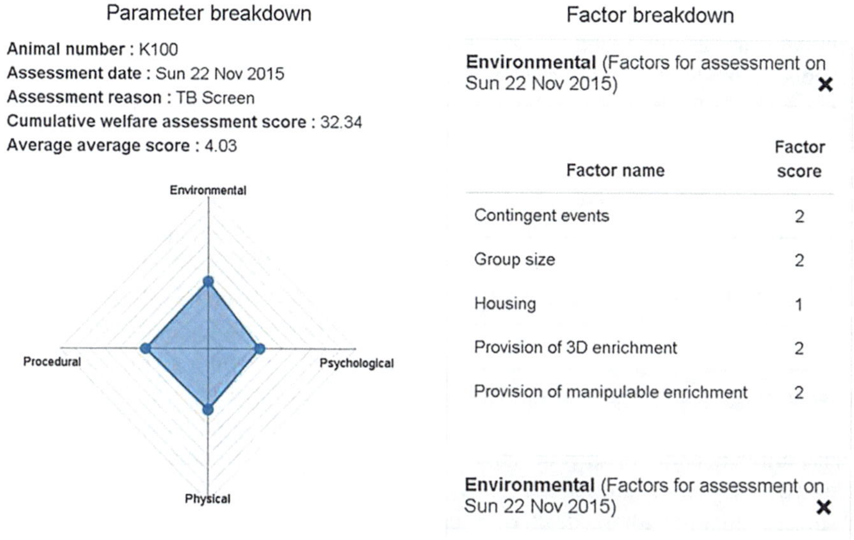

Parameter breakdown

Animal number : K100
Assessment date : Sun 22 Nov 2015
Assessment reason : TB Screen
Cumulative welfare assessment score : 32.34
Average average score : 4.03

Factor breakdown

Environmental (Factors for assessment on Sun 22 Nov 2015) ✗

Factor name	Factor score
Contingent events	2
Group size	2
Housing	1
Provision of 3D enrichment	2
Provision of manipulable enrichment	2

Environmental (Factors for assessment on Sun 22 Nov 2015) ✗

Fig. 2 Clicking on the CWAS at one point brings up the Grid for that moment in time and the detail of the factors contributing to that welfare assessment

if so, there must be clearly articulated reasons based on the harm:benefit justification about why nothing is done. While we wish to evaluate the animal's quality of life, the animal's perception of its welfare is not affected by the reason it is maintained (whether it is for experimental or breeding use, for example) or the cause of its suffering. Welfare assessment can be used to demonstrate the implications of research and the effects of the implementation of refinements to other users, funders, and ethical review bodies; but the outcome of monitoring and assessment must be improvements in welfare; welfare assessment is simply the tool to demonstrate the action is effective. The entire research team must accept responsibility in delivering animal welfare; monitoring is simply the tool that demonstrates whether this delivery is being achieved.

7 Considerations Around End of Life of Primates

Decisions on euthanasia must be reached swiftly and appropriate actions taken promptly if suffering is to be prevented. Animals used in experiments may need to be killed for a variety of reasons. The UK Animals (Scientific Procedures) Act requires that at the end of a series of regulated procedures, the animals used must be killed, except in certain specified circumstances. Waiting until the animal dies should not be the endpoint of the experiment and is unacceptable.

In addition to using euthanasia to apply an experimental humane endpoint, animals may also be killed if their health gives cause for concern and they fail to respond to treatment, if they have reached the end of their breeding life, if they are unwanted stock, or if tissues and blood are required. Euthanasia, if carried out competently, may be less traumatic and easier on the animal than those parts of a procedure that may have caused pain, suffering or distress. For additional information, there are published guidelines which refer to the animal welfare concerns when carrying out euthanasia. See Close et al. (1996, 1997), British Veterinary Association (2016), American Veterinary Medical Association (2001), and chapter "Euthanasia and other fates for laboratory animals" in the Universities Federation for Animal Welfare handbook (Wolfensohn 2010a). Staff who euthanize nonhuman primates must be competent in the chosen method (for example giving an intravenous injection), and it is essential that they receive good training, adequate supervision, and any necessary compassionate support. Performing euthanasia can be the most difficult part of the procedure for the researcher, particularly if they have worked daily with an animal and developed a close working rapport with it. Indeed, coping with the death of research animals can take a toll on the emotions of those involved (Pekow 1994; Halpern-Lewis 1996). For further discussion on the ethics of killing animals which have an inherent value either to themselves or to others, see Nuffield Council on Bioethics (2005).

The definition of death is "an animal shall be regarded as continuing to live until the permanent cessation of the circulation or the destruction of the brain" (Animals Scientific Procedures Act 1986 Section 1(4)). Therefore, any method of euthanasia must ensure that one or both criteria are met. The person carrying out the killing must be competent to do so without causing distress to the animals involved. Therefore, training and competence must be ensured by local management and a register of those deemed competent should be kept and regularly updated.

There are a number of different methods of euthanasia available and Table 3 lists the points that should be considered when choosing a particular method (Wolfensohn 2010a), which should be selected after discussion with veterinary staff and with due regard for the collection of scientific data of adequate quality.

Table 3 Points that should be considered when choosing a particular method of euthanasia (from Wolfensohn 2010a)

Death must occur without producing pain
The time required to produce loss of consciousness must be as short as possible
The time required to produce death must be as short as possible
The method must be reliable and nonreversible
There must be minimal psychological stress on the animal
There must be minimal psychological stress to the operators and any observers
It must be safe for personnel carrying out the procedure
It must be compatible with the requirements of the experiment
It must be compatible with any requirement to carry out histology on the tissues
Any drugs used should be readily available and have minimum abuse potential
The method should be economically acceptable
The method should be simple to carry out with little room for error

The most common methods for killing primates include an overdose of an anesthetic, such as sodium pentobarbital, which is administered intravenously. For larger primates, such as *M. mulatta,* it may be necessary to sedate the individual before attempting any intravenous injection. In the UK, all these drugs are prescription-only medicines and therefore must only be used under veterinary direction. Methods of sedation will vary depending on the size of the primate, the housing environment, the primate's underlying state of health, and whether the animal has been trained to take medication or receive injections (Wolfensohn 2010b). Many experimental procedures require gaining access to the individual animal and since primates are intelligent they can readily be trained to cooperate (see Bloomsmith et al. 2022), even when they are living in social groups (Honess et al. 2005). Training is encouraged by the use of positive reinforcement, usually as a favorite food or drink, and rhesus monkeys also respond to being trained to cooperate using voice commands (Reinhardt et al. 1995; Sauceda and Schmidt 2000; Reinhardt 2003). They can be trained to present their hindquarters for injection and this avoids the need to catch them or to use a "restraint" cage in which the back is pulled gently forward or the front pushed gently backwards. Most animals find the use of restraint cages very stressful and it is much better to spend time training the monkey to present its hindquarters for injection of the sedative, which can then be carried out without stress to animal or handler (Wolfensohn and Finnemore 2006). Other methods used to catch laboratory primates include net-catching (Luttrell et al. 1994), but this is a significant stress factor and most animals will not respond well to this method of capture. A useful discussion of the welfare implications of the various techniques is contained in Rennie and Buchanan-Smith (2006). Whatever the method of capture, macaques should be safely restrained and their body weight supported when they are carried.

When primates are euthanized, every effort should be made to make full use of their tissues and blood, particularly if this will minimize the number of other animals used in experiments (Jennings and Prescott 2009). This requires good mechanisms for communication to match supply and demand both within and between scientific institutes. Establishing tissue banks and data exchange networks is one means of coordinating, optimizing, reducing, and refining primate use. In the USA, Primate Info Net (http://pin.primate.wisc.edu/) achieves this objective and in Europe, EUPRIM-NET http://www.euprim-net.eu/ offers a tissue and gene bank.

8 Incorporation of Humane Endpoints in Management

The researcher is the person who is the most aware of the scientific objectives of the study, but endpoints should be decided in consultation with the veterinary surgeon, animal care staff, and the animal welfare and ethical review committee. The researcher should have in-depth knowledge of previous studies that have been conducted using the primate model, either from their own efforts or from knowing the relevant literature. The expertise of the entire team will assist in deciding how to monitor the animals and to ensure that the appropriate people are present to humanely end the animals' lives or otherwise stop the experiment when the end point will be reached. A clear chain of command and management structure needs to be established and communicated to all relevant personnel so that the person who will decide on the termination of the primate's life is available and informed. Animals need to be monitored with sufficient frequency, as guided by thorough knowledge of the normal behavior and physiology of the species, and the use of appropriate monitoring and assessment tools as described earlier.

There should be proactive planning of the study when considering the research design and use of humane endpoints, It may be necessary to use a pilot study to generate and validate an improved endpoint (for example, detection of a biomarker) which is better for the animal's welfare by being implemented earlier before the welfare deteriorates. Comparing data with prior studies using a previously accepted endpoint should be encouraged and then included in methodological publication. The use of humane endpoints should be monitored and recorded throughout the experiment, and reviewed and changed as required, and this information should always be included when publishing the results of the study.

Developments in noninvasive technologies, such as imaging, and 24-h remote monitoring of behavior, and movement detection by automated recording are becoming more sophisticated. Use of these technologies and physiological monitoring via biotelemetry may enable pain and distress to be avoided by identifying nonclinical humane endpoint criteria that occur prior to observable suffering or clinical manifestations of a condition.

References

American Veterinary Medical Association (2001) American Veterinary Medical Association report of the AMVA panel on euthanasia. J Am Vet Med Assoc 218(5):669–696. https://doi.org/10.2460/javma.2001.218.669

Animal Science Committee (2017) Review of harm-benefit analysis in the use of animals in research. Home Office

Bailey J (2022) Arguments against using nonhuman primates in research. In: Robinson LM, Weiss A (eds) Nonhuman primate welfare: from history, science, and ethics to practice. Springer, Cham, pp 547–574

Bayne K, Hau J, Morris T (2022) The welfare impact of regulations, policies, guidelines and directives and nonhuman primate welfare. In: Robinson LM, Weiss A (eds) Nonhuman primate welfare: from history, science, and ethics to practice. Springer, Cham, pp 629–646

Bloomsmith M, Perlman J, Franklin A, Martin AL (2022) Training research primates. In: Robinson LM, Weiss A (eds) Nonhuman primate welfare: from history, science, and ethics to practice. Springer, Cham, pp 517–546

Boly M, Seth AK, Wilke M et al (2013) Consciousness in humans and non-human animals: recent advances and future directions. Front Psychol 4:625. https://doi.org/10.3389/fpsyg.2013.00625

British Veterinary Association (2016) British Veterinary Association Euthanasia of animals guidelines. https://www.bva.co.uk/media/2981/bva_guide_to_euthanasia_2016.pdf. Accessed 16 June 2021

Brønstad A, Newcomer CE, Decelle T et al (2016) Current concepts of harm–benefit analysis of animal experiments–report from the AALAS–FELASA working group on harm–benefit analysis, part 1. Lab Anim 50(1 Suppl):1–20. https://doi.org/10.1177/0023677216642398

Canadian Council on Animal Care (1998) Canadian Council on Animal Care Guidelines on choosing an appropriate endpoint in experiments using animals for research and testing.

https://ccac.ca/Documents/Standards/Guidelines/Appropriate_endpoint.pdf. Accessed 16 June 2021

Carstens E, Moberg GP (2000) Recognizing pain and distress in laboratory animals. ILAR J 41(2):62–71. https://doi.org/10.1093/ilar.41.2.62

Chatfield K, Morton D (2018) The use of non-human primates in research. In: Schroeder D, Cook J, Hirsch F, Fenet S, Muthuswamy V (eds) Ethics dumping. SpringerBriefs in research and innovation governance. Springer, Cham. https://doi.org/10.1007/978-3-319-64731-9_10. Accessed 16 June 2021

Close B, Banister K, Baumans V et al (1996) Recommendations for euthanasia of experimental animals: part 1. Lab Anim 30(4):293–316. https://doi.org/10.1258/002367796780739871

Close B, Banister K, Baumans V et al (1997) Recommendations for euthanasia of experimental animals: part 2. Lab Anim 31(1):1–32. https://doi.org/10.1258/002367797780600297

Davies G, Golledge H, Hawkins P, Rowland A, Wolfensohn S, Smith J, Wells D (2017) Review of harm-benefit analysis in the use of animals in research. Report of the Animals in Science Committee Harm-Benefit Analysis Sub-group. https://assets.publishing.service.gov.uk/government/uploads/system/uploads/attachment_data/file/675002/Review_of_harm_benefit_analysis_in_use_of_animals_18Jan18.pdf. Accessed 16 June 2021

European Commission (2010) Directive 2010/63/EU of the European Parliament and of the Council of 22 September 2010 on the protection of animals used for scientific purposes. https://eur-lex.europa.eu/LexUriServ/LexUriServ.do?uri=OJ:L:2010:276:0033:0079:en:PDF. Accessed 16 June 2021

Gilbert C, Wolfensohn S (2011) Veterinary ethics and the use of animals in research: are they compatible? In: Wathes C, Corr S, May S et al (eds) Veterinary and animal ethics: Proceedings of the first International Conference on Veterinary and Animal Ethics. Wiley-Blackwell, Oxford, pp 155–173

Gleerup KB, Forkman B, Lindegaard C, Andersen PH (2015) An equine pain face. Vet Anaesth Analg 42(1):103–114. https://doi.org/10.1111/vaa.12212

Halpern-Lewis JG (1996) Understanding the emotional experiences of animal research personnel. Contemp Top Lab Anim Sci 35(6):58–60

Hendriksen C, Cussler K, Morton D (2011) Use of humane endpoints to minimize suffering. In: Howard B, Nevalainen T, Peratta G (eds) The COST manual of laboratory animal care and use. CRC Press, Boca Raton, FL, pp 333–354

Home Office (2014) Guidance on the operation of the animals (Scientific Procedures) Act 1986. https://www.gov.uk/guidance/research-and-testing-using-animals. Accessed 16 June 2021

Home Office (2015) The harm-benefit analysis process: New project licence applications. Advice Note: 05/2015. Animals in Science Regulation Unit, Home Office, London. https://assets.publishing.service.gov.uk/government/uploads/system/uploads/attachment_data/file/487914/Harm_Benefit_Analysis__2_.pdf. Accessed 16 June 2021

Honess PE, Wolfensohn S (2010a) The extended welfare assessment grid: a matrix for the assessment of welfare and cumulative suffering in experimental animals. Altern Lab Anim 38(3):205–212. https://doi.org/10.1177/026119291003800304

Honess PE, Wolfensohn S (2010b) Welfare of exotic animals in captivity. In: Tynes VV (ed) Behavior of exotic pets. Wiley-Blackwell, Oxford, p 215

Honess PE, Marin C, Brown AP, Wolfensohn SE (2005) Assessment of stress in non-human primates: application of the neutrophil activation test. Anim Welf 14(4):291–295

Jennings M, Prescott MJ (2009) Refinements in husbandry, care and common procedures for non-human primates. Lab Anim 43(Suppl 1):1–47. https://doi.org/10.1258/la.2008.007143

Justice WSM, O'Brien MF, Szyszka O et al (2017) Adaptation of the animal welfare assessment grid (AWAG) for monitoring animal welfare in zoological collections. Vet Rec 181(6):143–143. https://doi.org/10.1136/vr.104309

Langford DJ, Bailey AL, Chanda ML et al (2010) Coding of facial expressions of pain in the laboratory mouse. Nat Methods 7:447–449. https://doi.org/10.1038/nmeth.1455

Lloyd MH, Foden BW, Wolfensohn SE (2008) Refinement: promoting the 3Rs in practice. Lab Anim 42(3):284–293. https://doi.org/10.1258/la.2007.007045

Luttrell L, Acker L, Urben M, Reinhardt V (1994) Training a large troop of rhesus macaques to co-operate during catching: analysis of time investment. Anim Welf 3:135–140

Lutz CK, Baker KC (2022) Using behavior to assess primate welfare. In: Robinson LM, Weiss A (eds) Nonhuman primate welfare: from history, science, and ethics to practice. Springer, Cham, pp 171–206

Main DCJ, Whay HR, Green LE, Webster AJF (2003) Preliminary investigation into the use of expert opinion to compare the overall welfare of dairy cattle farms in different farm assurance schemes. Anim Welf 12(4):565–569

Mellor DJ (2012) Affective states and the assessment of laboratory-induced animal welfare impacts. ALTEX 444–449

Morton DB (2000) A systematic approach for establishing humane endpoints. ILAR J 41(2):80–86. https://doi.org/10.1093/ilar.41.2.80

Morton DB, Griffiths PH (1985) Guidelines on the recognition of pain, distress and discomfort in experimental animals and an hypothesis for assessment. Vet Rec 116(16):431–436. https://doi.org/10.1136/vr.116.16.431

Nuffield Council on BioEthics (2005) The ethics of research involving animals. https://www.nuffieldbioethics.org/publications/animal-research. Accessed 16 June 2021

Organisation for Economic Cooperation and Development (2000) Guidance document on the recognition, assessment, and use of clinical signs as humane endpoints for experimental animals used in safety evaluation. https://www.ncbi.nlm.nih.gov/books/NBK32660/. Accessed 16 June 2021

Pekow CA (1994) Suggestions from research workers for coping with research animal death. Lab Anim (NY) 23:28–29

Reinhardt V (2003) Working with rather than against macaques during blood collection. J Appl Anim Welf Sci 6(3):189–197. https://doi.org/10.1207/s15327604jaws0603_04

Reinhardt V, Liss C, Stevens C (1995) Restraint methods of laboratory non-human primates: a critical review. Anim Welf 4(3):221–238

Rennie AE, Buchanan-Smith HM (2006) Refinement of the use of non-human primates in scientific research. Part III: refinement of procedures. Anim Welf 15(3):239–261

Russell W, Burch R (1959) The principles of humane experimental technique. Methuen & Co, London

Sauceda R, Schmidt MG (2000) Refining macaque handling and restraint techniques. Lab Anim 29:47–49

Sebel PS, Bowdle TA, Ghoneim MM et al (2004) The incidence of awareness during anesthesia: a multicenter United States study. Anesth Analg 99(3):833–839. https://doi.org/10.1213/01.ane.0000130261.90896.6c

Smith E, Delargy M (2005) Locked-in syndrome. BMJ 330(7488):406–409. https://doi.org/10.1136/bmj.330.7488.406

Sotocina SG, Sorge RE, Zaloum A et al (2011) The rat grimace scale: a partially automated method for quantifying pain in the laboratory rat via facial expressions. Mol Pain 7:1–10. https://doi.org/10.1186/1744-8069-7-55

t'Hart BA, Laman JD, Kap YS (2022) An unexpected symbiosis of animal welfare and clinical relevance in a refined nonhuman primate model of human autoimmune disease. In: Robinson LM, Weiss A (eds) Nonhuman primate welfare: from history, science, and ethics to practice. Springer, Cham, pp 591–612

Wallace J (2000) Humane endpoints and cancer research. ILAR J 41(2):87–93. https://doi.org/10.1093/ilar.41.2.87

Wolfensohn S (2010a) Euthanasia and other fates for laboratory animals. In: Hubrecht RC, Kirkwood J (eds) The UFAW handbook on the care and management of laboratory and other research animals, 8th edn. Wiley-Blackwell, Oxford, pp 219–226

Wolfensohn SE (2010b) Old World monkeys. In: Hubrecht RC, Kirkwood J (eds) The UFAW handbook on the care and management of laboratory and other research animals, 8th edn. Wiley-Blackwell, Oxford, pp 592–617

Wolfensohn S (2020) Too cute to kill? The need for objective measurements of quality of life. Animals 10(6):1054. https://doi.org/10.3390/ani10061054

Wolfensohn S, Finnemore P (2006) Refinements in primate husbandry: a DVD training resource

Wolfensohn S, Honess PE (2007) Laboratory animal, pet animal, farm animal, wild animal: which gets the best deal? Anim Welf 16:117–123

Wolfensohn SE, Lloyd MH (2013) Handbook of laboratory animal management and welfare, 4th edn. Wiley-Blackwell, Oxford

Wolfensohn S, Sharpe S, Hall I, Lawrence S, Kitchen S, Dennis M (2015) Refinement of welfare through development of a quantitative system for assessment of lifetime experience. Anim Welf 24:139–149. https://doi.org/10.7120/09627286.24.2.139

Wolfensohn S, Shotton J, Bowley H, Davies S, Thompson S, Justice W (2018) Assessment of welfare in zoo animals: towards optimum quality of life. Animals 8(7):110. https://doi.org/10.3390/ani8070110

Part IV

Individual Differences, Application, and Improvement of Nonhuman Primate Welfare

Primate Personality and Welfare

Lauren M. Robinson and Alexander Weiss

Abstract

Like most species, primates display stable individual differences in their behavior, perceptions, cognition, and emotions. In this chapter, we examine the relationship between these traits and different measures of welfare. To this end, we first review the work on the measurement, validation, and structure of personality in nonhuman primates. We then review research on the association between personality traits and physiological, behavioral, and psychological measures of primate welfare. Finally, we introduce possible applications of personality measures and discuss new directions that this field could take. We conclude that, while no one approach will meet the welfare needs of all individuals of a particular primate species, to move primate welfare and care forward it is vital to increase our understanding of personality and welfare.

Keywords

Temperament · Well-being · Psychometrics · Variation · Care

L. M. Robinson (✉)
Wolf Science Center, University of Veterinary Medicine Vienna, Vienna, Austria

Language Research Center, Georgia State University, Atlanta, GA, USA

A. Weiss
School of Philosophy, Psychology and Language Sciences, Department of Psychology, University of Edinburgh, Edinburgh, UK

Kyoto University, Wildlife Research Center, Kyoto, Japan

Scottish Primate Research Group, Fife, UK

© Springer Nature Switzerland AG 2023
L. M. Robinson, A. Weiss (eds.), *Nonhuman Primate Welfare*,
https://doi.org/10.1007/978-3-030-82708-3_17

1 Primate Personality and Welfare

Individuals within any species of nonhuman primate vary in traits related to temperament or personality. Some of these traits appear to be related to better or poorer welfare, and so understanding how and which personality traits impact welfare can aid us in meeting, and possibly exceeding, the standard of care for captive primates. Because the study of personality and welfare is at a fairly early stage, the goal of this chapter is to introduce the findings of this work and their implications. Thus, we first describe what is meant by personality and how it can be measured. We then review the literature on associations between personality and welfare. Finally, we discuss ways in which measures of personality can be used to improve welfare, highlight limitations of this literature, and note some promising future directions.

2 Personality

In the human and the animal literature, personality traits are defined as stable individual differences in ways of behaving, feeling, and thinking (McCrae and Costa 1997; Pervin and John 1999; Gosling 2001). Personality traits, then, are characteristics of the individual and not the situation or transient mood states.

Even if one applies this definition of what a personality trait is, as we are sure the reader can imagine, the English language alone contains a large number of words that refer to the personalities of humans and/or animals. This fact was established in a seminal 1936 study by Gordon Allport and Henry S. Odbert, who combed through Webster's New International Dictionary and found 17,953 words that could be used to describe peoples' personalities. If each of these words described a unique aspect of an individual's personalities, it would be difficult to conduct personality research unless researchers limited themselves to a manageable number, and then one would be missing out on all of the other terms. Fortunately, and we expect that this will come as no surprise to the reader, there is overlap between many of these terms that can be used to describe people. For instance, individuals who are described as "fragile" are very likely also described as "sad" and less likely to be described as "resilient".

In addition to testing whether personality traits were characteristics of the individual being measured (Kenrick and Funder 1988 for a review), early studies of human personality examined the overlap between personality traits to determine just how many domains were needed to adequately describe a person (Goldberg 1993; John et al. 2008 for a review). Although there is still some disagreement (e.g., Ashton and Lee 2007), a general consensus was reached that five domains—Neuroticism, Extraversion, Openness, Agreeableness, and Conscientiousness—were sufficient (Digman 1990; McCrae and Costa 2008). With these questions (mostly) behind it, human personality research has flourished and turned to other questions, including whether personality is associated with life outcomes, including physical and mental health and well-being (see, e.g., Strickhouser et al. 2017).

Research into nonhuman primate personality no doubt benefited from this extensive human literature. Researchers up until the last years of the previous century, for instance, tested whether measures of personality traits in primates were reliable, stable, related to behavior, heritable, and/or examined the structure of these measures (Crawford 1938; Hebb 1946, 1949; Chamove et al. 1972; Buirski et al. 1973, 1978; Stevenson-Hinde and Zunz 1978; Stevenson-Hinde et al. 1980a, b; Buirski and Plutchik 1991; Gold and Maple 1994; King and Figueredo 1997; Capitanio 1999; Weiss et al. 2000). Building on these successes, researchers studying primate personality have begun to move forward to other questions, including whether animal personality traits are associated with individual differences in health and well-being.

3   Measuring Personality

This volume includes chapters devoted to different ways to measure welfare in primates (see, for example, Lutz and Baker 2022; Capitanio et al. 2022; Gartner 2022). Although many of the lessons apply to personality, we thought it would be a good idea to summarize how one goes about measuring personality in primates. The answer to this question turns out to differ depending on who you ask. This is because different disciplines, and even researchers within disciplines, often use different methods. These methods typically fall under one of two broad categories: behavioral coding and ratings (Freeman and Gosling 2010).

Behavioral coding involves recording (coding) what an individual primate does. Of course, it is not possible to record *everything* that a primate does, so researchers who use this method have to decide in advance which behaviors to focus on and then devise a set of operational definitions, known as an ethogram. Depending on the questions being asked, behaviors may be recorded for their duration, frequency, intensity, and/or latency (Martin and Bateson 2007).

The methods used to assess personality via behavioral coding differ with regard to the context in which behaviors are observed. One method involves coding behaviors that are freely expressed (Freeman and Gosling 2010). For example, van Hooff (1970) observed 25 semi-free ranging chimpanzees for 53 naturally occurring behaviors, such as touching, pouting, and grooming. The other behavioral coding method involves recording behaviors that are expressed in response to one or more behavioral tests (Fairbanks and Jorgensen 2011). A behavioral test can involve, for instance, introducing the animal to a stimulus such as an unfamiliar object (e.g., human toys, kitchen instruments, anything that is not commonly found in their environment) or person (Capitanio et al. 2017). Another kind of behavioral test involves placing that animal in a novel environment (Perals et al. 2017). For example, Johnson et al. (2015) recorded the behaviors of olive (*Papio anubis*) and yellow baboons (*P. cynocephalus*) who were put into individual cages and presented with a mirror and two novel objects (a plastic truck and a plastic bear). The ethogram included 73 behaviors relating to frequency and duration of aggression, abnormal behavior, and interactions with each object.

In its most common form, measuring personality by ratings, also known as "observer ratings" or "trait ratings", involves asking people who are familiar with the individual animals to complete a questionnaire (Wemelsfelder et al. 2000; Gosling and Vazire 2002; Freeman and Gosling 2010; Carter et al. 2012a; Freeman et al. 2013). Researchers, in this instance, can use an existing questionnaire or design their own. Using an existing questionnaire may be a good option if a questionnaire exists for the species under study or for a similar species. For example, the Emotions Profile Index (Plutchik 1962, 1980), Madingley Questionnaire (Stevenson-Hinde and Zunz 1978), and the Chimpanzee Personality Questionnaire (King and Figueredo 1997) have all been used to assess personality in nonhuman primates. If there is no existing survey for the given species or a researcher is interested in studying traits that they do not think are captured by existing questionnaires, then they can modify an existing questionnaire or develop a new one.

3.1 Reliability

Regardless of whether one uses behavioral coding or trait ratings, it is important to establish the reliability of one's measures. Which method of estimating reliability one uses depends on how one measures personality, the research question and design, and practical considerations. When personality is assessed via behavioral codings, and especially when behavioral tests are used, reliability is often assessed by collecting the measure on two or more occasions and then computing the repeatability of the measure (Boake 1989; Nakagawa and Schielzeth 2010). When personality is assessed via ratings, reliability is often assessed by obtaining measures from multiple raters and then computing the interrater reliability (Shrout and Fleiss 1979). Both types of reliability estimates are a form of intraclass correlation and so indicate the proportion of the variation in the measure that is attributable to differences between the animals. As such, it is possible (and even desirable) to estimate reliability by using a research design in which, for example, multiple coders are recording the behavior of the animal in response to multiple applications of a behavioral test or where multiple raters rate each animal on multiple occasions. In either of these situations, it is possible to compute intraclass correlations that indicate the degree to which the measure generalizes over different observers/raters and/or different occasions (Shavelson et al. 1989; Hernández-Lloreda and Colmenares 2006).

3.2 Validity

In addition to ensuring that one's personality measures are reliable, it is important to demonstrate that they are measuring what one thinks they are measuring. To do so, one first needs to develop hypotheses regarding what measures should and should not be correlated with the trait of interest (Campbell and Fiske 1959; Carter et al. 2012b, 2013). For example, if one wanted to validate a measure of fearfulness, one might expect that it would be correlated with a tendency to avoid risky situations and not a tendency to avoid grooming conspecifics, because grooming is probably more indicative of sociability. One would then collect data on their new measure of fearfulness, other purported measures of fearfulness, and measures of sociability, and perhaps other traits, too. If the new measure of fearfulness is valid, it should be highly correlated with other measures of fearfulness, but only negligibly (or uncorrelated) with the other personality traits (Campbell and Fiske 1959).

3.3 Structure

Oftentimes, researchers who wish to study personality will collect data on multiple measures: questionnaires used in rating studies typically include multiple items, multiple behavioral tests may be administered, multiple measures may be taken from a single behavioral test, or multiple behaviors may be included in an ethogram. Like the dictionary terms investigated by Allport and Odbert (1936), these measures are bound to not be independent of one another. As such, it is useful and important to examine the underlying structure of these measures. To do so, researchers can use data reduction, which is a family of statistical analyses that includes principal components analysis and different kinds of factor analyses (Gorsuch 1983). These analyses are useful in that they inform the researcher which variables can be grouped together to form more reliable measures, that is, composites or scales.

It is beyond the scope of this chapter to go into detail about how to conduct data reduction analyses and there are various guides to conducting these analyses (e.g., Costello and Osborne 2005; Budaev 2010). However, some words of caution on these guides are warranted. It has been our observation that some people have been misled by these guides. That is not to say that these guides are "wrong". Instead, they are often, by necessity, oversimplified. Unlike many statistical techniques, in factor analysis, there are many nuances and calls for judgment at every step, and what constitutes good judgment is borne from experience. We therefore advise prospective users of these methods to not exclusively rely on these guides, but to familiarize themselves with the literature, including specialist texts (e.g., Gorsuch 1983). We would also advise consulting somebody with experience in data reduction and to build up one's own experience, perhaps by using simulated data or data with known properties.

4 Personality and Welfare

Because this book contains a section dedicated to the different ways that one can measure welfare, we will move straight into a review of the associations between personality and welfare. However, before we do so, we thought it would be worth highlighting the need to consider welfare as a psychological construct. That is, these measures are based on definitions of welfare that focus on the experiences of individual animals and quantify welfare by measuring signs and signals that an animal uses to show how well it is coping (Broom 1991, 2001). These signs may include biological function, emotional experience, and the relationships between them (Hemsworth et al. 2015), as well as other ones discussed elsewhere in this volume. Different measures are thus manifestations of some underlying welfare state and can thus be used to "check" one another. In other words, by incorporating multiple measures in our studies, we can check the convergent and discriminant validities of any purported measure of welfare (see our discussion of validity in the context of personality measures).

So why would we expect to find associations between personality and primate welfare? First, stable traits related to how an individual behaves in, perceives, and reacts to the world will influence whether and what kind of mischief they can get into and whether and how they react to and experience bad *and* good things that happen to them. Second, there is a large literature showing that personality is related to outcomes in humans that would be considered welfare outcomes if they were measured in animals. Personality traits in humans are related to subjective well-being (see reviews by DeNeve and Cooper 1998; Steel et al. 2008; Anglim et al. 2020), coping (Carver and Connor-Smith 2010), and better mental and physical health (Strickhouser et al. 2017).

4.1 Ratings

King and Landau (2003) developed a four-item subjective well-being questionnaire. Each item was an attempt to measure an aspect of well-being that had been identified in studies of human happiness. The interrater reliability of ratings on these items by keepers and volunteers was satisfactory and the items loaded on a single subjective well-being factor, which was stable over time (King and Landau 2003). After establishing that subjective well-being *can* be measured in chimpanzees (*Pan troglodytes*), King and Landau, as part of the same study, tested whether subjective well-being was related to personality and behavior. Chimpanzees with higher subjective well-being scores were more likely to engage in submissive behaviors. Moreover, higher dominance, extraversion, and dependability (later termed "conscientiousness") were independently related to higher subjective well-being. The latter findings are notable in that they are broadly consistent with findings from studies of humans (again, see reviews by DeNeve and Cooper 1998; Steel et al. 2008; Anglim et al. 2020). One exception, however, was that, instead of there being a negative association between emotionality (later termed "neuroticism") and subjective well-being, King and Landau found a positive association between dominance, a personality factor seemingly absent in humans (King and Weiss 2011; Weiss 2018), and subjective well-being (King and Landau 2003).

The discrepancy between some personality and subjective well-being relationships in King and Landau's study and those in studies of humans may have been attributable to the items that made up the emotionality and dominance factors. Just three items related to human neuroticism—excitable, stable (reversed), and unemotional (reversed)—loaded on emotionality. On the other hand, five items, including, perhaps crucially, fearful, timid, and cautious, related to human neuroticism, had negative loadings on dominance. Two studies, each involving an independent sample of chimpanzees, supported this explanation. In the first study, Weiss et al. (2009) measured personality using an expanded version of the questionnaire, which included more items related to neuroticism (Weiss 2017). They found that, of the six personality factors, higher dominance, extraversion, agreeableness, and openness, and lower neuroticism, were related to higher subjective well-being. In the second study, Robinson et al. (2017) found associations between higher extraversion and lower neuroticism, and higher subjective well-being. Personality and subjective well-being in this study were measured using the same questionnaire as in Weiss et al. (2009).

Studies of other nonhuman primates have also revealed evidence for associations between personality and subjective well-being ratings. These studies mostly relied on the same expanded personality questionnaire used by Weiss et al. (2009) and the subjective well-being items devised by King and Landau (2003). Overall, the findings of these studies are broadly consistent with those of the chimpanzee studies described above. In zoo-housed orangutans (*Pongo pygmaeus and P. abelii*), higher subjective well-being was positively correlated with extraversion and agreeableness and negatively correlated with neuroticism; subjective well-being was not associated with orangutan dominance or intellect (Weiss et al. 2006). In semi-free-ranging rhesus macaques (*Macaca mulatta*) that lived on Cayo Santiago, subjective well-being was positively correlated with confidence and friendliness, and negatively correlated with anxiety, but not with the openness, dominance, or activity factors of this species (Weiss et al. 2011). A later study, one of laboratory-housed infant rhesus macaques, found that lower anxiety and higher dominance were associated with higher subjective well-being (Simpson et al. 2019). In a study of zoo-housed brown capuchin monkeys (*Sapajus apella*), both bivariate correlations and models in which the authors adjusted for sex, age, and the presence of motor stereotypies, higher subjective well-being was associated with higher sociability and assertiveness, but not neuroticism, openness, or attentiveness (Robinson et al. 2016). Finally, a study of common marmosets (*Callithrix jacchus*) housed in a laboratory facility found that subjective well-being was positively correlated with sociability and negatively correlated with neuroticism (Inoue-Murayama et al. 2018).

In addition to testing for associations between personality and welfare, as measured using the subjective well-being scale, some of the studies above have examined these associations using a scale designed specifically to assess welfare. Robinson et al. designed a 12-item (2016, 2017) and a 16-item (2021) welfare survey based on the five contributors to animal quality of life: social relationships, mental stimulation, physical health, stress, and control of one's physical and social environment (McMillan 2005). They tested this questionnaire in chimpanzees, rhesus macaques, and brown capuchin monkeys. Robinson et al. found that, in all three species, the welfare ratings were highly intercorrelated and that a single component loaded on the welfare and subjective well-being items. Brown capuchins rated as being higher on welfare and subjective well-being were rated as being higher in sociability, higher in assertiveness, and lower in neuroticism and attentiveness (Robinson et al. 2016). Chimpanzees rated as being higher in welfare and subjective well-being were rated as being higher in extraversion and openness, and lower in neuroticism (Robinson et al. 2017). Rhesus macaques rated as being higher in welfare and subjective well-being were rated as being higher in confidence, openness, dominance, and friendliness (Robinson et al. 2021).

4.2 Behavior

One study that examined the relationships between personality traits and behaviors was conducted by Vandeleest et al. (2011). Although they did not find main effects for personality, they did find a significant interaction effect between personality and rearing: rhesus macaques that were reared indoors were at greater risk of stereotypies if they were rated as being more gentle or more nervous.

A pair of studies found evidence that appeared to contradict Vandeleest et al.'s findings. These studies used behavioral tests, behavioral observations, and ratings to measure the degree to which rhesus macaques that were housed indoors were gentle, active, and likely to make contact with a novel object (Gottlieb et al. 2013). The first of these studies found that macaques that were less gentle, more active, and more likely to touch a novel object were more likely to display stereotypies (Gottlieb et al. 2013). The second showed that, among indoor-housed Chinese and Indian rhesus macaques, individuals that did not touch a novel food item in a behavioral test ("inhibited" monkeys) were less likely to develop stereotypies (Gottlieb et al. 2015). To explain the latter findings, the authors suggested that the monkeys that did touch the novel object were those that used stereotypies as a coping mechanism (cf. Ijichi et al. 2013 cited in Gottlieb et al. 2015).

A recent study of rhesus macaques that were housed indoors at the Washington National Research Center supported the notion that stereotypies are coping mechanisms that are related to certain personality traits. In this study, Peterson et al. (2017) examined the reactions of rhesus macaques to a human intruder test. Compared to macaques that did not engage in self-injurious behaviors, those that engaged in these behaviors displayed less threatening behavior and less anxious behavior during the test. These monkeys also showed a lower overall number of behavioral events. Evidence to support the notion that stereotypies are personality-linked coping methods has also emerged in zoo-housed chimpanzees. For instance, one study found that animals rated as being higher in neuroticism displayed more self-directed behaviors, such as rough scratching (Herrelko et al. 2012).

4.3 Health and Physiology

Personality traits, and particularly those related to social interactions, are associated with physical health outcomes in nonhuman primates. A study of personality ratings and veterinary records revealed that golden snub-nosed monkeys (*Rhinopithecus roxellana*) rated as less agreeable had more frequent and longer lasting illnesses (Jin et al. 2013). Another study of veterinary records, which also controlled for sex, observed behaviors related to welfare, and rank, found that rhesus macaques rated as more confident and more anxious accumulated fewer injuries throughout their lives (Robinson et al. 2018).

Two long-term follow-up studies have found associations between personality and longevity. The first study found that zoo-housed western lowland gorillas (*Gorilla gorilla gorilla*) rated as being higher in extraversion were less likely to have died over an 18.5-year follow-up period (Weiss et al. 2013). The second study, which was of zoo-housed chimpanzees found that males who were rated as higher in agreeableness were less likely to die over the follow-up period as well as a possible association between openness and longevity in females (Altschul et al. 2018).

Given that the diets of captive nonhuman primates are controlled and that these primates receive regular veterinary checkups, the findings on personality and health may very well come about because personality traits are related to the functioning of the immune systems of these animals. Studies of rhesus macaques support this possibility. A study by Maninger et al. (2003), for example, found that rhesus macaques rated as more sociable had an increased tetanus-specific antibody response. A study by Capitanio et al. (1999) found that, among rhesus macaques experimentally inoculated with the Simian Immunodeficiency Virus, ratings of sociability were positively associated with immune system response and negatively associated with viral load. These findings mirrored those of a nonexperimental study of humans who had contracted the human immunodeficiency virus (O'Cleirigh et al. 2007).

Along with the main effects of personality on health, there is evidence that personality may buffer individuals against stressors. Boyce et al. (1998) found that rhesus macaques rated as highly behaviorally inhibited were more prone to being injured during periods of group confinement. Another study, this one by Gottlieb et al. (2018), examined whether temperament was related to diarrhea in rhesus macaques and whether temperament traits moderated the association between exposure to a known stressor (relocation; Davenport et al. 2008) and diarrhea. Monkeys that were more nervous, gentle, vigilant, and confident were at greater risk of diarrhea and had diarrhea more often. Risk of having diarrhea was also associated with being relocated more often and this association was greater for animals that were less nervous, less gentle, and more confident (Gottlieb et al. 2018).

Associations between personality and cortisol have also been examined in nonhuman primate species. Byrne and Suomi (2002) found that personality ratings were associated with cortisol in tufted capuchin monkeys (*Sapajus apella*) in much the same way as behavioral observations. Specifically, they found that cortisol reactivity was positively associated with the traits aggressive, confident, effective, curious, and opportunistic, and positively related to the traits apprehensive, fearful, insecure, submissive, and tense. In chimpanzees, Anestis (2005) found that individuals who used coalitions more during aggressive interactions received more grooming and had their play offers accepted more frequently, that is, chimpanzees with a "smart" personality type, had higher urinary cortisol levels. In a later study, Anestis et al. (2006) found that chimpanzees that used coalitions more often during aggressive interactions had a greater stress response to sedation than those lower in this personality style. Finally, a study of common marmosets by Inoue-Murayama et al. (2018) found that individuals rated as being higher in sociability had higher levels of cortisol, which was measured in hair.

In addition to these relationships between personality traits and physiological variables that are collected invasively, recent evidence suggests that there is also a relationship between personality and hair loss (alopecia). In perhaps the only study published at this time, Coleman et al. (2017) found that rhesus monkeys whose temperaments were characterized as being either more inhibited or more anxious on the basis of behavioral observations and tests were less likely to have alopecia.

5   Applications

In humans and in nonhuman primates, there thus appear to be consistent associations between individual differences in personality and in outcomes related to physical and psychological well-being, including welfare. The major findings to come out of these areas of research are that (a) the gregarious, active, and socially connected fare better than the shy, inactive, and socially disconnected; and (b) the fearful, emotional, and unstable fare more poorly than the fearless, even-tempered, and stable. We would like to think that these findings can be put to use in such a way as to maximize individual primate welfare. Unfortunately, although there have been recommendations for how personality may be used to improve primate welfare (e.g., Tetley and O'Hara 2012; Gartner and Weiss 2018), little attention has been paid to how the personality and welfare relationship can be used in this regard. It seems to us that one major means of improving welfare stem from these results.

Speaking from experience, upon entering a primate facility, one is often told to avoid a certain animal because that animal is reactive to, fearful of, or aggressive to newcomers. This information is often part of the "institutional memory" of a captive facility. If these data were systematized, say by asking staff to rate individuals or by behavioral observations, and digitized, facilities would have some idea of which animals might be more (or less) at risk of poor health, injuries, or low well-being. Individual animals deemed "at-risk" could be closely monitored, and staff could intervene early and minimize the adverse impact of any problems that might arise. More "resilient" animals, on the other hand, that may not need human intervention or checks as often, are likely to be identified, too, and these animals would benefit by having less unnecessary contact with caretakers. This system might benefit especially those primates transferred from one facility to another; staff at the new facility would have a snapshot of what these animals are like. Moreover, by transferring the "institutional memory" to an "institutional cloud," additional ratings or observations over time could be used to update animals' personality profiles. Likewise, if detailed pedigrees are available, it would be possible estimate the levels of different personality traits of individuals who have not been reported on yet and even those individuals who have not been born (Lynch and Walsh 1998).

In addition to assisting with the daily monitoring of individual animals, these data could generate new findings. For example, because the relationships between personality and welfare measures are not perfect, along with primates whose welfare is exactly where one would expect it to be given their personality, there will be individuals whose welfare is better or worse than expected. Identifying these individuals and determining whether these deviations are attributable to measurement error or are systematic may yield insights into how differences between facilities impact welfare. For example, one could identify whether institutions that use some kind of innovative housing have primates that experience better (or poorer) welfare, on average, than they would expect in their animals.

6 Future Improvements

The area of primate welfare and well-being research has yielded multiple tantalizing findings. However, there are ways that the field could be improved.

First, the research to date has focused on a handful of primate species, and many of the studies have been based in zoos. For obvious reasons, this limits the generalizability of the findings. The solution to this problem is to expand the research to other species of primates and primates kept in research facilities, and those in the field.

Another shortcoming is that most studies of personality and welfare are cross-sectional. As such, we are limited with respect to understanding the direction of causation (Cook and Campbell 1979). Although longitudinal studies cannot exclude the possibility that poor welfare causes personality, and it could not rule out the possibility that a third variable is causing both, it would be a good first step.

A third shortcoming is that we do not know the extent to which the "file-drawer problem" (Rosenthal 1979) plagues personality and welfare research. That is to say, to what extent is the current state of knowledge biased by the tendency of journals to not report nonsignificant findings and even for authors to not submit such papers to journals? If, as we suspect, it is considerable in some areas, then the strength of some of the relationships may be overstated at best and null at worst. Addressing this will mean preregistering studies and changing the publication culture. At the very least, researchers should submit reports on their findings, whether null or otherwise, to preprint servers, such as BioRxiV.

There are also opportunities for research that have, so far as we know, not yet been explored. For one, there has been, so far as we are aware, just one study that even hinted at whether personality traits are related to cognitive bias tests (see Bethell and Pfefferle 2022). In that study, Bateson and Nettle (2015) found that, among three chimpanzees, the individual with the highest rank was less pessimistic. This is consistent with the findings of rating-based studies of chimpanzees that we discussed earlier (King and Landau 2003; Weiss et al. 2009; Robinson et al. 2017).

Additional studies in this area will undoubtedly advance our understanding of personality and of cognitive bias in nonhuman primates.

In addition, there has been a call in studies of personality and health in humans to consider long-term causal models to better understand the mechanisms by which personality traits end of helping or hindering us (Friedman et al. 2014). There has also been a call to examine the associations between lower-levels of personality and health (Weiss and Deary 2020). With increasing sample sizes, better measures, and aging populations, it should eventually be possible to conduct similar studies in nonhuman primates.

7 Conclusions

If anything, the findings described here suggest that we cannot expect a "one size fits all" approach to primate welfare to work. Moreover, the links between mostly stable personality traits and different measures of primate welfare suggest that the best way to improve welfare may be to work toward attending to individual personalities and their concomitant needs. Our understanding of personality's relationship with welfare has increased, but there is much left to be done if we are to maximize the potential of this field to increase our knowledge of the welfare of nonhuman primates and humans alike.

References

Allport GW, Odbert HS (1936) Trait-names: a psycho-lexical study. Psychol Monogr 47(1):i–171. https://doi.org/10.1037/h0093360

Altschul DM, Hopkins WD, Herrelko ES et al (2018) Personality links with lifespan in chimpanzees. eLife 7:e33781. https://doi.org/10.7554/eLife.33781

Anestis SF (2005) Behavioral style, dominance rank, and urinary cortisol in young chimpanzees (*Pan troglodytes*). Behaviour 142(9–10):1245–1268. https://doi.org/10.1163/156853905774539418

Anestis SF, Bribiescas RG, Hasselschwert DL (2006) Age, rank, and personality effects on the cortisol sedation stress response in young chimpanzees. Physiol Behav 89(2):287–294. https://doi.org/10.1016/j.physbeh.2006.06.010

Anglim J, Horwood S, Smillie LD et al (2020) Predicting psychological and subjective well-being from personality: a meta-analysis. Psychol Bull 146(4):279–323. https://doi.org/10.1037/bul0000226

Ashton MC, Lee K (2007) Empirical, theoretical, and practical advantages of the HEXACO model of personality structure. Personal Soc Psychol Rev 11(2):150–166. https://doi.org/10.1177/1088868306294907

Bateson M, Nettle D (2015) Development of a cognitive bias methodology for measuring low mood in chimpanzees. PeerJ 3:e998. https://doi.org/10.7717/peerj.998

Bethell EJ, Pfefferle D (2022) Cognitive bias tasks: a new set of approaches to assess welfare in nonhuman primates. In: Robinson LM, Weiss A (eds) Nonhuman primate welfare: from history, science, and ethics to practice. Springer, Cham, pp 207–230

Boake CRB (1989) Repeatability: its role in evolutionary studies of mating behavior. Evol Ecol 3:173–182. https://doi.org/10.1007/BF02270919

Boyce WT, O'Neill-Wagner P, Price CS et al (1998) Crowding stress and violent injuries among behaviorally inhibited rhesus macaques. Health Psychol 17(3):285–289. https://doi.org/10.1037/0278-6133.17.3.285

Broom DM (1991) Animal welfare: concepts and measurement. J Anim Sci 69(10):4167–4175. https://doi.org/10.2527/1991.69104167x

Broom DM (2001) Coping, stress and welfare. In: Coping with challenge: welfare in animals including humans. Proceedings of the Dahlem Conference, Dahlem University Press, Berlin, pp 1–9

Budaev SV (2010) Using principal components and factor analysis in animal behaviour research: caveats and guidelines. Ethology 116(5):472–480. https://doi.org/10.1111/j.1439-0310.2010.01758.x

Buirski P, Plutchik R (1991) Measurement of deviant behavior in a Gombe chimpanzee: relation to later behavior. Primates 32:207–211. https://doi.org/10.1007/BF02381177

Buirski P, Kellerman H, Plutchik R et al (1973) A field study of emotions, dominance, and social behavior in a group of baboons (*Papio anubis*). Primates 14:67–78. https://doi.org/10.1007/BF01730516

Buirski P, Plutchik R, Kellerman H (1978) Sex differences, dominance, and personality in the chimpanzee. Anim Behav 26(1):123–129. https://doi.org/10.1016/0003-3472(78)90011-8

Byrne G, Suomi SJ (2002) Cortisol reactivity and its relation to homecage behavior and personality ratings in tufted capuchin (*Cebus apella*) juveniles from birth to six years of age. Psychoneuroendocrinology 27(1–2):139–154. https://doi.org/10.1016/S0306-4530(01)00041-5

Campbell DT, Fiske DW (1959) Convergent and discriminant validation by the multitrait-multimethod matrix. Psychol Bull 56(2):81–105. https://doi.org/10.1037/h0046016

Capitanio JP (1999) Personality dimensions in adult male rhesus macaques: prediction of behaviors across time and situation. Am J Primatol 47(4):299–320. https://doi.org/10.1002/(SICI)1098-2345(1999)47:4<299::AID-AJP3>3.0.CO;2-P

Capitanio JP, Mendoza SP, Baroncelli S (1999) The relationship of personality dimensions in adult male rhesus macaques to progression of simian immunodeficiency virus disease. Brain Behav Immun 13(2):138–154. https://doi.org/10.1006/brbi.1998.0540

Capitanio JP, Blozis SA, Snarr J et al (2017) Do "birds of a feather flock together" or do "opposites attract"? Behavioral responses and temperament predict success in pairings of rhesus monkeys in a laboratory setting. Am J Primatol 79(1):e22464. https://doi.org/10.1002/ajp.22464

Capitanio JP, Vandeleest J, Hannibal DL (2022) Physiological measures of welfare. In: Robinson LM, Weiss A (eds) Nonhuman primate welfare: from history, science, and ethics to practice. Springer, Cham, pp 231–254

Carter AJ, Marshall HH, Heinsohn R, Cowlishaw G (2012a) Evaluating animal personalities: do observer assessments and experimental tests measure the same thing? Behav Ecol Sociobiol 66(1):153–160. https://doi.org/10.1007/s00265-011-1263-6

Carter AJ, Marshall HH, Heinsohn R, Cowlishaw G (2012b) How not to measure boldness: novel object and antipredator responses are not the same in wild baboons. Anim Behav 84(3):603–609. https://doi.org/10.1016/j.anbehav.2012.06.015

Carter AJ, Feeney WE, Marshall HH et al (2013) Animal personality: what are behavioural ecologists measuring? Biol Rev Camb Philos Soc 88(2):465–475. https://doi.org/10.1111/brv.12007

Carver CS, Connor-Smith J (2010) Personality and coping. Annu Rev Psychol 61:679–704. https://doi.org/10.1146/annurev.psych.093008.100352

Chamove AS, Eysenck HJ, Harlow HF (1972) Personality in monkeys: factor analyses of rhesus social behaviour. Q J Exp Psychol 24(4):496–504. https://doi.org/10.1080/14640747208400309

Coleman K, Lutz CK, Worlein JM et al (2017) The correlation between alopecia and temperament in rhesus macaques (*Macaca mulatta*) at four primate facilities. Am J Primatol 79(1):e22504. https://doi.org/10.1002/ajp.22504

Cook TD, Campbell DT (1979) Quasi-experimentation: design and analysis issues for field settings. Houghton Mifflin, Boston

Costello AB, Osborne JW (2005) Best practices in exploratory factor analysis: Four recommendations for getting the most from your analysis. Pract Assess 10:7

Crawford MP (1938) A behavior rating scale for young chimpanzees. J Comp Psychol 26(1):79–92. https://doi.org/10.1037/h0054503

Davenport MD, Lutz CK, Tiefenbacher S et al (2008) A rhesus monkey model of self-injury: effects of relocation stress on behavior and neuroendocrine function. Biol Psychiatry 63(10):990–996. https://doi.org/10.1016/j.biopsych.2007.10.025

DeNeve KM, Cooper H (1998) The happy personality: a meta-analysis of 137 personality traits and subjective well-being. Psychol Bull 124(2):197–229. https://doi.org/10.1037/0033-2909.124.2.197

Digman JM (1990) Personality structure: emergence of the five-factor model. Annu Rev Psychol 41:417–440. https://doi.org/10.1146/annurev.ps.41.020190.002221

Fairbanks LA, Jorgensen MJ (2011) Objective behavioral tests of temperament in nonhuman primates. In: Weiss A, King JE, Murray L (eds) Personality and temperament in nonhuman primates. Springer, New York, pp 103–127

Freeman HD, Gosling SD (2010) Personality in nonhuman primates: a review and evaluation of past research. Am J Primatol 72(8):653–671. https://doi.org/10.1002/ajp.20833

Freeman HD, Brosnan SF, Hopper LM et al (2013) Developing a comprehensive and comparative questionnaire for measuring personality in chimpanzees using a simultaneous top-down/bottom-up design. Am J Primatol 75(10):1042–1053. https://doi.org/10.1002/ajp.22168

Friedman HS, Kern ML, Hampson SE, Duckworth AL (2014) A new life-span approach to conscientiousness and health: combining the pieces of the causal puzzle. Dev Psychol 50(5):1377–1389. https://doi.org/10.1037/a0030373

Gartner MC (2022) Questionnaires and their use in primate welfare. In: Robinson LM, Weiss A (eds) Nonhuman primate welfare: from history, science, and ethics to practice. Springer, Cham, pp 255–266

Gartner MC, Weiss A (2018) Studying primate personality in zoos: implications for the management, welfare and conservation of great apes. Int Zoo Yearb 52(1):79–91. https://doi.org/10.1111/izy.12187

Gold KC, Maple TL (1994) Personality assessment in the gorilla and its utility as a management tool. Zoo Biol 13(5):509–522. https://doi.org/10.1002/zoo.1430130513

Goldberg LR (1993) The structure of phenotypic personality traits. Am Psychol 48(1):26–34. https://doi.org/10.1037/0003-066X.48.1.26

Gorsuch RL (1983) Factor analysis. L. Erlbaum Associates, Hillsdale, NJ

Gosling SD (2001) From mice to men: what can we learn about personality from animal research? Psychol Bull 127(1):45–86. https://doi.org/10.1037/0033-2909.127.1.45

Gosling SD, Vazire S (2002) Are we barking up the right tree? Evaluating a comparative approach to personality. J Res Pers 36(6):607–614. https://doi.org/10.1016/S0092-6566(02)00511-1

Gottlieb DH, Capitanio JP, McCowan BJ (2013) Risk factors for stereotypic behavior and self-biting in rhesus macaques (*Macaca mulatta*): animal's history, current environment, and personality. Am J Primatol 75(10):995–1008. https://doi.org/10.1002/ajp.22161

Gottlieb DH, Maier A, Coleman K (2015) Evaluation of environmental and intrinsic factors that contribute to stereotypic behavior in captive rhesus macaques (*Macaca mulatta*). Appl Anim Behav Sci 171:184–191. https://doi.org/10.1016/j.applanim.2015.08.005

Gottlieb DH, Del Rosso L, Sheikhi F et al (2018) Personality, environmental stressors, and diarrhea in rhesus macaques: an interactionist perspective. Am J Primatol 80(12):e22908. https://doi.org/10.1002/ajp.22908

Hebb DO (1946) Emotion in man and animal: an analysis of the intuitive processes of recognition. Psychol Rev 53(2):88–106. https://doi.org/10.1037/h0063033

Hebb DO (1949) Temperament in chimpanzees: I. Method of analysis. J Comp Physiol Psychol 42(3):192–206. https://doi.org/10.1037/h0056842

Hemsworth PH, Mellor DJ, Cronin GM, Tilbrook AJ (2015) Scientific assessment of animal welfare. N Z Vet J 63(1):24–30. https://doi.org/10.1080/00480169.2014.966167

Hernández-Lloreda MV, Colmenares F (2006) The utility of generalizability theory in the study of animal behaviour. Anim Behav 71(4):983–988. https://doi.org/10.1016/j.anbehav.2005.04.023

Herrelko ES, Vick S-J, Buchanan-Smith HM (2012) Cognitive research in zoo-housed chimpanzees: influence of personality and impact on welfare. Am J Primatol 74(4):828–840. https://doi.org/10.1002/ajp.22036

Ijichi CL, Collins LM, Elwood RW (2013) Evidence for the role of personality in stereotypy predisposition. Anim Behav 85(6):1145–1151. https://doi.org/10.1016/j.anbehav.2013.03.033

Inoue-Murayama M, Yokoyama C, Yamanashi Y, Weiss A (2018) Common marmoset (*Callithrix jacchus*) personality, subjective well-being, hair cortisol level and AVPR1a, OPRM1, and DAT genotypes. Sci Rep 8:10255. https://doi.org/10.1038/s41598-018-28112-7

Jin J, Su Y, Tao Y et al (2013) Personality as a predictor of general health in captive golden snub-nosed monkeys (*Rhinopithecus roxellana*). Am J Primatol 75(6):524–533. https://doi.org/10.1002/ajp.22127

John OP, Naumann LP, Soto CJ (2008) Paradigm shift to the integrative big five trait taxonomy. In: John OP, Robins RW, Pervin LA (eds) Handbook of personality: theory and research. Guilford Press, New York, pp 114–158

Johnson Z, Brent L, Alvarenga JC et al (2015) Genetic influences on response to novel objects and dimensions of personality in Papio baboons. Behav Genet 45(2):215–227. https://doi.org/10.1007/s10519-014-9702-6

Kenrick DT, Funder DC (1988) Profiting from controversy: lessons from the person-situation debate. Am Psychol 43(1):23–34. https://doi.org/10.1037/0003-066X.43.1.23

King JE, Figueredo AJ (1997) The five-factor model plus dominance in chimpanzee personality. J Res Pers 31(2):257–271. https://doi.org/10.1006/jrpe.1997.2179

King JE, Landau VI (2003) Can chimpanzee (*Pan troglodytes*) happiness be estimated by human raters? J Res Pers 37(1):1–15. https://doi.org/10.1016/S0092-6566(02)00527-5

King JE, Weiss A (2011) Personality from the perspective of a primatologist. In: Weiss A, King JE, Murray L (eds) Personality and temperament in nonhuman primates. Springer, New York, pp 77–99

Lutz CK, Baker KC (2022) Using behavior to assess primate welfare. In: Robinson LM, Weiss A (eds) Nonhuman primate welfare: from history, science, and ethics to practice. Springer, Cham, pp 171–206

Lynch M, Walsh B (1998) Genetics and analysis of quantitative traits. Sinauer, Sunderland, MA

Maninger N, Capitanio JP, Mendoza SP, Mason WA (2003) Personality influences tetanus-specific antibody response in adult male rhesus macaques after removal from natal group and housing relocation. Am J Primatol 61(2):73–83. https://doi.org/10.1002/ajp.10111

Martin P, Bateson P (2007) Measuring behaviour: an introductory guide, 3rd edn. Cambridge University Press, Cambridge

McCrae RR, Costa PT (1997) Personality trait structure as a human universal. Am Psychol 52(5):509–516. https://doi.org/10.1037/0003-066X.52.5.509

McCrae RR, Costa PT (2008) The five factor theory of personality. In: John OP, Robins RW, Pervin LA (eds) Handbook of personality: theory and research. The Guildford Press, New York, pp 159–181

McMillan FD (2005) Mental wellness: the concept of quality of life in animals. In: McMillan FD (ed) Mental health and well-being in animals. Blackwell Publishing, New York

Nakagawa S, Schielzeth H (2010) Repeatability for Gaussian and non-Gaussian data: a practical guide for biologists. Biol Rev 85(4):935–956. https://doi.org/10.1111/j.1469-185X.2010.00141.x

O'Cleirigh C, Ironson G, Weiss A, Costa PT (2007) Conscientiousness predicts disease progression (CD4 number and viral load) in people living with HIV. Health Psychol 26(4):473–480. https://doi.org/10.1037/0278-6133.26.4.473

Perals D, Griffin AS, Bartomeus I, Sol D (2017) Revisiting the open-field test: what does it really tell us about animal personality? Anim Behav 123:69–79. https://doi.org/10.1016/j.anbehav.2016.10.006

Pervin LA, John OP (1999) Handbook of personality: theory and research. The Guildford Press, New York

Peterson EJ, Worlein JM, Lee GH et al (2017) Rhesus macaques (*Macaca mulatta*) with self-injurious behavior show less behavioral anxiety during the human intruder test. Am J Primatol 79(1):e22569. https://doi.org/10.1002/ajp.22569

Plutchik R (1962) The emotions: facts, theories and a new model. Random House, New York

Plutchik R (1980) Emotion: a psychoevolutionary synthesis. Harper & Row, New York

Robinson LM, Waran NK, Leach MC et al (2016) Happiness is positive welfare in brown capuchins (*Sapajus apella*). Appl Anim Behav Sci 181:145–151. https://doi.org/10.1016/j.applanim.2016.05.029

Robinson LM, Altschul DM, Wallace EK et al (2017) Chimpanzees with positive welfare are happier, extraverted, and emotionally stable. Appl Anim Behav Sci 191:90–97. https://doi.org/10.1016/j.applanim.2017.02.008

Robinson LM, Coleman K, Capitanio JP et al (2018) Rhesus macaque personality, dominance, behavior, and health. Am J Primatol 80(2):e22739. https://doi.org/10.1002/ajp.22739

Robinson LM, Waran NK, Handel I, Leach MC (2021) Happiness, welfare, and personality in rhesus macaques (*Macaca mulatta*). Appl Anim Behav Sci 236:105268. https://doi.org/10.1016/j.applanim.2021.105268

Rosenthal R (1979) The file drawer problem and tolerance for null results. Psychol Bull 86(3):638–641. https://doi.org/10.1037/0033-2909.86.3.638

Shavelson RJ, Webb NM, Rowley GL (1989) Generalizability theory. Am Psychol 44(6):922–932. https://doi.org/10.1037/0003-066X.44.6.922

Shrout PE, Fleiss JL (1979) Intraclass correlations: uses in assessing rater reliability. Psychol Bull 86(2):420–428. https://doi.org/10.1037//0033-2909.86.2.420

Simpson EA, Robinson LM, Paukner A (2019) Infant rhesus macaque (*Macaca mulatta*) personality and subjective well-being. PLoS One 14(12):e0226747. https://doi.org/10.1371/journal.pone.0226747

Steel P, Schmidt J, Shultz J (2008) Refining the relationship between personality and subjective well-being. Psychol Bull 134(1):138–161. https://doi.org/10.1037/0033-2909.134.1.138

Stevenson-Hinde J, Zunz M (1978) Subjective assessment of individual rhesus monkeys. Primates 19(3):473–482. https://doi.org/10.1007/BF02373309

Stevenson-Hinde J, Stillwell-Barnes R, Zunz M (1980a) Individual differences in young rhesus monkeys: consistency and change. Primates 21(4):498–509. https://doi.org/10.1007/BF02373838

Stevenson-Hinde J, Stillwell-Barnes R, Zunz M (1980b) Subjective assessment of rhesus monkeys over four successive years. Primates 21(1):66–82. https://doi.org/10.1007/BF02383825

Strickhouser JE, Zell E, Krizan Z (2017) Does personality predict health and well-being? A metasynthesis. Health Psychol 36(8):797–810. https://doi.org/10.1037/hea0000475

Tetley C, O'Hara S (2012) Ratings of animal personality as a tool for improving the breeding, management and welfare of zoo mammals. Anim Welf 21(4):463–476. https://doi.org/10.7120/09627286.21.4.463

van Hooff JARAM (1970) A component analysis of the structure of the social behaviour of a semi-captive chimpanzee group. Experientia 26(5):549–550. https://doi.org/10.1007/BF01898505

Vandeleest JJ, McCowan BJ, Capitanio JP (2011) Early rearing interacts with temperament and housing to influence the risk for motor stereotypy in rhesus monkeys (*Macaca mulatta*). Appl Anim Behav Sci 132(1–2):81–89. https://doi.org/10.1016/j.applanim.2011.02.010

Weiss A (2017) Exploring factor space (and other adventures) with the hominoid personality questionnaire. In: Personality in nonhuman animals. Springer, Cham, pp 19–38

Weiss A (2018) Personality traits: a view from the animal kingdom. J Pers 86(1):12–22. https://doi.org/10.1111/jopy.12310

Weiss A, Deary IJ (2020) A new look at neuroticism: should we worry so much about worrying? Curr Dir Psychol Sci 29(1):92–101. https://doi.org/10.1177/0963721419887184

Weiss A, King JE, Figueredo AJ (2000) The heritability of personality factors in chimpanzees (*Pan troglodytes*). Behav Genet 30(3):213–221. https://doi.org/10.1023/A:1001966224914

Weiss A, King JE, Perkins L (2006) Personality and subjective well-being in orangutans (*Pongo pygmaeus* and *Pongo abelii*). J Pers Soc Psychol 90(3):501–511. https://doi.org/10.1037/0022-3514.90.3.501

Weiss A, Inoue-Murayama M, Hong K-W et al (2009) Assessing chimpanzee personality and subjective well-being in Japan. Am J Primatol 71(4):283–292. https://doi.org/10.1002/ajp.20649

Weiss A, Adams MJ, Widdig A, Gerald MS (2011) Rhesus macaques (*Macaca mulatta*) as living fossils of hominoid personality and subjective well-being. J Comp Psychol 125(1):72–83. https://doi.org/10.1037/a0021187

Weiss A, Gartner MC, Gold KC, Stoinski TS (2013) Extraversion predicts longer survival in gorillas: an 18-year longitudinal study. Proc Biol Sci 280(1752):20122231. https://doi.org/10.1098/rspb.2012.2231

Wemelsfelder F, Hunter EA, Mendl MT, Lawrence AB (2000) The spontaneous qualitative assessment of behavioural expressions in pigs: first explorations of a novel methodology for integrative animal welfare measurement. Appl Anim Behav Sci 67(3):193–215. https://doi.org/10.1016/s0168-1591(99)00093-3

Sociality, Health, and Welfare in Nonhuman Primates

Brianne A. Beisner, Darcy L. Hannibal, Jessica J. Vandeleest, and Brenda McCowan

Abstract

Herein we provide a review of the importance of social relationships on health and welfare of nonhuman primates. Social relationships are a key component of life for most primates. Sociality, however, comes with costs and benefits to health and welfare. When examining these benefits and costs we can look at two main types of primate relationships: dominance relationships/social status and affiliative relationships/social bonds. Over millions of years of evolution, many primate species experienced selection for larger brains (relative to body size) and living in cohesive social groups, which has resulted in a social life that is fundamentally linked to their welfare. These two evolutionary characteristics lead to complexity in both the dominance relationships and social bonds that develop in many primate societies. Individual dominance rank may be associated with differential risk of injury, exposure to stress, and other welfare-related outcomes because dominance relationships govern competitive interactions over access to resources and mating opportunities. Similarly, the affiliative social bonds between primates can also impact welfare because such interactions offer a way to cope with social stress yet increase risk of disease transmission via contact. Zoos, sanctuaries, research facilities, and other institutions that house nonhuman primates in captivity must take into consideration such social

---

B. A. Beisner (✉) · D. L. Hannibal · J. J. Vandeleest
Department of Population Health & Reproduction, School of Veterinary Medicine, University of California Davis, Davis, CA, USA
e-mail: brianne.aminta.beisner@emory.edu

B. McCowan
Department of Population Health & Reproduction, School of Veterinary Medicine, University of California Davis, Davis, CA, USA

Neuroscience and Behavior Unit, California National Primate Research Center, University of California Davis, Davis, CA, USA

relationships in their management strategies and procedures to promote good welfare. Here we discuss these social relationships in relation to individual health, fitness, and well-being.

Keywords

Social bonds · Aggression · Hierarchy · Health · Affiliation · Grooming

1 Introduction

Animal behavior is largely driven by three major selection pressures that animals are evolved to cope with: obtaining enough food to eat, finding mates for reproduction, and avoiding predation (e.g., *Papio cynocephalus*: Bertram 1978; Darwin 1871; Clutton-Brock 1974; Bradbury and Vehrencamp 1976; Kay et al. 1988; Janson and Goldsmith 1995). Primates are distinguished from other orders of mammals by their social, behavioral, and cognitive adaptations for coping with these three selective pressures (Byrne and Whiten 1988; Cheney and Seyfarth 1990; Dunbar 1998). Therefore, in captivity, a beneficial social environment buffers against stress and provides an appropriate context to engage social adaptations, while an absent or poor social environment lacks these and can even result in physical injury or abnormal changes to both behavioral and biological systems (e.g., digestive, immune, cardiovascular, endocrine) that impact welfare (e.g., Harlow and Harlow 1962; Clarke et al. 1995; Kempes et al. 2008; Gilbert and Baker 2011; Beisner et al. 2012; Gottlieb et al. 2013; Capitanio and Cole 2015). Deleterious effects of a poor social environment can be transient and reversible, longer-lasting, or permanent depending on several key and interacting factors (e.g., sex, personality, length, and scale of change, whether experienced during growth and development, etc.; Suomi et al. 1973; Coe 1993; Capitanio et al. 1998; Schapiro et al. 2000; Vandeleest et al. 2011; Gottlieb et al. 2013; McCowan et al. 2016). Because a species-appropriate social environment is the most important enrichment that can be provided for good welfare in captive primates, research facilities are mandated by national, international, and professional institutions to provide socialization in captivity (NC3Rs 2006; The Commission of European Communities 2007; European Commission 2010; National Research Council 2011; U.S. Department of Agriculture 2013; Association for Assessment and Accreditation of Laboratory Animal Care Internatonal 2015; Office of Laboratory Animal Welfare 2015).

Animal welfare has both physical and psychological components, and these components are often intertwined (e.g., poor psychological welfare may be associated with poor physical welfare). For primates, psychological welfare has taken center stage since the 1985 amendment to the Animal Welfare Act requiring "a physical environment adequate to promote the psychological well-being of primates." (Novak and Suomi 1988, Sec. 43(a)(2)(A)). Given primates' adaptations for social complexity and intelligence, it is not surprising that social housing is regarded to be the best form of enrichment for captive primates (Lutz and Novak

2005). Indeed, research facilities endeavor to pair house their research subjects (e.g., macaques) whenever possible. Furthermore, some research facilities, as well as most sanctuaries and zoos, provide more species-appropriate social housing (e.g., mixed-sex, age-graded groups for baboons, capuchins, chimpanzees, and macaques; male-female pairs with immature offspring for monogamous species such as owl monkeys and titi monkeys) which goes further to meet primates' psychological well-being. Given this chapter's emphasis on the welfare implications of social status and social bonds, we focus here on animals living in social groups, specifically addressing the physical and/or psychological welfare issues that may arise due to forming typical dominance and affiliative relationships with their group mates. Despite primates' (particularly monkeys and apes) clear need for social housing, social relationships are not without cost; navigating social relationships includes risks of aggression, harassment, trauma, and psychosocial stress, all of which have implications for both psychological and physical well-being. Finally, because the overwhelming majority of welfare research on primates in captivity is based on members of the genus *Macaca*, much of the literature is based on macaque models; however, where possible, we include examples from other primate species as well.

Primates evolved to live in cohesive social groups because of the fitness benefits from cooperative defense of food resources (Wrangham 1980), reduced risk of predation or infanticide (van Schaik 1983, 1989; Sterck et al. 1997), and/or avoiding the costs of dispersal for the philopatric sex (Isbell 2004). And although group-living comes with costs (e.g., within group competition for resources and increased risk of exposure to diseases and parasites: Alexander 1974; Isbell 1991; Chapman and Chapman 2000; Vitone et al. 2004), primates are adapted to develop and maintain different types of complex and clearly defined social relationships that mitigate the costs of group-living. For example, dominance relationships govern priority of access to defensible food resources (Whitten 1983; van Schaik and van Noordwijk 1988; Barton and Whiten 1993), which minimize the frequency of aggressive competitive interactions. Additionally, beneficial grooming activity may be exchanged for tolerance at feeding sites (Kapsalis and Berman 1996; Chancellor and Isbell 2009). These behaviors are exhibited across a variety of habitats and tend to persist even in the captive environment where predation is absent and food is abundant (Brennan and Anderson 1988; Deutsch and Lee 1991).

Therefore, primates' social behavioral interactions and relationships constitute a major part of primate daily life (e.g., Dunbar 1991; Lehmann et al. 2007). As such, a wealth of research has shown that social relationships impact a wide range of physical and mental aspects of health and well-being (for summary see McCowan et al. 2016; Hannibal et al. 2017). We divide social relationships into two main categories for the purposes of understanding individual health and welfare: social status or dominance rank and affiliative social bonds. Below we summarize the beneficial and costly implications of both types of social relationships on welfare in nonhuman primates and highlight where such relationships may intersect with management approaches.

2 Social Status, Dominance Relationships, and Welfare

Many, if not most, primate societies are organized around a dominance hierarchy, including most members of the taxonomic groups representing lemurs, New World monkeys, Old World monkeys, and apes (Smuts et al. 1987; Campbell et al. 2011). Dominance relationships are generally maintained via regular, but usually noninjurious, agonistic interactions. These relationships serve to mitigate conflict among members of a social group by allowing individuals to predict the outcome of potential contests (Bernstein and Gordon 1974; Rowell 1974). Both dominant and subordinate animals can use this relationship information to avoid potential injury from a more serious physical contest. Therefore, a certain level of aggression, and even injury (e.g., among macaques and baboons), is normal in stable social groups, although what can be considered normal in terms of frequency and severity of aggression depends on the species (Ruehlmann et al. 1988; Alford et al. 1995; Byrne et al. 1996; McCowan et al. 2008; Beisner et al. 2012).

2.1 Potential Costs of Being Subordinate

The relative costs and benefits of high versus low status among social living animals have been extensively studied, particularly with respect to agonistic interactions over access to high-quality, defensible food resources: (Whitten 1983; Saito 1996; Isbell et al. 1999; Stahl and Kaumanns 2003; Scott and Lockard 2006), reproductive success and/or access to mating opportunities (Harcourt 1987; Cowlishaw and Dunbar 1991; Pusey et al. 1997; van Noordwijk and van Schaik 1999), and psychosocial stress and stress-related disease (Creel 2001; Abbott et al. 2003; Sapolsky 2005). Both psychosocial stress and agonistic and/or competitive interactions have implications for welfare (whereas there is no evidence to suggest that welfare is reduced by absence of mating opportunities), thus we explore primarily the first two topics here. The picture that has emerged from this body of literature is that subordinates generally experience greater costs than dominants (see section on *Psychosocial Stress and Welfare* for details), suggesting that the frequency and type of welfare issues related to social status are greater for subordinates than dominants. However, in a given social group, subordinates do not always experience greater psychosocial stress or reduced access to food resources. For example, among studies of stress and stress-related disease in primates and other animals, whether evidence of stress is greater in subordinates or dominants depends upon species' social structure and the type of health measure examined (Sapolsky 2005; Creel et al. 2013; Marmot and Sapolsky 2014; Habig and Archie 2015; Vandeleest et al. 2016). In other words, being subordinate does not automatically imply poor welfare, nor does being dominant automatically imply good welfare.

2.2 Resource Competition, Aggressive Interactions, and Welfare

A consequence of having dominance relationships is that subordinates can be competitively excluded from important resources, such as high-quality food (defined by Schoener 1971 as providing a high net energy gain) that is clumped in distribution and worth monopolizing, as is true with most fruit and seed foods (Boccia et al. 1988; Stahl and Kaumanns 2003; Scott and Lockard 2006). Further, dominance relationships must be maintained through occasional or regular agonistic interactions that can range from displacements (a subordinate animal relinquishes their spot to an approaching dominant), to mild threats (facial and/or vocal aggressive signal with no physical contact), to attacks or bites. Both social processes can potentially influence the welfare of subordinates in captive settings and we discuss these in more detail below.

Generally, when food resources are high-quality and distributed in discrete pockets, such as monkey chow being provided in a single location, dominant animals can monopolize the food resource and reduce subordinates' access to this food (e.g., Boccia et al. 1988). Clumped and desirable resources also increase the frequency and intensity of aggression as animals become increasingly likely to resort to aggression to access or defend these resources (Barton et al. 1996; Goldberg et al. 2001). Long-term, pervasive exclusion from food has the potential to result in serious nutritional and caloric deficiencies in subordinate animals if not managed appropriately. Providing food in such a large abundance that there is enough food to satiate all individuals is one strategy (e.g., observed at multiple National Primate Research Centers) to ensure that subordinates do not suffer from poor nutritional status or body condition. Indeed, the impact of dominance on feeding behavior (e.g., relative food intake, feeding rate) is greatest when the group receives the least amount of food relative to overall group nutritional needs (Deutsch and Lee 1991). However, even under conditions of abundant food (where malnutrition is rarely an issue), discrepancies in nutritional status can arise since most primates are energy maximizers (Sensu Schoener 1971) and are adapted to eat more calories than needed to store fat for future calorie deficits. For example, in a social group of captive rhesus macaques given abundant monkey chow in a single location, higher-ranking females show higher body fat than lower-ranking females (Small 1981). Such rank-related differences in body fat and body weight can further impact onset of puberty and potentially reproductive success (Fairbanks and McGuire 1984; Bercovitch and Strum 1993; Zehr et al. 2005), which may be of interest to facilities with breeding programs, such as zoos or research centers. Furthermore, providing overly abundant food resources runs the risk of causing obesity in captive animals. Obesity is a common problem among animals in captivity (Leigh 1994; Schwitzer and Kaumanns 2001; Videan et al. 2007), suggesting that alternative feeding regimes might be considered to maximize welfare of all social group members.

A potentially more effective feeding strategy to promote good welfare for both subordinate and dominant individuals is to spread food out over a large area and/or provide alternative food resources, thereby discouraging or preventing contest competition over food. For example, providing foraging materials or vegetation can reduce rates of aggression in captive primates (Chamove et al. 1982; Beisner and Isbell 2011), suggesting that competitive interactions over food are reduced. Widely distributing food has similar beneficial effects (Boccia et al. 1988). Although widely distributed and/or alternative food resources mimic wild conditions, it is worth noting that many studies of wild groups find little to no difference between dominants and subordinates in nutritional and/or caloric intake (Majolo et al. 2012; Heesen et al. 2013). It may be that dominant animals in wild groups cannot competitively exclude subordinates from all food resources all the time when food patches vary in number, size, and seasonal availability/abundance. Another method that has the potential to mitigate feeding competition in captivity is the use of automated feeders that portion food dispensed according to information in each animal's electronic chip (e.g., RFID chip implanted beneath the skin on an animal's hand: Wilson et al. 2008; Ethun et al. 2015). However, these systems are costly and a similar setup and monitoring may be important to replicate the surprisingly low theft rate of food by dominants reported by these investigators.

Another consequence of dominance relationships that impacts animal welfare is injury from social aggression. The simple acts of exercising one's dominance status (e.g., to obtain a resource) and regular maintenance of dominance relationships can place individuals, and particularly subordinate animals, at risk of injury from social aggression. Subordinates are at greatest risk of injury from social aggression in species where subordinates are (1) frequent targets of redirected aggression by dominants, (2) frequently excluded competitively from resources, and (3) regularly subject to physical contests with dominants. For example, among wild female baboons (Archie et al. 2014) and captive rhesus macaques (McCowan unpublished data; Fig. 1), subordinate animals are injured more frequently than dominant animals.

2.2.1 Addressing Welfare Concerns Related to Aggression and Trauma

While some amount of aggression and injury is considered normal and expected for some species, such as macaques, chimpanzees, and capuchins (Ruehlmann et al. 1988; Alford et al. 1995; Byrne et al. 1996), occasionally the frequency and/or severity of aggressive interactions reaches a level that is unhealthy for both individuals targeted by the aggression and for the social group (e.g., Oates-O'Brien et al. 2010). Determining whether an individual should be removed from the social group based on either receipt or instigation of aggressive incidents; however, is not always obvious. These decisions involve not only consideration of the animal's current social situation but also what future social opportunities are available and whether they offer improvement over current conditions. Among macaques, for example, it is difficult to introduce an adult female into an existing social group (and possible, though dangerous, to introduce an adult male). Therefore, the housing options for a macaque that has been removed from its social group generally include (a) being pair-housed indoors or, if no suitable partners are available, being singly housed, and (b) occasionally, being included in a new group formation, but these formations are typically infrequent and labor-intensive to manage until well-established.

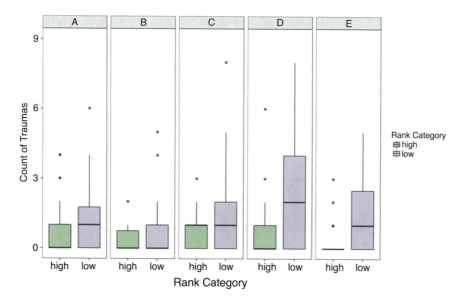

Fig. 1 Trauma rate for high versus low rank across five captive groups (**A, B, C, D,** and **E**) of rhesus macaques (McCowan unpublished data). According to Poisson regression models of trauma rate, low-ranking animals are 1.2 to 5.5 more likely to receive trauma than high ranked animals (**A**: $p = 0.36$; **B**: $p = 0.05$; **C**: $p = 0.06$; **D**: $p < 0.001$; **E**: $p < 0.001$)

Such dramatic change in social housing conditions can negatively impact welfare and must be balanced against the current welfare concerns in the animal's current social group. For example, relocating a macaque from an outdoor social group to indoor pair housing or single housing can diminish the animals' psychological welfare by reducing the opportunity to form a variety of complex and well-differentiated social relationships with a variety of conspecifics. Relocation from group housing to pair or single housing has other negative impacts, such as increasing the incidence of abnormal behavior such as stereotypies and self-injurious behavior (Suomi et al. 1971; Anderson and Chamove 1980; Eaton et al. 1994; Fontenot et al. 2006; Gottlieb et al. 2013). Further, the welfare impacts of relocation to single housing are also demonstrated in physiological changes, with alterations to cortisol and leukocyte levels that can take 1–5 months to reach a new set point which affects both the animal's welfare as well as its utility as a research subject in research institutions (Capitanio et al. 1998). Returning subjects to a social group after several months of absence or introducing them to a new group as part of a new social group formation have similarly disruptive effects on physiology that take weeks or months to normalize (Gust et al. 1991, 1993). When group membership is unstable and unpredictable, the altered physiology of individuals can be so dramatic, and welfare can be so compromised, that it can hasten the progress to disease states (Cohen et al. 1992; Gordon et al. 1992; Gust et al. 1992, 1994; Capitanio et al. 1998). Finally, if removal of an animal is deemed necessary, it is important to remember that its effect on the net welfare of the individual's former social group is difficult to anticipate because individuals that serve a keystone role in the group can have a disproportionate impact on social dynamics (good or bad) with emergent properties that are only detectable once the keystone is absent (Modlmeier et al. 2014). For example, individuals that connect two otherwise-separate clusters of the social group or individuals that are one of a few highly connected "hubs" in a social network are likely to hold keystone roles. Although social connections are expected to "rewire" after removal of any individual, the ability to fully recover depends on how critical the individuals were to the global social structure of the group. For instance, natural systems, such as social groups, often exhibit scale-free structure in their connectivity in which few individuals are the well-connected hubs that unite all remaining members of the group. While such networks are resilient to removal of random individuals, they are extremely vulnerable to removal of a well-connected hub because the pre-existing social structure (such as a dominance hierarchy) may not have enough redundant information among remaining members to allow social ties to reconnect (Albert et al. 2000; Whitacre 2010). As with food competition, management strategies and procedures can be adjusted to reduce wounding and aggression, thereby reducing the need to consider whether social relocation is necessary and improving both individual-level and group-level welfare. Strategies for reducing aggression and wounding involve keeping animals occupied (e.g., structural enrichment, foraging enrichment: Chamove et al. 1982), reducing opportunities for aggressive competition over resources (e.g., change feeding practices to minimize feeding competition, including alternate food sources: Beisner and Isbell 2008), spreading food over a large area, or providing smaller units of a high-quality food resource (Mathy and Isbell 2001), and providing opportunities for subordinates to avoid or escape aggression from dominant individuals (Erwin et al. 1976; Estep and Baker 1991).

2.3 Psychosocial Stress and Welfare

A less obvious potential impact of social status on welfare is via psychosocial stress and diseases related to chronic stress. The experience of acute, manageable stress is a fact of everyday living and, in fact, can lead to resilience in the face of future stressors (Lyons and Parker 2007). Chronic stress, however, has been linked to many poor psychological and physical outcomes, including increased susceptibility to infectious disease, cardiovascular disease, depression, and hippocampal dysfunction among others (McEwen 1998; Cohen et al. 2007). Glucocorticoid (GC) concentration is the most commonly investigated measure of stress, partially because when GCs are chronically released they can suppress the immune system and can leave individuals vulnerable to a variety of diseases and infectious agents (Sapolsky et al. 2000). However, GCs represent only one aspect of the stress response, and systems other than the hypothalamic-pituitary-adrenal (HPA) axis are also affected by chronic stress (see Capitanio et al. 2022). For example, stress causes activation of the sympathetic nervous system. Repeated activation of the sympathetic nervous system can lead to cardiovascular "wear and tear" and an increased risk of atherosclerosis and cardiovascular disease (Shively and Day 2015).

Perhaps the most familiar hypothesis regarding the impact of social status on physiological health is that individuals exposed to chronic psychosocial stress may show greater incidence of stress-related disease and thus poorer health and welfare (Sapolsky 1982; Dhabhar 2014; Habig and Archie 2015). Typically, it is subordinate animals that experience chronic psychosocial stress as they tend to experience more frequent and unpredictable aggression and have fewer social outlets to help them cope (see below). Although elevated levels of glucocorticoids have been reported in both high-ranked and low-ranked individuals in many primate species, the literature points to a greater number of potential health costs for low-status individuals than for high-status individuals, such as higher basal levels of GCs (Sapolsky 1982), poorer cardiovascular function (Sapolsky and Share 1994; Shively and Clarkson 1994), and greater risk of infection (Cohen et al. 1997; Foerster et al. 2015).

The social dynamics of rank acquisition and maintenance strongly influence which group members experience greater stress. Subordinates are exposed to greater psychosocial stress in societies where (a) their daily activities are frequently disrupted (e.g., unpredictable displacement aggression or disrupted feeding by dominants: Ray and Sapolsky 1992; Sapolsky and Spencer 1997), (b) they have reduced access to social support for coping with stress (e.g., grooming partners: Keverne et al. 1984; Sapolsky and Ray 1992; Abbott et al. 2003), and (c) resource inequity is great, with subordinates being competitively excluded from resources (Sapolsky 2005; Cavigelli and Caruso 2015). These factors can be ameliorated in captivity via the same management strategies for reducing feeding competition, aggression, and social trauma (see above) by ensuring that there is enough physical space and variation in structural enrichment (e.g., vertical perching, visual barriers) for subordinates to avoid stressful interactions with dominants (Erwin 1977; Fairbanks et al. 1978) and that important resources, such as food, water, enrichment, and desirable resting substrates are plentiful enough and/or distributed in such a way that subordinates can access these (Boccia et al. 1988; Scott and Lockard 2006). Further, for some taxa whose natural social structure involves complicated fission-fusion dynamics, such as chimpanzees, the risk of aggression, and injury to subordinates (and presumably the incidence of psychosocial stress) may be reduced through active management of social housing to emulate fission-fusion by regularly shuffling group membership (see Morimura et al. 2022).

Evidence of chronic stress in subordinates comes from a variety of measures, species, and environments. Subordinates may show higher basal levels of GCs (Sapolsky 1982; Yodyingyuad et al. 1985; Gust et al. 1993; Ostner et al. 2008), although this pattern is not universally true (Table 1; e.g., Bercovitch and Clarke 1995; Smith and French 1997; Setchell et al. 2008). Further, subordinates may experience increased susceptibility to disease and infectious agents to due immune

Table 1 Studies of glucocorticoid levels by dominance rank in nonhuman primate species

Species	GC pattern	Sex	Study condition	Sample medium	Citations
Lemur catta	Dom > Subord	f	Wild	Feces	Cavigelli (1999)
Callithrix jacchus	Dom > Subord	f	Captive	Urine	Saltzman et al. (1998); as found in Abbott et al. (2003)
Saguinus oedipus	Dom > Subord	m, f	Captive	Urine	Ziegler et al. (1995); as found in Abbott et al. (2003)
Saimiri sciurius	Dom > Subord	m	Captive		Coe et al. (1979)
Macaca fuscata	Dom > Subord	m	Wild	Feces	Barrett et al. (2002)
Pan troglodytes	Dom > Subord	m	Wild	Urine	Muller and Wrangham (2004)
Saimiri sciurius	Subord > Dom	m	Captive	Blood	Manogue (1975)
Macaca assamensis	Subord > Dom	m	Wild	Feces	Ostner et al. (2008)
Macaca fascicularis	Subord > Dom	f	Captive	Urine	Abbott et al. (2003)
Macaca mulatta	Subord > Dom	f	Captive		Gust et al. (1993)
Miopithecus talapoin	Subord > Dom	f	Captive	Urine	Yodyingyuad et al. (1985), Abbott et al. (2003)
Papio anubis	Subord > Dom	m	Wild	Blood	Sapolsky (1983)
Papio cynocephalus	Subord > Dom	m	Wild	Feces	Gesquiere et al. (2011)
Callithrix kuhli	Dom = Subord	f	Captive		Smith and French (1997)
Macaca mulatta	Dom = Subord	m	Captive	Blood	Bercovitch and Clarke (1995)
Microcebus murinus	Dom = Subord	m	Captive		Perrett (1992)
Mandrillus sphinx	Dom = Subord	f	Semicaptive	Feces	Setchell et al. (2008)
Papio hamadryas ursinus	Dom = Subord	f	Wild	Feces	Weingrill et al. (2004)

suppression (Cohen et al. 1997; Foerster et al. 2015) and poorer cardiovascular function (Adams et al. 1985; Sapolsky and Mott 1987; Sapolsky and Share 1994; Shively and Clarkson 1994). Although elevated GC levels are generally interpreted as evidence of chronic stress, it is important to be cautious when using GC concentration to assess animal welfare. Chronically high levels of GCs can alter the regulation of the HPA axis which can also lead to a blunting of the cortisol response after exposure to long-term social stress. Under such conditions of long-term stress, it is expected that an animal would have low levels of GCs even though they may continue to experience high levels of stress (Capitanio et al. 1998). Therefore, other indicators of welfare (e.g., avoidance of social interactions, anxiety and abnormal behaviors, parasite load, biomarkers of inflammation, evidence of stress-related disease) should also be assessed. Given the number of potential negative health outcomes from exposure to chronic stress, establishing surveillance methods to detect individuals experiencing chronic stress and stress-related disease is obviously important, but also challenging.

One potentially effective health surveillance strategy, targeted at subordinates, may be to collect biological samples during regular animal health checks, such as during annual or semiannual health round-ups when animals are thoroughly examined by veterinarians. If behavioral observation data are available, biological sample collection and/or processing may be targeted at individuals showing evidence of exposure to chronic stress (e.g., being frequent targets of aggression by conspecifics) and/or lack of ability to cope with social stress (e.g., few or no grooming or huddling partners; few locations to escape or avoid dominant individuals). Examples of biological sample collection and processing might include drawing blood samples with which to measure biomarkers of inflammation (e.g., C-reactive protein: Vandeleest et al. 2016) or glucocorticoids, getting a urine sample to measure catecholamines, or obtaining a fecal sample to test for gastrointestinal infection (e.g., *Shigella*: Balasubramaniam et al. 2016). Repeated physiological evidence of stress may warrant increased monitoring for health decline.

If an animal is found to be experiencing chronic psychosocial stress (especially if no evidence of other disease is found), we suggest that, given the importance of the social environment delineated above, the best course of action is to continue to monitor them for behavioral and/or physical signs of further welfare issues. For example, subordinates with elevated GCs but who also groom and huddle frequently with others in their social group (such as adult offspring, mother, or other close affiliates) are demonstrating an ability to cope with the stresses of their social environment (see *Social Bonds and Welfare* section below). So long as there is no evidence of serious infection, poor cardiovascular function, psychological disorders such as depression, or other disease, the benefits of group living still likely outweigh the costs, and social relocation (being moved to a new enclosure and social setting, such as indoor pair housing) is likely not necessary. Importantly, social relocation is also stressful, particularly if the new social situation is dramatically different from the animal's current social group, such as relocation from group housing to pair housing (Capitanio et al. 1998). In other words, the psychosocial stresses of the new social environment, as well as the relocation process, should be weighed against the psychosocial stress an animal is experiencing in its social group when deciding whether social relocation is necessary. In contrast, if behavioral observation of a subordinate with elevated GCs reveals that group members repeatedly aggressively target the animal, and/or the animal appears to have no social affiliates to help it cope with this social stress, social relocation may offer greater benefits than costs. For species that evolved to live in cohesive social groups, the benefits of remaining in a social group, on average, outweigh the costs of sociality even for subordinates experiencing psychosocial stress—evidence for this comes from both field work, such as the finding that subordinate male baboons experiencing chronic psychosocial stress yet choose to remain in their social group, as well as decades of captive research on comparing group-housed monkeys to single- and pair-housed monkeys in research facilities (e.g., Gust et al. 1992; Clarke et al. 1995; Baker et al. 2012).

2.4 Potential Costs of High Rank

High social status generally comes with numerous advantages, ranging from greater access to high-quality food resources and mating opportunities, to lower risk of social trauma, to lower incidence of chronic stress, and its associated health issues (see above). Thus, many of the welfare concerns that we have for subordinate animals do not typically apply to dominant animals. However, dominance is not without cost, and managers of captive social groups should not assume that there are no welfare concerns for high-ranking individuals. There are circumstances where dominant animals have higher GC levels, greater parasitism, or reduced immune function than subordinates (Manogue 1975; Hausfater and Watson 1976; Abbott et al. 2003; Melfi and Poyser 2007; Habig and Archie 2015). For example, high-ranking males (particularly alpha males) may show higher levels of GCs than low-ranking males (Manogue 1975; Cavigelli 1999; Barrett et al. 2002; Muller and Wrangham 2004; Gesquiere et al. 2011). Most of these reports come from wild populations of primates and are likely applicable to captive settings as well. Importantly, these costs are generally not interpreted as evidence of chronic stress, but rather are thought to reflect greater energetic demands of either acquiring/maintaining high rank or of engaging in extensive reproductive effort (Cavigelli 1999; Barrett et al. 2002; Muller and Wrangham 2004).

The greater energetic demands of dominance are thought to be achieved, in part, by reducing energetic investment in other bodily systems, such as immune function, which can have downstream consequences. Consistent with this is the finding that parasitism is greater in dominant individuals than subordinates in nearly every population examined (but see Foerster et al. 2015; Habig and Archie 2015). For example, high-ranking individuals exhibit greater parasite richness (e.g., *Macaca mulatta*: MacIntosh et al. 2012), higher prevalence for certain parasite species, (*P. cynocephauls* Hausfater and Watson 1976; e.g., *M. fuscatta*: MacIntosh et al. 2012) greater helminth burden (e.g., *Pan troglodytes*: Muehlenbein and Watts 2010), higher tick loads (e.g., *P. cynocephalus*: Akinyi et al. 2013), and higher prevalence of SIV infection (*Cercocebus atys*: Santiago et al. 2005). This evidence suggests that welfare surveillance of high-ranking individuals might focus on parasitism. Such parasite surveillance might be incorporated into annual or semiannual health examinations; fecal samples could be collected from high-ranking individuals and examined for evidence of parasites such as helminths or bacterial pathogens. In addition, veterinarians could administer regular antiparasite treatments for high-ranking individuals. The increased risk for parasitism puts high-ranking animals at health risk because ectoparasites such as ticks may carry diseases (tick-borne hemoprotozoan *Babesia microti*: Akinyi et al. 2013) and gastrointestinal parasites such as nematodes place an additional energetic burden on the host animal by sapping calories and nutrients that would otherwise support host growth and homeostasis.

Notably there may be rare occasions when dominant individuals do experience greater psychosocial stress than subordinates. For example, constant changes to party membership in fission-fusion societies may prevent dominants from observing others' political interactions in their absence, which may cause some instability in the hierarchy and heightened psychosocial stress for those who need to continually reassert their dominance (Muller and Wrangham 2004). Thus, captive housing options for fission-fusion species such as chimpanzees, bonobos, and spider monkeys, might consider whether mimicking the constantly changing subgroup composition of wild groups, which has been successfully implemented with chimpanzees at Kumamoto Sanctuary in Japan (Morimura et al. 2022), may cause dominant individuals to unnecessarily reassert their dominance more often, which may contribute to stress.

Cooperatively breeding species, where reproduction is reduced or suppressed in subordinates, are an interesting exception to the trade-offs model. Among cooperative breeders, such as marmosets and tamarins as well as many nonprimates, dominants show higher GC levels than subordinates (Ziegler et al. 1995; Saltzman et al. 1998; Creel 2001), but this likely reflects suppression of reproduction in subordinates rather than status-related variation in psychosocial stress levels (Smith and French 1997; Abbott et al. 2003).

2.5 Rank Stability and Certainty

The impact of social status on welfare goes beyond the effect of high versus low rank and includes the relative predictability or certainty of one's status relationships. Within a stable hierarchy, evidence of rank instability or ambiguity is more stressful for dominants, who stand to drop in rank, than it is for subordinates, who stand to rise in rank. In adjacently ranked pairs of male baboons whose dominance interactions were highly ambiguous in direction, only the more dominant males (who were about to lose rank) showed elevated basal levels of GCs (Sapolsky 1992). However, evidence of rank instability or ambiguity need not come from direct interactions or adjacently ranked pairs. Recent work on multiple social groups of captive rhesus macaques used a novel network approach to examine each individual's "fit" within the group hierarchy, a measure called dominance certainty (Vandeleest et al. 2016). High-ranking macaques whose dominance relationships are more ambiguous, on average, have higher levels of biomarkers of inflammation (C-reactive protein, IL-6, and TNF-alpha) than adult females and other males, whereas these same biomarker levels are much lower in high-ranking macaques with high certainty about their rank compared to all other adult group members (Vandeleest et al. 2016).

The welfare implications of these elevations in stress due to rank instability or ambiguity are likely minimal so long as the individual is experiencing brief rank instability (e.g., a rank reversal eventually occurs, and the new rank relationship becomes well-established). Occasional bouts of extreme stress are a normal part of social life, and brief elevations in GCs and/or biomarkers of inflammation likely do not pose significant health risks, and in fact could lead to brief enhancements of immunity (Dhabhar 2014). However, perpetual rank instability or ambiguity, and the associated psychosocial stress, would be cause for concern, and in captive populations, human intervention to remove the recipient or instigator of aggression might be necessary. Further investigation of individuals with greater ambiguity in their dominance relationships (but with no observed rank reversals) has also shown that this ambiguity is associated with more frequent injury from social aggression (Beisner and McCowan 2014) and may be symptomatic of other health or welfare risks. Regular behavioral observations of dominance interactions over an extended period offer the best possibility of detecting such long-term instabilities in rank relationships, and therefore identifying which individuals may be at risk of poor welfare.

2.5.1 Group-Level Aspects of the Status-Welfare Relationship

Up to this point we have focused on individual health and welfare. However, there are group-level dynamics that are relevant to both individual welfare and group-level welfare. When animals are socially housed, the distribution of the costs and benefits of sociality differs across individuals, and it is not always possible to achieve the same level of welfare for all group members. Further, single individuals can have a disproportionate impact on the welfare of the rest of the group via their impact on group dynamics such as conflict policers who quell fights among group members and thereby reduce rates of severe aggression and trauma in the group (Flack et al. 2005b; McCowan et al. 2011; von Rohr et al. 2012; Beisner and McCowan 2013).

Social stability and its association with group-level welfare have perhaps been best studied in captive macaques. Macaques are overwhelmingly the most commonly maintained nonhuman primates at captive research facilities (Smith 2012; Lankau et al. 2014), especially rhesus macaques. Group-housed rhesus macaques are highly aggressive toward one another and severe aggression often results in wounding of one or more individuals (Thierry et al. 2004; McCowan et al. 2008), which compromises the welfare of injured individuals. In extreme circumstances, social instability can result in social overthrow, with equally extreme impact on welfare as it typically leads to severe injuries of multiple group members and often relocation to indoor housing. Individual-level welfare can therefore be improved, for most group members, by managing macaque social groups in such a way that preserves long-term group-level stability.

Although the end result of uncorrected social instability in macaques is severe aggression, aggression and trauma rates are not reliable measures of a group's social stability. Evidence of a fully connected acyclic subordination signaling network with

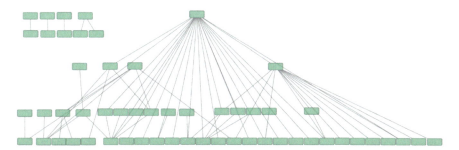

Fig. 2 Subordination signaling (i.e., peaceful silent-bared-teeth displays) network for a captive group of rhesus macaques. Ovals represent individuals and arrows between them represent observed peaceful silent-bared-teeth signals. Arrows are drawn from the signaler to the receiver

multiple tiers (i.e., a network in which all animals participate and where no signals go against the hierarchy; Fig. 2) is associated with stability in captive rhesus macaque groups (Fushing et al. 2014) and this pattern may also be present in other macaque societies in which subordination signals play a critical role in communicating status relationships (although an acyclic pattern was not found in a group of pigtailed macaques *M. nemestrina*: Flack and Krakauer 2006, nor in a group of Barabary macaques *M. sylvanus*: Preuschoft 1992). Individuals that receive formal subordination signals (e.g., peaceful silent-bared-teeth signals) from many other group members have more social power, but this power is distinct from social rank even though individuals with high social power tend to have high rank (Flack and Krakauer 2006; Flack and de Waal 2007; Beisner and McCowan 2014; Beisner et al. 2016). By having a large number of group members that all signal their subordinate status to a few powerful dominant individuals, those powerful dominants have the ability to intervene to stop the fights of other group members with little risk of retaliation or injury (Flack et al. 2005a; Beisner and McCowan 2013). This is known as conflict policing, and the presence of one or more conflict policers in a social group offers a significant welfare advantage to all members of the social group via reduced exposure to social instability and its associated costs, such as social stress, deleterious aggression, and injury (Beisner et al. 2012; von Rohr et al. 2012; Beisner and McCowan 2013). Thus, social stability in macaque groups stems from two interconnected behavioral systems: (1) a skewed distribution of social power across group members (communicated via subordination signals) in which a few high-ranking individuals (typically adult males) have very high social power (Flack and Krakauer 2006; Beisner et al. 2016) while most other group members have very little social power, and (2) conflict policing by powerful individuals to stop group fights (Flack et al. 2005a, b; Beisner and McCowan 2013; Beisner et al. 2016). Monitoring group stability requires regular behavioral observation of both subordination signaling, to estimate social power of individuals at or near the top of the subordination signaling network, and conflict policing behavior.

Although conflict policing is not necessarily linked to subordination signaling in other species, it is a stabilizing mechanism, primarily performed by dominant males (Beisner and McCowan 2013; Ehardt and Bernstein 1992; macaques: Flack et al.

2005a; hamadryas baboons *P. hamadryas*: Kummer 1967; Petit and Thierry 1994; mountain gorillas *Gorilla beringei beringei*: Sicotte 1995; orangutans *Pongo pygmaeus*: Tajiama and Kurotori 2010, e.g., chimpanzees: von Rohr et al. 2012). The clear sex bias in the performance of conflict policing, combined with the finding that low male:female adult sex ratio is associated with reduced policing behavior and social instability (Beisner et al. 2012), indicates that maintaining an adequate number of dominant adult males is important to preventing social instability and overthrow. If sufficient behavioral observation effort cannot be performed due to time constraints, sex ratio can at a minimum be maintained so that the dominant males are not overwhelmed by too many individuals to monitor (Beisner et al. 2012; von Rohr et al. 2012). However, even occasional observations of group conflicts may reveal important information, such as whether certain animals intervene in a manner that limits aggression or whether certain individuals exacerbate aggression. Knowing the identities of individuals that are either good interveners or the instigators/agitators of aggression can inform efforts to maintain the social stability of populations.

Take-home messages regarding social status and welfare

- Most primates evolved to function within complex social environments, and a rich social environment produces a more physically and psychologically healthy primate than an inadequate social environment, regardless of social status.
- Subordinates experience a greater range of potential welfare issues than dominants including greater risk of receiving social aggression and trauma, being excluded from critical resources, being exposed to psychosocial stress and its associated disease states, and having fewer avenues for coping with psychosocial stress, such as social grooming. However, subordinate status alone does not lead to poor welfare in the social group and individuals should not be removed simply because of the downsides associated with low social status.
- Management strategies to improve welfare of subordinates across this range of potential issues are largely the same: (a) provide visual barriers or other structural enrichment that allows subordinates to escape harassment from dominants, and (b) provide critical resources (e.g., food, shelter, water) in sufficient abundance, distribute widely, and/or provide alternative resources to reduce aggressive and stressful competition for access to these resources.
- Both behavioral observation and health surveillance strategies can provide key information about the welfare status of individuals. Behavior observation is critical for identifying individual's approximate social status in the group, as well as identifying whether they have friends and allies to help them cope with stress and whether they hold a keystone position in the social group. Biological samples obtained during regular health examinations are critical for identifying individuals experiencing disease states (e.g., parasitic infection; poor cardiovascular function) that are not observable behaviorally.

3 Social Bonds and Welfare

Affiliative social interactions play a key role in the daily lives of primates, and individuals develop close social bonds with one or more members of their group. Both affiliative interactions and strong social bonds can positively impact health and welfare in a variety of species in both captive and wild settings. For example, there are benefits to giving and receiving grooming (Boccia et al. 1989; Shutt et al. 2007) and having consistent, long-standing social partners (Silk et al. 2010). Further, there may be benefits to one's network position or the group-level patterning of affiliative interactions (Tung et al. 2015; Balasubramaniam et al. 2016). Yet there are also potential costs of affiliative interactions and relationships because having many different close affiliates increase an individual's exposure to communicable diseases, which is a commonly cited cost of group living (MacIntosh et al. 2012). We discuss these costs and benefits below.

3.1 Affiliative Interactions and Stress Physiology

Affiliative interactions, such as social grooming (distinct from self-grooming), are linked to stress and physiological indicators of welfare. Grooming is a known mechanism for coping with social tension and has been found to have physiological benefits in both captive and wild populations. Among captive populations of macaques, heart rate decreases while receiving grooming from conspecifics (and is lower than during matched control periods and self-grooming) (Boccia et al. 1989; Aureli et al. 1999). In addition, individuals who receive frequent grooming (*M. mulatta*: Gust et al. 1993), and in some cases those that frequently groom others (*M. sylvanus*: Shutt et al. 2007), have lower GC levels, suggesting that both receiving and giving grooming can serve to help individuals cope with social stress. In fact, a study of talapoin monkeys showed that both giving and receiving grooming increased β-endorphin production, an endogenous opioid, which is a biomarker of good psychological well-being (Keverne et al. 1989).

Given the psychological value of social grooming in many primate species, the presence of social grooming in captive social groups is one positive indicator of well-being. If systematic behavioral observation reveals that certain individuals rarely, if ever, participate in social grooming, it may be worth investigating whether these individuals are experiencing psychosocial stress and/or have insufficient avenues for coping with stress (see section on social status and stress above). It is worth noting, however, that due to personality differences, some individuals may not want to participate in social grooming very often.

3.2 Strong Social Bonds

In addition to grooming interactions, the presence of a particular social partner (e.g., a mate, offspring, or "friend") can reduce the impact of a stressful event through social buffering (Kikusui et al. 2006). Even auditory contact with a pairmate has been shown to buffer against a stressor (Rukstalis and French 2005). Importantly, not all conspecifics can buffer individuals from stress: there needs to be a strong connection like that of mother-infant, pairmate, or preferred social partner. Affiliative interactions are often used to assess the strength of a social bond (Silk et al. 2003). In the wild, social bonds offer individuals a coping mechanism for dealing with social stressors that are a normal part of social life for both males and females. For example, during periods of social instability, such as alpha male turn over, female chacma baboons (*P. ursinus*) with small, focused grooming networks (Wittig et al. 2008) showed less dramatic increases in GCs than females with more diverse grooming networks. Furthermore, females who narrowed their grooming networks in response to social instability reduced the extent of the rise in GC levels. Similarly, members of a captive group of pigtailed macaques also narrowed their social networks (e.g., smaller mean degree centrality) when conflict policers were experimentally removed (Flack et al. 2006).

Even during socially stable periods, the strength and patterning of social bonds can mitigate stress. In wild Barbary macaques faced with frequent receipt of social aggression, males with strong social bonds have lower fecal GC levels (Young et al. 2014), and female chacma baboons have lower GC levels during months when their grooming networks are small and focused compared to months with a greater diversity of grooming partners (Crockford et al. 2008).

Social bonds can further impact physiological state, susceptibility to disease, and fitness, as the presence, patterning, and quality of social bonds have been linked to a variety of health and welfare metrics. For example, gut microbiome composition is linked to many different health outcomes and disease states in humans and primates (colitis: McKenna et al. 2008; obesity: Turnbaugh et al. 2009; immune function: Hooper et al. 2012), and research in wild baboons shows that close grooming partners have a similar gut microbiome composition (Tung et al. 2015). Thus, one's close grooming partners stand to significantly impact aspects of health that are linked to host-microbe interactions. Furthermore, differential gene expression is predicted by social connectedness and rank (Runcie et al. 2013). Finally, strong social bonds, such as those measured via grooming relationships and proximity, are linked to survival and reproductive success in baboon females. Both infant survival (Silk et al. 2003, 2009) and adult female longevity (Silk et al. 2010) are enhanced by having strong social bonds, especially with close kin such as mothers and daughters.

The absence of key social bonds can also impact health. Work on social isolation and loneliness in macaques has identified a class of individuals that appear to experience loneliness—individuals who, despite initiating as many social interactions as highly sociable individuals, rarely succeed in translating these attempts into actual interaction with others (Capitanio et al. 2014). In humans, being socially isolated increases the risk of physical and mental illness as well as mortality (House et al. 1988; Cacioppo et al. 2010; Holt-Lunstad et al. 2010). Thus, evidence of social isolation or loneliness in socially housed animals is noteworthy because it has the potential to cause or contribute to poor welfare.

3.3 Social Network Position and Structure and Welfare

The impact of affiliative interactions and social bonds on welfare goes beyond that of direct ties. An individual's social network position, which incorporates direct and indirect ties, as well as the patterning of social relationships in the group, can also significantly influence health and welfare. Disease transmission is viewed as one of the primary costs of group living (Alexander 1974; Freeland 1976). Both theoretical and empirical evidence suggest that greater social connectedness puts individuals at greater risk of contact-mediated transmission of pathogens (Corner et al. 2003; Christley et al. 2005; MacIntosh et al. 2012; Drewe and Perkins 2015). Among primates, for example, high-ranking female Japanese macaques are both the most central in the grooming network and also show higher prevalence of nematode parasite infection (MacIntosh et al. 2012). And in brown spider monkeys *Ateles hybridus*, well-connected individuals in the physical contact network show higher gastrointestinal parasite species richness than less-connected individuals (Rimbach et al. 2015). Further evidence from nonprimates also illustrates the potential costs of holding a highly central network position (for a review see Drewe and Perkins 2015).

Despite the increased risk of contact-mediated pathogen transmission that socially well-connected individuals face, there is evidence that social network position and structure can mitigate infection risk. Modularity is a structural feature of the network that describes the extent of community structure or local connectedness. In highly modular networks there exist subsets of highly connected nodes (i.e., a module) which are not well connected to nodes outside the module. Disease can spread rapidly among members of the same module (within a social network that has modular structure), but disease transmission between modules is limited because infections tend to die out before spreading to other modules (Salathe and Jones 2010; Griffin and Nunn 2012). Work by Griffin and Nunn (2012) demonstrates that the presence of modular network structure in large primate social groups is associated with lower parasite richness (i.e., smaller number of parasite species).

There are other aspects of an individual's position in their network that can mitigate infection risk such as access to social support. Just as the presence of a pairmate or preferred social partner can socially buffer one against psychosocial and/or environmental stressors, so can one's network centrality. Sociopositive interactions such as grooming are important ways of alleviating stress, suggesting that being well-connected in your social network may decrease susceptibility to infection (as opposed to increasing risk of infection through contact transmission). In captive social groups of rhesus macaques, for example, individuals with more direct and especially indirect connections in their grooming and huddling social networks were less susceptible to *Shigella* infection (Balasubramaniam et al. 2016). In other words, social buffering can include the effects of having a well-connected social circle of both primary and secondary connections to elicit support.

3.4 Welfare Implications of Affiliative Interactions and Social Bonds

The value of social bonds and affiliative interactions is apparent when considering how animals cope with the daily stressors they experience in captive settings, such as disruptions due to routine colony management (e.g., health checks, feeding, and cleaning). Therefore, social housing is the first step toward reaping the welfare benefits of affiliative interactions and social bonds because the presence of conspecifics gives captive animals an opportunity to develop beneficial relationships. Furthermore, negotiating social relationships, both through affiliative and aggressive interactions, stimulates normal cognitive and behavioral functions and allows captive primates to meet the social and socio-ecological challenges they are evolved to cope with (Hinde 1976; van Schaik and van Hooff 1983; Dunbar 1988, 1998; Thierry et al. 2000).

In circumstances where animals must be temporarily separated from their group, such as for veterinary care, it may be beneficial to bring along a preferred social partner or pair-mate of the targeted animal. For example, the time it takes to heal after severe injury or after surgery may be accelerated with the presence of a social partner due to reduced social stress. However, the presence of a preferred social partner is not without difficulty, as the social partner takes up additional space in the veterinary hospital, may interfere with treatment of the ill or injured animal (e.g., picking at stiches and bandages), and will also experience separation from the social group (Flack et al. 2005b; McCowan et al. 2011; Beisner et al. 2016).

Alternatively, veterinarians might favor home-cage treatments for less serious illness or injury to keep animals in their social group rather than temporarily relocating them in the hospital. Keeping animals in their normal social environment for the duration of veterinary care has the added benefit of avoiding removal of key individuals that uphold the stability of the group.

Take-home messages regarding affiliative relationships and welfare
- Affiliative interactions such as grooming offer physiological benefits including lower heart rate, increased endorphin production, and reduced GC levels.
- Having strong social bonds can allow social buffering against the negative effects of stressful conditions and improve longevity and survival.
- An animal's position in the network and the higher-order structure of the network affect welfare by influencing both disease transmission and susceptibility to infection. Animals with more social connections, both direct and indirect, have greater potential exposure to infect others, yet by having more social connections they may also be socially buffered against infection by preventing immune suppression and thereby lowering susceptibility.
- Colony managers can promote good welfare by monitoring whether animals have access to social partners to allow them to cope with daily stressors. Further, during particularly stressful situations, such as transport to the hospital for veterinary care, technicians and veterinarians should consider whether home-cage treatment or allowing access to a preferred partner in the hospital is possible.

4 Conclusions/Summary

Primates evolved to live in cohesive social groups, and they exhibit complex social, behavioral, and cognitive adaptations. A species-appropriate social environment is the most valuable form of enrichment for promoting good welfare in captive primates. Such environments provide an appropriate context to engage social adaptations, buffer against stress, and, for research subjects, ensure that subjects exhibit normal ranges of behavioral, biological, and physiological functioning that enhance external validity of biomedical research conducted on those animals. The costs and benefits of sociality vary across individuals, and two primary domains of social relationships contribute to this balance: social status (a.k.a. dominance) and affiliative relationships. We have the ability leverage our knowledge of the relationships between sociality and health and welfare to maximize welfare for each animal by implementing preventative management strategies (e.g., reducing competitive interactions by changing distribution of food) and adopting systematic surveillance methods to detect evidence of individuals experiencing poor welfare. Ultimately, good welfare benefits not only the captive animals themselves, but also the various human stakeholders that stand to benefit from these captive populations of primates.

References

Abbott DH, Keverne EB, Bercovitch FB et al (2003) Are subordinates always stressed? A comparative analysis of rank differences in cortisol levels among primates. Horm Behav 43 (1):67–82. https://doi.org/10.1016/s0018-506x(02)00037-55

Adams MR, Kaplan JR, Clarkson TB, Koritnik DR (1985) Ovariectomy, social status, and atherosclerosis in cynomolgus monkeys. Arteriosclerosis 5(2):192–200. https://doi.org/10.1161/01.atv.5.2.192

Akinyi MY, Tung J, Jeneby M et al (2013) Role of grooming in reducing tick load in wild baboons (*Papio cynocephalus*). Anim Behav 85(3):559–568. https://doi.org/10.1016/j.anbehav.2012.12.012

Albert R, Jeong H, Barabasi AL (2000) Error and attack tolerance of complex networks. Nature 406:378–381. https://doi.org/10.1038/35019019

Alexander RD (1974) The evolution of social behavior. Annu Rev Ecol Syst 5:324–382. https://doi.org/10.1146/annurev.es.05.110174.001545

Alford PL, Bloomsmith MA, Keeling ME, Beck TF (1995) Wounding aggression during the formation and maintenance of captive, multimale chimpanzee groups. Zoo Biol 14 (4):347–359. https://doi.org/10.1002/zoo.1430140406

Anderson JR, Chamove AS (1980) Self-aggression and social aggression in laboratory-reared macaques. J Abnorm Psychol 89(4):539–550. https://doi.org/10.1037/0021-843X.89.4.539

Archie EA, Altmann J, Alberts SC (2014) Costs of reproduction in a long-lived female primate: injury risk and wound healing. Behav Ecol Sociobiol 68(7):1183–1193. https://doi.org/10.1007/s00265-014-1729-4

Association for Assessment and Accreditation of Laboratory Animal Care Internatonal (2015) Position statements: social housing. https://www.aaalac.org/accreditation-program/position-statements/

Aureli F, Preston SD, de Waal FBM (1999) Heart rate responses to social interactions in free-moving rhesus macaques (*Macaca mulatta*): A pilot study. J Comp Psychol 113(1):59–65. https://doi.org/10.1037/0735-7036.113.1.59

Baker KC, Bloomsmith MA, Oettinger BC et al (2012) Benefits of pair housing are consistent across a diverse population of rhesus macaques. Appl Anim Behav Sci 137(3–4):148–156. https://doi.org/10.1016/j.applanim.2011.09.010

Balasubramaniam KN, Beisner BA, Vandeleest JJ et al (2016) Social buffering and contact transmission: network connections have beneficial and detrimental effects on Shigella infection risk among captive rhesus macaques. Peer J 4:e2630. https://doi.org/10.7717/peerj.2630

Barrett GM, Shimizu K, Bardi M et al (2002) Endocrine correlates of rank, reproduction, and female-directed aggression in male Japanese macaques (*Macaca fuscata*). Horm Behav 42(1):85–96. https://doi.org/10.1006/hbeh.2002.1804

Barton RA, Whiten A (1993) Feeding competition among female olive baboons, *Papio anubis*. Anim Behav 46(4):777–789. https://doi.org/10.1006/anbe.1993.1255

Barton RA, Byrne RW, Whiten A (1996) Ecology, feeding competition and social structure in baboons. Behav Ecol Sociobiol 38(5):321–329

Beisner B, Isbell LA (2008) Ground substrate affects activity budgets and hair loss in outdoor captive groups of rhesus macaques (*Macaca mulatta*). Am J Primatol 70(12):1160–1168. https://doi.org/10.1002/ajp.20615

Beisner BA, Isbell LA (2011) Factors affecting aggression among females in captive groups of rhesus macaques (*Macaca mulatta*). Am J Primatol 73(11):1152–1159. https://doi.org/10.1002/ajp.20982

Beisner BA, McCowan B (2013) Policing in nonhuman primates: partial interventions serve a prosocial conflict management function in rhesus macaques. PLoS One 8(10):e77369. https://doi.org/10.1371/journal.pone.0077369

Beisner BA, McCowan B (2014) Signaling context modulates social function of silent bared teeth displays in rhesus macaques (*Macaca mulatta*). Am J Primatol 76(2):111–121. https://doi.org/10.1002/ajp.22214

Beisner BA, Jackson ME, Cameron A, McCowan B (2012) Sex ratio, conflict dynamics and wounding in rhesus macaques (*Macaca mulatta*). Appl Anim Behav Sci 137(3–4):137–147. https://doi.org/10.1016/j.applanim.2011.07.008

Beisner BA, Hannibal DL, Finn KR et al (2016) Social power, conflict policing, and the role of subordination signals in rhesus macaque society. Am J Phys Anthropol 160(1):102–112. https://doi.org/10.1002/ajpa.22945

Bercovitch FB, Clarke AS (1995) Dominance rank, cortisol concentrations, and reproductive maturation in male rhesus macaques. Physiol Behav 58(2):215–221. https://doi.org/10.1016/0031-9384(95)00055-N

Bercovitch FB, Strum SC (1993) Dominance rank, resource availability, and reproductive maturation in female savanna baboons. Behav Ecol Sociobiol 33:313–318. https://doi.org/10.1007/bf00172929

Bernstein I, Gordon TP (1974) The function of aggression in primate societies. Am Sci 62:304–311

Bertram BCR (1978) Living in groups: predators and prey. In: Krebs JR, Davies NB (eds) Behavioural ecology: an evolutionary approach. Blackwell, Oxford, pp 64–96

Boccia ML, Laudenslager M, Reite M (1988) Food distribution, dominance, and aggressive behaviors in bonnet macaques. Am J Primatol 16(2):123–130. https://doi.org/10.1002/ajp.1350160203

Boccia ML, Reite M, Laudenslager M (1989) On the physiology of grooming in a pigtail macaque. Physiol Behav 45(3):667–670. https://doi.org/10.1016/0031-9384(89)90089-9

Bradbury JW, Vehrencamp SL (1976) Social organization and foraging in emballonurid bats. Behav Ecol Sociobiol 1:337–381. https://doi.org/10.1007/BF00299284

Brennan J, Anderson JR (1988) Varying responses to feeding competition in a group of rhesus monkeys (*Macaca mulatta*). Primates 29:353–360. https://doi.org/10.1007/bf02380958

Byrne RW, Whiten A (1988) Machiavellian intelligence. Oxford University Press, Oxford

Byrne GD, Abbott KM, Suomi SJ (1996) Reorganization of dominance rank among adult males in a captive group of tufted capuchins (*Cebus apella*). Lab Primate Newsl 35:1–4

Cacioppo JT, Hawkley LC, Thisted RA (2010) Perceived social isolation makes me sad: 5-year cross-lagged analyses of loneliness and depressive symptomatology in the Chicago Health, Aging, and Social Relations Study. Psychol Aging 25(2):453–463. https://doi.org/10.1037/a0017216

Campbell CJ, Fuentes A, MacKinnon KC et al (2011) Primates in perspective. Oxford University Press, Oxford

Capitanio JP, Cole SW (2015) Social instability and immunity in rhesus monkeys: the role of the sympathetic nervous system. Philos Trans R Soc B Biol Sci 370(1669):20140104. https://doi.org/10.1098/rstb.2014.0104

Capitanio JP, Mendoza SP, Lerche NW (1998) Individual differences in peripheral blood immunological and hormonal measures in adult male rhesus macaques (*Macaca mulatta*): Evidence for temporal and situational consistency. Am J Primatol 44(1):29–41. https://doi.org/10.1002/(sici)1098-2345(1998)44:1<29::aid-ajp3>3.0.co;2-z

Capitanio JP, Mendoza SP, Lerche NW, Mason WA (1998) Social stress results in altered glucocorticoid regulation and shorter survival in simian acquired immune deficiency syndrome. Proc Natl Acad Sci U S A 95(8):4714–4719. https://doi.org/10.1073/pnas.95.8.4714

Capitanio JP, Hawkley LC, Cole SW, Cacioppo JT (2014) A behavioral taxonomy of loneliness in humans and rhesus monkeys (*Macaca mulatta*). PLoS One 9(10):e110307. https://doi.org/10.1371/journal.pone.0110307

Capitanio JP, Vandeleest J, Hannibal DL (2022) Physiological measures of welfare. In: Robinson LM, Weiss A (eds) Nonhuman primate welfare: from history, science, and ethics to practice. Springer, Cham, pp 231–254

Cavigelli SA (1999) Behavioural patterns associated with faecal cortisol levels in free-ranging female ring-tailed lemurs, *Lemur catta*. Anim Behav 57(4):935–944. https://doi.org/10.1006/anbe.1998.1054

Cavigelli SA, Caruso MJ (2015) Sex, social status and physiological stress in primates: the importance of social and glucocorticoid dynamics. Philos Trans R Soc B Biol Sci 370(1669):20140103. https://doi.org/10.1098/rstb.2014.0103

Chamove AS, Anderson JR, Morgan-Jones SC, Jones SP (1982) Deep woodchip litter: hygiene, feeding, and behavioral enhancement in eight primate species. Int J Study Anim Probl 3(4):308–318

Chancellor RL, Isbell LA (2009) Food site residence time and female competitive relationships in wild gray-cheeked mangabeys (*Lophocebus albigena*). Behav Ecol Sociobiol 63(10):1447–1458. https://doi.org/10.1007/s00265-009-0805-7

Chapman CA, Chapman LJ (2000) Constraints on group size in red colobus and red-tailed guenons: Examining the generality of the ecological constraints model. Int J Primatol 21:565–585. https://doi.org/10.1023/A:1005557002854

Cheney D, Seyfarth R (1990) Attending to behaviour versus attending to knowledge: examining monkeys' attribution of mental states. Anim Behav 40(4):742–753. https://doi.org/10.1016/S0003-3472(05)80703-1

Christley RM, Pinchbeck GL, Bowers RG et al (2005) Infections in social networks: using network analysis to identify high-risk individuals. Am J Epidemiol 162(10):1024–1031. https://doi.org/10.1093/aje/kwi308

Clarke AS, Czekala NM, Lindburg DG (1995) Behavioral and adrenocortical responses of male cynomolgus and lion-tailed macaques to social stimulation and group formation. Primates 36:41–56. https://doi.org/10.1007/bf02381914

Clutton-Brock TH (1974) Primate social organisation and ecology. Nature 250:539–542. https://doi.org/10.1038/250539a0

Coe CL (1993) Psychosocial factors and immunity in nonhuman-primates: a review. Psychosom Med 55(3):298–308. https://doi.org/10.1097/00006842-199305000-00007

Coe CL, Mendoza SP, Levine S (1979) Social status constrains the stress response in the squirrel monkey. Physiol Behav 23(4):633–638. https://doi.org/10.1016/0031-9384(79)90151-3

Cohen S, Kaplan JR, Cunnick JE et al (1992) Chronic social stress, affiliation, and cellular immune-response in nonhuman primates. Psychol Sci 3(5):301–304. https://doi.org/10.1111/j.1467-9280.1992.tb00677.x

Cohen S, Line S, Manuck SB et al (1997) Chronic social stress, social status, and susceptibility to upper respiratory infections in nonhuman primates. Psychosom Med 59(3):213–221. https://doi.org/10.1097/00006842-199705000-00001

Cohen S, Janicki-Deverts D, Miller GE (2007) Psychological stress and disease. JAMA 298 (14):1685–1687. https://doi.org/10.1001/jama.298.14.1685

Corner LAL, Pfeiffer DU, Morris RS (2003) Social-network analysis of *Mycobacterium bovis* transmission among captive brushtail possums (*Trichosurus vulpecula*). Prev Vet Med 59 (3):147–167. https://doi.org/10.1016/S0167-5877(03)00075-8

Cowlishaw G, Dunbar RIM (1991) Dominance rank and mating success in male primates. Anim Behav 41(6):1045–1056. https://doi.org/10.1016/S0003-3472(05)80642-6

Creel S (2001) Social dominance and stress hormones. Trends Ecol Evol 16(9):491–497. https://doi.org/10.1016/S0169-5347(01)02227-3

Creel S, Dantzer B, Goymann W, Rubenstein DR (2013) The ecology of stress: effects of the social environment. Funct Ecol 27(1):66–80. https://doi.org/10.1111/j.1365-2435.2012.02029.x

Crockford C, Wittig RM, Whitten PL et al (2008) Social stressors and coping mechanisms in wild female baboons (*Papio hamadryas ursinus*). Horm Behav 53(1):254–265. https://doi.org/10.1016/j.yhbeh.2007.10.007

Darwin C (1871) The descent of man and selection in relation to sex. John Murray, London

Deutsch JC, Lee PC (1991) Dominance and feeding competition in captive rhesus monkeys. Int J Primatol 12:615–628. https://doi.org/10.1007/bf02547673

Dhabhar FS (2014) Effects of stress on immune function: the good, the bad, and the beautiful. Immunol Res 58(2–3):193–210. https://doi.org/10.1007/s12026-014-8517-0

Drewe JA, Perkins SE (2015) Disease transmission in animal social networks. In: Krause J, James R, Croft DP (eds) Animal social networks. Oxford University Press, Oxford, pp 95–110

Dunbar RIM (1988) Primate social systems. Chapman & Hall, London

Dunbar RIM (1991) Functional significance of social grooming in primates. Folia Primatol 57:121–131. https://doi.org/10.1159/000156574

Dunbar RIM (1998) The social brain hypothesis. Evol Anthropol 6(5):178–190. https://doi.org/10.1002/(SICI)1520-6505(1998)6:5%3C178::AID-EVAN5%3E3.0.CO;2-8

Eaton GG, Kelley ST, Axthelm MK et al (1994) Psychological well-being in paired adult female rhesus (*Macaca mulatta*). Am J Primatol 33(2):89–99. https://doi.org/10.1002/ajp.1350330204

Ehardt CL, Bernstein IS (1992) Conflict intervention behavior by adult male macaques: structural and functional aspects. In: Harcourt AH, De Waal FBM (eds) Coalitions and alliances in humans and other animals. Oxford University Press, Oxford, pp 83–111

Erwin J (1977) Factors influencing aggressive-behavior and risk of trauma in pigtail macaque (*Macaca nemestrina*). Lab Anim Sci 27:541–547

Erwin J, Anderson B, Erwin N et al (1976) Aggression in captive pigtail monkey groups: effects of provision of cover. Percept Mot Skills 42(1):319–324. https://doi.org/10.2466/pms.1976.42.1.319

Estep DQ, Baker SC (1991) The effects of temporary cover on the behavior of socially housed stumptailed macaques (*Macaca arctoides*). Zoo Biol 10(6):465–472. https://doi.org/10.1002/zoo.1430100605

Ethun K, Dicker S, Hughes B et al (2015) Use of novel automated feeders to control obesity of socially housed rhesus macaques. J Am Assoc Lab Anim Sci 54:661

European Commission (2010) Directive 2010/63/EU of the European Parliament and of the Council of 22 September 2010 on the protection of animals used for scientific purposes OJL276/33

Fairbanks LA, McGuire MT (1984) Determinants of fecundity and reproductive success in captive vervet monkeys. Am J Primatol 7(1):27–38. https://doi.org/10.1002/ajp.1350070106

Fairbanks LA, McGuire TM, Kerber W (1978) Effects of group-size, composition, introduction technique and cage apparatus on aggression during group formation in rhesus-monkeys. Psychol Rep 42(1):327–333. https://doi.org/10.2466/pr0.1978.42.1.327

Flack JC, de Waal FBM (2007) Context modulates signal meaning in primate communication. Proc Natl Acad Sci 104(5):1581–1586. https://doi.org/10.1073/pnas.0603565104

Flack JC, Krakauer DC (2006) Encoding power in communication networks. Am Nat 168(3):E87–E102. https://doi.org/10.1086/506526

Flack JC, de Waal FBM, Krakauer DC (2005a) Social structure, robustness, and policing cost in a cognitively sophisticated species. Am Nat 165(5):E126–E139. https://doi.org/10.1086/429277

Flack JC, Krakauer DC, de Waal FBM (2005b) Robustness mechanisms in primate societies: a perturbation study. Proc R Soc B Biol Sci 272(1568):1091–1099. https://doi.org/10.1098/rspb.2004.3019

Flack JC, Girvan M, de Waal FBM, Krakauer DC (2006) Policing stabilizes construction of social niches in primates. Nature 439:426–429. https://doi.org/10.1038/nature04326

Foerster S, Kithome K, Cords M, Monfort SL (2015) Social status and helminth infections in female forest guenons (*Cercopithecus mitis*). Am J Phys Anthropol 158(1):55–66. https://doi.org/10.1002/ajpa.22764

Fontenot MB, Wilkes MN, Lynch CS (2006) Effects of outdoor housing on self-injurious and stereotypic behavior in adult male rhesus macaques (*Macaca mulatta*). J Am Assoc Lab Anim Sci 45(5):35–43

Freeland WJ (1976) Pathogens and the evolution of primate sociality. Biotropica 8(1):12–24. https://doi.org/10.2307/2387816

Fushing H, Jordà Ò, Beisner B, McCowan B (2014) Computing systemic risk using multiple behavioral and keystone networks: The emergence of a crisis in primate societies and banks. Int J Forecast 30(3):797–806. https://doi.org/10.1016/j.ijforecast.2013.11.001

Gesquiere LR, Learn NH, Simao MCM et al (2011) Life at the top: rank and stress in wild male baboons. Science 333(6040):357–360. https://doi.org/10.1126/science.1207120

Gilbert MH, Baker KC (2011) Social buffering in adult male rhesus macaques (*Macaca mulatta*): effects of stressful events in single vs. pair housing. J Med Primatol 40(2):71–78. https://doi.org/10.1111/j.1600-0684.2010.00447.x

Goldberg JL, Grant JWA, Lefebvre L (2001) Effects of the temporal predictability and spatial clumping of food on the intensity of competitive aggression in the Zenaida dove. Behav Ecol 12 (4):490–495. https://doi.org/10.1093/beheco/12.4.490

Gordon TP, Gust DA, Wilson ME et al (1992) Social separation and reunion affects immune-system in juvenile rhesus monkeys. Physiol Behav 51(3):467–472. https://doi.org/10.1016/0031-9384(92)90166-y

Gottlieb DH, Capitanio JP, McCowan BJ (2013) Risk factors for stereotypic behavior and self-biting in rhesus macaques (*Macaca mulatta*): animal's history, current environment, and personality. Am J Primatol 75(10):995–1008. https://doi.org/10.1002/ajp.22161

Griffin RH, Nunn CL (2012) Community structure and the spread of infectious disease in primate social networks. Evol Ecol 26:779–800. https://doi.org/10.1007/s10682-011-9526-2

Gust DA, Gordon TP, Wilson ME et al (1991) Formation of a new social group of unfamiliar female rhesus monkeys affects the immune and pituitary adrenocortical systems. Brain Behav Immun 5 (3):296–307. https://doi.org/10.1016/0889-1591(91)90024-5

Gust DA, Gordon TP, Wilson ME et al (1992) Removal from natal social group to peer housing affects cortisol levels and absolute numbers of T cell subsets in juvenile rhesus monkeys. Brain Behav Immun 6(2):189–199. https://doi.org/10.1016/0889-1591(92)90018-j

Gust D, Gordon TP, Hambright MK, Wilson ME (1993) Relationship between social factors and pituitary-adrenocortical activity in female rhesus monkeys (*Macaca mulatta*). Horm Behav 27 (3):318–331. https://doi.org/10.1006/hbeh.1993.1024

Gust DA, Gordon TP, Brodie AR, McClure HM (1994) Effect of a preferred companion in modulating stress in adult female rhesus monkeys. Physiol Behav 55(4):681–684. https://doi.org/10.1016/0031-9384(94)90044-2

Habig B, Archie EA (2015) Social status, immune response and parasitism in males: a meta-analysis. Philos Trans R Soc B Biol Sci 370(1669):20140109. https://doi.org/10.1098/rstb.2014.0109

Hannibal D, Bliss-Moreau E, Vandeleest JJ et al (2017) Laboratory rhesus macaque social housing and social changes: implications for research. Am J Primatol 79(1):e22528. https://doi.org/10.1002/ajp.22528

Harcourt AH (1987) Dominance and fertility among female primates. J Zool Soc Lond 213(3):471–487. https://doi.org/10.1111/j.1469-7998.1987.tb03721.x

Harlow HF, Harlow MK (1962) Social deprivation in monkeys. Sci Am 207:136–146

Hausfater G, Watson DF (1976) Social and reproductive correlates of parasite ova emissions by baboons. Nature 262:688–689. https://doi.org/10.1038/262688a0

Heesen M, Rogahn S, Ostner J, Schülke O (2013) Food abundance affects energy intake and reproduction in frugivorous female Assamese macaques. Behav Ecol Sociobiol 67:1053–1066. https://doi.org/10.1007/s00265-013-1530-9

Hinde RA (1976) Interactions, relationships, and social structure. Man 11(1):1–17

Holt-Lunstad J, Smith TB, Layton JB (2010) Social relationships and mortality risk: a meta-analytic review. PLoS Med 7(7):e1000316. https://doi.org/10.1371/journal.pmed.1000316

Hooper LV, Littman DR, Macpherson AJ (2012) Interactions between the microbiota and the immune system. Science 336(6086):1268–1273. https://doi.org/10.1126/science.1223490

House JS, Landis KR, Umberson D (1988) Social relationships and health. Science 241(4865):540–545. https://doi.org/10.1126/science.3399889

Isbell LA (1991) Contest and scramble competition: patterns of female aggression and ranging behavior among primates. Behav Ecol 2(2):143–155. https://doi.org/10.1093/beheco/2.2.143

Isbell LA (2004) Is there no place like home? Ecological bases of female dispersal and philopatry and their consequences for the formation of kin groups. In: Chapais B, Berman CM (eds) Kinship and behavior in primates. Oxford University Press, Oxford, pp 71–108

Isbell LA, Pruetz JD, Lewis M, Young TP (1999) Rank differences in ecological behavior: a comparative study of patas monkeys (*Erythrocebus patas*) and vervets (*Cercopithecus aethiops*). Int J Primatol 20:257–272. https://doi.org/10.1023/A:1020574504017

Janson CH, Goldsmith ML (1995) Predicting group size in primates: foraging costs and predation risks. Behav Ecol 6(3):326–336. https://doi.org/10.1093/beheco/6.3.326

Kapsalis E, Berman C (1996) Models of affiliative relationships among free-ranging rhesus monkeys (*Macaca mulatta*). 2. Testing predictions for three hypothesized organizing principles. Behaviour 133(15/16):1235–1263

Kay RF, Plavcan JM, Glander KE, Wright PC (1988) Sexual selection and canine dimorphism in New World monkeys. Am J Phys Anthropol 77(3):385–397. https://doi.org/10.1002/ajpa.1330770311

Kempes MM, Gulickx MMC, van Daalen HJC et al (2008) Social competence is reduced in socially deprived rhesus monkeys (*Macaca mulatta*). J Comp Psychol 122(1):62–67. https://doi.org/10.1037/0735-7036.122.1.62

Keverne EB, Eberhart JA, Yodyingyuad U, Abbott DH (1984) Social influences on sex differences in the behaviour and endocrine state of talapoin monkeys. Prog Brain Res 61:331–347. https://doi.org/10.1016/S0079-6123(08)64445-3

Keverne EB, Martensz ND, Tuite B (1989) Beta-endorphin concentrations in cerebrospinal fluid of–monkeys are influenced by grooming relationships. Psychoneuroendocrinology 14(1–2):155–161. https://doi.org/10.1016/0306-4530(89)90065-6

Kikusui T, Winslow JT, Mori Y (2006) Social buffering: relief from stress and anxiety. Philos Trans R Soc B Biol Sci 361(1476):2215–2228. https://doi.org/10.1098/rstb.2006.1941

Kummer H (1967) Tripartite relations in hamadryas baboons. In: Altmann SA (ed) Social communication among primates. University of Chicago Press, Chicago, pp 113–121

Lankau EW, Turner PV, Mullan RJ, Galland GG (2014) Use of nonhuman primates in research in North America. J Am Assoc Lab Anim Sci 53(3):278–282

Lehmann J, Korstjens AH, Dunbar RIM (2007) Group size, grooming and social cohesion in primates. Anim Behav 74(6):1617–1629. https://doi.org/10.1016/j.anbehav.2006.10.025

Leigh SR (1994) Relations between captive and noncaptive weights in anthropoid primates. Zoo Biol 13(1):21–43. https://doi.org/10.1002/zoo.1430130105

Lutz CK, Novak MA (2005) Environmental enrichment for nonhuman primates: Theory and application. ILAR J 46(2):178–191. https://doi.org/10.1093/ilar.46.2.178

Lyons DM, Parker KJ (2007) Stress inoculation-induced indications of resilience in monkeys. J Trauma Stress 20(4):423–433. https://doi.org/10.1002/jts.20265

MacIntosh AJJ, Jacobs A, Garcia C et al (2012) Monkeys in the middle: parasite transmission through the social network of a wild primate. PLoS One 7(12):e51144. https://doi.org/10.1371/journal.pone.0051144

Majolo B, Lehmann J, Vizioli AD, Schino G (2012) Fitness-related benefits of dominance in primates. Am J Phys Anthropol 147(4):652–660. https://doi.org/10.1002/ajpa.22031

Manogue KR (1975) Dominance status and adrenocortical reactivity to stress in squirrel monkeys (*Saimiri sciureus*). Primates 14:457–463. https://doi.org/10.1007/BF02382742

Marmot M, Sapolsky RM (2014) Of baboons and men: Social circumstances, biology, and the social gradient in health. In: Weinstein M, Lane MA (eds) Sociality, hierarchy, health: comparative biodemography: a collection of papers. The National Academies Press, Washington, DC, pp 365–388

Mathy JW, Isbell LA (2001) The relative importance of size of food and interfood distance in eliciting aggression in captive rhesus macaques (*Macaca mulatta*). Folia Primatol 72 (2):268–277. https://doi.org/10.1159/000049948

McCowan B, Anderson K, Heagarty A, Cameron A (2008) Utility of social network analysis for primate behavioral management and well-being. Appl Anim Behav Sci 109(2–4):396–405. https://doi.org/10.1016/j.applanim.2007.02.009

McCowan B, Beisner BA, Capitanio JP et al (2011) Network stability is a balancing act of personality, power, and conflict dynamics in rhesus macaque societies. PLoS ONE 6(8): e22350. https://doi.org/10.1371/journal.pone.0022350

McCowan B, Beisner BA, Bliss-Moreau E et al (2016) Connections matter: social networks and lifespan health in primate translational models. Front Psychol 7:433. https://doi.org/10.3389/fpsyg.2016.00433

McEwen BS (1998) Stress, adaptation, and disease: allostasis and allostatic load. Ann N Y Acad Sci 840(1):33–44. https://doi.org/10.1111/j.1749-6632.1998.tb09546.x

McKenna P, Hoffmann C, Minkah N et al (2008) The macaque gut microbiome in health, lentiviral infection, and chronic enterocolitis. PLoS Pathog 4(2):e20. https://doi.org/10.1371/journal.ppat.0040020

Melfi V, Poyser F (2007) Trichuris burdens in zoo-housed *Colobus guereza*. Int J Primatol 28:1449–1456. https://doi.org/10.1007/s10764-007-9206-9

Modlmeier AP, Keiser CN, Watters JV et al (2014) The keystone individual concept: an ecological and evolutionary overview. Anim Behav 89:53–62. https://doi.org/10.1016/j.anbehav.2013.12.020

Morimura N, Hirata S, Matsuzawa T (2022) Challenging cognitive enrichment: examples from caring for the chimpanzees in the Kumamoto Sanctuary, Japan and Bossou, Guinea. In: Robinson LM, Weiss A (eds) Nonhuman primate welfare: from history, science, and ethics to practice. Springer, Cham, pp 489–516

Muehlenbein MP, Watts DP (2010) The costs of dominance: testosterone, cortisol and intestinal parasites in wild male chimpanzees. BioPsychoSocial Med 4:21. https://doi.org/10.1186/1751-0759-4-21

Muller MN, Wrangham RW (2004) Dominance, cortisol, and stress in wild chimpanzees (*Pan troglodytes*). Behav Ecol Sociobiol 55:332–340. https://doi.org/10.1007/s00265-003-0713-1

National Research Council (2011) Guide for the care and use of laboratory animals, 8th edn. The National Academies Press, Washington, DC

NC3Rs (2006) NC3Rs guidelines: Primate accomodations, care and use. National Center for the Replacement, Refinement and Reduction of Animals in Research, London

Novak MA, Suomi SJ (1988) Psychological well-being of primates in captivity. Am Psychol 43 (10):765–773. https://doi.org/10.1037/0003-066X.43.10.765

Oates-O'Brien RS, Farver TB, Anderson-Vicino KC et al (2010) Predictors of matrilineal overthrows in large captive breeding groups of rhesus macaques (*Macaca mulatta*). J Am Assoc Lab Anim Sci 49(2):196–201

Office of Laboratory Animal Welfare (2015). http://grants.nih.gov/grants/olaw/faqs.htm#1655

Ostner J, Heistermann M, Schülke O (2008) Dominance, aggression, and physiological stress in wild male Assamese macaques (*Macaca assamensis*). Horm Behav 54(5):613–619. https://doi.org/10.1016/j.yhbeh.2008.05.020

Perret M (1992) Environmental and social determinants of sexual function in the male lesser mouse lemur (*Microcebus murinus*). Folia Primatol 59(1):1–25. https://doi.org/10.1159/000156637

Petit O, Thierry B (1994) Aggressive and peaceful interventions in conflicts in Tonkean macaques. Anim Behav 48(6):1427–1436. https://doi.org/10.1006/anbe.1994.1378

Preuschoft S (1992) "Laughter" and "smile" in Barbary macaques (*Macaca sylvanus*). Ethology 91 (3):220–236. https://doi.org/10.1111/j.1439-0310.1992.tb00864.x

Pusey A, Williams J, Goodall J (1997) The influence of dominance rank on the reproductive success of female chimpanzees. Science 277(5327):828–831. https://doi.org/10.1126/science.277.5327.828

Ray JC, Sapolsky RM (1992) Styles of male social behavior and their endocrine correlates among high-ranking wild baboons. Am J Primatol 28(4):231–250. https://doi.org/10.1002/ajp.1350280402

Rimbach R, Bisanzio D, Galvis N et al (2015) Brown spider monkeys (*Ateles hybridus*): a model for differentiating the role of social networks and physical contact on parasite transmission dynamics. Philos Trans R Soc Lond B Biol Sci 370(1669):20140110. https://doi.org/10.1098/rstb.2014.0110

Rowell TE (1974) The concept of social dominance. Behav Biol 11(2):131–154. https://doi.org/10.1016/S0091-6773(74)90289-2

Ruehlmann TE, Bernstein IS, Gordon TP, Balcaen P (1988) Wounding patterns in three species of captive macaques. Am J Primatol 14(2):125–134. https://doi.org/10.1002/ajp.1350140203

Rukstalis M, French JA (2005) Vocal buffering of the stress response: exposure to conspecific vocalizations moderates urinary cortisol excretion in isolated marmosets. Horm Behav 47 (1):1–7. https://doi.org/10.1016/j.yhbeh.2004.09.004

Runcie DE, Wiedmann RT, Archie EA et al (2013) Social environment influences the relationship between genotype and gene expression in wild baboons. Philos Trans R Soc Lond B Biol Sci 368(1618):20120345. https://doi.org/10.1098/rstb.2012.0345

Saito C (1996) Dominance and feeding success in female Japanese macaques, *Macaca fuscata*: effects of food patch size and inter-patch distance. Anim Behav 51(5):967–980. https://doi.org/10.1006/anbe.1996.0100

Salathe M, Jones JH (2010) Dynamics and control of diseases in networks with community structure. PLoS Comput Biol 6(4):e1000736. https://doi.org/10.1371/journal.pcbi.1000736

Saltzman W, Schultz-Darken NJ, Wegner FH et al (1998) Suppression of cortisol levels in subordinate female marmosets: reproductive and social contributions. Horm Behav 33 (1):58–74. https://doi.org/10.1006/hbeh.1998.1436

Santiago ML, Range F, Keele BF et al (2005) Simian immunodeficiency virus infection in free-ranging sooty mangabeys (*Cercocebus atys atys*) from the Taï Forest, Côte d'Ivoire: implications for the origin of epidemic human immunodeficiency virus type 2. J Virol 79 (19):12515–12527. https://doi.org/10.1128/jvi.79.19.12515-12527.2005

Sapolsky RM (1982) The endocrine stress-response and social status in the wild baboon. Horm Behav 16(3):279–292. https://doi.org/10.1016/0018-506X(82)90027-7

Sapolsky RM (1983) Endocrine aspects of social instability in the olive baboon (*Papio anubis*). Am J Primatol 5(4):365–379. https://doi.org/10.1002/ajp.1350050406

Sapolsky RM (1992) Cortisol concentrations and the social significance of rank instability among wild baboons. Psychoneuroendocrinology 17(6):701–709. https://doi.org/10.1016/0306-4530(92)90029-7

Sapolsky RM (2005) The influence of social hierarchy on primate health. Science 308 (5722):648–652. https://doi.org/10.1126/science.1106477

Sapolsky RM, Mott GE (1987) Social subordinance in wild baboons is associated with suppressed high density lipoprotein-cholesterol concentrations: the possible role of chronic social stress. Endocrinology 121(5):1605–1610. https://doi.org/10.1210/endo-121-5-1605

Sapolsky RM, Ray JC (1992) Styles of male social behavior and their endocrine correlates among high-ranking wild baboons. Am J Primatol 28(4):231–250. https://doi.org/10.1002/ajp.1350280402

Sapolsky RM, Share LJ (1994) Rank-related differences in cardiovascular function among wild baboons: role of sensitivity to glucocorticoids. Am J Primatol 32(4):261–268. https://doi.org/10.1002/ajp.1350320404

Sapolsky RM, Spencer E (1997) Insulin-like growth factor I is suppressed in socially subordinate male baboons. Am J Physiol 273(4):R1346–R1351. https://doi.org/10.1152/ajpregu.1997.273.4.R1346

Sapolsky RM, Romero LM, Munck AU (2000) How do glucocorticoids influence stress responses? Integrating permissive, suppressive, stimulatory, and preparative actions. Endocr Rev 21 (1):55–89. https://doi.org/10.1210/edrv.21.1.0389

Schapiro SJ, Nehete PN, Perlman JE, Sastry KJ (2000) A comparison of cell-mediated immune responses in rhesus macaques housed singly, in pairs, or in groups. Appl Anim Behav Sci 68 (1):67–84. https://doi.org/10.1016/s0168-1591(00)00090-3

Schoener TW (1971) Theory of feeding strategies. Annu Rev Ecol Syst 2:369–404. https://doi.org/10.1146/annurev.es.02.110171.002101

Schwitzer C, Kaumanns W (2001) Body weights of ruffed lemurs (*Varecia variegata*) in European zoos with reference to the problem of obesity. Zoo Biol 20(4):261–269. https://doi.org/10.1002/zoo.1026

Scott J, Lockard JS (2006) Captive female gorilla agonistic relationships with clumped defendable food resources. Primates 47:199–209. https://doi.org/10.1007/s10329-005-0167-3

Setchell JM, Smith T, Wickings EJ, Knapp LA (2008) Factors affecting fecal glucocorticoid levels in semi-free-ranging female mandrills (*Mandrillus sphinx*). Am J Primatol 70(11):1023–1032. https://doi.org/10.1002/ajp.20594

Shively CA, Clarkson TB (1994) Social status and coronary artery atherosclerosis in female monkeys. Arterioscler Thromb 14(5):721–726. https://doi.org/10.1161/01.ATV.14.5.721

Shively CA, Day SM (2015) Social inequalities in health in nonhuman primates. Neurobiol Stress 1:156–163. https://doi.org/10.1016/j.ynstr.2014.11.005

Shutt K, MacLarnon A, Heistermann M, Semple S (2007) Grooming in Barbary macaques: better to give than to receive? Biol Lett 3(3):231–233. https://doi.org/10.1098/rsbl.2007.0052

Sicotte P (1995) Interposition in conflicts between males in bimale groups of mountain gorillas. Folia Primatol 65(1):14–24. https://doi.org/10.1159/000156871

Silk JB, Alberts SC, Altmann J (2003) Social bonds of female baboons enhance infant survival. Science 302(5648):1231–1234. https://doi.org/10.1126/science.1088580

Silk JB, Beehner JC, Bergman TJ et al (2009) The benefits of social capital: close social bonds among female baboons enhance offspring survival. Proc R Soc B Biol Sci 276(1670):3099–3104. https://doi.org/10.1098/rspb.2009.0681

Silk JB, Beehner JC, Bergman TJ et al (2010) Strong and consistent social bonds enhance the longevity of female baboons. Curr Biol 20(15):1359–1361. https://doi.org/10.1016/j.cub.2010.05.067

Small M (1981) Body fat, rank, and nutritional status in a captive group of rhesus macaques. Int J Primatol 2:91–95. https://doi.org/10.1007/BF02692303

Smith DG (2012) Taxonomy of nonhuman primates used in biomedical research. In: Abee C, Mansfield K, Tardif S, Morris T (eds) Nonhuman primates in biomedical research: biology and management, vol 1. Elsevier, New York, pp 57–85

Smith TE, French JA (1997) Social and reproductive conditions modulate urinary cortisol excretion in black tufted-ear marmosets (*Callithrix kuhli*). Am J Primatol 42(4):253–267. https://doi.org/10.1002/(sici)1098-2345(1997)42:4%3C253::aid-ajp1%3E3.0.co;2-w

Smuts BB, Cheney D, Seyfarth R et al (1987) Primate societies. University of Chicago Press, Chicago

Stahl D, Kaumanns W (2003) Food competition in captive female sooty mangabeys (*Cercocebus torquatus atys*). Primates 44(3):203–216. https://doi.org/10.1007/s10329-002-0012-x

Sterck EHM, Watts DP, van Schaik CP (1997) The evolution of female social relationships in nonhuman primates. Behav Ecol Sociobiol 41:291–309. https://doi.org/10.1007/s002650050390

Suomi SJ, Harlow HF, Kimball SD (1971) Behavioral effects of prolonged partial social isolation in the rhesus monkey. Psychol Rep 29(3):1171–1177. https://doi.org/10.2466/pr0.1971.29.3f.1171

Suomi SJ, Collins ML, Harlow HF (1973) Effects of permanent separation from mother on infant monkeys. Dev Psychol 9(3):376–384. https://doi.org/10.1037/h0034896

Tajiama T, Kurotori H (2010) Nonaggressive interventions by third parties in conflict among captive Bornean orangutans (*Pongo pygmaeus*). Primates 51(2):179–182. https://doi.org/10.1007/s10329-009-0180-z

The Commission of European Communities (2007) Commission recommendation of 18 June 2007 on guidelines for the accommodation and care of animals used for experimental and other scientific purposes. Off J Eur Union 30.7.2007:L197/1–L197/89. https://eur-lex.europa.eu/LexUriServ/LexUriServ.do?uri=OJ:L:2007:197:0001:0089:EN:PDF

Thierry B, Iwaniuk AN, Pellis SM (2000) The influence of phylogeny on the social behaviour of macaques (Primates: Cercopithedicdae, genus *Macaca*). Ethology 106(8):713–728. https://doi.org/10.1046/j.1439-0310.2000.00583.x

Thierry B, Singh M, Kaumanns W (2004) Macaque societies: a model for the study of social organization. Cambridge University Press, Cambridge

Tung J, Barreiro LB, Burns MB et al (2015) Social networks predict gut microbiome composition in wild baboons. Elife 4:e05224. https://doi.org/10.7554/eLife.05224

Turnbaugh PJ, Hamady M, Yatsunenko T et al (2009) A core gut microbiome in obese and lean twins. Nature 457:480–484. https://doi.org/10.1038/nature07540

U.S. Department of Agriculture (2013) Animal Welfare Act and Animal Welfare Regulations

van Noordwijk M, van Schaik CP (1999) The effects of dominance rank and group size on female lifetime reproductive success in wild long-tailed macaques, *Macaca fascicularis*. Primates 40:105–130. https://doi.org/10.1007/BF02557705

van Schaik CP (1983) Why are diurnal primates living in groups? Behaviour 87(1/2):120–144

van Schaik CP (1989) The ecology of social relationships amongst female primates. In: Standen V, Foley RA (eds) Comparative socioecology: the behavioural ecology of humans and other mammals. Blackwell Scientific Publications, Oxford, pp 195–218

van Schaik CP, van Hooff JARAM (1983) On the ultimate causes of primate social systems. Behaviour 85(1/2):91–117

van Schaik CP, van Noordwijk M (1988) Scramble and contest in feeding competition among female long-tailed macaques (*Macaca fascicularis*). Behaviour 105(1/2):77–98. https://doi.org/10.1163/156853988X00458

Vandeleest JJ, McCowan BJ, Capitanio JP (2011) Early rearing interacts with temperament and housing to influence the risk for motor stereotypy in rhesus monkeys (*Macaca mulatta*). Appl Anim Behav Sci 132(1–2):81–89. https://doi.org/10.1016/j.applanim.2011.02.010

Vandeleest JJ, Beisner BA, Hannibal DL et al (2016) Decoupling social status and status certainty effects on health in macaques: a network approach. PeerJ 4:e2394. https://doi.org/10.7717/peerj.2394

Videan EN, Fritz J, Murphy J (2007) Development of guidelines for assessing obesity in captive chimpanzees (*Pan troglodytes*). Zoo Biol 26(2):93–104. https://doi.org/10.1002/zoo.20122

Vitone ND, Altizer S, Nunn CL (2004) Body size, diet and sociality influence the species richness of parasitic worms in anthropoid primates. Evol Ecol Res 6(2):183–199 von Rohr CR, Koski SE, Burkart JM et al (2012) Impartial third-party interventions in captive chimpanzees: a reflection of community concern. PLoS ONE 7(3):e32494. https://doi.org/10.1371/journal.pone.0032494

Weingrill T, Gray DA, Barrett L, Henzi SP (2004) Fecal cortisol levels in free-ranging female chacma baboons: relationship to dominance, reproductive state and environmental factors. Horm Behav 45(4):259–269. https://doi.org/10.1016/j.yhbeh.2003.12.004

Whitacre JM (2010) Degeneracy: a link between evolvability, robustness and complexity in biological systems. Theor Biol Med Model 7:6. https://doi.org/10.1186/1742-4682-7-6

Whitten PL (1983) Diet and dominance among female vervet monkeys (*Cercopithecus aethiops*). Am J Primatol 5(2):139–159. https://doi.org/10.1002/ajp.1350050205

Wilson ME, Fisher J, Fischer A et al (2008) Quantifying food intake in socially housed monkeys: social status effects on caloric consumption. Physiol Behav 94(4):586–594. https://doi.org/10.1016/j.physbeh.2008.03.019

Wittig RM, Crockford C, Lehmann J et al (2008) Focused grooming networks and stress alleviation in wild female baboons. Horm Behav 54(1):170–177. https://doi.org/10.1016/j.yhbeh.2008.02.009

Wrangham R (1980) An ecological model of female-bonded primate groups. Behaviour 75(3/4):262–300

Yodyingyuad U, de la Riva C, Abbott DH et al (1985) Relationship between dominance hierarchy, cerebrospinal fluid levels of amine transmitter metabolites (5-hydroxyindole acetic acid and homovanillic acid) and plasma cortisol in monkeys. Neuroscience 16(4):851–858. https://doi.org/10.1016/0306-4522(85)90099-5

Young C, Majolo B, Heistermann M et al (2014) Responses to social and environmental stress are attenuated by strong male bonds in wild macaques. Proc Natl Acad Sci U S A 111(51):18195–18200. https://doi.org/10.1073/pnas.1411450111

Zehr JL, Van Meter PE, Wallen K (2005) Factors regulating the timing of puberty onset in female rhesus monkeys (*Macaca mulatta*): role of prenatal androgens, social rank, and adolescent body weight. Biol Reprod 72(5):1087–1094. https://doi.org/10.1095/biolreprod.104.027755

Ziegler TE, Scheffler G, Snowdon CT (1995) The relationship of cortisol levels to social environment and reproductive functioning in female cotton-top tamarins, *Saguinus oedipus*. Horm Behav 29(3):407–424. https://doi.org/10.1006/hbeh.1995.1028

Research Benefits of Improving Welfare in Captive Primates

Steven J. Schapiro and Jann Hau

Abstract

The maintenance of high levels of welfare in captive nonhuman primates is essential to research. Behavioral management techniques, incorporating socialization strategies, environmental enrichment procedures, and positive reinforcement training techniques typically result in high levels of species-appropriate behaviors and low levels of abnormal behaviors. Additionally, these techniques can yield physiological and immunological response patterns indicative of the suitability of nonhuman primate subjects for use in biomedical and other types of research projects. Similarly, subject selection procedures that account for relevant characteristics of the nonhuman primates (disease status, species, temperament, etc.) are also likely to positively influence data quality. Behavioral management procedures and subject selection strategies typically result in fewer confounding influences on experimental data, resulting in less problematic interindividual variation in studies that employ appropriate behavioral management techniques. The implementation of behavioral management refinements results in enhanced welfare for the subjects, higher quality data, more reliable and robust results, and potentially, a reduction in the number of subjects required for research projects. Using positive reinforcement training techniques that allow socially housed, appropriately selected, nonhuman primates living in enriched environments to participate in research procedures is critical, if not imperative, to the collection of reliable and valid data, the foundation of all types of scientific investigations.

S. J. Schapiro (✉)
Department of Experimental Medicine, University of Copenhagen, København, Denmark

Department of Comparative Medicine, The University of Texas MD Anderson Cancer Center, Bastrop, TX, USA
e-mail: sschapir@mdanderson.org

J. Hau
Department of Experimental Medicine, University of Copenhagen, København, Denmark

Keywords

Behavioral management · Socialization · Environmental enrichment · Positive reinforcement training · Validity · Refinement

1 Introduction

Managers of animals in captivity are ethically obligated to ensure that the animals in their care live a good life, in conditions that promote a high state of welfare (Schapiro 2017, other chapters this volume). The present chapter discusses the welfare of captive nonhuman primates, but rather than focusing on the ways in which particular management manipulations, e.g., socially housing animals, affect general welfare-related behavioral measures (e.g., time spent engaged in social behaviors), we will focus more on the importance of the welfare of the primates as a fundamental prerequisite for the generation of reliable and valid scientific data. We will identify specific conditions, manipulations, and/or refinements that are likely to result in good welfare and therefore good data; and conditions and/or manipulations that are likely to result in compromised welfare, and therefore, potentially compromised data (Schapiro et al. 2000). Throughout the chapter, we will discuss several other welfare- and data-related issues (e.g., animal characteristics, including temperament; gentle handling) that are associated with ensuring the utility of research findings, and hence, the value of the scientific contributions made by the animals.

Because this chapter emphasizes the implications of enhanced welfare on the collection of reliable and valid data, much of the information contained herein will be derived from studies of nonhuman primates living in laboratories rather than from nonhuman primates living in the wild, in sanctuaries, or in zoos. However, most of the findings and implications of the cited work will be applicable to the collection of data from nonhuman primates in most captive settings.

The majority of this chapter will be organized around the concept of behavioral management: a system that focuses on refinements to the captive environment that are intended to influence and enhance the welfare of the animals (Coleman et al. 2012). We will discuss the ways in which socialization strategies, environmental enrichment plans, and positive reinforcement training techniques affect behavioral and physiological measures associated with animal welfare (Schapiro et al. 1993). We will especially emphasize how socialization, enrichment, and training influence the suitability of nonhuman primates as research models; and even more specifically, how these types of manipulations and refinements affect the dependent variables that are likely to be measured in research projects involving captive nonhuman primates (Schapiro 2002). For example, if nonhuman primates are used for immunodeficiency virus research, where immunological variables (lymphocyte subsets, cytokines, proliferation assays, etc.) are of great interest, we will discuss how behavioral management strategies (e.g., single vs. pair vs. group housing; Schapiro et al. 2000), independent of the experimental manipulations, are related to changes in immunological parameters.

To put it rather simply (but quite accurately), "happy animals make good science" (Poole 1997), and the present chapter is intended to address why competent and kind care, and attention to the enhancement of primate welfare, is vital for the production of high-quality science.

It is important to note that the classification of specific behaviors exhibited by nonhuman primates as abnormal (for review of behavioral assessment see Lutz and Baker 2022) is vital for assessing welfare and can be accomplished in two ways (Erwin and Deni 1979; Novak et al. 2017). The first is to consider behaviors that are observed in captive nonhuman primates, but are never observed in their wild conspecifics, as abnormal. Repetitive locomotion (pacing) and self-injurious behavior might be two reasonable examples. The second approach is to consider behaviors that occur in captivity at significantly different (higher or lower) frequencies than they do in the wild as abnormal, with higher levels of self-grooming as a prime example. Virtually all primates self-groom in the wild, but relatively few groom themselves bald. Most ethograms for studying captive primates include both types of classifications of abnormal behaviors (Honess 2017). It is necessary to consider the potential effects of abnormal behavior patterns on the integrity of scientific data for all species of laboratory animals (Garner 2005), not just for nonhuman primates.

Without detailed observations and analyses, many of the more subtle behavioral abnormalities, especially those related to changes in behavioral frequencies, which might have clinical implications, would not be recognized. However, such behavioral alterations may be important and adversely affect the value of nonhuman primate models for certain types of investigations, including the development and testing of drugs that affect the central nervous system (Shively 2017). For example, it has been demonstrated that people suffering from anxiety or depression interpret ambiguous events more negatively than healthy people, a condition referred to as cognitive bias (Holmes et al. 2008; in relation to primates: Bethell and Pfefferle 2022). Singly housed rhesus monkeys seem to display a similar response, as Bethell et al. (2012) demonstrated, i.e. a negative shift in cognitive bias following mildly stressful events compared to periods following feeding enrichment. These data emphasize the importance of attempting to make certain that nonhuman primate subjects are minimally stressed, and healthy, when they are used for biomedical research.

2 Welfare, Behavioral Management, and Scientific Data

As other chapters in this volume and numerous other studies have demonstrated, nonhuman primates that (1) perform many abnormal behaviors, (2) have heightened glucocorticoid levels, and/or (3) have "diminished" immune responses (among other detrimental effects) are likely to be experiencing compromised welfare (Schapiro 2002; Novak et al. 2017). In many cases, we have a very good understanding of the types of captive environments and management programs that (1) are likely to result in these kinds of problematic responses and (2) are likely to prevent the development of these kinds of responses. When nonhuman primates are subjects in scientific investigations, it is of the utmost importance, from a data quality perspective, that they be maintained in environments and conditions that enhance, rather than compromise, welfare. Compromised welfare is likely to be associated with a high potential for the introduction of uncontrolled, confounding factors into the experimental procedures, potentially affecting the interpretation and validity of the data (Hopkins and Latzman 2017; Shively 2017). Animals that are overly fearful or stressed, or that are depressed and experiencing other psychological or somatic problems, may make it difficult, if not impossible, to accurately identify the true effects of an experimental treatment, and erroneous conclusions may be drawn from the study (Shively et al. 2005; Shively 2017). In other words, variability in welfare across subjects and/or treatments may result in increased interindividual variation in experimental responses, potentially resulting in studies that are less statistically powerful (more prone to certain types of statistical errors), while requiring more subjects (Capitanio 2017). Efforts to enhance the welfare of nonhuman primate subjects, through behavioral management refinements, should result in lower interindividual variation (Schapiro et al. 2005; Lambeth et al. 2006), addressing one of the 3Rs (for review see Prescott 2022) by leading to a reduction in the number of subjects necessary for many studies (Schapiro et al. 2005; although see Capitanio's 2017 discussion of the value of individual differences in primate research).

As mentioned previously, we have some reasonably good understanding of the types of captive environments and management plans that are likely to result in compromised welfare, and therefore, compromised data. For over 50 years, since Harry Harlow published his social deprivation work with rhesus monkeys (Harlow and Harlow 1962), it has been well known that infant monkeys need to be reared in social circumstances to develop normally (Suomi et al. 1975) and to be appropriate subjects for scientific investigations. Similarly, more recent work has shown that nonhuman primates living in nonenriched environments, in which they have large quantities of empty time, relatively few opportunities to perform species-typical behaviors, and little or no control over what happens to them, also experience diminished welfare (Schapiro and Bloomsmith 1995; Xie et al. 2014). And finally, nonhuman primates that do not participate in positive reinforcement training programs are likely to exhibit behavioral and physiological responses that differ from those that have the chance to participate, through training, in aspects of their care and research experiences (Bloomsmith et al. 2022; Lambeth et al. 2006; Schapiro et al. 2017).

Not only do we have ideas related to the types of management plans that are likely to result in compromised data, we also have an ever-expanding understanding of the types of captive environments and management plans that are likely to result in enhanced welfare, and therefore, subjects that are likely to yield valuable data (Schapiro 2017). Again, Harlow's work (Harlow and Harlow 1962) taught us how *not* to rear infant macaques, if they were going to be used in scientific research projects. Nonhuman primate subjects are now reared and housed socially (McGrew 2017; Truelove et al. 2017), so as to provide data that are useful. Similarly, most nonhuman primates used in research live in enriched environments (e.g., structurally complex environments with multiple opportunities to perform species-typical behaviors; Coleman et al. 2012; Schapiro et al. 2014, see also Kemp 2022; Morimura et al. 2022) that seek to minimize the "empty time" experienced by the animals. Additionally, many nonhuman primates now participate in positive reinforcement training programs (Baker 2016) in which the animals voluntarily cooperate in a variety of husbandry, veterinary, and research procedures (Coleman et al. 2008; Magden et al. 2013; Schapiro et al. 2014; Magden et al. 2016; Schapiro et al. 2018).

2.1 Socialization and Data

There are considerable data available that demonstrate that early social deprivation of nonhuman primates results in a variety of lifelong effects on multiple parameters likely to diminish the suitability of socially deprived individuals as viable models for human conditions (Laudenslager et al. 2013; Brunelli et al. 2014). Given the substantial literature available on the effects of social restriction in nonhuman primates, including chapters in this volume, there is little need to go into detail concerning this issue.

Relatively few nonhuman primates live solitarily in the wild, and a social partner in captivity provides many opportunities for variable, species-typical interactions (Schapiro et al. 1996), one of the objective measures of good welfare. There are many examples of single housing being associated with stress responses, and altered values for the dependent variables that typically comprise the data in many different types of investigations (Schapiro et al. 2000; Schapiro 2002; Doyle et al. 2008). Having said this, there are, however, a few other lines of inquiry, specifically some neuroscience investigations, that can be conducted with nonhuman primates that take part in, e.g., cognitive tests and/or are housed singly, and that appear unstressed by living and/or working alone (Slater et al. 2016). Additionally, numerous studies have demonstrated that social rearing results not only in enhanced welfare, but in enhanced research suitability as well, with most socially reared animals exhibiting values for many important parameters that are more representative of the species' norms (Rommeck et al. 2009, 2011). Perhaps more importantly, comparisons of singly housed to socially housed older animals that were reared socially when young indicate that social housing positively affects the utility of these animals as research subjects (Benton et al. 2013). Schapiro et al. (2000) compared single-, pair-, and group-housed adult rhesus monkeys on a variety of immunological responses relevant for immunodeficiency virus vaccine research, identifying numerous statistically significant differences across housing conditions. For most parameters, the socially housed subjects appeared to yield more relevant data; data that are indicative of both greater scientific utility/translatability and enhanced welfare.

The stability of social groupings is also important for the collection of data that are valid and reliable. Capitanio et al. (1998) have shown that male rhesus macaques living in social groups that were manipulated to be unstable were more likely to be adversely affected by immunodeficiency viruses than were male rhesus that lived in stable social groups. Therefore, unstable group composition is another social factor that contributes to reduced welfare and hence, scientific data that may be of limited value. Again, unless one is interested in studying the effects of group instability, maintaining research subjects in stable social groups is important for promoting subject health and for obtaining reliable and valid data. This work by Capitanio and colleagues demonstrates that the composition, and especially the stability, of social groupings can introduce confounds that can affect the interpretation of research findings.

Shively and colleagues (Shively et al. 2005; Shively 2017) have identified "depression" in captive cynomolgus monkeys (*Macaca fasciularis*), operationalized as a set of very specific postures and behaviors, which is likely to be indicative of diminished welfare, and appears to also be affected by social group instability, among other factors (Shively et al. 2005). Shively and her group have clearly demonstrated that there are many data- and research-relevant implications of including depressed monkeys in non-depression-related research (Shively 2017). These involve well-documented physiological and neurobiological differences between depressed and non-depressed monkeys associated with the Hypothalamus-Pituitary-Adrenal axis (Shively 2002), ovarian function (Shively et al. 2005), atherosclerosis (Shively et al. 2009), and hippocampal volume (Willard et al. 2009), to name just a few. This work highlights the importance of using welfare-related behavioral criteria (depressed posture and behavior) when selecting subjects for scientific studies. The selection of subjects should be performed carefully to prevent the introduction of depression as a confounding factor in investigations of other independent variables.

2.2 Enrichment and Data

While a reasonable quantity of data exist that connect environmental enrichment with enhanced welfare, there are relatively few studies in the literature that directly connect environmental enrichment practices with the reliability and validity of scientific data from nonhuman primates (Schapiro et al. 1998; Bayne 2005; Weed and Raber 2005; Bayne and Wurbel 2014). Among those studying rodents, some question whether environmental enrichment is beneficial for achieving reliable and valid data (Eskola et al. 1999; Baumans et al. 2010). Environmental enrichment practices that promote the performance of appropriate levels of species-typical (normal) behaviors positively impact nonhuman primate welfare (Schapiro et al. 1998; Coleman et al. 2012) and seem likely to enhance the quality of the data collected from these animals.

2.3 Positive Reinforcement Training and Data

Positive reinforcement training (reviewed by Bloomsmith et al. 2022), a behavioral management technique that allows captive nonhuman primates to (1) exert some control over aspects of their environment by making meaningful choices concerning the way they are treated and (2) participate in different procedures, has been shown to result in behavioral and physiological changes that suggest enhanced welfare (Schapiro et al. 2001, 2005; Laule et al. 2003; Lambeth et al. 2006; Magden et al. 2013). More importantly, positive reinforcement training can result in data that are less variable across individuals, which may mean that fewer potential confounds may influence the analysis and interpretation of the findings.

Training nonhuman primates using positive reinforcement training techniques generally enhances the welfare of captive nonhuman primates (Laule et al. 2003; Magden et al. 2013), but more importantly for this chapter, the training enhances the utility of the data attainable from trained subjects. Training nonhuman primates to participate in research procedures (1) decreases procedure-related stress for the subjects and for the people; (2) allows for the analysis of parameters that might not be measurable in untrained subjects; and again, most importantly for this chapter, (3) minimizes variability between subjects due to minimizing potential confounds related to trained behaviors. Melanie Graham's diabetes-related research program at the University of Minnesota is an extremely relevant example of the value of training macaques to participate in mildly invasive research protocols (Graham et al. 2012; Graham 2017). Physical and chemical restraint of nonhuman primates affects many of the physiological parameters important for the assessment of treatment effects in diabetes-related studies. By training nonhuman primate subjects to participate in drug administration and biological sampling procedures, Graham and colleagues (Graham et al. 2012; Graham 2017) were able to minimize the potential confounding effects of restraint on their data.

Lambeth et al. (2006) have also found that positive reinforcement training can influence general physiological parameters (i.e., complete blood counts, chemistry, immune responses), and especially immunological parameters, that are assessed in many research programs. In this study, the method of administering anesthetic to chimpanzees *Pan troglodytes* (voluntary present for the anesthetic injection versus a nonvoluntary injection) affected these stress-sensitive and research-relevant variables. Subjects trained to voluntarily receive the anesthetic injection had significantly lower mean total white blood cell counts and glucose levels than did subjects that were nonvoluntarily anesthetized. Additionally, many physiological and immunological parameters differed when blood samples obtained from chimpanzees voluntarily, using the blood sleeve and no anesthesia, were compared to blood samples that were obtained from chimpanzees after they had been anesthetized (Schapiro et al. 2005). These results demonstrate that positive reinforcement training can significantly improve the animals' utility as biomedical models. Positive reinforcement training, as a component of a comprehensive behavioral management program, is now practiced at most primate facilities in the USA (Baker 2016) and is an integral part of Primtrain (www.primtrain.eu), a European transnational primate center initiative.

It is important to note that in most successful positive reinforcement training programs, food or fluid control is rarely, if ever, necessary to enhance the motivation of nonhuman primate subjects. Neither the macaques at the University of Minnesota (Graham 2017) nor the chimpanzees at the National Center for Chimpanzee Care (Schapiro et al. 2018), for example, are food- or fluid-deprived. While food or fluid control is a complicated issue (Prescott et al. 2010; Westlund 2012), such procedures are likely to have welfare consequences and seem unlikely to yield experimental data of the highest quality.

3 Nonhuman Primate Subject Characteristics and Their Effects on Data

The nonhuman primates that are maintained at most primate research facilities and that are used in most scientific investigations are, in general, outbred animals that exhibit considerably more interindividual variation than the genetically defined rodent strains that are used in the vast majority (nonhuman primates comprise less than 0.5% of animals that are used in research) of studies involving research animals. Such genetic variation typically results in variation in how animals respond to both (1) stressors in the environment that affect measures of welfare and (2) experimental treatments, and potentially reduce statistical power, thus necessitating the use of larger sample sizes to test hypotheses. This is contrary to one of the 3Rs (reduction) and should be avoided when possible. However, the consequences of increased interindividual variation may not be all bad (Capitanio 2017), and a recent article by Vallender and Miller (2013) suggests that the genetic variability of nonhuman primates makes them particularly useful as translational research models. The development of personalized biological therapies, combined with advancements in functional genomics and studies of genetic variation in nonhuman primates (Hopkins and Latzman 2017), allow for the establishment of better, targeted models and techniques to evaluate therapies for human diseases.

Different species of nonhuman primates are differentially affected by captive management procedures in general, and especially by behavioral management procedures (Schapiro 2017). Socialization, enrichment, and training techniques that work for rhesus macaques (*Macaca mulatta*; Gottlieb et al. 2017) may not be as successful for African green monkeys (*Chlorocebus* spp.; Jorgensen 2017) or owl monkeys (*Aotus*; Williams and Ross 2017). Even closely related species, such as rhesus, cynomolgus, and bonnet macaques (*M. radiata*) differ with respect to their (hormonal) responsiveness to certain stressors (Clarke et al. 1988; Gottlieb et al. 2017; Honess 2017). Similarly, titi (*Callicebus* spp.) and squirrel (*Saimiri* spp.) monkeys show very different hormonal responses to stressors (Mendoza and Mason 1984) with titi monkeys being considerably less adaptable to some captive management protocols (Poole 1997), especially single housing.

3.1 Pathogens and Data

Although nonhuman primates can be reasonably conveniently obtained from high-quality, microbiologically defined (Specific-Pathogen-Free, SPF) breeding colonies, the situation is considerably different from the situation for laboratory rodents. SPF rodents are routinely used in research, and there are relatively few welfare problems reported. SPF nonhuman primates, however, must be carefully "derived" and subsequently housed, with strict attention paid to the welfare consequences and pathogen-related consequences of the derivation techniques and housing strategies employed. Animals that are pathogen-free, but behaviorally abnormal, may not be particularly useful for research. Derivation techniques that account for both favorable welfare and pathogen outcomes exist and have been successfully employed by our group (Schapiro et al. 1994; Schapiro and Bernacky 2012) and by others (Hilliard and Ward 1999; Budda et al. 2013).

Nonhuman primates that are not SPF, and that enter research colonies or projects, could be carriers of clinically silent or dormant infections. Different species of nonhuman primates are natural hosts for several exogenous retroviruses, including gibbon-ape leukemia virus, simian sarcoma virus, simian T-cell lymphotropic virus, simian immunodeficiency virus, simian type D retrovirus, and simian foamy virus (Wolf et al. 2010). These viruses may establish persistent infections, the effects of which can range from high levels of pathogenesis to no pathogenesis, depending on various host, virus, and environmental factors. Latent or subclinical infections are common, and stress associated with research procedures may stimulate virus activation and consequently, clinical disease. Lerche and Osborn (2003) reviewed the adverse effects of undetected retroviral infections on toxicological research and found that such infections occur and may result in the loss of experimental subjects and statistical power, due to increased morbidity and mortality. Toxicological findings were discovered to be confounded by virus-associated abnormalities, such as pathological lesions and alteration of physiologic parameters. Macaques are hosts to a range of herpes viruses that cause persistent, latent, lifelong infections. Primary infection of macaques with normal immune responses is typically subclinical, with associated morbidity and mortality being quite rare. However, among affected animals, immune modulation or suppression can result in disease, and even death (Simmons 2010). Immunosuppressive drugs, like tacrolimus, and irradiation may also reactivate viruses (Mahalingam et al. 2010). For these reasons, SPF primates are highly desirable subjects for many nonhuman primate research projects, with behaviorally normal SPF primates the most desirable, and, in some cases, the only suitable, subjects.

3.2 Temperament and Data

Recently, considerable effort has been devoted to studying the temperament, or personality, of animals, particularly nonhuman primates (Capitanio et al. 2011; Freeman et al. 2013; Coleman and Pierre 2014). While many of these investigations originally focused on theoretical issues associated with temperament (Freeman et al. 2013; Latzman et al. 2017), more recent publications have emphasized applied aspects of understanding nonhuman primate temperament (Robinson et al. 2016, Capitanio 2017; Coleman 2017; reviewed by Robinson and Weiss 2022). One applied area has revolved around selecting subjects identifying those animals that are best suited for particular types of research protocols; for instance, animals that score highly on exploratory temperament dimensions may be more appropriate for certain studies compared with animals that do not (Capitanio 2017). Some of the motivation for this approach has been to enhance welfare by not selecting individuals to be used in projects and procedures that they may be ill-equipped to deal with. An example of this would be to avoid individuals exhibiting a nervous temperament for infectious disease studies, because these individuals show evidence of glucocorticoid desensitization and have physiological characteristics with a potential impact on their inflammatory responses to infection. Another aspect motivating the application of temperament assessments for subject selection has been to minimize interindividual variation by matching temperaments across control and experimental subjects. This work suggests quite clearly that evaluating temperament can enhance both the welfare of nonhuman primate subjects and the utility of the data which they "produce" (Capitanio 2017; Coleman 2017).

4   Acclimation and Data Quality

Capture and confinement of wild nonhuman primates is associated with severe stress, as demonstrated in some detail by studies of newly captured African green monkeys (*C. aethiops*; Else 1985), which are more stress-prone than, for example, baboons (*Papio* spp.). Recently captured African green monkeys exhibited a wide range of abnormal biological characteristics when maintained in captivity, including persistently high cortisol levels, immunosuppression, enlarged adrenal glands, gastric ulcers, and hippocampal damage (Uno et al. 1989; Suleman et al. 1999, 2000, 2004). An investigation of fluctuations in hematological values of 50 wild-caught African green monkeys during habituation to captivity demonstrated that adaptation periods of at least several months appear to be necessary (Kagira et al. 2007). The large variations observed between individuals on many parameters suggest that within-subjects designs, employing comparisons to baseline values for each individual, may be required to minimize interindividual variance when African green monkeys are used in research projects. This confirms earlier findings (Else 1985) that complete adjustment to captivity took a minimum of one year. Additionally, the establishment of stable breeding groups of adult African green monkeys was initially difficult due to fighting among females (Else 1985). These findings clearly demonstrate that wild-caught African green monkeys show stress-induced biological abnormalities during the early phases of their adaptation to captivity that are so severe that they should not be considered suitable as animal models in biomedical research until they have spent approximately one year in captivity. Consequently, from an ethical point of view, the use of wild-caught African green monkeys in biomedical research is criticizable.

While the stress associated with capture from the wild and relocation to captive conditions is likely to be considerably more severe than the stress associated with transport and relocation from one captive setting to another, there is evidence from multiple nonhuman primate species that such relocation-related manipulations are also likely to adversely affect the quality of research data. Chimpanzees (Schapiro et al. 2012; Neal Webb et al. 2018), macaques (Capitanio et al. 2005; Fernström et al. 2008; Koban et al. 2010; Nehete et al. 2017; Shelton et al. 2019), owl monkeys

(Williams et al. 2010), and squirrel monkeys (Williams et al. 2010) all display behavioral and/or physiological changes as a function of relocation to different/new captive conditions. It would not be appropriate in any of these circumstances to begin a study immediately upon the animals' arrival in their new research setting. In general, acclimation periods, ranging from 6 weeks to 6 months, which involve the implementation of numerous behavioral management techniques (e.g., enrichment, socialization, training), should be required prior to study enrollment and the initiation of data collection. Even social separations for periods as short as three hours can result in changes in immunological responses (e.g., proliferation responses to concanavalin A in juvenile squirrel monkeys; Friedman et al. 1991). Efforts must be made to restore, maintain, and enhance the welfare of nonhuman primates, prior to their use in research projects, when moving from one facility to another. Such efforts should include behavioral conditioning (see above) at the facility of origin (Fernandez et al. 2017), any intermediate facilities, and of course, at the facility at which the research is taking place (Schapiro et al. 2014).

5   Animal Handling and Data Quality

Nonhuman primates are naturally fearful of people and kind, gentle, confident handling is crucial to avoid unnecessarily stressing the animals. Even routine husbandry procedures may cause nervousness and increases in heart rate for up to two hours after the event (Line et al. 1989). When possible, minor procedures should be carried out with (1) minimal disturbance and restraint and (2) the use of positive reinforcement training techniques, because the animals are likely to respond with a stress response to restraint and/or unfamiliar surroundings/procedures. The cortisol response of single-caged adult female rhesus monkeys during venipuncture in a restraint apparatus was compared with the cortisol response of 10 paired and five single-caged adult female rhesus monkeys during venipuncture in the home-cage. Results demonstrated that in-home-cage venipuncture offers a methodological refinement for research protocols that require blood collection from undisturbed animals (Reinhardt et al. 1990). Using positive reinforcement training to allow nonhuman primate subjects to participate in blood sampling procedures (Schapiro et al. 2005; Lambeth et al. 2006; Coleman et al. 2008; Graham et al. 2012; Reamer et al. 2014; Graham 2017; Magden 2017) is simply the next refinement in animal handling.

6   Multiple Protocols, Re-Use, and Data Quality

Nonhuman primates often are used in multiple research projects during their lifetime. Protocols early in the nonhuman primates' life may influence the animals' mental and physical health, and the quality of the data they generate later in life. Hau and Schapiro (2007) advocated the importance of maintaining records for individual nonhuman primates that contain not just their medical and experimental protocol histories, but also critical aspects of their psychological histories. Such psychological factors might include the nonhuman primates' social housing experiences, personality/temperament assessments, any special behavioral tendencies they may exhibit, and the numbers and types of studies in which they have been used (Neal Webb et al. 2019b). These data will facilitate the process of monitoring individual cumulative severity (lifetime experience, Wolfensohn et al. 2015) and the potential welfare effects of using nonhuman primates in experimental protocols. The extended welfare assessment grid, as conceived and developed by Honess and Wolfensohn (Wolfensohn 2022; Honess and Wolfensohn 2010), is designed to facilitate ethical oversight, and to draw attention to the temporal and cumulative effects on welfare that may be overlooked during the planning and implementation of research projects when nonhuman primate subjects are used in multiple (sequential) protocols.

In addition, maintaining the welfare of research veteran and/or aged nonhuman primates requires a considerable amount of special attention. Nonhuman primates develop many of the same diseases and maladies that older humans do, including type 2 diabetes (Reamer et al. 2014; Graham 2017), obesity, arthritis (Magden et al. 2013), and dementia (Darusman et al. 2013, 2019; Edler et al. 2017). Consequently, special care must be provided to aging nonhuman primates to maintain their welfare and scientific value, including the use of acupuncture, laser therapy, implantable cardiac monitors, and weight management (Magden et al. 2013, 2016; Lambeth et al. 2013; Neal Webb et al. 2019a).

7 Conclusions

A focus on the maintenance of high levels of welfare among captive nonhuman primates is essential to generate reliable and robust research data when they are used as models in biomedical research. Behavioral management techniques, incorporating socialization strategies, environmental enrichment procedures, and positive reinforcement training techniques typically result in high levels of species-appropriate behaviors and low levels of abnormal behaviors. Additionally, these techniques can also yield physiological and immunological response patterns indicative of the suitability of nonhuman primate subjects for use in biomedical and other types of research projects. Similarly, subject selection procedures that account for relevant characteristics of the nonhuman primates (disease status, species, temperament, etc.) are also likely to positively influence data quality. Behavioral management procedures and subject selection strategies typically result in fewer confounding influences on experimental data, resulting in less problematic interindividual variation in studies that employ appropriate behavioral management techniques. The implementation of behavioral management refinements results in enhanced welfare for the subjects, higher quality data, and eventually, an increase in scientific robustness of the studies and a potential for reduction in the number of subjects required for research projects. Using positive reinforcement training techniques that allow socially housed, appropriately selected, nonhuman primates living in enriched environments to participate in research procedures is critical, if not imperative, to the collection of reliable and valid data, the foundation of all types of scientific investigations.

References

Baker KC (2016) Survey of 2014 behavioral management programs for laboratory primates in the United States. Am J Primatol 78(7):780–796. https://doi.org/10.1002/ajp.22543

Baumans V, van Loo PLP, Pham TM (2010) Standardisation of environmental enrichment for laboratory mice and rats: utilisation, practicality and variation in experimental results. Scand J Lab Anim Sci 37(2):1010–1014

Bayne K (2005) Potential for unintended consequences of environmental enrichment for laboratory animals and research results. ILAR J 46(2):129–139. https://doi.org/10.1093/ilar.46.2.129

Bayne K, Wurbel H (2014) The impact of environmental enrichment on the outcome variability and scientific validity of laboratory animal studies. Rev Sci Tech 33(1):273–280. https://doi.org/10.20506/rst.33.1.2282

Benton CG, West MW, Hall SM et al (2013) Effect of short-term pair housing of juvenile rhesus macaques (*Macaca mulatta*) on immunologic parameters. J Am Assoc Lab Anim Sci 52(3):240–246

Bethell EJ, Pfefferle D (2022) Cognitive bias tasks: a new set of approaches to assess welfare in nonhuman primates. In: Robinson LM, Weiss A (eds) Nonhuman primate welfare: from history, science, and ethics to practice. Springer, Cham, pp 207–230

Bethell EJ, Holmes A, MacLarnon A, Semple S (2012) Evidence that emotion mediates social attention in rhesus macaques. PLoS ONE 7(8):e44387. https://doi.org/10.1371/journal.pone.0044387

Bloomsmith M, Perlman J, Franklin A, Martin AL (2022) Training research primates. In: Robinson LM, Weiss A (eds) Nonhuman primate welfare: from history, science, and ethics to practice. Springer, Cham, pp 517–546

Brunelli RL, Blake J, Willits N et al (2014) Effects of a mechanical response-contingent surrogate on the development of behaviors in nursery-reared rhesus macaques (*Macaca mulatta*). J Am Assoc Lab Anim Sci 53(5):464–471

Budda ML, Ely JJ, Doan S et al (2013) Evaluation of reproduction and raising offspring in a nursery-reared SPF Baboon (*Papio hamadryas anubis*) colony. Am J Primatol 75(8):798–806. https://doi.org/10.1002/ajp.22136

Capitanio JP (2017) Variation in biobehavioral organization. In: Schapiro SJ (ed) Handbook of primate behavioral management. Taylor & Francis, Boca Raton, FL, pp 55–74

Capitanio JP, Mendoza SP, Lerche NW, Mason WA (1998) Social stress results in altered glucocorticoid regulation and shorter survival in simian acquired immune deficiency syndrome. Proc Natl Acad Sci U S A 95(8):4714–4719. https://doi.org/10.1073/pnas.95.8.4714

Capitanio JP, Mendoza SP, Mason WA, Maninger N (2005) Rearing environment and hypothalamic-pituitary-adrenal regulation in young rhesus monkeys (*Macaca mulatta*). Dev Psychobiol 46(4):318–330. https://doi.org/10.1002/dev.20067

Capitanio JP, Mendoza SP, Cole SW (2011) Nervous temperament in infant monkeys is associated with reduced sensitivity of leukocytes to cortisol's influence on trafficking. Brain Behav Immun 25(1):151–159. https://doi.org/10.1016/j.bbi.2010.09.008

Clarke AS, Mason WA, Moberg GP (1988) Differential behavioral and adrenocortical responses to stress among three macaque species. Am J Primatol 14(1):37–52. https://doi.org/10.1002/ajp.1350140104

Coleman K (2017) Individual differences in temperament and behavioral management. In: Schapiro SJ (ed) Handbook of primate behavioral management. Taylor & Francis, Boca Raton, FL, pp 95–114

Coleman K, Pierre PJ (2014) Assessing anxiety in nonhuman primates. ILAR J 55(2):333–346. https://doi.org/10.1093/ilar/ilu019

Coleman K, Pranger L, Maier A et al (2008) Training rhesus macaques for venipuncture using positive reinforcement techniques: a comparison with chimpanzees. J Am Assoc Lab Anim Sci 47(1):37–41

Coleman K, Bloomsmith MA, Crockett CM et al (2012) Behavioral management, enrichment, and psychological well-being of laboratory nonhuman primates. In: Abee C, Mansfield K, Suzette T, Morris T (eds) Nonhuman primates in biomedical research: biology and management, vol 1, 2nd edn. Academic, London, pp 149–176

Darusman HS, Sajuthi D, Kalliokoski O et al (2013) Correlations between serum levels of beta amyloid, cerebrospinal levels of tau and phospho tau, and delayed response tasks in young and aged cynomolgus monkeys (*Macaca fascicularis*). J Med Primatol 42(3):137–146. https://doi.org/10.1111/jmp.12044

Darusman HS, Agungpriyono DR, Kusumaputri VA et al (2019) Granulovacuolar degeneration in brains of senile cynomolgus monkeys. Front Aging Neurosci 11:50. https://doi.org/10.3389/fnagi.2019.00050

Doyle LA, Baker KC, Cox LD (2008) Physiological and behavioral effects of social introduction on adult male rhesus macaques. Am J Primatol 70(6):542–550. https://doi.org/10.1002/ajp.20526

Edler MK, Sherwood CC, Meindl RS et al (2017) Aged chimpanzees exhibit pathologic hallmarks of Alzheimer's disease. Neurobiol Aging 59:107–120. https://doi.org/10.1016/j.neurobiolaging.2017.07.006

Else JG (1985) Captive propagation of vervet monkeys (*Cercopithecus aethiops*) in harems. Lab Anim Sci 35(4):373–375

Erwin J, Deni R (1979) Strangers in a strangeland: abnormal behaviors or abnormal environment? In: Erwin J, Maple TL, Mitchell G (eds) Captivity and behavior: primates in breeding colonies, laboratories, and zoos. Van Nostrand Reinhold, New York, pp 1–28

Eskola S, Lauhikari M, Voipio HM et al (1999) Environmental enrichment may alter the number of rats needed to achieve statistical significance. Scand J Lab Anim Sci 26(3):134–144

Fernandez L, Griffiths M-A, Honess PE (2017) Providing behaviorally manageable primates for research. In: Schapiro SJ (ed) Handbook of primate behavioral management. Taylor & Francis, Boca Raton, FL, pp 481–494

Fernström AL, Sutian W, Royo F et al (2008) Stress in cynomolgus monkeys (*Macaca fascicularis*) subjected to long-distance transport and simulated transport housing conditions. Stress 11(6):467–476. https://doi.org/10.1080/10253890801903359

Freeman HD, Brosnan SF, Hopper LM et al (2013) Developing a comprehensive and comparative questionnaire for measuring personality in chimpanzees using a simultaneous top-down/bottom-up design. Am J Primatol 75(10):1042–1053. https://doi.org/10.1002/ajp.22168

Friedman EM, Coe CL, Ershler WB (1991) Time-dependent effects of peer separation on lymphocyte proliferation responses in juvenile squirrel monkeys. Dev Psychobiol 24(3):159–173. https://doi.org/10.1002/dev.420240303

Garner JP (2005) Stereotypies and other abnormal repetitive behaviors: potential impact on validity, reliability, and replicability of scientific outcomes. ILAR J 46(2):106–117. https://doi.org/10.1093/ilar.46.2.106

Gottlieb DH, Coleman K, Prongay K (2017) Behavioral management of *Macaca* spp. (except *M. fascicularis*). In: Schapiro SJ (ed) Handbook of primate behavioral management. Taylor & Francis, Boca Raton, FL, pp 279–304

Graham ML (2017) Positive reinforcement training and research. In: Schapiro SJ (ed) Handbook of primate behavioral management. Taylor & Francis, Boca Raton, FL, pp 187–200

Graham ML, Rieke EF, Mutch LA et al (2012) Successful implementation of cooperative handling eliminates the need for restraint in a complex non-human primate disease model. J Med Primatol 41(2):89–106. https://doi.org/10.1111/j.1600-0684.2011.00525.x

Harlow HF, Harlow MK (1962) Social deprivation in monkeys. Sci Am 207:136–146

Hau J, Schapiro SJ (2007) The welfare of non-human primates. In: Kaliste E (ed) The welfare of laboratory animals, vol 2. Kluwer, Boston, pp 291–314

Hilliard JK, Ward JA (1999) B-virus specific-pathogen-free breeding colonies of macaques (*Macaca mulatta*): retrospective study of seven years of testing. Comp Med 49(2):144–148

Holmes A, Nielsen MK, Green S (2008) Effects of anxiety on the processing of fearful and happy faces: an event-related potential study. Biol Psychol 77(2):159–173. https://doi.org/10.1016/j.biopsycho.2007.10.003

Honess PE (2017) Behavioral management of long-tailed macaques (*Macaca fascicularis*). In: Schapiro SJ (ed) Handbook of primate behavioral management. Taylor & Francis, Boca Raton, FL, pp 305–338

Honess PE, Wolfensohn S (2010) The Extended Welfare Assessment Grid: a matrix for the assessment of welfare and cumulative suffering in experimental animals. Altern Lab Anim 38(3):205–212. https://doi.org/10.1177/026119291003800304

Hopkins WD, Latzman RD (2017) Future research with captive chimpanzees in the United States: integrating scientific programs with behavioral management. In: Schapiro SJ (ed) Handbook of primate behavioral management. Taylor & Francis, Boca Raton, FL, pp 141–155

Jorgensen MJ (2017) Behavioral management of Chlorocebus spp. In: Schapiro SJ (ed) Handbook of primate behavioral management. Taylor & Francis, Boca Raton, FL, pp 339–365

Kagira JM, Ngotho M, Thuita JK et al (2007) Hematological changes in vervet monkeys (*Chlorocebus aethiops*) during eight months' adaptation to captivity. Am J Primatol 69(9):1053–1063. https://doi.org/10.1002/ajp.20422

Kemp C (2022) Enrichment. In: Robinson LM, Weiss A (eds) Nonhuman primate welfare: from history, science, and ethics to practice. Springer, Cham, pp 451–488

Koban TL, Schapiro SJ, Kusznir T et al (2010) Effects of international transit and relocation on cortisol values in cynomolgus macaques (*Macaca fascicularis*). Am J Primatol 72(Suppl):51. https://doi.org/10.1002/ajp.20862

Lambeth SP, Hau J, Perlman JE et al (2006) Positive reinforcement training affects hematologic and serum chemistry values in captive chimpanzees (*Pan troglodytes*). Am J Primatol 68(3):245–256. https://doi.org/10.1002/ajp.20148

Lambeth S, Schapiro SJ, Bernacky B, Wilkerson G (2013) Establishing "quality of life" parameters using behavioural guidelines for humane euthanasia of captive non-human primates. Anim Welf 22(4):429–435. https://doi.org/10.7120/09627286.22.4.429

Latzman RD, Patrick CJ, Freeman HD et al (2017) Etiology of triarchic psychopathy dimensions in chimpanzees (*Pan troglodytes*). Clin Psychol Sci 5(2):341–354. https://doi.org/10.1177/2167702616676582

Laudenslager ML, Natvig C, Corcoran CA et al (2013) The influences of perinatal challenge persist into the adolescent period in socially housed bonnet macaques (*Macaca radiata*). Dev Psychobiol 55(3):316–322. https://doi.org/10.1002/dev.21030

Laule GE, Bloomsmith MA, Schapiro SJ (2003) The use of positive reinforcement training techniques to enhance the care, management, and welfare of primates in the laboratory. J Appl Anim Welf Sci 6(3):163–173. https://doi.org/10.1207/S15327604JAWS0603_02

Lerche NW, Osborn KG (2003) Simian retrovirus infections: potential confounding variables in primate toxicology studies. Toxicol Pathol 31(Suppl):103–110. https://doi.org/10.1080/01926230390174977

Line SW, Morgan KN, Markowitz H, Strong S (1989) Influence of cage size on heart rate and behavior in rhesus monkeys. Am J Vet Res 50(9):1523–1525

Lutz CK, Baker KC (2022) Using behavior to assess primate welfare. In: Robinson LM, Weiss A (eds) Nonhuman primate welfare: from history, science, and ethics to practice. Springer, Cham, pp 171–206

Magden ER (2017) Positive reinforcement training and health care. In: Schapiro SJ (ed) Handbook of primate behavioral management. Taylor & Francis, Boca Raton, FL, pp 201–216

Magden ER, Haller RL, Thiele EJ et al (2013) Acupuncture as an adjunct therapy for osteoarthritis in chimpanzees (*Pan troglodytes*). J Am Assoc Lab Anim Sci 52(4):475–480

Magden ER, Sleeper MM, Buchl SJ et al (2016) Use of an implantable loop recorder in a chimpanzee (*Pan troglodytes*) to monitor cardiac arrhythmias and assess the effects of acupuncture and laser therapy. Comp Med 66(1):52–58

Mahalingam R, Traina-Dorge V, Wellish M et al (2010) Latent simian varicella virus reactivates in monkeys treated with tacrolimus with or without exposure to irradiation. J Neurovirol 16(5):342–354. https://doi.org/10.3109/13550284.2010.513031

McGrew K (2017) Pairing strategies for cynomolgus macaques. In: Schapiro SJ (ed) Handbook of primate behavioral management. Taylor & Francis, Boca Raton, FL, pp 255–264

Mendoza SP, Mason WA (1984) Rambunctious *Saimiri* and reluctant *Callicebus*: systemic contrasts in stress physiology. Am J Primatol 6(4):415. https://doi.org/10.1002/ajp.1350060413

Morimura N, Hirata S, Matsuzawa T (2022) Challenging cognitive enrichment: examples from caring for the chimpanzees in the Kumamoto Sanctuary, Japan and Bossou, Guinea. In: Robinson LM, Weiss A (eds) Nonhuman primate welfare: from history, science, and ethics to practice. Springer, Cham, pp 489–516

Neal Webb SJ, Hau J, Schapiro SJ (2018) Captive chimpanzee (*Pan troglodytes*) behavior as a function of space per animal and enclosure type. Am J Primatol 80(3):e22749. https://doi.org/10.1002/ajp.22749

Neal Webb SJ, Hau J, Lambeth SP, Schapiro SJ (2019a) Differences in behavior between elderly and non-elderly captive chimpanzees and the effects of the social environment. J Am Assoc Lab Anim Sci 58(6):783–789. https://doi.org/10.30802/AALAS-JAALAS-19-000019

Neal Webb SJ, Hau J, Schapiro SJ (2019b) Relationships between captive chimpanzee (*Pan troglodytes*) welfare and voluntary participation in behavioural studies. Appl Anim Behav Sci 214:102–109. https://doi.org/10.1016/j.applanim.2019.03.002

Nehete PN, Shelton KA, Nehete BP et al (2017) Effects of transportation, relocation, and acclimation on phenotypes and functional characteristics of peripheral blood lymphocytes in rhesus monkeys (*Macaca mulatta*). PLoS ONE 12(12):e0188694. https://doi.org/10.1371/journal.pone.0188694

Novak MA, Hamel AF, Ryan AM et al (2017) The role of stress in abnormal behavior and other abnormal conditions such as hair loss. In: Schapiro SJ (ed) Handbook of primate behavioral management. Taylor & Francis, Boca Raton, FL, pp 75–94

Poole T (1997) Happy animals make good science. Lab Anim 31(2):116–124. https://doi.org/10.1258/002367797780600198

Prescott MJ (2022) Using primates in captivity: research, conservation, and education. In: Robinson LM, Weiss A (eds) Nonhuman primate welfare: from history, science, and ethics to practice. Springer, Cham, pp 57–78

Prescott MJ, Brown VJ, Flecknell PA et al (2010) Refinement of the use of food and fluid control as motivational tools for macaques used in behavioural neuroscience research: report of a Working Group of the NC3Rs. J Neurosci Methods 193(2):167–188. https://doi.org/10.1016/j.jneumeth.2010.09.003

Reamer LA, Haller RL, Thiele EJ et al (2014) Factors affecting initial training success of blood glucose testing in captive chimpanzees (*Pan troglodytes*). Zoo Biol 33(3):212–220. https://doi.org/10.1002/zoo.21123

Reinhardt V, Cowley D, Scheffler J et al (1990) Cortisol response of female rhesus monkeys to venipuncture in homecage versus venipuncture in restraint apparatus. J Med Primatol 19(6):601–606

Robinson LM, Weiss A (2022) Primate personality and welfare. In: Robinson LM, Weiss A (eds) Nonhuman primate welfare: from history, science, and ethics to practice. Springer, Cham, pp 387–402

Robinson LM, Waran NK, Leach MC et al (2016) Happiness is positive welfare in brown capuchins (*Sapajus apella*). Appl Anim Behav Sci 181:145–151. https://doi.org/10.1016/j.applanim.2016.05.029

Rommeck I, Gottlieb DH, Strand SC, McCowan B (2009) The effects of four nursery rearing strategies on infant behavioral development in rhesus macaques (*Macaca mulatta*). J Am Assoc Lab Anim Sci 48(4):395–401

Rommeck I, Capitanio JP, Strand SC, McCowan BJ (2011) Early social experience affects behavioral and physiological responsiveness to stressful conditions in infant rhesus macaques (*Macaca mulatta*). Am J Primatol 73(7):692–701. https://doi.org/10.1002/ajp.20953

Schapiro SJ (2002) Effects of social manipulations and environmental enrichment on behavior and cell-mediated immune responses in rhesus macaques. Pharmacol Biochem Behav 73(1):271–278. https://doi.org/10.1016/S0091-3057(02)00779-7

Schapiro SJ (ed) (2017) Handbook of primate behavioral management. Taylor & Francis, Boca Raton, FL

Schapiro SJ, Bernacky BJ (2012) Socialization strategies and disease transmission in captive colonies of nonhuman primates. Am J Primatol 74(6):518–527. https://doi.org/10.1002/ajp.21001

Schapiro SJ, Bloomsmith MA (1995) Behavioral effects of enrichment on singly-housed, yearling rhesus monkeys: An analysis including three enrichment conditions and a control group. Am J Primatol 35(2):89–101. https://doi.org/10.1002/ajp.1350350202

Schapiro SJ, Bloomsmith MA, Kessel AL, Shively CA (1993) Effects of enrichment and housing on cortisol response in juvenile rhesus monkeys. Appl Anim Behav Sci 37(3):251–263. https://doi.org/10.1016/0168-1591(93)90115-6

Schapiro SJ, Lee-Parritz DE, Taylor LL et al (1994) Behavioral management of specific pathogen-free rhesus macaques: group formation, reproduction, and parental competence. Lab Anim Sci 44(3):229–234

Schapiro SJ, Bloomsmith MA, Porter LM, Suarez SA (1996) Enrichment effects on rhesus monkeys successively housed singly, in pairs, and in groups. Appl Anim Behav Sci 48(3–4):159–171. https://doi.org/10.1016/0168-1591(96)01038-6

Schapiro SJ, Nehete PN, Perlman JE et al (1998) Effects of dominance status and environmental enrichment on cell-mediated immunity in rhesus macaques. Appl Anim Behav Sci 56(2–4):319–332. https://doi.org/10.1016/S0168-1591(97)00087-7

Schapiro SJ, Nehete PN, Perlman JE, Sastry KJJ (2000) A comparison of cell-mediated immune responses in rhesus macaques housed singly, in pairs, or in groups. Appl Anim Behav Sci 68(1):67–84. https://doi.org/10.1016/s0168-1591(00)00090-3

Schapiro SJ, Perlman JE, Boudreau BA (2001) Manipulating the affiliative interactions of group-housed rhesus macaques using positive reinforcement training techniques. Am J Primatol 55(3):137–149. https://doi.org/10.1002/ajp.1047

Schapiro SJ, Perlman JE, Thiele E, Lambeth S (2005) Training nonhuman primates to perform behaviors useful in biomedical research. Lab Anim (NY) 34(5):37–42. https://doi.org/10.1038/laban0505-37

Schapiro SJ, Lambeth SP, Jacobsen KR et al (2012) Physiological and welfare consequences of transport, relocation, and acclimatization of chimpanzees (*Pan troglodytes*). Appl Anim Behav Sci 137(3–4):183–193. https://doi.org/10.1016/J.APPLANIM.2011.11.004

Schapiro SJ, Coleman K, Akinyi MY et al (2014) Nonhuman primate welfare in the research environment. In: Bayne K, Turner PV (eds) Laboratory animal welfare. Academic Press, San Diego, pp 197–212

Schapiro SJ, Brosnan SF, Hopkins WD et al (2017) Collaborative research and behavioral management. In: Schapiro SJ (ed) Handbook of primate behavioral management. Taylor & Francis, Boca Raton, FL, pp 243–254

Schapiro SJ, Magden ER, Reamer LA et al (2018) Behavioral training as part of the health care program. In: Thompson G, Weichbrod R, Thomas X (eds) Management of animal care and use programs in research, teaching, and testing, 2nd edn. Taylor & Francis, Boca Raton, FL, pp 771–792

Shelton KA, Nehete BP, Chitta S et al (2019) Effects of transportation and relocation on immunologic measures in cynomolgus macaques (*Macaca fascicularis*). J Am Assoc Lab Anim Sci 58(6):774–782. https://doi.org/10.30802/AALAS-JAALAS-19-000007

Shively CA (2002) Depression and coronary artery atherosclerosis and reactivity in female cynomolgus monkeys. Psychosom Med 64(5):699–706. https://doi.org/10.1097/01.PSY.0000021951.59258.C7

Shively CA (2017) Depression in captive nonhuman primates: theoretical underpinnings, methods, and application to behavioral management. In: Schapiro SJ (ed) Handbook of primate behavioral management. Taylor & Francis, Boca Raton, FL, pp 115–125

Shively CA, Register TC, Friedman DP et al (2005) Social stress-associated depression in adult female cynomolgus monkeys (*Macaca fascicularis*). Biol Psychol 69(1):67–84. https://doi.org/10.1016/j.biopsycho.2004.11.006

Shively CA, Musselman DL, Willard SL (2009) Stress, depression, and coronary artery disease: modeling comorbidity in female primates. Neurosci Biobehav Rev 33(2):133–144. https://doi.org/10.1016/j.neubiorev.2008.06.006

Simmons JH (2010) Herpesvirus infections of laboratory macaques. J Immunotoxicol 7:102–113

Slater H, Milne AE, Wilson B et al (2016) Individually customisable non-invasive head immobilisation system for non-human primates with an option for voluntary engagement. J Neurosci Methods 269:46–60. https://doi.org/10.1016/j.jneumeth.2016.05.009

Suleman MA, Yole D, Wango E et al (1999) Peripheral blood lymphocyte immunocompetence in wild African green monkeys (*Cercopithecus aethiops*) and the effects of capture and confinement. In Vivo (Brooklyn) 13(1):25–27

Suleman MA, Wango E, Farah IO, Hau J (2000) Adrenal cortex and stomach lesions associated with stress in wild male African green monkeys (*Cercopithecus aethiops*) in the post-capture period. J Med Primatol 29(5):338–342. https://doi.org/10.1034/j.1600-0684.2000.290505.x

Suleman MA, Wango E, Sapolsky RM et al (2004) Physiologic manifestations of stress from capture and restraint of free-ranging male African green monkeys (*Cercopithecus aethiops*). J Zoo Wildl Med 35(1):20–25. https://doi.org/10.1638/01-025

Suomi SJ, Eisele CD, Grady SA, Harlow HF (1975) Depressive behavior in adult monkeys following separation from family environment. J Abnorm Psychol 84(5):576–578. https://doi.org/10.1037/h0077066

Truelove MA, Martin AL, Perlman JE et al (2017) Pair housing of macaques: a review of partner selection, introduction techniques, monitoring for compatibility, and methods for long-term maintenance of pairs. Am J Primatol 79(1):e22485. https://doi.org/10.1002/ajp.22485

Uno H, Tarara R, Else J et al (1989) Hippocampal damage associated with prolonged and fatal stress in primates. J Neurosci 10(9):1705–1711. https://doi.org/10.1523/JNEUROSCI.09-05-01705.1989

Vallender EJ, Miller GM (2013) Nonhuman primate models in the genomic era: a paradigm shift. ILAR J 54(2):154–165. https://doi.org/10.1093/ilar/ilt044

Weed JL, Raber JM (2005) Balancing animal research with animal well-being: establishment of goals and harmonization of approaches. ILAR J 46(2):118–128. https://doi.org/10.1093/ilar.46.2.118

Westlund K (2012) Questioning the necessity of food– and fluid regimes: Reply to Prescott and colleagues' response. J Neurosci Methods 204(1):210–213. https://doi.org/10.1016/j.jneumeth.2011.10.022

Willard SL, Friedman DP, Henkel CK, Shively CA (2009) Anterior hippocampal volume is reduced in behaviorally depressed female cynomolgus macaques. Psychoneuroendocrinology 34(10):1469–1475. https://doi.org/10.1016/j.psyneuen.2009.04.022

Williams L, Ross CN (2017) Behavioral management of neotropical species: *Aotus, Callithrix*, and *Saimiri*. In: Schapiro SJ (ed) Handbook of primate behavioral management. Taylor & Francis, Boca Raton, FL, pp 409–434

Williams LE, Nehete PN, Schapiro SJ, Lambeth SP (2010) Effects of relocation on immunological measures in two captive nonhuman primate species: squirrel monkeys and owl monkeys. Am J Primatol 72(Suppl 1):28. https://doi.org/10.1002/ajp.20862

Wolf RF, Eberle R, White GL (2010) Generation of a specific-pathogen-free baboon colony. J Am Assoc Lab Anim Sci 49(6):814–820

Wolfensohn S (2022) Humane end points and end of life in primates used in laboratories. In: Robinson LM, Weiss A (eds) Nonhuman primate welfare: from history, science, and ethics to practice. Springer, Cham, pp 369–386

Wolfensohn S, Sharpe S, Hall I et al (2015) Refinement of welfare through development of a quantitative system for assessment of lifetime experience. Anim Welf 24(2):139–149. https://doi.org/10.7120/09627286.24.2.139

Xie L, Zhou Q, Liu S et al (2014) Effect of living conditions on biochemical and hematological parameters of the cynomolgus monkey. Am J Primatol 76(11):1011–1024. https://doi.org/10.1002/ajp.22285

Enrichment

Caralyn Kemp

Abstract

Enrichment is a vital component of husbandry and housing practices for captive primates. Primates in enriched environments are better equipped to cope with the challenges of captivity, make for more reliable research subjects, and are physically and psychologically healthier compared to animals in unenriched environments. Enrichment is more than just "giving animals toys." At best practice, the application of enrichment occurs as part of a well-thought-out program with set goals and consideration of the five main types of enrichment (social, physical, food-based, sensory, and cognitive), and is modified using an evidence-based approach. This latter point is particularly important as without assessment it is not possible to determine if an enrichment item is actually enriching the lives of the target animal or animals. In this chapter, I discuss what enrichment is and is not, the principles for developing effective enrichment programs, and a range of ways enrichment can be applied easily and cheaply. Enrichment should be assessed for effectiveness, with results, whether positive or negative, published to help inform the primate care community in making more appropriate decisions when designing and applying enrichment.

Keywords

Enrichment · Animal welfare · Assessment · Animal husbandry

C. Kemp (✉)
School of Environmental and Animal Sciences, Unitec Institute of Technology, Auckland, New Zealand
e-mail: ckemp2@unitec.ac.nz

1 Introduction

The provision of stimuli to enhance the captive environment for primates is essential for their physical and psychological wellbeing (Shepherdson et al. 1998). These stimuli are commonly known as "enrichment," but, as will be discussed, enrichment is more than just the provision of "toys" and other items to make an environment seem complex and interesting. Given the topic of this book, I am specifically referring to primates and enrichment for primate species; however, the guidelines and principles outlined in this chapter are relevant to captive animals in general and can be applied widely regardless of species. The use of enrichment traditionally stems from an identified need to combat abnormal, stereotypic, and repetitive behaviors in captive animals (Swaisgood and Shepherdson 2006). However, at best practice, it is a husbandry principle, aiming to enhance the quality of the captive environment (Shepherdson et al. 1998) and facilitate natural behaviors (Hosey 2005). Ultimately, prevention of the development of abnormal behaviors is better than treatment. Thus, the application of enrichment stimuli should aim to reduce the likelihood that negative and unwanted behaviors will arise by encouraging species-appropriate behavior, meeting physical and psychological needs, and adding an element of control and predictability to the captive environment. Enrichment does not just improve behavioral diversity; enriched primates have fewer health complications compared to animals in unenriched environments, which leads to lower medical upkeep, and enriched primates are more likely to breed successfully and produce healthy offspring (Newberry 1995; Ventura and Buchanan-Smith 2003). Furthermore, enrichment demonstrates to the general public that an active approach to the improvement of the welfare of animals in our care is taking place (Kutska 2009).

Enrichment is often viewed as an "add-on." That is, it is something to be considered, created, and applied when care staff have spare time. Yet, enrichment has been the cornerstone of the improvements we have seen in the housing of primates in all types of facilities (zoos, research laboratories, and breeding facilities) in recent decades. Without enrichment, primates would still be housed in barren cages. To suggest that enrichment is not a vital component of husbandry is underestimating its importance and value. Effective enrichment does not need to be complex or expensive. In fact, enrichment items should be easy to set up, apply, remove, and/or clean as necessary to encourage continued use. Simple, cost-effective items can be applied for the purpose of enrichment and have a large impact on animal welfare. However, it is not just a matter of throwing in some "toys" and assuming that this will be sufficient. Instead, an enrichment program with carefully set goals should be designed. Without goals, there is no guarantee that the animals and their environments are being enriched appropriately. It is important to note that any stimulus or item used for the purpose of enriching the captive environment should not be called "enrichment" unless evidence has demonstrated that it is used by the animal or animals in the manner in which it was intended and achieves an appropriate goal (Newberry 1995; Hosey et al. 2013). The overuse of the term "enrichment" has led to a widespread belief that enrichment is commonly actively applied in captive settings and is occurring simply because a stimulus is named as such. This is obviously a misnomer; giving an item the name of enrichment does not make it enrichment. Stimuli can only be considered as *alleged* enrichment until an assessment has confirmed the desired effect (Clark et al. 2005; Clark and King 2008; Hosey et al. 2013). Good intentions and assumptions are not enough. Assessment of how the primates use the stimuli will help to get the best out of the enrichment, and these positive outcomes should inspire further application (see Sect. 6).

In this chapter I examine the aims of enrichment, the development of enrichment programs, safety considerations, and the five main different types of enrichment (social, physical, food-based, sensory, and cognitive). Novel approaches to enrichment are discussed, including the use of comparative psychology experiments to provide more complex challenges than can otherwise be designed using more naturalistic and traditional approaches. The chapter concludes with a discussion on the assessment of the effectiveness of enrichment items.

2 Aims of Enrichment

Enrichment is used to create a complex captive environment (Buchanan-Smith 2010). More specifically, the use of enrichment should improve or maintain a primate's physical and psychological health, provide opportunities for a diversity of species-specific behaviors to be displayed, increase the utilization of the environment, prevent or reduce the frequency of abnormal and unwanted behaviors such as stereotypies (Mason et al. 2007), and increase an individual's ability to cope with the challenges of captivity (Jennings and Prescott 2009; Young 2003). Enrichment increases the options for choice; choice is linked to control and predictability, which are important for welfare (Buchanan-Smith 2010; Gottlieb et al. 2013; Videan et al. 2005).

Enrichment provides opportunities for primates to make choices about how, when, and where they will interact with their environment (Buchanan-Smith 2010). It is important to note that choice within the enrichment context requires the application of multiple stimuli within the same category of enrichment. That is, having a choice of nesting areas, for example, where one area has a soil substrate, another is a fire hose hammock, and a third can be made by the primate from browse, is not the same as having multiple soil-based nesting areas and multiple browse- and hammock-based nests. By providing options, we are asking the primate to show us what they prefer. Preference testing is an ideal method to determine what an individual would choose if given the choice (Buchanan-Smith 2010) and creates an opportunity to determine consumer-demand (Schapiro and Lambeth 2007). It is important to note that the primate's preferences can be moderated by the presence of other options and factors (Bateson 2004; Hubrecht 2010; Buchanan-Smith 2010). The preference is not ultimate; it is only based on what is available.

The most common aim for enrichment is the generation of a behavioral change in the target animal or animals (Hosey et al. 2013). This is a very broad aim and can result in the application of unsuitable and ineffective stimuli. Therefore, the specifics of these aims need to be clarified. For example, care staff for an orangutan set a goal

Table 1 Questions to consider when determining the most appropriate item to achieve any given goal

Question	Example
Why	Why do these animals need enriching? Why is this behavior occurring/not occurring?
When	When do these animals need to be enriched? When is an undesirable behavior most likely to occur? When was enrichment last given to this animal or animals?
How	How will the enrichment be applied? How does the enrichment need to look (naturalistic versus artificial)? How many items will be needed? How long should the enrichment be made available to the animal or animals? How many animals are in the group? How will the enrichment be assessed?
Where	Where should I apply the enrichment? Where does an undesirable behavior occur?
Who	Who is receiving the enrichment? Who might monopolize the enrichment? Who might be adversely affected by this enrichment?
What	What type of enrichment will encourage natural behaviors? What enrichment is suitable for this environment? What are the potential risks of this stimulus? What should this item achieve? What are the physical requirements of the animal or animals?

to increase the time the orangutan spends locomoting each day. They set up climbing ropes in the animal's enclosure. However, across the first fortnight, the orangutan was observed pacing more than she had done so prior to the introduction of the ropes and does not use the new climbing features. Given the goal was to increase the time the orangutan spent locomoting, the rope can be considered as successful enrichment, as pacing is a form of locomotion. If the goal had been to encourage an increase in climbing-based locomotion, then the rope would be deemed as ineffective. This is why having specific goals, and avoiding vague and broad aims, is important.

Consideration of these details in the target behavior will help to set achievable, measurable goals. When we begin to think about applying stimuli, it is easy to become caught up in the "what"—*What will I give to these animals?* However, there are other questions which should be asked first; the answers generated will help to guide the goals and determine what stimulus or stimuli will be most appropriate (Table 1). When these questions are asked, enrichment is considered in more holistic terms and allows for the development of an enrichment program.

3 Enrichment Program

Developing an enrichment program helps to fulfill the overall aims described above, reduce the likelihood that stimuli become boring, and ensures that the various types of enrichment (see Sect. 5) are covered. An enrichment program aids in the application of stimuli using a pre-determined roster, rather than applying items ad libitum. The rotation of items can be on a weekly, fortnightly, monthly, or even yearly basis, and is dependent on the type of enrichment. For example, food-based enrichment will most likely have a weekly or fortnightly rotation, while a pile of snow may only be applied once a year (in areas where snow is not part of the normal environment).

Variability in access to an enrichment device is the most effective way to maintain long-term interest (Csatádi et al. 2008; Kuczaj et al. 2002). The enrichment program also aids in the adjustment of goals over time for both groups of animals and individuals; these goals need to be evaluated and updated throughout an animal's lifetime, according to its needs (Coleman and Novak 2017).

The S.P.I.D.E.R. Framework (Set Goals—Plan—Implement—Document—Evaluate—Readjust; Colahan and Breder 2003), developed at Disney's Animal Kingdom, is a step-by-step method which can be incorporated within an enrichment program and aids in the development of appropriate and effective enrichment stimuli. This tool guides the user through the process of considering which behavior or behaviors to target, setting goals for increasing or decreasing the prevalence of performance of the behavior or behaviors, planning the creation and implementation of enrichment, applying the enrichment item, recording how an animal uses the item, and evaluating the effectiveness of the item in achieving the original goal. Using this, or a similar, method, the application of enrichment becomes a thoroughly thought-out process, rather than a haphazard approach.

An enrichment program should ideally be designed by a committee representing a variety of agendas and viewpoints. Care staff/keepers, veterinarians, behavioral researchers, welfare officers, and nutritionists will all have important inputs into the type of enrichment needed. Brain-storming and the development or use of established, goal-setting tools, forms and databases will help this process (e.g., Disney n.d.). Table 2 lists questions and considerations the committee should consider while designing an enrichment program.

While the specifics may change between species, there are general principles of successful enrichment programs (American Society of Primatologists 2016). These include awareness of the species' natural history, the housing facility, safety concerns (Sect. 4), appropriate goals, the needs of individual animals, and evaluation of effectiveness (Sect. 6). The species' normal behavior, activity levels, ecological niche, natural foraging behavior, social system, environment (e.g., terrestrial or arboreal), and sensory system will help to determine which stimuli are suitable (Lutz and Novak 2005). Due to overlaps in natural history, some enrichment may be applicable to multiple primate taxa. However, generalizations cannot always be made and success with one species or even group may not translate to another (Jennings and Prescott 2009; Lutz and Novak 2005; Maple and Finlay 1989; National Research Council 2011).

Enrichment programs also need to consider the housing facility. Many zoos prefer the use of naturalistic enrichment when their exhibits are designed to resemble a primate's natural habitat. While animals require complex environments, there is limited evidence to suggest that this environment needs to be naturalistic (Jacobson et al. 2017); however, zoo visitors certainly prefer naturalistic enclosures when viewing animals, and their assessment of welfare is influenced by this setting (Davey 2006; Ross et al. 2012; Lukas and Ross 2014; Razal and Miller 2019). This preference has certainly had a large influence on the development of more natural-appearing enclosures in zoos. Primate research and breeding facilities are less likely to be affected by these viewpoints. However, it is important to note that

Table 2 Questions to consider when developing an enrichment program

Questions	Considerations
What are the overarching goals?	Ideally, enrichment should provide opportunities for animals to display their full behavioral repertoire and prevent the development of abnormal behavior. However, enrichment is most commonly applied retrospectively.
What items should be considered for use?	
What are the goals for individual items?	
Should items go through a wider approval process?	If a committee represents only one or two viewpoints (e.g., care staff only), items can be assessed for their suitability and appropriateness with an application form. The form should detail: • The species in question • The behavior the enrichment is designed to encourage • The type of item • Estimated set-up time and financial costs • Safety concerns • Improvements to previously rejected or questioned items • Staff from whom approval is being sought
Where will items be placed?	
Do additional items need to be purchased or created?	• How will new items be generated? • Is there a budget? • Redundancy in the case of overlap with already established items
What are the safety considerations for each item (see Sect. 4)?	
What are the ideal behaviors the species should be exhibiting?	
What behaviors need to be targeted by enrichment to try to increase or decrease their frequency?	All components of the behavior need to be considered. For example, eating is the end result of foraging for food. Foraging may include physical and visual searching of the environment and solving tasks to gain access to the food. The process of eating and the mechanical requirements will be dependent on the type of food.
How will enrichment items be assessed and documented to determine their effectiveness for both the general goals and for individual animals (see Lutz and Novak 2005)? The committee should generate a general enrichment analysis datasheet which can be used for any item.	

(continued)

Table 2 (continued)

Questions	Considerations
What are the size limitations?	
How might husbandry (e.g., ease of cleaning) be affected by the application of the enrichment?	

How will a roster for the enrichment be created and what needs to be considered in the formulation of the roster?	• The number and type of enrichment items available
	• How many items should be available to the primate(s) at any one time
	• The number of animals
	• The use and goals of the items
	• When items should be applied and how long the animal(s) should be exposed to them
	• The number of food-based items within the schedule and the prevalence of additional sugary treats
	• The needs of the animal/s
	• The interest in the items. Highly desirable items need to be numerous in order to avoid hoarding by a single animal (Honess and Marin 2006)
	• The permanency of the items. For example:
	  – Structural enrichment and housing items are long-term items and would not be alternated more often than every few weeks
	  – Food-based items are short-term and would need to be changed daily
	  – Destructible items, such as cardboard boxes, should be removed after a day or two at the most
	  – The ratio of enrichment items (especially if it is specialized, such as a feeding item) to number of animals within the group
	• The accessibility to items. This will depend on:
	  – The social structure of the group—high ranking primates of species with rigid social structures are more likely to have access to desirable enrichment items than low ranking animals (Bloomstrand et al. 1986)
	  – The location of items—enrichment placed up high may not be easily accessible to older individuals with limited mobility

(continued)

Table 2 (continued)

Questions	Considerations
How will the individuals within a group or collection be catered for in the program?	• Who gets the enrichment? • Are there any animals within the group or collection which often display signs of stress and stereotypy? They may require more enrichment than other individuals. • Consideration of individual responses as different animals may not benefit from each enrichment item in the same way (e.g., see Buchanan-Smith 2010). • What is the composition of the primate group/collection and how might these factors affect the ability to engage with enrichment items? Considerations should include age and sex (Coleman et al. 2012), mobility (Waitt et al. 2010), disability (Fig. 1), disease, injuries old or new, and use (research or human-animal encounter animals), as well as multi-species enclosures.
What emergency procedures will be put in place in the case of adverse reactions to enrichment items?	
How will a reduction in interest in enrichment items over time be determined and how will enrichment fatigue be combated?	Decline in interaction with enrichment items can occur over a short—a single day (e.g., Crockett et al. 1989)—or long—weeks or months (e.g., Kessel and Brent 1998)—period.
How often should the goals be assessed and updated?	

Fig. 1 Catering to different physical abilities is important. These puzzle balls are manageable for this lemur (*Lemur catta*), despite it missing a forelimb, to manipulate by rolling the ball so the food inside falls out. This type of enrichment provides an appropriate food-based challenge and avoids placing food in simple bowls. Photo credit unless otherwise stated: C. Kemp

enrichment items which may not be suitable for display areas can be used behind the scenes in the animals' night dens. These areas, being typically small and unstimulating, may require more enrichment, especially when the animals' main living areas are inaccessible due to cleaning, refurbishment, or nighttime lock up.

3.1 Designing an Enrichment Program

A thorough and laid-out program will help ensure that multiple aspects of behavioral and ecological needs are considered and covered. An enrichment program helps develop overarching goals; individual stimuli are then determined to help achieve these goals while having smaller aims of their own. For example, a program may determine that enrichment is needed to encourage activity, reduce the animal's weight, occupy its time, and provide opportunities for natural foraging behavior (see example in Table 3). Multiple stimuli may be applied to achieve these goals and it is unlikely that any one item will achieve all four. In this example, two items are considered: a climbing structure and a puzzle feeder. Individually, they have particular goals. The aim of the climbing structure is to provide the opportunity for the primate to utilize its arboreal skills, while the puzzle feeder makes accessing food more challenging. If the puzzle feeder is placed high on the climbing structure, then the primate is forced to move off the ground to access it, thereby encouraging activity and helping in the goal of weight loss. The puzzle feeder helps to occupy the primate's time and by being placed arboreally provides the opportunity for a tree-dwelling primate to perform a more natural foraging behavior. This combination results in an accumulative effect for the achievement of the original goals.

4 Safety Considerations

The potential dangers of any enrichment item should be carefully considered and assessed prior to incorporation into an enrichment program (Hare et al. 2008; Young 2003). First, the species' size, temperament, behavior, ecological needs, and strength need to be taken into consideration. What is suitable for chimpanzees, for example, may not be suitable for tamarin species. Small, dexterous fingers may get caught in small holes that larger digits will not be able to access. Chimpanzees are also more capable of breaking items than tamarins and so durability needs to be considered. Table 4 lists other safety considerations which need to be addressed before any stimulus is applied.

While physical safety concerns are often well considered, it is easy to overlook potentially detrimental psychological effects (see Hoy et al. 2010), including fear and boredom. Primates who show extreme avoidance or fear toward new and unfamiliar objects, food, or situations (neophobia) may respond aversively to seemingly benign, novel stimuli. This may be especially important when enrichment items are introduced to an individual or group for the first time. Although some primates may overcome their hesitancy toward some novel stimuli with repeated

Table 3 Example of a simple enrichment program

Enrichment Program

Version number: 3		Version date: 24/5/18		Program designed by: Keeping staff (SK, MH)	
Species: Orangutan (*Pongo pygmaeus abelii*)		Number of animals: 3		Sex: 1:2	Ages: 1.5 – 34 years
Known concerns:	Adult male overweight Adult female spending a lot of time on the ground, inactive				
Enrichment goals:	Increase activity levels Encourage foraging off the ground Improve health associated with weight Occupy more of the animals' time between feeding schedules Reduce incidences of abnormal behavior Avoid development of abnormal behavior in infant orangutan				
Current enrichment:	Type:	Item:	Approved by:	Last assessed:	Next assessment:
	Social:	Adult male and female housed together with infant female offspring. Male given time each day (approx. 2hrs) to himself.	Keepers	02/04/17	2019
	Physical:	Outdoor enclosure: 　Climbing structures approx. 2m high max. (timber frame, wooden platforms, firehose hammocks, ropes, fiberglass poles) 　Cotton sheets 　Nesting material (straw) Indoor enclosure: 　Concrete platforms 　Rope 　Cotton sheets 　Cardboard boxes	Keepers, curator, welfare scientist	15/11/17	2020
	Food:	Rotational diet (see food plan) Popcorn piñatas (1x week) Frozen pinecone with honey and oats (1x week) Treat board (1x week) Browse (5x week)	Keepers	Never	01/11/2018
	Sensory:	Wooden texture	Keepers	Never	TBD

Outcome of assessments since last version of document:	Cognitive: Novel:	Grass Heat lamp Frozen stimuli Not currently implemented Snow (1x year)	Keepers, curator		16/2/18 2021
		Snow to be continued as novel enrichment (previously implemented 3 times) Higher climbing structures required Activity levels low in adult orangutans Orangutans using wooden platforms more than firehose hammocks			
Planned new enrichment:	Social Physical Food Sensory Cognitive Novel	☐ Yes ☒ No ☒ Yes ☐ No ☒ Yes ☐ No ☐ Yes ☒ No ☒ Yes ☐ No ☐ Yes ☒ No			
	Type:	Item, schedule of use, and intended implementation date:	Approved by:		Assessment plan:
	Physical:	Climbing structure approx. 10m high max. Permanent. 30/9/2018	Keepers, curator, head of Life Sciences, welfare scientist		Daily behavioral observations for 2 weeks from implementation, then one day per month for a year (volunteers). Also, weekly weighing.
	Food/ Cognitive:	Puzzle feeders x5 (to be placed at highest points of climbing structures). Daily. 30/9/2018	Keepers, curator, welfare scientist, nutritionist		Daily records of length of time engaging with item for 2 weeks from implementation, then one day per month for a year (volunteers).
Aim of new enrichment:		See Documents 8.2 and 8.3			
Safety concerns:		Height of new climbing structure may increase the likelihood of injuries from falls Animals may ingest cotton threads from sheets Potential for fingers to be caught in puzzle feeders Small cuts from pinecones Risk of diabetes from high sugar treats Risk of injury due to fights when animals housed together			
Incident reports:		None since V2.0 of this document			

Table 4 Safety concerns to address before applying enrichment items

Concern	Example
Choking hazards	Small items or items with removable pieces
Contagions	Bacteria spread by sharing enrichment items between housing without cleaning
Toxicity	Paints
Weaponry	Parts or whole items which can be used as weapons, particularly small, heavy objects
Sharps	Staples in magazines, edges of items, exposed nails
Strangling hazards	Loose thread, loops, large holes, gaps
Length	Short pieces of hose which can only be secured at one end or long items which may be used to escape open-air enclosures
The individual animal	Temperament, age (Reinhardt 1990; Lutz and Novak 2005), sex (Novak et al. 1993; Parks and Novak 1993), disabilities, and the past experiences of the animals

exposure, other individuals may continue to exhibit signs of fear. Understanding which features of enrichment items may trigger such responses will be essential to avoid compromising a primate's welfare.

It is possible to incorporate acclimatization (habituation) toward enrichment items as part of a training program, slowly introducing the item and positively rewarding and reinforcing any interactions (personal experience; also, Brendan Host, personal communication). This is not an area which has received much attention in the literature. There may be some concerns that training an animal to use an enrichment device negates the potential benefits and removes the option of choice. However, the intention of this process is to demonstrate to the animal, in a safe environment with a trainer the animal trusts, that an item poses no threat and can be engaging. The goal with this method is to later provide opportunities for the animal to choose to interact with the item on their own, once they no longer show aversion in training sessions. Ultimately, this method desensitizes individuals to perceived "aversive" stimuli by associating positive experiences with interactions between the animal and the stimulus. This is similar to the process of social learning, in which individuals watch how conspecifics respond toward new stimuli (e.g., Addessi and Visalberghi 2006). However, it is important to note that this training method differs from the more common approach of associating new stimuli with food. Care staff will often place desired food items on new objects, to encourage the animal to approach. This then creates an association between the item and food; the value of the item lessens in the absence of food. From personal experience, once the food has been consumed, it is common for animals to lose interest in non-food-based enrichment items which were originally combined with food to elicit exploration. Therefore, it is recommended that non-food-based enrichment items are presented to the individual or group without food, to allow the animal or animals to choose how they want to interact with, or avoid, the stimulus.

Boredom is another issue which needs to be considered for safety reasons (Buchanan-Smith 2010). When primates lose interest in an item, they may begin to interact with it in a manner that was not intended. Given their dexterous fingers and strong jaws, primates will often pick pieces off the item or try to pull it apart. This can result in damaged investments or small pieces which an animal may try to eat and possibly choke on.

It is essential to always observe animals with any new enrichment items to ensure that there are no dangers in the use of the stimuli and to be able to intervene quickly if necessary. Some enrichment items are suitable for long-term use (e.g., durable and structural enrichment) while others should be removed within a short-time span (e.g., destructible items). Items which are repeatedly used should be inspected regularly for points of weakness (e.g., areas which have been chewed on) or potential hazards (e.g., loose threads). The Shape of Enrichment database has a list of safety concerns identified by care staff worldwide for different enrichment items and the species for which they are used (The Shape of Enrichment n.d.).

5 Types of Enrichment

Enrichment is typically divided into two types: social and inanimate (Lutz and Novak 2005). Within the inanimate division, there are four general categories (Bloomsmith et al. 1991; Buchanan-Smith 2010; Keeling et al. 1991): physical (also known as environmental and/or structural), food-based, sensory, and cognitive (including training). Enrichment can also be categorized as occupational (Coleman et al. 2012), novel (Britt 1998a; Csatádi et al. 2008; Paquette and Prescott 1988), manipulative, cooperative, and emotional (Morris et al. 2011). It is likely that any given enrichment item falls into multiple categories. For example, a frozen food item is both food-based enrichment and sensory enrichment.

5.1 Social Enrichment

Most primates are social, or at the very least classified as semi-social (e.g., orangutans, *Pongo spp.*) (de Waal 1991). Group housing is one of the most important aspects of captive primate welfare and is essential for normal development (Ventura and Buchanan-Smith 2003), the promotion of typical species-specific behavior (Bourgeois and Brent 2005; Lutz and Novak 2005; Leonardi et al. 2010), and reducing the occurrence of abnormal behaviors (Sackett et al. 1982; Lutz and Novak 2005). Furthermore, social housing can also result in individuals engaging with other forms of enrichment more often than do individually-housed primates (e.g., macaques: Line et al. 1991; Novak et al. 1993).

Creating opportunities for social interactions, such as allogrooming, huddling, and play (Fig. 2), promotes physical and psychological health (Hutchins and Barash 1976; Reinhardt 1989). These interactions help to develop and maintain social bonds. Group-housed primates show a resistance to stress and stressful events, as

Fig. 2 Social housing allows primates to engage in play behavior (*left*) and huddling for warmth and physical contact (*right*)

well as better recovery from aversive experiences (Young et al. 2014) and an improved ability to cope with change, compared to individually-housed primates (Levine et al. 1978; Coe et al. 1978; Mendoza 1978). This phenomenon is known as social buffering (see Gust et al. 1994; Kikusui et al. 2006).

Although social housing of primates is generally the norm in modern zoos, research facilities can be hesitant due to limited space and concerns with issues around research protocols and potential conflicts between animals (Baker et al. 2007; Baker 2016). Conflicts can be managed to allow the welfare benefits of social housing (Hartner et al. 2001; Wolfensohn 2004). Guidelines for the introduction of animals to each other can be found on the National Centre for the Replacement, Refinement and Reduction of Animals in Research's macaque website (n.d.). Prior to putting animals together during the introductory phase, it can be beneficial to scatter enrichment items and food around the housing. These may act as distractions so that the primates' attention is on these stimuli, rather than the unfamiliar animal.

5.2 Physical Enrichment

The basis of effective physical enrichment is in the design of the facility. Thus, with modern improvements to enclosure design due to both a better understanding of primate welfare and a human preference for naturalistic environments for captive animals (e.g., Melfi et al. 2004), physical enrichment is a basic standard and is typically in-built. However, in any environment, whether zoo, research, private ownership, or breeding facility, it is the use of the space that is most important (Chamove 1989). Primates can move in all three dimensions and so it is important that the use of space is maximized (Fig. 3).

Structural enrichment may be permanent or temporary, moving or stationary, natural or non-natural, but should ideally incorporate all of these possibilities (Fig. 4). Providing several options will allow individuals to choose how to use their environment. Indoor facilities can use cage mesh and room ceilings to suspend objects while outdoor housing must build from the ground up. Items should be

Fig. 3 An outdoor orangutan/siamang exhibit (*left*) uses up to 10 m of vertical height with platforms, tall poles, and ropes, while indoor rhesus macaque (*Macaca mulatta*) housing (*right*) also utilizes the vertical space with climbing structures provided at different elevations (Credit right picture: J. Nightingale)

securely fixed, or at least of a suitable weight and size so that the animals are unlikely to be able to lift them up and throw them. Furthermore, the animals using the environment need to be taken into consideration. For example, older animals tend to prefer more stable structures compared to younger individuals (Bryant et al. 1988; Kopecky and Reindhert 1991; Lehman and Lessnau 1992; Dexter and Bayne 1994; Reinhardt and Reinhardt 2008); gorillas typically stay close to the ground and require large, sturdy climbing frames due to their size; social groups need physical barriers which create hiding spots. Physical barriers are also useful in allowing display animals privacy from visitors when desired.

Durable items which are easy to clean are ideal for structural enrichment. A combination of wooden platforms and poles, nesting opportunities, fiberglass, polyvinyl chloride (PVC) and other plastics, firehose and rope, as well as the use of natural fibers, such as hessian (Fig. 5), create textures, shapes, and colors, to increase the sensational experience of the captive environment (see Sect. 5.4). These items can be easily sourced. For example, fire stations often have lengths of old firehose they are willing to give away (and have even been known to help hang; personal experience); for smaller primates, in particular callitrichids, tea towels can be used as nesting material; plastic horse jumping blocks and saddle carriers, which can be bought online, are durable items which add bright colors to the environment (Fig. 6). The use of these items is flexible and they can be left free-standing or attached to other items within the housing (Fig. 6).

Physical enrichment can also incorporate climatic gradients and other environmental features. This is easier for outdoor housing, but there is a reduced control over exposure to the elements. For indoor housing, one option is to provide primates with access to natural sunlight through skylights and windows. The use of water features can also be enriching, especially for primates who are known to swim (e.g., macaques) or wade (e.g., orangutans). A tub of water is a popular enrichment item for macaques, and can be used for water foraging, bathing, and diving, and be further varied with the use of a non-toxic children's bubble bath. Water features in an

Fig. 4 Lemur exhibit, with substrate and climbing opportunities including low logs (*top left*), stable platforms under heat lamps with small compartments to hide food items (*top right*), non-rigid rope (*bottom left*), and braided firehose (*bottom right*)

outdoor exhibit (Fig. 7) may also help to establish microclimates and produce sounds (e.g., from waterfalls) which may help to reduce noise from care staff and visitors. However, this needs verification as it is also possible that the primates find the constant noise from the waterfall to be irritating. While moats are popular features for primate enclosures, care needs to be taken with deep water moats as primates have been known to drown in them.

Although primates will continue to use the structures in their housing, these structures will eventually no longer offer any particular challenge and the unchanging scenery may elicit boredom. Having a selection of items which can be changed, introduced, and removed will help to keep the captive environment varied and novel (Csatádi et al. 2008), as the primates would experience in the wild. For example,

Fig. 5 Hessian used to wrap around branches and tied with strips of callico provide a suitable, and more naturalistic, gripping surface for marmosets (*Callithrix jacchus*), while PVC tubing creates hiding spots and play tunnels. Covering the outside of the tubing with a non-slippery fabric will add greater usable surface

changing the way ropes are connected will create new climbing courses and areas to access. Large logs can occasionally be rearranged to develop new food hiding spots. Objects moved between housings can also bring with them exciting new smells from other animals (see Sect. 5.4.4). On a cautionary note, there is a fine balance in how often environments should be altered and how many structures should be changed at any given time. Change is linked to unpredictability, which is known to lead to stress responses in primates (Buchanan-Smith 2010). Given the limited research on the effect of environmental change on stress responses (Fairhurst et al. 2011) in primates, a cautious approach needs to be taken and only some aspects of the habitat should be altered at any one time.

5.3 Food-Based Enrichment

Food-based enrichment is probably the most readily used form of transitory enrichment. It is popular because it tends to produce an immediate response from the primate, is typically easy to prepare, and can be produced in a variety of different ways. There are two ways in which food can be enriching: when the food is presented in an enriching manner (e.g., scatter feeds, puzzle feeders) and when the food itself is the enrichment (e.g., treats, sensory foods) (Table 5). Due to the short-term nature of food, this form of enrichment is unlikely to have a lasting effect on reducing behavioral problems (Lutz and Farrow 1996). However, as animals spend much of their time in the wild foraging and eating, finding methods to prolong feeding periods in captivity can go a long way toward better emulating wild behavior (Morimura 2007; Reinhardt and Roberts 1997). Presenting a primate's daily food rations using enrichment methods can be stimulating, encourages activity, and occupies their time. Indeed, studies have shown that primates prefer to work for

Fig. 6 Use of colorful, plastic horse blocks and saddle carriers to provide climbing structures, both stable and moving, and allow for hierarchal positioning and play

their food when not overtly hungry (contrafreeloading: Jensen 1963; e.g., Markowitz 1982; Inglis et al. 1997; Jones and Pillay 2004).

Food-based enrichment should not be in addition to the daily feed—it should be incorporated into the normal feeding regime. Enrichment is the alternative to placing food in bowls. Bowls should not be used (ban the bowl!), except in extreme cases

Fig. 7 Chimpanzees sit on sun-warmed rocks by a waterfall in an outdoor exhibit

(i.e., injury or disability), as they provide no enrichment value and make it too easy for the animals to access the food. Furthermore, food programs should consider the enrichment as well as the nutritional value of each item. Typically, food-based enrichment is seen as an addition to the daily diet and so treats and high sugar foods are given to "enrich" the animals' feeding experience; this should be discouraged as obesity and diabetes can be common problems for primates in captivity (Bauer et al. 2011; Videan et al. 2007). Instead, consider how foods can be stimulating by presenting them as whole pieces (e.g., an entire apple rather than apple pieces), very small pieces for which the primates must spend time searching (e.g., seeds in a deep substrate; Fig. 8), or food in objects which provide a challenge to access (e.g., a puzzle feeder which requires a tool to manipulate a food item to the exit point). A mixture of these approaches will create complexity in the feeding regimen.

Although primates need to be fed daily, highly challenging food-based enrichment items do not need to be provided every day. Research has found that the application of these devices increases natural foraging through substrates for up to 2 days after the removal of the devices (Gottlieb et al. 2011). Indeed, one of the most effective ways to encourage natural foraging and occupy the primates for an extended period of time is through scatter feeding. The distribution of a primate's food across multiple surfaces and heights will promote long-term activity and reduce competition in socially-housed animals. Scatter feeding can be done with both large and small food items. A deep substrate (at least 2.5 cm/1 inch), such as straw and wood-shavings (Fig. 9) or bark, will require primates to seek out their food (Chamove et al. 1982). If it is not possible to provide a deep substrate throughout

Table 5 Examples of food enrichment for primates, presented for their own merit (enrichment foods) or for their access to be an enriching experience (presentation)

Types of presentation	Examples	Enrichment foods	Reasoning
Scattered	Large items placed around housing at different levels. Small pieces (e.g., oats, lentils, dried peas, rice) thrown into substrate.	Browse	Leafy browse can provide naturalistic food items while looking attractive. Hang for increased foraging difficulty. Research has found that the addition of browse can increase activity levels (Dishman et al. 2009).
Puzzle feeders and mazes	Food inside objects which the primates must manipulate in order to access. Hang for increased difficulty.	Cooked pasta	A source of carbohydrates and easily chewable by primates with worn teeth (French and Fite 2005).
Termite mounds (Fig. 8)	Food smears placed inside artificial termite mound with access holes—provide chimpanzees with browse they must strip in order to work into usable tools for dipping.	Live insects (Fig. 8)	Some primates may relish the opportunity to catch live prey. The insects can also be put in small plastic jars with either a loose lid or stoppered with straw.
Piñatas	Papier mâché items in a variety of shapes and sizes, with or without holes. Can be painted in bright colors with non-toxic children's paint. Glue should be made from boiled water and flour. Hang for increased difficulty.	Ice blocks	Use plain water or heavily diluted cordial and add food items such as fruit pieces. Can be presented in a variety of shapes. Can also freeze a carabiner at one end to use to hang the item later.
Magazines and phonebooks	Food items can be placed between the pages. Can be hung.	Edible flowers	Adds color and novelty to the diet.
Cardboard rolls and boxes (Fig. 8)	Rolls and boxes can be stuffed with straw and wood wool (excelsior) with a few treats—fold up the ends. Can be frozen for later.	Popcorn	Unsalted plain popcorn (without butter) can be used in a variety of ways (e.g., stuck to smears, in phonebooks) and is a relatively healthy treat.
Pinecones	Smear the pinecone with peanut butter, honey, etc., and roll in a mix of oats, dried fruits, popcorn, etc. Can be frozen for later.	Jelly	Novel texture compared to other food substances. Requires effort to obtain due to slippery and wobbly characteristics.
Toys (Fig. 8)	Dog toys, including Kongs and balls, can be filled with small treats or smeared. They can be presented freely or hung.	Bamboo	Requires either excellent jaw strength or perseverance to consume.
Wood boards and logs	Drill holes into the boards and logs which can either be wide enough for a finger or narrow for twig-only access. Can be hung.	Smears	Examples of smears include: blended or cooked fruits, jams, peanut butter, and yogurt. Can be smeared on housing surfaces or other enrichment items.

(continued)

Table 5 (continued)

Types of presentation	Examples	Enrichment foods	Reasoning
Hollowed coconuts	Empty coconuts can be hung in exhibits and used as naturalistic feeding devices. Drill holes in them for food to pass out with manipulation (alternative to plastic balls).	Whole foods	As food is often presented in small pieces to reduce competition in social groups, whole food items can provide occasional novelty. Ensuring the quantity is enough for the group size, all animals should be able to access the food.
Cage-top	Food is placed on the top of caging to encourage climbing and hanging to feed (see Britt 1998b).		

the housing facility, filling buckets, tubs, and children's paddling pools with substrate are suitable alternatives.

Food enrichment is also an excellent way to present a variety of tastes, textures, and sensations (sensory experiences). Cold foods are particularly popular, especially in warm environments. Moreover, cold foods are easy to make and store and require a lot of perseverance by the primate, which is excellent for time consumption (Fig. 10). Jello (known as "jelly" outside the U.S. jelly) is another example of food enrichment which provides a different physical sensation and challenge for primates to typical primate food. However, avoid jello products high in sugar.

5.4 Sensory Enrichment

Animals experience the world around them through a variety of sensory modalities. However, as humans consider themselves visual beings (Hosey et al. 2013), enrichment often focuses on the visual modality. Indeed, the enrichment provided in on-display housing in zoos is typically chosen based on its visual appeal to the visiting public (McPhee et al. 1998; also, personal experience). Yet, to create an immersive and complex captive environment, enrichment needs to cater to the sensory ecology of the primate species. Engaging their other senses will create a richer experience to captive living. Primates communicate using visual, auditory, tactile, and olfactory cues (Liebal et al. 2013), and so it stands to reason that they can perceive stimuli through these modalities as well.

5.4.1 Visual Enrichment

This form of stimulus is already well-catered within Sect. 5.2 (physical enrichment). Stimuli should encompass a variety of shapes and colors, although the latter is likely to be less important for nocturnal (e.g., owl monkeys, *Aotus* spp.) and dichromatic species (e.g., New World monkeys). Other forms of visual enrichment can include

Fig. 8 Cardboard rolls can be used to create food parcels (*top left*) and even frozen (*bottom right*). Dog toys (*top right*) can create fun puzzles with food inserted or smeared (but they should be checked regularly for signs of chewing). Mealworms (*middle*) and crickets are easily bred and can be given to primates in a myriad of ways and provide opportunities for presenting live prey. An artificial termite mound (*bottom left*) with holes can be used by chimpanzees to exhibit natural behavior

televisions and videos, still images (magazines), and shiny objects, such as mirrors (Lambeth and Bloomsmith 1992; de Groot and Cheyne 2016) and CDs. Not all human forms of visual entertainment will be suitable for nonhuman primates and careful testing should be trialed before long-term exposure.

Fig. 9 A juvenile rhesus macaque foraging for small food items in deep substrate

Mirrors allow animals to view themselves, down corridors, or other social groups (Lambeth and Bloomsmith 1992). Access to a mirror should be dependent on the species, as large primates may break the glass if given direct access. Instead, they can be given indirect access, such as a mirror on the outside of a window with a directional control knob inside the housing (see, for example, the set-up at the Centre for Macaques; Medical Research Centre 2016). Alternatively, using reflective, shatter-proof plastic mirrors is an option for direct access.

Televisions and videos are potentially good sources of easily-applied enrichment as the variability and transitory nature of the images should reduce the likelihood of habituation (Ogura and Matsuzawa 2012). While various primate species have been noted to spend time watching videos (e.g., Bloomsmith and Lambeth 2000; Brent and Stone 1996; personal observation), watching videos does not promote species-typical behavior and may not reduce the prevalence of abnormal behaviors (Platt and Novak 1997; Bloomsmith and Lambeth 2000; although see Ogura and Matsuzawa 2012). If video-based stimulation is being considered as an enrichment tool, it is important to test what images the primates prefer—for example, brightly-colored, noisy cartoons, animal videos, or shows with human actors (Ogura and Matsuzawa 2012). The ability to choose to engage with a television with the option of an on/off switch the primates can access will reduce the likelihood of moving images and high-pitched sounds causing stress.

Colors can also be applied to otherwise dull enrichment items. For example, piñatas and papier mâché items can be painted using non-toxic children's paint; food dyes can be added to ice blocks; and some primates may even use paints as directed

Fig. 10 Examples of frozen enrichment items. Top left: silicone and plastic molds of a variety of shapes and sizes can be used to present frozen enrichment. Top right: a Christmas tree mold has been used to freeze carrot, pumpkin, and cucumber. Bottom left: cylindrical molds create a more long-lasting frozen treat, with broccoli, eggplant, carrot, capsicum, and tomato pieces. Bottom right: a pinecone has been used as a novel method for presenting oats and popcorn using smeared honey before being frozen

by keepers or on their housing walls when given free reign. They are also known to mix colors to preference (personal observation). However, care should be considered when applying color as studies have shown that some colors, such as red, can elicit undesirable responses in rhesus macaques (e.g., Humphrey 1971; Humphrey and Keeble 1975). Other studies have found a preference for green and blue stimuli in chimpanzees and gorillas (e.g., Fritz et al. 1997; Wells et al. 2008).

5.4.2 Tactile Enrichment

Textures are already incorporated into other forms of enrichment. Examples include the hair of a conspecific during a grooming session (social enrichment), the difference between a wooden platform and a plastic one (physical enrichment), or frozen fruit compared to smeared fruit compared to fresh whole fruit (food-based enrichment). Tactile enrichment can also take the form of bedding, including straw and leaves for making nests (important for great apes), hammocks, sheets (popular with orangutans), tea towels (more suitable for small primates such as marmosets), nesting boxes or hollows, and hessian, which can be used to reduce the slipperiness of other textures, or as bags for hiding treats or using as covers.

5.4.3 Auditory Enrichment

This form of sensory stimuli may be the least utilized (Farmer et al. 2011). What research has been done has had largely non-favorable findings. The most commonly used forms of auditory enrichment are human music and natural habitat sounds (e.g., waterfalls). While these sounds may be considered enriching and are sometimes used to try to mask unpleasant background noises, research suggests that some types of music can cause aggression, elicit no increase in positive behaviors (e.g., Hanbury et al. 2009; Howell et al. 2003; Wells 2009), and even increase stereotypic behavior (Robbins and Margulis 2014). Indeed, when given a choice, chimpanzees, tamarins, and marmosets will often choose silence over music (Richardson et al. 2006). Wallace et al. (2017) found no evidence that music was either enriching or had a negative effect on the welfare of group-housed chimpanzees; however, the apes were more likely to exit the area when songs with a high beat count per minute were played. These results demonstrate that care staff should be cautious when attempting to use music as enrichment, and the genre of music needs to be considered. However, Snowdon and Teie (2010) found that tamarins (*Saguinus oedipus*) exhibited calm behaviors when exposed to music which was based on their vocalizations.

Alternatives to contrived sounds include noise makers and rattles, which primates can use to produce sounds, as well as the vocalizations of other animals. The latter may occur naturally within the primate's environment (i.e., from nearby animals) or can be deliberately applied through playback experiments. In howler monkeys (*Alouatta caraya*), for example, playbacks of other howler vocalizations increased calling from subject animals and other associated behaviors (Farmer et al. 2011). Vocalizations could, therefore, be a form of positive auditory enrichment.

Some primates also detect food (live insects) using auditory cues (e.g., aye-ayes, *Daubentonia madagascariensis*; Charles-Dominique 1977; Erickson et al. 1998; MacKinnon and MacKinnon 1980; Niemitz 1979). While auditory cues are likely very important for nocturnal species, it is also thought that chimpanzees also use auditory cues when hunting prey (Milton 2000). While research on this topic is limited, there is room for exploring the use of prey-based auditory enrichment for primates.

5.4.4 Olfactory Enrichment

Olfactory enrichment refers to the application of scents or scented material (Swaisgood and Shepherdson 2006). The items which can be used to increase the olfactory experience of captive primates are numerous and include food smells, infused and essential oils, artificial odors, animal derivatives, and commercial lures. This form of enrichment is underutilized for primates (Clark and King 2008), perhaps due to the long-held belief that primates are not olfactory-driven; this may reflect our limited understanding of this sensory modality in ourselves let alone other primate species (Hübener and Laska 2001; Barton 2006). The species most commonly given olfactory-based enrichment tend to be those known to perform scent

Fig. 11 Flavoring essences (*left*) can be diluted in water to between 1:1000 and 1:10, depending on their strength, the primate species, and any previous indications of interest in the odor. The odor can then be sprayed on to objects in the environment or soak a cotton ball in the liquid and place it inside a tea infuser before hanging the infuser on the outside of the caging (*right*)

marking, such as lemurs and marmosets. Although the olfactory bulb does reduce in size along the primate line from prosimians to apes, that does not mean that olfactory ability disappears or is not important. In fact, some primate species can detect some odors at lower concentrations than rats and dogs, two species which are highly sensitive to odors (Danilova and Hellekant 2000, 2002; Hellekant et al. 1997; Hudson et al. 1992; Laska and Hudson 1993; Laska et al. 1996, 2000, 2007; Laska and Freyer 1997). In particular, ethanol-based cues may be important for primate foraging (Dudley 2000). Specifically, primates may use the smell of the ethanol in the fruit to find and select the ripest pieces (Dominy 2004).

Odors can be presented in a variety of ways: sprayed on surfaces, as crushed or whole plants (e.g., dried or fresh herbs and spices), and in bodies of water. Bedding material and other substrates can also be rotated between species and groups to share animal scents. Fecal matter and urine (with permission from veterinarians) from disease-free animals can also be used for olfactory enrichment, either diluted or in their original forms.

Most odors can be diluted (e.g., oils and essences; Fig. 11); an odor does not have to be fully concentrated to produce a behavioral reaction or encourage exploration. Diluted odors should be initially used; the concentration can then be incrementally raised if no detection or investigation of the odor is observed. It is also recommended that one should not apply odors in too many locations or use airborne dispersers in a housing facility because, in the event of an adverse reaction, rapid removal of the source of the odor may be required. Alternatively, liquid odors can be presented on the outside of caging by soaking a cotton ball and placing it inside a hanging tea infuser (Kemp and Kaplan 2012; Fig. 11). This will give primates more control and avoid the odor if desired. It is important to note that the animals may be able to detect the odor from a distance and so physical proximity may not be needed for the odor to be impactful.

5.5 Cognitive Enrichment

Despite primates having large, complex brains which are capable of a myriad of impressive behaviors, and, in the wild, solving problems requiring various mental skills and processes, cognitive enrichment is often overlooked (Clark 2017; Meehan and Mench 2007). Still, other types of enrichment encompass elements of cognitive stimulation. Puzzle feeders, for example, provide something of a mental and physical challenge as the animal must manipulate the object to get the food to the exit hole (Fig. 12). However, primates learn the best techniques quickly and so the challenge posed by some puzzle feeders will diminish accordingly. Complex puzzle feeders, with movable parts and segments which can be swapped to create new challenges can reduce this problem and will be a more effective form of enrichment in the long-term.

Fig. 12 Examples of relatively simplistic puzzle feeders, which can utilize readily available dog toys such as Kongs (*top left*) and naturalistic items including hollowed-out bamboo (*top right*). Large, clean petrol canisters may provide more of a challenge with small exit holes (*bottom left*) while food cages (*bottom right*) can be used to house large pieces of food items, which may require a stick to help access, or even frozen foods. When puzzle feeders cannot be used, placing food items on the top of caging, requiring the primate to work through the bars, is a viable alternative which achieves a similar purpose

Still, primates are capable of navigation, tool use, and cooperative activities which they are rarely able to express in the captive environment (Meehan and Mench 2007). Devising enrichment items to target these skills can be challenging for care staff. Opportunities for the development of suitable challenges may come from comparative psychology and neuroscience research. Research tasks often incorporate cognitive skills—the use of touchscreen computers or joysticks (e.g., Fagot and Paleressompoulle 2009; Martin et al. 2014, 2017; Morimura and Matsuzawa 2001; Platt and Novak 1997; Washburn and Rumbaugh 1992), problem solving (e.g., Gunhold et al. 2014), cooperative efforts for goal achievement (e.g., Suchak et al. 2016), and hidden object-object permanence paradigms (e.g., Call 2007; Mendes and Huber 2004)—but these activities have yet to make it into mainstream enrichment programs. It is possible that technology may make some facilities uneasy due to their artificial nature (Carter et al. 2015) or because these apparatuses are expensive, yet the benefits to both captive primates and human viewers could outweigh these disadvantages. Indeed, there is evidence that these research activities can provide forms of enrichment for captive animals (e.g., Fagot and Bonté 2010; Fagot et al. 2014), as well as being engaging for visitors and staff (Webber et al. 2017). However, further research is needed.

There is a bias toward food-based cognitive enrichment; while food may encourage primates to interact with items and technologies, it does not have to be the main motivator. For example, research has found macaques will perform cognitive tasks for images of other macaques (Deaner et al. 2005). Computer technology has recently been introduced into the orangutan exhibit at Melbourne Zoo, Australia (Webber 2018) in which the only reward is from the interaction with the system. In this case, the orangutans do not touch any screens or get food rewards; instead, images are directed into their exhibit using Microsoft Kinect, which tracks the primates' movements and detects when they touch the projections. This creates a large touch-enabled surface without allowing the orangutans access to the fragile equipment. The system also allows for interactions to occur from outside the exhibit, potentially creating positive human–animal interactions and an enriching experience for zoo visitors.

Positive reinforcement (reward-based) training may also provide cognitive stimulation for captive primates (Buchanan-Smith 2010; Hosey et al. 2013; Laule and Desmond 1998). Positive-reinforcement training is becoming increasingly commonplace in many facilities as a way to encourage captive primates to behave in desired ways and to reduce the stress of husbandry procedures and research activities (Prescott and Buchanan-Smith 2003, 2007; Prescott et al. 2005; Kemp et al. 2017). While there is evidence that reward-based training may be effective in reducing stereotypies in primates (Bourgeois and Brent 2005), there is still some debate in the literature as to whether it constitutes a form of enrichment (Baker et al. 2010; Hare and Sevenich 2001; Melfi 2013; Westlund 2014). It is certainly not a replacement for environmental enrichment and may only constitute an enriching experience during the period in which the behavior being targeted is learned by the animal (Melfi 2013).

6 Assessment

It is important that stimuli are assessed to determine if they are achieving the goals set out prior to their application. New items should be assessed upon initial application, but repeatedly used items should also have regular "check-ups" to determine if habituation and boredom have occurred (Kuczaj et al. 2002). Not all enrichment is equally enriching (Galef 1999; Mason et al. 1998; Mellen and Sevenich MacPhee 2001; Morgan et al. 1998) and given that resources (time, labor, and financial costs) for enrichment are limited, maximizing the outcome for the effort put in is ideal. Furthermore, if we cannot be assured that our efforts have lasting benefits and improve animal welfare, we are undermining the quality of our care for captive primates. As Galef (1999, p. 279) pointed out:

> enrichment programs based on unscientific belief systems or unscientific methods must be counter productive in the end. Good will toward animals plus professional judgement is simply not enough. We need to undertake research on the efficacy of whatever enrichment procedures we propose to implement. If we do not, we are not meeting our moral obligations, either to the animals... or to the public that asks that we treat our animals as humanely as we can[.]

The assessment of how the primate or primates interact with and utilize any given stimulus results in data which helps to validate the effort and resources involved. Enrichment is an investment: staff time in the development and application of items, as well as financial costs, need to be outweighed by the benefits of applying enrichment. Analyzing the enrichment's effectiveness to achieve the set goal or goals will help fine-tune the enrichment program so that the results are maximized. This will help to determine if a stimulus should continue to be used, if a stimulus requires modifications to improve its enrichment value, and how to modify the item to best meet the goal. Assessment will also help to rapidly determine if a stimulus is having an adverse effect on an individual or group of animals (Bayne 2005).

When assessing the effectiveness of a stimulus, it is important to remember that enrichment is experienced by the individual, even when applied in a group setting. It is the individual animal which benefits or does not benefit, not the group. A group of primates does not interact with a stimulus; the individuals does so. Therefore, it is important that any assessment examines individual responses to the stimulus.

There are four major ways in which enrichment use can be measured (Coleman et al. 2012): (1) physiological responses—these can include cortisol, heart rate, and immune function; this method is likely to be more suitable for research primates than for zoo animals or those in breeding facilities, (2) indirect results—if an item is applied to encourage locomotion in an overweight primate, then evidence of the effectiveness of the item will be measurable through physical condition. Note that indirect measures such as these should only be used when the specific aim of the enrichment was to target a physical state. While indirect measures provide an easy measure of effectiveness, there are potential problems: data will need to be collected over a long period of time and adverse behavioral responses may go unnoticed. (3) Behavioral responses—as the aim of enrichment is typically to increase the frequency and variety of species-typical behaviors while reducing the levels of abnormal behaviors, behavioral assessment is highly practical and can be conducted fairly easily. It is important, though, that any assessment of behavior is more than just a did/did not interact with the item dichotomy as avoidance of an enrichment item is still a behavioral response and may not necessarily be indicative of aversion. Considering a variety of behaviors will more likely result in determining the effect of the stimulus. (4) Preference—choice tests can be used to "ask" the primate with which enrichment item it would prefer to interact (e.g., Mehrkam and Dorey 2015). For further discussion on assessing and measuring welfare, see Part II of this book.

Documentation and assessment of primate reactions to enrichment items should ideally consider short- and long-term responses. This will help to determine how quickly animals habituate to items and will provide indications of how long any item should remain accessible. Items may be kept interesting by their frequent removal and withholding, if research shows that the animal or animals lose interest after short periods of time (hours, days). Preferably, most enrichment should be removed before animals become bored with them. For enrichment used to deliver food items, animals may show a lack of engagement, even when food is still present, if the item is proving too much of a challenge. Documenting this response can help to determine whether a simpler challenge is required to help the animal gain confidence in using the enrichment and to encourage its continued use. Documentation also allows for the possibility of publishing results. The application of effective enrichment is still a developing science; publishing successful and null results of enrichment trials will help the primate care community make more appropriate decisions when designing enrichment programs.

7 Conclusion

Although establishing an enrichment program may seem time-consuming, costly, and require additional staff resources, the benefits are numerous. Moreover, simple, cheap approaches can have a large positive impact on welfare. The enticement to express a rich behavioral repertoire will have flow-on effects regardless of the facility type; zoo visitors will be more impressed by active and engaging animals and researchers will be able to place more stock in the reliability of their data. Ultimately, choice and control by the primates will yield the most effective enrichment. Careful planning of enrichment will reduce risks and have long-term benefits.

References

Addessi E, Visalberghi E (2006) How social influences affect food neophobia in captive chimpanzees: a comparative approach. In: Matsuzawa T, Tomonaga M, Tanaka M (eds) Cognitive development in chimpanzees. Springer, Tokyo, pp 246–264

American Society of Primatologists (2016) Introduction to environmental enrichment for primates

Baker KC (2016) Survey of 2014 behavioral management programs for laboratory primates in the United States. Am J Primatol 78(7):780–796. https://doi.org/10.1002/ajp.22543

Baker KC, Weed JL, Crockett CM, Bloomsmith MA (2007) Survey of environmental enhancement programs for laboratory primates. Am J Primatol 69(4):377–394. https://doi.org/10.1002/ajp.20347

Baker KC, Bloomsmith MA, Neu K et al (2010) Positive reinforcement training as enrichment for singly housed rhesus macaques (*Macaca mulatta*). Anim Welf 19(3):307–313

Barton RA (2006) Olfactory evolution and behavioral ecology in primates. Am J Primatol 68(6):545–558. https://doi.org/10.1002/ajp.20251

Bateson M (2004) Mechanisms of decision-making and the interpretation of choice tests. Anim Welf 13(Suppl 1):115–120

Bauer SA, Arndt TP, Leslie KE et al (2011) Obesity in rhesus and cynomolgus macaques: a comparative review of the condition and its implications for research. Comp Med 61(6):514–526

Bayne K (2005) Potential for unintended consequences of environmental enrichment for laboratory animals and research results. ILAR J 46(2):129–139. https://doi.org/10.1093/ilar.46.2.129

Bloomsmith MA, Lambeth SP (2000) Videotapes as enrichment for captive chimpanzees (*Pan troglodytes*). Zoo Biol 19(6):541–551. https://doi.org/10.1002/1098-2361(2000)19:6<541::AID-ZOO6>3.0.CO;2-3

Bloomsmith M, Brent L, Schapiro S (1991) Guidelines for developing and managing an environmental enrichment program for nonhuman primates. Lab Anim Sci 41(4):372–377

Bloomstrand M, Riddle K, Alford P, Maple TL (1986) Objective evaluation of a behavioral enrichment device for captive chimpanzees (*Pan troglodytes*). Zoo Biol 5(3):293–300. https://doi.org/10.1002/zoo.1430050307

Bourgeois SR, Brent L (2005) Modifying the behaviour of singly caged baboons: evaluating the effectiveness of four enrichment techniques. Anim Welf 14(1):71–81

Brent L, Stone AM (1996) Long-term use of televisions, balls, and mirrors as enrichment for paired and singly caged chimpanzees. Am J Primatol 39(2):139–145. https://doi.org/10.1002/(SICI)1098-2345(1996)39:2<139::AID-AJP5>3.0.CO;2-#

Britt A (1998a) Environmental enrichment for apes: a literature review. In: Field DA (ed) Guidelines for environmental enrichment, association of British wild animal keepers. Association of British Wild Animal Keepers, New York, pp 233–247

Britt A (1998b) Encouraging natural feeding behavior in captive-bred black and white ruffed lemurs (*Varecia variegata variegata*). Zoo Biol 17(5):379–392. https://doi.org/10.1002/(sici)1098-2361(1998)17:5<379::aid-zoo3>3.3.co;2-o

Bryant CE, Rupniak NM, Iversen SD (1988) Effects of different environmental enrichment devices on cage stereotypies and autoaggression in captive cynomolgus monkeys. J Med Primatol 17(5):257–267

Buchanan-Smith HM (2010) Environmental enrichment for primates in laboratories. Adv Sci Res 5(1):41–56. https://doi.org/10.5194/asr-5-41-2010

Call J (2007) Apes know that hidden objects can affect the orientation of other objects. Cognition 105(1):1–25. https://doi.org/10.1016/j.cognition.2006.08.004

Carter M, Webber S, Sherwen S (2015) Naturalism and ACI: augmenting zoo enclosures with digital technology. In: Proceedings of the 12th International Conference on Advances in Computer Entertainment Technology. ACM Press, New York, pp 1–5

Chamove AS (1989) Cage design reduces emotionality in mice. Lab Anim 23(3):215–219. https://doi.org/10.1258/002367789780810608

Chamove AS, Anderson JR, Morgan-jones SC, Jones SP (1982) Deep woodchip litter: hygiene, feeding, and behavioral enhancement in eight primate species. Int J Study Anim Probl 3 (4):308–318

Charles-Dominique P (1977) Ecology and behaviour of nocturnal primates: prosimians of equatorial West Africa. Columbia University Press, New York

Clark FE (2017) Cognitive enrichment and welfare: current approaches and future directions. Anim Behav Cogn 4(1):52–71. https://doi.org/10.12966/abc.05.02.2017

Clark FE, King AJ (2008) A critical review of zoo-based olfactory enrichment. In: Clark F, King AJ (eds) Chemical signals in vertebrates 11. Springer, New York, pp 391–398

Clark FE, Melfi V, Mitchell H (2005) Wake up and smell the enrichment: a critical review of an olfactory enrichment study. In: Proceedings of the Seventh International Conference on Environmental Enrichment, Wildlife Conservation Society, New York City. Wildlife Conservation Society, New York, pp 178–185

Coe CL, Mendoza SP, Smotherman WP, Levine S (1978) Mother-infant attachment in the squirrel monkey: adrenal response to separation. Behav Biol 22(2):256–263. https://doi.org/10.1016/S0091-6773(78)92305-2

Colahan H, Breder C (2003) Primate training at Disney's Animal Kingdom. J Appl Anim Welf Sci 6(3):235–246. https://doi.org/10.1207/S15327604JAWS0603_08

Coleman K, Novak MA (2017) Environmental enrichment in the 21st century. ILAR J 58 (2):295–307. https://doi.org/10.1093/ilar/ilx008

Coleman K, Bloomsmith MA, Crockett CM et al (2012) Behavioral management, enrichment, and psychological well-being of laboratory nonhuman primates. In: Abee CR, Mansfield K, Tardiff S, Morris T (eds) Nonhuman primates in biomedical research: biology and management, vol 1, 2nd edn. Academic Press, London, pp 149–176

Crockett CM, Bielitzki J, Carey A, Velez A (1989) Kong ® Toys as enrichment devices for singly-caged macaques. Lab Primate Newsl 28(2):21–22

Csatádi K, Leus K, Pereboom JJM (2008) A brief note on the effects of novel enrichment on an unwanted behaviour of captive bonobos. Appl Anim Behav Sci 112(1–2):201–204. https://doi.org/10.1016/j.applanim.2007.09.001

Danilova V, Hellekant G (2000) The taste of ethanol in a primate model: II. Glossopharyngeal nerve response in *Macaca mulatta*. Alcohol 21(3):259–269. https://doi.org/10.1016/S0741-8329(00)00094-X

Danilova V, Hellekant G (2002) Oral sensation of ethanol in a primate model III: responses in the lingual branch of the trigeminal nerve of *Macaca mulatta*. Alcohol 26(1):3–16. https://doi.org/10.1016/S0741-8329(01)00178-1

Davey G (2006) Relationships between exhibit naturalism, animal visibility and visitor interest in a Chinese zoo. Appl Anim Behav Sci 96(1–2):93–102. https://doi.org/10.1016/j.applanim.2005.04.018

de Groot B, Cheyne SM (2016) Does mirror enrichment improve primate well-being? Anim Welf 25(2):163–170. https://doi.org/10.7120/09627286.25.2.163

de Waal FBM (1991) The social nature of primates. In: Novak MA, Petto AJ (eds) Through the looking glass. American Psychological Association, Washington, DC, pp 69–77

Deaner RO, Khera AV, Platt ML (2005) Monkeys pay per view: adaptive valuation of social images by rhesus macaques. Curr Biol 15(6):543–548. https://doi.org/10.1016/j.cub.2005.01.044

Dexter S, Bayne KAL (1994) Results of providing swings to individually housed rhesus monkeys (*Macaca mulatta*). Lab Primate Newsl 33(2):9–12

Dishman DL, Thomson DM, Karnovsky NJ (2009) Does simple feeding enrichment raise activity levels of captive ring-tailed lemurs (*Lemur catta*)? Appl Anim Behav Sci 116:88–95. https://doi.org/10.1016/j.applanim.2008.06.012

Disney (n.d.) Animal enrichment. http://www.animalenrichment.org. Accessed 28 June 2021

Dominy NJ (2004) Fruits, fingers, and fermentation: the sensory cues available to foraging primates. Integr Comp Biol 44(4):295–303. https://doi.org/10.1093/icb/44.4.295

Dudley R (2000) Evolutionary origins of human alcoholism in primate frugivory. Q Rev Biol 75 (1):3–15. https://doi.org/10.1086/393255

Erickson CJ, Nowicki S, Dollar L, Goehring N (1998) Percussive foraging: stimuli for prey location by aye-ayes (*Daubentonia madagascariensis*). Int J Primatol 19:111–122. https://doi.org/10.1023/A:1020363128240

Fagot J, Bonté E (2010) Automated testing of cognitive performance in monkeys: use of a battery of computerized test systems by a troop of semi-free-ranging baboons (*Papio papio*). Behav Res Methods 42(2):507–516. https://doi.org/10.3758/BRM.42.2.507

Fagot J, Paleressompoulle D (2009) Automatic testing of cognitive performance in baboons maintained in social groups. Behav Res Methods 41(2):396–404. https://doi.org/10.3758/BRM.41.2.396

Fagot J, Gullstrand J, Kemp C et al (2014) Effects of freely accessible computerized test systems on the spontaneous behaviors and stress level of Guinea baboons (*Papio papio*). Am J Primatol 76 (1):56–64. https://doi.org/10.1002/ajp.22193

Fairhurst GD, Frey MD, Reichert JF et al (2011) Does environmental enrichment reduce stress? An integrated measure of corticosterone from feathers provides a novel perspective. PLoS ONE 6 (3):e17663. https://doi.org/10.1371/journal.pone.0017663

Farmer HL, Plowman A, Leaver L (2011) Auditory enrichment for captive black howler monkeys (*Alouatta caraya*); efficacy, keeper opinions and its future in husbandry practices. Paper presented at 10th International Conference on Environmental Enrichment, Portland, Oregon

French JA, Fite FE (2005) Marmosets and tamarins (callitrichids). National Institutes of Health Office of Laboratory Animal Welfare, Bethesda, MD

Fritz J, Howell SM, Schwandt ML (1997) Colored light as environmental enrichment for captive chimpanzees (*Pan troglodytes*). Lab Primate Newsl 36(2):1–4

Galef BG Jr (1999) Environmental enrichment for laboratory rodents: animal welfare and the methods of science. J Appl Anim Welf Sci 2(4):267–280. https://doi.org/10.1207/s15327604jaws0204_2

Gottlieb DH, Ghirardo S, Minier DE et al (2011) Efficacy of 3 types of foraging enrichment for rhesus macaques (*Macaca mulatta*). J Am Assoc Lab Anim Sci 50(6):888–894

Gottlieb DH, Capitanio JP, McCowan BJ (2013) Risk factors for stereotypic behavior and self-biting in rhesus macaques (*Macaca mulatta*): Animal's history, current environment, and personality. Am J Primatol 75(10):995–1008. https://doi.org/10.1002/ajp.22161

Gunhold T, Whiten A, Bugnyar T (2014) Video demonstrations seed alternative problem-solving techniques in wild common marmosets. Biol Lett 10(9):20140439. https://doi.org/10.1098/rsbl.2014.0439

Gust DA, Gordon TP, Brodie AR, McClure HM (1994) Effect of a preferred companion in modulating stress in adult female rhesus monkeys. Physiol Behav 55(4):681–684. https://doi.org/10.1016/0031-9384(94)90044-2

Hanbury DB, Fontenot MB, Highfill LE et al (2009) Efficacy of auditory enrichment in a prosimian primate (*Otolemur garnettii*). Lab Anim 38(4):122–125. https://doi.org/10.1038/laban0409-122

Hare VJ, Sevenich M (2001) Is it training or is it enrichment? In: Proceedings of the Fourth International Conference on Environmental Enrichment, pp 40–47

Hare V, Rich B, Worley K (2008) Enrichment gone wrong! The Shape of Enrichment, Inc., San Diego, CA. https://theshapeofenrichmentinc.wildapricot.org/resources/Documents/hare_2008.pdf

Hartner M, Hall J, Penderghest J, Clark LP (2001) Group-housing subadult male cynomolgus macaques in a pharmaceutical environment. Lab Anim 30(8):53–57. https://doi.org/10.1038/5000167

Hellekant G, Danilova V, Roberts T, Ninomiya Y (1997) The taste of ethanol in a primate model: I. Chorda tympani nerve response in *Macaca mulatta*. Alcohol 14(5):473–484. https://doi.org/10.1016/S0741-8329(96)00215-7

Honess PE, Marin CM (2006) Enrichment and aggression in primates. Neurosci Biobehav Rev 30 (3):413–436. https://doi.org/10.1016/j.neubiorev.2005.05.002

Hosey G (2005) How does the zoo environment affect the behaviour of captive primates? Appl Anim Behav Sci 90(2):107–129. https://doi.org/10.1016/j.applanim.2004.08.015

Hosey G, Melfi V, Pankhurst S (2013) Zoo animals: behaviour, management, and welfare, 2nd edn. Oxford University Press, Oxford

Howell S, Schwandt M, Fritz J et al (2003) A stereo music system as environmental enrichment for captive chimpanzees. Lab Anim 32(10):31–36. https://doi.org/10.1038/laban1103-31

Hoy JM, Murray PJ, Tribe A (2010) Thirty years later: enrichment practices for captive mammals. Zoo Biol 29(3):303–316. https://doi.org/10.1002/zoo.20254

Hübener F, Laska M (2001) A two-choice discrimination method to assess olfactory performance in pigtailed macaques, *Macaca nemestrina*. Physiol Behav 72(4):511–519. https://doi.org/10.1016/S0031-9384(00)00447-9

Hubrecht RC (2010) Enrichment: animal welfare and experimental outcomes. In: Hubrecht RC, Kirkwood J (eds) The UFAW handbook on the care and management of laboratory and other research animals, 8th edn. Wiley-Blackwell, Oxford, pp 136–146

Hudson R, Laska M, Ploog D (1992) A new method for testing perceptual and learning capacities in unrestrained small primates. Folia Primatol 59(1):56–60. https://doi.org/10.1159/000156643

Humphrey N (1971) Colour and brightness preferences in monkeys. Nature 229(5287):615–617. https://doi.org/10.1038/229615a0

Humphrey NK, Keeble GR (1975) Interactive effects of unpleasant light and unpleasant sound. Nature 253:346–347. https://doi.org/10.1038/253346a0

Hutchins M, Barash DP (1976) Grooming in primates: implications for its utilitarian function. Primates 17:145–150. https://doi.org/10.1007/BF02382848

Inglis IR, Forkman B, Lazarus J (1997) Free food or earned food? A review and fuzzy model of contrafreeloading. Anim Behav 53(6):1171–1191. https://doi.org/10.1006/anbe.1996.0320

Jacobson SL, Hopper LM, Shender MA et al (2017) Zoo visitors' perceptions of chimpanzee welfare are not affected by the provision of artificial environmental enrichment devices in a naturalistic exhibit. J Zoo Aquarium Res 5(1):56–61. https://doi.org/10.19227/jzar.v5i1.250

Jennings M, Prescott MJ (2009) Refinements in husbandry, care and common procedures for non-human primates. Lab Anim 43(Suppl 1):1–47. https://doi.org/10.1258/la.2008.007143

Jensen GD (1963) Preference for bar pressing over "freeloading" as a function of number of rewarded presses. J Exp Psychol 65(5):451–454. https://doi.org/10.1037/h0049174

Jones M, Pillay N (2004) Foraging in captive hamadryas baboons: implications for enrichment. Appl Anim Behav Sci 88:101–110. https://doi.org/10.1016/J.APPLANIM.2004.03.002

Keeling ME, Alford PL, Bloomsmith MA (1991) Decision analysis for developing programs of psychological well-being: a bias-for-action approach. In: Novak MA, Petto AJ (eds) Through the looking glass: issues of psychological well-being in captive nonhuman primates. American Psychological Association, Washington, DC, pp 57–65

Kemp C, Kaplan G (2012) Olfactory cues modify and enhance responses to visual cues in the common marmoset (*Callithrix jacchus*). J Primatol 01:102–114. https://doi.org/10.4172/2167-6801.1000102

Kemp C, Thatcher H, Farningham D et al (2017) A protocol for training group-housed rhesus macaques (*Macaca mulatta*) to cooperate with husbandry and research procedures using positive reinforcement. Appl Anim Behav Sci 197:90–100. https://doi.org/10.1016/j.applanim.2017.08.006

Kessel AL, Brent L (1998) Cage toys reduce abnormal behavior in individually housed pigtail macaques. J Appl Anim Welf Sci 1(3):227–234. https://doi.org/10.1207/s15327604jaws0103_3

Kikusui T, Winslow JT, Mori Y (2006) Social buffering: relief from stress and anxiety. Philos Trans R Soc B Biol Sci 361:2215–2228. https://doi.org/10.1098/rstb.2006.1941

Kopecky J, Reindhert V (1991) Comparing the effectiveness of PVC swings versus PVC perches as environmental enrichment objects for caged female rhesus macaques (*Macaca mulatta*). Lab Primate Newsl 30(2):5–6

Kuczaj S, Lacinak T, Otto F et al (2002) Keeping environmental enrichment enriching. Int J Comp Psychol 15(2):127–137. https://doi.org/10.46867/C4XK5N

Kutska D (2009) Variation in visitor perceptions of a polar bear enclosure based on the presence of natural vs. un-natural enrichment items. Zoo Biol 28(4):292–306. https://doi.org/10.1002/zoo.20226

Lambeth SP, Bloomsmith MA (1992) Mirrors as enrichment for captive chimpanzees (*Pan troglodytes*). Lab Anim Sci 42(3):261–266

Laska M, Freyer D (1997) Olfactory discrimination ability for aliphatic esters in squirrel monkeys and humans. Chem Senses 22(4):457–465. https://doi.org/10.1093/chemse/22.4.457

Laska M, Hudson R (1993) Assessing olfactory performance in a New World primate, *Saimiri sciureus*. Physiol Behav 53(1):89–95. https://doi.org/10.1016/0031-9384(93)90015-8

Laska M, Alicke T, Hudson R (1996) A study of long-term odor memory in squirrel monkeys (*Saimiri sciureus*). J Comp Psychol 110(2):125–130. https://doi.org/10.1037/0735-7036.110.2.125

Laska M, Seibt A, Weber A (2000) 'Microsmatic' primates revisited: olfactory sensitivity in the squirrel monkey. Chem Senses 25(1):47–53. https://doi.org/10.1093/chemse/25.1.47

Laska M, Bautista RMR, Hofelmann D et al (2007) Olfactory sensitivity for putrefaction-associated thiols and indols in three species of non-human primate. J Exp Biol 210(Pt 23):4169–4178. https://doi.org/10.1242/jeb.012237

Laule G, Desmond T (1998) Positive reinforcement training as an enrichment strategy. In: Shepherdson D, Mellen J, Hutchins M (eds) Second nature: environmental enrichment for captive animals. Smithsonian Institution Press, Washington, DC, pp 302–313

Lehman SM, Lessnau RG (1992) Pickle barrels as enrichment objects for rhesus macaques. Lab Anim Sci 42(4):392–397

Leonardi R, Buchanan-Smith HM, Dufour V et al (2010) Living together: behavior and welfare in single and mixed species groups of capuchin (*Cebus apella*) and squirrel monkeys (*Saimiri sciureus*). Am J Primatol 72(1):33–47. https://doi.org/10.1002/ajp.20748

Levine S, Coe CL, Smotherman WP, Kaplan J (1978) Prolonged cortisol elevation in the infant squirrel monkey after reunion with mother. Physiol Behav 20(1):7–10. https://doi.org/10.1016/0031-9384(78)90194-4

Liebal K, Waller BM, Burrows AM, Slocombe KE (2013) Primate communication: a multimodal approach. Cambridge University Press, Cambridge

Line SW, Markowitz H, Morgan KN, Strong S (1991) Effects of cage size and environmental enrichment on behavioral and physiological responses of rhesus macaques to the stress of daily events. In: Novak MA, Petto AJ (eds) Through the looking glass. American Psychological Association, Washington, DC, pp 160–179

Lukas KE, Ross SR (2014) Naturalistic exhibits may be more effective than traditional exhibits at improving zoo-visitor attitudes toward African apes. Anthrozoös 27(3):435–455. https://doi.org/10.2752/175303714X14023922797904

Lutz CK, Farrow RA (1996) Foraging device for singly housed longtailed macaques does not reduce stereotypies. Contemp Top Lab Anim Sci 35(3):75–78

Lutz CK, Novak MA (2005) Environmental enrichment for nonhuman primates: theory and application. ILAR J 46(2):178–191. https://doi.org/10.1093/ilar.46.2.178

MacKinnon J, MacKinnon K (1980) The behavior of wild spectral tarsiers. Int J Primatol 1:361–379. https://doi.org/10.1007/BF02692280

Maple TL, Finlay TW (1989) Applied primatology in the modern zoo. Zoo Biol 8(S1):101–116. https://doi.org/10.1002/zoo.1430080511

Markowitz H (1982) Behavioral enrichment in the zoo. Van Nostrand Reinhold, New York

Martin CF, Biro D, Matsuzawa T (2014) The arena system: a novel shared touch-panel apparatus for the study of chimpanzee social interaction and cognition. Behav Res Methods 46:611–618. https://doi.org/10.3758/s13428-013-0418-y

Martin CF, Biro D, Matsuzawa T (2017) Chimpanzees spontaneously take turns in a shared serial ordering task. Sci Rep 7:14307. https://doi.org/10.1038/s41598-017-14393-x

Mason GJ, McFarland D, Garner JP (1998) A demanding task: assessing the needs of captive animals. Anim Behav 55:1071–1075

Mason G, Clubb R, Latham N, Vickery S (2007) Why and how should we use environmental enrichment to tackle stereotypic behaviour? Appl Anim Behav Sci 102(3–4):163–188. https://doi.org/10.1016/j.applanim.2006.05.041

McPhee ME, Foster JS, Sevenich M, Saunders CD (1998) Public perceptions of behavioral enrichment: assumptions gone awry. Zoo Biol 17(6):525–534. https://doi.org/10.1002/(SICI)1098-2361(1998)17:6<525::AID-ZOO6>3.0.CO;2-W

Medical Research Centre (2016) Observing observers at the MRC Centre for Macaques. https://youtu.be/kWI3bMmzDkY?list=PLSus4fp7v7sSw-yEyjtn9OwlJk4wu0-HF. Accessed 28 June 2021

Meehan CL, Mench JA (2007) The challenge of challenge: can problem solving opportunities enhance animal welfare? Appl Anim Behav Sci 102(3–4):246–261. https://doi.org/10.1016/j.applanim.2006.05.031

Mehrkam LR, Dorey NR (2015) Preference assessments in the zoo: keeper and staff predictions of enrichment preferences across species. Zoo Biol 34(5):418–430. https://doi.org/10.1002/zoo.21227

Melfi V (2013) Is training zoo animals enriching? Appl Anim Behav Sci 147(3–4):299–305. https://doi.org/10.1016/j.applanim.2013.04.011

Melfi VA, McCormick W, Gibbs A (2004) A preliminary assessment of how zoo visitors evaluate animal welfare according to enclosure style and the expression of behavior. Anthrozoös 17(2):98–108

Mellen J, Sevenich MacPhee M (2001) Philosophy of environmental enrichment: past, present, and future. Zoo Biol 20(3):211–226. https://doi.org/10.1002/zoo.1021

Mendes N, Huber L (2004) Object permanence in common marmosets (*Callithrix jacchus*). J Comp Psychol 118(1):103–112. https://doi.org/10.1037/0735-7036.118.1.103

Mendoza S (1978) The physiological response to group formation in adult male squirrel monkeys. Psychoneuroendocrinology 3(3–4):221–229. https://doi.org/10.1016/0306-4530(78)90012-4

Milton K (2000) Quo vadis? Tactics of food search and group movements in primates and other animals. In: Boinski S, Garber PA (eds) On the move: how and why animals travel in groups. University of Chicago Press, Chicago, pp 375–417

Morgan KN, Line SW, Markowitz H (1998) Zoos, enrichment, and the skeptical observer: the practical value of assessment. In: Shepherdson DJ, Mellen JD, Hutchins M (eds) Second nature: environmental enrichment for captive animals. Smithsonian Institution Press, Washington, DC, pp 153–171

Morimura N (2007) Note on effects of a daylong feeding enrichment program for chimpanzees (*Pan troglodytes*). Appl Anim Behav Sci 106(1–3):178–183. https://doi.org/10.1016/j.applanim.2006.06.015

Morimura N, Matsuzawa T (2001) Memory of movies by chimpanzees (*Pan troglodytes*). J Comp Psychol 115(2):152–158. https://doi.org/10.1037/0735-7036.115.2.152

Morris CL, Grandin T, Irlbeck NA (2011) Companion Animals Symposium: environmental enrichment for companion, exotic, and laboratory animals. J Anim Sci 89(12):4227–4238. https://doi.org/10.2527/jas.2010-3722

National Centre for the Replacement, Refinement and Reduction of Animals in Research (n.d.) The Macaque Website. https://macaques.nc3rs.org.uk/. Accessed 28 June 2021

National Research Council (2011) Guide for the care and use of laboratory animals, 8th edn. The National Academies Press, Washington, DC

Newberry RC (1995) Environmental enrichment: increasing the biological relevance of captive environments. Appl Anim Behav Sci 44(2–4):229–243. https://doi.org/10.1016/0168-1591(95)00616-Z

Niemitz C (1979) Outline of the behavior of *Tarsius bancanus*. In: Doyle GA, Martin RD (eds) The study of prosimian behavior. Academic Press, New York, pp 631–660

Novak MA, Musante A, Munroe H et al (1993) Old, socially housed rhesus monkeys manipulate objects. Zoo Biol 12(3):285–298. https://doi.org/10.1002/zoo.1430120306

Ogura T, Matsuzawa T (2012) Video preference assessment and behavioral management of single-caged Japanese macaques (*Macaca fuscata*) by movie presentation. J Appl Anim Welf Sci 15(2):101–112. https://doi.org/10.1080/10888705.2012.624887

Paquette D, Prescott J (1988) Use of novel objects to enhance environments of captive chimpanzees. Zoo Biol 7(1):15–23. https://doi.org/10.1002/zoo.1430070103

Parks KA, Novak MA (1993) Observations of increased activity and tool use in captive rhesus monkeys exposed to troughs of water. Am J Primatol 29(1):13–25. https://doi.org/10.1002/ajp.1350290103

Platt DM, Novak MA (1997) Video stimulation as enrichment for captive rhesus monkeys (*Macaca mulatta*). Appl Anim Behav Sci 52(1–2):139–155. https://doi.org/10.1016/S0168-1591(96)01093-3

Prescott MJ, Buchanan-Smith HM (2003) Training nonhuman primates using positive reinforcement techniques: guest editors' introduction. J Appl Anim Welf Sci 6(3):157–161. https://doi.org/10.1207/S15327604JAWS0603_01

Prescott MJ, Buchanan-Smith HM (2007) Training laboratory-housed non-human primates, part I: a UK survey. Anim Welf 16:288–303

Prescott MJ, Bowell VA, Buchanan-Smith HM (2005) Training laboratory-housed non-human primates, part 2: resources for developing and implementing training programmes. Anim Technol Welf 4:133–148

Razal CB, Miller LJ (2019) Examining the impact of naturalistic and unnaturalistic environmental enrichment on visitor perception of naturalness, animal welfare, and conservation. Anthrozoös 32(1):141–153. https://doi.org/10.1080/08927936.2019.1550289

Reinhardt V (1989) Behavioral responses of unrelated adult male rhesus monkeys familiarized and paired for the purpose of environmental enrichment. Am J Primatol 17(3):243–248. https://doi.org/10.1002/ajp.1350170305

Reinhardt V (1990) Social enrichment for laboratory primates: a critical review. Lab Primate Newsl 29(3):7–11

Reinhardt V, Reinhardt A (2008) Environmental enrichment and refinement for nonhuman primates kept in research laboratories: a photographic documentation and literature review, 3rd edn. Animal Welfare Institute, Washington, DC

Reinhardt V, Roberts A (1997) Effective feeding enrichment for non-human primates: a brief review. Anim Welf 6:265–272

Richardson AS, Lambeth SP, Schapiro SJ (2006) Control over the auditory environment: a study of music preference in captive chimpanzees (*Pan troglodytes*). Int J Primatol 27:423

Robbins L, Margulis SW (2014) The effects of auditory enrichment on gorillas. Zoo Biol 33(3):197–203. https://doi.org/10.1002/zoo.21127

Ross SR, Melber LM, Gillespie KL, Lukas KE (2012) The impact of a modern, naturalistic exhibit design on visitor behavior: a cross-facility comparison. Visit Stud 15(1):3–15. https://doi.org/10.1080/10645578.2012.660838

Sackett GP, Tripp R, Grady S (1982) Rhesus monkeys reared in isolation with added social, nonsocial and electrical brain stimulation. Ann Ist Super Sanita 18(2):203–213

Schapiro SJ, Lambeth SP (2007) Control, choice, and assessments of the value of behavioral management to nonhuman primates in captivity. J Appl Anim Welf Sci 10(1):39–47. https://doi.org/10.1080/10888700701277345

Shepherdson DJ, Mellen JD, Hutchins M (1998) Tracing the path of environmental enrichment in zoos. In: Shepherdson DJ, Mellen JD, Hutchins M (eds) Second nature: environmental enrichment for captive animals environmental enrichment for captive animals. Smithsonian Institution Press, Washington, DC, pp 1–12

Snowdon CT, Teie D (2010) Affective responses in tamarins elicited by species-specific music. Biol Lett 6(1):30–32. https://doi.org/10.1098/rsbl.2009.0593

Suchak M, Eppley TM, Campbell MW et al (2016) How chimpanzees cooperate in a competitive world. Proc Natl Acad Sci U S A 113(36):10215–10220. https://doi.org/10.1073/pnas.1611826113

Swaisgood R, Shepherdson D (2006) Environmental enrichment as a strategy for mitigating stereotypies in zoo animals: a literature review and meta-analysis. In: Mason G, Rushen J (eds) Stereotypic animal behaviour: fundamentals and applications to welfare. CABI, Wallingford, pp 256–285

The Shape of Enrichment (n.d.) Safety database. https://theshapeofenrichmentinc.wildapricot.org/. Accessed 28 June 2021

Ventura R, Buchanan-Smith HM (2003) Physical environmental effects on infant care and development in captive *Callithrix jacchus*. Int J Primatol 24:399–423. https://doi.org/10.1023/A:1023061502876

Videan EN, Fritz J, Schwandt ML et al (2005) Controllability in environmental enrichment for captive chimpanzees (*Pan troglodytes*). J Appl Anim Welf Sci 8(2):117–130. https://doi.org/10.1207/s15327604jaws0802_4

Videan EN, Fritz J, Murphy J (2007) Development of guidelines for assessing obesity in captive chimpanzees (*Pan troglodytes*). Zoo Biol 26(2):93–104. https://doi.org/10.1002/zoo.20122

Waitt CD, Bushmitz M, Honess PE (2010) Designing environments for aged primates. Lab Primate Newsl 49(3):5–9

Wallace EK, Altschul DM, Körfer K et al (2017) Is music enriching for group-housed captive chimpanzees (*Pan troglodytes*)? PLoS ONE 12(3):e0172672. https://doi.org/10.1371/journal.pone.0172672

Washburn DA, Rumbaugh DM (1992) Testing primates with joystick-based automated apparatus: lessons from the language research center's computerized test system. Behav Res Methods Instrum Comput 24(2):157–164. https://doi.org/10.3758/BF03203490

Webber S (2018) Kinecting with orang-utans: Microsoft Centre for social natural user interfaces at the University of Melbourne. https://socialnui.unimelb.edu.au/research/zoos/. Accessed 19 Jan 2018

Webber S, Carter M, Smith W, Vetere F (2017) Interactive technology and human–animal encounters at the zoo. Int J Hum Comput Stud 98:150–168. https://doi.org/10.1016/j.ijhcs.2016.05.003

Wells DL (2009) Sensory stimulation as environmental enrichment for captive animals: a review. Appl Anim Behav Sci 118(1–2):1–11. https://doi.org/10.1016/j.applanim.2009.01.002

Wells DL, McDonald CL, Ringland JE (2008) Color preferences in gorillas (*Gorilla gorilla gorilla*) and chimpanzees (*Pan troglodytes*). J Comp Psychol 122(2):213–219. https://doi.org/10.1037/0735-7036.122.2.213

Westlund K (2014) Training is enrichment—and beyond. Appl Anim Behav Sci 152:1–6. https://doi.org/10.1016/j.applanim.2013.12.009

Wolfensohn S (2004) Social housing of large primates: methodology for refinement of husbandry and management. Altern Lab Anim 32(Suppl 1A):149–151. https://doi.org/10.1177/026119290403201s24

Young RJ (2003) Environmental enrichment for captive animals. Blackwell Science, Malden, MA

Young C, Majolo B, Heistermann M et al (2014) Responses to social and environmental stress are attenuated by strong male bonds in wild macaques. Proc Natl Acad Sci U S A 111(51):18195–18200. https://doi.org/10.1073/pnas.1411450111

Challenging Cognitive Enrichment: Examples from Caring for the Chimpanzees in the Kumamoto Sanctuary, Japan and Bossou, Guinea

Naruki Morimura, Satoshi Hirata, and Tetsuro Matsuzawa

Abstract

We must ensure the welfare of captive chimpanzees. One way to do so is by building environments that enable chimpanzees to express evolved cognitive abilities and skills. These environments must therefore include "cognitive enrichment" that resemble daily challenges that chimpanzees in the wild must meet if they are to survive and reproduce. In the Kumamoto Sanctuary of Kyoto University, Japan, we introduced fission–fusion emulation, a dynamic group management system in which the spatiotemporal cohesion within a group in terms of space, group size, and group membership was changed by human caretakers. Kumamoto Sanctuary also instituted a new experimental system that balances the needs of human experimenters and the chimpanzee participants by ensuring that the chimpanzees do not experience stress during experimental procedures. Moreover, because conservation has become an increasingly important consideration, the cognitive challenges at Kumamoto Sanctuary are designed to allow captive chimpanzees to engage in decision making on a daily basis. Conservation activities also need to consider the needs of local people and their chimpanzee neighbors. A plantation project in a fragmented habitat with anthropogenic activity can be regarded as environmental enrichment on a large scale, in chimpanzees' natural habitat. These enrichment designs, one in Kumamoto Sanctuary and one in chimpanzees' natural habitat, not only maximize chimpanzee welfare, but also enable chimpanzees to express their natural behaviors to the fullest extent possible.

N. Morimura (✉) · S. Hirata
Kumamoto Sanctuary of Wildlife Research Center, Kyoto University, Kumamoto, Japan
e-mail: morimura.naruki.5a@kyoto-u.ac.jp

T. Matsuzawa
Division of the Humanities and Social Sciences, California Institute of Technology, Pasadena, CA, USA

Keywords

Cognitive enrichment · Conservation · Chimpanzee · Decision making · Agency · Behavioral freedom

1   Introduction

Distinct behavioral and cognitive characteristics in primates are considered products of cumulative evolutionary changes over time (e.g., de Waal and Ferrari 2009). As part of their daily lives, primates are exposed to situations that require problem-solving via cognitive abilities, such as physical or spatial cognition, memory, planning, making and using tools, engaging in cooperative or competitive social interactions, and so forth (e.g., Menzel 1973; Goodall 1986; Schmelz and Call 2016). This suggests that welfare considerations must include and focus on providing environments with characteristics that allow primate species to fully express cognitive competencies acquired through the evolutionary process (Hughes and Duncan 1988; Morimura 2006; Brydges and Braithwaite 2008). Thus, the daily challenges that animals face, ultimately for survival and reproduction, are an important component of the environmental enrichment of captive primates; this is known as cognitive enrichment (Matsuzawa 2006; Morimura 2006; Clark 2011).

When considering what the cognitive challenges in primates are, debates about the evolution of (human) intelligence can provide many insights into their nature. The leading hypothesis, which is related to the evolutionary increase in brain size, suggests that dietary and/or social complexities were major selective pressures in the evolution of primate cognition (e.g., MacLean et al. 2014). Relative brain size in primates, for example, is greater in frugivores than in folivores (Clutton-Brock and Harvey 1980). This theory posits that dietary complexity, such as the variety and spatiotemporal distribution of fruits, was the primary driver of primate cognitive evolution (Zuberbühler and Janmaat 2010; Barton 2012). The neocortex ratio, another index of brain size used as a proxy for cognition, was also found to be positively correlated with species' typical group size (Barton 1996; Dunbar 1998). This has led to social intelligence theory, which proposes that social complexity (e.g., increased social group size) was the major selective pressure in cognitive evolution (Barton 2006; Shultz and Dunbar 2007).

Notably, there has been no evidence suggesting that either hypotheses alone is enough to fully explain cognitive evolution. In fact, both ecological and social factors have been associated with different brain volume indices (Dunbar and Shultz 2007). An example of this is that both the percentage of fruit in the diet and social group size correlate positively with the neocortex ratio in primates (Barton 1996). Such studies suggest that the challenges that captive primates are asked to face in an environmental enrichment program can include foraging, exploring their living space, object manipulation, gathering information about the environment using different modalities, and in their social life with conspecifics and/or heterospecifics.

Environmental enrichment has generally been categorized into five types: food-based, physical, sensory, social, and cognitive (e.g., Brent 2001; Hosey et al. 2009; see also Kemp 2022). As mentioned earlier, animals' cognitive skills could be challenged by the demands required in each context of food-based, physical, sensory, and social enrichment programs. For example, modifications in group size (social enrichment) can be used to challenge evolved cognitive skills that deal with increasing social complexity (e.g., Whiten and Erdal 2012).

In this chapter, we first describe the recent movement to rescue retired laboratory chimpanzees (*Pan troglodytes*) in Japan. The cessation of invasive biomedical research using chimpanzees at a pharmaceutical company created a surplus of 79 laboratory chimpanzees. Of these chimpanzees, 22 were isolated in a laboratory when the sanctuary was launched in 2007. Social enrichment has been one of the major strategies to enhance the recovery of retired laboratory chimpanzees. However, there are major differences between the social life of captive and wild chimpanzees, especially in terms of spatiotemporal social dynamics. Therefore, we will spend some time discussing how cognitive enrichment can be used to tackle the limitations of captive environments to benefit the well-being of chimpanzees.

2   Kumamoto Sanctuary: The First Sanctuary for Retired Laboratory Chimpanzees in Japan

As of June 1st, 2019, according to an open-source database (the Great Ape Information Network or GAIN 2009), there are 305 captive chimpanzees living in Japan. Next to the USA, which has 1478 chimpanzees listed in the ChimpCare website (2009), Japan has the second largest population of captive chimpanzees in the world. Between 1925 and the end of 2013, 501 chimpanzees were relocated from Africa, and from zoos in the USA and Europe, to Japan, and up until the 1980s most of these chimpanzees were wild caught (Watanuki et al. 2014). According to studbook databases, over 80% of the chimpanzees imported into the USA (Ely et al. 2005), Western Europe (Carlsen 2009), and Japan (Morimura et al. 2011a, b) are of the West African subspecies (*P. t. verus*). These figures suggest that importing chimpanzees has had a large negative impact on wild chimpanzee populations, and especially those in West Africa. All the chimpanzee facilities, including research institutes, zoos, and sanctuaries, therefore, must take responsibility for the future of chimpanzees, and especially Western chimpanzees, both in captivity and in the wild.

The lack of laws or guidelines prohibiting the use of laboratory chimpanzees in severely invasive biomedical research in Japan allowed the use of captive chimpanzees living in Japan in behavioral, genetic, and biomedical experiments until the late 2000s. It was only in 2006, after protests by and denunciation from researchers belonging to Support for African/Asian Great Apes (SAGA 1998), that invasive laboratory studies on chimpanzees in Japan ended. Indeed, the last biomedical institute that used chimpanzees, owned by Sanwa Kagaku Kenkyusyo Co. Ltd., was transformed into the largest chimpanzee sanctuary in Japan, with the donation of 79 retired chimpanzees, facilities, and skilled staff. The sanctuary, renamed the

Fig. 1 Panoramic view of Kumamoto Sanctuary for chimpanzees and bonobos. Note: The first and second buildings (BLDG1 and 2, respectively) are for all-male groups. The fifth building (BLDG5) are for mixed-sex groups of chimpanzees and bonobos. The fourth building (BLDG4) is an emergency space. The third building is not shown as it was demolished in 2012

Kumamoto Sanctuary (KS) (Fig. 1), has been run by Kyoto University since 2011 following a transitionary period that began in 2007 (Morimura et al. 2011a). By 2013, there were no more chimpanzees in biomedical institutes throughout Japan as the last three chimpanzees had been relocated to KS.

Of the 305 chimpanzees in Japan, 53 currently live in KS. After the sanctuary's launch, the care of chimpanzees changed drastically. These changes came about largely because of the introduction of well-organized environmental enrichment programs. The annual variety of food items, for instance, increased from approximately 10 to 71 items, and a weekly feeding enrichment program was introduced (Fig. 2a). Caretakers also celebrate the birthday of each chimpanzee and Japanese holidays by providing the chimpanzees with special meals decorated with fresh fruits and vegetables (Fig. 2b).

Social enrichment was also a major target during the transition phase of the sanctuary in 2007 as 22 of the 79 retired laboratory chimpanzees were previously isolated in indoor rooms. Wild chimpanzees are complex and flexible social beings that form parties based on fission-fusion dynamics (Nishida and Hiraiwa-Hasegawa 1987; Aureli et al. 2008), which is defined by variation in spatiotemporal cohesion of individual membership in a group over time. By May 2008, all the chimpanzees,

Fig. 2 (**a**) The weekly schedule of the enrichment program consisted of paper-wrapped feeder (Mon.), fire horse puzzle feeder (Tue.), fresh green (Wed.), cardboard box gift (Thu.), honey-juice dipping (Fri.). (**b**) The 11-year-old birthday party for Hatsuka

except for one blind female (Hirata et al. 2017), had been returned to some type of group, including all-male, single-male and multi-female, and multi-male/multi-female groups.

3 Behavior that Is Typical in Captivity Can Be Atypical in the Wild

Matsuzawa (2006) pointed out that to maximize the welfare of captive chimpanzees, laboratory work should be founded on efforts to provide environmental enrichment that allows chimpanzees to express their full behavioral repertoires within time budgets that are close to those observed in wild chimpanzees. Thus, when developing environmental enrichments, understanding the differences between behaviors that are typical in captivity but atypical in the wild is essential (for review of behavioral measures of primate welfare, see Lutz and Baker 2022).

As with many other nonhuman primates, chimpanzees form enduring social bonds (Aureli et al. 2008). In addition to the aforementioned fission-fusion dynamics that characterize their social groups (Aureli et al. 2008), wild chimpanzees travel daily in woodlands, which are anywhere from 6 to 70 km^2 in size (e.g., Humle 2011a; Samson and Hunt 2012). Chimpanzees have a patrilineal society in which females disperse into neighboring communities shortly after reaching sexual maturity (Pusey et al. 1997). In contrast, most captive chimpanzees are characterized by a relatively "static" life in a social group that consists of the same group members, in the same enclosure, for numerous years or, in the worst case, until the end of their lives. Since 2008, in an effort to reproduce the complex social life of captive chimpanzees, KS introduced fission-fusion emulation (FFE), a group management system in which the spatiotemporal cohesion within a group in terms of space, group size, and group membership were changed by human caretakers (Fig. 3).

Prior to the transitional running of the sanctuary in 2007, most males had been separated from females as part of a moratorium on reproduction that had been ongoing since 2000. Lethal aggressive interactions are known among wild male chimpanzees (Wilson et al. 2014a). Consequently, multi-male group formations have a higher risk of severe fights than multi-female groups. As such, most males were still singly-housed in the sanctuary in 2007 whereas the isolated females joined new social groups. We needed to address the problem of how to introduce these males into social groups.

In the wild, adult male chimpanzees have been known to form equitable, long-lasting social bonds (Mitani 2009). Moreover, although all-male groups of chimpanzees had not been observed in the wild, males have been known to occasionally form all-male parties through fission-fusion dynamics (e.g., Lanjouw 2002; Hockings 2011). Therefore, we expected that forming all-male groups through FFE would lead to the formation of male bonds and prevent the escalation of aggressive interactions. In fact, it was the only solution that might enable the chimpanzees to return to their social lives.

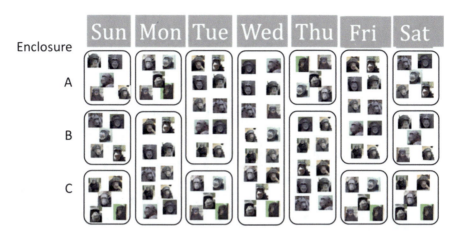

Fig. 3 Schema of the fission-fusion emulation for 15 males at BLDG1 of KS. (**a–c**) indicate the place of enclosure. ab, bc, and abc indicate that two or three enclosures were connected by opening separation doors

Our attempt to form male bonds proceeded as follows. Eight isolated males (9–34 years old in 2008) at the second KS building (BLDG2) were the first target for the all-male group formation. This allowed us to not only end their social isolation but also to expand the amount of space available for the new social group as the group would be able to use the rooms in which the isolated males were kept. The dynamic group management by FFE, moreover, promoted variation in the size and membership of each group by changing the connection pattern of indoor rooms and outdoor cages (Fig. 4a). The isolated males had lived for several years in these rooms with visual or tactile contact with their neighbors through the iron bars. This previous experience allowed us to form an all-male group of familiar unrelated individuals in short introductory steps by connecting the isolated cages one at a time. Another all-male group of six individuals was formed following the same procedure. All individuals at BLDG2 returned to their social lives by forming two all-male groups of eight and six individuals (Fig. 4b).

There were fifteen adult males (16–41 years old in 2011) at the first KS building (BLDG1) consisting of thirteen familiar individuals and two unfamiliar individuals who were relocated from a local zoo in 2008. Figure 5 shows the group's history with FFE. The largest all-male group in the sanctuary was first formed by individuals from three of the existing all-male groups (Teramoto et al. 2007). This introduction was conducted based on individual familiarity among males. As a first step, two males that were familiar with each other were paired; then a third individual who was familiar with one male of the pair and unfamiliar with the other member was introduced into the group; the male who was familiar with both males intervened in the affiliative interaction between the two unfamiliar males. More males were introduced to this group until it reached 15 males. FFE consisted of the daily formation of 2–3 groups of 2–13 individual males and was introduced in June

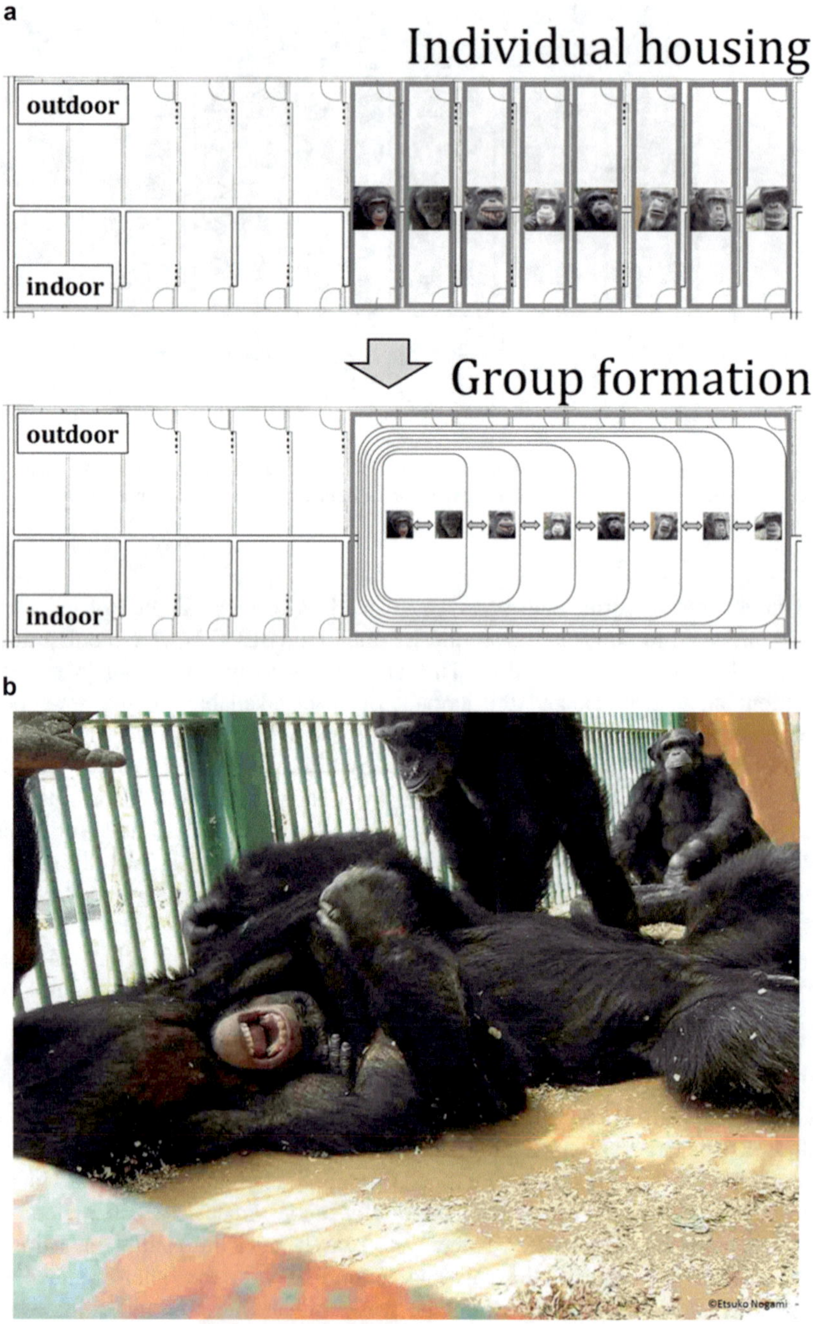

Fig. 4 (**a**) Group formation of eight males at BLDG2 in 2008. These males were housed individually before the formation. During introduction, one male was added at a time, forming a pair, triad, tetrad, and so on. Introduction was from the most subordinate to the most dominant individuals. (**b**) Social play in an all-male group of eight chimpanzees

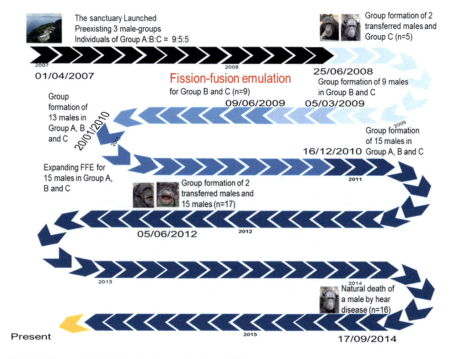

Fig. 5 History of the fission-fusion emulation at KS

2009. The 15 individuals were united for the first time on December 16th, 2010; then, the entire group was formed once every 2 weeks. The weekly schedule of FFE at BLDG1 (Fig. 3) was as follows: three groups of five individuals for 2 days, two groups of five and ten individuals for 4 or 5 days, and one group of 15 individuals once every 2 weeks.

To evaluate the risk of aggressions during the formation of all-male groups, the number of medical care events in the pre-FFE period in 2003–2007 and post-FFE period in 2008–2015 were compared (Fig. 6). Medical care in KS was categorized as either medication or operation under anesthesia. The number of operations did not increase in the post-FFE compared to the pre-FFE period (Mann-Whitney U test: $z = -0.88, P = 0.38$), whereas the medication rate increased (Mann-Whitney U test: $z = -2.64, P < 0.01$). The number of medical care events peaked for 3 years after the start of the formation of the 15-male group in 2008 and then declined to baseline levels. These findings suggest that FFE enabled these chimpanzees to adapt to social living. Therefore, FFE successfully reduced the number of adult males kept individually at KS to zero.

We also evaluated the social relationships among the all-male group of 15 individuals at BLDG1 that emerged under FFE. We were especially interested in the question of whether changes in the spatiotemporal cohesion in FFE strengthened social bonds. The formation of social bonds was analyzed based on 161 h of observational data collected in 2012. The results showed that social behavior

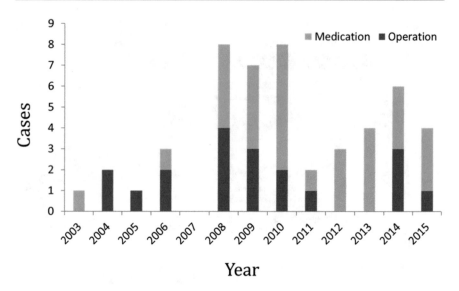

Fig. 6 Temporal distribution of medical care events of the all-male group of 15 chimpanzees

(number ± SEM) was predominantly affiliative (Fig. 7; 19.2 ± 2.1%) and that antagonistic interactions were rarely observed (0.2 ± 0.0%). Wild chimpanzees have been known to display non-random patterns of grooming within their groups (Nishida and Hosaka 1996). Grooming in the all-male group at BLDG1 was observed in 63 of 105 dyads. The analysis of grooming equality (± SEM) (Mitani 2009) within pairs showed that individuals within several pairs of the group groomed each other equally. Moreover, the frequency of social behavior (i.e., the social bond) for each dyad was positively related to grooming equality (Fig. 8; $r_{80} = 0.50$, $t = 5.18$, $P < 0.01$). Primates typically interact repeatedly with conspecifics in relatively stable social groups (Smuts et al. 1987). The repeated social association of male chimpanzees at BLDG1 was positively related to the strength of dyads' social bonds in the all-male group (Fig. 9; $r_{208} = 0.75$, $t = 16.11$, $P < 0.01$).

The group member with whom each individual had the strongest bond was evaluated by determining the partner with whom they had the most frequent social behaviors. The strongest bond per week block (± SEM) was 10.9 ± 0.9. Next, the average number of partners with whom individuals engaged in social behavior, i.e., the number of social bonds, was compared. The number (± SEM) of social bonds was 3.3 ± 0.3. Individuals in the all-male group did not interact with group members simply as a function of increasing member numbers. The value of the strongest bond and the number of social bonds were significantly correlated (Fig. 10; $r_{30} = 0.79$, $t = 6.95$, $P < 0.01$).

In total, these results showed that the male chimpanzees at BLDG1 formed enduring social bonds under FFE. The development of a stronger social bond with a given individual was associated with having stronger bonds with other individuals, too (Fig.10). This finding shows that, even though the quality and

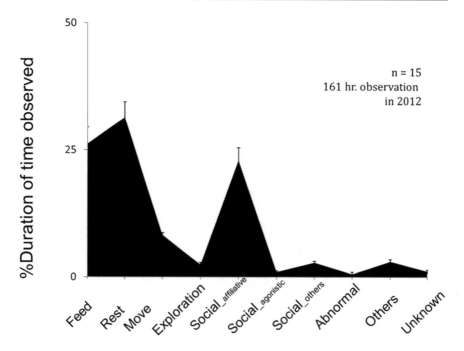

Fig. 7 Time allocation of 10 behavioral repertoires of the all-male group of 15 chimpanzees

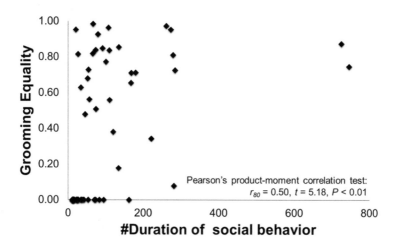

Fig. 8 Correlation between social bond and grooming equality

strength of social bonds differs among dyads, the social bonds that develop between dyads do not conflict with the development of social bonds with other individuals within a group. Thus, captive male chimpanzees at KS could strengthen social bonds in dyads under FFE. Moreover, a follow-up study that focused on grooming, a measure of the strength of social relationships in primates (Fedurek and Dunbar

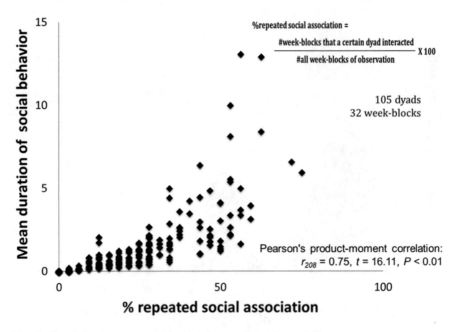

Fig. 9 Correlation between social bond and repeated social association

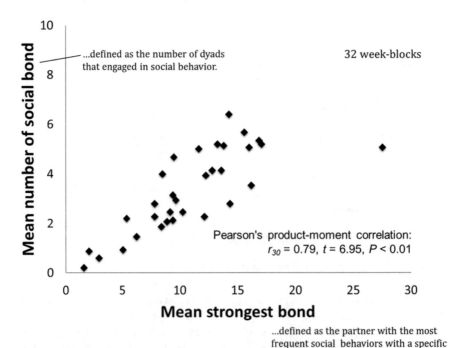

Fig. 10 Correlation between the strongest bond and the number of social bonds

Fig. 11 Grooming in the all-male group of 15 chimpanzees

2009) that was performed 2 years after the introduction of FFE showed that the original 15 males engaged in grooming regardless of their differences in early rearing history (Levé et al. 2016). This follow-up study supported the view that captive male chimpanzees formed stable social bonds under dynamic social cohesion through FFE (Fig. 11).

4 The Agency of Chimpanzees Living in Captivity

Chimpanzees can recognize themselves in a mirror or in live-video footage (e.g., Hirata 2007; Anderson 2015). They are also capable of agency, that is, they can manipulate an object based on their recognition of the object as a representation of an action generated by the self (Kaneko and Tomonaga 2011). Cognitive challenges related to agency have focused on welfare issues, such as animals' ability to control their environment by applying cognitive skills (e.g., Young 2003; Buchanan-Smith and Badihi 2012). Permitting control over aversive events from the physical environment also improves animal welfare (Hanson et al. 1976). In humans, the perception that one can control his or her environment plays an important role in the process through which a person sets challenging goals for themselves (e.g., Bandura 1993). In contrast, exposure to uncontrollable aversive events, which instills learned helplessness, results in a reduction in overall activity and in instrumental responses that would eliminate other aversive events (Overmier and Seligman 1967).

In general, the behavior of wild animals in captivity is known to be affected by limitations in the available resources, which may lead to skews in kinship, age, dominance rank, ranging behavior, predation risk, infectious disease, etc. (Clubb and Mason 2003). Wild chimpanzees, for example, while foraging, travel daily within their home range, because their feeding places are spread out. In contrast, captive chimpanzees have limited feeding places and/or foraging opportunities because of facilities' need to control costs and maintain work efficiency. Given the ecological and evolutionary impact of dietary constraints (e.g., Aiello and Wheeler 1995; Fish and Lockwood 2003; Allen and Kay 2012), limitations in foraging can distort chimpanzees' decision-making processes. Moreover, wild chimpanzees can choose which individuals to approach during fission-fusion dynamics. In contrast, for the captive chimpanzees at KS, human caretakers planned and instituted group formation under FFE (Fig. 12). In a captive environment, the needs and resources of the chimpanzees compete with the needs of the human caretakers, scientists, and other staff. Consequently, these limitations that are imposed by captivity may result in differences in the underlying decision-making processes of wild and captive chimpanzees (Chamove and Anderson 1989).

These differences in decision-making processes between captive and wild chimpanzees can degrade the "cognitive quality of life" (e.g., Kiang et al. 2016) of captive chimpanzees. Wild chimpanzees often express their complex cognitive

DATE	George	Goro	Kanao	Lennon	Mikota	Minato	Mizuo	Naoya	Norihei	Norio	Satoru	Shiro	Takabo	Toon	Shoubou	Musashi
2016/10/1	C	C	C	A	A	A	B	A	B	A	C	B	A	B	B	C
2016/10/2	C	C	C	AB	AB	AB	C	AB	AB	AB	AB	AB	AB	AB	C	C
2016/10/3	C	B	B	B	C	A	C	A	A	B	A	A	C	B	C	C
2016/10/4	BC	BC	A	A	A	BC	BC	A	BC	A	BC	BC	BC	BC	BC	BC
2016/10/5	ABC	ABC	ABC	ABC	ABC	ABC	ABC	ABC	ABC	ABC	ABC	ABC	ABC	ABC	G	G
2016/10/6	C	A	B	A	A	B	C	B	A	C	C	B	A	B	C	C
2016/10/7	C	A	A	C	A	B	C	B	A	A	B	C	C	B	C	C
2016/10/8	C	C	C	AB	AB	AB	C	AB	AB	AB	AB	AB	AB	AB	C	C
2016/10/9	C	C	C	A	A	A	B	A	B	A	C	B	A	B	C	B
2016/10/10	B	C	C	C	B	A	B	A	A	C	A	A	B	C	B	B
2016/10/11	BC	BC	A	A	A	BC	BC	A	BC	A	BC	BC	BC	BC	BC	BC
2016/10/12	C	A	A	A	A	B	B	B	C	A	C	C	C	B	C	C
2016/10/13	C	A	B	A	A	B	C	B	A	C	C	B	A	B	C	C
2016/10/14	C	A	A	C	A	B	C	B	A	A	B	C	C	B	C	C
2016/10/15	C	C	C	AB	AB	AB	C	AB	AB	AB	AB	AB	AB	AB	C	C
2016/10/16	C	C	C	A	A	A	B	A	B	A	C	B	A	B	B	C
2016/10/17	C	B	B	B	C	A	C	A	A	B	A	A	C	B	C	C
2016/10/18	BC	BC	A	A	A	BC	BC	A	BC	A	BC	BC	BC	BC	BC	BC
2016/10/19	ABC	ABC	ABC	ABC	ABC	ABC	ABC	ABC	ABC	ABC	ABC	ABC	ABC	ABC	G	G
2016/10/20	C	A	B	A	A	B	C	B	A	C	C	B	A	B	C	C
2016/10/21	C	A	A	C	A	B	C	B	A	A	B	C	C	B	C	C
2016/10/22	C	C	C	AB	AB	AB	C	AB	AB	AB	AB	AB	AB	AB	C	C
2016/10/23	C	C	C	A	A	A	B	A	B	A	C	B	A	B	C	B
2016/10/24	B	C	C	C	B	A	B	A	A	C	A	A	B	C	B	B

Fig. 12 The latest version of the fission-fusion emulation in the all-male group of 15 chimpanzees as of October 2016

Fig. 13 Jire taking care of the carcass of Jodoamon

abilities related to decision-making processes through their social interactions with conspecifics. When humans play a major role in the social lives of captive chimpanzees by managing their social affairs, captive chimpanzees lose the opportunity to confront cognitive challenges. Furthermore, chimpanzees exhibit enduring mother-infant bonds, which may even persist after the death of an infant. There have been reports from Bossou, Guinea, for example, of mothers who continued to carry and care for their infant's body for a month or more after their death (Biro et al. 2010). Therefore, in captive situations, even though it may seem wise to take a prematurely born chimpanzee infant from their mother for urgent medical purposes, we should be aware that the mother would have been deprived of the cognitive challenges (the infant's death) in a highly species-typical context (Fig. 13).

5 Sharing Resources in Captive Conditions

Cognitive challenges that arise in captive conditions are at least partly attributable to competition between chimpanzees and humans over resources. Some of these challenges resemble those encountered by chimpanzees in the wild. For example, community density in individuals per km^2 in KS and Bossou are 50,000 and 1.5, respectively. The atypical captive condition may result in escalations of lethal aggression (Wilson et al. 2014a). The difficulties in adapting to antagonistic interactions provide a reason for sometimes prioritizing the demands of humans over those of chimpanzees.

After the launch of KS, to attempt to balance the demands for human safety and the chimpanzees' cognitive needs, KS instituted a new experimental system. In 2011 and 2012, by renovating the biomedical facilities in KS, two new huge outdoor cages (W:D:H [m] = 14:26:4 and 26:18:12; Fig. 14) were constructed for the WISH (Web for the Integrated Studies of the Human Mind) project (Matsuzawa 2015). In addition, six bonobos (*Pan paniscus*) from two zoos in the USA were relocated to KS in 2012 (Kano and Hirata 2015). Computerized touch panels for cognitive tasks were installed in these outdoor cages. Previous cognitive tasks using these devices had been conducted in an experimental booth by separating a chimpanzee participant from its group. However, the new system allows chimpanzees or bonobos to voluntarily participate in cognitive experiments in the outdoor cage (Fig. 15). For this to work, we had to devise an experimental setup that could be attractive for participants. Additionally, it is sometimes necessary to separate a chimpanzee from its group, such as when the experiment takes place in an indoor experimental booth (W:D:H [m] = 1.5:3:2), for example, when a human experimenter conducts the task in a face-to-face setup called the "participant-observation procedure" (Fig. 16; Matsuzawa 2009). In "participant-observation" studies, the experimenter needs to communicate with the chimpanzee to keep it motivated. This approach balances the needs of human experimenters and the chimpanzee participants and forms the basis of the research performed at KS. The fact that the KS system prioritizes chimpanzees' needs has enabled us to develop a new comparative method to study emotions in such a way that the chimpanzees do not experience stress during the experimental procedures (Kano et al. 2016).

The new connection cage system between the KS buildings was installed in 2011. Prior to this, the chimpanzees and bonobos at KS lived in three different buildings. The connection cage system allowed the chimpanzees and bonobos to travel between buildings through an iron-mesh bridge, which was 154 m long, 60 cm wide, and 90 cm tall (Fig. 16). Each group of chimpanzees that lived in the three KS buildings had a weekly training routine that encouraged the use of the connection cage. The cage system also included three small outdoor cages (W:D:H [m] = 2:3:3 and 4:8:4) attached to the iron-mesh bridge. These outdoor cages were designed as places where the chimpanzees could rest. The natural vegetation surrounding the cage system also provided an opportunity for chimpanzees to feed on their preferred plants at any time.

6 The Wild Model: Sharing Resources in a Natural Habitat

Captive animals can be categorized into four major groups: laboratory, farm, companion, and zoo animals (Young 2003). Wild species, such as primates, that are housed in laboratories and zoos are unique in that their captive welfare status can be evaluated by comparing their behavior to their wild counterparts. The comparison with wild conspecifics can be helpful especially in cases in which captive and wild endangered species face common threats related to human activity. For example, the reduction in the behavioral freedom of chimpanzees that results from conflicts with

Fig. 14 Huge outdoor cages for the WISH project constructed in 2012 (**a**) and 2013 (**b**)

Fig. 15 Chimpanzees (**a**) and bonobos (**b**) voluntarily participating in a computerized cognitive task

Fig. 16 Panoramic view of the connection cage system

humans is an ongoing problem both in captivity and in the wild. Chimpanzees have been listed in the endangered or critically endangered Red List by the International Union for the Conservation of Nature and Natural Resources (IUCN) since 1996 (Oates 2006; Humle et al. 2016; Kühl et al. 2017). Most wild chimpanzees face not only ecological, but also intense anthropogenic pressures, including logging, agriculture, and the bushmeat trade (Kormos et al. 2003). As a consequence, chimpanzees' habitats have been fragmented, which has led to the isolation of local populations (Humle and Kormos 2011). This suggests that conservation activities need to balance the needs of local people and their chimpanzee neighbors. These considerations can also provide insights into the welfare of captive chimpanzees.

In Bossou, which is located 4 km west of the Nimba Mountains, the only World Natural Heritage site (UNESCO) in Guinea, a group of chimpanzees (*P. t. verus*) has been studied since 1976, when the first long-standing field study of wild chimpanzees in West Africa was established by Yukimaru Sugiyama (Matsuzawa et al. 2011). The Bossou chimpanzees (Fig. 17), which numbered seven in the beginning of 2019, have coexisted over many generations with the local Manon people who live in a nearby village. With the local human population growth in Bossou, crop raiding by chimpanzees has become an increasingly serious source of conflict between the chimpanzees and the villagers (Hockings et al. 2010). Because of the lack of female migration since the start of the field study and a flu-like

Fig. 17 Bossou chimpanzees: Foaf (left), Yo (center), and Fana (right)

Fig. 18 Schematic map of the Green Corridor project

epidemic that occurred in November 2003 (Matsuzawa et al. 2004; Humle 2011b), this isolated population is in danger of extinction. In a conservation effort to save this population, a tree plantation effort in the savanna—the Green Corridor project (Fig. 18)—was started in 1997 to promote individual interchanges between wild chimpanzee groups that lived in Bossou and the Nimba Mountains (Hirata et al. 1998; Matsuzawa and Kourouma 2008; Morimura et al. 2011b). The plantation was established in 2014 (Morimura et al. 2014) and the saplings have since grown and become small forest patches over 5 m high (Fig. 19).

Currently, bushfire from livestock farming in the dry season destroys part of Green Corridor plantation area once every 2–3 years (Fig. 20). Without consensus among local people for chimpanzee conservation, the completion and maintenance of the Green Corridor is difficult. The progress of the Green Corridor represents the

Fig. 19 Plantation trees with height over 5 m

Fig. 20 Local people have been fighting bushfires in the plantation area of Green Corridor almost every year

local population's tolerance toward chimpanzees in Bossou. Reforestation of this area requires that humans agree to prioritize the needs of the Bossou chimpanzees and to abandon agriculture and economic resources at least in the nearby protected area. Ultimately, conservation activities demand that humans relinquish some freedom to protect endangered species.

This philosophy of conservation has been growing within animal welfare. Prioritizing the demands of captive animals, for example, can raise serious concerns in reintroduction programs for endangered species (chimpanzees: Humle et al. 2011; orangutans *Pongo* spp.: Wilson et al. 2014b). A single chimpanzee experiences captive life according to welfare considerations and then experiences natural life under conservation protection in the course of reintroduction projects. To ensure the survival and reproduction of reintroduced chimpanzees, their captive environment must include ways in which the chimpanzees can learn behavioral strategies and cognitive skills for controlling and exploring the environment. Additionally, a plantation in a natural habitat, such as the Green Corridor project, can be regarded as a large-scale version of environmental enrichment in the natural habitat. Reforestation is expected to enhance the natural behavior of wild chimpanzees to the greatest extent possible. The issues common to welfare and conservation advance a new perspective on how to reduce differences between wild and captive environments for chimpanzees and other endangered species. Therefore, welfare and conservation can complement one another to provide more opportunities for fostering the decision-making processes that are necessary if individual animals are to cope with environmental challenges (Meehan and Mench 2007).

7 The Behavioral Freedom of Chimpanzees: Cognitive Enrichment as a Bridge Between Welfare and Conservation

The future contribution of cognitive enrichment is likely to be an in-depth pursuit for expanding the behavioral freedom of wild animals in captivity. Animal welfare is grounded in the notion of freedom. This is demonstrated by the emphasis, early in the animal welfare movement, of the five freedoms (Broom 1988; Hosey et al. 2009) from the Brambell Report (1965). After over 50 years of debate, the recognition of its importance has been growing. This growth is partly fueled by the better understanding of animal minds and the expanding public awareness of welfare and conservation issues (e.g., Matsuzawa et al. 2006; Penning et al. 2009; Arcus Foundation 2015; Barongi et al. 2015). Even after the broad implementation of environmental enrichment in captive animals, it has been rare for captive animals to have much control over their environment in daily life. As long as captive animals are taken care of by humans, captive animals have no choice but to express their behavioral freedom in ways that their human caretakers permit.

The behavioral freedom of captive animals, then, differs from human freedom, which has evolutionary basis and is rooted in anatomical, genetic, physiological, behavioral, and cognitive characteristics that are not granted or dictated to us by other members of society. Human freedom is therefore focused on autonomy

(Beauchamp and Wobber 2014), perception and recognition in spatiotemporal dimensions, self-other representations, and verbal and non-verbal communication. Of nonhuman species, chimpanzees are the closest relatives of humans (The Chimpanzee Sequencing and Analysis Consortium 2005). Therefore, we can acknowledge that chimpanzees share aspects of some or all of these competencies with humans. Cognitive features of behavioral freedom may have evolved not only in humans, but also in chimpanzees and other nonhuman animals (Dennett 2003).

Because of the cognitive competencies of chimpanzees and bonobos described above, KS recognizes behavioral freedom as being independent of the care and management provided by humans. As it is difficult to provide captive chimpanzees and bonobos with the same amount of space that their natural habitats offer, we provide as an alternative the opportunity for these apes to increase the degree of freedom in decision-making. As we are responsible for taking care of these chimpanzees and bonobos throughout their captive lives, we must design cognitive challenges that allow them to, as much as possible, make decisions about group formation in the FFE. To allow captive chimpanzees' commitment to the decision-making process in social management with FFE, humans must prioritize chimpanzees' demands by giving up some control. This care management system can be applicable to a reintroduction program of rescued wild chimpanzees and bonobos. In this way, cognitive enrichment plays a key role in the advance of welfare by connecting it with conservation. The notions of behavioral freedom can be applied to all types of captive animals. Moreover, welfare improvements developed for wild species may lead to the improvement of the welfare of other captive animals, such as farm species.

8   Conclusion

Primate welfare research can enhance and enrich the lives of zoo, companion, laboratory, and farm animals. Cognitive enrichment is an important component of captive animal welfare. It focuses on the degree of behavioral freedom in daily decision-making. During a reintroduction program, a chimpanzee experiences a captive life that is guided by welfare considerations. Cognitive enrichment provides challenges that aid in the development of behavioral strategies and cognitive skills needed for survival and reproduction. After reintroduction, the individual experiences natural life under conservation protection while practicing what it learned through cognitive enrichment. With the increase in the number of endangered species that results from the destruction of natural habitats by humans, captive facilities will have to assume a greater burden to sustain captive and wild populations. Projects bridging captivity and wild habitats, such as reintroduction programs, will have to increasingly enhance the degree of behavioral freedom by reducing the differences between life in captive and wild environments.

Acknowledgements   The research described in this chapter was supported by Grants from MEXT KAKENHI (#24000001 and #16H06283 to TM), JSPS KAKENHI (# 23650135 and #15K12048 to NM; 26245069, 16H06301, 16H06283, and 18H05524 to SH), JSPS-LGP-U04 (PWS) and Great

Ape Information Network to SH and MT, the JSPS Core-to-Core Program, A. Advanced Research Networks to TM, Toyota Environmental Activities Grant Program (G2013-036 and G2015-004 to TM), and The Mitsui & Co. Environment Fund (K18-0098 to NM). Thanks are due to Gen'ichi Idani, Masaki Tomonaga, Shinya Yamamoto, Fumihiro Kano, Yumi Yamanashi, Etsuko Nogami, Kazuyo Nasu, Mori Yusuke, Toshifumi Udono, Hirosuke Uesaka, Migaku Teramoto, and other colleagues in the laboratory. Thanks are also due to Mafory Bangougra, Soumah. A. Gaspar, Yukimaru Sugiyama, Gen Yamakoshi, Gaku Ohashi, Boniface Zogbila, Jiles Doré, Gouanou Zogbila, Henry Camara, Vincent Traoré, Lawé Goweibé, Dagouka Samy, Jacqueline Labila, Bouna Zogbila, Remy Touréand, and other colleagues at the Institut de Recherche Environnementale de Bossou and the Direction Générale de la Recherche Scientifique et Innovation Technologique in Guinea.

References

Aiello LC, Wheeler P (1995) The expensive-tissue hypothesis: the brain and the digestive system in human and primate evolution. Curr Anthropol 36(2):199–221

Allen KL, Kay RF (2012) Dietary quality and encephalization in platyrrhine primates. Proc R Soc B Biol Sci 279(1729):715–721. https://doi.org/10.1098/rspb.2011.1311

Anderson JR (2015) Mirror self-recognition: a review and critique of attempts to promote and engineer self-recognition in primates. Primates 56(4):317–326. https://doi.org/10.1007/s10329-015-0488-9

Arcus Foundation (2015) State of the apes: industrial agriculture and ape conservation. Cambridge University Press, Cambridge, UK

Aureli F, Schaffner CM, Boesch C, Bearder SK, Call J, Chapman CA, Connor R, Di Fiore A, Dunbar RIM, Henzi SP, Holekamp K, Korstjens AH, Layton R, Lee P, Lehmann J, Manson JH, Ramos-Fernandez G, Strier KB, van Schaik CP (2008) Fission-fusion dynamics: new research frameworks. Curr Anthropol 49(4):627–654. https://doi.org/10.1086/586708

Bandura A (1993) Perceived self-efficacy in cognitive development and functioning. Educ Psychol 28(2):117–148. https://doi.org/10.1207/s15326985ep2802_3

Barongi R, Fisken FA, Parker M, Gusset M (eds) (2015) Committing to conservation: the World Zoo and Aquarium Conservation Strategy. WAZA Executive Office, Gland, Switzerland

Barton RA (1996) Neocortex size and behavioural ecology in primates. Proc R Soc B Biol Sci 263 (1367):173–177. https://doi.org/10.1098/rspb.1996.0028

Barton RA (2006) Primate brain evolution: integrating comparative, neurophysiological, and ethological data. Evol Anthropol 15(6):224–236. https://doi.org/10.1002/evan.20105

Barton RA (2012) Embodied cognitive evolution and the cerebellum. Philos Trans R Soc Lond Ser B Biol Sci 367(1599):2097–2107. https://doi.org/10.1098/rstb.2012.0112

Beauchamp TL, Wobber V (2014) Autonomy in chimpanzees. Theor Med Bioeth 35(2):117–132. https://doi.org/10.1007/s11017-014-9287-3

Biro D, Humle T, Koops K, Sousa C, Hayashi M, Matsuzawa T (2010) Chimpanzee mothers at Bossou, Guinea carry the mummified remains of their dead infants. Curr Biol 20(8):R351–R352. https://doi.org/10.1016/j.cub.2010.02.031

Brambell Report (1965) Report of the technical committee to enquire into the welfare of animals kept under intensive livestock husbandry systems. Her Majesty's Stationery Office, London

Brent L (2001) The care and management of captive chimpanzees. American Society of Primatologists, San Antonio

Broom DM (1988) Needs, freedoms and the assessment of welfare. Appl Anim Behav Sci 19 (3–4):384–386. https://doi.org/10.1016/0168-1591(88)90023-8

Brydges NM, Braithwaite VA (2008) Measuring animal welfare: what can cognition contribute? Annu Rev Biomed Sci 10:T91–T103. https://doi.org/10.5016/1806-8774.2008.v10pT91

Buchanan-Smith HM, Badihi I (2012) The psychology of control: effects of control over supplementary light on welfare of marmosets. Appl Anim Behav Sci 137(3–4):166–174. https://doi.org/10.1016/j.applanim.2011.07.002

Carlsen F (2009) European studbook for the chimpanzee (*Pan troglodytes*) Copenhagen Zoo, Frederiksberg, Denmark

Chamove AS, Anderson J (1989) Examining environmental enrichment. In: Segal EF (ed) Housing, care and psychological well-being of captive and laboratory primates. Noyes Publication, New Jersey, pp 183–202

ChimpCARE (2009) Project ChimpCARE. https://www.lpzoo.org/science-project/project-chimpcare/. Accessed 24 June 2021

Clark FE (2011) Great ape cognition and captive care: can cognitive challenges enhance well-being? Appl Anim Behav Sci 135(1–2):1–12. https://doi.org/10.1016/j.applanim.2011.10.010

Clubb R, Mason G (2003) Captivity effects on wide-ranging carnivores. Nature 425:473–474. https://doi.org/10.1038/425473a

Clutton-Brock TH, Harvey PH (1980) Primates, brains and ecology. J Zool 190(3):309–323. https://doi.org/10.1111/j.1469-7998.1980.tb01430.x

de Waal FBM, Ferrari PF (eds) (2009) The primate mind: built to connect with other minds. Harvard University Press, Cambridge, MA

Dennett DC (2003) Freedom evolves. Viking, New York

Dunbar RI (1998) The social brain hypothesis. Evol Anthropol 6(5):178–190. https://doi.org/10.1002/(SICI)1520-6505(1998)6:5<178::AID-EVAN5>3.0.CO;2-8

Dunbar RI, Shultz S (2007) Understanding primate brain evolution. Philos Trans R Soc Lond Ser B Biol Sci 362(1480):649–658. https://doi.org/10.1098/rstb.2006.2001

Ely JJ, Dye B, Frels WI, Fritz J, Gagneux P, Khun HH, Switzer WM, Lee DR (2005) Subspecies composition and founder contribution of the captive U.S. chimpanzee (*Pan troglodytes*) population. Am J Primatol 67(2):223–241. https://doi.org/10.1002/ajp.20179

Fedurek P, Dunbar RI (2009) What does mutual grooming tell us about why chimpanzees groom? Ethology 115(6):566–575. https://doi.org/10.1111/j.1439-0310.2009.01637.x

Fish JL, Lockwood CA (2003) Dietary constraints on encephalization in primates. Am J Phys Anthropol 120(2):171–181. https://doi.org/10.1002/ajpa.10136

Goodall J (1986) The chimpanzees of Gombe: patterns of behaviour. Harvard University Press, Cambridge, MA

Great Ape Information Network (2009) National Bio Resource Project. https://shigen.nig.ac.jp/gain/index.jsp. Accessed 24 June 2021

Hanson JD, Larson ME, Snowdon CT (1976) The effects of control over high intensity noise on plasma cortisol levels in rhesus monkeys. Behav Biol 16(3):333–340. https://doi.org/10.1016/S0091-6773(76)91460-7

Hirata S (2007) A note on the responses of chimpanzees (*Pan troglodytes*) to live self-images on television monitors. Behav Proc 75(1):85–90. https://doi.org/10.1016/j.beproc.2007.01.005

Hirata S, Morimura N, Matsuzawa T (1998) Green passage plan (tree-planting project) and environmental education using documentary videos at Bossou: a progress report. Pan Afr News 5(2):18–20. https://doi.org/10.5134/143371

Hirata S, Hirai H, Nogami E, Morimura N, Udono T (2017) Chimpanzee Down syndrome: a case study of trisomy 22 in a captive chimpanzee. Primates 58(2):267–273. https://doi.org/10.1007/s10329-017-0597-8

Hockings KJ (2011) The crop-raiders of the sacred hill. In: Matsuzawa T, Humle T, Sugiyama Y (eds) The chimpanzees of Bossou and Nimba. Springer, Tokyo, pp 211–220

Hockings KJ, Yamakoshi G, Kabasawa A, Matsuzawa T (2010) Attacks on local persons by chimpanzees in Bossou, Republic of Guinea: long-term perspectives. Am J Primatol 72(10):887–896. https://doi.org/10.1002/ajp.20784

Hosey G, Melfi V, Pankhurst S (2009) Zoo animals: behaviour, management, and welfare. Oxford University Press, Oxford

Hughes BO, Duncan IJH (1988) The notion of ethological 'need', models of motivation and animal welfare. Anim Behav 36(6):1696–1707. https://doi.org/10.1016/S0003-3472(88)80110-6

Humle T (2011a) Location and ecology. In: Matsuzawa T, Humle T, Sugiyama Y (eds) The chimpanzees of Bossou and Nimba. Springer, Tokyo, pp 13–21

Humle T (2011b) The 2003 epidemic of a flu-like respiratory disease at Bossou. In: Matsuzawa T, Humle T, Sugiyama Y (eds) The chimpanzees of Bossou and Nimba. Springer, Tokyo, pp 325–333

Humle T, Kormos R (2011) Chimpanzees in Guinea and in West Africa. In: Matsuzawa T, Humle T, Sugiyama Y (eds) The chimpanzees of Bossou and Nimba. Springer, Tokyo, pp 393–401

Humle T, Colin C, Laurans M, Raballand E (2011) Group release of sanctuary chimpanzees (*Pan troglodytes*) in the Haut Niger National Park, Guinea, West Africa: ranging patterns and lessons so far. Int J Primatol 32:456–473. https://doi.org/10.1007/s10764-010-9482-7

Humle T, Maisels F, Oates JF, Plumptre A, Williamson EA (2016) *Pan troglodytes*. The IUCN Red List of Threatened Species 2016: e.T15933A17964454. Accessed 06 Oct 2016

Kaneko T, Tomonaga M (2011) The perception of self-agency in chimpanzees (*Pan troglodytes*). Proc R Soc B Biol Sci 278(1725):3694–3702. https://doi.org/10.1098/rspb.2011.0611

Kano F, Hirata S (2015) Great apes make anticipatory looks based on long-term memory of single events. Curr Biol 25(19):2513–2517. https://doi.org/10.1016/j.cub.2015.08.004

Kano F, Hirata S, Deschner T, Behringer V, Call J (2016) Nasal temperature drop in response to a playback of conspecific fights in chimpanzees: a thermo-imaging study. Physiol Behav 155:83–94. https://doi.org/10.1016/j.physbeh.2015.11.029

Kemp C (2022) Enrichment. In: Robinson LM, Weiss A (eds) Nonhuman primate welfare: from history, science, and ethics to practice. Springer, Cham, pp 451–488

Kiang A, Weinberg VK, Cheung KHN, Shugard E, Chen J, Quivey JM, Yom SS (2016) Long-term disease-specific and cognitive quality of life after intensity-modulated radiation therapy: a cross-sectional survey of nasopharyngeal carcinoma survivors. Radiat Oncol 11(1):127. https://doi.org/10.1186/s13014-016-0704-9

Kormos R, Boesch C, Bakarr MI, Butynski TM (eds) (2003) West African chimpanzees: status survey and conservation action plan. IUCN, Gland

Kühl HS, Sop T, Williamson EA et al (2017) The critically endangered western chimpanzee declines by 80%. Am J Primatol 79(9):e22681. https://doi.org/10.1002/ajp.22681

Lanjouw A (2002) Behavioural adaptations to water scarcity in Tongo chimpanzees. In: Boesch C, Hohmann G, Marchant LF (eds) Behavioural diversity in chimpanzees and bonobos. Cambridge University Press, Cambridge, pp 52–60

Levé M, Sueur C, Petit O, Matsuzawa T, Hirata S (2016) Social grooming network in captive chimpanzees: does the wild or captive origin of group members affect sociality? Primates 57:73–82. https://doi.org/10.1007/s10329-015-0494-y

Lutz CK, Baker KC (2022) Using behavior to assess primate welfare. In: Robinson LM, Weiss A (eds) Nonhuman primate welfare: from history, science, and ethics to practice. Springer, Cham, pp 171–206

MacLean EL et al (2014) The evolution of self-control. Proc Natl Acad Sci U S A 111(20):E2140–E2148. https://doi.org/10.1073/pnas.1323533111

Matsuzawa T (2006) Sociocognitive development in chimpanzees: a synthesis of laboratory work and fieldwork. In: Matsuzawa T, Tomonaga M, Tanaka M (eds) Cognitive development in chimpanzees. Springer, Tokyo, pp 3–33

Matsuzawa T (2009) The chimpanzee mind: in search of the evolutionary roots of the human mind. Anim Cogn 12:S1–S9. https://doi.org/10.1007/s10071-009-0277-1

Matsuzawa T (2015) Interdisciplinary approach to understanding the human mind: The WISH project and the unit at Kyoto University. Research Activities 5(3):18

Matsuzawa T, Kourouma M (2008) The Green Corridor Project: long-term research and conservation in Bossou, Guinea. In: Wrangham R, Ross E (eds) Science and conservation in African forests: the benefits of long-term research. Cambridge University Press, New York, pp 201–212

Matsuzawa T, Humle T, Koops K, Biro D, Hayashi M, Sousa C, Mizuno Y, Kato A, Yamakoshi G, Ohashi G, Sugiyama Y, Kourouma M (2004) Wild chimpanzees at Bossou-Nimba: deaths through a flu-like epidemic in 2003 and the green-corridor project (Japanese with English abstract). Primate Res 20(1):45–55. https://doi.org/10.2354/psj.20.45

Matsuzawa T, Tomonaga M, Tanaka M (2006) Cognitive development in chimpanzees. Springer, Tokyo

Matsuzawa T, Humle T, Sugiyama Y (eds) (2011) The chimpanzees of Bossou and Nimba. Springer, Tokyo

Meehan CL, Mench JA (2007) The challenge of challenge: can problem solving opportunities enhance animal welfare? Appl Anim Behav Sci 102(3–4):246–261. https://doi.org/10.1016/j.applanim.2006.05.031

Menzel E (1973) Chimpanzee spatial memory organization. Science 182(4115):943–945. https://doi.org/10.1126/science.182.4115.943

Mitani JC (2009) Male chimpanzees form enduring and equitable social bonds. Anim Behav 77(3):633–640. https://doi.org/10.1016/j.anbehav.2008.11.021

Morimura N (2006) Cognitive enrichment in chimpanzees: an approach of welfare entailing an animal's entire resources. In: Matsuzawa T, Tomonaga M, Tanaka M (eds) Cognitive development in chimpanzees. Springer, Tokyo, pp 368–391

Morimura N, Idani G, Matsuzawa T (2011a) The first chimpanzee sanctuary in Japan: an attempt to care for the "surplus" of biomedical research. Am J Primatol 73(3):226–232. https://doi.org/10.1002/ajp.20887

Morimura N, Ohashi G, Gaspard SA, Matsuzawa T (2011b) Bush fire control using arbors in Green Corridor project at Bossou. Pan Africa News 18(1):10–11

Morimura N, Ohashi G, Matsuzawa T (2014) A survey of the savanna vegetation in Bossou, Guinea. Pan Africa News 21(2):22–24

Nishida T, Hiraiwa-Hasegawa M (1987) Chimpanzees and bonobos: Cooperative relationships among males. In: Smuts BB, Cheney DL, Seyfarth RM, Wrangham RW, Struhsaker TT (eds) Primate societies. University of Chicago Press, Chicago, pp 165–177

Nishida T, Hosaka K (1996) Coalition strategies among adult male chimpanzees of the Mahale Mountains, Tanzania. In: McGrew WC, Marchant L, Nishida T (eds) Great ape societies. Cambridge University Press, London, pp 114–134

Oates J (2006) Is the chimpanzee, *Pan troglodytes*, an endangered species? It depends on what "endangered" means. Primates 47:102–112. https://doi.org/10.1007/s10329-005-0149-5

Overmier JB, Seligman MEP (1967) Effects of inescapable shock upon subsequent escape and avoidance responding. J Comp Physiol Psychol 63(1):28–33. https://doi.org/10.1037/h0024166

Penning M, McG-Reid G, Koldewey H, Dick G, Andrews B, Arai K, Garratt P, Gendron S, Lange J, Tanner K, Tonge S, Van den Sande P, Warmolts D, Gibson C (eds) (2009) Turning the tide: a global aquarium strategy for conservation and sustainability. World Association of Zoos and Aquariums, Bern, Switzerland

Pusey AE, Williams JM, Goodall J (1997) The influence of dominance rank on the reproductive success of female chimpanzees. Science 277(5327):828–831. https://doi.org/10.1126/science.277.5327.828

SAGA (1998) Support for African/Asian Great Apes. http://www.saga-jp.org/indexe.html. Accessed 24 June 2021

Samson DR, Hunt KD (2012) A thermodynamic comparison of arboreal and terrestrial sleeping sites for dry-habitat chimpanzees (*Pan troglodytes schweinfurthii*) at the Toro-Semliki Wildlife Reserve, Uganda. Am J Primatol 74(9):811–818. https://doi.org/10.1002/ajp.22031

Schmelz M, Call J (2016) The psychology of primate cooperation and competition: a call for realigning research agendas. Philos Trans R Soc B 371(1686):20150067. https://doi.org/10.1098/rstb.2015.0067

Shultz S, Dunbar RIM (2007) The evolution of the social brain: anthropoid primates contrast with other vertebrates. Proc R Soc B Biol Sci 274(1624):2429–2436. https://doi.org/10.1098/rspb.2007.0693

Smuts BB, Cheney DL, Seyfarth RM, Wrangham RW, Struhsaker TT (1987) Primate societies. University of Chicago Press, Chicago

Teramoto M, Mori Y, Nagano K, Hayasaka I, Kutsukake N, Ikeda K, Hasegawa T (2007) Successful formation and maintenance of all-male groups in captive chimpanzees (Japanese with English abstract). Primate Res 23(1):33–43. https://doi.org/10.2354/psj.23.33

The Chimpanzee Sequencing and Analysis Consortium (2005) Initial sequence of the chimpanzee genome and comparison with the human genome. Nature 437:69–87. https://doi.org/10.1038/nature04072

Watanuki K, Ochiai T, Hirata S, Morimura N, Tomonaga M, Idani G, Matsuzawa T (2014) Review and long-term survey of the status of captive chimpanzees in Japan in 1926–2013 (Japanese with English abstract). Primate Res 30(1):147–156. https://doi.org/10.2354/psj.30.009

Whiten A, Erdal D (2012) The human socio-cognitive niche and its evolutionary origins. Philos Trans R Soc Lond Ser B Biol Sci 367(1599):2119–2129. https://doi.org/10.1098/rstb.2012.0114

Wilson ML, Boesch C, Fruth B, Furuichi T, Gilby IC, Hashimoto C, Hobaiter CL, Hohmann G, Itoh N, Koops K, Lloyd JN, Matsuzawa T, Mitani JC, Mjungu DC, Morgan D, Muller MN, Mundry R, Nakamura M, Pruetz J, Pusey AE, Riedel J, Sanz C, Schel AM, Simmons N, Waller M, Watts DP, White F, Wittig RM, Zuberbühler K, Wrangham RW (2014a) Lethal aggression in *Pan* is better explained by adaptive strategies than human impacts. Nature 513:414–417. https://doi.org/10.1038/nature13727

Wilson HB, Meijaard E, Venter O, Ancrenaz M, Possingham HP (2014b) Conservation strategies for orangutans: reintroduction versus habitat preservation and the benefits of sustainably logged forest. PLoS ONE 9(7):e102174. https://doi.org/10.1371/journal.pone.0102174

Young RJ (2003) Environmental enrichment for captive animals. Blackwell Science, Oxford, UK

Zuberbühler K, Janmaat K (2010) Foraging cognition in nonhuman primates. In: Platt ML, Ghazanfar AA (eds) Primate neuroethology. Oxford University Press, New York, pp 64–83

Training Research Primates

Mollie Bloomsmith, Jaine Perlman, Andrea Franklin, and Allison L. Martin

Abstract

The development of positive reinforcement animal training methods has been an important refinement in the care of primates living in research settings. Using operant conditioning techniques, research primates have been taught to cooperate with veterinary, husbandry, and research procedures. Successfully trained behaviors include moving to locations on cue, body examination behaviors, biological sample collection, cooperating with restraint, and promoting social housing. Classical conditioning techniques have also been used, particularly counter-conditioning, to reduce fear. This chapter describes these training accomplishments and reviews the scientific evidence for the enhancement of animal welfare associated with training. Animal training can improve the quality of the science being conducted with primate subjects and increase the ease with which people can work with the animals. Opportunities to make further progress are discussed, including more incorporation of applied behavior analysis, and thoughts on how best to develop training programs and select trainers. The use of positive reinforcement training is continuing to grow in the primate research community, and its full potential has not yet been achieved.

Keywords

Animal welfare · Applied behavior analysis · Classical conditioning · Operant conditioning · Positive reinforcement training · Training program

M. Bloomsmith (✉) · J. Perlman · A. Franklin
Yerkes National Primate Research Center, Emory University, Atlanta, GA, USA
e-mail: mabloom@emory.edu

A. L. Martin
Yerkes National Primate Research Center, Emory University, Atlanta, GA, USA

Department of Psychological Science, Kennesaw State University, Kennesaw, GA, USA

1 Introduction

Over the last two decades, there have been major changes in how we care for nonhuman primates (hereafter, "primates") living in research facilities, and developments in animal training methods have been an important aspect of this change. There is a rising emphasis on using nonaversive training techniques, such as positive reinforcement, to encourage participation of research primates with husbandry and research procedures. This is in contrast to the use of more traditional handling techniques that often relied on physical restraint or coercion (Laule et al. 2003). The transition to the use of positive reinforcement training (PRT) methods is not yet complete, but those working in the laboratory animal community have begun to recognize the power this approach has to improve the welfare of primates, the quality of care, and the quality of the biomedical research in which the animals participate. This movement is paralleled by similar changes in animal handling methods taking place in zoos and animal sanctuaries. PRT is becoming best practice for handling research primates. Within laboratory animal science, refinements refer to modifications of procedures to minimize animal pain and distress, and to enhance animal welfare (Russell and Burch 1959). PRT is a significant refinement.

2 Regulatory Support for the Use of PRT Methods

There is increased attention to animal training in regulatory and guideline documents in the USA and around the world. Within the United Kingdom, Home Office guidelines, the Medical Research Council, and other groups recommend training primates to cooperate with procedures to reduce adverse impact on them (Prescott and Buchanan-Smith 2007). Similarly, the European Directive on the protection of animals used for scientific purposes emphasizes training programs that encourage animal cooperation with scientific procedures to benefit the animals, the animal care staff, and the quality of the science (Council of Europe 2010). In a 2010 report in the USA, the Office of Laboratory Animal Welfare of the National Institutes of Health supported the tenet that PRT may help in reducing stress and concluded that when safe and feasible, nonhuman primates should be afforded PRT opportunities (OLAW 2010). In 2011, *The Guide for the Care and Use of Laboratory Animals* (National Research Council 2011) further supported the use of PRT techniques. *The Guide* is commonly followed by institutions worldwide as a part of maintaining their AAALAC International accreditation, which includes a large percentage of the facilities housing research primates. *The Guide* recommends using stepwise shaping techniques for voluntary cooperation with procedures, training animals for voluntary movement, using habituation techniques to reduce the stress, and training animals to remain immobile to either eliminate the need for, or reduce the stress associated with, restraint (National Research Council 2011).

3   Basic Training Concepts

Positive reinforcement training uses operant conditioning techniques to teach animals to voluntarily cooperate with procedures that are a routine part of life for research primates. The fundamental principle of operant conditioning is that behavior is determined by its consequences such that what follows the actions of an animal will affect the frequency with which those actions are repeated in the future (Pierce and Cheney 2017). When a consequence increases the probability or rate of a behavior, it is considered a reinforcer. When a consequence decreases the probability or rate of a behavior, it is considered a punisher (Pierce and Cheney 2017). In a typical PRT sequence, the animal is presented with a stimulus (e.g., verbal cue, "open"). When the animal shows the desired behavior (opening his mouth), the behavior is reinforced (by the consequence of being given a grape) and this process increases the chance that the animal will open his mouth again when the word "open" is spoken in the future. Animal training is teaching, and it requires a subject who actively chooses to participate. With an entirely positive approach, if the animal chooses not to participate, she is not coerced into participating and there is no negative consequence to her. The provision of choice is considered an important component in captive animal welfare (Laule and Whittaker 2007; Washburn et al. 1991). However, given that the animal can choose to not participate, it is up to the trainer to arrange contingencies that engage the animal in the training session to ensure success. A good trainer keeps the subject engaged by carefully observing and responding to the subject's behavioral reactions and works through any resistance to the behavior at the animal's pace. Ensuring success requires considering factors such as scheduling training during times the subjects and nearby animals are less likely to be anticipating meals or procedures, and adjusting so the monkeys are less likely to be distracted by nearby activities or social dynamics. Using preferred foods and selecting individual reinforcement preferences are also important to increasing training engagement (see below for further description; Martin et al. 2018).

The two most basic operant conditioning concepts that should be understood in the context of animal training are positive reinforcement and negative reinforcement (see Ramirez 1999 for a more in-depth discussion of training terminology). Both positive reinforcement and negative reinforcement increase the likelihood that the preceding behavior will recur. When using positive reinforcement, the primate receives something (e.g., food) after displaying the desired behavior following a cue (e.g., verbal signal). For example, if the desired behavior is for a monkey to move to the front of her cage, when she does this, she receives a food reward. The timely delivery of food reinforces the behavior of moving to the cage front. When using negative reinforcement training (NRT), the likelihood of a behavior being repeated is increased by removing an aversive stimulus. So, to achieve the same behavior of moving to the cage front, the "squeeze-back" wall of the cage is moved slowly toward her, she moves away from it and toward the front of the cage. When she reaches the cage front, the squeeze back is released. Both techniques result in the same behavior—the animal moving to the front of her cage—but the two methods involve different experiences for the animal. In the PRT example, the monkey is earning something she likes (the food), and in the NRT example, she is avoiding something she does not (the moving squeeze wall). The impact of these training processes may have important welfare implications, and the differing effects of these training procedures have been studied in other species. For example, horses trained with PRT showed more exploratory behavior, less discomfort behaviors, and participated more in training sessions than those trained with NRT (Hendriksen et al. 2011; Innes and McBride 2008). Combining PRT with NRT led to horses that showed behaviors associated with improved welfare compared to when only NRT was used (Warren-Smith and McGreevy 2007). Further studies with primates are needed to assess welfare implications of using positive and negative reinforcement techniques.

Behaviors conditioned through operant conditioning are generally considered voluntary behaviors, but some of the responses we wish to change would be considered reflexive (e.g., fear response). Changing these reflexive behaviors requires classical conditioning in which a behavior is changed through the pairing of stimuli (Pierce and Cheney 2017). Counter-conditioning is a common application of classical conditioning in which a stimulus that elicits one response (e.g., food that elicits excitement) is repeatedly paired with a stimulus that elicits an incompatible response (e.g., a needle that elicits fear) to change the response elicited by the initial stimulus (e.g., to change the fear response to the needle). This concept was initially developed by Wolpe (1958), and the resulting technique of systematic desensitization (pairing relaxation with increasing degrees of a stimulus that elicits fear) is commonly used to modify fear in a variety of settings, including human phobias and posttraumatic stress disorders (Chance 2003; Powell et al. 2017). In applied animal training contexts, the terms "counter-conditioning" and "systematic desensitization" are both used to describe this process. Over time, this pairing process causes the negative event to slowly lose its ability to adversely influence behavior (Chance 2003). Counter-conditioning can also involve reinforcing a competing response (such as sitting calmly) to interfere with a fearful response (Callen and Boyd 1990; Wolpe 1958). For example, in a study by Clay et al. (2009), rhesus macaques were desensitized to several items, including a pole used by animal care staff to check the functioning of the automatic watering device. Using a counter-conditioning technique, a food treat was paired with the insertion of the pole into the cage in stages, and desensitization effectively reduced stress-related behaviors originally associated with the stimulus.

Habituation, in contrast to the active process of desensitization, is a passive process in which simple exposure to a stimulus results in a reduced reflexive response to it (Webster 1994). Strictly, habituation and desensitization cannot be separated because while desensitization may be taking place, habituation is as well. Habituation can be useful in reducing a monkey's initial fearfulness of an object or place by just extending exposure to that item. An example of habituation is helping a primate adjust to a transfer box (a small enclosure used to move a monkey to a different location) by putting it near or inside his home enclosure, so he can inspect it, touch it, and even move in and out of it on his own, thus reducing any initial reluctance. Habituation is not appropriate to use if the animal is extremely fearful.

4 First Steps in Positive Reinforcement Training

Whether PRT is applied to teach the primate to enter a transport box or to shift into an outside play yard, the starting point of most PRT is establishing that the individual will accept hand feeding (Owen and Amory 2011). If the primate is fearful and does not accept food from a human, desensitization can reduce or eliminate this (Clay et al. 2009). Once food is accepted, the next step is creating an association between an initially neutral stimulus (e.g., the sound of a "clicker") and a primary reinforcer (e.g., food). This establishes communication whereby the sound of the clicker (now established as a secondary reinforcer) signals to the animal that food will be delivered. The click becomes a tool for the human to communicate that the behavior that the primate displayed was correct. Next, the trainer uses the training technique of shaping (breaking the behavior down into small steps or "successive approximations"), reinforcing each step toward a behavior goal. The speed at which training occurs depends on trainer experience (Perlman et al. 2012; Young and Cipreste 2004), animal temperament (Coleman et al. 2005; Coleman 2012), the preference for the reinforcement being used (Martin et al. 2018; Clay et al. 2009b), any competing social or environmental distractions, and the types of behaviors being trained.

5 How Positive Reinforcement Training Is Used in the Research Environment

5.1 Training for Movement

Teaching primates to touch a target is a common first step because targets can provide a way to efficiently move and position an animal. For example, a stationary target (e.g., a piece of polyvinyl chloride pipe, clipped to the enclosure) can be used to identify the location the trainer wants the animal to go to. When animals are housed in groups, these targets can be used to keep individuals in particular locations, allowing trainers to work with other animals without interference (Kemp et al. 2017; Schapiro et al. 2001), or to direct individuals to a location for shifting or separation. Alternatively, a mobile target that the trainer holds (e.g., a long, disposable paper stick) is a more dynamic option. By training an animal to move his/her body toward the mobile target, the trainer can get access to different body parts for inspection or treatment (e.g., the upper leg for injection) (Gillis et al. 2012).

Primates are trained for a variety of behaviors related to movement to facilitate their care. Many are regularly shifted within their home enclosures, or between their home enclosures and play cages, outdoor yards, testing cages, holding areas, or transport boxes. PRT is useful in establishing consistent and cooperative movement responses by the animals (Bloomsmith et al. 1998; Veeder et al. 2009; Owen and Amory 2011). The primates move toward the trainer or a target, and their movement is then reinforced. The trainer may prompt with verbal cues ("shift") and visual cues (movement of the arm from one side to the other) which together signal "I want you to move over to the next run/into the box/play cage." Consistent responses by the animals make it easier to offer access to play or activity cages or to outside spaces.

5.2 Positive Reinforcement Training and Social Housing

PRT is used in research settings to promote primate welfare via social housing. Social housing promotes welfare in captive primates (National Research Council 2011), and PRT facilitates being able to conduct research while animals remain socially housed. "Cooperative feeding" is used to reduce aggression during feeding times (Bloomsmith et al. 1994), whereby the dominant animal's behavior of allowing the subordinate to take food is reinforced, and the subordinate's behavior of taking food in the presence of the dominant is reinforced. It can be used in situations from large primate groups housed in outdoor compounds, to pairs of monkeys living in indoor caging and can make it possible to continue to socially house primates who fight when food is delivered but who are otherwise compatible. This technique can also be applied to promote social and affiliative behaviors (Schapiro et al. 2001; Cox 1987; Desmond et al. 1987), and it may be beneficial when introducing unfamiliar primates to one another (Laule and Desmond 1990). A more advanced technique called "collaborative training" involves training animals to work together toward a common goal by rewarding their mutual efforts (e.g., two individuals are rewarded for working together to move an object), which may strengthen social relationships (Laule and Whittaker 2007).

Somewhat ironically, using PRT to train primates to temporarily separate from group members into individual cages (as needed for some research procedures) helps continue social housing. Reliable and fast social separation that can be used for procedures (e.g., biological sample collection) makes it unnecessary to individually house monkeys for these studies (Truelove et al. 2017; Martin et al. 2018). Steps for social separation involve training individuals to move to targets in different cages and to remain at the targets while a cage divider is used to separate the animals. Training may also eliminate the need for social separation during research tasks. For example, Kemp et al. (2017) trained groups of monkeys to sit and stay at targets during research testing procedures. With 61 out of 65 monkeys successfully trained, this demonstrated that the complexities of social groupings can be managed so individuals can participate in research testing while remaining within their social group.

5.3 Training for Biological Sample Collection

Many types of biomedical and behavioral research require collecting biological samples, and primates have been trained to cooperate with a huge number of these procedures. Primates voluntarily participate in conscious blood collection from the saphenous vein in the leg and from the cephalic vein in the arm (Coleman et al. 2008; Laule et al. 1996), and via capillaries in fingers and toes (Reamer et al. 2014). With

PRT, the collection of bodily fluids such as urine (Bloomsmith et al. 2015; Kelley and Bramblett 1981; Laule et al. 1996; Perlman et al. 2010; Stone et al. 1994), semen (VandeVoort et al. 1993; Perlman et al. 2003), saliva (Lutz et al. 2000; Kaplan et al. 2012; Behringer et al. 2014), and vaginal secretions (Bloomsmith et al. 2013) have all been accomplished without anesthesia or restraint. Many of these collections can be conducted while the animals remain in their social groups by stationing group members so that they do not interfere with the procedures (Kemp et al. 2017). It may be difficult to imagine primates repeatedly cooperating with blood collections or injections because of the accompanying discomfort, but with proper preparation, they can.

5.4 Training for Other Research Procedures

Administration of medications and the collection of physiological and biological measurements can also be obtained via PRT training. Voluntary injections for anesthesia, antibiotics, or vaccinations (Lambeth et al. 2006; Schapiro et al. 2005; Videan et al. 2005), and subcutaneous injections (Perlman et al. 2004; Schapiro et al. 2005) can facilitate research. As described above, time must be taken to ensure the positive experience associated with the procedure is more salient to the animal than the discomfort.

Primates can also be trained to accept oral medications by targeting individuals to the front of the enclosure, providing them with preferred fluids and foods in syringes or cups, and slowly replacing those foods or fluids with the medication. Primates can be trained to position and hold in place during ultrasound and other diagnostic examinations (e.g., X-rays), and while using monitoring units (e.g., loop recorders that measure cardiac patterns) (Schapiro et al. 2018). Ideally, the one animal that is being treated can remain in view of his/her social group during the procedures (Drews et al. 2011).

5.5 Training for Restraint

Restraint devices such as the squeeze back of a cage and chair restraint (see next paragraph) are used for research that requires alert monkeys to remain still, during which time the researcher needs access to certain areas of the monkey's body (e.g., leg, back) and the ability to control the animal's position and movement. Since these periods of restraint are stressful (Ruys et al. 2004; Shirasaki et al. 2013), the refinement of using PRT represents a significant shift in research animal management. Because most of the research monkeys who live in cages will probably have the squeeze-back mechanism used at times, it would be prudent to prepare them all for this process. PRT and desensitization can be applied to teach calm behavior while the squeeze back is slowly engaged.

There are generally two types of restraint chairs in use (see Fig. 1; and also McMillan et al. 2017). With the "open chair," the animal is removed from their home

Fig. 1 Examples of "box" (1a) and "open" (1b) restraint chairs for nonhuman primates. Photo credits: Yerkes National Primate Research Center, Emory University

cage with a pole that is attached to a neck collar, guided into the chair, and then secured via the collar. In the "box chair," the animal is free to move its arms and legs, and to turn around (unless additional restraints are added). With the box chair, the animal is typically shifted from the home cage into a transfer cage and then into the fully enclosed device. The monkey's head is positioned, and a neck yoke is secured. Body parts are accessible though doors on the box. The open chair typically requires a "pole and collar" method to transfer a monkey into the chair. McMillan et al. (2014) described an approach to prepare rhesus macaques for this using PRT with minimal amounts of NRT, which required an average of 17 weeks. A different approach by Bliss-Moreau and Moadab (2016) used both PRT and NRT to teach macaques to enter a box chair, and the training was completed in 14 days. Unfortunately, neither of these studies assessed physiological measures of stress, so whether these approaches impact stress at that level is not yet addressed. Nevertheless, the methods show promising changes in behavior.

Technology is being used to facilitate training and to prepare monkeys for studies that use restraint chairs. In one study, rhesus who had access to a button pressing task that provided fluid reinforcement progressed more quickly through the training regimen than those without such experience (Tulip et al. 2017). This simple experience helped predict which subjects were appropriate for the study which required restraint. Ponce et al. (2016) reported on technology utilizing microcontrollers coupled to a water reward system and touch and proximity sensors to train monkeys to enter a box-type primate chair and position their heads. This "smart chair" responded to the animal and trains the monkeys to enter the chair without humans in the room. The monkeys began entering the chair within hours, had completed the training process within 14–21 sessions, requiring fewer human hours than with their previous methods. Although this training relied on water control for the subjects, it would be interesting to adapt it to see if the same level of success could be attained without fluid restriction.

5.6 Moving Away from the Need for Restraint

In some situations, PRT can eliminate the need for restraint. This is responsive to *The Guide* (National Research Council 2011, p. 29) which states that "Alternatives to physical restraint should be considered." For example, when blood samples were obtained from conscious monkeys by using the squeeze-back mechanism as compared to using chair restraint, there were changes in physiological measures (e.g., plasma glucose, cortisol) indicating reduced stress with the squeeze back in the home cage (Shirasaki et al. 2013). In some cases, it may be possible to completely avoid the use of physical restraint. In one laboratory, comprehensive changes were made such that research sample collection, veterinary care, and husbandry were all converted to be trained primarily with PRT (Graham et al. 2012). The people in the laboratory trained monkeys on hand feeding and drinking, shifting into a transport box to obtain weight, and presenting a limb for vascular access port use (for intravenous access for large volume blood collection and intravenous injection), blood collection (heel stick), drug administration (subcutaneous and intramuscular), and basic physical examination. The training consisted mostly of PRT and resulted in almost completely replacing chair restraint with home cage sample collection; the need for chemical restraint was negligible for the trained cohort. (See later section for a discussion of how this enhanced science.)

5.7 Training Cooperation with Veterinary Care

Other behaviors trained with PRT can promote medical management of primates. For example, body examination behaviors where the primate presents limbs, hands, feet, ears, eyes, back, sides, open their mouths, etc., on cue, help facilitate their health care. Health problems such as ear infections and eye trauma can be treated (Magden 2017), rectal or tympanic temperatures can be assessed, and wounds treated. Being able to readily apply topical treatments reduces the need to remove animals from their social groups and reduces the use of anesthesia (Schapiro et al. 2018; Magden 2017). Presentation of body parts can also facilitate pain management via acupuncture treatment (Magden et al. 2013). Training for presentation of the perineum allowed the rapid detection of pinworm infestations (Schapiro et al. 2005), and training for swabbing nasal passages allowed testing for infection (Hanley et al. 2015).

Primates have been trained to cooperate with ultrasound examination needed for the evaluation of prenatal development (Savastano et al. 2003) and with treatment using a nebulizer to manage respiratory problems (Gresswell and Goodman 2011). Some primates have been trained to cooperate with blood pressure measurement (Zoo Atlanta 2011). Finally, weight management in captive primates is a common clinical concern necessitating diets and monitoring. McKinley et al. (2003) used PRT to train marmosets get on a scale for weighing, and other primates can be trained to enter a box containing a scale or to step onto a scale.

6   Is There a Role for Negative Reinforcement?

As should be obvious by now, PRT's value for improving welfare and developing calm and willing subjects for research is well established, but there are practical drawbacks to using only PRT. As we will discuss in a later section, PRT typically requires significantly more time to train animals on a particular behavior than when NRT is used, and developing proficient trainers requires a good deal of learning, as they develop an understanding of animal behavior, of training terminology, training options, and the problem-solving skills that are an essential component of training. For the sake of personnel efficiency, in many cases NRT is used with PRT even as programs are evolving toward making maximal use of PRT (Perlman et al. 2012; Prescott and Buchanan-Smith 2007). Continuing the use of NRT may be needed during this transition period as the laboratory animal community embraces the increasing use of PRT. For now, the reality is that NRT is probably used at almost all research facilities for some behaviors, for certain animals, and/or at some times. We believe it is important to make deliberate decisions about the appropriateness of using NRT within your own facility, with the goal of minimizing its use. Examples of aversive stimuli that are commonly used in NRT are noises (e.g., clapping of hands or banging) to get an animal to move, use of the squeeze back mechanism, a person entering an enclosure to get the monkey to move into a transfer box or tunnel, and threatening a primate with a dart gun to get the animal to present for an injection. Novel stimuli (e.g., a distinctive bucket) are sometimes aversive and therefore can serve as negative reinforcers and have been used in shifting primates (Wergard et al. 2015).

There have been few evaluations comparing the use of PRT and NRT in primates, and we encourage more such studies to be conducted. Wergard et al. (2015) compared rhesus macaques that were trained to move into a section of their enclosure and to tolerate a gate being closed to confine them. They compared two methods—PRT and combined reinforcement training. The second method included presenting novel objects as aversive stimuli (which the monkeys avoided by moving into the desired area) and then positively reinforcing the monkeys' movement. None of the monkeys trained with PRT completed the behavior in the allotted time, while 10 of the 12 monkeys in the combination method were trained. They also found that the combination method had no negative effect on the monkeys' responses toward the trainer. The monkeys trained with PRT exclusively sought contact with their trainer more, while those monkeys who experienced the combination method showed a nonstatistically significant tendency to seek this contact. The monkeys did not show overt signs of anxiety during the combination method training sessions. Another study found that animal temperament may play a role in how primates respond to different training methods. Hannibal et al. (2013) reported that, although rhesus with certain personality measures responded more quickly to the NRT training method, they also showed more human-directed aggression and fear. They concluded that, although training times maybe faster using NRT for monkeys with certain temperaments, due to human safety and animal welfare concerns, it was better to exclusively rely on PRT (Hannibal et al. 2013).

We suggest certain guidelines that can be followed when NRT is being used, that will help to more quickly transition into a PRT model as soon as practical. These are (1) to use PRT to accomplish as much of the task as possible; and when NRT must be introduced, (2) to apply a minimal amount of the aversive stimulus, (3) to remove the aversive stimulus as soon as the needed behavior is offered, and (4) to immediately follow up the desired behavior with PR. The overall goal is to eliminate the use of NR as soon as is feasible. As an example of this, Bliss-Moreau et al. (2013) used PR and NR when training macaques to enter an enclosed restraint chair and found that the NR could be gradually eliminated for all of the subjects. We applied our guidelines for a study requiring the collection of vaginal fluid samples from female rhesus macaques. We needed the monkeys to come to the front of their enclosures, turn around so that their hindquarters faced the trainer, and hold this position while a cotton swab was inserted for sample collection. A date was set as to when reliable performance would be needed so the study could begin. We conducted 40 training sessions with each monkey using only PRT. At this point, some were reliably performing the behavior, but for those who were not, we began using NRT. The trainer gave the cue for the behavior, and if the monkey did not respond, the squeeze-back mechanism was unlocked and slowly moved forward. To ensure we were using the minimal magnitude of the aversive stimulus, as soon as the subject performed the needed behavior, the squeeze back was released and the trainer took her hand off it. Furthermore, when the subject showed the behavior, the trainer immediately delivered a positive reinforcer. In the end, subjects who were not fully cooperating after the first 40 training sessions were cooperating by the time they had an average of 14 more sessions using the combination of PRT and NRT. We have used similar approaches training other behaviors with similar success (McMillan et al. 2014). Incorporating minimal NRT into PRT sessions can help to balance the welfare benefits of PRT with the practical need to conduct research.

7 Training to Improve Animal Welfare

This section provides examples of how PRT improves the welfare of primates. An underlying principle of PRT is that it gives primates control over some events, fosters exploration, enhances performance on cognitive tasks, and may reduce stress responses (Hanson et al. 1976; Mineka et al. 1986; Roma et al. 2006; Washburn et al. 1991). These outcomes are all related to improved well-being (Novak and Suomi 1988).

Training reduces stress experienced by research primates by diminishing the need for physical restraint and/or the use of anesthesia, so the stress that accompanies those procedures is encountered less often (Graham et al. 2012). Another way that

PRT reduces stress is by changing the animal's experience with procedures such that they become less stressful. For example, chimpanzees who voluntarily presented for the injection of an anesthetic agent had significantly reduced physiological measures of stress when compared to chimpanzees who were anesthetized via chemical darts (Lambeth et al. 2006). Cynomolgus macaques given daily PRT sessions showed decreased stress compared to controls by hormonal, hematological, and cardiovascular measures (Koban et al. 2005). A measurement of acute, chronic stress—thymus involution—was significantly reduced in cynomolgus macaques who received primarily PRT for research procedures when compared to subjects who were handled using conventional methods (Graham et al. 2012). And cynomolgus macaques who experienced refined restraint techniques and positive socialization with humans showed less fear to care staff, as well as lower heart rate and blood pressure values with less between-animal variation, thus improving the sensitivity of those measures (described in Prescott and Lidster 2017).

Physical restraint and conventional handling methods restrict an animal's ability to make choices and exert control over their environment. By reducing their use through training, we also provide animals with more opportunities for choice and control. Control over one's environment is considered to be a basic component of welfare, and providing animals with control over some elements of their daily routines can improve behavioral and physiological measures of welfare (Buchanan-Smith and Badihi 2012; Hanson et al. 1976; Mineka et al. 1986). Choices are based on an individual's past experiences with the available options, and the nature of past consequences likely impacts the emotional valence of the choice. In a training context, choices based on a history with PRT-only contingencies are likely to produce the most welfare benefit as animals can choose whether to participant in training or not, and no negative consequences result from the choice not to participate. However, if aversive consequences have been used, an animal's choice might be analogous to us choosing between paying taxes and going to jail. The animal may still choose, but the choice is not without stress. Nonetheless, even when the eventual outcome is the same, animals prefer choice conditions over no-choice conditions (Catania and Sagvolden 1980; Perdue et al. 2014; Suzuki 1999; Voss and Homzie 1970), further indicating the importance of choice in welfare considerations.

An additional mechanism by which PRT improves welfare is by facilitating social housing as described above (Sect. 5.2, Positive Reinforcement Training and Social Housing). PRT can also be used to reduce behavioral problems such as stereotyped behavior (i.e., repetitive, invariant, and seemingly without function, such as pacing; see Lutz and Baker 2022). Training can reduce the frequency of stereotyped behaviors during times when a trainer is working with the primates (Bloomsmith et al. 1997; Morgan et al. 1993). However, behavioral data collected outside of training sessions show conflicting findings; two studies showed no generalized effect so stereotypies were not reduced when the trainer was absent (Baker et al. 2009; Bloomsmith et al. 1997), while three studies reported a generalized decrease in stereotypies (Bourgeois and Brent 2005; Coleman and Maier 2010; Pomerantz and Terkel 2009). When PRT was used to train a behavior incompatible with pacing (e.g., remaining still), stereotyped behavior was reduced outside of training times

(Bourgeois and Brent 2005). PRT may also prevent the development of stereotyped behavior, as has been found with young nursery-reared rhesus monkeys (Brunelli et al. 2009). Combining PRT with other approaches (e.g., social group manipulations, drug therapy) reduced the extreme stereotyped behavior in a male bonobo (Prosen and Bell 2000). While there seems to be promise in using PRT as a treatment and perhaps as a preventative measure, it is important to note that no study found that PRT eliminated stereotyped behaviors.

Self-injurious behavior (self-directed hitting, biting, and so on that can cause physical harm) has also been addressed through PRT. One chimpanzee was treated with a combination of enrichment, socialization, drug therapy, and PRT (Bourgeois et al. 2007), and at the time of publication, he had not self-injured for 2 years. The PRT component involved training him to perform a behavior incompatible with the self-injurious behavior and desensitizing him to arousing stimuli. Prosen and Bell (2000) used a similar combination approach to successfully treat a bonobo who exhibited self-injurious behaviors.

Counter-conditioning techniques are used to reduce fearful behavior. Although fear has not been fully studied in relation to primate welfare, it is used as a measurement of welfare in domestic species. For example, fear of humans is an essential component for the assessment of the "human–animal relationship" in farmed animals (Waiblinger et al. 2006). Fearful behavior in rhesus macaques was reduced using PRT when individuals who showed fear to routine husbandry tasks (e.g., removing food from a cage, spraying water from a hose) were desensitized to these experiences (Clay et al. 2009).

8 Positive Reinforcement Training Helps People Working in Research Settings

Successful animal training programs benefit the people working with animals, too. First, although PRT typically requires more time than training the same behavior with more traditional techniques, it can result in time savings over the long term. Procedures such as moving primates to different areas (e.g., locking them outside while indoor spaces are cleaned), transporting them between enclosures, and collecting biological samples can be completed much more quickly once primates have been trained. Several studies have documented these time savings. McKinley et al. (2003) found a savings of 91% of the time required to weigh marmosets once they were trained. Veeder et al. (2009) trained a group of mangabeys to move from one area of the enclosure to another for cleaning, saving 46 min a day of animal care staff time, and the total training time required was recouped in fewer than 35 days. Research procedures can have similar time savings. Graham et al. (2010) reported that monkeys trained to present their vascular access ports had procedures completed in less than 5 min in their home cage as compared to chemical restraint for the same procedures, which required a minimum of 30 min, or using chair restraint, which required a minimum of 10 min. They suggested that such training is especially relevant for procedures that require repeated accesses during a short period.

Second, job satisfaction can be improved by implementing animal training. Training is a challenging collaboration with animals. Many people working in animal care, veterinary care, and research roles clearly derive personal satisfaction from training the monkeys with whom they work. This is bound to be especially true when the training diminishes stress for the animals and/or for staff. If people are not aware of alternatives to forcing an animal to complete a procedure, they may feel badly for the animal, but also feel as though they have no choice. Once individuals learn to perform the same procedure in a way that is better for the animal, they are relieved and take pride in their accomplishment. Becoming a competent animal trainer is an important skill that people can take with them as their careers develop and may help them qualify them for other jobs. PRT can also improve the job experience for people by reducing monkey aggression toward humans (Minier et al. 2011) and reducing the risk of injury for people and animals (Graham et al. 2010).

The widespread implementation of positive reinforcement training techniques in the laboratory setting also has the potential to improve public perception of primate research. Using PRT affords another opportunity to educate the public about how those working in the primate research community are concerned about the welfare of each individual primate in their care. There is a growing movement for increasing transparency and strengthening public engagement within the biomedical research community (e.g., Animal Research Tomorrow 2020), and being able to demonstrate how primates are taken care of and prepared for their role in research should be a part of this.

9 Positive Reinforcement Training Improves Science

PRT approaches to animal handling can enhance the quality of science by improving animal welfare. Prescott and Lidster (2017) review evidence in support of the notion that "good welfare equals good science," (p. 152) which highlights the relationship between animal welfare and the quality of science. They conclude that "...in all cases every effort should be made to minimize unnecessary harm because animals with compromised welfare have disturbed behavior, physiology and immunology." (p. 152) (also Schapiro and Hau 2022). The reliability and repeatability of scientific findings are increased by furthering the welfare of the animal subjects (Graham et al. 2012; Prescott and Lidster 2017).

As described in the animal welfare section above, many studies have demonstrated how PRT can reduce stress experienced by primates. This stress can be an experimental confound to many research studies, so PRT benefits science by minimizing this confound, which improves data quality and facilitates better interpretation of the data. This may be especially important for studies of immunology, cardiovascular health, or other studies of stress-sensitive measures. Substantial reductions in stress attributed to no longer needing to restrain subjects may decrease variation in the animal model and could reduce the number of animals necessary for meaningful results. Graham et al. (2012) make the case that one benefit of PRT in research is that it is more similar to medical management in humans by eliminating restraint, keeping the environment familiar, dynamic, and social, and providing positive relationships with caregivers.

PRT also increases the ease of performing certain procedures with primate subjects and, in some cases, makes it possible to collect data that could otherwise not be collected. For example, chimpanzees involved in a long-term study of aging were trained for urine collection needed for hormonal analysis. Voluntary urine collection was deemed superior to blood collection methods because it was noninvasive, fast, and easier to collect frequently and over a long period of time (Bloomsmith et al. 2015). As Graham et al. (2012) point out, cooperation from monkeys makes it possible to collect more complete experimental data, as well as producing data that are more reliable, in both in vivo and laboratory evaluations.

10   Advancing Animal Training Through Behavior Analysis

As should be clear from this chapter, there has been tremendous progress in training primates in a research setting, but there is always room for improvement. One path to further growth is to better utilize a field of psychology called applied behavior analysis (ABA). Practitioners of this approach, called behavior analysts, build on principles of classical and operant conditioning and study how to gain control over behavioral problems in humans. ABA is best known for its success in addressing problem behaviors in children with autism and other developmental disabilities. Several arguments for the inclusion of ABA techniques have been published (Bloomsmith et al. 2007; Maple and Segura 2015; Martin 2017). The application of ABA offers the opportunity to determine the cause of a behavior and to treat that behavior according to its function, as seen in studies involving functional analysis of human behavior. Recently, some experimental techniques developed in ABA have been applied to the study of captive animals.

10.1   Functional Analysis

In human clinical settings, behavior analysts have developed a functional analysis technique that allows them to identify the underlying reinforcer, or function, of an undesirable behavior (e.g., social attention from a parent might be inadvertently reinforcing head-banging behavior in a child with a developmental disorder). Identifying and altering the consequences that reinforce these problems can ameliorate that problem behavior. During a functional analysis, an experimenter provides the subject with putative reinforcers (e.g., social attention, escape from a task or demand) contingent on the occurrence of the problem behavior. The rate of problem behavior in these test conditions is compared with its rate in a control condition in which there are no demands placed on the subject and she/he is given response-independent social attention and/or tangible items (Iwata et al. 1982). By analyzing which contingencies result in higher rates of the problem behavior, the function of the behavior can be determined. Functional analysis techniques have been used to understand the environmental contingencies maintaining self-injury (e.g., Iwata et al. 1982), aggression (e.g., Derby et al. 1994; Mace et al. 1986; Thompson et al. 1998), stereotypies (e.g., Derby et al. 1994; Goh et al. 1995), food refusal (e.g., Munk and Repp 1994; Piazza et al. 2003), and other problem behaviors in developmentally disabled children and adults (see Hanley et al. 2003 for a review), and to successfully treat those behaviors (Didden et al. 1997; Ingram et al. 2005; Newcomer and Lewis 2004).

Functional analysis was adapted to determine the function of feces-throwing and spitting in a male chimpanzee (Martin et al. 2011). The results showed that these behaviors were highest during the positive reinforcement condition, indicating that the combination of social attention by humans and access to fruit juice were the strongest reinforcers for these behaviors. A treatment system was designed that combined extinction contingencies (no longer providing attention or juice) for the target behaviors (i.e., feces-throwing and spitting were ignored) with an alternate, more appropriate way for the subject to gain access to these reinforcers (holding on to a plastic ring). This intervention was effective in reducing the problem behaviors to a small fraction of their previous levels (Martin et al. 2011). In zoos, functional analysis-based techniques have been used to successfully treat hair pulling, hand biting, and foot biting in olive baboons (Dorey et al. 2009), and human-directed aggression in a black-and-white ruffed lemur (Farmer-Dougan 2014). Given the success of this technique, its use is drawing more attention, with several recent publications advocating for an increase in research in this area (Edwards and Poling 2011; Farhoody 2012; Maple and Segura 2015).

Despite the success of functional analysis, its use has limitations in animal settings. The previous studies (Dorey et al. 2009, 2012; Farmer-Dougan 2014; Martin et al. 2011) all tested and identified reinforcers that were provided by humans. When behaviors are maintained by human-controlled reinforcers, functional assessments are relatively easy since the human experimenter can differentially deliver those reinforcers in the experimental conditions. However, it is unlikely that social reinforcement from humans is the reinforcer for most problem behaviors seen in research primates. These behaviors are probably also maintained by "automatic" (nonsocial) reinforcement, such as an increase or decrease in arousal (see Martin 2017 for discussion), or by social reinforcement provided by other monkeys. These reinforcers are much harder to vary systematically as one would in a standard functional assessment, so further adaptations of functional analysis will be needed to test these additional functions.

10.2 Preference Assessment

Preference assessment offers an empirical approach to determining stimuli that may work as positive reinforcers (Fernandez et al. 2004; Clay et al. 2009). Although it may seem that people familiar with the training subject would know the best reinforcers for that person or animal, studies show that staff opinion is not very reliable (Gaalema et al. 2011; Green et al. 1988; Mehrkam and Dorey 2015). There are several ways to conduct preference assessments, and a popular method, multiple stimulus without replacement (DeLeon and Iwata 1996), involves the subject being presented with an array of items, and the subject chooses items one at a time. Preference testing can be completed quickly, just prior to a training session, since preferences may be dynamic.

After preferred items are identified, they can be further evaluated to determine their effectiveness as reinforcers. That is, when a food item is highly preferred, will it improve the speed of acquisition or the strength of reinforcement (Hutson and Mourik 1981; Vicars et al. 2014)? More preferred foods resulted in higher levels of engagement during training sessions with rhesus macaques (Martin et al. 2018). So, it seems that preference testing can be used to help boost the efficiency of PRT sessions.

Functional analysis and preference testing are just some of the techniques used within ABA that may be useful in captive animal environments. Maple and Segura (2015) have recommended partnerships between universities and zoos to expand the application of ABA in zoos. Such partnerships offer unique opportunities for collaboration to increase the well-being of animals in zoos and should serve as an example for primates in research facilities.

11 Developing Animal Training Programs

11.1 Program Structure

As the value of PRT has become clearer to those working with research primates, there is growing interest in establishing formal animal training programs. Such programs may include dedicated personnel, education programs, practices to increase consistency among trainers, documentation systems, and evaluations of trained animals. More comprehensive descriptions of how to set up training programs have been published (Perlman et al. 2012; Prescott et al. 2005), so here we will illustrate a few key elements. Animal training programs can be structured in three ways: project-based, section-based, or facility-wide (Perlman et al. 2012; Whittaker et al. 2008), each with its own strengths and weaknesses.

11.2 Project-Based Approach

This approach usually involves a small number of animals and a small number of staff members. Typically, specific objectives are set out, such as a veterinary technician teaching the monkeys in one study to present their ears for a blood sampling procedure. Advantages of this approach are conferred to the small number of animals and to the individuals who conduct the training. The project-based approach can serve as a catalyst for other training projects and may develop into a section-wide program if others working with the first individual begin training. Drawbacks are that relatively few animals and people are involved. Additionally, without much supervisory oversight there may not be sufficient time and other resources devoted to the project. Training techniques and protocols may also not be adequately developed. Lastly, the project may be stopped prematurely if the primary trainer stops or leaves his or her position.

11.3 Section-Wide Approach

The section-wide approach is when an entire department or research unit is involved, usually supervised by a manager and including multiple staff working on training objectives with all the animals in their area. For example, all members of a research laboratory might coordinate to train their primates to cooperate with research procedures. With involvement of a manager, typically oversight is improved. Trainers are better supported in learning needed skills, having the time to train, and consistency is increased across trainers. Weaknesses of this approach are that it is limited to one work unit, so the trained behaviors may not be used by others working with the same animals (e.g., a veterinary technician may not ask a monkey to present for an injection, even though the research staff has trained the monkey). Additionally, the program may not continue if the manager leaves his or her position.

11.4 Facility-Wide Approach

The most comprehensive approach is facility-wide, which is implemented throughout the institution and has support from multiple levels of management and multiple departments. There may be people involved from animal care, veterinary staff, colony management, behavioral management, research groups, administration, occupational health and safety, facility or operations management, and regulatory oversight groups (e.g., Institutional Animal Care and Use Committee). For example, a research facility might have a centralized training committee with representatives from these groups who develop broad training objectives that involve most or all the animals at the institution. Advantages of such an approach are that more animals, people, and research projects will experience the benefits of the training, and typically, more resources are devoted to the program. Such an approach creates a framework of safety, education, consistency, and communication that is very valuable. Drawbacks of a facility-wide training program are largely the significant costs and personnel time required to support such a program.

11.5 Personnel Considerations

How should animal training programs in research facilities be staffed? Individuals need to be assessed for their potential to serve as animal trainers, given appropriate educational opportunities to learn animal training techniques and terminology, and afforded proper oversight and support. Schapiro et al. (2018) suggest that patience and consistency are important characteristics to look for in personnel who serve as trainers, and we agree. Patience is important when training is not progressing as quickly as the trainer might prefer. Consistency is important in getting an animal to understand what you are asking him/her to do, as well as facilitating multiple people training the same primates. In addition, a deep understanding of classical and operant conditioning is fundamental. When training is ineffective, it is important for the trainer to be able to discern what behaviors are being reinforced, or not, as these can signal why training may not be going well. Openness to new ideas, a high motivation to train, and a willingness to try different techniques are also important. New ideas in PRT are constantly being published, and an openness to them can make a trainer more successful. Additionally, being open-minded to the critiques of other trainers is a valuable characteristic. Finally, as with most endeavors, attitude is everything. It is important to keep a positive attitude and optimistic outlook when training an animal. Frustration and low morale can be manifested during training sessions in ways that trainers may not even be aware. If a trainer doubts that training will be successful, there is almost a 100% guarantee that it will not be. When a negative attitude gets in the way of training, patience and openness to new ideas begin to wane. To maintain a positive outlook, it is important to celebrate the small gains, too, particularly as they usually lead to larger ones.

Trainers will develop expertise by having the proper educational background and continuing education in animal training. The most relevant educational background will include formal coursework in animal learning and conditioning or ABA. When this is lacking, it may be possible for trainers to take courses at a local college or through an online program. Continuing education options include reading books and articles about training and the training accomplishments of others (see Table 1 for a list of resources), completing a certification program (e.g., Karen Pryor Academy 2021; Animal Behavior Institute 2017; ILLIS ABC Animal Behavior Consulting 2018), or attending a workshop (see American Society of Primatologists 2020). Some facilities have arranged short internships for new trainers so they can observe an established program at another institution. Hiring consultants can be useful to review training programs, provide instruction on training techniques, and give feedback to animal trainers. We also suggest that those who are training within an institution work together and establish facility-wide goals through a training committee. Trainers may also review IACUC protocols for animal training opportunities as studies are being started. Lastly, plan discussion groups that bring laboratories or departments together to troubleshoot and discuss training techniques for challenging behaviors (Perlman et al. 2012).

Table 1 Suggested readings for an introduction to PRT

Graham, ML (2017) Positive reinforcement training and research. In: Schapiro SJ (ed) Handbook of primate behavioral management. CRC Press, Boca Raton, pp 189–197.
Laule GE, Bloomsmith MA, Schapiro SJ (2003) The use of positive reinforcement training techniques to enhance the care, management, and welfare of laboratory primates. J Appl Anim Welf Sci 6(3):163–174.
Magden ER (2017) Positive reinforcement training and health care. In: Schapiro SJ (ed) Handbook of primate behavioral management. CRC Press, Boca Raton, pp 201–213.
Martin AL (2017) The primatologist as a behavioral engineer. Am J Primatol 79(1): 1–10.
Perlman JE, Bloomsmith MA, Whittaker MA et al (2012) Implementing positive reinforcement animal training programs at primate laboratories. Appl Anim Behav Sci 137(3):114–126.
Prescott MJ, Buchanan-Smith HM, Rennie AE (2005) Training of laboratory-housed non-human primates in the UK. Anthrozoös 18(3):288–303.
Pryor K (2002) Don't shoot the dog! The new art of teaching and training. Ringpress Books Ltd, Lydney.
Ramirez K (1999) Animal training: successful animal management through positive reinforcement. Chicago.
Schapiro SJ, Magden ER, Reamer LA et al (2018) Behavioral training as part of the health care program. In: Weichbrod RH, Thompson GAH, Norton JN (eds) Management of animal care and use programs in research, education, and testing. CRC Press, Boca Raton, pp 772–788.

12 Moving Forward

We conclude that PRT for primates in a research setting is effective, enhances primate welfare, and improves the quality of biomedical research. These outcomes are well worth the investment of time, money, and personnel education needed to facilitate a PRT program. The use of PRT is continuing to grow in the primate research community, but its full potential has yet to be achieved. For this movement to continue, several things need to take place. Existing techniques need to be applied more broadly within veterinary care programs and within biomedical research programs. More research facilities should hire animal trainers. Furthermore, continuing educational opportunities need to be made more widely available for people who want and need to become competent trainers. Finally, the increased use of technology to support and enhance animal training, and to improve its efficiency, will be an important step in the direction of realizing the potential of PRT.

Behavioral scientists should play key roles in continuing to foster the use of positive approaches to training. Those who understand animal learning and conditioning need to be leading this process, and they should conduct more quantitative evaluations of the effects of training on a range of dependent measures related to welfare. Specialists in ABA will be important as they adapt their techniques to the research setting and teach others to correctly apply this paradigm to modify the behavior of primates. Those scientists who conduct biomedical research with primates also have an important role as they incorporate the time and money required for PRT into grant proposals and study plans from the beginning. They should also assess the effects of PRT on the dependent measures in their research programs. Some have found benefits to their research, but if there are unintended or undesirable effects, those should be documented and dealt with so that the research progresses with PRT best practices firmly in place.

References

American Society of Primatologists (2020) Introduction to animal training techniques. https://asp.org/welfare/animal-training/. Accessed 16 June 2021

Animal Behavior Institute (2017) Animal behavior Institute's Animal Training & Enrichment Certificate Program. https://www.animaledu.com/Programs/Animal-Training-Enrichment?d=1. Accessed 16 June 2021

Animal Research Tomorrow (2020) Basel Declaration. https://animalresearchtomorrow.org/. Accessed 16 June 2021

Baker KC, Bloomsmith M, Neu K et al (2009) Positive reinforcement training moderates only high levels of abnormal behavior in singly housed rhesus macaques. J Appl Anim Welf Sci 12 (3):236–252. https://doi.org/10.1080/10888700902956011

Behringer V, Stevens JM, Hohmann G et al (2014) Testing the effect of medical positive reinforcement training on salivary cortisol levels in bonobos and orangutans. PLoS One 9(9):e108664. https://doi.org/10.1371/journal.pone.0108664

Bliss-Moreau E, Moadab G (2016) Variation in behavioral reactivity is associated with cooperative restraint training efficiency. J Am Assoc Lab Anim Sci 55(1):41–49

Bliss-Moreau E, Theil JH, Moadab G (2013) Efficient cooperative restraint training with rhesus macaques. J Appl Anim Welf Sci 16(2):98–117. https://doi.org/10.1080/10888705.2013.768897

Bloomsmith MA, Laule GE, Alford PL et al (1994) Using training to moderate chimpanzee aggression during feeding. Zoo Biol 13(6):557–566. https://doi.org/10.1002/zoo.1430130605

Bloomsmith MA, Lambeth SP, Stone AM et al (1997) Comparing two types of human interaction as enrichment for chimpanzees. Am J Primatol 42(2):96. https://doi.org/10.1002/(SICI)1098-2345(1997)42:2<96::AID-AJP3>3.0.CO;2-T

Bloomsmith MA, Stone AM, Laule GE (1998) Positive reinforcement training to enhance the voluntary movement of group-housed chimpanzees within their enclosures. Zoo Biol 17(4):333–341. https://doi.org/10.1002/(SICI)1098-2361(1998)17:4<333::AID-ZOO6>3.0.CO;2-A

Bloomsmith MA, Marr MJ, Maple TL (2007) Addressing nonhuman primate behavioral problems through the application of operant conditioning: is the human treatment approach a useful model? Appl Anim Behav Sci 102(3–4):205–222. https://doi.org/10.1016/j.applanim.2006.05.028

Bloomsmith M, Franklin A, Neu K et al (2013) Training time required to collect a variety of biological samples using primarily positive reinforcement training methods. Am J Primatol 75 (S1):65. https://doi.org/10.1002/ajp.22188

Bloomsmith M, Neu K, Franklin A et al (2015) Positive reinforcement methods to train chimpanzees to cooperate with urine collection. J Am Assoc Lab Anim Sci 54(1):66–69

Bourgeois SR, Brent L (2005) Modifying the behaviour of singly caged baboons: evaluating the effectiveness of four enrichment techniques. Anim Welf 14(1):71–81

Bourgeois SR, Vazquez M, Brasky K (2007) Combination therapy reduces self-injurious behavior in a chimpanzee (*Pan troglodytes troglodytes*): a case report. J Appl Anim Welf Sci 10 (2):123–140. https://doi.org/10.1080/10888700701313454

Brunelli RL, Gottlieb D, Holcomb K et al (2009) Effects of positive reinforcement training on infant behavioral development in nursery-reared rhesus macaques (*Macaca mulatta*). Am J Primatol 71(S1):74. https://doi.org/10.1002/ajp.20733

Buchanan-Smith HM, Badihi I (2012) The psychology of control: effects of control over supplementary light on welfare of marmosets. Appl Anim Behav Sci 137(3–4):166–174. https://doi.org/10.1016/j.applanim.2011.07.002

Callen EJ, Boyd TL (1990) Examination of a backchaining/counterconditioning process during the extinction of conditioned fear. Behav Res Ther 28(4):261–271. https://doi.org/10.1016/0005-7967(90)90077-v

Catania AC, Sagvolden T (1980) Preference for free choice over forced choice in pigeons. J Exp Anal Behav 34(1):77–86. https://doi.org/10.1901/jeab.1980

Chance P (2003) Learning and behavior. Wadsworth, Belmont

Clay AW, Bloomsmith MA, Marr MJ (2009) Habituation and desensitization as methods for reducing fearful behavior in singly housed rhesus macaques. Am J Primatol 71(1):30–39. https://doi.org/10.1002/ajp.20622

Clay AW, Bloomsmith MA, Marr MJ et al (2009b) Systematic investigation of the stability of food preferences in captive orangutans: implications for positive reinforcement training. J Appl Anim Welf Sci 12(4):306–313. https://doi.org/10.1080/10888700903163492

Coleman K (2012) Individual differences in temperament and behavioral management practices for nonhuman primates. Appl Anim Behav Sci 137(3–4):106–113. https://doi.org/10.1016/j.applanim.2011.08.002

Coleman K, Maier A (2010) The use of positive reinforcement training to reduce stereotypic behavior in rhesus macaques. Appl Anim Behav Sci 124(3–4):142–148. https://doi.org/10.1016/j.applanim.2010.02.00

Coleman K, Tully LA, McMillan JL (2005) Temperament correlates with training success in adult rhesus macaques. Am J Primatol 65(1):63–71. https://doi.org/10.1002/ajp.20097

Coleman K, Pranger L, Maier A et al (2008) Training rhesus macaques for venipuncture using positive reinforcement techniques: a comparison with chimpanzees. J Am Assoc Lab Anim Sci 47(1):37–41

Council of Europe (2010) Directive 2010/63/EU of the European parliament and of the council of 22 September on the protection of animals used for scientific purposes. Off J Eur Union L276:33–79

Cox C (1987) Increase in the frequency of social interactions and the likelihood of reproduction among drills. In: proceedings of the American Association of Zoological Parks and Aquariums [AAZPA] Western Regional Conference, Wheeling, WV, pp 321–328

DeLeon IG, Iwata BA (1996) Evaluation of a multiple-stimulus presentation format for assessing reinforcer preferences. J Appl Behav Anal 29(4):519–533. https://doi.org/10.1901/jaba.1996.29-519

Derby KM, Wacker DP, Peck S et al (1994) Functional analysis of separate topographies of aberrant behavior. J Appl Behav Anal 27(2):267–278. https://doi.org/10.1901/jaba.1994.27-267

Desmond T, Laule G, McNary J (1987) Training to enhance socialization and reproduction in drills. In: Proceedings of the American Association of Zoological Parks and Aquariums [AAZPA] Western Regional Conference, Wheeling, WV, pp 435–441

Didden R, Duker PC, Korzilius H (1997) Meta-analytic study on treatment effectiveness for problem behaviors with individuals who have mental retardation. Am J Ment Retard 101(4):387–399

Dorey NR, Rosales-Ruiz J, Smith R et al (2009) Functional analysis and treatment of self-injury in a captive olive baboon. J Appl Behav Anal 42(4):785–794. https://doi.org/10.1901/jaba.2009.42-785

Dorey NR, Tobias JS, Udell MA et al (2012) Decreasing dog problem behavior with functional analysis: linking diagnoses to treatment. J Vet Behav Clin Appl Res 7(5):276–282. https://doi.org/10.1016/j.jveb.2011.10.002

Drews B, Harmann LM, Beehler LL et al (2011) Ultrasonographic monitoring of fetal development in unrestrained bonobos (*Pan paniscus*) at the Milwaukee County Zoo. Zoo Biol 30(3):241–253. https://doi.org/10.1002/zoo.20304

Edwards TL, Poling A (2011) Animal research in the journal of applied behavior analysis. J Appl Behav Anal 44(2):409–412. https://doi.org/10.1901/jaba.2011.44-409

Farhoody P (2012) A framework for solving behavior problems. Vet Clin Exotic Anim Pract 15(3): 399–411. https://doi.org/10.1016/j.cvex.2012.06.002

Farmer-Dougan V (2014) Functional analysis of aggression in a black-and-white ruffed lemur (*Varecia variegata variegata*). J Appl Anim Welf Sci 17(3):282–293. https://doi.org/10.1080/10888705.2014.917029

Fernandez EJ, Dorey N, Rosales-Ruiz J (2004) A two-choice preference assessment with five cotton-top tamarins (*Saguinus oedipus*). J Appl Anim Welf Sci 7(3):163–169. https://doi.org/10.1207/s15327604jaws0703_2

Gaalema DE, Perdue BM, Kelling AS (2011) Food preference, keeper ratings, and reinforcer effectiveness in exotic animals: the value of systematic testing. J Appl Anim Welf Sci 14(1): 33–41. https://doi.org/10.1080/10888705.2011.527602

Gillis TE, Janes AC, Kaufman MJ (2012) Positive reinforcement training in squirrel monkeys using clicker training. Am J Primatol 74(8):712–720. https://doi.org/10.1002/ajp.22015

Goh HL, Iwata BA, Shore BA et al (1995) An analysis of the reinforcing properties of hand mouthing. J Appl Behav Anal 28(3):269–283. https://doi.org/10.1901/jaba.1995.28-269

Graham ML (2017) Positive reinforcement training and research. In: Schapiro SJ (ed) Handbook of primate behavioral management. CRC Press, Boca Raton, pp 189–197

Graham ML, Mutch LA, Rieke EF et al (2010) Refinement of vascular access port placement in nonhuman primates: complication rates and outcomes. Comp Med 60(6):479–485

Graham ML, Rieke EF, Mutch LA et al (2012) Successful implementation of cooperative handling eliminates the need for restraint in a complex non-human primate disease model. J Med Primatol 41(2):89–106. https://doi.org/10.1111/j.1600-0684.2011.00525.x

Green CW, Reid DH, White LK et al (1988) Identifying reinforcers for persons with profound handicaps: staff opinion versus systematic assessment of preferences. J Appl Behav Anal 21(1): 31–43. https://doi.org/10.1901/jaba.1988.21-31

Gresswell C, Goodman G (2011) Case study: training a chimpanzee (*Pan troglodytes*) to use a nebulizer to aid the treatment of airsacculitis. Zoo Biol 30(5):570–578. https://doi.org/10.1002/zoo.20388

Hanley GP, Iwata BA, McCord BE (2003) Functional analysis of problem behavior: a review. J Appl Behav Anal 36(2):147–185. https://doi.org/10.1901/jaba.2003.36-147

Hanley PW, Barnhart KF, Abee CR et al (2015) Methicillin-resistant *Staphylococcus aureus* prevalence among captive chimpanzees, Texas, USA, 2012. Emerg Infect Dis 21(12):2158–2160. https://doi.org/10.3201/eid2112.142004

Hannibal D, Minier D, Capitanio J et al (2013) Effect of temperament on the behavioral conditioning of individual rhesus monkeys. Am J Primatol 75(Suppl 1):66. https://doi.org/10.1002/ajp.22188

Hanson JD, Larson ME, Snowdon CT (1976) The effects of control over high intensity noise on plasma cortisol levels in rhesus monkeys. Behav Biol 16:333–340. https://doi.org/10.1016/S0091-6773(76)91460-7

Hendriksen P, Elmgreen K, Ladewig J (2011) Trailer-loading of horses: is there a difference between positive and negative reinforcement concerning effectiveness and stress-related signs? J Vet Behav Clin Appl Rev 6(5):261–266. https://doi.org/10.1016/j.jveb.2011.02.007

Hutson GD, van Mourik SC (1981) Food preferences of sheep. Aust J Exp Agric 21(113):575–582. https://doi.org/10.1071/EA9810575

ILLIS ABC Animal Behavior Consulting (2018) Courses. https://illis.se/en/courses-menu/?fbclid=IwAR1e1I24_Lf4pC9AplFk4hiTw0zDcEyaeHHVYgaJzYr5GMEWjYrlQwbx6Ds. Accessed 16 June 2021

Ingram K, Lewis-Palmer T, Sugai G (2005) Function-based intervention planning: comparing the effectiveness of FBA function-based and non-function-based intervention plans. J Posit Behav Interv 7(4):224–236. https://doi.org/10.1177/10983007050070040401

Innes L, McBride S (2008) Negative versus positive reinforcement: an evaluation of training strategies for rehabilitated horses. Appl Anim Behav Sci 112(3–4):357–368. https://doi.org/10.1016/j.applanim.2007.08.011

Iwata BA, Dorsey MF, Slifer KJ et al (1982) Toward a functional analysis of self-injury. Anal Interv Dev Disabil 2(1):3–20. https://doi.org/10.1901/jaba.1994.27-197

Kaplan G, Pines MK, Rogers LJ (2012) Stress and stress reduction in common marmosets. Appl Anim Behav Sci 137(3–4):175–182. https://doi.org/10.1016/j.applanim.2011.04.011

Karen Pryor Academy (2021) Dog trainer professional. https://karenpryoracademy.com/courses/dog-trainer-professional/. Accessed 16 June 2021

Kelley TM, Bramblett CA (1981) Urine collection from vervet monkeys by instrumental conditioning. Am J Primatol 1(1):95–97. https://doi.org/10.1002/ajp.1350010112

Kemp C, Thatcher H, Farningham D et al (2017) A protocol for training group-housed rhesus macaques (*Macaca mulatta*) to cooperate with husbandry and research procedures using positive reinforcement. Appl Anim Behav Sci 197:90–100. https://doi.org/10.1016/j.applanim.2017.08.006

Koban T, Miyamoto M, Donmoyer G et al (2005) Effects of positive reinforcement training on cortisol, hematology and cardiovascular parameters in cynomolgus macaques (*Macaca fascicularis*). In: Abstracts of Seventh International Conference on Environmental Enrichment, Wild Conservation Society, Bronx, NY, p 233

Lambeth SP, Hau J, Perlman JE et al (2006) Positive reinforcement training affects hematologic and serum chemistry values in captive chimpanzees (*Pan troglodytes*). Am J Primatol 68(3):245–256. https://doi.org/10.1002/20148

Laule G, Desmond T (1990) Use of positive behavioral techniques in primates for husbandry and handling. In: Proceedings of the American Association of Zoo Veterinarians Annual Conference, South Padre Island, TX, pp 269–273

Laule G, Whittaker M (2007) Enhancing nonhuman primate care and welfare through the use of positive reinforcement training. J Appl Anim Welf Sci 10(1):31–38. https://doi.org/10.1080/10888700701277311

Laule GE, Thurston RH, Alford PL et al (1996) Training to reliably obtain blood and urine samples from a young diabetic chimpanzee (*Pan troglodytes*). Zoo Biol 15(6):587–591. https://doi.org/10.1002/(SICI)1098-2361(1996)15:6<587::AID-ZOO4>3.0.CO;2-7

Laule GE, Bloomsmith MA, Schapiro SJ (2003) The use of positive reinforcement training techniques to enhance the care, management, and welfare of primates in the laboratory. J Appl Anim Welf Sci 6(3):163–174. https://doi.org/10.1207/s15327604jaws0603_02

Lutz CK, Baker KC (2022) Using behavior to assess primate welfare. In: Robinson LM, Weiss A (eds) Nonhuman primate welfare: from history, science, and ethics to practice. Springer, Cham, pp 171–206

Lutz CK, Tiefenbacher S, Jorgensen MJ et al (2000) Techniques for collecting saliva from awake, unrestrained, adult monkeys for cortisol assay. Am J Primatol 52(2):93–99. https://doi.org/10.1002/1098-2345(200010)52:2<93::AID-AJP3>3.0.CO;2-B

Mace FC, Page TJ, Ivancic MT et al (1986) Analysis of environmental determinants of aggression and disruption in mentally retarded children. Appl Res Ment Retard 7(2):203–221. https://doi.org/10.1016/0270-3092(86)90006-8

Magden ER (2017) Positive reinforcement training and health care. In: Schapiro SJ (ed) Handbook of primate behavioral management. CRC Press, Boca Raton, pp 201–213

Magden ER, Haller RL, Thiele EJ et al (2013) Acupuncture as an adjunct therapy for osteoarthritis in chimpanzees (*Pan troglodytes*). J Am Assoc Lab Anim Sci 52(4):475–480

Maple TL, Segura VD (2015) Advancing behavior analysis in zoos and aquariums. Behav Anal 38(1):77–91. https://doi.org/10.1007/s40614-014-0018-x

Martin AL (2017) The primatologist as a behavioral engineer. Am J Primatol 79(1):e22500. https://doi.org/10.1002/ajp.22500

Martin AL, Bloomsmith MA, Kelley ME et al (2011) Functional analysis and treatment of human directed undesirable behavior exhibited by a captive chimpanzee. J Appl Behav Anal 44 (1):139–143. https://doi.org/10.1901/jaba.2011.44-139

Martin AL, Franklin AN, Perlman JE et al (2018) Systematic assessment of food item preference and reinforcer effectiveness: enhancements in training laboratory housed rhesus macaques. Behav Process 157:445–452. https://doi.org/10.1016/j.beproc.2018.07.002

McKinley J, Buchanan-Smith HM, Bassett L et al (2003) Training common marmosets (*Callithrix jacchus*) to cooperate during routine laboratory procedures: ease of training and time investment. J Appl Anim Welf Sci 6(3):209–220. https://doi.org/10.1207/s15327604jaws0603_06

McMillan JL, Perlman JE, Galvan A et al (2014) Refining the pole-and-collar method of restraint: emphasizing the use of positive training techniques with rhesus macaques (*Macaca mulatta*). J Am Assoc Lab Anim Sci 53(1):61–66

McMillan JL, Bloomsmith MA, Prescott MJ (2017) An international survey of approaches to chair restraint of nonhuman primates. Comp Med 67(5):442–451

Mehrkam LR, Dorey NR (2015) Preference assessments in the zoo: keeper and staff predictions of enrichment preferences across species. Zoo Biol 34(5):418–430. https://doi.org/10.1002/zoo.21227

Mineka S, Gunnar M, Champoux M (1986) Control and early socioemotional development: infant rhesus monkeys reared in controllable versus uncontrollable environments. Child Dev 57 (5):1241–1256. https://doi.org/10.2307/1130447

Minier DE, Tatum L, Gottlieb DH (2011) Human-directed contra-aggression training using positive reinforcement with single and multiple trainers for indoor-housed rhesus macaques. Appl Anim Behav Sci 132(3–4):178–186. https://doi.org/10.1016/j.applanim.2011.04.009

Morgan L, Howell SM, Fritz J (1993) Regurgitation and reingestion in a captive chimpanzee. Lab Anim 22:42–45

Munk DD, Repp AC (1994) Behavioral assessment of feeding problems of individuals with severe disabilities. J Appl Behav Anal 27(2):241–250. https://doi.org/10.1901/jaba.1994.27-241

National Research Council (2011) Guide for the care and use of laboratory animals, 8th edn. National Academies Press, Washington, DC

Newcomer LL, Lewis TJ (2004) Functional behavioral assessment: An investigation of assessment reliability and effectiveness of function-based interventions. J Emot Behav Disord 12 (3):168–181. https://doi.org/10.1177/10634266040120030401

Novak MA, Suomi SJ (1988) Psychological well-being of primates in captivity. Am Psychol 43 (10):765–773. https://doi.org/10.1037/0003-066X.43.10.765

Owen Y, Amory JR (2011) A case study employing operant conditioning to reduce stress of capture for red-bellied tamarins (*Saguinus labiatus*). J Appl Anim Welf Sci 14(2):124–137. https://doi.org/10.1080/10888705.2011.55162

Perdue BM, Evans TA, Washburn DA, Rumbaught DM, Beran MJ (2014) Do monkeys choose to choose? Learn Behav 42(2):164–175. https://doi.org/10.3758/s13420-014-0135-0

Perlman JE, Bowsher TR, Braccini SN et al (2003) Using positive reinforcement training techniques to facilitate the collection of semen in chimpanzees (*Pan troglodytes*). Am J Primatol 60(Suppl 1):77–78. https://doi.org/10.1002/ajp.10085

Perlman JE, Thiele E, Whittaker MA et al (2004) Training chimpanzees to accept subcutaneous injections using positive reinforcement training techniques. Am J Primatol 62(S1):96. https://doi.org/10.1002/ajp.20029

Perlman JE, Horner V, Bloomsmith MA et al (2010) Positive reinforcement training, social learning, and chimpanzee welfare. In: Lonsdorf EV, Ross SR, Matsuzawa T (eds) The mind of the chimpanzee: ecological and experimental perspectives. University of Chicago Press, Chicago, pp 320–331

Perlman JE, Bloomsmith MA, Whittaker MA et al (2012) Implementing positive reinforcement animal training programs at primate laboratories. Appl Anim Behav Sci 137(3–4):114–126. https://doi.org/10.1016/j.applanim.2011.11.003

Piazza CC, Patel MR, Gulotta CS et al (2003) On the relative contributions of positive reinforcement and escape extinction in the treatment of food refusal. J Appl Behav Anal 36(3):309–324. https://doi.org/10.1901/jaba.2003.36-309

Pierce WD, Cheney CD (2017) Behavior analysis and learning: a biobehavioral approach, 6th edn. Routledge, New York

Pomerantz O, Terkel J (2009) Effects of positive reinforcement training techniques on the psychological welfare of zoo-housed chimpanzees (*Pan troglodytes*). Am J Primatol 71(8):687–695. https://doi.org/10.1002/ajp.20703

Ponce CR, Genecin MP, Perez-Melara G et al (2016) Automated chair-training of rhesus macaques. J Neurosci Methods 263:75–80. https://doi.org/10.1016/j.jneumeth.2016.01.024

Powell RA, Honey PL, Symbaluk DG (2017) Introduction to learning and behavior, 5th edn. Cengage, Boston

Prescott MJ, Buchanan-Smith HM (2007) Training laboratory-housed non-human primates, Part I: a UK survey. Anim Welf 16(1):21–36

Prescott MJ, Lidster K (2017) Improving quality of science through better animal welfare: the NC3Rs strategy. Lab Anim 46(4):152–156. https://doi.org/10.1038/laban.1217

Prescott MJ, Buchanan-Smith HM, Rennie AE (2005) Training of laboratory-housed non-human primates in the UK. Anthrozoös 18(3):288–303. https://doi.org/10.2752/089279305785594153

Prosen H, Bell B (2000) A psychiatrist consulting at the zoo (the therapy of Brian bonobo). In: The apes: Challenges for the 21st Century-conference proceedings, Brookfield, IL, 10–13 May, 2000, pp 161–164

Pryor K (2002) Don't shoot the dog! The new art of teaching and training. Ringpress Books Ltd, Lydney

Ramirez K (1999) Animal training: successful animal management through positive reinforcement. Shedd Aquarium, Chicago

Reamer LA, Haller RL, Thiele EJ et al (2014) Factors affecting initial training success of blood glucose testing in captive chimpanzees (*Pan troglodytes*). Zoo Biol 33(3):212–220. https://doi.org/10.1002/zoo.21123

Report to Office of Extramural Research Acting Director on Office of Laboratory Animal Welfare (OLAW) Site Visits to Chimpanzee Facilities (2010) NOT-OD-10-121. National Institutes of Health, Bethesda

Roma RG, Champoux M, Suomi SJ (2006) Environmental control, social context, and individual differences in behavioral and cortisol responses to novelty in infant rhesus monkeys. Child Dev 77(1):118–131. https://doi.org/10.1111/j.1467-8624.2006.00860.x

Russell WMS, Burch RL (1959) The principles of humane experimental technique. Methuen, London

Ruys JD, Mendoza SP, Capitanio JP et al (2004) Behavioral and physiological adaptation to repeated chair restraint in rhesus macaques. Physiol Behav 82(2–3):205–213. https://doi.org/10.1016/j.physbeh.2004.02.031

Savastano G, Hanson A, McCann C (2003) The development of an operant conditioning training program for New World primates at the Bronx Zoo. J Appl Anim Welf Sci 6(3):247–261. https://doi.org/10.1207/S15327604JAWS0603_09

Schapiro SJ, Hau J (2022) Benefits of improving welfare in captive primates. In: Robinson LM, Weiss A (eds) Nonhuman primate welfare: from history, science, and ethics to practice. Springer, Cham, pp 433–450

Schapiro SJ, Perlman JE, Boudreau BA (2001) Manipulating the affiliative interactions of group-housed rhesus macaques using positive reinforcement training techniques. Am J Primatol 55(3):137–149. https://doi.org/10.1002/ajp.1047

Schapiro SJ, Perlman JE, Thiele E et al (2005) Training nonhuman primates to perform behaviors useful in biomedical research. Lab Anim 34(5):37–42. https://doi.org/10.1038/laban0505-37

Schapiro SJ, Magden ER, Reamer LA et al (2018) Behavioral training as part of the health care program. In: Weichbrod RH, Thompson GAH, Norton JN (eds) Management of animal care and use programs in research, education, and testing. CRC Press, Boca Raton, pp 772–788

Shirasaki Y, Yoshioka N, Kanazawa K et al (2013) Effect of physical restraint on glucose tolerance in cynomolgus monkeys. J Med Primatol 42(3):165–168. https://doi.org/10.1111/jmp.12039

Stone AM, Bloomsmith MA, Laule GE et al (1994) Documenting positive reinforcement training for chimpanzee urine collection. Am J Primatol 33(3):242. https://doi.org/10.1002/ajp.1350330303

Suzuki S (1999) Selection of forced- and free-choice by monkeys (*Macaca fascicularis*). Percept Mot Skills 88(1):242–250. https://doi.org/10.2466/PMS.88.1.242-250

Thompson RH, Fisher WW, Piazza CC et al (1998) The evaluation and treatment of aggression maintained by attention and automatic reinforcement. J Appl Behav Anal 31(1):103–116. https://doi.org/10.1901/jaba.1998.31-103

Truelove MA, Martin AL, Perlman JE et al (2017) Pair housing of macaques: a review of partner selection, introduction techniques, monitoring for compatibility, and methods for long-term maintenance of pairs. Am J Primatol 79(1):1–15. https://doi.org/10.1002/ajp.22485

Tulip J, Zimmermann JB, Farningham D et al (2017) An automated system for positive reinforcement training of group-housed macaque monkeys at breeding and research facilities. J Neurosci Methods 285:6–18. https://doi.org/10.1016/j.jneumeth.2017.04.015

VandeVoort CA, Neville LE, Tollner TL et al (1993) Noninvasive semen collection from an adult orangutan. Zoo Biol 12(3):257–265. https://doi.org/10.1002/zoo.1430120303

Veeder CL, Bloomsmith MA, McMillan JL et al (2009) Positive reinforcement training to enhance the voluntary movement of group-housed sooty mangabeys (*Cercocebus atys atys*). J Am Assoc Lab Anim Sci 48(2):192–195

Vicars SM, Miguel CF, Sobie JL (2014) Assessing preference and reinforcer effectiveness in dogs. Behav Process 103:75–83. https://doi.org/10.1016/j.beproc.2013.11.006

Videan EN, Fritz J, Murphy J et al (2005) Training captive chimpanzees to cooperate for an anesthetic injection. Lab Anim 34(5):43–48. https://doi.org/10.1038/laban0505-43

Voss SC, Homzie MJ (1970) Choice as a value. Psychol Rep 26(3):912–914. https://doi.org/10.2466/pr0.1970.6.3.912

Waiblinger S, Boivin X, Pedersen V et al (2006) Assessing the human–animal relationship in farmed species: a critical review. Appl Anim Behav Sci 101(3–4):185–242. https://doi.org/10.1016/j.applanim.2006.02.001

Warren-Smith AK, McGreevy PD (2007) The use of blended positive and negative reinforcement in shaping the halt response of horses (*Equus caballas*). Anim Welf 16(4):481–488

Washburn DA, Hopkins WD, Rumbaugh DM (1991) Perceived control in rhesus monkeys (*Macaca mulatta*): enhanced video-task performance. J Exp Psychol Anim Behav Process 17(2):123–129. https://doi.org/10.1037//0097-7403.17.2.123

Webster J (1994) Animal welfare: a cool eye towards Eden. Blackwell Science, Oxford

Wergard EM, Temrin H, Forkman B et al (2015) Training pair-housed rhesus macaques (*Macaca mulatta*) using a combination of negative and positive reinforcement. Behav Process 113:51–59. https://doi.org/10.1016/j.beproc.2014.12.008

Whittaker M, Perlman J, Laule G (2008) Facing real world challenges: keeping behavioral management programs alive and well. In: Hare HJ, Kroshko JE (eds) Proceedings of the Eighth International Conference on Environmental Enrichment, Vienna, Austria. The shape of enrichment, San Diego, pp 87–89

Wolpe J (1958) Psychotherapy by reciprocal inhibition. Stanford University Press, Stanford

Young RJ, Cipreste CF (2004) Applying animal learning theory: training captive animals to comply with veterinary and husbandry procedures. Anim Welf 13(2):225–232

Zoo Atlanta (2011) Groundbreaking gorilla blood pressure procedure. https://www.youtube.com/watch?v=-kNLXz5wUGs. Accessed 16 June 2021

Part V

Biomedical Research, Ethics, and Legislation Surrounding Nonhuman Primate Welfare

Arguments Against Using Nonhuman Primates in Research

Jarrod Bailey

Abstract

Most people oppose using nonhuman primates for research from an ethical standpoint because it can cause pain and suffering to the subjects and usually leads to their death. Some accept it, often reluctantly, because they believe it is only conducted when necessary, and that it results in medical progress, and therefore human benefit. Here, I outline an argument that nonhuman primate experiments are unnecessary, misleading, are mostly counter-productive due to wide-ranging biological differences between species; and that there are other humane and superior alternatives that researchers could, and should, be making much more use of instead. Nonhuman primate experiments have failed in many areas, including drug testing, and research into HIV/AIDS, neurological function, and diseases such as Parkinson's, stroke, and others. Yet, they persist, possibly due to convention, habit, and/or vested interests, and humans pay the price with failed attempts at understanding human diseases and finding treatments for them, while the nonhuman primates pay the price in laboratories. A more critical appreciation of the harms and benefits involved, which are greater and lesser, respectively, than commonly portrayed, would expedite an overdue shift away from nonhuman primate experiments that would benefit all primates—including humans.

Keywords

Ethics · Drug testing · Toxicology · Harm/benefit · Translatability · Alternatives

J. Bailey (✉)
Animal Free Research UK, London, UK
e-mail: Jarrod@animalfreeresearchuk.org

1 Introduction

I have spent more than a quarter of a century as a scientific researcher. I spent my student days studying genetics/molecular biology, my days in the laboratory investigating premature birth, and more recently I have been working as a researcher/writer critiquing the use of animals in science and promoting humane alternatives. Throughout this time, my opposition to the use of animals in harmful experiments has grown, both ethically and scientifically. This chapter presents some of the evidence on which my opposition is based, including my own work. At the same time, it is not a simple statement of evidence: I have been forthright in stating my opinions and conclusions. While I remain confident in them, I naturally encourage the reader to take on board other evidence and points of view, to come to their own conclusions.

Subjecting animals, and in particular nonhuman primates, to scientific experiments is highly controversial. Those who practise, fund, and advocate such experiments claim that these experiments are indispensable for medical progress; yet public opinion polls over the years demonstrate clear and growing opposition to this work, often in spite of professed human benefit. A UK poll conducted on behalf of the government in 2014 found that just 37% of people agreed with the use of animals in "all types of research", with only 16% accepting experiments on monkeys, even if they clearly benefitted people (Ipsos MORI 2014). Across Europe, a 2009 poll revealed that 79% of people supported a new law to "prohibit all experiments on animals which do not relate to serious or life-threatening human conditions", with 84% agreeing this "should prohibit all experiments causing severe pain or suffering to any animal" (BUAV/YouGov plc 2009); and in 2010, 56% of Europeans agreed that scientists should not experiment on large animals like dogs and monkeys for the improvement of human health and well-being (TNS Opinion and Social 2010). In the U.S., various polls consistently show a majority of the public opposing animal experiments, and the proportion is increasing over time (see Merkley et al. 2018). Such public objection has had little impact on nonhuman primate experiments, however, which continue to take place in the tens of thousands each year. Recent statistics reveal more than 72,000 nonhuman primates were used in experiments annually, including around 64,000 in the U.S. (United States Department of Agriculture (USDA)—Animal and Plant Health Inspection Service (APHIS) 2014), and 6000 across the EU (European Commission 2013), including around 2400 in the UK (UK Home Office 2020). Many more nonhuman primates are housed in labs and used for breeding, but not used experimentally, in any one year, and many more will be involved in experiments across the rest of the world (Taylor et al. 2008; Knight 2008). Until 2011, and despite significant public objection, invasive chimpanzee research was still funded in the U.S., with some 1000 or so individuals being housed in American laboratories (Institute of Medicine and National Research Council 2011).

These 72,000 nonhuman primates are used in experiments that range from testing new drugs and chemicals for information about human safety, to the creation of "models" to help understand diseases that affect people, and to help develop new therapies for these diseases (see Sect. 2, "Uses of Nonhuman Primates in Research" below). These experiments indisputably cause pain and suffering, which can be substantial and severe (see Sect. 4.5, "Ethical Argument" below). This chapter investigates two related claims by those who advocate nonhuman primate research (see, for example, Jennings et al. 2016; Verdier et al. 2015; Phillips et al. 2014; Tardif et al. 2013; Vallender and Miller 2013; Chan 2013; Bateson 2011). The first is that suffering is mitigated and largely negligible, or at worst, and rarely, "moderate". The second is that, where suffering does occur, it is justified by human benefits that *depend* on that research in the form of new treatments for human diseases, or at least steps toward an improved understanding of those diseases, and how they may be tackled in the future. They also claim that such experiments are a last resort, when no alternative approach is possible; that the welfare of those nonhuman primates is important, that their suffering is kept to a minimum; and that data derived from those experiments are frequently relevant to humans, particularly given the close evolutionary relationship of nonhuman primates to humans, especially relative to, for example, rodents.

My chief argument is that using nonhuman primate experiments to inform human biology, disease, and responses to drugs and chemicals is not only a current and historical failure, but that it can only ever be counterproductive due to intractable, myriad, and notable biological interspecies differences. I also argue that opposition to, and questioning of, nonhuman primate research is growing, even within a largely defensive and conservative scientific community; and that there are many and varied, superior, less inhumane, more human-relevant research and testing methods that should be adopted much more widely, and which would expedite our understanding of human diseases and the quest for treatments for them. Finally, and only relatively briefly owing to guidelines on chapter length (though in no way suggesting the ethical argument is subordinate to the scientific one), I argue that the suffering of nonhuman primates in experiments is frequently much greater than is acknowledged, and that the standard of care for the nonhuman primates is often found wanting, or even extremely poor.

2 Uses of Nonhuman Primates in Research

2.1 Drug/Product Testing

Before drugs are tested in humans for the first time, it is obviously important to ensure that they are going to be safe for clinical trial volunteers, and likely to be efficacious. Regulatory agencies, such as the U.S. Food and Drug Administration (FDA) and the European Medicines Agency (EMA), require evidence from the manufacturers to this end. Much of these data are from preclinical trials—including animal tests—which establish the pharmacokinetics of the new drug (how it is absorbed and distributed around the body, metabolized, and excreted) and how toxic it may be. These animal tests are required to be conducted in at least two species: one rodent (typically rats or mice), and one nonrodent (typically dogs or monkeys) (e.g., European Parliament 2004; U.S. Congress 1938). This requirement is in itself an acknowledgment of differences in drug effects, toxicity, and response between species, conducted in the hope that one species might detect effects not revealed by the other (Hasiwa et al. 2011) (see Sect. 3, "Empirical Evidence of Benefits" below).

2.2 Disease Modeling/Basic Research

It is probably true to say that, if a human disease exists, or if a biological system needs more understanding, attempts to model/investigate it in nonhuman primates will have been made. Common areas of research have included: Alzheimer's and Parkinson's diseases; cardiovascular diseases including stroke; neurological research; infectious diseases, including viral hepatitis and HIV/AIDS; and respiratory diseases. In many cases, nonhuman primates do not naturally suffer from the human disease being investigated, so efforts are made to create "models" that have similar symptoms. As we will see, the presence of a few similar symptoms in a nonhuman primate does not translate to human relevance.

3 Scientific Argument: Empirical Evidence of Benefits

Unfortunately, for animals involved in experiments (and for humans relying on biomedical research), and despite many calls for it, there remains little retrospective critical assessment of animal experimentation (Pound and Nicol 2018; Balls 2018). This is surprising for several reasons: first, critical reflection, and a willingness to change direction based on it, is core to the scientific method; science should constantly ask questions of itself, seeking improvement with a determination to use the best methods to achieve a goal, even if this means changing tack (Balls 2018). If a scientific model is not sufficiently reflective of what is being modeled (i.e., [something in] humans), then alternative, more relevant methods must be used. Secondly, this *particularly* applies to animal research due to the harm involved, and perhaps, especially, to nonhuman primate research, in light of the acknowledged pain, suffering, and death it involves. If harms are great, this approach cannot be used habitually and without powerful justification.

3.1 Drug Testing

While there has been some progress in critiquing nonhuman primate research lately, there remains a vigorous defense of nonhuman primate research by those who conduct and profit from it (Bateson 2011; Benabid et al. 2015a, b; Editorial 2008; SCHEER 2017; VandeBerg and Zola 2005; Weatherall 2006). Nonhuman primate research therefore continues not only in the face of mounting and formidable evidence against it, but also a stark absence of valid, robust supportive evidence. In drug testing, for example—which involves almost 80% of nonhuman primates used in science in the UK (UK Home Office 2020)—there remains, after many decades, no published evidence demonstrating its necessity, save the occasional publication based on scant data and/or weak, incomplete statistics (Bailey and Balls 2019). This is supported by the Toxicology Working Group of the UK Parliament's House of Lords Select Committee on Animals in Scientific Procedures, which stated that the use of two species in toxicity testing was not scientifically justifiable, but actually acknowledged problematic species differences, both between nonhuman animals, and animals and humans (House of Lords 2002).

An example of an unsuitable report proffered by advocates of animal drug-testing examined only sensitivity (toxicities correctly identified by animal tests), but not specificity (absence of toxicities correctly identified by animal tests) (Olson et al. 2000), yet *both* are essential to assess the evidential weight they provide to human toxicity (Matthews 2008). Another study did include both measures, but used incorrect definitions (Schein et al. 1970). This has not prevented that study from being used as evidence to support animal tests (e.g., Greaves et al. 2004), even though, when correct definitions were used, the evidence demonstrated that the animal tests did not contribute any predictive value for humans (Matthews 2008). Meanwhile, the International Conference for Harmonization (ICH) acknowledged that nonhuman primate pharmacokinetic data "...can differ from humans as much as other species" (International Conference for Harmonisation of Technical Requirements for Registration of Pharmaceuticals for Human Use 2005); developmental toxicity data from nonhuman primates correspond to human data only 50% of the time, less even than results from more evolutionarily distant species (Bailey 2008a); for the prediction of drug-induced liver injury, primates are less predictive than rodents, the latter having failed to predict up to 51% of the effects in humans (Spanhaak et al. 2008); and single-dose toxicity tests have been scientifically discredited (Robinson et al. 2008). One high-profile example of the failure of nonhuman primate tests is the drug Theralizumab or TGN1412: cynomolgus macaques showed it to be safe, even at a dose 500 times higher than the dose in humans that almost killed six healthy clinical trial volunteers (UK Department of Health 2006).

While it is acknowledged that data to perform analyses are scarce, due to concerns over privacy and commercial interests (Hasiwa et al. 2011), it is disappointing that, after so many years of animal-based tests, large-scale animal testing persists with little or no published scientific basis. For example, it has been admitted there is an "inbuilt prejudice" that nonhuman primate toxicity data are more predictive of liver toxicity in humans than are such data in other species, and there is "little evidence" to support this assumption (Foster 2005). The pharmaceutical industry has historically been unwilling to conduct and publish its own analyses, or to share even anonymized data to facilitate third-party analyses, therefore my colleagues and I addressed this issue by publishing several papers (Bailey et al. 2013; 2014; 2015; Bailey 2014b; Bailey and Balls 2019). We reported our own analyses of what published data were commercially available, assessing the contribution of evidential weight by animal tests for or against the toxicity of drugs in humans. We showed that, generally, the most commonly used species (mice, rats, rabbits, dogs, and nonhuman primates) cannot be considered fit for purpose. Inter alia, but most importantly, we showed that the absence of toxicity of a new drug in any species provides little or no evidential weight to the probability of a similar lack of toxicity—or "safety"—in any other species, including humans. In other words, when a new drug is found to be "safe" in animal tests, this provides essentially no evidential weight to the same being true in humans. Nonhuman primates (almost exclusively cynomolgus macaques *Macaca fascicularis*) seemed the worst species of all for human prediction. This is crucial, because the critical observation to allow a new drug to proceed to human trials is the *absence* of toxicity in animals. To quantify: Suppose a new drug in testing has a prior probability (based on prior information, such as similarity to other drugs, data from in vitro or in silico tests, and so on) of causing no adverse effects in humans of 70%. A subsequent negative test in nonhuman primates increases this probability of no adverse effects in humans to just 70.4%. While our analyses suggested toxicity in animals was also likely to be present in humans, this was very variable, with no clear pattern in terms of types of toxic effects or types of drugs, so could not be considered particularly consistent or reliable. To illustrate: the range of positive likelihood ratios (the statistical metric gauging the degree of evidential weight provided by the nonhuman primate tests to likelihood of human toxicity) was relatively high (605). This indicated that some tests (for some types of toxicity) were providing evidential weight for human toxicity, but others were not. The median value, however, was 9.39, which is lower than the acknowledged "diagnostic" value for a useful test of 10 (Grimes and Schulz 2005).

This should have serious implications for the pharmaceutical industry and its regulators. Many drugs that *are* toxic to humans are not detected in animal tests, resulting in risk of harm to clinical trial participants, and to consumers of drugs that do reach market. In fact, adverse drug reactions and drug failure rate have increased significantly over the past two decades—92% of drugs that appear safe and effective in animal tests go on to fail in human trials, and many (up to half in any 1 year) are subsequently withdrawn or relabeled due to adverse reactions not detected in the animal tests (e.g., BioSpace 2012). It is thus increasingly clear that animal testing of human drugs is not sufficiently predictive for humans, and this is especially true in light of the array of human-relevant methods now available to science for this purpose (see Sect. 4.6, "What Can Be Done?—Alternative Methods" below). Combined with the level of public concern over animal experiments, and the high ethical costs of doing so, testing drugs in animals cannot be justified ethically or scientifically. At the very least, it is incumbent on the pharmaceutical industry and its regulators—for too long virtually silent about this matter—to take on board these concerns, and to expedite a more predictive testing regime. To their credit, efforts are being made (U.S. Food and Drug Administration 2021; Tagle 2019), though many argue that much more could be done, and more quickly.

3.2 Disease Models/Basic Research

As for nonhuman primate use in drug testing, nonhuman primate experimentation to further our understanding of human diseases, and to realize treatments for them, has similarly been subject to very little retrospective analysis and critical reflection in spite of calls for this (Animal Procedures Committee 2003). Yet, claims of the essential nature of nonhuman primate experiments continue to be made by those who conduct, support, fund, and profit from them (e.g., SCHER 2009; The Boyd Group 2002; Tardif et al. 2013; Vallender and Miller 2013; Chan 2013), in the face of published studies outlining how poorly data from animals, including nonhuman primates, translate to humans. For example, just 0.3% of animal research publications report results that translate to human benefit (Lindl et al. 2005).

An examination of more than 25,000 papers from the best medical journals over a five-year period revealed that 101 of them showed novel therapeutic or preventive promises, of which 64 involved animal studies: 20 years later, just 19 of the 101 papers (19%) had resulted in a positive clinical trial (0.075% of the total number of papers examined); only five of the 27 technologies had been licensed for use (18.5% of the reported technologies); and just one had shown extensive clinical advantages (3.7% of technologies, or 0.004% of the total papers examined) (Contopoulos-Ioannidis et al. 2003; Crowley 2003). Crucially, this single study did not depend on the animal research, but instead was based on rational drug design (Cushman and Ondetti 1999). A review of the extent to which animal data correlate with humans concluded that animal tests fail to reliably predict effects in humans, and that many animal experiments are of poor quality (for example, with respect to randomization and blinding, which minimize bias) (Perel et al. 2007). Four of six interventions studied in animals failed to predict human outcomes, and in two cases they predicted benefit when the treatment was ineffective and harmful to humans (Perel et al. 2007). Some animal studies were conducted concurrently or even after human studies had demonstrated a treatment's efficacy, and there was a lack of communication between animal researchers and those conducting clinical trials, and the results of animal studies are not being incorporated into human research—the key justification for conducting them in the first place (Perel et al. 2007).

Regarding nonhuman primates, failures have been reported in many areas. Below I highlight the more prominent cases.

3.2.1 HIV/AIDS

Macaque use in HIV/AIDS research is considered by some scientists a failure and of questionable human relevance (da Silva and Richtmann 2006; Tonks 2007; D'Souza et al. 2004; Johnston 2000; Taylor 2006; Hu 2005; Lévy 2005; Tonini et al. 2005). Many, if not all of some 100 different types of HIV vaccines have been tested in nonhuman primates with positive results, yet none has provided protection or therapeutic benefit in humans—the case at the time of an extensive review (Bailey 2008b), and still heading toward 2020—more than a decade later. Many critical observations have been made of macaque HIV experiments, which exhibit major differences from HIV-positive people. These include, "efficacy of HIV-1 based vaccines cannot be directly evaluated in the SIV [Simian Immunodeficiency Virus] model" (Hu 2005); "...the crucial role of human testing in the development of any vaccine...human immune system variability or virus diversity can't really be mimicked by any of the currently used laboratory animal models" (Grant 2009);

"What is not known is how studies in monkeys will translate into humans" (Haynes and Burton 2017), and more.

3.2.2 Alzheimer's Disease

Years of effort have failed to create a "good" Alzheimer's disease (AD) animal model (Emerich et al. 2000; Conner and Tuszynski 1999; Lindholm 1997; Snowdon et al. 1997), and little progress has been made in understanding the pathologies associated with the disease (Pippin et al. 2019; Pistollato et al. 2016). For example, plaques and tangles in the brain are the hallmark of AD in humans but not in nonhuman primates (St George-Hyslop and Westaway 1999). Humans and great apes alone possess a particular type of projection neuron in the anterior cingulate cortex of the brain, which is severely affected in the degenerative process of AD (Nimchinsky et al. 1999). Also, a neuropeptide (molecules that help neurons communicate with each other), galanin, which regulates the function of the basal forebrain, differs in its chemoanatomic organization (i.e., is present in different types of neuron at different levels) across species (Mufson et al. 1998), which may exacerbate cholinergic cellular dysfunction in AD. Such differences could critically affect the human relevance of nonhuman primate research into AD.

Many experts in the field concluded some time ago that human, not animal, research was the way forward (Pippin et al. 2019; Pistollato et al. 2016). Human clinical research, epidemiological studies, and in vitro techniques have given rise to hypotheses for AD and important findings such as the AD-associated decrease in choline acetyltransferase (ChAT) activity, and links with the presenilin 1, presenilin 2, and APOE-e4 genes, and vitamin B12/folate deficiency and high fat/high cholesterol diets (Greek and Greek 2003). None relied on the use of nonhuman primates or other animals. In fact, nonhuman primate use can also harm humans. The once much-vaunted AD "vaccine" AN-1792 (AIP-001) dramatically slowed brain damage in an AD mouse model; further, it "was well tolerated when tested in several animal species, including monkeys" (Sibal and Samson 2001; Young 2002), yet clinical trials were suspended following inflammation of the central nervous system and ischemic strokes in 15 participants (Steinberg 2002). Several recent papers, including some in high-profile journals such as Nature and Science, have noted the failure, and been critical of, animal models of AD (Chakradhar 2018; Reardon 2018; Golde et al. 2018; Pistollato et al. 2016), and a book chapter summarized the sorry state of affairs comprehensively (Pippin et al. 2019) .

3.2.3 Stroke

Differences in responses to ischemic injury (injuries resulting from inadequate blood supply) exist between species (even between strains of mice). Extrapolation of data from one species to another with regard to stroke modeling is thus difficult or impossible (Huang et al. 2000). Decades of research, much in nonhuman primates, have produced more than 4000 reports involving more than 1300 successful interventions, including more than 700 for acute ischemic stroke, none of which has led to significant human benefit (O'Collins et al. 2006; Macleod and Registrar 2005; Macleod et al. 2004). Some consider animal models of stroke a failed paradigm, and have argued for more human research (Wiebers et al. 1990; Editorial 1990), and/or been critical of animal models or even lamented that animal models cannot be translated to humans (Molinari 1988; Neff 1989; Johnston 2006; Johnson and Goldstein 2003; Editorial 2006; Sena et al. 2010; Feuerstein and Chavez 2009). While more recent reviews of the failure of animal experiments in stroke research include other contributory factors, such as the design of both animal and human trials, and choice of animal species for particular experiments (Herson and Traystman 2014; Fluri et al. 2015), the predictive value of using animals remains an acknowledged issue.

3.2.4 Parkinson's Disease

Parkinson's disease (PD) is often studied in nonhuman primates in whom PD-like symptoms have been induced using neurotoxic chemicals (e.g., see Sect. 4.5.1, "Direct Harms" section below). The literature is replete with caveats concerning their reliability, human relevance, and applicability to PD. For example, many candidate therapeutic compounds have shown great promise and efficacy in nonhuman primates and murine models of PD, but none has proven to be efficacious in humans (Kouroupi et al. 2020). It is estimated that more than 2300 studies of PD have been conducted using nonhuman primates since 1990, predominantly to test new therapies, yet none of the new therapies showing promise have translated to clinical benefit in the form of a neuroprotective or disease-modifying drug (Konnova and Swanberg 2018; Duty and Jenner 2011). There is also poor translational value of experimental data generally from animal models (including nonhuman primates) to clinical studies (Rõlova et al. 2020).

Furthermore, significant gaps exist in the knowledge of causative events and factors, and pathological mechanisms, despite decades of research, much of it in animals, including thousands of nonhuman primates (Kouroupi et al. 2020; Duty and Jenner 2011). Animal models, including many nonhuman primate models, largely fail to recapitulate the human disease (Kouroupi et al. 2020; Outeiro et al. 2021; Konnova and Swanberg 2018; Duty and Jenner 2011), and this is likely to be due to notable species differences between humans and nonhuman primates, including "age-related alterations, brain size, gene-environment interactions, and genotype–phenotype relationships" (Sittig et al. 2016). For example, nonhuman primate models of PD in which the animals' brains have been injected with the neurotoxin MPTP to induce some PD-like symptoms show functional recovery, unlike human PD patients, and there is also a lack of animal models that show any of the nonmotor symptoms seen in humans with PD (Konnova and Swanberg 2018).

One often-claimed success of nonhuman primate research is deep brain stimulation (DBS) for the control of PD-related tremors. It involves implanting an electrode into the brain, to stimulate specific deep-brain areas, such as the subthalamic nucleus (STN) part of the brain's basal ganglia. Advocates of nonhuman primate research have long insisted that DBS of the STN could not have been developed without, and "owes everything to", this research (Benabid et al. 2015a, b). I published two detailed rebuttals (Bailey 2015a, b), showing, in contrast to these claims, that DBS of the STN "owes everything" to everything *but* nonhuman primate research. In brief, the macaque model of PD was not reported until 1983. Yet, the STN—targeted in DBS for PD therapy—was linked to movement disorders in the 1920s by human observations; the basal ganglia were being surgically targeted in patients to alleviate movement disorders in the 1940s; the suppression of movement disorder symptoms using electrode stimulation has been reported since the 1960s; and human DBS has been used since the 1970s, including in the basal ganglia, to control tremor. It is clear that monkey experiments on the 1980s reignited interest in DBS and the STN with regard to PD, but significant human data and experience had provided a huge weight of evidence in the decades prior to that.

The failure, in many circumstances, or at the very least the lack of necessity, of nonhuman primate experiments is not surprising when one appreciates how poor research involving the chimpanzee—the most closely related species, genetically, to humans—proved to be. Chimpanzees were used in biomedical research in the U.S. until very recently despite calls to terminate it being resisted by chimpanzee researchers for many years (VandeBerg and Zola 2005; Institute of Medicine and National Research Council 2011). The U.S. National Institutes of Health (NIH) eventually—under pressure from animal welfare groups—charged the U.S. Institute of Medicine (IOM) with conducting an evaluation. Much of my work critiqued chimpanzee research as part of a campaign to end it. This work, which I submitted and presented to the IOM as part of its inquiry, showed that chimpanzee research had failed in many areas (Bailey et al. 2007; Knight 2007), including HIV/AIDS (Bailey 2008b), hepatitis C (Bailey 2010a, b), and cancer (Bailey 2009), due in large part to myriad genetic differences between humans and chimpanzees (Bailey 2011). The IOM concluded in 2011 that chimpanzee research was *not* scientifically necessary, contrary to the claims of chimpanzee researchers (Institute of Medicine and National Research Council 2011). Its only exception was a disputed "evenly-split committee" regarding a narrow area of work into hepatitis C vaccines. However, NIH committed to no further funding of new invasive research on chimpanzees, to retiring the chimpanzees it supports to permanent sanctuaries, and the U.S. Fish and Wildlife Service declared that it will give the same endangered status and protections to chimpanzees in captivity as it does their free-living cousins in Africa. In the U.S., federally funded invasive research involving chimpanzees has ended.

One must ask the obvious question: if chimpanzees—our closest genetic relative, and therefore the species most likely to be reflective of, and predictive for, humans—cannot serve as good models in biomedical research, how can any other species that is further removed from humans genetically serve as a good model?

4 Reasons for Failure

We now know, increasingly, *why* animal experimentation is—and can only ever be—a failed paradigm. Some of the reasons are summarized in this section. More details may be found in published reviews of mine, detailing genetic differences between species (Bailey 2011, 2014a), as well as other confounding factors such as stress and anesthesia (Bailey 2017, 2018).

4.1 Genetic Differences Between Humans and Nonhuman Primates

The nonhuman primate species most commonly used in science are the rhesus (*Macaca mulatta*) and cynomolgus macaque (*Macaca fascicularis*—the long-tailed or crab-eating macaque). These species are around 90–93% genetically similar to humans (Gibbs et al. 2007). Yet, the 7–10% difference means that many biological differences manifest, confounding their use in research (Bailey 2014a). Old World monkeys like macaques diverged from a common ancestor between 14 and 35 million years ago (Raaum et al. 2005; Glazko and Nei 2003; Kulski et al. 2004; Kumar and Hedges 1998; Perelman et al. 2011), providing ample opportunity for differences to accrue, leading to major variation even within species. This means that regional populations of monkeys may show stark biological differences from each other, confounding their use as model organisms for human biology (Kanthaswamy et al. 2013; Sturt 1984). Six rhesus macaque subspecies have been noted, displaying a variety of biological differences that can impact susceptibility to malarial parasites, SIV infection and pathology, and the metabolism of drugs (Groves 2001) (see Sect. 4.4). If such differences can make it impossible to extrapolate data between nonhuman primates within the same genus, or even between populations of the same species, how can data be extrapolated from nonhuman primates to humans?

Notable differences appear at all levels of genetics and gene expression, that is, major genomic rearrangements (associated with, for instance, developmental disabilities, cancers, multiple sclerosis, and more); different types and prevalence of mobile and repetitive DNA elements (affecting gene expression and the immune system, and associated with cancers, hemophilias, and AD); gene complement (many hundreds of genes have been lost and gained between different primate species, many affecting immune function and drug metabolism); and differences in coding sequences within genes, which may affect gene function (Bailey 2011, 2014a, b). In addition, even where genes are shared by nonhuman and human primates, they are frequently expressed differently, and so hundreds or even thousands of such differences exist in all major organs and tissues, associated with many and varied diseases (Bailey 2011, 2014a, b). Further, gene expression is affected by other factors that differ between species, including epigenetic influences (in turn mediated by many environmental factors), transcription-factor differences (which regulate hundreds or thousands of other genes), micro-RNAs and small-interfering RNAs (also regulating the expression of hundreds or thousands of genes), and RNA-splicing mechanisms. Variability in these factors, and the power of the consequences, is illustrated by their association with many and varied diseases and disorders (see Bailey 2011, 2014a).

4.2 Consequences of Genetic Differences for Nonhuman Primate Use in Drug Tests

These genetic differences lead to major functional differences that seriously affect the utility of using nonhuman primates to inform human medicine (Bailey 2011, 2014a, b). One of the classes of genes most affected by interspecies differences is that involved in drug metabolism (how drugs are absorbed into the body, distributed around it, metabolized and excreted by it, and how those drugs and/or their associated metabolic products may be toxic). There is—surprisingly—a paucity of data, but it is clear that even very small genetic differences can have substantial effects (Bailey 2014a). Differences include the activity, levels, and presence of enzymes. It is acknowledged that species differences in cytochrome P450 genes pose serious problems in interpreting animal data for humans and must be done "with some caution" for some types, while "more caution" and even "major problems" are associated with other types (Guengerich 1997). These differences may be at the root of the empirical problems in drug testing described above, and the consequent record level of failure of new drugs as they progress from animal to human trials.

4.3 Consequences of Genetic Differences for Nonhuman Primate Use in Neuroscience

There has also been a startling lack of comparative study of the suitability of nonhuman primates to inform human neuroscience. While this has improved in recent years, species differences discovered to date should have resulted in a move away from using nonhuman primates (Bailey 2011, 2014a). In the brain, many differences between primate species are known. Thousands of genetic differences have been identified between the brains of humans and even their closest relative, the chimpanzee (Bailey 2011, 2014a). Many genes, even when common to some species, are expressed in one but not the other, or at different levels, and this is the case for many brain regions investigated to date. Many differences are associated with brain and central nervous system development, maintenance, structure, function, and/or with neurological diseases that are often studied in nonhuman primates. Species differences in brain architecture are associated with functional differences in sensory perception, visceral functions, higher-order cognitive functions, and emotional and reward-related behaviors. Further, various imaging technologies have revealed other consequences of altered gene expression between species, involving smaller scale and specialized morphology, including organization of the cortex ("laminar organization, cellular specialization, and structural association"), connectivity within and between various brain regions, aging, visual and auditory pathways, and the impact of this on various sensory and motor functions, awareness, emotions, perception; all of which are likely to confound translational research (Bailey 2011, 2014a).

While one analysis of brain connectivity suggests just over half of brain regions may be "similarly wired" between humans and macaques (Li et al. 2013); this means that almost half are not. Augmented by another study showing "major differences", especially in the inferior parietal, polar, and medial prefrontal cortices, implying differences in high-order cognition, emotional and reward-related behaviors, visceral functions, and sensory processing, these studies lead to concerns over the suitability of using macaques to inform human neuroscience (Li et al. 2013). Some researchers have noted the growing discrepancy between data from invasive, electrode-based nonhuman primate brain research and human data from fMRI imaging, predicting, for instance (though somewhat controversially), that the nonhuman primate model will undoubtedly "break down" at some point, as science pushes toward "higher-level processes such as consciousness, learning, and decision making" (Boynton 2011). Genetic differences are the basis of several other crucial differences in human and nonhuman primate neuroscientific investigations (see Bailey and Taylor 2016).

4.4 Stress Associated with Laboratory Environment and Experimentation

Stress associated with laboratory life and experimentation is widely acknowledged to affect the welfare of animals in labs, and is discussed below. Somewhat less appreciated, I argue, are its consequences specifically for the reliability and human relevance of data obtained from the animals, over and above unreliability due to inherent species differences. Overall, in all species examined, including humans, stress can perturb immune function and increase susceptibility to physical and mental disorders—though, as may be expected by now, the specifics may vary between species (Gurfein et al. 2012; Obernier and Baldwin 2006; Kurokawa et al. 2010; Wright 2011). Consequences for nonhuman primate data, specifically, have not been studied as much as might have been expected, but when one considers the effects in humans—hypertension, heart attacks, stroke, obesity, Alzheimer's disease, AIDS dementia complex (Raber 1998), poor responses to vaccines, increased susceptibility to infections, and accelerated progression of several diseases, and more, it is clear that experimental results from stressed nonhuman primates must be rendered unreliable (Bailey 2018).

The effects of stress on immune function may be particularly important, though, again, the specifics differ between species (Dhabhar 2009; Christian 2012; Sorenson et al. 2011). Human sufferers of post-traumatic stress disorder, for instance, show increased cytokine levels, impaired activity of natural killer cells, lower T-lymphocyte counts, and epigenetic modifications exerting a lifelong impact on immune and inflammatory function (Pace and Heim 2011); depressed adults show greater inflammatory responses to vaccinations (Glaser et al. 2003); and many more effects on immune function (e.g., Papathanassoglou et al. 2010; Nater et al. 2009; Kamezaki et al. 2012). Though I believe animal data translate very poorly to humans, it is generally acknowledged that this can be mitigated, to some degree and in some respects, by improved welfare (see Swan and Hickman 2014). The impact of stress is critical as it may exacerbate and compound crucial immune differences between humans and nonhumans—particularly as much animal experimentation involves infectious agents (Bailey 2011, 2014a). Many scientists have been aware of these effects, and cautioned against disregarding them (Mason et al. 1968; Roberts et al. 1995; Brenner et al. 1990), yet they often have been not reported or underreported in scientific papers (Reinhardt et al. 1995).

4.5 The Ethical Argument

Much of the objection to animal use in scientific experiments is that it is simply wrong and unethical to inflict pain and suffering on sentient beings, who cannot give consent, no matter any claimed benefits for mankind. This objection is particularly strong when it comes to animals that humans can empathize with, such as companion animals like dogs and cats, and species considered especially intelligent and evolutionarily close to humans, such as types of nonhuman primate.

4.5.1 Capacity to Suffer, and Direct Harms

Nonhuman primates are cognitively and emotionally advanced, and they suffer when they are captured from the wild, transported to laboratories and breeding centers, housed therein or in labs with restricted freedom and movement in alien environments, subjected to stressful routine procedures and experimental procedures, and so on (Bailey 2014a, b). Indeed, this is reflected in EU Directives, European Conventions (e.g., Council of Europe 1986) and in Animal Welfare Acts of EU member states, which stipulate, for example, that pain, suffering, and harm caused to animals must be ethically justifiable, and the choice of species must be carefully made and well justified, particularly with regard to "higher" species like dogs and nonhuman primates (European Commission 2010; Hasiwa et al. 2011) (and see chapter by Bayne et al., "The Welfare Impact of Regulations, Policies, Guidelines and Directives and Nonhuman Primate Welfare"). Pain and suffering in testing, specifically, are acknowledged by, for example, the Organization for Economic Co-operation and Development (OECD 2000) and the Nuffield Council on Bioethics (2005, p 66). These organizations accept many signs that animals are experiencing pain and/or distress and suffering, including for example, difficulty breathing, convulsions, coma, bleeding from any orifice, paralysis, and others. Often, pain- or distress-relieving drugs are withheld, due to concerns that they might alter the toxicity profile of the chemical being tested (Mena et al. 2010; Loepke and Soriano 2008). Overall, toxicity testing is regarded as imposing moderate to severe pain (Combes et al. 2002; National Research Council (NRC) (US) 1992, Table 4-3), and data from the U.S. and Canada indicate that regulatory testing procedures account for the vast majority of animals reported in the highest categories of pain and distress (Stephens et al. 2002).

Wild capture has been banned in many countries, including the UK (European Commission 2010), due to associated "significant distress, suffering and physical injury" (European Union Committee Sub-committee D (Environment and Agriculture) 2009; Williams et al. 2008; Cruelty Free International 2006). Conditions in breeding/holding facilities are often appalling (Cruelty Free International 2008, 2010). Despite bans and restrictions, wild-caught primates continue to be used to breed animals for export to labs in the U.S., which can cause serious and chronic welfare problems, and death (Honess et al. 2004; Fernström et al. 2008).

Once in the lab, many experiments cause pain and suffering (Tardif et al. 2013; Scientific Committee on Animal Health and Welfare 2002; NC3Rs 2018), often involving highly invasive, stressful procedures, including single caging for up to 10–13 years (Lutz et al. 2003; Griffis et al. 2013), the induction of movement disabilities such as tremor and rigidity, severing spinal nerves, drug infusion into the brain, organ removal, repeated tissue biopsies, exposure to nerve gases, infection with anthrax, and death due either to the experiments or deliberate killing at their end. Suffering in neuroscience experiments is often classified as "substantial" (European Commission 2010), involving brain lesions and damage, electrodes implanted in the brain, chair-based restraint, implanted eye coils, water deprivation, and implanted skull devices to prevent head movement.

Parkinson's research has high impact. Neurotoxins such as MPTP may be injected into nonhuman primates' brains to inflict damage, resulting in movement disorders, tremors, inability to feed, etc. (Henderson et al. 1998; Escola et al. 2002). Nonhuman primates may be housed singly throughout the experiments, during which time they have recordings taken from implanted brain electrodes with their heads restrained, and overall they experience "distress" and "devastating welfare costs" (Anon 2007; Escola et al. 2002; Henderson et al. 1998). In stroke research, surgery frequently involves removal of part of the skull (craniectomy), or removing an eye and using a drill to enter the skull through the eye socket, so that blood vessels can be blocked often while nonhuman primates are awake and restrained (Fukuda and del Zoppo 2003; Nudo et al. 2003; Gao et al. 2006).

As previously stated, most nonhuman primates are used to test the safety of new drugs (approaching 80% in the UK; UK Home Office 2020). Many of these tests cause suffering that can be severe and prolonged, with nonhuman primates receiving the test substance daily for up to 9 months (National Toxicology Program 2011; European Commission 2010). They may breathe the test substance via a mask, or orogastric gavage is often used for administration directly into the stomach, involving forcible restraint and therefore significant pain and distress (European Commission 2010; Maninger et al. 2010; Reinhardt et al. 1995) (and see below). Complications can include accidental administration of substances into the trachea and lungs, aspiration pneumonia, esophageal trauma or perforation, gastric distension, edematous lungs, hemothorax, and death (Balcombe et al. 2004). Restraint chairs may be used, causing psychological stress and distress, similarities to human post-traumatic stress disorder with repeated use (Suliman et al. 2009; Musazzi et al. 2018; Fenster et al. 2018; Compean and Hamner 2018; Garfin et al. 2018; Cohen et al. 2015; Maninger et al. 2010; Capaldo and Bradshaw 2011; Bradshaw et al. 2008; Ferdowsian et al. 2011), and inguinal hernia and rectal prolapse (Reinhardt et al. 1995).

4.5.2 Indirect Harms

Nonhuman primates housed in labs and undergoing experiments experience considerable stress and distress, which exacerbate suffering (see above). While all species experience stress, psychological and physiological problems may arise with recurrent stressors, and/or when stress becomes chronic and too difficult to cope with. This results in excessive wear and tear on the body, and may manifest in psychological trauma (distress) and altered physiological responses, which can be harmful (Maestripieri and Hoffman 2011). The point is that it is increasingly acknowledged that diverse species, from fish to nonhuman primates, experience pain, stress and distress, and/or depression and anxiety disorders (see Ferdowsian et al. 2011).

I discussed the inherent, often intractable, stress from laboratory life and its impact on the welfare of animals in laboratories, and on the reliability of scientific data from those animals, in two reviews to which I refer the reader for more detail and references (Bailey 2017, 2018). Briefly, animals in laboratories experience significant and repeated stress, which is unavoidable and is caused by many aspects of laboratory life. This is difficult to mitigate, and lack of significant desensitization/habituation can result in considerable psychological and physiological welfare problems mediated by neuroendocrine networks with numerous and pervasive effects. Psychological damage can be reflected in stereotypical behaviors, including repetitive pacing and circling, and even self-harm. Physical consequences include adverse effects on immune function, inflammatory responses, metabolism, and disease susceptibility and progression. Some of these effects are epigenetic, and therefore potentially transgenerational. These effects must have consequences for the reliability of experimental data and their extrapolation to humans, and this may not be recognized sufficiently among those who use animals in experiments.

4.6 What Can Be Done? Alternative Methods

Scientific evidence supporting the case for ending the use of nonhuman primates in harmful experiments is increasing, and is formidable enough to stand separate from the ethical case. Nonhuman primate experimentation involves significant harms, so should not continue. Science and ethics demand that an approach must be viewed critically and comparatively with alternatives. This is especially true when that approach causes direct harm (to nonhuman primates) and indirect harm (to humans seeking treatments for diseases). Yet, there is an extensive array of humane investigative techniques at our disposal, many of which could barely have been envisaged just a few decades or even years ago. These methods permit extremely powerful, human-specific, and humane experimentation, and also augment proven, humane, and human-relevant methods of research that have existed for many years, and which are the bedrock of biomedical progress.

The breadth, capacity, and potential of alternatives can be appreciated in any abstract compilations accompanying relevant worldwide congresses, for example, the Proceedings of World Congresses on Alternatives (https://altex.ch). Of particular relevance to nonhuman primates, alternatives in drug testing include growing human cells and tissues in the laboratory. While cells in a dish are different, to some degree, from cells in living people, human cells are likely to be more relevant to human biology than are animal cells, even in whole, living animals (Bailey 2014a). A drive toward their use in place of animal tests has thus been outlined in road maps, such as by the U.S. NRC, FDA, EPA, and others. Such human cell- and tissue-cultures are constantly being improved, for example using ethically derived stem cells (e.g., those not derived from destroyed embryos). These cultures have the potential to develop into the different types of specialized cells found throughout the human body, and 3D rather than 2D cultures, which help maintain faithful physiological function, and which permit the coculture of several types of cells in close proximity in the form of "mini-organs" or organoids, and better reflect in vivo functionality. One application of these technologies is the "organ/body on a chip" approach, in which several distinct 3D-cultures of human cells are grown on a plate, the size of a smartphone, with circulatory systems connecting them, better mimicking real-life systems. Companies such as TissUse, CN Bio Innovations, Emulate, and Kirkstall are leading the way, with 10-plus-organ chips now a reality, and a host of scientific publications supporting their validity (see, e.g., tissuse.com, cn-bio.com, kirkstall.com, for example (Ramachandran et al. 2015; Tsamandouras et al. 2017; Beilmann et al. 2018).

Many scientific bodies concur. In the U.S., for instance, the National Academy of Sciences (NAS), NRC, Environmental Protection Agency, and FDA have all stated, some years ago, that toxicology *must* transform from testing in animals toward such alternatives (National Research Council (NRC) (US) 2007). Microdosing also shows great promise to revolutionize drug development and predictive drug toxicology: by administering approximately 1% of the estimated pharmacological dose of a potential new drug to human volunteers, important and predictive human-specific data can be obtained (Burt et al. 2018; Burt and Combes 2018), with no need for animal tests. By sampling and analyzing the subject's blood and other bodily fluids, and use of clinical imaging techniques, valuable human-specific properties of the potential new drug can be identified (Burt et al. 2018; Burt and Combes 2018). In combination with in silico (computer-based) data analysis and modeling, this information can greatly aid decisions to proceed to later stage clinical trials that determine human efficacy and toxicity (Fuloria et al. 2013; Lappin et al. 2006, 2013; Burt et al. 2018; Burt and Combes 2018).

Disease research also is more advanced than ever, using imaging techniques and DNA-based technologies to drill down to the very causes and pathology of human diseases. For example, "micro-lungs" are used to research and test new drugs for lung disorders such as asthma. Around 400 mini lungs can be obtained from just one tissue sample, which can also be used by cosmetic companies to test the safety of aerosols, such as hairspray (BéruBé 2013; Barker 2014). Microbrains are being used to research human brain development and brain disorders (Lancaster et al. 2013; Lancaster and Knoblich 2014), and have the potential to augment other alternatives in neuroscience, which presently uses many nonhuman primates. I discussed the scope of these methods, which provide powerful means to elucidate human brain function without recourse to nonhuman primate experiments in a review (Bailey and Taylor 2016).

Examples of methods discussed include transcranial magnetic stimulation (TMS), functional magnetic resonance imaging (fMRI), electrocorticography (ECoG), intracranial electroencephalography (EEG), magnetoencephalography (MEG), single unit/microelectrode recordings, cortico-cortical evoked potentials (CCEP), diffusion tensor imaging (DTI), transcranial direct-current stimulation (tDCS), postmortem human brain dissection, and others. I argue that these approaches make the nonhuman primate work unnecessary, particularly when one considers the associated advantages of species specificity. In light of these methods—which are constantly improving and providing more detailed data—it is perplexing why those who do similar research using nonhuman primates maintain that much of this type of research is the preserve of their field. When one considers the poor translation of nonhuman primate data to humans, largely based on genetic differences already discussed here, as well as other confounding factors mentioned above, the scientific argument against nonhuman primate use, and supporting human studies, is strong. Given the ethical costs of using nonhuman primates, as well as the opposition of much of the public to it, at the very least the onus is on those who use nonhuman primates to make their case.

4.7 Summary and Conclusion

Nonhuman primate research continues, with little or no reduction in numbers of animals used over time, based on an *assumption* that it has human relevance, benefits humans, and provides a means of obtaining important data that cannot be obtained in any other way. Yet, in the year 2021, and following decades of nonhuman primate research, there is, in my estimation, no evidence (at least nothing published that withstands critical scrutiny) supporting a scientific rationale for it and this must be balanced against the costs to the nonhuman primates. The only possible, valid conclusion must be that nonhuman primate experimentation continues out of habit and to serve vested interests—to support those who benefit from it, either financially or professionally, and arguably out of a resolve not to be proved wrong or to concede. To compound matters, any substantive critique of nonhuman primate research is resisted and ignored, save for the occasional "in house" self-affirmation exercise (e.g., Bailey et al. 2008; Bailey and Taylor 2009; Balls 2018; Bateson 2011; Greek and Hansen 2011; Knight 2012; SCHEER 2017; VandeBerg and Zola 2005; Weatherall 2006). Far from doing what ethics demand and searching for reason to move away from nonhuman primate research, too many within the industry, instead, seek reasons to defend it. Their reasons may be superficial and specious; some even seek to find *new* avenues to use nonhuman primates, and to denigrate proposed alternatives which, if they are not perfect, are brushed aside. These issues within science have been acknowledged and criticized for some time. As long ago as 1964, Platt, writing in Science, urged a "Strong inference" attitude to scientific research, and decried a resistance within science to ask what was being done wrong, and to test, disprove, and exclude hypotheses. This had led to scientists being shackled to particular approaches, and to habitually being method, instead of problem, oriented (Platt 1964). In 1974, physicist Feynman similarly denounced what he called "Cargo cult science" (which may be paraphrased as "It has worked at times in the past, so it doesn't need to be disproved"), and advocated science with integrity, in which researchers asked tough questions of their approaches and attempted to invalidate them, embracing all available information (not simply whatever suited them), and eliminating methods that were not good enough (Feynman 1974). For all these reasons, and others, nonhuman primate experiments, never validated in any way, persist (e.g., Bailey et al. 2008; Bailey 2015a, b; Bailey and Taylor 2009; Balls 2018; Bateson 2011; Benabid et al. 2015a, b; Greek and Hansen 2011; Knight 2012; SCHEER 2017; Taylor 2018; VandeBerg and Zola 2005; Weatherall 2006).

I have argued here, with relatively few examples from the voluminous published and peer-reviewed evidence, that the need for, and benefits from, nonhuman primate research is greatly overstated, if not absent. This is certainly the case with regard to what it has contributed to science and medicine, which, while exaggerated, is of academic interest only, and cannot therefore be used to support any contemporary or future need for it. It also, however, applies to speculations of future necessity. Simultaneously, alternatives to nonhuman primate research are underplayed, and their power, the significance of what they have contributed to medicine, and what they can contribute now, and in the future, are overlooked while focus falls on their shortcomings.

The lack of translation of nonhuman primate experiments to humans is inherent, and due to intractable, substantial biological differences, borne of myriad and widespread genetic and gene expression differences. These differences affect the biological systems that matter most for nonhuman primate research: the immune system and systems relating to the metabolism of drugs. These differences are further confounded by the serious consequences of unavoidable chronic stress experienced in labs, anesthesia, sex differences, and more. In fact, we know that any one individual of *any* species cannot reliably model any other individual of the same species. Many drugs are likely to work in a minority of people; different people are affected differently by adverse effects, by infectious agents, by diseases, and so on. It must therefore be folly to claim that experimental data from one species can inform another. There may be some examples of nonhuman primate data that correlate with human data, but this cannot be used to imply or infer general value: all the research that did not translate to humans must be taken into account and weighed against the few examples that do.

The case, then, that research would be much more successful if science moved away from using nonhuman primates, is powerful. When the ethical aspects—the "harm" or "costs"—are considered too, it becomes overwhelming. And the public, internationally, agrees. It is long overdue for those who cause pain and suffering to nonhuman primates in the name of science to embrace a less inhumane, and more human-relevant, line of inquiry. At the very least, I hope this chapter will lead some nonhuman primate researchers to pause for more, and more critical, reflection, and to embrace alternatives to animal use with alacrity and positivity. Nonhuman primates, *and humanity*, will benefit.

References

Animal Procedures Committee (2003) Review of cost-benefit assessment in the use of animals in research
Anon (2007) R (BUAV) v Secretary of State for the Home Department [2007] EWHC 1964 (Admin) (High Court) and [2008] EWCA Civ 417 (Court of Appeal)
Bailey J (2008a) Developmental toxicity testing: protecting future generations? Altern Lab Anim 36 (6):718–721. https://doi.org/10.1177/026119290803600618
Bailey J (2008b) An assessment of the role of chimpanzees in AIDS vaccine research. Altern Lab Anim 36(4):381–428. https://doi.org/10.1177/026119290803600403
Bailey J (2009) An examination of chimpanzee use in human cancer research. Altern Lab Anim 37 (4):399–416. https://doi.org/10.1177/026119290903700410
Bailey J (2010a) An assessment of the use of chimpanzees in hepatitis C research past, present and future: 1. Validity of the chimpanzee model. Altern Lab Anim 38(5):387–418. https://doi.org/10.1177/026119291003800501
Bailey J (2010b) An assessment of the use of chimpanzees in hepatitis C research past, present and future: 2. Alternative replacement methods. Altern Lab Anim 38(6):471–494. https://doi.org/10.1177/026119291003800602
Bailey J (2011) Lessons from chimpanzee-based research on human disease: the implications of genetic differences. Altern Lab Anim 39(6):527–540. https://doi.org/10.1177/026119291103900608
Bailey J (2014a) Monkey-based research on human disease: the implications of genetic differences. Altern Lab Anim 42(5):287–317. https://doi.org/10.1177/026119291404200504
Bailey J (2014b) A response to the ABPI's letter to the use of dogs in predicting drug toxicity in humans. Altern Lab Anim 42(2):149–153. https://doi.org/10.1177/026119291404200208
Bailey J (2015a) Letter to the editor. Altern Lab Anim 43(3):206–207. https://doi.org/10.1177/026119291504300310
Bailey J (2015b) Letter to the editor. Altern Lab Anim 43(6):428–431. https://doi.org/10.1177/026119291504300612
Bailey J (2017) Does the stress inherent to laboratory life and experimentation on animals adversely affect research data. Altern Lab Anim 45(6):299–301. https://doi.org/10.1177/026119291704500605
Bailey J (2018) Does the stress of laboratory life and experimentation on animals adversely affect research data? A critical review. Altern Lab Anim 46(5):291–305. https://doi.org/10.1177/026119291804600501
Bailey J, Balls M (2019) Recent efforts to elucidate the scientific validity of animal-based drug tests by the pharmaceutical industry, pro-testing lobby groups, and animal welfare organisations. BMC Med Ethics 20(1):16. https://doi.org/10.1186/s12910-019-0352-3
Bailey J, Taylor K (2009) The SCHER report on non-human primate research - biased and deeply flawed. Altern Lab Anim 37(4):427–435. https://doi.org/10.1177/026119290903700412
Bailey J, Taylor K (2016) Nonhuman primates in neuroscience research: the case against its scientific necessity. Altern Lab Anim 44(1):43–69. https://doi.org/10.1177/026119291604400101
Bailey J, Balcombe J, Capaldo T (2007) Chimpanzee research: an examination of its contribution to biomedical knowledge and efficacy in combating human diseases. https://releasechimps.org/resources/publication/chimpanzee-research. Accessed 18 June 2019
Bailey J, Capaldo T, Conlee K et al (2008) Experimental use of nonhuman primates is not a simple problem. Nat Med 14:1011–1012. https://doi.org/10.1038/nm1008-1011b
Bailey J, Thew M, Balls M (2013) An analysis of the use of dogs in predicting human toxicology and drug safety. Altern Lab Anim 41(5):335–350. https://doi.org/10.1177/026119291304100504
Bailey J, Thew M, Balls M (2014) An analysis of the use of animal models in predicting human toxicology and drug safety. Altern Lab Anim 42(3):181–199. https://doi.org/10.1177/026119291404200306
Bailey J, Thew M, Balls M (2015) Predicting human drug toxicity and safety via animal tests: can any one species predict drug toxicity in any other, and do monkeys help? Altern Lab Anim 43 (6):393–403. https://doi.org/10.1177/026119291504300607
Balcombe JP, Barnard ND, Sandusky C (2004) Laboratory routines cause animal stress. Contemp Top Lab Anim Sci 43(6):42–51

Balls M (2018) Why are validated alternatives not being used to replace animal tests? Altern Lab Anim 46(1):1–3. https://doi.org/10.1177/026119291804600105

Barker C (2014) "Human lung" model developed to replace animal testing. https://www.cosmeticsdesign-europe.com/Article/2013/11/15/Human-lung-model-developed-to-replace-animal-testing. Accessed 18 June 2019

Bateson P (2011) Review of research using non-human primates. Available https://wellcomeacuk/sites/default/files/wtvm052279_1pdf. Accessed 18 June 2019

Beilmann M, Boonen H, Czich A et al (2018) Optimizing drug discovery by investigative toxicology: current and future trends. ALTEX 36(2):289–313. https://doi.org/10.14573/altex.1808181

Benabid AL, Delong M, Hariz M (2015a) Letter to the editor. Altern Lab Anim 43(3):205–206. https://doi.org/10.1177/026119291504300309

Benabid AL, Delong M, Hariz M (2015b) Letter to the editor. Altern Lab Anim 43(6):427–428. https://doi.org/10.1177/026119291504300611

BéruBé KA (2013) Medical waste tissues - breathing life back into respiratory research. Altern Lab Anim 41(6):429–434. https://doi.org/10.1177/026119291304100604

BioSpace (2012) Annual R&D general metrics study highlights new success rate drugs and cycle time data. https://www.biospace.com/article/releases/-b-pharmaceutical-benchmarking-forum-b-annual-r-and-d-general-metrics-study-highlights-new-success-rate-and-cycle-time-data-/. Accessed 26 Oct 2021

Boynton GM (2011) Spikes, BOLD, attention, and awareness: a comparison of electrophysiological and fMRI signals in V1. J Vis 11(5):12. https://doi.org/10.1167/11.5.12

Bradshaw GA, Capaldo T, Lindner L et al (2008) Building an inner sanctuary: complex PTSD in chimpanzees. J Trauma Dissociation 9(1):9–34. https://doi.org/10.1080/15299730802073619

Brenner GJ, Cohen N, Ader R et al (1990) Increased pulmonary metastases and natural killer cell activity in mice following handling. Life Sci 47(20):1813–1819. https://doi.org/10.1016/0024-3205(90)90283-w

BUAV/YouGov plc (2009) Opinion poll on animal experiments, available on request from Cruelty Free International, London. Sample Size: 7139, Fieldwork: 24th February - 4th March 2009

Burt T, Combes RD (2018) The use of imaging, biomonitoring and microdosing in human volunteers to improve safety assessments and clinical development. In: Balls M, Combes R, Worth A (eds) The history of alternative test methods in toxicology. Academic Press, Cambridge, MA

Burt T, Vuong LT, Baker E et al (2018) Phase 0, including microdosing approaches: applying the three Rs and increasing the efficiency of human drug development. Altern Lab Anim 46(6):335–346. https://doi.org/10.1177/026119291804600603

Capaldo T, Bradshaw GA (2011) The bioethics of chimpanzee psychological wellbeing: Psychiatric injury and duty of care. Animals and Society Institute Animals and Society Institute Policy Paper. https://www.animalsandsociety.org/public-policy/policy-papers/the-bioethics-of-great-ape-well-being/. Accessed 18 June 2019

Chakradhar S (2018) Treatments that made headlines in 2018. Nat Med 24:1785–1787. https://doi.org/10.1038/s41591-018-0292-3

Chan AWS (2013) Progress and prospects for genetic modification of nonhuman primate models in biomedical research. ILAR J 54(2):211–223. https://doi.org/10.1093/ilar/ilt035

Christian LM (2012) Psychoneuroimmunology in pregnancy: immune pathways linking stress with maternal health, adverse birth outcomes, and fetal development. Neurosci Biobehav Rev 36(1):350–361. https://doi.org/10.1016/j.neubiorev.2011.07.005

Cohen BE, Edmondson D, Kronish IM (2015) State of the art review: depression, stress, anxiety, and cardiovascular disease. Am J Hypertens 28(11):1295–1302. https://doi.org/10.1093/ajh/hpv047

Combes R, Schechtman L, Stokes WS, Blakey D (2002) The International Symposium on Regulatory Testing and Animal Welfare: Recommendations on Best Scientific Practices for Subchronic/Chronic Toxicity and Carcinogenicity Testing. ILAR Journal V43 Supplement. https://academic.oup.com/ilarjournal/article/43/Suppl_1/S112/756645. Accessed 18 June 2019

Compean E, Hamner M (2018) Post-traumatic stress disorder with secondary psychotic features (PTSD-SP): diagnostic and treatment challenges. Prog Neuro-Psychopharmacol Biol Psychiatry 88:265–275. https://doi.org/10.1016/j.pnpbp.2018.08.001

Conner JM, Tuszynski MH (1999) Cholinergic lesions as a model of Alzheimer's disease. In: Emerich DF (ed) Central nervous system diseases: innovative animal models from lab to clinic. Humana Press, Totowa, NJ, pp 65–80

Contopoulos-Ioannidis DG, Ntzani E, Ioannidis JP (2003) Translation of highly promising basic science research into clinical applications. Am J Med 114(6):477–484. https://doi.org/10.1016/s0002-9343(03)00013-5

Council of Europe (1986) European Convention for the Protection of Vertebrate Animals used for Experimental and Other Scientific Purposes.http://conventions.coe.int/treaty/en/treaties/html/123.htm. Accessed 18 June 2019

Crowley WFJ (2003) Translation of basic research into useful treatments: how often does it occur? Am J Med 114(6):503–505. https://doi.org/10.1016/S0002-9343(03)00119-0

Cruelty Free International (2006) Monkey business in Vietnam. Available on request

Cruelty Free International (2008) Cambodia: the trade in primates for research. Available on request

Cruelty Free International (2010) Mauritius: the trade in primates for research. Available on request

Cushman DW, Ondetti MA (1999) Design of angiotensin-converting enzyme inhibitors. Nat Med 5:1110–1113. https://doi.org/10.1038/13423

D'Souza MP, Allen M, Sheets R et al (2004) Current advances in HIV vaccines. Curr HIV/AIDS Rep 1(1):18–24. https://doi.org/10.1007/s11904-004-0003-1

da Silva LJ, Richtmann R (2006) Vaccines under development: Group B streptococcus, herpes-zoster, HIV, malaria and dengue. J Pediatr (Rio J) 82(3 Suppl):S115–S124. https://doi.org/10.2223/jped.1476

Dhabhar FS (2009) Enhancing versus suppressive effects of stress on immune function: implications for immunoprotection and immunopathology. Neuroimmunomodulation 16(5):300–317. https://doi.org/10.1159/000216188

Duty S, Jenner P (2011) Animal models of Parkinson's disease: a source of novel treatments and clues to the cause of the disease. Br J Pharmacol 164(4):1357–1391. https://doi.org/10.1111/j.1476-5381.2011.01426.x

Editorial (1990) Relevance of animal models to stroke. Stroke 21(7):1091–1092. https://doi.org/10.1161/01.STR.21.7.1091

Editorial (2006) Neuroprotection: the end of an era? Lancet 368(9547):1548. https://doi.org/10.1016/s0140-6736(06)69645-1

Editorial (2008) When less is not more. Nat Med 14:791–791. https://doi.org/10.1038/nm0808-791

Emerich DF, Dean RL, Sanberg PR (2000) Central nervous system diseases: innovative animal models from lab to clinic. Humana Press, Totowa, NJ

Escola L, Michelet T, Douillard G et al (2002) Disruption of the proprioceptive mapping in the medial wall of parkinsonian monkeys. Ann Neurol 52(5):581–587. https://doi.org/10.1002/ana.10337

European Commission (2010) Directive 2010/63/EU of the European Parliament and of the Council of 22 September 2010 on the protection of animals used for scientific purposes. https://eur-lex.europa.eu/LexUriServ/LexUriServ.do?uri=OJ:L:2010:276:0033:0079:EN:PDF. Accessed 18 June 2019

European Commission (2013) Seventh report on the statistics on the number of animals used for experimental and other scientific purposes in the member states of the European Union. Report from the Commission to the Council and the European Parliament. https://eur-lex.europa.eu/LexUriServ/LexUriServ.do?uri=COM:2013:0859:FIN:EN:PDF. Accessed 18 June 2019

European Parliament (2004) Directive 2004/27/EEC of the European Parliament and the Council of 31 March 2004, on the Community code relative to medicinal products for human use. https://ec.europa.eu/health//sites/health/files/files/eudralex/vol-1/dir_2004_27/dir_2004_27_en.pdf. Accessed 18 June 2019

European Union Committee Sub-committee D (Environment and Agriculture) (2009) Inquiry into the Revision of the Directive on the protection of animals used for scientific purposes: summary of evidence submitted by the Animal Procedures Committee. https://publications.parliament.uk/pa/ld200809/ldselect/ldeucom/164/164i.pdf. Accessed 18 June 2019

Fenster RJ, Lebois LAM, Ressler KJ et al (2018) Brain circuit dysfunction in post-traumatic stress disorder: from mouse to man. Nat Rev Neurosci 19(9):535–551. https://doi.org/10.1038/s41583-018-0039-7

Ferdowsian HR, Durham DL, Kimwele C et al (2011) Signs of mood and anxiety disorders in chimpanzees. PLoS One 6(6):e19855. https://doi.org/10.1371/journal.pone.0019855

Fernström AL, Sutian W, Royo F et al (2008) Stress in cynomolgus monkeys (*Macaca fascicularis*) subjected to long-distance transport and simulated transport housing conditions. Stress 11(6):467–476. https://doi.org/10.1080/10253890801903359

Feuerstein GZ, Chavez J (2009) Translational medicine for stroke drug discovery: the pharmaceutical industry perspective. Stroke 40(3 Suppl):S121–S125. https://doi.org/10.1161/strokeaha.108.535104

Feynman R (1974) Cargo cult science. Eng Sci 37(7):10–13

Fluri F, Schuhmann MK, Kleinschnitz C (2015) Animal models of ischemic stroke and their application in clinical research. Drug Des Dev Ther 9:3445–3454. https://doi.org/10.2147/dddt.s56071

Foster JR (2005) Spontaneous and drug-induced hepatic pathology of the laboratory beagle dog, the cynomolgus macaque and the marmoset. Toxicol Pathol 33(1):63–74. https://doi.org/10.1080/01926230590890196

Fukuda S, del Zoppo GJ (2003) Models of focal cerebral ischemia in the nonhuman primate. ILAR J 44(2):96–104. https://doi.org/10.1093/ilar.44.2.96

Fuloria NK, Fuloria S, Vakiloddin S (2013) Phase zero trials: a novel approach in drug development process. Ren Fail 35(7):1044–1053. https://doi.org/10.3109/0886022x.2013.810543

Gao H, Liu Y, Lu S et al (2006) A reversible middle cerebral artery occlusion model using intraluminal balloon technique in monkeys. J Stroke Cerebrovasc Dis 15(5):202–208. https://doi.org/10.1016/j.jstrokecerebrovasdis.2006.05.010

Garfin DR, Thompson RR, Holman EA (2018) Acute stress and subsequent health outcomes: A systematic review. J Psychosom Res 112:107–113. https://doi.org/10.1016/j.jpsychores.2018.05.017

Gibbs RA, Rogers J, Katze MG et al (2007) Evolutionary and biomedical insights from the rhesus macaque genome. Science 316(5822):222–234. https://doi.org/10.1126/science.1139247

Glaser R, Robles TF, Sheridan J et al (2003) Mild depressive symptoms are associated with amplified and prolonged inflammatory responses after influenza virus vaccination in older adults. Arch Gen Psychiatry 60(10):1009–1014. https://doi.org/10.1001/archpsyc.60.10.1009

Glazko GV, Nei M (2003) Estimation of divergence times for major lineages of primate species. Mol Biol Evol 20(3):424–434. https://doi.org/10.1093/molbev/msg050

Golde TE, DeKosky ST, Galasko D (2018) Alzheimer's disease: the right drug, the right time. Science 362(6420):1250–1251. https://doi.org/10.1126/science.aau0437

Grant B (2009) HIV vax testers react to Thai trial. The Scientist. https://www.the-scientist.com/the-nutshell/hiv-vax-testers-react-to-thai-trial-43871. Accessed 18 June 2019

Greaves P, Williams A, Eve M (2004) First dose of potential new medicines to humans: how animals help. Nat Rev Drug Discov 3(3):226–236. https://doi.org/10.1038/nrd1329

Greek CR, Greek JS (2003) Specious science: how genetics and evolution reveal why medical research on animals harms humans. Continuum, New York

Greek R, Hansen LA (2011) An analysis of the Bateson Review of research using nonhuman primates. Medicoleg Bioeth 1:3–22. https://doi.org/10.2147/MB.S25938

Griffis CM, Martin AL, Perlman JE et al (2013) Play caging benefits the behavior of singly housed laboratory rhesus macaques (*Macaca mulatta*). J Am Assoc Lab Anim Sci 52(5):534–540

Grimes DA, Schulz KF (2005) Refining clinical diagnosis with likelihood ratios. Lancet 365(9469):1500–1505. https://doi.org/10.1016/s0140-6736(05)66422-7

Groves CP (2001) Primate taxonomy (Smithsonian series in comparative evolutionary biology). Smithsonian Books, Washington, DC

Guengerich FP (1997) Comparisons of catalytic selectivity of cytochrome P450 subfamily enzymes from different species. Chem Biol Interact 106(3):161–182. https://doi.org/10.1016/s0009-2797(97)00068-9

Gurfein BT, Stamm AW, Bacchetti P et al (2012) The calm mouse: an animal model of stress reduction. Mol Med 18(1):606–617. https://doi.org/10.2119/molmed.2012.00053

Hasiwa N, Bailey J, Clausing P et al (2011) Critical evaluation of the use of dogs in biomedical research and testing in Europe. ALTEX 28(4):326–340. https://doi.org/10.14573/altex.2011.4.326

Haynes BF, Burton DR (2017) Developing an HIV vaccine. Science 355(6330):1129–1130. https://doi.org/10.1126/science.aan0662

Henderson JM, Annett LE, Torres EM et al (1998) Behavioural effects of subthalamic nucleus lesions in the hemiparkinsonian marmoset (*Callithrix jacchus*). Eur J Neurosci 10(2):689–698. https://doi.org/10.1046/j.1460-9568.1998.00077.x

Herson PS, Traystman RJ (2014) Animal models of stroke: translational potential at present and in 2050. Future Neurol 9(5):541–551. https://doi.org/10.2217/fnl.14.44

Honess PE, Johnson PJ, Wolfensohn SE (2004) A study of behavioural responses of non-human primates to air transport and re-housing. Lab Anim 38(2):119–132. https://doi.org/10.1258/002367704322968795

House of Lords (2002) Select committee on animals in scientific procedures. Volume I - report. https://publications.parliament.uk/pa/ld200102/ldselect/ldanimal/150/150.pdf. Accessed 18 June 2019

Hu SL (2005) Non-human primate models for AIDS vaccine research. Curr Drug Targets Infect Disord 5(2):193–201. https://doi.org/10.2174/1568005054201508

Huang J, Mocco J, Choudhri TF et al (2000) A modified transorbital baboon model of reperfused stroke. Stroke 31(12):3054–3063. https://doi.org/10.1161/01.str.31.12.3054

Institute of Medicine and National Research Council (2011) Chimpanzees in biomedical and behavioral research: Assessing the necessity. The National Academies Press, Washington, DC. https://doi.org/10.17226/13257

International Conference for Harmonisation of Technical Requirements for Registration of Pharmaceuticals for Human Use (2005) ICH Harmonised Tripartite Guideline: Detection of toxicity to reproduction for medicinal products & toxicity to male fertility S5(R2). https://www.ich.org/products/guidelines/safety/safety-single/article/detection-of-toxicity-to-reproduction-for-medicinal-products-toxicity-to-male-fertility.html. Accessed 18 June 2019

Ipsos MORI (2014) Attitudes to animal research in 2014 – a report by Ipsos MORI for the Department for Business, Innovation & Skills. https://www.ipsos-mori.com/researchpublications/publications/1695/Attitudes-to-animal-research-in-2014.aspx. Accessed 18 June 2019

Jennings CG, Landman R, Zhou Y et al (2016) Opportunities and challenges in modeling human brain disorders in transgenic primates. Nat Neurosci 19:1123–1130. https://doi.org/10.1038/nn.4362

Johnson EJ, Goldstein D (2003) Medicine. Do defaults save lives? Science 302(5649):1338–1339. https://doi.org/10.1126/science.1091721

Johnston MI (2000) The role of nonhuman primate models in AIDS vaccine development. Mol Med Today 6(7):267–270. https://doi.org/10.1016/s1357-4310(00)01724-x

Johnston SC (2006) Translation: case study in failure. Ann Neurol 59(3):447–448. https://doi.org/10.1002/ana.20783

Kamezaki Y, Katsuura S, Kuwano Y et al (2012) Circulating cytokine signatures in healthy medical students exposed to academic examination stress. Psychophysiology 49(7):991–997. https://doi.org/10.1111/j.1469-8986.2012.01371.x

Kanthaswamy S, Ng J, Satkoski Trask J et al (2013) The genetic composition of populations of cynomolgus macaques (*Macaca fascicularis*) used in biomedical research. J Med Primatol 42(3):120–131. https://doi.org/10.1111/jmp.12043

Knight A (2007) The poor contribution of chimpanzee experiments to biomedical progress. J Appl Anim Welf Sci 10(4):281–308. https://doi.org/10.1080/10888700701555501

Knight A (2008) 127 million non-human vertebrates used worldwide for scientific purposes in 2005. Altern Lab Anim 36(5):494–496

Knight A (2012) A critique of the Bateson Review of research using non-human primates. AATEX 17(2):53–62

Konnova EA, Swanberg M (2018) Animal models of Parkinson's disease. In: Stoker TB, Greenland JC (eds) Parkinson's disease: pathogenesis and clinical aspects. Codon Publications, Brisbane, Australia, pp 83–106

Kouroupi G, Antoniou N, Prodromidou K, Taoufik E, Matsas R (2020) Patient-derived induced pluripotent stem cell-based models in Parkinson's disease for drug identification. Int J Mol Sci 21(19):7113. https://doi.org/10.3390/ijms21197113

Kulski JK, Anzai T, Shiina T et al (2004) Rhesus macaque class I duplicon structures, organization, and evolution within the alpha block of the major histocompatibility complex. Mol Biol Evol 21(11):2079–2091. https://doi.org/10.1093/molbev/msh216

Kumar S, Hedges SB (1998) A molecular timescale for vertebrate evolution. Nature 392:917–920. https://doi.org/10.1038/31927

Kurokawa K, Kuwano Y, Tominaga K et al (2010) Brief naturalistic stress induces an alternative splice variant of SMG-1 lacking exon 63 in peripheral leukocytes. Neurosci Lett 484(2):128–132. https://doi.org/10.1016/j.neulet.2010.08.031

Lancaster MA, Knoblich JA (2014) Organogenesis in a dish: Modeling development and disease using organoid technologies. Science 345(6194):1247125. https://doi.org/10.1126/science.1247125

Lancaster MA, Renner M, Martin CA et al (2013) Cerebral organoids model human brain development and microcephaly. Nature 501:373–379. https://doi.org/10.1038/nature12517

Lappin G, Kuhnz W, Jochemsen R et al (2006) Use of microdosing to predict pharmacokinetics at the therapeutic dose: experience with 5 drugs. Clin Pharmacol Ther 80(3):203–215. https://doi.org/10.1016/j.clpt.2006.05.008

Lappin G, Noveck R, Burt T (2013) Microdosing and drug development: past, present and future. Expert Opin Drug Metab Toxicol 9(7):817–834. https://doi.org/10.1517/17425255.2013.786042

Lévy Y (2005) Therapeutic HIV vaccines: an update. Curr HIV/AIDS Rep 2:5–9. https://doi.org/10.1007/s11904-996-0002-5

Li L, Hu X, Preuss TM et al (2013) Mapping putative hubs in human, chimpanzee and rhesus macaque connectomes via diffusion tractography. NeuroImage 80:462–474. https://doi.org/10.1016/j.neuroimage.2013.04.024

Lindholm D (1997) Models to study the role of neurotrophic factors in neurodegeneration. J Neural Transm Suppl 49:33–42. https://doi.org/10.1007/978-3-7091-6844-8_4

Lindl T, Voelkel M, Kolar R (2005) Animal experiments in biomedical research. An evaluation of the clinical relevance of approved animal experimental projects: No evident implementation in human medicine within 10 years. ALTEX 22(3):143–151

Loepke AW, Soriano SG (2008) An assessment of the effects of general anesthetics on developing brain structure and neurocognitive function. Anesth Analg 106(6):1681–1707. https://doi.org/10.1213/ane.0b013e318167ad77

Lutz C, Well A, Novak M (2003) Stereotypic and self-injurious behavior in rhesus macaques: a survey and retrospective analysis of environment and early experience. Am J Primatol 60(1):1–15. https://doi.org/10.1002/ajp.10075

Macleod M, Registrar S (2005) What can systematic review and meta-analysis tell us about the experimental data supporting stroke drug development. Int J Neuroprot Neuroregen 1(3):201.

Archived at https://web.archive.org/web/20060819132325/http://www.ijnn.org/dl/vol1/iss2/IJNN_Vol1_Iss2.pdf

Macleod MR, O'Collins T, Howells DW et al (2004) Pooling of animal experimental data reveals influence of study design and publication bias. Stroke 35(5):1203–1208. https://doi.org/10.1161/01.str.0000125719.25853.20

Maestripieri D, Hoffman CL (2011) Chronic stress, allostatic load, and aging in nonhuman primates. Dev Psychopathol 23(4):1187–1195. https://doi.org/10.1017/s0954579411000551

Maninger N, Mason W, Ruys J et al (2010) Acute and chronic stress increase DHEAS concentrations in rhesus monkeys. Psychoneuroendocrinology 35(7):1055–1062. https://doi.org/10.1016/j.psyneuen.2010.01.006

Mason JW, Wool MS, Wherry FE et al (1968) Plasma growth hormone response to avoidance sessions in the monkey. Psychosom Med 30(5):760–773

Matthews RA (2008) Medical progress depends on animal models - doesn't it? J R Soc Med 101(2):95–98. https://doi.org/10.1258/jrsm.2007.070164

Mena MÁ, Perucho J, Rubio I et al (2010) Studies in animal models of the effects of anesthetics on behavior, biochemistry, and neuronal cell death. J Alzheimers Dis 22(Suppl 3):43–48. https://doi.org/10.3233/jad-2010-100822

Merkley R, Pippin JJ, Joffe AR (2018) A survey to understand public opinion regarding animal use in medical training. Altern Lab Anim 46(3):133–143. https://doi.org/10.1177/026119291804600308

Molinari GF (1988) Why model strokes? Stroke 19(10):1195–1197. https://doi.org/10.1161/01.STR.19.10.1195

Mufson EJ, Kahl U, Bowser R et al (1998) Galanin expression within the basal forebrain in Alzheimer's disease. Comments on therapeutic potential. Ann N Y Acad Sci 863:291–304. https://doi.org/10.1111/j.1749-6632.1998.tb10703.x

Musazzi L, Tornese P, Sala N et al (2018) What acute stress protocols can tell us about PTSD and stress-related neuropsychiatric disorders. Front Pharmacol 9:758. https://doi.org/10.3389/fphar.2018.00758

Nater UM, Whistler T, Lonergan W et al (2009) Impact of acute psychosocial stress on peripheral blood gene expression pathways in healthy men. Biol Psychol 82(2):125–132. https://doi.org/10.1016/j.biopsycho.2009.06.009

National Research Council (NRC) (US) (1992) Pain and distress in laboratory animals. National Academy Press, Washington DC. https://doi.org/10.17226/1542

National Research Council (NRC) (US) (2007) Toxicity testing for the 21st century: a vision and a strategy. http://dels.nas.edu/resources/static-assets/materials-based-on-reports/reports-in-brief/Toxicity_Testing_final.pdf. Accessed 18 June 2019

National Toxicology Program (2011) Specifications for the conduct of studies to evaluate the toxic and carcinogenic potential of chemical, biological and physical agents in laboratory animals for the National Toxicology Program (NTP). http://ntp.niehs.nih.gov/ntp/Test_Info/FinalNTP_ToxCarSpecsJan2011.pdf. Accessed 18 June 2019

NC3Rs (2018) The welfare of non-human primates. https://www.nc3rs.org.uk/welfare-non-human-primates. Accessed 18 June 2019

Neff S (1989) Clinical relevance of stroke models. Stroke 20(5):699–701. https://doi.org/10.1161/str.20.5.699b

Nimchinsky EA, Gilissen E, Allman JM et al (1999) A neuronal morphologic type unique to humans and great apes. Proc Natl Acad Sci U S A 96(9):5268–5273. https://doi.org/10.1073/pnas.96.9.5268

Nudo RJ, Larson D, Plautz EJ et al (2003) A squirrel monkey model of poststroke motor recovery. ILAR J 44(2):161–174. https://doi.org/10.1093/ilar.44.2.161

Nuffield Council on Bioethics (2005) The ethics of research involving animals. http://nuffieldbioethicsorg/wp-content/uploads/The-ethics-of-research-involving-animals-full-reportpdf. Accessed 18 June 2019

O'Collins VE, Macleod MR, Donnan GA et al (2006) 1,026 experimental treatments in acute stroke. Ann Neurol 59(3):467–477. https://doi.org/10.1002/ana.20741

Obernier JA, Baldwin RL (2006) Establishing an appropriate period of acclimatization following transportation of laboratory animals. ILAR J 47(4):364–369. https://doi.org/10.1093/ilar.47.4.364

OECD OFEC-oAD (2000) Guidance document on the recognition, assessment, and use of clinical signs as humane endpoints for experimental animals used in safety evaluation (ENV/JM/MONO (2000)7). http://www.oecd.org/officialdocuments/publicdisplaydocumentpdf/?cote=ENV/JM/MONO(2000)7&docLanguage=En. Accessed 18 June 2019

Olson H, Betton G, Robinson D et al (2000) Concordance of the toxicity of pharmaceuticals in humans and in animals. Regul Toxicol Pharmacol 32(1):56–67. https://doi.org/10.1006/rtph.2000.1399

Outeiro TF, Heutink P, Bezard E, Cenci AM (2021) From iPS cells to rodents and nonhuman primates: filling gaps in modeling Parkinson's disease. Mov Disord 36(4):832–841. https://doi.org/10.1002/mds.28387

Pace TWW, Heim CM (2011) A short review on the psychoneuroimmunology of post-traumatic stress disorder: from risk factors to medical comorbidities. Brain Behav Immun 25(1):6–13. https://doi.org/10.1016/j.bbi.2010.10.003

Papathanassoglou ED, Giannakopoulou M, Mpouzika M et al (2010) Potential effects of stress in critical illness through the role of stress neuropeptides. Nurs Crit Care 15(4):204–216. https://doi.org/10.1111/j.1478-5153.2010.00363.x

Perel P, Roberts I, Sena E et al (2007) Comparison of treatment effects between animal experiments and clinical trials: systematic review. BMJ 334(7586):197. https://doi.org/10.1136/bmj.39048.407928.be

Perelman P, Johnson WE, Roos C et al (2011) A molecular phylogeny of living primates. PLoS Genet 7(3):e1001342. https://doi.org/10.1371/journal.pgen.1001342

Phillips KA, Bales KL, Capitanio JP et al (2014) Why primate models matter. Am J Primatol 76(9):801–827. https://doi.org/10.1002/ajp.22281

Pippin JJ, Cavanaugh SE, Pistollato F (2019) Animal research for Alzheimer disease: Failures of science and ethics. In: Herrmann K, Jayne K (eds) Animal experimentation: working towards a paradigm change. Brill, Boston, MA, pp 480–516

Pistollato F, Ohayon EL, Lam A et al (2016) Alzheimer disease research in the 21st century: past and current failures, new perspectives and funding priorities. Oncotarget 7(26):38999–39016. https://doi.org/10.18632/oncotarget.9175

Platt JR (1964) Strong inference: certain systematic methods of scientific thinking may produce much more rapid progress than others. Science 146(3642):347–353. https://doi.org/10.1126/science.146.3642.347

Pound P, Nicol CJ (2018) Retrospective harm benefit analysis of pre-clinical animal research for six treatment interventions. PLoS ONE 13(3):e0193758. https://doi.org/10.1371/journal.pone.0193758

Raaum RL, Sterner KN, Noviello CM et al (2005) Catarrhine primate divergence dates estimated from complete mitochondrial genomes: concordance with fossil and nuclear DNA evidence. J Hum Evol 48(3):237–257. https://doi.org/10.1016/j.jhevol.2004.11.007

Raber J (1998) Detrimental effects of chronic hypothalamic-pituitary-adrenal axis activation. From obesity to memory deficits. Mol Neurobiol 18(1):1–22. https://doi.org/10.1007/bf02741457

Ramachandran SD, Schirmer K, Münst B et al (2015) In vitro generation of functional liver organoid-like structures using adult human cells. PLoS ONE 10(10):e0139345. https://doi.org/10.1371/journal.pone.0139345

Reardon S (2018) Frustrated Alzheimer's researchers seek better lab mice. Nature 563:611–612. https://doi.org/10.1038/d41586-018-07484-w

Reinhardt V, Liss C, Stevens C (1995) Restraint methods of laboratory non-human primates: a critical review. Anim Welf 4(3):221–238

Roberts RA, Soames AR, James NH et al (1995) Dosing-induced stress causes hepatocyte apoptosis in rats primed by the rodent nongenotoxic hepatocarcinogen cyproterone acetate. Toxicol Appl Pharmacol 135(2):192–199. https://doi.org/10.1006/taap.1995.1223

Robinson S, Delongeas JL, Donald E et al (2008) A European pharmaceutical company initiative challenging the regulatory requirement for acute toxicity studies in pharmaceutical drug development. Regul Toxicol Pharmacol 50(3):345–352. https://doi.org/10.1016/j.yrtph.2007.11.009

Rõlova T, Lehtonen Š, Goldsteins G, Kettunen P, Koistinaho J (2020) Metabolic and immune dysfunction of glia in neurodegenerative disorders: focus on iPSC models. Stem Cells 39:256–265

SCHEER (2017) Scientific Committee on Health, Environmental and Emerging Risks. The need for non-human primates in biomedical research, production and testing of products and devices (update 2017). https://ec.europa.eu/health/scientific_committees/non-human-primates-testing_en. Accessed 18 June 2019

Schein PS, Davis RD, Carter S et al (1970) The evaluation of anticancer drugs in dogs and monkeys for the prediction of qualitative toxicities in man. Clin Pharmacol Ther 11:3–40

SCHER (2009) Scientific Committee on Health and Environmental Risks. The need for non-human primates in biomedical research, production and testing of products and devices. http://ec.europa.eu/health/ph_risk/committees/04_scher/docs/scher_o_110.pdf. Accessed 18 Jan 2019

Scientific Committee on Animal Health and Animal Welfare (2002) The welfare of non-human primates used in research. European Commission, Health and Consumer Protection Directorate-General. https://ec.europa.eu/food/sites/food/files/safety/docs/sci-com_scah_out83_en.pdf. Accessed 18 June 2019

Sena ES, van der Worp HB, Bath PM et al (2010) Publication bias in reports of animal stroke studies leads to major overstatement of efficacy. PLoS Biol 8(3):e1000344. https://doi.org/10.1371/journal.pbio.1000344

Sibal LR, Samson KJ (2001) Nonhuman primates: a critical role in current disease research. ILAR J 42(2):74–84. https://doi.org/10.1093/ilar.42.2.74

Sittig LJ, Carbonetto P, Engel KA, Krauss KS, Barrios-Camacho CM, Palmer AA (2016) Genetic background limits generalizability of genotype-phenotype relationships. Neuron 91(6):1253–1259. https://doi.org/10.1016/j.neuron.2016.08.013

Snowdon DA, Greiner LH, Mortimer JA et al (1997) Brain infarction and the clinical expression of Alzheimer disease. The Nun Study. JAMA 277(10):813–817. https://doi.org/10.1001/jama.1997.03540340047031

Sorenson M, Janusek L, Mathews H (2011) Psychological stress and cytokine production in multiple sclerosis: correlation with disease symptomatology. Biol Res Nurs 15(2):226–233. https://doi.org/10.1177/1099800411425703

Spanhaak S, Cook D, Barnes J et al (2008) Species concordance for liver injury. BioWisdom, Cambridge, UK. Available from Instem Scientific. https://www.instem.com, on request

St George-Hyslop PH, Westaway DA (1999) Alzheimer's disease. Antibody clears senile plaques. Nature 400:116–117. https://doi.org/10.1038/22006

Steinberg D (2002) Companies halt first Alzheimer vaccine trial investigators are looking into what inflamed patients' brains. Scientist 16(7):22–23

Stephens ML, Conlee K, Alvino G et al (2002) I - Possibilities for refinement and reduction: Future improvements within regulatory testing. ILAR J 43(Suppl):S74–S79. https://doi.org/10.1093/ilar.43.suppl_1.s74

Sturt E (1984) Analysis of linkage and association for diseases of genetic aetiology. Stat Med 3(1):57–72. https://doi.org/10.1002/sim.4780030108

Suliman S, Mkabile SG, Fincham DS et al (2009) Cumulative effect of multiple trauma on symptoms of post-traumatic stress disorder, anxiety, and depression in adolescents. Compr Psychiatry 50(2):121–127. https://doi.org/10.1016/j.comppsych.2008.06.006

Swan MP, Hickman DL (2014) Evaluation of the neutrophil-lymphocyte ratio as a measure of distress in rats. Lab Anim (NY) 43(8):276–282. https://doi.org/10.1038/laban.529

Tagle DA (2019) The NIH microphysiological systems program: developing in vitro tools for safety and efficacy in drug development. Curr Opin Pharmacol 48:146–154. https://doi.org/10.1016/j.coph.2019.09.007

Tardif SD, Coleman K, Hobbs TR et al (2013) IACUC review of nonhuman primate research. ILAR J 54(2):234–245. https://doi.org/10.1093/ilar/ilt040

Taylor K (2006) Still dying of ignorance? 25 years of failed primate AIDS research. Available from Cruelty Free International on request

Taylor KD (2018) Harms versus benefits: a practical critique of utilitarian calculations. In: Linzey A, Linzey C (eds) The ethical case against animal experiments. Oxford Centre for Animal Ethics, University of Illinois Press, Champaign, IL, pp 148–159

Taylor K, Gordon N, Langley G et al (2008) Estimates for worldwide laboratory animal use in 2005. Altern Lab Anim 36(3):327–342. https://doi.org/10.1177/026119290803600310

The Boyd Group (2002) The use of non-human primates in research and testing. https://web.archive.org/web/20061029121536/http://www.boyd-group.demon.co.uk/. Accessed 18 June 2019

TNS Opinion & Social (2010) Special Eurobarometer 340/ Wave 73.1, Science and Technology Report, 61. http://ec.europa.eu/public_opinion/archives/ebs/ebs_340_en.pdf. Accessed 18 June 2019

Tonini T, Barnett S, Donnelly J et al (2005) Current approaches to developing a preventative HIV vaccine. Curr Opin Investig Drugs 6(2):155–162

Tonks A (2007) Quest for the AIDS vaccine. BMJ 334(7608):1346–1348. https://doi.org/10.1136/bmj.39240.416968.AD

Tsamandouras N, Kostrzewski T, Stokes CL et al (2017) Quantitative assessment of population variability in hepatic drug metabolism using a perfused three-dimensional human liver microphysiological system. J Pharmacol Exp Ther 360(1):95–105. https://doi.org/10.1124/jpet.116.237495

U.S. Congress (1938) Federal Food, Drug and Cosmetics Act. https://legcounsel.house.gov/Comps/Federal%20Food,%20Drug,%20And%20Cosmetic%20Act.pdf. Accessed 18 June 2019

U.S. Food and Drug Administration (2021) Advancing new alternative methodologies at FDA. https://www.fda.gov/media/144891/download. Accessed 26 Oct 2021

UK Department of Health (2006) Expert scientific group on phase one clinical trials. https://webarchive.nationalarchives.gov.uk/20130105090249/http://www.dh.gov.uk/en/Publicationsandstatistics/Publications/PublicationsPolicyAndGuidance/DH_063117. Accessed 18 June 2019

UK Home Office (2020) Statistics of Scientific Procedures on Living Animals: Great Britain 2019. https://www.gov.uk/government/statistics/statistics-of-scientific-procedures-on-living-animals-great-britain-2019. Accessed 2021

United States Department of Agriculture (USDA)—Animal and Plant Health Inspection Service (APHIS) (2014) Animals used in research. http://www.aphis.usda.gov/animal_welfare/downloads/7023/Animals%20Used%20In%20Research%202013.pdf. Accessed 18 June 2019

Vallender EJ, Miller GM (2013) Nonhuman primate models in the genomic era: a paradigm shift. ILAR J 54(2):154–165. https://doi.org/10.1093/ilar/ilt044

VandeBerg JL, Zola SM (2005) A unique biomedical resource at risk. Nature 437:30–32. https://doi.org/10.1038/437030a

Verdier JM, Acquatella I, Lautier C et al (2015) Lessons from the analysis of nonhuman primates for understanding human aging and neurodegenerative diseases. Front Neurosci 9:64. https://doi.org/10.3389/fnins.2015.00064

Weatherall D (2006) The use of non-human primates in research. The Academy of Medical Sciences, London, UK. https://acmedsci.ac.uk/file-download/34785-nhpdownl.pdf. Accessed 18 June 2019

Wiebers DO, Adams HPJ, Whisnant JP (1990) Animal models of stroke: are they relevant to human disease? Stroke 21(1):1–3. https://doi.org/10.1161/01.STR.21.1.1

Williams PT, Poole MJ, Katos AM et al (2008) A new device for the capture and transport of small nonhuman primates in scientific research. Lab Anim 37:116–119. https://doi.org/10.1038/laban0308-116

Wright R (2011) Epidemiology of stress and asthma: from constricting communities and fragile families to epigenetics. Immunol Allergy Clin N Am 31(1):19–39. https://doi.org/10.1016/j.iac.2010.09.011

Young E (2002) Alzheimer's vaccine trial suspended. https://www.newscientist.com/article/dn1820-alzheimers-vaccine-trial-suspended/. Accessed 18 June 2019

The Indispensable Contribution of Nonhuman Primates to Biomedical Research

Stefan Treue and Roger Lemon

Abstract

Research in nonhuman primates is one of the cornerstones of modern biomedical research. This approach, while quantitatively contributing only a very small component to biomedical research, has allowed important and fundamental insights into our understanding of the physiology and anatomy of healthy organisms as well as the causes and mechanisms of disease in humans and other animals. In recent times, much progress has been achieved in advancing both the welfare of nonhuman primates (and other species) in research, as well as the quality and quantity of scientific insights possible by carefully combining responsible animal research with human and nonanimal methods and approaches. Despite these advances, there is a growing need and responsibility for individual scientists, their institutions, and scientific organizations to engage in a transparent dialogue with society about the scientific benefits and the ethical justification of animal research. Without such fact-based discourse, there is the risk that the broad consensus across human societies that some animal research (including research using nonhuman primates) is necessary and justified to ensure scientific and medical progress will erode.

Keywords

Animal research · Bioethics · Medical research · Basic research · Responsibility

---

S. Treue (✉)
Cognitive Neuroscience Laboratory, German Primate Center, Göttingen, Germany
e-mail: treue@gwdg.de

R. Lemon
Institute of Neurology, University College London, London, UK
e-mail: r.lemon@ucl.ac.uk

© Springer Nature Switzerland AG 2023
L. M. Robinson, A. Weiss (eds.), *Nonhuman Primate Welfare*,
https://doi.org/10.1007/978-3-030-82708-3_24

Biomedical research, the effort to fundamentally understand human and nonhuman physiology and anatomy, has revolutionized our society. Unraveling the causality of physiological processes underlying the many dimensions of our daily experience and applying this knowledge toward medical diagnosis and therapy has freed us to see diseases as malfunctions of physiological processes and find science-based therapies. It has also contributed to an understanding of what makes us human and how our behavior and experiences are evolutionary adaptations to our physical and social environment that link us to other species, in the present and from the past.

The recognition that all organisms on this planet are genetically related through a joint evolutionary history allows us to turn case studies on individual species into insights about the processes of life on earth as a whole. It not only explains the fascinating similarities between fundamental processes in humans and other animals, but also offers an interpretation for species differences, where they exist.

Understanding humans as members of the animal kingdom challenges our view of mankind as a privileged species. While science has identified a number of features and capacities unique to humans, or at least uniquely developed in humans, it has also eroded the view that humans are fundamentally different from any other life form on earth.

Three important conclusions can be drawn from these developments:

1. Studying animal species, both in detail and in comparison with humans, offers fundamental insights into human physiology, behavior, and health.
2. As we strive to improve equality and fairness for all humans by developing more extensive rights and legal protection, we need to consider our moral responsibility toward animals that can experience pain and distress and ensure that they are protected from unnecessary and avoidable harms.
3. Given the importance of science for our society, that society needs to ensure that the general public has the basic scientific literacy and the skills to interpret scientific information to make informed, fact-based political and societal decisions.

In this chapter we will discuss these issues in the context of the use of nonhuman primates in biomedical research. The focus will be on the scientific basis for this approach and the need for science communication. The need and ethical justification of responsible animal research, including the use of nonhuman primates, have been recognized in all research-intensive societies worldwide, but an in-depth discussion of ethical issues is beyond the scope of this chapter.

1 The Role of Nonhuman Primates in Biomedical Research

Animal research represents a minute fraction of the animals used and killed, voluntarily and involuntarily, by human societies. In European countries for every single research animal used, about 200–300 animals are killed for human consumption. Of the research animals used, more than 80% are rodents and less than 0.1% are nonhuman primates. Thus, we consume about 500,000 animals for every nonhuman primate used in research (Roelfsema and Treue 2014).

Despite the small fraction of nonhuman primates in animal research, in the past they have made and for the future they can be expected to make central and indispensable contributions to biomedical research (Mitchell et al. 2018). The quantitatively largest contributions, about three quarters of the nonhuman primates used in the EU, are toxicological studies, the production of antibodies, and blood products and other legally required tests of new substances or medical devices (ALURES Statistical EU Database 2021). Here, particularly for potential drugs for the treatment of complex diseases, the close physiological similarity of the human and nonhuman primate organ systems is the reason that tests in nonhuman primates are considered essential and often the last mandated safeguard before substances move to clinical testing in human populations. In its report, "Working to reduce the use of animals in scientific research" (UK Government 2014), the UK Government states: "the development of monoclonal antibody therapies over the last 20 years has completely transformed our ability to treat diseases including breast and other cancers, rheumatoid arthritis and multiple sclerosis. The development of this technology would not have been possible without the use of animals both in developing the fundamental elements of the technology and in producing the medicines used to treat patients." Similarly, the EU's Scientific Committee on Health, Environmental and Emerging Risks (SCHEER) concludes "appropriate use of nonhuman primates remains essential in some areas of biomedical and biological research and for the safety assessment of pharmaceuticals" (SCHEER 2017; Epstein and Vermeire 2017).

The following table documents the broad range of nonhuman primate contributions to medical progress.

Select Health Breakthroughs Involving Nonhuman Primates
Strategies for restoring lost motor function following spinal cord injury
The development of an innovative surgical procedure called deep brain stimulation to combat the effects of Parkinson's disease
A cure for hepatitis C infection
The development of an Ebola vaccine, which is currently being tested in humans
Stem cell therapies, which are being developed and tested to combat several diseases including macular degeneration, spinal cord injury, heart disease, and ALS (also known as motor neuron disease or Lou Gehrig's disease)
Gene therapy advancements for mitochondrial-linked disorders which can cause diabetes, deafness, blindness, dementia, or epilepsy
Information on the impacts of Zika infection and new approaches for combatting the disease
New approaches for combatting HIV infection through the study of SIV and SHIV (the primate versions of HIV)
Research that led to the successful human clinical trials and FDA approval of belatacept, the first new transplant drug since 1999
Gene therapy advancements for combatting Huntington's disease
New genetic findings related to alcohol dependence
New avenues for the treatment of obesity, including childhood obesity
Hormone therapy for cognitive issues related to aging

Source: Americans for Medical Progress (https://www.amprogress.org)

A quantitatively much smaller fraction of nonhuman primates in animal research (about one quarter, ALURES Statistical EU Database 2021) is used for research that is often classified as "basic/fundamental" research or "applied/translational/medical" research. This reflects a misleading division, since biomedical research is a continuum, stretching from studies of fundamental physiological processes to an understanding of the principles of disease and the development of therapies (Basel Declaration, www.basel-declaration.org). The range of such research with nonhuman primates has provided the basis for the examples of ground-breaking medical discoveries and progress listed in the table above, but has also been the strategic focus of very vocal and sometimes violent activism from well-funded groups opposed to any animal-based biomedical research (Speaking of Research 2021). The two most important areas of nonregulatory research with nonhuman primates are infectious diseases and the neurosciences. Less prominent contributions are made in other areas, such as ophthalmology and (xeno)transplantation (VandeBerg and Williams-Blangero 1997; Newsome and Stein-Aviles 1999; Carlsson et al. 2004; Roska and Sahel 2018; Sato and Sasaki 2018; Picaud et al. 2019; Feng et al. 2020). The following table highlights Nobel Prize-winning research that involved nonhuman primates.

Nobel Prize Winning Research Involving Nonhuman Primates		
Year	Discovery	Scientist(s)
1928	Pathogenesis of typhus	Nicolle
1951	Development of yellow fever vaccine	Theiler
1954	Culture of poliovirus that led to development of vaccine	Enders, Weller, Robbins
1975	Interaction between tumor viruses and genetic material	Baltimore, Dulbecco, Temin
1976	New mechanisms for the origin and dissemination of diseases	Blunberg, Gajdusek
1981	Processing of visual information by the brain	Sperry, Hubel, Wiesel
1988	Discoveries of important principles for drug treatment	Black, Elion
2003	Discoveries concerning magnetic resonance imaging (MRI)	Lauterbur, Mansfield
2008	Discovery of human immunodeficiency virus (HIV)	Barre-Sinoussi, Montagnier
2015	Development of drugs used to treat malaria and roundworm parasites	Campbell, Omaru, Tu
2020	Discovery of the hepatitis C virus	Alter, Houghton, Rice

Adapted from: Animal Roles in Medical Discoveries, the American Association for Laboratory Animal Science Foundation (https://www.aalasfoundation.org)

2 Categorical Challenges to the Scientific Value of Research on Nonhuman Primates

Despite the broad and fundamental contributions made across biomedical research by studies involving nonhuman primates, organizations opposed to any animal research and a small number of publications in peer-reviewed scientific journals have argued against the necessity of research in nonhuman primates for past and future scientific progress (Bailey and Taylor 2016; Lauwereyns 2018). Broadly, their claims can be grouped into the following arguments:

1. Knowledge gained from research in animals (and therefore from nonhuman primates) cannot be transferred to humans as there are categorical differences between humans and animals that make such a translation impossible.
2. There is no added benefit of performing research in nonhuman primates compared to studying the same questions in rodents.
3. The scientific knowledge gained from research in nonhuman primates could have been achieved by human studies.
4. Research involving nonhuman primates is only upheld because researchers working with nonhuman primates have lost the ability to recognize that it is scientifically misguided.

Given that the claims against nonhuman primate research are advanced from various angles, some claims are incompatible, such as the second and first, as the second acknowledges the benefits of animal research that the first denies. Additionally, some claims, such as the first, are not specific to nonhuman primates. Furthermore, there are bioethical positions that argue against the use of *any* animal species in biomedical research. The latter goes beyond the scope of this chapter and will therefore not be addressed.

Claim 1 is particularly widespread amongst organizations fundamentally opposed to animal research and typically takes the specific form of "Penicillin is toxic for species x, chocolate is toxic for species y, aspirin is toxic for species z, etc.— therefore the toxicology of any substance for humans cannot be assessed from animal studies". Despite its wide use, this conclusion does not follow from the examples of toxic effects of some substances in some species. In its pre-Darwinian approach, the claim ignores the biological reason for the similarities between different animal species and humans. It fails to recognize that understanding why some substances are toxic only for some species helps science understand the toxic mechanism per se, leading to a better understanding of the risks associated with a given substance and even with other related substances (Baker 2011; Carbone 2012; Ergorul and Levin 2013; Barré-Sinoussi and Montagutelli 2015).

Claim 2 argues almost opposite to Claim 1, in that it proposes that rodents are biomedical models for humans of equal suitability as nonhuman primates, despite the much greater evolutionary distance between rodents and humans. Examples of Claim 2 can be found in Lauwereyns (2018), where the author argues that neuroscience research into high-level cognitive functions, such as the neural mechanism of decision-making, and into the brain's motor control system (e.g., for the

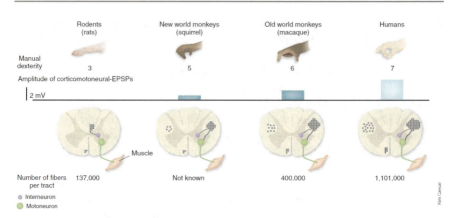

Fig. 1 Schematic diagram illustrating species differences in the involvement of the corticospinal tract in motor control. In rodents, there are no direct connections between corticospinal axons and the cervical motoneurons that innervate forelimb muscles, and indirect brainstem and spinal pathways, here indicated by a spinal interneuron, mediate cortical input to motoneurons. Direct cortico-motoneuronal (CM) connections with motoneurons are present in some primates, which also exhibit an increase in the size and number of corticospinal fibers. CM excitatory effects are relatively weak in the New World squirrel monkey but of moderate size in the Old World macaque monkey, represented here by the size of the CM excitatory postsynaptic potential (EPSP) recorded from hand muscle motoneurons. Indirect estimates suggest that CM effects are strongest in humans. In rats, most corticospinal fibers are crossed and travel in the dorsal columns, while in primates the majority are also crossed but descend in the dorsolateral funiculus. There are also species differences in the proportion of uncrossed fibers which descend ipsilaterally. The evolution of the corticospinal tract is closely correlated with the development of skilled hand use, as indicated here by increase in the index of dexterity from rat to monkey to human (reproduced from Courtine et al. 2007)

development of neuroprosthetic devices) could be shifted from nonhuman primates to rodents, as rodents also have to be adept at decision-making and motor control. This argument for moving research from nonhuman primates to rodents is certainly valid if the aim of the research is to understand decision-making and motor control in rodents. It might also be valid if the aim is to understand fundamental aspects of decision-making and motor control shared across many mammals. If on the other hand, the goal is to have the best chance to learn about human decision-making and motor control from animal studies, a combined approach is essential. This means that such processes need to be studied in animal species evolutionary close as well as further away from humans to dissect fundamental aspects of these processes from those aspects that have developed only (or predominantly) in primates. For example, in motor control there are fundamental differences between the corticospinal control system in rodents, nonhuman primates, and humans, as depicted in Fig. 1, which is likely to reflect the much greater importance of skilled hand function in primates than in rodents. Hand function is devastated by many different neurological diseases, and patients prioritize the recovery of hand function over other bodily functions (Anderson 2004).

Rather than following along Claim 2 to replace nonhuman primate research with research in rodents, the ethically and scientifically most responsible approach is to carefully combine research in multiple species with studies in humans, as well as other nonanimal methods. This leverages each approach's strengths and compensates for the various weaknesses, ensuring the best science with the least animal use (Carbone 2012).

Examples of Claim 3 can be found in Bailey and Taylor (2016), where the authors argue that, in the case of neurophysiological recordings of awake rhesus monkeys (a very small fraction, ca. 5%, of nonhuman primate research), nonhuman primate approaches have been rendered redundant by noninvasive and invasive approaches in humans. Unfortunately, the essence of this claim boils down to a combination of a false antithesis with a wide-ranging conspiracy theory (see below). The false antithesis is that invasive and noninvasive methods used in humans can serve as a replacement to experiments in nonhuman primates. In fact, as argued above, the nonhuman primate and human approaches are complementary and thus their careful combination allows insights not possible by a single-species approach, given the limitations of research in each species by itself. Correspondingly, there is no doubt as to the enormous contributions that noninvasive techniques in humans, such as transcranial magnetic stimulation (TMS), electroencephalography (EEG), magnetoencephalography (MEG), and functional magnetic resonance imaging (fMRI) have made to the understanding of the human brain. Neuroscientists involved in nonhuman primate research also make use of these noninvasive techniques (see Fig. 2 below, from Lemon 2018), alongside their studies in nonhuman primates that, although invasive, allow understanding the bases of the noninvasive techniques. But these scientists view such noninvasive methods as *complementary* to the research in animals, rather than as alternatives to ensure the ethically and scientifically most responsible approach outlined above (Lemon 2018).

In their attempt to discredit nonhuman primate research, Bailey and Taylor (2016) ignore three key points:

1. These noninvasive techniques could never have been developed, and cannot be further improved, without an understanding of the underlying brain properties, almost all of which are derived from animal studies.
2. As these techniques develop, they require validation with further animal studies, including some in primates. For example, some approaches, which are required to understand brain circuits and causal interactions between neurons, such as optogenetic or pharmacological activation/inactivation, are not yet safe and reliable enough to be used in humans.
3. These noninvasive techniques cannot address the fundamental mechanisms that operate at the single neuron level because this is beyond the resolution of such methods. For example, the otherwise very useful brain imaging techniques of functional magnetic resonance imaging, electroencephalography or magnetoencephalography are fundamentally unsuitable for measuring single neuron activity (Logothetis 2008). This is one of the reasons why single unit recording, which has been so successful in nonhuman primate studies, is increasingly being adopted in patient studies. But this invasive approach entails considerable limitations given that electrode placement in people is solely driven by clinical needs and often involves studying of pathological tissue (see below).

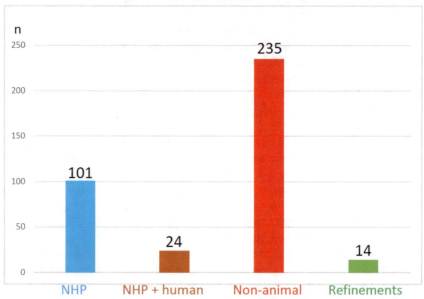

Fig. 2 These data demonstrate that neuroscientists who use nonhuman primates in their research, not only publish papers involving nonhuman primates (cyan, 101 papers) but also carry out comparative studies involving both nonhuman primates and humans (brown, 24) and a large amount of research which does not involve any animal models (red, 235). These alternative studies complemented their primate work, and involved human volunteers and patients as well as computer models. Finally, these same scientists also carried out research on new techniques to refine research involving nonhuman primates (green, 14) (data from papers published by UK scientists 2014-2017; Lemon 2018)

Bailey and Taylor (2016) claim that microelectrode recordings in human patients could be used to replace those made in experimental monkeys. This claim is not shared by leading proponents of the important technique of recordings in humans. For example, the neurosurgeon Izthak Fried states in his introduction to a major textbook on the subject of human microelectrode recording (our italics):

> You cannot choose just any question and hope to record from the relevant neurons. The sites of recordings will always be completely determined by the clinical imperative and thus will be fixed in locations that cannot be altered. *The question you elect to explore has to be grounded in animal physiology and has to build on this knowledge.* Yet, the question must also be relevant and unique to the human condition. In particular, *you need to take what we know from nonhuman primate neurophysiology to the next level, the human level.* Single neuron human neurophysiology is a small field between animal neurophysiology and human functional neuroimaging and other noninvasive methods customarily used in cognitive neuroscience. (Fried 2014)

Similarly, the neurosurgeon Adam Marmelak points out:

> Human MER [microelectrode recording] offers a unique and unprecedented opportunity to study how the brain works. Research efforts along these lines should be applauded and encouraged as they are likely to lead to profound and potentially revolutionary insight into how the human brain functions and may indeed lead to novel treatments for a variety of brain diseases. However, due to the unique ethical and practical considerations in these studies, a strict adherence to ethical principles with a pragmatic approach to experiment design and execution are needed to ensure success in this endeavor. (Mamelak 2014)

Some of the considerations Marmelak is referring to include the fact that such recordings can only be carried out in severely ill patients (e.g., those suffering from epilepsy, tetraplegia, or Parkinson's disease), in agreement with the ethical standards of the Helsinki Declaration (2013). In addition, such recordings, while clearly of great value to human neuroscience, are not without risk and other complications. As the neurosurgeon George Ojemann has put it:

> ...it is the author's view that since microelectrode recording is invasive, and associated with some risk of significant damage to the region of recording, it should be restricted to tissue that would be subject to surgical injury independent of any microelectrode recording. Thus recording should be restricted to tissue that would be resected as part of a therapeutic procedure independent of any considerations of microelectrode recording or along the track or at the target of electrodes placed for clinical reasons independent of microelectrode recording. Moreover, the extent of surgical exposure and anesthetic techniques utilized for an operation should be based only on clinical considerations and not microelectrode recording. (Ojemann 2014)

Bailey and Taylor appear to misinterpret and misquote published studies using such recordings. For example, when considering the problem that in most cases recordings can only be made from brain tissue that is compromised by the disease process, they state: "However, it has been argued that any differences between epileptic and non-epileptic brains may be gross in nature, and therefore that there is no evidence that single neuron activity is affected" (Bailey and Taylor 2016, p. 49). This statement is referenced to Ojemann (2014), who writes "There is evidence that even at the level of gross anatomy there are widespread differences between these patients and other populations, changes that extend beyond any gross pathology (McDonald et al. 2008), though whether these changes alter single neuron activity is unknown".

Again, Bailey and Taylor write "any potential differences are mitigated by the recording of activity in tissue away from the epileptogenic focus (i.e., with no epileptiform activity)." (Bailey and Taylor 2016, p. 49). What Ojemann (2014) actually writes is "Epilepsy effects in single neuron recordings can be mitigated to some extent by limiting recording to tissue that does not have electrocorticographic evidence of epileptiform activity, and neurons that do not show 'bursting' activity, but that does not mitigate any unknown effects of the more widespread 'reorganization' of cognitive processes." (Ojemann 2014).

Again, Bailey and Taylor cite Ojemann (2014) when they state "Further, investigation is not restricted to areas of the brain around sampling sites, and the use of other sites (such as the sensorimotor cortex, for example) is generally accepted if there is informed consent by the patient, as well as Institutional Review Board approval". Ojemann actually points out:

> The ideal recording sites based on current models of the brain regions involved in different cognitive processes are not necessarily the areas that would be surgically exposed for purely diagnostic or therapeutic reasons and, indeed, might be areas that would generally be spared in any resection, such as sensorimotor cortex or areas considered "eloquent" for language. While recordings from such areas have been reported (Goldring and Ratcheson 1972), it is the author's view that since microelectrode recording is invasive, and associated with some risk of significant damage to the region of recording, it should be restricted to tissue that would be subject to surgical injury independent of any microelectrode recording. Thus recording should be restricted to tissue that would be resected as part of a therapeutic procedure independent of any considerations of microelectrode recording or along the track or at the target of electrodes placed for clinical reasons independent of microelectrode recording.... However, when opportunities for microelectrode recording that meet these conditions exist, every effort should be made to utilize them, with the patient's informed consent that includes issues of risk and comfort, and after Institutional Review Board approval. (Ojemann 2014)

In summary, there are good clinical and ethical reasons why neurosurgeons cannot and do not adopt experimental approaches that might be of interest, or indeed record anywhere in the brain of their choosing. They include the definite risks associated with microelectrode recording in sick patients (Nakajima et al. 2011), restricting such recordings in accordance with the ethical criteria of the Helsinki Declaration. And once again, intracranial recordings in human patients are not a substitute for understanding brain mechanisms using nonhuman primates but are complementary to the work in animals and depend on advances from it, as pointed out above.

Both Lauwereyns (2018) and Bailey and Taylor (2016) argue for a complete replacement of nonhuman primate research on scientific grounds by claiming that studies in rodents and/or humans can fully replace neurophysiological studies in nonhuman primates. At the same time (a) such nonhuman primate research is performed in many dozens of labs, (b) each experiment conducted by these labs has to undergo an independent evaluation that includes a specific determination as to whether the study can be replaced by nonprimates, or even nonanimal approaches (see also Bayne et al. 2022), (c) is funded based on careful and competitive peer-review by many scientific grant agencies and through government funding, despite its relatively high cost, (d) multiple commissions and other bodies have made substantial and carefully judged evidence-based assessments as to the continuing need for basic neuroscience research involving nonhuman primates and its irreplaceability (EU-Directive 2010/63/EU 2010; Weatherall 2006; SCHEER 2017), and (e) neuroscientists involved in research in nonhuman primates also make use of noninvasive techniques (Fig. 2). In the light of this, Lauwereyns' and Bailey and Taylor's claim require a major conspiracy, not only of the scientists involved, but also for a large and diverse group of other experts and government agencies, to ignore the supposed rodent and human alternatives to nonhuman primate-based approaches.

3 A Nuanced Approach to Nonhuman Primate Research

As shown above, claims of science-based categorical arguments against nonhuman primate biomedical research are in disagreement with basic scientific and methodological facts. These claims also run counter to a broad consensus in the scientific community and the societal consensus expressed in national laws and international directives. But this should not distract the scientific community from the real challenge, namely maintaining a continuous momentum toward the best possible combination of high-value science with optimal animal welfare (see Schapiro and Hau 2022, for a review of the benefits of high welfare for better quality science). Here we want to highlight one focus for each of the two sides of this combination, i.e., science and animal welfare.

Within the last decade, there has been recognition across many fields of experimental research that more has to be done to ensure science of the highest possible value by enhancing its reliability and reproducibility. While this "reproducibility crisis" has been recognized across many fields, a lack of reproducibility of animal-based research has the additional consequence of negatively affecting the harm/benefit tradeoff inherit in any experimental study involving animals: that is, are the harms experienced by animals in a scientific study justified by the scientific advances and benefits gained from that study? It is therefore of critical importance in the context of this chapter that scientists, scientific institutions, and scientific organizations make every effort to ensure best scientific practice in biomedical research. This includes (a) adequate teaching and training of scientists (e.g., by FELASA, the Federation of European Laboratory Animal Science Associations), (b) the identification and removal of pressures or incentives that discourage best scientific practices (Pusztai et al. 2013), and (c) support for transparency in method, data, and result presentation and sharing (Percie du Sert et al. 2020).

On the animal welfare side, the last decades have seen a wave of methodological refinements that enhance the welfare of nonhuman primates in biomedical research (see Schapiro and Hau 2022; Bayne et al. 2022 for more details). This has been particularly extensive in neurophysiology research with awake behaving nonhuman primates. Here, developments include improvements (a) in surgical and analgesic procedures and approaches (Oliveira and Dimitrov 2008), (b) in the design of and the care for cranial implants (Lanz et al. 2013), (c) in the technology and the extensive use of noninvasive imaging (Logothetis et al. 1999; Milham et al. 2020), and (d) in the animals' training, handling, enrichment, and housing techniques (Prescott and Bowell 2005; Prescott et al. 2010; Calapai et al. 2017; Berger et al. 2018).

While these developments have continually improved the harm/benefit balance for such research, the focus of the public debate about animal research in general or nonhuman primate research in particular, and on a possible abolishment of such research, has distracted the public, politicians, and the scientific community from much more important topics in this context. We would argue that these topics include the need for the development of objective, scientific, rather than anthropomorphic, measures of animal welfare, the need to identify the most efficient use of limited resources to get the biggest 3R-effects, and the political debate around variations in animal welfare standards across continents. It should be remembered that scientific collaboration and interaction is the best way to continuously inform and improve these standards.

4 The Need for Transparent Science Communication About Responsible Nonhuman Primate Research

The need and ethical justification of responsible animal research, including the use of nonhuman primates, have been recognized in all research-intensive societies worldwide. This consensus has led to legal and regulatory frameworks, such as the Directive 2010/63/EU of the European Union on the protection of animals used for scientific purposes (EU-Directive 2010/63/EU 2010) and the resulting national animal protection laws, which are built on the broad consensus across science, politics, and society that a certain amount of research on animals (including nonhuman primates) is necessary and justifiable (see Bayne et al. 2022 for more details).

Unfortunately, the scientific community has widely ignored its responsibility to contribute to the scientific literacy needed for a science-based public debate about how, when, and why responsible animal research is and should be conducted. Some of this responsibility has been addressed by national information initiatives, such as the UK's Understanding Animal Research (www.understandinganimalresearch.org.uk), France's Gircor (www.recherche-animale.org), the United States' Foundation for Biomedical Research (fbresearch.org), and Germany's Tierversuche verstehen (tierversuche-verstehen.de). Additional efforts have come from Europe-wide initiatives such as the European Animal Research Association (http://eara.eu/en/) as well as extensive websites of Primate Centers (German Primate Center (Germany) www.dpz.eu/en/home.html; National Primate Research Centers (USA) nprcresearch.org/primate/; Primate Research Institute Kyoto University (Japan) www.pri.kyoto-u.ac.jp). But there is still widespread fear of triggering damaging and potentially threatening activism against scientific institutes and individual scientists when they begin to openly communicate about animal research. The UK's Concordat on Openness on Animal Research (concordatopenness.org.uk), which has served as a model for several other European countries (including Belgium, France, Germany, Portugal, and Spain), has offered a highly successful approach to address these concerns. While these initiatives have started to shift an emotionally charged public debate, dominated by groups opposed to any animal-based biomedical research, toward a more fact-based discourse, these efforts need to be accompanied by corresponding efforts by scientific institutes and individual scientists.

In summary, carefully regulated and responsible animal research in nonhuman primates provides a cornerstone for biomedical research. This is recognized worldwide across a wide range of stakeholders in science and society. While contributing only a very small quantitative component to biomedical research as a whole, studies in nonhuman primates often provide a critical link between the research results achieved through rodent studies and those from human or nonanimal approaches. To ensure that this important component of biomedical research receives the continued societal support it needs and deserves, scientists working with nonhuman primates need to conduct their studies with the utmost care and engage in a continuous exchange with society about the "why" and "how" of the scientific need and the ethical justification of this important method.

References

ALURES Statistical EU Database (2021). https://ec.europa.eu/environment/chemicals/lab_animals/alures_en.htm and https://ec.europa.eu/environment/chemicals/lab_animals/reports_en.htm. Accessed June 2021

Anderson KD (2004) Targeting recovery: priorities of the spinal cord-injured population. J Neurotrauma 21(10):1371–1383. https://doi.org/10.1089/neu.2004.21.1371

Bailey J, Taylor K (2016) Non-human primates in neuroscience research: The case against its scientific necessity. Altern Lab Anim 44(1):43–69. https://doi.org/10.1177/026119291604400101

Baker M (2011) Inside the minds of mice and men. Nature 475:123–128. https://doi.org/10.1038/475123a

Barré-Sinoussi F, Montagutelli X (2015) Animal models are essential to biological research: issues and perspectives. Future Sci OA 1(1):FSO63. https://doi.org/10.4155/fso.15.63

Bayne K, Hau J, Morris T (2022) The welfare impact of regulations, policies, guidelines and directives and nonhuman primate welfare. In: Robinson LM, Weiss A (eds) Nonhuman primate welfare: from history, science, and ethics to practice. Springer, Cham, pp 629–646

Berger M, Calapai A, Stephan V, Niessing M, Burchardt L, Gail A, Treue S (2018) Standardized automated training of rhesus monkeys for neuroscience research in their housing environment. J Neurophysiol 119(3):796–807. https://doi.org/10.1152/jn.00614.2017

Calapai A, Berger M, Niessing M, Heisig K, Brockhausen R, Treue S, Gail A (2017) A cage-based training, cognitive testing and enrichment system optimized for rhesus macaques in neuroscience research. Behav Res Methods 49(1):35–45. https://doi.org/10.3758/s13428-016-0707-3

Carbone L (2012) The utility of basic animal research. Hastings Cent Rep 42(s1):S12–S15. https://doi.org/10.1002/hast.101

Carlsson HE, Schapiro SJ, Farah I, Hau J (2004) Use of primates in research: a global overview. Am J Primatol 63(4):225–237. https://doi.org/10.1002/ajp.20054

Courtine G, Bunge MB, Fawcett JW, Grossman RG, Kaas JH, Lemon R, Maier I, Martin J, Nudo RJ, Ramon-Cueto A, Rouiller EM, Schnell L, Wannier T, Schwab ME, Edgerton VR (2007) Can experiments in nonhuman primates expedite the translation of treatments for spinal cord injury in humans? Nat Med 13(5):561–566. https://doi.org/10.1038/nm1595

Epstein MM, Vermeire T (2017) An Opinion on non-human primates testing in Europe. Drug Discov Today Dis Models 23:5–9. https://doi.org/10.1016/j.ddmod.2017.09.001

Ergorul C, Levin LA (2013) Solving the lost in translation problem: improving the effectiveness of translational research. Curr Opin Pharmacol 13(1):108–114. https://doi.org/10.1016/j.coph.2012.08.005

EU-Directive 2010/63/EU (2010) Directive 2010/63/EU of the European Parliament and of the Council of 22 September 2010 on the protection of animals used for scientific purposes. http://data.europa.eu/eli/dir/2010/63/oj. Accessed June 2021

Feng G, Jensen FE, Greely HT, Okano H, Treue S, Roberts A, Fox JG, Caddick S, Poo MM, Newsome W, Morrison JH (2020) Opportunities and limitations of genetically modified nonhuman primate models for neuroscience research. Proc Nat Acad Sci U S A 117(39):24022–24031. https://doi.org/10.1073/pnas.2006515117

Fried I (2014) Open letter to a beginning researcher in the field of human single neuron investigations. In: Fried IRU, Cerf M, Kreiman G (eds) Single neuron studies of the human brain. MIT Press, Cambridge, MA, pp vii–viii

Lanz F, Lanz X, Scherly A, Moret V, Gaillard A, Gruner P, Hoogewoud HM, Belhaj-Saif A, Loquet G, Rouiller EM (2013) Refined methodology for implantation of a head fixation device and chronic recording chambers in non-human primates. J Neurosci Methods 219(2):262–270. https://doi.org/10.1016/j.jneumeth.2013.07.015

Lauwereyns J (2018) Rethinking the three R's in animal research. Palgrave MacMillan, New York

Lemon R (2018) Applying the 3Rs to neuroscience research involving nonhuman primates. Drug Discov Today 23(9):1574–1577. https://doi.org/10.1016/j.drudis.2018.05.002

Logothetis NK (2008) What we can do and what we cannot do with fMRI. Nature 453:869–878. https://doi.org/10.1038/nature06976

Logothetis NK, Guggenberger H, Peled S, Pauls J (1999) Functional imaging of the monkey brain. Nat Neurosci 2:555–562. https://doi.org/10.1038/9210

Mamelak AN (2014) Ethical and practical considerations for human microelectrode recording studies. In: Fried IRU, Cerf M, Kreiman G (eds) Single neuron studies of the human brain. MIT Press, Cambridge, MA, pp 27–42

Milham M, Petkov CI, Margulies DS, Schroeder CE, Basso MA, Belin P, Fair DA, Fox A, Kastner S, Mars RB, Messinger A, Poirier C, Vanduffel W, Van Essen DC, Alvand A, Becker Y, Ben Hamed S, Benn A, Bodin C, Boretius S, Cagna B, Coulon O, El-Gohary SH, Evrard H, Forkel SJ, Friedrich P, Froudist-Walsh S, Garza-Villarreal EA, Gao Y, Gozzi A, Grigis A, Hartig R, Hayashi T, Heuer K, Howells H, Ardesch DJ, Jarraya B, Jarrett W, Jedema HP, Kagan I, Kelly C, Kennedy H, Klink PC, Kwok SC, Leech R, Liu X, Madan C, Madushanka W, Majka P, Mallon A-M, Marche K, Meguerditchian A, Menon RS, Merchant H, Mitchell A, Nenning K-H, Nikolaidis A, Ortiz-Rios M, Pagani M, Pareek V, Prescott M, Procyk E, Rajimehr R, Rautu I-S, Raz A, Roe AW, Rossi-Pool R, Roumazeilles L, Sakai T, Sallet J, García-Saldivar P, Sato C, Sawiak S, Schiffer M, Schwiedrzik CM, Seidlitz J, Sein J, Shen Z-m, Shmuel A, Silva AC, Simone L, Sirmpilatze N, Sliwa J, Smallwood J, Tasserie J, Thiebaut de Schotten M, Toro R, Trapeau R, Uhrig L, Vezoli J, Wang Z, Wells S, Williams B, Xu T, Xu AG, Yacoub E, Zhan M, Ai L, Amiez C, Balezeau F, Baxter MG, Blezer ELA, Brochier T, Chen A, Croxson PL, Damatac CG, Dehaene S, Everling S, Fleysher L, Freiwald W, Griffiths TD, Guedj C, Hadj-Bouziane F, Harel N, Hiba B, Jung B, Koo B, Laland KN, Leopold DA, Lindenfors P, Meunier M, Mok K, Morrison JH, Nacef J, Nagy J, Pinsk M, Reader SM, Roelfsema PR, Rudko DA, Rushworth MFS, Russ BE, Schmid MC, Sullivan EL, Thiele A, Todorov OS, Tsao D, Ungerleider L, Wilson CRE, Ye FQ, Zarco W, Y-d Z (2020) Accelerating the evolution of nonhuman primate neuroimaging. Neuron 105(4):600–603. https://doi.org/10.1016/j.neuron.2019.12.023

Mitchell AS, Thiele A, Petkov CI, Roberts A, Robbins TW, Schultz W, Lemon R (2018) Continued need for non-human primate neuroscience research. Curr Biol 28(20):R1186–R1187. https://doi.org/10.1016/j.cub.2018.09.029

Nakajima T, Zrinzo L, Foltynie T, Olmos IA, Taylor C, Hariz MI, Limousin P (2011) MRI-guided subthalamic nucleus deep brain stimulation without microelectrode recording: Can we dispense with surgery under local anaesthesia? Sterotact Funct Neurosurg 89(5):318–325. https://doi.org/10.1159/000330379

Newsome WT, Stein-Aviles JA (1999) Nonhuman primate models of visually based cognition. ILAR J 40(2):78–91. https://doi.org/10.1093/ilar.40.2.78

Ojemann G (2014) Fifty-plus years of human single neuron recordings: a personal perspective. In: Fried IRU, Cerf M, Kreiman G (eds) Single neuron studies of the human brain. MIT Press, Cambridge, MA, pp 7–16

Oliveira LMO, Dimitrov D (2008) Surgical techniques for chronic implantation of microwire arrays in rodents and primates. In: Nicolelis MA (ed) Methods for neural ensemble recordings. Taylor & Francis, New York, pp 21–46

Percie du Sert N et al (2020) The ARRIVE guidelines 2.0: updated guidelines for reporting animal research. PLoS Biol 18(7):e3000410. https://doi.org/10.1371/journal.pbio.3000410

Picaud S, Dalkara D, Marazova K, Goureau O, Roska B, Sahel JA (2019) The primate model for understanding and restoring vision. Proc Natl Acad Sci U S A 116(52):26280–26287. https://doi.org/10.1073/pnas.1902292116

Prescott MJ, Bowell VA (2005) Training laboratory-housed non-human primates, part 2: Resources for developing and implementing training programmes. Anim Technol Welf 4:133–148

Prescott MJ, Brown VJ, Flecknell PA, Gaffan D, Garrod K, Lemon RN, Parker AJ, Ryder K, Schultz W, Scott L, Watson J, Whitfield L (2010) Refinement of the use of food and fluid control as motivational tools for macaques used in behavioural neuroscience research: Report of a Working Group of the NC3Rs. J Neurosci Methods 193(2):167–188. https://doi.org/10.1016/j.jneumeth.2010.09.003

Pusztai L, Hatzis C, Andre F (2013) Reproducibility of research and preclinical validation: problems and solutions. Nat Rev Clin Oncol 10(12):720–724. https://doi.org/10.1038/nrclinonc.2013.171

Roelfsema PR, Treue S (2014) Basic neuroscience research with nonhuman primates: a small but indispensable component of biomedical research. Neuron 82(6):1200–1204. https://doi.org/10.1016/j.neuron.2014.06.003

Roska B, Sahel JA (2018) Restoring vision. Nature 557:359–367. https://doi.org/10.1038/s41586-018-0076-4

Sato K, Sasaki E (2018) Genetic engineering in nonhuman primates for human disease modeling. J Hum Genet 63:125–131. https://doi.org/10.1038/s10038-017-0351-5

Schapiro SJ, Hau J (2022) Benefits of improving welfare in captive primates. In: Robinson LM, Weiss A (eds) Nonhuman primate welfare: from history, science, and ethics to practice. Springer, Cham, pp 433–450

Scientific Committee on Health Environmental and Emerging Risks (SCHEER) (2017) The need for non-human primates in biomedical research, production and testing of products and devices. https://ec.europa.eu/health/sites/health/files/scientific_committees/scheer/docs/followup_cons_primates_en.pdf

Speaking of Research (2021) Animal Rights Extremism. https://speakingofresearch.com/extremism-undone/ar-extremism/. Accessed June 2021

UK Government (2014) Working to reduce the use of animals in scientific research. https://assets.publishing.service.gov.uk/government/uploads/system/uploads/attachment_data/file/277942/bis-14-589-working-to-reduce-the-use-of_animals-in-research.pdf. Accessed June 2021

VandeBerg JL, Williams-Blangero S (1997) Advantages and limitations of nonhuman primates as animal models in genetic research on complex diseases. J Med Primatol 26(3):113–119. https://doi.org/10.1111/j.1600-0684.1997.tb00042.x

Weatherall D (2006) The use of non-human primates in research (Sponsored by the Academy of Medical Sciences, the Royal Society, the Medical Research Council, The Wellcome Trust), London

An Unexpected Symbiosis of Animal Welfare and Clinical Relevance in a Refined Nonhuman Primate Model of Human Autoimmune Disease

Bert A. 't Hart, Jon D. Laman, and Yolanda S. Kap

Abstract

Aging Western populations are confronted with an increasing prevalence of chronic inflammatory and degenerative diseases for which adequate treatments are lacking. One of the major hurdles in therapy development is the poor translation of disease concepts, often developed in rodent disease models, into effective treatments for the patient. Reasons for the high failure rate of promising drug candidates are unforeseen toxicity and lack of efficacy. Essential elements of human disease are apparently lacking in the current preclinically used animal models. Results obtained in a generic nonhuman primate model of human autoimmunity, the marmoset experimental autoimmune encephalomyelitis (EAE) model, are discussed to emphasize the claim that primates are essential complementary models that can help to bridge the wide translational gap between mouse and man.

Keywords

Animal model · Neuroinflammation · Multiple sclerosis · 3R's · Validity

---

B. A. 't Hart (✉)
Department of Immunobiology, Biomedical Primate Research Centre, Rijswijk, The Netherlands

Department of Neuroscience, University Medical Center Groningen, University of Groningen, Groningen, The Netherlands

Department of Anatomy and Neuroscience, Amsterdam University Medical Center, Amsterdam, The Netherlands

J. D. Laman
Department of Neuroscience, University Medical Center Groningen, University of Groningen, Groningen, The Netherlands

Y. S. Kap
Department of Immunobiology, Biomedical Primate Research Centre, Rijswijk, The Netherlands

1 Introduction

Animal models have an important role in the translational research of human disease. Although many aspects of the disease process can be investigated in cell or tissue cultures, most scientists are convinced that research into the complex connections and interactions of these processes requires live animals (Barre-Sinoussi and Montagutelli 2015). Nevertheless, the use of animal models in preclinical research of human disease is the subject of increasing debate. Some opponents in the public debate even claim that animal models are completely irrelevant and therefore are unethical.

When considering the relevance of animal models for translational research into the pathogenesis and treatment of human disease, a classical aphorism by the statistician George Box is worth mentioning: *Essentially all models are wrong, but some are useful* (Box and Draper 1987). In the context of a discussion on the relevance of a certain animal model for human disease, the aphorism can be interpreted as: *the relevance of the model depends to a large extent on its intended use*. In our field of expertise, which is neuroimmunology in general and multiple sclerosis (MS) in particular, a plethora of potentially useful animal models exists, including *Caenorhabditis elegans* worms, *Drosophila* flies, *Brachydanio rerio* fish, mice, rats, and primates. Each of these models has provided important information on pathogenic mechanisms in MS, but none of them faithfully replicates all pathological and clinical aspects of the human disease. It is therefore not surprising that the translation of the accumulated scientific knowledge into safe and effective treatments for the patient has been notoriously difficult. Apparently, essential aspects of the human disease are lacking in each of the available animal models.

The subject of this chapter is the (essential) role of primates in preclinical research. The discussion will be focused on a subgroup of diseases caused by the immune system, namely those in which the own body is attacked causing autoimmune disease. Experimental autoimmune encephalomyelitis (EAE) is one of the most intensively investigated autoimmune animal models and is used both as a specific model of the autoimmune neuroinflammatory disease MS and as a model of human autoimmune disease in general.

We will discuss that although the specific pathogen-free (SPF)-bred laboratory mouse is the gold standard in this research, unique aspects of primate EAE make it an essential complementary model that can help bridge the wide gap between the laboratory mouse and the patient. Specific attention will also be paid to welfare aspects of the primate EAE model, in particular the compliance with the 3R principles (Russell and Burch 1959).

2 Concise Phylogeny of Animal Models Used in Preclinical Immunology Research

The basic role of the human immune system is to protect the organism against infections and cancer (nonself), without causing harm to the organism (self), and to promote repair. This vital task involves a complex interplay of innate and adaptive immune functions, which are activated upon exposure to hostile intruders, while at the same time self is ignored (Nossal 1991). A fundamental modification of this dogma has been the discovery that the adaptive arm of the immune system is only activated when the innate arm recognizes danger (Matzinger 2002).

The nematode worm, *Caenorhabditis elegans* has an ancestral immune system via which it can recognize and combat viral, bacterial, and fungal infections (Ermolaeva and Schumacher 2014). The template of the worm immune system shows similarities with the innate arm of the human immune system. Consequently, *C. elegans* has been used to unravel principles of human innate immunity. The worm is a powerful model as its whole genome has been sequenced and annotated, and loss of function mutants of almost all genes are available. This, added to the neurological (only 320 neurons) and immunological (only innate immunity) simplicity of the worms creates a strong research tool for developing a deep understanding of the neural regulation of innate immunity and the innate immune regulation of neurological functions.

Drosophila is well equipped for the recognition and combating of infection by microorganisms as they have a capable innate immune system. The *Drosophila* system uses a set of germ-line encoded receptors together with effector cells and molecules, which have evolved into the essential factors of the human innate immune system (Hoffmann et al. 1999; Janeway and Medzhitov 2002). However, just like *C. elegans*, *Drosophila* lacks adaptive immune functions executed by T- and B-lymphocytes (Langenau and Zon 2005) and is therefore incomplete models of human immunity.

Zebra fish have both innate and adaptive immunity, which enable them not only to recognize and combat infections, but also to store information on previous pathogen exposures in memory cells. The latter capacity enables a faster and more effective response upon subsequent exposures to the same microorganisms. The basic templates of the fish and human immune system are remarkably similar (Langenau and Zon 2005).

For many years, the mouse has been the elected animal model of human immunology as many similarities exist both in the architecture as well as the functioning of the innate and adaptive immune systems (Davis 2008). As, by far, the greatest majority of fundamental discoveries in immunology were done in mice, it would be ridiculous to downscale the relevance of the mouse for our current understanding of the human immune system. However, despite the many similarities, there are also essential differences between the immune systems of mice and man, such as complement functions and the ratio between neutrophils and lymphocytes in blood, to give a few examples (Mestas and Hughes 2004). Moreover, recent studies showed that, due to the SPF breeding conditions, the immune systems of standard laboratory mice are essentially immature and lack effector memory cells (Beura et al. 2016; Abolins et al. 2017).

Nonhuman primates are the closest living relatives of man. This evolutionary proximity is reflected in the high immunological similarity between humans and nonhuman primates, as expressed in the highly polymorphic genes that encode molecules involved in antigen presentation and recognition (Bontrop et al. 1995). Moreover, captive colonies of nonhuman primates in research centers such as the BPRC (Rijswijk, Netherlands) are bred and raised under conventional conditions, where they are exposed to similar and often the same types of pathogens as humans are exposed to (www.bprc.nl). Work from our group shows that the pathogen-educated nonhuman primate immune system harbors potentially auto-aggressive effector memory T cells, which, upon in vivo activation, can turn on pathogenic mechanisms leading to features of MS pathology that are not seen in other animal models ('t Hart et al. 2011; 't Hart 2016) (Fig. 1). The observation that in vivo activation of these autoaggressive effector memory T cells can be achieved with relatively mild adjuvants, such as incomplete Freund's adjuvant (IFA), has formed the basis for a set of atypical EAE models which are not only more animal friendly than the classical models based on strong bacterial adjuvants, but also clinically more relevant ('t Hart et al. 2011).

3 Multiple Sclerosis (MS)

Multiple sclerosis (MS) is an autoimmune neuroinflammatory disease that selectively affects the human central nervous system (CNS). The cause of the disease is not known. Genome-wide association studies and the beneficial effects of therapies targeting immune functions indicate an important role of the immune system in the initiation and/or perpetuation of the disease (Sospedra and Martin 2005; Sawcer et al. 2011). Indeed, once established, chronic disease development is driven by the synergy of autoreactive T and B cells specific for components of the myelin sheaths that wrap around axons (Sospedra and Martin 2005). Also, the trigger of the pathogenic autoimmune reactions is not known, but it could be an interplay of genetic and microbial factors or a dysregulated response to autoantigens released from an idiopathic lesion within the CNS (Stys et al. 2012).

Mouse EAE models have shaped our current understanding of immunopathogenic mechanisms (Steinman 2014). However, despite the vast body of accumulated knowledge, there remain open questions for which we have no satisfactory answer yet. Accumulating evidence indicates that nonhuman primate EAE models can help bridging the gap between mouse EAE and MS.

A poorly understood phenomenon in MS is the heterogeneous clinical course. In the majority of MS patients (±85%), the disease initially follows a relapsing-remitting course, where episodes of neurological dysfunction (relapse) alternate with recovery (remission). In most patients, the relapsing-remitting course of the disease transits after a variable time into a secondary progressive course. During the latter course, recovery no longer occurs and neurological functions worsen

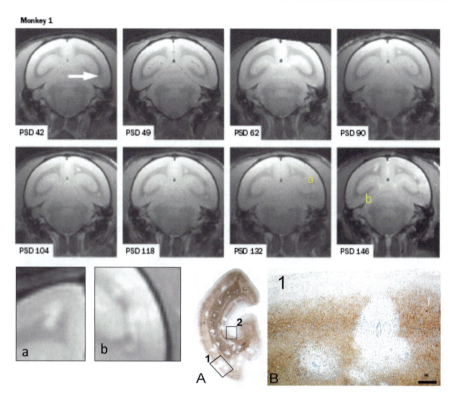

Fig. 1 Pathological characterization of marmoset EAE induced with rhMOG/CFA. Following a single inoculation with recombinant human MOG (residues 1–125) emulsified with complete Freund's adjuvant (CFA), T2-weighted brain MR images were made (psd = postsensitization day). Lesion development, visible as hyperintense spots, is disseminated in time and space and initially confined to the white matter. The white arrow in the image at psd 42 points to the first formed lesion. In late stage disease (psd 132 and 146), lesions seem to colonize also the cortical gray matter. Inserts **a** and **b** show late lesions at higher magnification. The histological pictures (A, B) show PLP staining of an EAE brain from the same model albeit another monkey. Figure composed of parts of figures published in 't Hart et al. (2004a) and Jagessar et al. (2015)

progressively. In a minority of patients (± 15%), the disease is progressive from the onset, and is referred to as primary progressive disease. The factor(s) that underlie the transition of relapsing-remitting to secondary progressive disease are unknown; the cause of primary progressive MS is not known either (Steinman and Zamvil 2016). The available therapies for relapsing-remitting MS do not show a relevant beneficial effect in progressive disease, indicating that relapsing-remitting and secondary progressive MS may be driven by different pathogenic mechanisms.

According to the prevailing concept, MS is an autoimmune disease which is elicited when a genetically predisposed individual encounters an environmental trigger. However, despite decades of intensive research in patients and animal models, an environmental trigger of MS has not been identified. Demographic studies indicate that the thus far elusive trigger may be infection with a virus or bacterium, which is encountered around the age of 15. The infection more frequently leads to MS in moderate climate areas than around the equator. Moreover, people migrating from a high-risk to a low-risk region before the age of 15 adopt the risk of their new environment, whereas people migrating after age 15 keep the risk of their country of origin.

The genetic risk is dominated by the major histocompatibility complex (MHC) class II genomic region, which is a cluster of highly polymorphic genes encoding molecules expressed on professional antigen-presenting cells (APC), via which antigens are presented to CD4+ T cells. However, the dominant subset of T cells present in established MS lesions is not CD4+ but CD8+, and depletion of CD4+ T cells with anti-CD4 monoclonal antibody (mAb) did not reduce disease activity (van Oosten et al. 1997). Moreover, treatment of RRMS patients with ustekinumab (another mAb) against the shared p40 subunit of interleukin (IL)-12 and -23, two sister cytokines engaged in the skewing of CD4+ T cells toward a proinflammatory profile (Th1 and Th17), was clinically ineffective (Segal et al. 2008). This does not preclude, however, a pathogenic role of CD4+ T cells early in the disease, i.e., before the diagnosis MS has been made. The question which pathogenic roles autoaggressive CD4+ and CD8+ T cells subsets exert is subject of intensive research.

Epstein Barr Virus (EBV) is the most important infectious risk factor for MS. Overall, a conservative estimate indicates that the relative risk of developing MS is 15 for people with evidence of asymptomatic EBV infection at adolescent age and even 30 for those having experienced symptomatic infection, i.e., infectious mononucleosis (Thacker et al. 2006). By contrast, a negative risk factor for developing MS has been linked to a minority of the adult population (<10%) who have not encountered EBV infection (Pakpoor et al. 2013). These are striking ratios for a disease in which the strongest genetic factors (the presence of the HLA-DRB1*1501, –DRB5*0101, -DQB1*0602 alleles) confer a relative risk of 3–4 (Hoppenbrouwers and Hintzen 2011). However, the mechanisms underlying the association between EBV infection and enhanced MS risk are poorly understood. An explanation for the paradox between the high EBV infection prevalence in the healthy population (90%) and the low prevalence of MS (0.1%) eludes us as well.

The poor translation of scientific concepts into effective treatments for MS patients is probably the best illustration that essential elements of MS are lacking in currently used animal models. Accumulating evidence presented in the next paragraph indicates that several of these elements are present in the well-established EAE model in common marmosets (*Callithrix jacchus*). As argued elsewhere, we believe that more investment should be made in a (reverse translational) analysis of the reasons why promising treatments failed in clinical trials ('t Hart et al. 2014). With this information in hand, the translational relevance of the currently used rodent and nonhuman primate EAE models can then be improved.

4 Translational Relevance of the Marmoset EAE Model

Nonhuman primate species used for the modeling of human autoimmune disease includes the larger rhesus and cynomolgus macaques (*Macaca mulatta* and *M. fascicularis*) and the small-bodied common marmoset. In our hands, the EAE models in both macaque species are rather acute, more closely resembling acute postinfectious demyelinating diseases, such as acute disseminated encephalomyelitis, while the model in marmosets more closely resembles chronic MS (Brok et al. 2001; 't Hart et al. 2005a). Marmoset EAE is therefore often the model of choice, while the macaque EAE models are used to test the efficacy of drugs that are inactive in marmosets, or for experiments requiring larger volumes of blood.

Marmosets provide translationally relevant models for a variety of clinical conditions, including (age-associated) autoimmune-mediated inflammatory diseases (AIMID) ('t Hart et al. 2012, 2013). Marmosets are nonprotected, small-bodied nonhuman primates (weighing 300–400 g at adult age), which have as their natural habitat the Amazon rainforest. They breed well in captivity, giving birth to one or two pairs of nonidentical twin pairs or triplets per year. Twin siblings often develop as bone marrow chimeras due to the sharing of the placental bloodstream (Haig 1999). As the immune systems of twin siblings are educated in the same thymic and bone marrow compartments, they are not only allotolerant, but also immunologically highly comparable. This is an important advantage for preclinical therapy studies as these can be set up in twins, where one sibling receives an experimental treatment and the other a relevant control preparation. Despite common marmosets' small body-size, it is nevertheless possible to perform immunological studies by using methods specially designed for working with small blood volumes (Jagessar et al. 2013b).

The outbred, pedigreed, purpose-bred marmoset colony at the BPRC is housed under conventional conditions in partly outdoor enclosures (Bakker et al. 2015). The monkeys are thus exposed to similar environmental factors as humans are exposed to. Marmosets also harbor chronic latent infections with herpesviruses related to the ones that humans are infected with and which have been implicated in the pathogenesis of MS (see below). Thus, just like humans, marmosets have a *"pathogen-educated"* immune system, which contains auto-aggressive effector memory T cells that mediate the high immune reactivity of marmosets against human CNS myelin ('t Hart et al. 2015).

The original EAE model was established by sensitization of marmosets against CNS myelin from an MS patient formulated with a suitable adjuvant (CFA), which elicited a neuroinflammatory disease that approximates MS in clinical and pathological presentation ('t Hart et al. 1998). A noticeable difference between marmoset and mouse EAE models is that in the former demyelinated lesions are present in the white and gray matter of the brain and spinal cord, while in the latter lesions are confined to the white matter of the spinal cord. This aspect of marmoset EAE has enabled an in-depth analysis of the histological correlates of brain lesions detectable with magnetic resonance imaging (MRI), the most frequently used imaging method in MS. We observed that essentially all MS white matter lesion types are also present

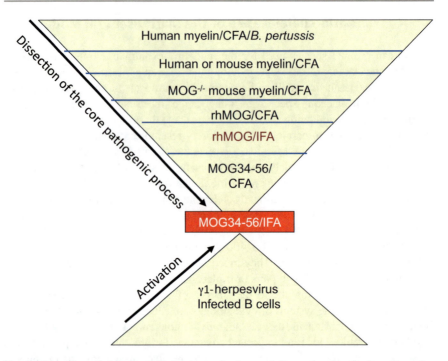

Fig. 2 Dissection of the core pathogenic mechanism and its mode of activation. Step-wise refinement of the marmoset EAE model induced by sensitization against MS myelin in CFA was performed on the guidance of clinical and pathological characteristics. The minimal induction requirement is a peptide of 23 residues emulsified with mineral oil. The activation of this core-pathogenic process appears to involve B cells infected with the EBV-related γ1-herpesvirus CalHV3

in the marmoset EAE model ('t Hart et al. 1998; Blezer et al. 2007). Later research demonstrated that this was also the case for the lesions in the cortical gray matter (Merkler et al. 2006; Kap et al. 2011; Dunham et al. 2017b). Moreover, brain MRI could be used as a clinically relevant read-out in immunotherapy studies in the model ('t Hart et al. 2006).

One focus of our research has been the dissection of the core pathogenic mechanism, as this should be the optimal target of therapy. To achieve this, we used the stepwise refinement procedure depicted in Fig. 2, as reviewed in 't Hart et al. (2009).

As a first step, we showed that autoimmunity against CNS myelin glycoprotein MOG is dispensable for EAE initiation, but essential for the evolution to progressive disease (Jagessar et al. 2008). A similar critical role of MOG was found in the Biozzi mouse EAE model (Smith et al. 2005). MOG has an essential role in the regulation of tolerance and autoimmunity against myelin (Garcia-Vallejo et al. 2014). As a normally glycosylated protein, MOG is tolerogenic as it binds the C-type lectin receptor DC-SIGN, which relays inhibitory signals for DC maturation to dendritic cells. Alteration of the normal glycosylation, for example under the inflammatory conditions present in MS lesions, makes MOG strongly immunogenic ('t Hart and Weissert 2016).

Next, we showed that sensitization of marmosets against a recombinant protein that encompasses the extracellular domain MOG (residues 1–125) of human MOG activates two nonoverlapping pathogenic mechanisms, which respectively mediate the initiation and the progression phase of the EAE model ('t Hart et al. 2011). The **initiation mechanism** involves T-helper 1 (Th1) cells recognizing a specific fragment (epitope) of the immunizing MOG protein, namely residues 24–36, which is presented via an invariant MHC class II allele, Caja-DRB*W1201. Moreover, antibodies binding a conformational epitope of the MOG molecule are induced. Both factors seem to act synergistically the Th1 cells induce inflammation and the antibodies elicit demyelination. This synergistic mechanism essentially replicates mouse EAE models. The clinical relevance of the initiation pathway for the EAE model was confirmed by the beneficial effect of therapeutic antibodies targeting the formation of proinflammatory Th1/Th17 cells (Brok et al. 2002; 't Hart et al. 2005b) or B cells (Boon et al. 2001; Kap et al. 2010; Kap et al. 2014).

We then found that the EAE **progression mechanism** involves the activation of autoaggressive CD8+CD56+ cytotoxic effector memory T cells specific for the epitope MOG40-48, which is presented by the invariant MHC class Ib allele Caja-E (Kap et al. 2008; Jagessar et al. 2012). This pathway has no known correlate in mouse EAE models. T cells driving the progression pathway require B cells infected with the EBV-related lymphocryptovirus CalHV3 for their activation. This was deduced from the discrepant clinical effects between mAbs against the pan B-cell marker CD20 and the B-cell growth and differentiation factors BlyS and APRIL (Jagessar et al. 2013a). It is noteworthy that this distinction has not been found in SPF mouse EAE models, while a similar paradoxical effect has been observed in MS clinical trials (Barun and Bar-Or 2012; Kappos et al. 2014).

5 Mechanistic Basis of MS Risk Factors: Lessons from the Marmoset EAE Model

5.1 Predisposing Genes

The strongest genetic effect on the risk to develop MS is exerted by the HLA-DR2 locus. Strong candidate risk alleles are HLA-DRB1*1501/HLA-DRB5*0101/HLA-DQB1*0602 (Hoppenbrouwers and Hintzen 2011). Studies in mice expressing HLA-DRB1*1501 transgene show that this MHC class II specificity binds the immunodominant MOG34-56 peptide and activates proinflammatory CD4+ T cells capable of inducing CNS inflammation (Rich et al. 2004). A direct equivalent of this allele has not been found in the MHC of marmosets, which is indicated with the acronym Caja (from *Callithrix jacchus*); hence the finding could not be confirmed. Nevertheless, the marmoset model shows a similar pathogenic role of Caja-DRB*1201 restricted Th1 cells specific for another epitope, MOG24-36, in the initiation of EAE. Blockade of this EAE initiation mechanism with therapeutic mAbs, such as anti-CD20, anti-CD40 or ustekinumab (anti-IL-12p40), abrogated EAE development (Boon et al. 2001; Brok et al. 2002; Kap et al. 2010). As the results from immunotherapies targeting CD4+ T cells in RRMS have been disappointing thus far, the relevance of this subset in MS has been disputed (Lassmann and Ransohoff 2004). However, the negative results obtained during ongoing MS do not preclude that Th1 cells exert a pathogenic function early in the disease process, possibly even before the disease is diagnosed.

Genome-wide association studies (GWAS) are designed for the identification of genes that are differentially expressed between MS patients and healthy controls. Ubiquitously expressed invariant genes are usually not detected. This may explain why HLA-E, which comprises only two alleles (HLA-E*0101/E^R and HLA-E*0103/E^G), did not emerge as a dominant risk factor in MS. Studies in the marmoset demonstrated that the direct equivalent of HLA-E, called Caja-E, functions as the restriction element of core pathogenic autoaggressive cytotoxic T lymphocytes (CTL) specific for the epitope MOG40-48. Upon in vivo activation, these CTL were found to be capable of inducing essential pathological elements of RRMS and SPMS (Jagessar et al. 2010; Dunham et al. 2017a).

A non-MHC gene associated with enhanced MS risk is the receptor of IL-7 (CD127). A mAb against this receptor was found to exert a beneficial effect on marmoset EAE, but only in monkeys that developed fast-progressing EAE (Dunham et al. 2016).

In summary, the marmoset EAE model revealed that distinct pathogenic mechanisms are involved in the induction of brain pathology and the induction of neurological symptoms. Therapies targeting the former mechanism, which accurately replicates pathogenic mechanisms in rodent EAE models, frequently failed to reproduce promising effects observed in mouse EAE when they were tested in the clinic. Data obtained thus far show that the latter mechanism, which is novel and has no equivalent in rodents, better represents the situation in MS. The new atypical EAE model in which this mechanism has a central pathogenic role ('t Hart et al. 2017) offers new unmet opportunities for therapy development.

5.2 Infections

The family herpesviridae comprises eight members (indicated HHV) that are known to cause disease in humans. A role of three of these HHV in MS pathogenesis is supported by marmoset EAE models.

HHV5/cytomegalovirus (CMV) is a β-herpes virus that causes usually asymptomatic infections in at least 60% of the adult human population; the infection prevalence can be >90% in high-risk groups, such as AIDS patients and offspring of mothers infected during pregnancy. The virus is viewed as a driving factor behind the aging of the immune system, in particular via the induction of oligoclonal expansion of potentially pathogenic T cells ('t Hart et al. 2013). Latent CMV infection is controlled by HLA-E restricted CD8+ CTL, which also expresses markers of natural killer (NK) cells (Moretta et al. 2003). Marmosets are naturally infected with a simian CMV or can be experimentally infected with human CMV. (Nigida et al. 1975, 1979), whether CMV has a pathogenic role in MS is debated (Vanheusden et al. 2015). Based on the specificity for a mimicry epitope shared between MOG and the UL86 antigen of CMV (Brok et al. 2007), the restriction by Caja-E, and the expression of the NK cell marker CD56 (Jagessar et al. 2012), we tentatively placed the CTL that drives the EAE progression pathway in the repertoire of anti-CMV effector memory T cells. Although a high proportion of mice in nature are normally infected with mouse CMV, SPF laboratory mice are not infected by the virus.

HHV4/Epstein Barr Virus (EBV) is a γ1-herpesvirus that causes usually asymptomatic infections in about 90% of the healthy adult population (Bar-Or et al. 2020). However, only a small minority of B cells actually contain the virus (Khan et al. 1996). The geographical latitude effect on MS has been attributed to the age at which children are infected with EBV. Around the equator, children are infected before the age of two, usually without clear clinical consequences. Exposure to the virus in adolescence can induce infectious mononucleosis, which is characterized by oligoclonal expansion of B cells, strong activation of antiviral T cells, and flu-like symptoms. It has been difficult to prove a causal relation between EBV infection and MS as the difference in infection prevalence between MS patients (100%) and the healthy population (>90%) is small. Nevertheless, seronegativity for EBV has been associated with a low-to-absent risk of developing MS (Pakpoor et al. 2013). On the other hand, a history of infectious mononucleosis (IM) has been reported to increase the risk of developing MS by at least twofold, when compared to individuals infected with EBV earlier in life (Ascherio and Munger 2015). Finally, in a case study in one secondary progressive MS patient, remission could be achieved by the infusion of cytotoxic T cells designed to clear the host of EBV-infected B cells (Pender et al. 2014). This is the first clear indication that EBV-infected B cells may have a core pathogenic role in progressive MS.

Mice infected with the murine gammaherpesvirus-68 are used as a model of human EBV infection for therapy development (Marquez and Horwitz 2015). However, this virus belongs to the group of γ2-herpesviruses, which have no known pathogenic role in MS. The model should therefore be deemed as suboptimal.

Marmosets are naturally infected with the EBV-related lymphocryptovirus callithrichine herpesvirus 3 (CalHV3) (Cho et al. 2001), but marmoset B cells can also be infected ex vivo with an EBV laboratory strain (95-8). Immunotherapy studies support a crucial pathogenic role of CalHV3-infected B cells most likely in the recruitment of the autoaggressive CTL from the anti-CMV repertoire (Jagessar et al. 2013a). The role of EBV/CalHV3 infection seems to be protection of the proteolysis sensitive MOG40–48 epitope against fast degradation by the serine protease cathepsin G in the endolysosomal compartment of B cells so that it can be cross-presented via Caja-E to the autoaggressive CTL. The protection mechanism involves citrullination of essential arginine residues and association of the peptide with autophagosomes ('t Hart et al. 2016).

HHV6A is a neurotropic β-herpesvirus that infects >90% of the human population (Clark 2004). Primary infections in immunocompetent individuals can result in neurological problems, such as meningitis and meningoencephalitis. The virus infects cells involved in MS, including CD4+ T cells and precursors of oligodendrocytes, the myelin-forming glial cells. The virus has been detected in brain tissue and CSF of MS patients. Marmosets infected with HHV6A develop signs of neuroinflammation and neurological problems (Leibovitch et al. 2013). Humanized SCID mice can also be infected with HHV6A, but obviously provide a highly artificial system (Reynaud and Horvat 2013).

HHV8/Kaposi's sarcoma-associated herpesvirus (KSHV) is a γ2-herpesvirus/rhadinovirus which has no known role in MS. Nevertheless, two animal studies hint at a possible pathogenic role in the disease. The murine herpesvirus 68 (MHV-68) infects mouse B cells, and for this reason, MHV-68 has been proposed as mouse model of EBV infection (Marquez and Horwitz 2015). However, MHV-68 is more closely related to HHV-8/KSHV than to EBV. A publication from the Oregon National Primate Center reported a spontaneous outbreak of MS-like disease in a colony of Japanese macaques, which was found to be associated with a thus far unknown simian rhadinovirus (Axthelm et al. 2011).

In conclusion, marmosets are susceptible to infections with three human herpesviruses, which all have been implicated in the initiation and/or course of MS. Marmosets are therefore a highly useful model for studies on the separate and interactive roles of these viruses in MS. The fact that these marmosets naturally infected with the EBV-related CalHV3, offers unique opportunities for translational research into the still poorly understood relation between EBV and MS.

6 Welfare Aspects

Aging societies are facing an increasing prevalence of chronic invalidating disorders of the central nervous system, such as Alzheimer's and Parkinson's, and MS. Despite substantially increased investments by the pharmaceutical industry, the output of successful new drugs for these disease remains disappointingly low (Kola and Landis 2004; Schafer and Kolkhof 2008). A main reason is the wide gap between animal models used in the pipeline selection of candidate drugs and the human disease.

The lack of valid preclinical animal models added to the increasing costs of animal research has stimulated the development of nonanimal models based on human-derived cells ranging from single cell cultures to complex multicellular systems, such as organs on a chip ('t Hart and Bajramovic 2008; Balls et al. 2019). Although the developments are promising and these models can be useful for the study of isolated pathological processes, we believe that the high complexity of neurological disorders such as MS cannot be adequately modeled without animals.

Important criteria in the selection of a valid animal model are: (i) whether the clinical and pathological presentations adequately replicate the human disease (face validity), (ii) whether disease mechanisms adequately replicate the human disease (construct validity), and (iii) whether pharmacological effects of a new drug are comparable between the model and the human disease. The data presented above illustrate the validity of the marmoset in the translational research and treatment of MS. Especially for the new generation of highly human-specific biological drugs, monoclonal antibodies for example, replacement by other species is often not an option (Chapman et al. 2007).

However, the two other Rs of the Russell and Burch triplet (1959) require special consideration. Marmosets are an outbred nonendangered species, which adapt well to captive conditions in moderate climate areas. The marmosets that we use for our research come from the purpose-bred and pedigreed colony that has been held for at least 30 years at the BPRC. Large investments have enabled the creation of optimal housing conditions of marmoset families (see www.bprc.nl). For a detailed description of our animal welfare policy in general and more specific information on housing, enrichment, and animal training, we refer readers to the institute's website: http://www.bprc.nl/en/welfare/. Marmosets selected for EAE experiments are moved to the experimental facility where in agreement with international standards, they are pair-housed in spacious indoor cages ($0.75 \times 0.70 \times 1.90$ m^3), enriched with sticks, branches, toys, and boxes that can be used for shelter.

Inevitably, the welfare of marmosets participating in EAE experiments is affected at different levels, including the procedures used for disease induction, stress or physical damage caused by the impairment or loss of neurological functions, and the procedures for collection of body fluids for immune monitoring. A large part of our research has been dedicated at achieving compliance of the marmoset EAE model with the 3R principles (Russell and Burch 1959), while keeping an eye on the clinical relevance of the model. Our work revealed a potential conflict among the 4R's (Relevance, Replacement, Reduction, and Refinement), which cannot easily be solved ('t Hart 2016).

The **Relevance** of an animal model for the preclinical efficacy screening of potential therapies, depends on whether essential clinical and pathological aspects of MS are reproduced in the EAE model (face validity) and whether the pathogenic mechanisms resemble those in the human disease (construct validity). The close similarity of the marmoset EAE model with MS implies that a certain amount of discomfort due to the loss of sensory and motor functions is inevitable. A potentially problematic factor is that marmosets in the experimental facility are pair housed in tall cages. To protect a motorically affected EAE marmoset against falling from a high altitude, separators are placed in the cage. Moreover, padded shelter is provided in the cage where a sick monkey can rest. Another important measure to minimize suffering is that the duration of the different levels of discomfort is maximized in a cumulative fashion (Jagessar et al. 2013b).

Considering the **Replacement** paradigm, it is important to stress that according to European legislation (EU directive 2010/63/EU; European Commission 2010), experiments in live nonhuman primates are only allowed when there is no other way to obtain the same information. Typically, the marmoset EAE model is used for the preclinical efficacy testing of new biological therapeutics, which, due to their high specificity cannot be tested in other animals. Importantly, usage of the model

Table 1 The effect of response variation on group size

Response to EAE	Response to treatment	Group size	Response to treatment	Group size
10/10	100%	4	80%	6
9/10	100%	5	80%	8
8/10	100%	6	80%	11
7/10	100%	8	80%	16
6/10	100%	10	80%	24
5/10	100%	12	80%	40

Shown is a power calculation of group size for a hypothetical experiment in the marmoset EAE model. The depicted example shows that the occurrence of nonresponders to EAE induction has a dramatic effect on group size even when 80% of the monkeys respond to the experimental treatment

for therapy evaluation requires deep understanding of the pathogenic mechanisms, which necessitates exploratory research in the model.

Compliance with the **Reduction** principle is achieved by using power analysis for calculating the minimum size of experimental groups needed for obtaining results that can be tested statistically (Cohen 1992). Moreover, techniques for the collection of more information from fewer animals have been developed, including live imaging, longitudinal immune monitoring, and biomarker analysis in body fluids ('t Hart et al. 2004b; Jagessar et al. 2013b). In addition, tissues collected at necropsy, including lymphoid organs, brain, and spinal cord, are intensively used for further analysis by histological and molecular biological techniques. As marmosets are an outbred species, heterogeneity in the response to EAE induction and to an experimental treatment should always be anticipated. Fortunately, the MHC class II (Caja-DRB*W1201) and class I (Caja-E) susceptibility alleles are invariant, but non-MHC genes, such as those encoding the IL-7 receptor, also appeared to exert a variable influence (Dunham et al. 2016). This variation can be dealt with to some extent by using bone marrow chimeric twins, which, as discussed above, are immunologically more comparable than nonrelated monkeys.

Considering **Refinement**, it is a central dogma in immunology that autoreactive T cells that have escaped negative selection in the thymus, and are therefore present in the healthy immune repertoire, are kept under strict control by potent regulatory cells (Bluestone and Abbas 2003; Peterson et al. 2008). Adjuvants are used for breaking such tolerance mechanisms and for the awakening of autoreactive T cells (Baxter 2007). The frequently used adjuvant CFA, which in rodent EAE models is combined with systemic administration of another adjuvant (*Bordetella pertussis*), is notorious for its serious adverse effects, of which the formation of necrotic granulomas at the injection sites is the most visible, albeit not the only, damage. The observation that the T cells that mediate EAE initiation and progression in marmosets can be activated in vivo by immunization with antigen in the much milder adjuvant IFA (discussed above) implies a major reduction of discomfort. However, these atypical EAE models are sensitive to variation in individual characteristics of the monkeys, such as their genetic background and history of infections. The inevitable consequence for these models is higher variation in the response to immunization and to the experimental treatment than observed in the more robust CFA-based models. Table 1 illustrates the impact of higher variation on the group size. The given example illustrates that an investment in one R (i.e., Refinement) can create a conflict with other Rs (i.e., Reduction).

A possible way out of this dilemma would be a different view on the design of studies involving precious animals, such as nonhuman primates. It has been argued by Bacchetti et al. that underpowered studies are not by definition irrelevant and can provide innovative data (Bacchetti et al. 2011, 2012). Above a certain sample size, the scientific or clinical value of each extra animal decreases, while the potential discomfort is the same for each added animal. In a recent marmoset EAE experiment comprising seven marmoset twins of which six developed EAE, we observed that only three twins responded to the experimental treatment, which was a novel mAb against the human IL-7 receptor. As the EAE course in these three responder twins clearly evolved faster than in the three nonresponder twins, we concluded that the treatment may have intervened in the process that accelerated EAE development (Dunham et al. 2016).

Of note, it is commonly observed in clinical trials that less than 100% of the participants respond to a tested treatment, which is usually attributed to heterogeneity of the pathogenic process. Even for a highly successful antirheumatic drug, the anti-TNFα mAb infliximab, which has been a trendsetting treatment for autoimmune inflammatory diseases, a response rate of 70 to 80% has been recorded (Maini and Taylor 2000).

7   Perspectives and Concluding Remarks

For many years, research in immunology has been concentrated on the adaptive arm of the immune system, i.e., the mechanism(s) used by T and B lymphocytes to distinguish infectious nonself from a species' self. The SPF-bred mouse has been at the center of all discoveries that shaped our current understanding of the system. Just two decades ago, interest in the role of the innate immune system was sparked by the discovery of evolutionary conserved pattern recognition receptors (PRR), such as Toll- and NOD-like receptors, with which immune cells recognize equally conserved pathogen-associated and cell damage-associated molecular patters (PAMPS and DAMPS) (Janeway and Medzhitov 2002). In addition, lectin-type receptors were identified on antigen-presenting cells that recognize carbohydrate structures, via which, self can be distinguished from nonself or altered self (e.g., on infected cells or cancer cells) ('t Hart and van Kooyk 2004; Geijtenbeek et al. 2004; Rabinovich et al. 2012). Research into the basic principles of innate immunity has involved, besides mice, other species, including invertebrates.

The current impressive body of immunological knowledge has enabled the development of treatments with satisfactory efficacy in RRMS. However, the list of failures, where the promising effects of new drugs in animal models could not be reproduced in patients, is much longer than the list of successes. There is growing awareness that the over-reliance of immunologists on a few well-defined SPF-bred and genetically homogeneous laboratory mouse strains hinders the development of better therapies for autoimmune diseases, cancer, and neurological diseases (Davis 2008).

As the causes of failure are usually not investigated, the predictive quality of the animal models currently used in preclinical research has not really changed. We have proposed elsewhere that lessons should be learned from failed clinical trials and that this knowledge should be used for elucidating why a given animal model has failed to predict efficacy of a promising treatment in the clinic ('t Hart et al. 2014). Unfortunately, this is rarely done.

We have used such a reverse translation approach for therapeutic biologicals that failed unexpectedly in RRMS clinical trials, namely the anti-IL-12p40 antibody ustekinumab (Segal et al. 2008) and atacicept, a chimeric construct combining IgG-Fc with the soluble TACI receptor of the B-cell cytokines BlyS and APRIL (Kappos et al. 2014).

Regarding ustekinumab, we discovered that the mAb is much more effective during EAE onset (Brok et al. 2002) than during established disease, although late stage treatment inhibited the activity and enlargement of lesions ('t Hart et al. 2005b). The explanation for this phenomenon could be that after a variable period of time, the autoimmune attack on the CNS transits from a mouse EAE like pathogenic mechanism, driven by the synergistic action of MHC class II-restricted Th1 cells and autoantibody, to an MS-like pathogenic mechanism, driven by MHC class Ib-restricted CD8+ CD56+ CTL, which seem to be absent in SPF mice (Kap et al. 2008).

Regarding atacicept, we discovered that capture of growth and differentiation factors, such as BlyS and APRIL, did induce depletion of B cells, but not of a small γ1-herpesvirus-infected fraction, which could be achieved with an anti-CD20 mAb that was effective in the clinic (ofatumumab). Indeed, survival of EBV-infected marmoset B lymphoblastoid cell lines in culture was not affected by the depletion of BlyS and APRIL (own unpublished data). This unexpected finding led us to the novel insight that immunotherapies targeting the small fraction of EBV-infected B cells ($<0.01\%$) may not only be effective, but also safe as nearly the entire B-cell compartment is left intact. In subsequent studies we analyzed why the virus-infected B-cell fraction is especially pathogenically relevant in MS ('t Hart et al. 2016). We discovered that EBV infection converts the destructive processing of the core pathogenic MOG34-56 peptide, which is a potential tolerance mechanism ('t Hart et al. 2016), into a productive processing and cross-presentation of the epitope, which is a potential autoimmune mechanism. This novel concept was recognized in an editorial in Science Translational Medicine as "*a new pathway by which infection triggers autoimmunity*" (Moore 2016).

In conclusion, we believe that the nonhuman primate is certainly not the translationally most relevant or the preclinically most useful model of human autoimmune disease. We also recognize that high costs and ethical constraints limit their use. This publication argues, however, that nonhuman primates are essential complementary models where the gap between mouse and man hinders progress in translational research. Moreover, when it comes to the development of innovative treatments, such as gene therapy (Goossens et al. 1999; Poliani et al. 2001; Bevaart et al. 2015) or stem cell therapy (Pluchino et al. 2009; Thiruvalluvan et al. 2016), nonhuman primate disease models have proven their usefulness.

Acknowledgments We would like to thank Mrs. Francisca van Hasselt (BPRC) for the artwork. The authors do not report conflicts of interest.

References

Abolins S, King EC, Lazarou L, Weldon L, Hughes L, Drescher P et al (2017) The comparative immunology of wild and laboratory mice, *Mus musculus* domesticus. Nat Commun 8:14811. https://doi.org/10.1038/ncomms14811

Ascherio A, Munger KL (2015) EBV and autoimmunity. Curr Top Microbiol Immunol 390 (Pt 1):365–385. https://doi.org/10.1007/978-3-319-22822-8_15

Axthelm MK, Bourdette DN, Marracci GH, Su W, Mullaney ET, Manoharan M et al (2011) Japanese macaque encephalomyelitis: a spontaneous multiple sclerosis-like disease in a nonhuman primate. Ann Neurol 70(3):362–373. https://doi.org/10.1002/ana.22449

Bacchetti P, Deeks SG, McCune JM (2011) Breaking free of sample size dogma to perform innovative translational research. Sci Transl Med 3(87):87ps24. https://doi.org/10.1126/scitranslmed.3001628

Bacchetti P, McCulloch C, Segal MR (2012) Being 'underpowered' does not make a study unethical (comment letter). Stat Med 31(29). https://doi.org/10.1002/sim.5451

Bakker J, Ouewerling B, Heidt PJ, Kondova I, Langermans JAM (2015) Advantages and risks of husbandry and housing changes to improve animal wellbeing in a breeding colony of common marmosets (*Callithrix jacchus*). J Am Assoc Lab Anim Sci 54(3):273–279

Balls M, Bailey J, Combes RD (2019) How viable are alternatives to animal testing in determining the toxicities of therapeutic drugs? Expert Opin Drug Metab Toxicol 15(12):985–987. https://doi.org/10.1080/17425255.2019.1694662

Bar-Or A, Pender MP, Khanna R, Steinman L, Hartung HP, Maniar T et al (2020) Epstein-Barr virus in multiple sclerosis: Theory and emerging immunotherapies. Trends Mol Med 26 (3):296–310. https://doi.org/10.1016/j.molmed.2019.11.003

Barre-Sinoussi F, Montagutelli X (2015) Animal models are essential to biological research: issues and perspectives. Future Sci OA 1(4):FSO63. https://doi.org/10.4155/fso.15.63

Barun B, Bar-Or A (2012) Treatment of multiple sclerosis with anti-CD20 antibodies. Clin Immunol 142(1):31–37. https://doi.org/10.1016/j.clim.2011.04.005

Baxter AG (2007) The origin and application of experimental autoimmune encephalomyelitis. Nat Rev Immunol 7(11):904–912. https://doi.org/10.1038/nri2190

Beura LK, Hamilton SE, Bi K, Schenkel JM, Odumade OA, Casey KA et al (2016) Normalizing the environment recapitulates adult human immune traits in laboratory mice. Nature 532 (7600):512–516. https://doi.org/10.1038/nature17655

Bevaart L, Aalbers CJ, Vierboom MP, Broekstra N, Kondova I, Breedveld E et al (2015) Safety, biodistribution, and efficacy of an AAV-5 vector encoding human interferon-Beta (ART-I02) delivered via intra-articular injection in rhesus monkeys with collagen-induced arthritis. Hum Gene Ther Clin Dev 26(2):103–112. https://doi.org/10.1089/humc.2015.009

Blezer EL, Bauer J, Brok HP, Nicolay K, 't Hart BA (2007) Quantitative MRI-pathology correlations of brain white matter lesions developing in a non-human primate model of multiple sclerosis. NMR Biomed 20(2):90–103. https://doi.org/10.1002/nbm.1085

Bluestone JA, Abbas AK (2003) Natural versus adaptive regulatory T cells. Nat Rev Immunol 3 (3):253–257. https://doi.org/10.1038/nri1032

Bontrop RE, Otting N, Slierendregt BL, Lanchbury JS (1995) Evolution of major histocompatibility complex polymorphisms and T-cell receptor diversity in primates. Immunol Rev 143:33–62. https://doi.org/10.1111/j.1600-065x.1995.tb00669.x

Boon L, Brok HP, Bauer J, Ortiz-Buijsse A, Schellekens MM, Ramdien-Murli S et al (2001) Prevention of experimental autoimmune encephalomyelitis in the common marmoset (*Callithrix jacchus*) using a chimeric antagonist monoclonal antibody against human CD40 is associated with altered B-cell responses. J Immunol 167(5):2942–2949. https://doi.org/10.4049/jimmunol.167.5.2942

Box GEP, Draper NR (1987) Empirical model-building and response surfaces. Wiley, Boca Raton, FL

Brok HP, Bauer J, Jonker M, Blezer E, Amor S, Bontrop RE et al (2001) Non-human primate models of multiple sclerosis. Immunol Rev 183:173–185. https://doi.org/10.1034/j.1600-065x.2001.1830114.x

Brok HPM, van Meurs M, Blezer E, Schantz A, Peritt D, Treacy G et al (2002) Prevention of experimental autoimmune encephalomyelitis in common marmosets using an anti-IL-12p40 monoclonal antibody. J Immunol 169(11):6554–6563. https://doi.org/10.4049/jimmunol.169.11.6554

Brok HPM, Boven L, van Meurs M, Kerlero de Rosbo N, Celebi-Paul L, Kap YS et al (2007) The human CMV-UL86 peptide 981-1003 shares a crossreactive T-cell epitope with the encephalitogenic MOG peptide 34-56, but lacks the capacity to induce EAE in rhesus monkeys. J Neuroimmunol 182(1–2):135–152. https://doi.org/10.1016/j.jneuroim.2006.10.010

Chapman K, Pullen N, Graham M, Ragan I (2007) Preclinical safety testing of monoclonal antibodies: the significance of species relevance. Nat Rev Drug Discov 6(2):120–126. https://doi.org/10.1038/nrd2242

Cho Y, Ramer J, Rivailler P, Quink C, Garber RL, Beier DR et al (2001) An Epstein-Barr-related herpesvirus from marmoset lymphomas. Proc Natl Acad Sci U S A 98(3):1224–1229. https://doi.org/10.1073/pnas.98.3.1224

Clark D (2004) Human herpesvirus type 6 and multiple sclerosis. Herpes 11(Suppl 2):112A–119A

Cohen J (1992) A power primer. Psychol Bull 112(1):155–159. https://doi.org/10.1037//0033-2909.112.1.155

Davis MM (2008) A prescription for human immunology. Immunity 29(6):835–838. https://doi.org/10.1016/j.immuni.2008.12.003

Dunham J, Lee LF, van Driel N, Laman JD, Ni I, Zhai W et al (2016) Blockade of CD127 exerts a dichotomous clinical effect in marmoset experimental autoimmune encephalomyelitis. J Neuroimmune Pharmacol 11(1):73–83. https://doi.org/10.1007/s11481-015-9629-6

Dunham J, van Driel N, Eggen BJ, Paul C, 't Hart BA, Laman JD et al (2017a) Analysis of the cross-talk of Epstein-Barr virus-infected B cells with T cells in the marmoset. Clin Transl Immunol 6(2):e127. https://doi.org/10.1038/cti.2017.1

Dunham J, Bauer J, Campbell GR, Mahad DJ, van Driel N, van der Pol SMA et al (2017b) Oxidative injury and iron redistribution are pathological hallmarks of marmoset experimental autoimmune encephalomyelitis. J Neuropathol Exp Neurol 76(6):467–478. https://doi.org/10.1093/jnen/nlx034

Ermolaeva MA, Schumacher B (2014) Insights from the worm: the *C. elegans* model for innate immunity. Semin Immunol 26(4):303–309. https://doi.org/10.1016/j.smim.2014.04.005

EU Directive 2010/63/EU of the European Parliament and of the Council of 22 September 2010 on the protection of animals used for scientific purposes. Available via https://eur-lex.europa.eu/649 legal-content/EN/TXT/?uricelex%3A32010L0063

Garcia-Vallejo JJ, Ilarregui JM, Kalay H, Chamorro S, Koning N, Unger WW et al (2014) CNS myelin induces regulatory functions of DC-SIGN-expressing, antigen-presenting cells via cognate interaction with MOG. J Exp Med 211(7):1465–1483. https://doi.org/10.1084/jem.20122192

Geijtenbeek TBH, van Vliet SJ, Engering A, 't Hart BA, van Kooyk Y (2004) Self- and nonself-recognition by C-type lectins on dendritic cells. Annu Rev Immunol 22:33–54. https://doi.org/10.1146/annurev.immunol.22.012703.104558

Goossens PH, Schouten GJ, 't Hart BA, Bout A, Brok HP, Kluin PM et al (1999) Feasibility of adenovirus-mediated nonsurgical synovectomy in collagen-induced arthritis-affected rhesus monkeys. Hum Gene Ther 10(7):1139–1149. https://doi.org/10.1089/10430349950018139

Haig D (1999) What is a marmoset? Am J Primatol 49(4):285–296. https://doi.org/10.1002/(SICI) 1098-2345(199912)49:4<285::AID-AJP1>3.0.CO;2-X

Hoffmann JA, Kafatos FC, Janeway CA, Ezekowitz RA (1999) Phylogenetic perspectives in innate immunity. Science 284(5418):1313–1318. https://doi.org/10.1126/science.284.5418.1313

Hoppenbrouwers IA, Hintzen RQ (2011) Genetics of multiple sclerosis. Biochim Biophys Acta 1812(2):194–201. https://doi.org/10.1016/j.bbadis.2010.09.017

Jagessar SA, Smith PA, Blezer E, Delarasse C, Pham-Dinh D, Laman JD et al (2008) Autoimmunity against myelin oligodendrocyte glycoprotein is dispensable for the initiation although essential for the progression of chronic encephalomyelitis in common marmosets. J Neuropathol Exp Neurol 67(4):326–340. https://doi.org/10.1097/nen.0b013e31816a6851

Jagessar SA, Kap YS, Heijmans N, van Driel N, van Straalen L, Bajramovic JJ et al (2010) Induction of progressive demyelinating autoimmune encephalomyelitis in common marmoset monkeys using MOG34-56 peptide in incomplete freund adjuvant. J Neuropathol Exp Neurol 69(4):372–385. https://doi.org/10.1097/NEN.0b013e3181d5d053

Jagessar SA, Heijmans N, Blezer EL, Bauer J, Blokhuis JH, Wubben JA et al (2012) Unravelling the T-cell-mediated autoimmune attack on CNS myelin in a new primate EAE model induced with MOG34-56 peptide in incomplete adjuvant. Eur J Immunol 42(1):217–227. https://doi.org/10.1002/eji.201141863

Jagessar SA, Fagrouch Z, Heijmans N, Bauer J, Laman JD, Oh L et al (2013a) The different clinical effects of anti-BLyS, anti-APRIL and anti-CD20 antibodies point at a critical pathogenic role of gamma-herpesvirus infected B cells in the marmoset EAE model. J Neuroimmune Pharmacol 8 (3):727–738. https://doi.org/10.1007/s11481-013-9448-6

Jagessar SA, Vierboom M, Blezer EL, Bauer J, 't Hart BA, Kap YS (2013b) Overview of models, methods, and reagents developed for translational autoimmunity research in the common marmoset (*Callithrix jacchus*). Exp Anim 62(3):159–171. https://doi.org/10.1538/expanim.62.159

Jagessar SA et al (2015) Immune profile of an atypical EAE model in marmoset monkeys immunized with recombinant human myelin oligodendrocyte glycoprotein in incomplete Freund's adjuvant. J Neuroinflammation 12:169. https://doi.org/10.1186/s12974-015-0378-5

Janeway CA Jr, Medzhitov R (2002) Innate immune recognition. Annu Rev Immunol 20:197–216. https://doi.org/10.1146/annurev.immunol.20.083001.084359

Kap YS, Smith P, Jagessar SA, Remarque E, Blezer E, Strijkers GJ et al (2008) Fast progression of recombinant human myelin/oligodendrocyte glycoprotein (MOG)-induced experimental autoimmune encephalomyelitis in marmosets is associated with the activation of MOG34-56-specific cytotoxic T cells. J Immunol 180(3):1326–1337. https://doi.org/10.4049/jimmunol.180.3.1326

Kap YS, van Driel N, Blezer E, Parren PW, Bleeker WK, Laman JD et al (2010) Late B-cell depletion with a human anti-human CD20 IgG1kappa monoclonal antibody halts the development of experimental autoimmune encephalomyelitis in marmosets. J Immunol 185 (7):3990–4003. https://doi.org/10.4049/jimmunol.1001393

Kap YS, Bauer J, van Driel N, Bleeker WK, Parren PW, Kooi EJ et al (2011) B-cell depletion attenuates white and gray matter pathology in marmoset experimental autoimmune encephalomyelitis. J Neuropathol Exp Neurol 70(11):992–1005. https://doi.org/10.1097/NEN.0b013e318234d421

Kap YS, van Driel N, Laman JD, Tak PP, 't Hart BA (2014) CD20+ B-cell depletion alters T-cell homing. J Immunol 192(9):4242–4253. https://doi.org/10.4049/jimmunol.1303125

Kappos L, Hartung H-P, Freedman MS, Boyko A, Radu EW, Mikol DD et al (2014) Atacicept in multiple sclerosis (ATAMS): a randomised, placebo-controlled, double-blind, phase 2 trial. Lancet Neurol 13(4):353–363. https://doi.org/10.1016/S1474-4422(14)70028-6

Khan G, Miyashita EM, Yang B, Babcock GJ, Thorley-Lawson DA (1996) Is EBV persistence in vivo a model for B-cell homeostasis? Immunity 5(2):173–179. https://doi.org/10.1016/s1074-7613(00)80493-8

Kola I, Landis J (2004) Can the pharmaceutical industry reduce attrition rates? Nat Rev Drug Discov 3(8):711–715. https://doi.org/10.1038/nrd1470

Langenau DM, Zon LI (2005) The zebrafish: a new model of T-cell and thymic development. Nat Rev Immunol 5(4):307–317. https://doi.org/10.1038/nri1590

Lassmann H, Ransohoff RM (2004) The CD4-Th1 model for multiple sclerosis: a critical [correction of crucial] re-appraisal. Trends Immunol 25(3):132–137. https://doi.org/10.1016/j.it.2004.01.007

Leibovitch E, Wohler JE, Cummings Macri SM, Motanic K, Harberts E, Gaitan MI et al (2013) Novel marmoset (*Callithrix jacchus*) model of human Herpesvirus 6A and 6B infections: immunologic, virologic and radiologic characterization. PLoS Pathog 9(1):e1003138. https://doi.org/10.1371/journal.ppat.1003138

Maini RN, Taylor PC (2000) Anti-cytokine therapy for rheumatoid arthritis. Annu Rev Med 51:207–229. https://doi.org/10.1146/annurev.med.51.1.207

Marquez AC, Horwitz MS (2015) The role of latently infected B cells in CNS autoimmunity. Front Immunol 6:544. https://doi.org/10.3389/fimmu.2015.00544

Matzinger P (2002) The danger model: a renewed sense of self. Science 296(5566):301–305. https://doi.org/10.1126/science.1071059

Merkler D, Böscke R, Schmelting B, Czéh B, Fuchs E, Brück W et al (2006) Differential macrophage/microglia activation in neocortical EAE lesions in the marmoset monkey. Brain Pathol 16(2):117–123. https://doi.org/10.1111/j.1750-3639.2006.00004.x

Mestas J, Hughes CC (2004) Of mice and not men: differences between mouse and human immunology. J Immunol 172(5):2731–2738. https://doi.org/10.4049/jimmunol.172.5.2731

Moore DJ (2016) Viral infection crosses up antigen presentation to drive autoimmunity. Sci Transl Med 8(349):349ec120. https://doi.org/10.1126/scitranslmed.aah4507

Moretta L, Romagnani C, Pietra G, Moretta A, Mingari MC (2003) NK-CTLs, a novel HLA-E-restricted T-cell subset. Trends Immunol 24(3):136–143. https://doi.org/10.1016/s1471-4906(03)00031-0

Nigida SM Jr, Falk LA, Wolfe LG, Deinhardt F, Lakeman A, Alford CA (1975) Experimental infection of marmosets with a cytomegalovirus of human origin. J Infect Dis 132(5):582–586. https://doi.org/10.1093/infdis/132.5.582

Nigida SM, Falk LA, Wolfe LG, Deinhardt F (1979) Isolation of a cytomegalovirus from salivary glands of white-lipped marmosets (*Saguinus fuscicollis*). Lab Anim Sci 29(1):53–60

Nossal GJ (1991) Molecular and cellular aspects of immunologic tolerance. Eur J Biochem 202(3):729–737. https://doi.org/10.1111/j.1432-1033.1991.tb16427.x

Pakpoor J, Disanto G, Gerber JE, Dobson R, Meier UC, Giovannoni G et al (2013) The risk of developing multiple sclerosis in individuals seronegative for Epstein-Barr virus: a meta-analysis. Mult Scler 19(2):162–166. https://doi.org/10.1177/1352458512449682

Pender MP, Csurhes PA, Smith C, Beagley L, Hooper KD, Raj M et al (2014) Epstein-Barr virus-specific adoptive immunotherapy for progressive multiple sclerosis. Mult Scler 20(11):1541–1544. https://doi.org/10.1177/1352458514521888

Peterson P, Org T, Rebane A (2008) Transcriptional regulation by AIRE: molecular mechanisms of central tolerance. Nat Rev Immunol 8(12):948–957. https://doi.org/10.1038/nri2450

Pluchino S, Gritti A, Blezer E, Amadio S, Brambilla E, Borsellino G et al (2009) Human neural stem cells ameliorate autoimmune encephalomyelitis in non-human primates. Ann Neurol 66(3):343–354. https://doi.org/10.1002/ana.21745

Poliani PL, Brok H, Furlan R, Ruffini F, Bergami A, Desina G et al (2001) Delivery to the central nervous system of a nonreplicative herpes simplex type 1 vector engineered with the interleukin 4 gene protects rhesus monkeys from hyperacute autoimmune encephalomyelitis. Hum Gene Ther 12(8):905–920. https://doi.org/10.1089/104303401750195872

Rabinovich GA, van Kooyk Y, Cobb BA (2012) Glycobiology of immune responses. Ann N Y Acad Sci 1253:1–15. https://doi.org/10.1111/j.1749-6632.2012.06492.x

Reynaud JM, Horvat B (2013) Animal models for human herpesvirus 6 infection. Front Microbiol 4:174. https://doi.org/10.3389/fmicb.2013.00174

Rich C, Link JM, Zamora A, Jacobsen H, Meza-Romero R, Offner H et al (2004) Myelin oligodendrocyte glycoprotein-35-55 peptide induces severe chronic experimental autoimmune encephalomyelitis in HLA-DR2-transgenic mice. Eur J Immunol 34(5):1251–1261. https://doi.org/10.1002/eji.200324354

Russell WMS, Burch RL (1959) The principles of humane experimental technique. Methuen, London

Sawcer S, Hellenthal G, Pirinen M, Spencer CC, Patsopoulos NA, Moutsianas L et al (2011) Genetic risk and a primary role for cell-mediated immune mechanisms in multiple sclerosis. Nature 476(7359):214–219. https://doi.org/10.1038/nature10251

Schafer S, Kolkhof P (2008) Failure is an option: learning from unsuccessful proof-of-concept trials. Drug Discov Today 13(21–22):913–916. https://doi.org/10.1016/j.drudis.2008.03.026

Segal BM, Constantinescu CS, Raychaudhuri A, Kim L, Fidelus-Gort R, Kasper LH et al (2008) Repeated subcutaneous injections of IL12/23 p40 neutralising antibody, ustekinumab, in patients with relapsing-remitting multiple sclerosis: a phase II, double-blind, placebo-controlled, randomised, dose-ranging study. Lancet Neurol 7(9):796–804. https://doi.org/10.1016/s1474-4422(08)70173-x

Smith PA, Heijmans N, Ouwerling B, Breij EC, Evans N, van Noort JM et al (2005) Native myelin oligodendrocyte glycoprotein promotes severe chronic neurological disease and demyelination in Biozzi ABH mice. Eur J Immunol 35(4):1311–1319. https://doi.org/10.1002/eji.200425842

Sospedra M, Martin R (2005) Immunology of multiple sclerosis. Annu Rev Immunol 23:683–747. https://doi.org/10.1146/annurev.immunol.23.021704.115707

Steinman L (2014) Immunology of relapse and remission in multiple sclerosis. Annu Rev Immunol 32:257–281. https://doi.org/10.1146/annurev-immunol-032713-120227

Steinman L, Zamvil SS (2016) Beginning of the end of two-stage theory purporting that inflammation then degeneration explains pathogenesis of progressive multiple sclerosis. Curr Opin Neurol 29(3):340–344. https://doi.org/10.1097/WCO.0000000000000317

Stys PK, Zamponi GW, van Minnen J, Geurts JJ (2012) Will the real multiple sclerosis please stand up? Nat Rev Neurosci 13(7):507–514. https://doi.org/10.1038/nrn3275

't Hart BA (2016) Primate autoimmune disease models; lost for translation? Clin Transl Immunol 5(12):e122. https://doi.org/10.1038/cti.2016.82

't Hart BA, Bajramovic J (2008) Non-human primate models of multiple sclerosis. Drug Discov Today Dis Model 5(2):97–104. https://doi.org/10.1016/j.ddmod.2008.06.001

't Hart BA, van Kooyk Y (2004) Yin-Yang regulation of autoimmunity by DCs. Trends Immunol 25(7):353–359. https://doi.org/10.1016/j.it.2004.04.006

't Hart BA, Weissert R (2016) Myelin oligodendrocyte glycoprotein has a dual role in T-cell autoimmunity against central nervous system myelin. MS J Exp Transl Clin 2:1–5. https://doi.org/10.1177/2055217316630999

't Hart BA, Bauer J, Muller HJ, Melchers B, Nicolay K, Brok H et al (1998) Histopathological characterization of magnetic resonance imaging-detectable brain white matter lesions in a primate model of multiple sclerosis: a correlative study in the experimental autoimmune encephalomyelitis model in common marmosets (*Callithrix jacchus*). Am J Pathol 153(2): P649–P663. https://doi.org/10.1016/S0002-9440(10)65606-4

't Hart BA, Vogels JT, Bauer J, Brok HPM, Blezer E (2004a) Non-invasive measurement of brain damage in a primate model of multiple sclerosis. Trends Mol Med 10(2):85–91. https://doi.org/10.1016/j.molmed.2003.12.008

't Hart BA, Laman JD, Bauer J, Blezer ED, van Kooyk Y, Hintzen RQ (2004b) Modelling of multiple sclerosis: lessons learned in a non-human primate. Lancet Neurol 3(10):589–597. https://doi.org/10.1016/s1474-4422(04)00879-8

't Hart BA, Bauer J, Brok HPM, Amor S (2005a) Non-human primate models of experimental autoimmune encephalomyelitis: variations on a theme. J Neuroimmunol 168(1–2):1–12. https://doi.org/10.1016/j.jneuroim.2005.05.017

't Hart BA, Brok HP, Remarque E, Benson J, Treacy G, Amor S et al (2005b) Suppression of ongoing disease in a nonhuman primate model of multiple sclerosis by a human-anti-human IL-12p40 antibody. J Immunol 175(7):4761–4768. https://doi.org/10.4049/jimmunol.175.7.4761

't Hart BA, Smith P, Amor S, Strijkers GJ, Blezer EL (2006) MRI-guided immunotherapy development for multiple sclerosis in a primate. Drug Discov Today 11(1–2):58–66. https://doi.org/10.1016/s1359-6446(05)03673-1

't Hart BA, Hintzen RQ, Laman JD (2009) Multiple sclerosis - a response-to-damage model. Trends Mol Med 15(6):235–244. https://doi.org/10.1016/j.molmed.2009.04.001

't Hart BA, Gran B, Weissert R (2011) EAE: imperfect but useful models of multiple sclerosis. Trends Mol Med 17(3):119–125. https://doi.org/10.1016/j.molmed.2010.11.006

't Hart BA, Abbott DH, Nakamura K, Fuchs E (2012) The marmoset monkey: a multi-purpose preclinical and translational model of human biology and disease. Drug Discov Today 17(21–22):1160–1165. https://doi.org/10.1016/j.drudis.2012.06.009

't Hart BA, Chalan P, Koopman G, Boots AM (2013) Chronic autoimmune-mediated inflammation: a senescent immune response to injury. Drug Discov Today 18(7–8):372–379. https://doi.org/10.1016/j.drudis.2012.11.010

't Hart BA, Jagessar SA, Kap YS, Haanstra KG, Philippens IHCHM, Serguera C et al (2014) Improvement of preclinical animal models for autoimmune-mediated disorders via reverse translation of failed therapies. Drug Discov Today 19(9):1394–1401. https://doi.org/10.1016/j.drudis.2014.03.023

't Hart BA, van Kooyk Y, Geurts JJ, Gran B (2015) The primate autoimmune encephalomyelitis model; a bridge between mouse and man (review). Ann Clin Transl Neurol 2(5):581–593. https://doi.org/10.1002/acn3.194

't Hart BA, Kap YS, Morandi E, Laman JD, Gran B (2016) EBV infection and multiple sclerosis: lessons from a marmoset model. Trends Mol Med 22(12):1012–1024. https://doi.org/10.1016/j.molmed.2016.10.007

't Hart BA, Dunham J, Faber BW, Laman JD, van Horssen J, Bauer J et al (2017) A B cell-driven autoimmune pathway leading to pathological hallmarks of progressive multiple sclerosis in the marmoset experimental autoimmune encephalomyelitis model. Front Immunol 8:804. https://doi.org/10.3389/fimmu.2017.00804

Thacker EL, Mirzaei F, Ascherio A (2006) Infectious mononucleosis and risk for multiple sclerosis: a meta-analysis. Ann Neurol 59(3):499–503. https://doi.org/10.1002/ana.20820

Thiruvalluvan A, Czepiel M, Kap YA, Mantingh-Otter I, Vainchtein I, Kuipers J et al (2016) Survival and functionality of human induced pluripotent stem cell-derived oligodendrocytes in a nonhuman primate model for multiple sclerosis. Stem Cells Transl Med 5(11):1550–1561. https://doi.org/10.5966/sctm.2016-0024

van Oosten BW, Lai M, Hodgkinson S, Barkhof F, Miller DH, Moseley IF et al (1997) Treatment of multiple sclerosis with the monoclonal anti-CD4 antibody cM-T412: results of a randomized, double-blind, placebo-controlled, MR-monitored phase II trial. Neurology 49(2):351–357. https://doi.org/10.1212/wnl.49.2.351

Vanheusden M, Stinissen P, 't Hart BA, Hellings N (2015) Cytomegalovirus: a culprit or protector in multiple sclerosis? Trends Mol Med 21(1):P16–P23. https://doi.org/10.1016/j.molmed.2014.11.002

Animal Welfare, Animal Rights, and a Sanctuary Ethos

Lori Gruen and Erika Fleury

Abstract

In this chapter, we will briefly examine the impact humans have on the other primates, particularly in the United States, and then explore different ethical frameworks that guide how we might exercise our responsibilities to nonhuman primates. We will examine the differences between an animal welfare position and an animal rights position. In closing we will advocate for the development of a sanctuary ethos that draws on both the welfare and the rights positions and urges us to re-examine our ethical agency in our relationships with other animals.

Keywords

Animal rights · Animal welfare · Entertainment · Ethics · Exploitation · Pets · Private ownership · Research · Retirement · Sanctuary · Speciesism

1 Ethics Beyond the Human

To deny that we have any responsibilities to animals other than human beings has been referred to as "speciesism" or "human exceptionalism." These terms highlight an ethically problematic prejudice that grants moral consideration based solely on species membership (see Gruen 2017; Ryder 2005; Singer 2002). Speciesism holds that humans are the only beings that do some things or have some capacity or capacities *and* that humans, by doing those things or having those capacities, are

L. Gruen (✉)
William Griffin Professor of Philosophy, Wesleyan University, Middletown, CT, USA
e-mail: lgruen@wesleyan.edu

E. Fleury
North American Primate Sanctuary Alliance, Oakland, CA, USA
e-mail: Efleury@primatesanctuaries.org

superior to species that do not do these things or have these capacities. The first of these claims raises largely empirical questions—what is the thing that only we do or have and are we really the only beings that do or have it? The second claim raises an evaluative or normative question—if we do discover a capacity that all and only humans share, does that make humans better, or more deserving of consideration, than others from an ethical point of view? (Gruen 2021).

Many capacities have been thought to distinguish humans from other animals, such as solving social problems, expressing emotions, developing culture, having relationships, laughing, playing, and having a sense of humor. As it turns out, none of these are unequivocally unique to humans. All animals living in socially complex groups solve problems that arise in their groups. Canids (Miklósi and Topál 2013) and primates (Stanford 2001) are particularly adept at it, yet even chickens (Bradshaw 1991, 1992; Schjelderup-Ebbe 1935) and horses (Williams 2004) are known to recognize large numbers of individuals in their social hierarchies and to maneuver within their social networks. Moreover, one way that nonhuman animals negotiate their social environments is by being attentive to the emotional states of conspecifics (Pennisi 2006). Many groups of animals develop skills that they pass along to others, which is one criterion of culture (Allen et al. 2013; Laland and Galef 2009). Social animals have complex relationships with others, including family and friends, and many animals engage in play and even play jokes on one another (Bekoff and Byers 1998; Bugnyar and Kotrschal 2002). Other primates have most of these capacities.

Yet there are differences between humans and animals. Only humans can read and write chapters such as this, and that is certainly a distinctive capacity. But not all humans can do so. Does that make those who cannot read or write any less important? Cheetahs can outrun humans and chimpanzees, but chimpanzees are stronger than humans and cheetahs. That makes both chimpanzees and cheetahs distinctive, but does that make either more important? All living beings are different from one another, as members of biological groups and as individuals. Given the tremendous variety of animal shapes, sizes, social structures, behaviors, and habitats, it seems odd to mark a divide between humans and all other animals for ethical reasons. Marking these marvelous differences does not obviously mean one whole species, the human species, with all the diversity within it, is better than all the other species.

In rejecting speciesism or human exceptionalism, we might instead focus not on how special we are, but rather on the type of people we want to be. Ethically conscientious people try to go through the world making the right sorts of choices, and doing good for, or at least minimizing harm to, other beings. In thinking about members of other species, such people consider whether their actions and choices are good for others and promote their well-being, or conversely, cause others harm. In so many contexts, the sad fact is that we rarely consider the impacts of our actions on others and end up causing them much harm, intentionally or not (for extended discussion see Gruen 2021).

2 Nonhuman Primates and Their Exploitation

Primates' lives are diverse, and this makes it difficult to generalize about what will promote their well-being. However, some generalizations can be made. For example, with a few exceptions, primates are quite social and require interactions with conspecifics to lead healthy lives (Sapolsky 2005). Compared to other mammals, primates have extended childhoods and stay with their mothers (and other parenting adults) for a year or more: chimpanzees and mountain gorillas wean at 48 months, woolly monkeys at 20 months, and olive baboons and long-tailed macaques at 14 months (Rowe 1996). During this extended childhood, individuals learn proper species behavior as well as survival skills. When primates are denied the opportunity to develop these skills, the negative effects shadow them for their entire lives. With limited early social opportunities, they develop stereotypies (repetitive, functionless behaviors) (Poirier and Bateson 2017) and serological evidence of stress (Jacobson et al. 2017).

Nonhuman primates' resemblance to humans, while remaining *not* human, has captivated people, and this human fascination has often led to the exploitation of nonhuman primates in various captive environments. Primates are held captive in the United States primarily in three industries: laboratory research, entertainment, and in the exotic pet trade. The exact numbers involved in these industries are difficult to obtain. Some species, for example, chimpanzees, are more closely tracked than other species, such as capuchins. In this case, the difference may be due to the fact that capuchin monkeys are not classified as endangered in the United States (USFWS 2018), and so trade in these species is less regulated. Here we briefly discuss the use of nonhuman primates in these three industries.

2.1 Laboratory Research

Most primates living in captivity in the United States are held in research laboratories, and the majority of laboratory-housed primates are New- or Old-World monkeys. A 2017 report by the Animal and Plant Health Inspection Service listed a total of 110,194 primates in research, including 34,369 which were held but were not currently involved in studies (APHIS 2018). The vast majority of primates held in laboratories are macaques, including rhesus and crab-eating. Other species commonly used in research are African green monkeys and New World monkeys such as titi monkeys, owl monkeys, marmosets, and tamarins (Carlsson et al. 2004; Phillips et al. 2014). At the time of this writing, 368 chimpanzees remain in laboratories (ChimpCARE n.d.), although they are no longer used in research (NIH 2018).

Improvements to laboratory primate welfare have been the focus of attention in recent decades, yet there is no way to completely eliminate the stress of laboratory captivity (Lahvis 2017b). Monkeys used in research may be captured in the wild and imported or they are bred in captivity. Wild captures destroy social groups, animals are frequently killed while struggling to free themselves, and individuals may perish during long international flights (Shukman and Piranty 2018). Commercial breeding avoids these difficulties.

According to federal regulations, rhesus macaques, the primate most frequently used in laboratory research (Conlee et al. 2004), can be kept in enclosures that provide just over 4 square feet, that is, just under 0.4 m^2, per animal, with a cage height of 30 inches, that is, just over 76 cm (APHIS 2019). Rhesus macaques are on average 20 inches (50 cm) tall (Rowe 1996), so they have little room to climb or move in these enclosures. To efficiently collect data, monkeys in biomedical laboratory research are almost always denied access to sunlight and fresh air. It is now common practice to socially housed monkeys in laboratories, or at least in pairs or quads, but this may not contribute to an individuals' well-being (Novak 2003). Depending on the relationships between the individuals housed together, the number of animals in a room, the stress levels, and vocalizations occurring, monkeys kept in such environments may experience significant distress (National Research Council 1992). Captive housing is just one aspect of laboratory research that impacts primate well-being and its scientifically confounding impact is often overlooked (Lahvis 2017a).

In the USA, all federally funded research must undergo review by an Institutional Animal Care and Use Committee (IACUC), a group that generally includes laboratory animal veterinarians, animal researchers, and someone from outside of the laboratory community. IACUCs are tasked with ensuring that protocols do not duplicate previous research, are not using animal models unnecessarily, and are making every effort to avoid or reduce animal suffering.

Even the most common laboratory procedures, such as a blood draw, may cause distress to the animal (Bush et al. 1977; Balcombe et al. 2004). More dramatic procedures, such as a darting with anesthesia, may provoke even more distress. Repeated exposures to such events may not dull an animals' reaction to them but will often instead make the animal react more strongly to these events. In other words, these events lead to sensitization and not habituation (Balcombe et al. 2004).

Primates in laboratories often exhibit symptoms of stress when they see other animals forced to undergo procedures, even if they themselves are spared (Balcombe et al. 2004). Many (if not all) laboratory procedures cause some level of distress, illness, or pain to the animals that undergo them, but the pursuit of improvements in human health is often the justification that permits laboratory research on nonhuman primates to continue (Friedman et al. 2017).

2.2 Entertainment

Although primates have acted the jester for centuries, playing an integral role in spectacles from circuses and live shows to films and television programs, primates are currently being used in entertainment less frequently. People are reconsidering the ethical permissibility of laughing and gawking at socially and emotionally sophisticated, intelligent primates which are trained, often through aversive methods, and forced to perform (Hevesi 2022; Newman 2014; Bodkin 2018;

Baeckler 2003). Additionally, technological advances provide realistic computerized apes and monkeys that are cheaper, safe to work with, and more readily accessible than live primate actors (Shaw-Williams 2016).

The number of primates used in entertainment are difficult to estimate, but, in the United States, it appears to be on the decline. For great apes in the United States, as recently as the 1990s, more than 100 chimpanzees and 35 orangutans were working for 40–50 trainers. A more recent survey in 2016 counted only 13 chimpanzees and 10 orangutans working for just five trainers (Ragan 2016). The situation is more dire throughout the world. In Indonesia, for instance, rhesus macaques and orangutans are regularly paraded around in chains, while dressed in clothing and make-up to delight tourists (BBC 2013).

For training purposes, primates are removed from their mothers often years before they would normally separate (Freeman and Ross 2014). Because primates generally become too aggressive and rebellious to handle as they age, trainers only can use primates in their first few years of life. Even then, to avoid injuries to humans, trainers often remove the primates' teeth, which disadvantages these animals for the rest of their lives (NAPSA n.d.-a; Baeckler 2003).

2.3 Zoos

Zoos exist in a middle ground between entertainment and conservation (Hosey 2022; Prescott 2022; Baker and Farmer 2022). Reputable zoos in the United States are accredited by the Association of Zoos and Aquariums (AZA). Worldwide zoo accreditation occurs under the World Association of Zoos and Aquariums. Zoos in the United Kingdom and Ireland are accredited by the British and Irish Association of Zoos and Aquariums. These organizations attempt to ensure that member institutions meet their standards of care and management. There are also unregulated roadside zoos in the United States and comparable facilities elsewhere that are not accredited by these organizations and often engage in practices that are more harmful to the animals and potentially dangerous to the public, such as nonprotected contact for photo opportunities.

Given that there are many types of zoos, there are no hard numbers about the populations of nonhuman primates currently living in zoos. ChimpCare, a project at the Lincoln Park Zoo in Chicago, Illinois, reports that there are 239 chimpanzees at AZA-accredited zoos, and 162 at unaccredited facilities, such as roadside zoos, but also pseudo-sanctuaries and breeders (ChimpCARE n.d.). However, chimpanzees are only one of many species of nonhuman primate living in zoos.

Life in an accredited zoo is likely to be one of the better options for a captive primate, as many zoos are working to improve animal welfare protocols. In the United States and across the globe, there are still ethically troubling aspects of zoo animal care. Zoos engage in selective breeding that often involves moving animals from one facility to another (AZA 2020), and moves can be stressful for the animals (National Research Council 2006). Most zoos will permit infants to be raised by their mothers and that can positively impact the well-being of the infants, the mothers, and their social groups. Well-run zoos provide regular medical care and enrichment programs that ensure that the primates' basic physical and psychological needs are met (AZA 2020). Although living a life on display can negatively impact well-being (Salas and Manteca 2017), there are zoos that permit primates in their care to have privacy and to escape visitors' prying eyes. However, the quality of zoos varies across countries and many long-lived primates in zoos are languishing in unnatural enclosures with little to stimulate their minds (McArthur 2017). Given there is little oversight to regulate the quality of care for animals in nonaccredited public attractions, the minimal standards of care negatively impact nonhuman primates (Rudloff 2017). Furthermore, some zoos worldwide, and many nonaccredited zoos in the United States, get animals from private breeders, which fund and encourage the primate pet trade (Green 1999).

2.4 The Pet Trade

The removal of infants from their mothers is rampant in the exotic pet industry (Hevesi 2022), where primate infants are removed from their mothers before they should be (Fleury 2013). When primates are unable to experience a typical infancy, the resulting stereotypies may include, but are not limited to, obsessive and compulsive actions such as rocking, repetitive banging, floating limb syndrome (where an animal's arm, leg or tail moves as if it is independent of the individual, which then often attacks it), aggression, depression, coprophagy, screaming, and self-injurious behaviors that include hair plucking, scratching, picking at injuries, insertion of foreign objects into bodily cavities, and other forms of mutilation, all which can cause infection and lead to further negative impacts on well-being (Marriner and Dickamer 1994; Mason 2006).

There is no federal law that regulates the private ownership of primates, and state restrictions vary. Although 29 states in the United States ban owning nonhuman primates as pets, nine allow for exceptions, depending on the animal's provenance or proof of legal ownership from another state. Primate pets may also be grandfathered in if they are already living in the state at the time the law was passed. Nine other states require some sort of permit to own a primate as a pet, and the remaining 12 states have some rules about animal care, but otherwise do not limit the purchase or housing of nonhuman primates (Paquette 2018). Oversight of these pets is little to nonexistent. As a result, there are likely to be nonhuman primates hidden in human homes and backyards throughout every state. The Humane Society of the United States estimates there are 15,000 primates living as pets in the United States (HSUS n.d.). The vast majority of these pets are monkeys, but they also include 25 chimpanzees (ChimpCARE n.d.) and one orangutan (Ragan 2016).

It is difficult to pinpoint the most troubling consequences of nonhuman primates living in human homes or backyards, but primate sanctuaries are frequently contacted regarding sanctuary housing for primates which have been confiscated or whose owners wish to voluntarily surrender them, suggesting that most of these pet owners regret their expensive purchase, lose interest in their pet, or do not have the space, funding, or time to adequately meet their pet's needs (RSCPA n.d.). These problems are likely the result of uneducated consumers who did not know how to care for primates and who usually do not know what they have gotten themselves into, as veterinarian Craig J. Blair explains:

> One of the largest brokers of pet monkeys in the eastern U.S. advertises on its web site that its employees spend up to two hours with new monkey owners educating them about everything they need to know about monkeys. I've been in exotic pet practice for 11 years and received training at the Cincinnati Zoo. I can say with certainty that two hours is barely enough time to realize that you will never know everything you need to know about monkeys (Blair 2005, p. 35).

3 Animal Welfare Concerns

Every animal, humans included, wants to avoid unnecessary pain. This sets a limit on what ethical agents should attend to when they are figuring out how to interact with other animals. Animal welfare was originally conceived as an absence of cruelty, and various organizations came into existence to expose cruelty and enact statutes to prevent it (Beers 2006). Animal welfare has since expanded to include broader considerations of the various ways that sentient beings have their basic interests violated when we cause them to suffer, emotionally, physically, or both (Palmer and Sandøe 2018). The many contexts in which primates and other animals are used today can cause suffering and negatively impact animal well-being.

In the United States, the first version of the Animal Welfare Act (AWA) was introduced in 1966 and it has since been amended multiple times to broaden its application (Gruen 2021). The AWA sets minimal standards of care that may keep animals alive, but it does not promote flourishing or a species-appropriate life for them. For example, AWA laboratory research regulations do not protect a research animal from experiencing pain. Instead, the Act asks researchers to "consider alternatives" if the procedure is "likely to produce pain to or distress" an animal (Favre 2002). If a procedure unexpectedly causes pain to the animal, the procedure is still acceptable under the AWA. If the research requires causing pain and there are no alternatives available, that too would be allowed under the AWA.

A 1985 amendment to the AWA sought specifically to do better for nonhuman primates by requiring that their psychological well-being be promoted. Some members of the research community fought this change, citing concerns that ranged from the cost of such changes to avoidance of increased aggression with larger numbers of socially housed primates (Gluck 2014), but regulations were nonetheless promulgated in 1991. Specifically, dealers, exhibitors, and research facilities were required to develop, document, and follow an appropriate plan for environment enhancement adequate to promote the psychological well-being of nonhuman primates. The plan must be in accordance with the currently accepted professional standards as cited in appropriate professional journals or reference guides, and as directed by the attending veterinarian (AL&HC 2015, np; Bayne et al. 2022).

Many believe that the AWA standards are not good enough (Frasch 2016). Accredited zoos often seek to go beyond the AWA standards and specifically focus on improving well-being. The World Zoo and Aquarium scientists have developed what they call the "Five Domains" model that includes nutrition, environment, physical health, behavior, and recognizing that well-being is a subjective state, the fifth domain is based on negative and positive experiences, or psychological states (WAZA n.d.). The "first four domains enable systematic consideration of a wide range of conditions that may give rise to a range of subjective experiences found within the fifth 'mental' domain. The net impact of all of these experiences is assessed as representing the animal's welfare status" (Gusset and Dick 2015). Some argue further that considerations of animal well-being should include a balance of positive over negative states, as well as the opportunity for animals to carry out natural behaviors, and to choose what matters most to them (Palmer and Sandøe 2018).

Of course, being free to express natural behaviors and make choices within unnatural environments may not be fully conducive to well-being. Not all harms can be avoided, such as when considering conflicts between wild animals and nearby humans. When conflicts of interest arise, as they often do in our world of limited resources, some interests will lose out. In speciesist societies like our own, it is the interests of other animals that usually are violated to promote human interests, and that is why some argue that animal welfare considerations do not go far enough.

4   Animal Rights

Advocates of animal rights grant that animal suffering is an important ethical consideration, but argue that it is not the only one. The difference between animal rights and animal welfare is that those concerned with animal welfare are not opposed to the use of animals, where those advocating animal rights believe animals have a right not to be used for human purposes (e.g., Regan 1985; Donaldson and Kymlicka 2011). Proponents of animal rights believe nonhuman animals are worthy of the respect that is granted to holders of human rights.

Animal rights proponents, like philosopher Tom Regan, argue that we should treat humans and other animals with the respect that they are due. For Regan, utilitarian considerations of weighing pleasures and pains, frustrations, and satisfaction should not be the basis of ethical action. He writes:

> The forlornness of the veal calf is pathetic, heart wrenching; the pulsing pain of the chimp with electrodes planted deep in her brain is repulsive; the slow, tortuous death of the raccoon caught in the leg-hold trap is agonizing. But what is wrong isn't the pain, isn't the suffering, isn't the deprivation. The fundamental wrong is the system that allows us to view animals as *our resources,* here for *us*... (italics in original, Regan 1985, p. 13).

According to Regan, all normal adult humans and other animals are what he calls "subjects of a life" (Regan 1985, p. 22) that have inherent worth and are due respect.

Subjects of a life are beings with relatively complex mental lives that include perceptions, desires, beliefs, memories, intentions, and at least a minimal sense of the future. Precisely who is a subject of a life is open to debate (for example, do octopuses have this sort of mental life? What about bats? Or humans who are in irreversible comas?), but the basic idea is that the lives of these individuals matter to them, "*equally,* whether they be human animals or not" (italics in original, Regan 1985, p. 23), that is what grounds their worth, and that is why they have rights.

Those who argue for animal rights are not necessarily seeking legal rights for other animals, but rather moral rights, rights that protect other animals in ways that go beyond the prevention of cruelty and suffering. But there are animal rights proponents who are seeking legal protections for primates. To date, these efforts have sought to grant legal rights to great apes given that we humans are great apes.

The Great Ape Project, which began with the publication of a book by that title in 1993, sought to establish that great apes, as subjects of meaningful lives, were deserving of rights. Through a collection of essays edited by Peter Singer and Paolo Cavalieri, the members of this project argue that, as members of a community of equals, there are basic rights that all great apes deserve:

1. Right to life—members of the community of equals, which include humans, may not be killed except in certain strictly defined circumstances such as self-defense.
2. Protection of individual liberty—members of the community of equals are not to be deprived of their liberty, and are entitled to immediate release where there has been no form of due process. The detention of great apes which have not been convicted of any crime or which are not criminally liable should be permitted only where it can be shown that the detention is in their own interests or is necessary to protect the public.
3. Prohibition of torture—the deliberate infliction of severe pain, on any great ape, whether wantonly or because of a perceived benefit to others should be prohibited (Singer and Cavalieri 1993, p. 4).

Building on some of these ideas, The Nonhuman Rights Project (NhRP), led by lawyer Steven Wise, filed writs of habeas corpus in New York state in 2013, seeking the release of privately owned chimpanzees to a sanctuary (Gorman 2013). The arguments hinge on the recognition of the chimpanzees' personhood, that is, their capacities for self-awareness and autonomous action. The NhRP built their case on research with chimpanzees that suggests they have a sense of self and they understand themselves as having interests that extend through time (NhRP 2013; Hirata et al. 2017; Silk et al. 2013). John Locke (1690) suggested the capacity to recognize oneself as having a past and future is to be a person. In our legal system, chimpanzees are not considered persons (no animals are) and thus animals cannot be the bearers of rights. Rather, animals are classified as property. The desire to change this classification is what motivates the NhRP. Given that the legal system only has two categories for distinguishing between beings (either person or property) classifying chimpanzees as legal persons are more accurate than considering them to be mere property, like a table or a cell phone.

Although no nonhuman animal species, including nonhuman primates, have been recognized as having legal rights, arguments for their moral rights have gained traction (Ballesteros 2017; Cetacean Rights 2010). The outcry that led the Institute of Medicine to explore whether we needed to continue research with chimpanzees, which ultimately led to the decision by the National Institutes of Health to end funding for chimpanzee research and retire government owned and supported chimpanzees to sanctuary (NIH 2015), was in part motivated by the view that great apes have significant moral status (Singer and Cavalieri 1993).

5 A Sanctuary Ethos

There is a perspective beyond animal welfare and animal rights that can be called a "sanctuary ethos." This is a way of engaging with other animals found in genuine sanctuaries, but the sanctuary ethos can also be developed in other sites of captivity. For example, the director of the Detroit Zoo, Ron Kagan, has been considering the implications of a sanctuary ethos for zoo settings (Kagan 2016). Sanctuary is more than a word; it is a place where the interests, choices, and well-being of animals come first; where their choices are taken seriously; a place where animals are treated with great care and respect. Care and respect in these settings are not just words. Renaming an institution of use with the word "care" is not the same as genuinely caring for and about nonhuman primates. There are facilities who wear the mantle of "sanctuary" but who engage in exploitive practices and are for-profit institutions where resident animals are used for profit by allowing public interactions and performances, and for whom animal welfare is not the first priority. At a bona fide sanctuary, care and respect are predicated on a recognition of the independent worth of another being, a recognition that they have their own interests, needs, personalities, and concerns that deserve to be attended to as their own, not for someone else's purpose. In sanctuaries, animals can interact with each other and caregivers when they decide to, exercise their bodies and minds as they want, and are free to choose how to spend their time and with whom.

In addition to providing the basic needs for well-being, including a healthy diet, clean air and water, and enough space, at reputable sanctuaries, animals are treated with dignity (Gruen 2014a; Kymlicka 2018; Nussbaum 2006). In sanctuaries, animals' lives are valued as theirs to live without our judgments and interference. Sanctuary residents are treated with respect and are cared for empathetically (Gruen 2014b). Recognizing the limits of our own ways of seeing the world and being open to learning how members of other species see the world helps promote animals' dignity, but also has the potential to expand our own ethical perceptions. Sanctuaries are sites of growth and care, for both humans and for nonhuman residents.

Accredited sanctuaries within the U.S. care for over 630 great apes (ChimpCARE n.d.) and many hundreds of monkeys. The number of sanctuary residents continue to grow as the use of primates is slowly phased out of entertainment, the pet trade, and laboratory research. In March of 2018, a turning point was reached in the U.S. when there were more chimpanzees living in sanctuaries than in laboratories (Grimm

2018). There had already been more chimpanzees in sanctuary than in zoos and in private ownership such as trainers and breeders (ChimpCARE n.d.). Although moving primates to sanctuary raises temporary concerns about their well-being, knowledgeable transport, quarantine, and acclimation procedures are in place to minimize temporary discomfort, and the end result of sanctuary retirement justifies any potential temporary inconvenience. Transporting primates to sanctuaries is a practice that has been relied upon for decades (Block 2018) and has been proven safe and reliable (NAPSA 2018).

Sanctuaries exist as a long-term bandage designed to patch up the problems wrought by the pet trade, by research, and by the entertainment industry. Of course, even in sanctuaries, monkeys and apes are still captive. Sanctuary life is not intended to be a replacement for living free of human control, but it exists as the best possible option for many animals. Institutions that house primates as means to an end (such as medical research or public education) have to balance their primary purpose with the well-being of the captive primates in their care. True sanctuaries exist for no purpose other than to provide for the animals. Primates are not used at sanctuaries, they are there to be cared for psychologically, medically, and socially. These facilities are not open to the public and do not transfer their residents unless it is in the individual animals' best interest (NAPSA n.d.-b). Reputable primate sanctuaries that are inspected and certified by international or national accrediting bodies provide a level of care, attention, respect, and freedom that cannot be provided at institutions in which primates are or were used.

6 Conclusion

Over the last few decades, concern about human use of nonhuman primates has grown, as has alarm about the destruction of primate habitats. We would venture to say that almost everyone is interested in promoting primate welfare, that is, to ensure that nonhuman primates held in laboratories, entertainment venues, zoos, and homes, are not treated cruelly, subjected to pain, or made to suffer. Despite that, current welfare standards are often inadequate and there are still too many nonhuman primates kept in conditions that are incompatible with their well-being. Animal rights advocates argue that nonhuman primates should not be held in captivity or used for human purposes, as this violates their right to freedom and respect. Given the fact that primates are already in captivity and cannot be moved to the wild, many have argued that they should be allowed to live out their lives in sanctuary. We have urged the importance of understanding sanctuary as a place that provides the highest level of care and one designed solely to promote the flourishing of the animals that can call it their home.

References

Allen J, Weinrich M et al (2013) Network-based diffusion analysis reveals cultural transmission of Lobtail feeding in humpback whales. Science 340(6131):485–488. https://doi.org/10.1126/science.1231976

Animal Legal & Historical Center (2015) Code of Federal Regulations. Title 9. Animals and Animal Products. https://www.animallaw.info/administrative/us-awa-awa-primate-regulations-subpart-d-primate-standards#s80. Accessed 3 Oct 2016

Association of Zoos & Aquariums (n.d.) About us. https://www.aza.org/about-us. Accessed 23 Oct 2016

Association of Zoos & Aquariums (AZA) (2020) The Accreditation Standards & Related Policies, 2018 Edition, pp 97–104. https://www.speakcdn.com/assets/2332/aza-accreditation-standards.pdf. Accessed 20 December 2019

Baeckler S (2003) Campaign to end the use of chimpanzees in entertainment. Testimony presented at a briefing co-hosted by the Chimpanzee Collaboratory and the Environmental Media Association, Los Angeles. http://www.eyesonapes.org/apes_in_entertainment/trainers/undercover/undercover_at_a_training_facility.pdf. Accessed 20 Oct 2016

Baker KR, Farmer HL (2022) The welfare of primates in zoos. In: Robinson LM, Weiss A (eds) Nonhuman primate welfare: from history, science, and ethics to practice. Springer, Cham, pp 79–96

Balcombe J, Barnard N et al (2004) Laboratory routines cause animal stress. Contemp Topics in Laboratory Animal Science 43(6):42–51.

Ballesteros C (2017) Elephants are Legal persons and deserve to be free, Group Claims in Court Petition. Newsweek. https://www.newsweek.com/animal-rights-elephants-human-rights-courts-709912. Accessed 21 July 2018

Bayne K, Hau J, Morris T (2022) The welfare impact of regulations, policies, guidelines and directives and nonhuman primate welfare. In: Robinson LM, Weiss A (eds) Nonhuman primate welfare: from history, science, and ethics to practice. Springer, Cham, pp 629–646

BBC News (2013) 'Abused' Indonesian monkeys taken off Jakarta Streets. BBC News.. https://www.bbc.co.uk/news/world-asia-24624847. Accessed 19 July 2018

Beers D (2006) For the prevention of cruelty: the history and legacy of animal rights activism in the United States. Swallow Press/Ohio University Press, Athens

Bekoff M, Byers J (eds) (1998) Animal play: evolutionary, comparative and ecological perspectives. Cambridge University Press, Cambridge

Blair C (2005) Pets or prisoners? Southsider Magazine. http://petmonkeyinfo.org/pets_or_prisoners.htm. Accessed 4 Mar 2013

Block K (2018) Tell NIH to send the last of the research chimpanzees to sanctuary. A Humane Nation. https://blog.humanesociety.org/2018/07/tell-nih-to-send-the-last-of-the-research-chimpanzees-to-sanctuary.html. Accessed 21 July 2018

Bodkin H (2018) Hollywood 'unethical' for starring chimpanzees and orangutans in films. The Telegraph. https://www.telegraph.co.uk/news/2018/01/18/hollywood-unethical-starring-chimpanzees-orangutans-films/. Accessed 4 Aug 2018

Bradshaw RH (1991) Discrimination of group members by laying hens *Gallus domesticus*. Behav Process 24(2):143–151. https://doi.org/10.1016/0376-6357(91)90006-1

Bradshaw RH (1992) Conspecific discrimination and social preference in the laying hen. Appl Anim Behav Sci 33(1):69–75. https://doi.org/10.1016/S0168-1591(05)80086-3

Bugnyar T, Kotrschal K (2002) Observational learning and the raiding of food caches in ravens, *Corvus corax*: is it 'tactical' deception? Anim Behav 64(2):185–195. https://doi.org/10.1006/anbe.2002.3056

Bush M, Custer R et al (1977) Physiologic measures of nonhuman primates during physical restraint and chemical immobilization. J Am Vet Med Assoc 171(9):866–869

Carlsson H, Schapiro S et al (2004) Use of primates in research: a global overview. Am J Primatol 63(4):225–237. https://doi.org/10.1002/ajp.20054

Cetacean Rights (2010) Declaration of Rights for Cetaceans: Whales and Dolphins. https://www.cetaceanrights.org. Accessed 21 July 2018

ChimpCARE (n.d.) Where are our amazing chimpanzees in the United States. http://www.chimpcare.org/map. Accessed 13 Nov 2018

Conlee K, Hoffeld E et al (2004) A demographic analysis of primate research in the United States. Altern Lab Anim 32(Suppl 1A):315–322. https://doi.org/10.1177/026119290403201s52

Donaldson S, Kymlicka W (2011) Zoopolis. Oxford University Press, New York

Favre D (2002) Overview of U.S. Animal Welfare Act. Michigan State University College of Law, Animal Legal & Historical Center. https://www.animallaw.info/article/overview-us-animal-welfare-act. Accessed 3 Oct 2016

Fleury E (2013) Monkey business: a history of nonhuman primate Rights. Createspace

Frasch P (2016) Gaps in US animal welfare law for laboratory animals: perspectives from an animal law attorney. ILAR J 57(3):285–292. https://doi.org/10.1093/ilar/ilw016

Freeman H, Ross S (2014) The impact of atypical early histories on pet or performer chimpanzees. PeerJ 2:3570. https://doi.org/10.7717/peerj.579

Friedman H, Haigwood H et al (2017) The critical role of nonhuman primates in medical research – white paper. Pathog Immun 2(3):352–365. https://doi.org/10.20411/pai.v2i3.186

Gluck J (2014) Moving beyond the welfare of psychological well-being for nonhuman primate: the case of chimpanzees. Theor Med Bioethics 35(2):105–116. https://doi.org/10.1007/s11017-014-9289-1

Gorman J (2013) Considering the humanity of nonhumans. The New York Times. https://www.nytimes.com/2013/12/10/science/considering-the-humanity-of-nonhumans.html. Accessed 21 July 2018

Green A, Center for Public Integrity (1999) Animal underworld: inside America's black market for rare and exotic species. Public Affairs, New York

Grimm D (2018) U.S. chimp retirement gains momentum, as famed pair enters sanctuary. Science. http://www.sciencemag.org/news/2018/03/us-chimp-retirement-gains-momentum-famed-pair-enters-sanctuary. Accessed 21 July 2018

Gruen L (2021) Ethics and animals: an introduction. Second Edition. Cambridge University Press, Cambridge

Gruen L (2014a) Dignity, captivity, and an ethics of sight. In: Gruen L (ed) The ethics of captivity. Oxford University Press, Oxford, pp 231–247

Gruen L (2014b) Entangled empathy: an alternative ethics for our relationships with animals. Lantern Press, New York

Gruen L (2017) The moral status of animals. Stanford Encyclopedia of Philosophy https://plato.stanford.edu/entries/moral-animal/

Gusset M, Dick G (2015) Editorial. World Association of Zoos and Aquariums (WAZA) Magazine

Hevesi R (2022) The welfare of primates kept as pets and entertainers. In: Robinson LM, Weiss A (eds) Nonhuman primate welfare: from history, science, and ethics to practice. Springer, Cham, pp 121–144

Hirata S, Fuwa K et al (2017) Chimpanzees recognize their own delayed self-image. R Soc Open Sci 4(8):170370. https://doi.org/10.1098/rsos.170370

Hosey G (2022) The history of primates in zoos. In: Robinson LM, Weiss A (eds) Nonhuman primate welfare: from history, science, and ethics to practice. Springer, Cham, pp 3–30

Humane Society of the United States, The (HSUS) (n.d.) Questions and answers about monkeys used in research. http://www.humanesociety.org/animals/monkeys/qa/questions_answers.html. Accessed 3 Oct 2016

Jacobson SL, Freeman HD et al (2017) Atypical experiences of captive chimpanzees (*Pan troglodytes*) are associated with higher hair cortisol concentrations as adults. R Soc Open Sci 4(12):170932. https://doi.org/10.1098/rsos.170932

Kagan R (2016) The show is over. https://www.humansandnature.org/the-show-is-over. Accessed 10 Nov2018

Kymlicka W (2018) Human rights without human supremacism. Can J Philos 48(6):763–792. https://doi.org/10.1080/00455091.2017.1386481

Lahvis G (2017a) Animal welfare: make animal models more meaningful. Nature 543:623. https://doi.org/10.1038/543623d

Lahvis G (2017b) Unbridle biomedical research from the laboratory cage. eLife 6:e27438. https://doi.org/10.7554/eLife.27438

Laland KN, Galef B (2009) The question of animal culture. Harvard University Press, Cambridge

Locke, J (1690) An essay concerning human understanding. ftp://ftp.dca.fee.unicamp.br/pub/docs/ia005/humanund.pdf. Accessed 4 Aug 2018

Marriner LM, Dickamer LC (1994) Factors influencing stereotyped behavior of primates in a zoo. Zoo Biol 13(3):267–275. https://doi.org/10.1002/zoo.1430130308

Mason GJ (2006) Stereotypic behaviour in captive animals: fundamentals and implications for welfare and beyond. In: Mason GJ, Rushen J (eds) Stereotypic animal behaviour. Fundamentals and applications to welfare, 2nd edn. CAB International, Wallingford, pp 325–356

McArthur J (2017) Captive. Lantern Publishing & Media, New York

Miklósi Á, Topál J (2013) What does it take to become 'best friends'? Evolutionary changes in canine social competence. Trends Cog Sci 17(6):287–294. https://doi.org/10.1016/j.tics.2013.04.005

National Institutes of Health (NIH) (2015) NIH will no longer support biomedical research on chimpanzees. https://www.nih.gov/about-nih/who-we-are/nih-director/statements/nih-will-no-longer-support-biomedical-research-chimpanzees. Accessed 23 Oct 2016

National Institutes of Health (NIH) (2018) Council of councils working group on assessing the safety of relocating at-risk chimpanzees. https://dpcpsi.nih.gov/sites/default/files/CoC_May_2018_WG_Report_508.pdf. Accessed 18 July 2018

National Research Council (1992) Recognition and alleviation of pain and distress in laboratory animals. The National Academies Press, Washington, DC, p 46. https://doi.org/10.17226/1542

National Research Council (US) Committee on Guidelines for the Humane Transportation of Laboratory Animals (2006) Guidelines for the humane transportation of research animals. National Academies Press (US), Washington, DC

Newman B (2014) No real animals aboard Hollywood Noah's ark. National Geographic. https://news.nationalgeographic.com/news/2014/03/140328-noah-animals-ark-movies-hollywood/. Accessed 4 Aug 2018

Nonhuman Rights Project, Inc (NhRP). v. Patrick C. Lavery (2013) Verified petition, State of New York, Supreme Court, County of Fulton. https://www.nonhumanrights.org/content/uploads/Petition-re-Tommy-Case-Fulton-Cty-NY.pdf. Accessed 4 Aug 2018

North American Primate Sanctuary Alliance (NAPSA) (2018) NAPSA's public comment to NIH Working Group on Assessing the Safety of Relocating At-Risk Chimpanzees. http://primatesanctuaries.org/wp-content/uploads/2018/08/NAPSAs-Public-Comment-to-NIH.pdf. Accessed 9 Aug 2018

North American Primate Sanctuary Alliance (NAPSA) (n.d.-a) Advocacy position statement: performing primates. http://primatesanctuaries.org/wp-content/uploads/2018/06/Advocacy-Statement-Entertainment.pdf. Accessed 21 July 2018

North American Primate Sanctuary Alliance (NAPSA) (n.d.-b) Advocacy position statement: True sanctuaries vs. pseudo-sanctuaries. http://primatesanctuaries.org/wp-content/uploads/2018/05/Advocacy-Statement-Pseudo-Sanctuaries.pdf. Accessed 21 July 2018

Novak M (2003) Housing for captive nonhuman primates: The balancing act. The development of science-based guidelines for laboratory animal care: Proceedings of the November 2003 International Workshop. https://www.ncbi.nlm.nih.gov/books/NBK25428/. Accessed 19 July 2018

Nussbaum M (2006) Frontiers of justice. Disability, nationality, species membership. The Belknap Press of the Harvard University Press, Cambridge, MA

Palmer C, Sandøe P (2018) Welfare. In: Gruen L (ed) Critical terms for animal studies. University of Chicago Press, Chicago, pp 424–438

Paquette N (2018) Primates as pets. NAPSA Workshop 2018, Gainesville, April 2018. http://primatesanctuaries.org/wp-content/uploads/2018/07/Paquette-Primates-as-Pets.pdf. Accessed 18 July 2018

Pennisi E (2006) Social animals prove their smarts. Science 312(5781):1734–1738. https://doi.org/10.1126/science.312.5781.1734

Phillips K, Bales K et al (2014) Why primate models matter. Am J Primatol 76(9):801–827. https://doi.org/10.1002/ajp.22281

Poirier C, Bateson M (2017) Pacing stereotypies in laboratory rhesus macaques: implications for animal welfare and the validity of neuroscientific findings. Neurosci Behav Rev 83:508–515. https://doi.org/10.1016/j.neubiorev.2017.09.010

Prescott MJ (2022) Using primates in captivity: research, conservation, and education. In: Robinson LM, Weiss A (eds) Nonhuman primate welfare: from history, science, and ethics to practice. Springer, Cham, pp 57–78

Ragan P (2016) Looking ahead. NAPSA Workshop 2016, Tacoma, September 2016

Regan T (1985) The case for animal Rights. In: Singer P (ed) Defense of animals. Basil Blackwell, New York, pp 13–26

Rowe N (1996) The pictorial guide to the living primates. Pogonias Press, Charlestown

Royal Society for the Prevention of Cruelty to Animals (RSPCA) (n.d.) Do You Give a Monkey's? https://www.rspca.org.uk/adviceandwelfare/pets/other/primates. Accessed 4 Aug 2018

Rudloff L (2017) Failure to launch: The lack of implementation and enforcement of the animal welfare act. Syracuse Law Rev 67:173–190. https://www.animallaw.info/sites/default/files/Rudloff_Failure%20to%20Launch.pdf. Accessed 19 July 2018

Ryder R (2005) All beings that feel pain deserve human rights. The Guardian. https://www.theguardian.com/uk/2005/aug/06/animalwelfare. Accessed 18 July 2018

Salas M, Manteca X (2017) Visitor effect on zoo animals. Zoo Animal Welfare Education Centre. https://www.zawec.org/media/com_lazypdf/pdf/Sheet%20ZAWEC%205.pdf. Accessed 4 Aug 2018

Sapolsky RM (2005) The influence of social hierarchy on primate health. Science 308(5722):648–652. https://doi.org/10.1126/science.1106477

Schjelderup-Ebbe T (1935) Social behavior of birds. In: Murchison C (ed) A handbook of social psychology. Worcester: Clark University Press, pp 947–972.

Shaw-Williams H (2016) Can CGI animals ever replace the real thing? Screen Rant. https://screenrant.com/jungle-book-cgi-animals-vfx-motion-capture/. Accessed 22 July 2018

Shukman D, Piranty S (2018) The secret trade in baby chimps. BBC News. https://www.bbc.co.uk/news/resources/idt-5e8c4bac-c236-4cd9-bacc-db96d733f6cf. Accessed 18 July 2018

Silk J, Brosnan S et al (2013) Chimpanzees share food for many reasons: the role of kinship, reciprocity, social bonds and harassment on food transfers. Anim Behav 85(5):941–947. https://doi.org/10.1016/j.anbehav.2013.02.014

Singer P (2002) Animal liberation. HarperCollins Publishers, New York

Singer P, Cavalieri C (1993) A declaration on great apes. In: Singer P, Cavalieri C (eds) The great ape project: equality beyond humanity. St. Martin's Press, New York, pp 4–7

Stanford C (2001) Significant others: the ape-human continuum and the quest for human nature. Basic Books, New York

United States Department of Agriculture, Animal and Plant Health Inspection Service (APHIS) (2018) Annual report animal usage by fiscal year. https://www.aphis.usda.gov/animal_welfare/downloads/reports/Annual-Report-Animal-Usage-by-FY2017.pdf?fbclid=IwAR05D8ZyB76yH1DxyD5cGqoZZpBU8buWqwuqH8XdaymTRuhlmxdOWUNoB80. Accessed 13 Nov 2018

United States Department of Agriculture, Animal and Plant Health Inspection Service (APHIS) (2019) USDA Animal Care, Animal Welfare Act and Animal Welfare Regulations. https://www.aphis.usda.gov/animal_welfare/downloads/bluebook-ac-awa.pdf. Accessed 17 Dec 2020

United States Fish and Wildlife Service (USFWS) (2018) Endangered species. https://www.fws.gov/endangered/. Accessed 18 July 2018

Williams C (2004) The basics of equine behavior. Rutgers New Jersey Agricultural Experiment Station. https://esc.rutgers.edu/fact_sheet/the-basics-of-equine-behavior. Accessed 18 July 2018

World Association of Zoos and Aquariums (WAZA) (n.d.) Positive animal welfare. http://www.waza.org/en/site/conservation/animal-welfare-1471340294/positive-animal-welfare. Accessed 19 July 2018

The Welfare Impact of Regulations, Policies, Guidelines, and Directives and Nonhuman Primate Welfare

Kathryn Bayne, Jann Hau, and Timothy Morris

Abstract

Many, but not all, countries and jurisdictions around the world have laws, regulations, policies, and other systems of oversight relating to the use of animals in science. There are variations in scope, scale, approach; in legal basis; in social and cultural perspectives; and in implementation. There is increasing convergence of the features of this oversight. In particular, there is increased emphasis on a wider scope of oversight to include all facets of animal use, in particular on ethical aspects and on standards and approaches to animal care and welfare. As the use of nonhuman primates in research requires special consideration of their welfare, many countries have adopted special requirements to achieve that objective. In addition, numerous professional organizations have published guidance documents designed to optimize the welfare of nonhuman primates in captive environments. The purpose of this chapter is to provide an introduction to the range of legislative approaches and to highlight associations that are impactful in this area.

Keywords

Regulations · Policies · Laws · Guidelines

K. Bayne (✉)
AAALAC International, Frederick, MD, USA
e-mail: kbayne@aaalac.org

J. Hau
Faculty of Health Sciences, Department of Experimental Medicine, University of Copenhagen and State Hospital, Copenhagen, Denmark
e-mail: jhau@sund.ku.dk

T. Morris
School of Veterinary Medicine and Science, University of Nottingham, Leicestershire, UK
e-mail: tmorris@timhmorris.co.uk

1 Introduction

There are well-recognized ethical challenges associated with captive housing of nonhuman primates which, at the moment, do not apply to the captive housing of many other species. In contrast to the majority of laboratory animals, nonhuman primates are basically wild animals, and although they may have been housed in captivity for a few generations, they remain undomesticated. Unlike most other species used in research, primates typically have a long lifespan, often spend years in captivity, and are frequently reused in several independent studies during the course of their lives (Boccia et al. 1995; van Vlissingen 1997; Carlsson et al. 2004). Captive production and housing of these intelligent and sentient animals for use in research are associated with close contact with people and some limitations compared to a free-ranging existence in the natural environment. Captive nonhuman primates are not able to exhibit their full behavioral repertoire and are prevented—and protected—from interacting freely with animals of other species. Taking control of the life of nonhuman primates thus obviously poses a moral issue and serious obligations for those responsible for their housing, care, and management while in captivity. The moral issue of providing captive nonhuman primates with a sufficient quality of life increases significantly when the nonhuman primates housed in captivity are used for research associated with pain, distress, suffering or lasting harm. However, increasing knowledge of the behavioral needs of nonhuman primates is reflected in modern guidelines advocating for group housing in large enclosures with ample foraging possibilities as well as furniture and fixtures allowing the animals to make full use of the space and flee vertically when frightened, with safe havens to escape unwanted attention from dominant animals, and enrichment items to expand their behavioral repertoire and engage their cognitive skills.

Housing and working with nonhuman primates in research environments are heavily regulated. Older regulations and guidelines were based on what has been termed a prescriptive "engineering" approach, while newer guidelines take a "performance" approach, which defines the desired outcome, but acknowledges that multiple methods may achieve that outcome. The performance approach relies on sound professional judgment and thus personnel competence and thorough organization of the nonhuman primate care and use program (Bayne 1998). The present chapter will focus on legislation and guidelines giving specific guidance of relevance to the behavioral management of nonhuman primates, updating previous reviews of this topic (e.g., Bayne and Morris 2012; Hau and Bayne 2017).

2 European Framework

Significant harmonization initiatives have been implemented within Europe that include a regulatory framework aimed at ensuring high quality research animal welfare. The most important legislations, and legislation-enforced guidelines, are the European Convention ETS 123 A (1986) and the European Directive (European

Union (EU) Directive 2010/63/EU), incorporating most of the accommodation and care guidelines from the European Convention ETS 123 Appendix A. Both these instruments have been revised within the last decade. According to the Directive's Annex, "care covers all aspects of the relationship between animals and man. Its substance is the sum of material and non-material resources provided by man to obtain and maintain an animal in a physical and mental state where it suffers least, and promotes good science. It starts from the moment the animal is intended to be used in procedures, including breeding or keeping for that purpose." The Directive further states that "...the use of non-human primates should be permitted only in those biomedical areas essential for the benefit of human beings, for which no other alternative replacement methods are yet available. Their use should be permitted only for basic research, the preservation of the respective non-human primate species or when the work, including xenotransplantation, is carried out in relation to potentially life-threatening conditions in humans or in relation to cases having a substantial impact on a person's day-to-day functioning, i.e., debilitating conditions." In considering the highly sensitive subject of research using great apes, the Directive states, "The use of great apes, as the closest species to human beings with the most advanced social and behavioral skills, should be permitted only for the purposes of research aimed at the preservation of those species and where action in relation to a life-threatening, debilitating condition endangering human beings is warranted, and no other species or alternative method would suffice in order to achieve the aims of the procedure."

Nonhuman primates are often reused in multiple protocols, and to minimize the impact on the individual research animal the EU Directive addresses the need to assess the cumulative impact of the entire experience of the research animal. This includes a retrospective analysis of the pain and suffering experienced by the animal after a project is finished. Lifetime experiences of the animals are a challenge to address qualitatively as well as quantitatively, but publications are beginning to appear in the literature addressing this complex issue (Honess and Wolfensohn 2010; Wolfensohn et al. 2015). Legislation requires that every nonhuman primate must have an individual health record containing all medical information, and Hau and Schapiro (2004) have advocated that health records should include a psychosocial profile listing information on the animal's housing history, social partners, dominance rank, compatibility with other animals, etc. When relevant, the file should include the animal's response to training and human contact as well as all of the experimental procedures that the animal has participated in during its lifetime.

The Directive recognizes that capturing nonhuman primates from the wild is both stressful for, and potentially injurious to, the individual animal and can be impactful on population conservation. To mitigate these consequences, the Directive stipulates that only F1 generation nonhuman primates (i.e., offspring of an animal which has been bred in captivity) or animals sourced from a self-sustaining colony may be used for research. The Directive encourages moving more exclusively toward the use of nonhuman primates derived from self-sustaining colonies.

The Directive's Annex III, which has been transposed into national legislation in the individual EU member states, lists fairly detailed requirements for the care and accommodation of laboratory animals, with species-specific guidelines for nonhuman primates. These include sections on health; breeding, and separation from the mother; enrichment; and handling, including training of the animals. Supplementing these general considerations for nonhuman primates, Annex III also contains additional detailed guidelines on the environment and its control; health; housing enrichment and care; training of personnel; and transport for marmosets and tamarins, squirrel monkeys, macaques, vervets, and baboons. Access to outdoor enclosures is important for the welfare of nonhuman primates and the newer guidelines emphasize this. The Directive contains an entire section on outdoor enclosures and also states that indoor enclosures, whenever possible, should be provided with windows, as they are a source of natural light and can provide environmental enrichment.

Prior to the most recent revision of the Directive, in 2009, the EU requested from its Scientific Committee on Health Environmental and Emerging Risks (SCHEER) an opinion on "The need for non-human primates in biomedical research, production and testing of products and devices." After the introduction of Directive 2010/63/EU, the EU Commission requested SCHEER to issue in 2017 an update on its 2009 Scientific Opinion (SCHEER 2017). This update highlighted both the many scientific approaches that could significantly contribute to the replacement, reduction, and refinement of nonhuman primate studies and tests, but also those issues that go beyond scientific rationale that prevents widespread adoption and development of alternatives for nonhuman primate laboratory use, with suggestions on how to overcome them. Through developing local policies and the mandatory ethics review process, this Opinion promotes the continuous raising of standards for scientific justification and animal care and use within the current EU legislative framework. This (SCHEER 2017) Opinion also recognized that tightening of the existing strict EU regulations for use of nonhuman primates may lead nonhuman primate research to transfer to other countries to the detriment of animal welfare. It recommended international cooperation to promote the international development of high standards for research and animal welfare and ethical use.

The welfare of nonhuman primates is also considered in circumstances of commercial trade. There are different conditions for commercial trade in primates depending on whether it is intra-EU trade or imports from outside the EU. For intra-EU trade, nonhuman primates can only be traded if the conditions set forth in Chapter II of the *Council Directive 92/65/EEC of 13 July* 1992 *laying down animal health requirements governing trade in and imports into the Community of animals, semen, ova, and embryos not subject to animal health requirements laid down in specific Community rules referred to in Annex A (I) to Directive 90/425/EEC (Council Directive 92/65/EEC)* are met: the premises they are going to are approved under Article 13 of *Council Directive 92/65/EEC*; they come from holdings that are approved by the competent authority in the country of origin; they show no sign of disease and come from holdings or areas which are not subject to any ban on health grounds; and they are accompanied by a veterinary certificate corresponding to the specimen in Part 1 of Annex E to *Council Directive 92/65/EEC*, completed by the approved veterinarian of the body, institute, or center of origin. Conditions for imports from non-EU countries are that nonhuman primates can only be imported if the conditions set forth in Chapter III of *Council Directive 92/65/EEC* are met: the premises the animals are going to must be approved under Article 13 of *Council Directive 92/65 EEC*; primates may have to be accompanied by a Convention on International Trade in Endangered Species (CITES) export certificate from the country of origin; they may have to have CITES import certificate from the country of destination; they must be accompanied by a veterinary certificate corresponding to the specimen in Part 1 of Annex E of *Council Directive 92/65 EEC* completed by the approved veterinarian of the body, institute or center of origin; 24 h prior to export from country of origin they must have been clinically examined by a veterinarian authorized by the competent authority in the country of origin; they must enter via a Border Inspection Post that handles live animals; the importer must ensure that Common Veterinary Entry Document is raised and part I is completed and returned to the BIP prior to entry; and if entering the UK they must go to a holding that is rabies approved and meets the conditions of *Council Directive 92/65/EEC*. The local institutional health and safety requirements relating to human contact with nonhuman primates are generally harmonized at a top level across the EU (Wood and Smith 1999). Some EU countries have instituted further oversight on third country breeding establishments outside the EU, for example the UK (Animal Procedures Committee 2007).

3 United States

The legal framework for the protection of research animals in the USA is a matrix of oversight provided primarily by the U.S. Department of Agriculture (USDA) and the National Institutes of Health (NIH) Office of Laboratory Animal Welfare (OLAW), though other agencies may play a role (e.g., Food and Drug Administration and Environmental Protection Agency) supplemented by a nonprofit international accrediting body (AAALAC International).

3.1 U.S. Department of Agriculture

The USDA is responsible for implementing the Animal Welfare Act, established by Congress, through the Animal Welfare Act Regulations (AWAR 1991). The Animal Welfare Act applies to all vertebrate species, as defined by the Secretary of Agriculture, though it excludes mice, rats, and birds used in research. Part 3 (standards) of the AWAR specifies humane handling, care, treatment, and transportation standards for several species, including nonhuman primates. Requirements for housing facilities, primary enclosures (including cage space), husbandry procedures (e.g., feeding, watering, sanitation), and care in transit are detailed. The AWARs also require the provision of environmental enhancement adequate to promote the psychological well-being of primates. Key aspects of enhancement that institutions must address are: (1) social grouping, (2) environmental enrichment, (3) use of restraint devices, and (4) special considerations. Exemptions can be granted for the following criteria: (1) The attending veterinarian determines that following the plan could adversely affect the clinical care of primates under medical treatment; this health-related exemption may remain in effect for a maximum of 30 days and then must be reviewed again by the attending veterinarian, or (2) The principal investigator provides a scientific justification that the environmental enhancement program would interfere with the objectives of the study. The Institutional Animal Care and Use Committee (IACUC) must approve these exceptions and review them at appropriate intervals, but at least annually.

Dealers, exhibitors, and research facilities must develop, document, and follow an appropriate plan for environmental enhancement adequate to promote the psychological well-being of nonhuman primates. The NIH's *Behavioral Management of Nonhuman Primates* (Appendix 2 of the *Guidelines for General Species Environmental Enrichment*, revised 2017) serves as one template plan for designing and implementing an enrichment program. The plan must be in accordance with currently accepted professional standards as cited in appropriate professional journals or reference guides and as directed by the attending veterinarian. At a minimum, the plan must address the social needs of nonhuman primate species known to exist in social groups in nature. Individual animals that are vicious, overaggressive, or debilitated should be individually housed. Nonhuman primates that are suspected of having a contagious disease must be isolated from healthy animals in the colony as determined by the attending veterinarian. Group-housed nonhuman primates must be determined to be compatible in accordance with generally accepted professional practices and actual observations, as directed by the attending veterinarian. Individually housed nonhuman primates must be able to see and hear members of their own or compatible species unless the attending veterinarian determines that this arrangement would endanger their health, safety, or well-being. Primary enclosures must be enriched by providing the animal the opportunity to express a diversity of species-typical behaviors. Environmental enrichment devices may include perches, swings, mirrors, manipulanda, and foraging or task-oriented feeding methods. Interaction with familiar and knowledgeable personnel is recommended, provided it is consistent with safety precautions. Special attention is required for infant and young juvenile nonhuman primates, those that exhibit signs of psychological distress, those entered in IACUC-approved research protocols that require restricted activity, and individually housed nonhuman primates without sensory contact with nonhuman primates of their own or compatible species. Great apes weighing more than 110 lb. (approximately 50 kg) must be provided additional opportunities to express species-typical behavior.

The AWARs specify that if a nonhuman primate must be maintained in a restraint device for an IACUC-approved protocol 2, such restraint must be for the minimum period possible. If the protocol requires more than 12 h of continuous restraint, the nonhuman primate must be provided the daily opportunity for at least 1 continuous hour of unrestrained activity, unless the IACUC approves an exception. Such an exception must be reviewed at least annually. The attending veterinarian may also exempt an individual nonhuman primate from participation in the environmental enhancement plan in consideration of its well-being. However, such an exemption must be documented and reviewed by the attending veterinarian every 30 days. All exemptions must be available for review by the USDA and federal funding agencies upon request and reported in the annual report to the USDA.

The USDA issued a policy manual, currently under review and at the time of this writing not available online, with the objective of clarifying certain aspects of the AWARs. Some policies apply specifically to nonhuman primates and must be considered in the care and use of these animals. For example, Policy #3, "Veterinary Care" includes a section on the reduction of canine teeth in nonhuman primates and references the American Veterinary Medical Association's position statement on the subject: "...The [American Veterinary Medical Association] is opposed to removal of canine teeth in captive nonhuman primates or exotic and wild (indigenous) carnivores, except when required for medical treatment or scientific research approved by an Institutional Animal Care and Use Committee. Reduction of canine teeth may be necessary to address medical and approved scientific research needs, or animal or human safety concerns. If reductions expose the pulp cavity, endodontic procedures must be performed by a qualified person. To minimize bite wounds, recommended alternatives to dental surgery include behavioral modification, environmental enrichment, and changes in group composition." Policy #7 addresses group classification of nonhuman primates, which then relates to the cage space made available to these different primates per the AWAR. Policy #18 addresses the need for a health certificate when transporting primates. Several other more general policies also have applicability to nonhuman primates.

3.2 Office of Laboratory Animal Welfare

OLAW oversees the care and use of research animals in U.S. Public Health Service-funded research (e.g., the National Institutes of Health). OLAW requires US-based institutions to conform with the U.S. Public Health Service Policy on Humane Care and Use of Laboratory Animals (2015) and the recommendations of the *Guide for the Care and Use of Laboratory Animals* (NRC 2011), several of which are specific to nonhuman primates, but the majority of which are more general in nature, and thus are also applicable to numerous other species. Of note, OLAW states in the Frequently Asked Questions located on their website that housing of primates in social groups or in pairs is the default system, and states that clear medical or scientific justification is required for any other method of housing: "Exemptions to the social housing requirement must be based on strong scientific justification approved by the IACUC or for a specific veterinary or behavioral reason. Lack of appropriate caging does not constitute an acceptable justification for exemption. When necessary, single housing of social animals should be limited to the minimum period necessary. When single housing is necessary, visual, auditory, olfactory, and (depending on the species) protected tactile contact with compatible conspecifics should be provided, if possible." In addition, OLAW sponsored the development and publication of a series of six booklets "that serve as an introduction to the subject of environmental enrichment for primates housed in a diversity of conditions" to assist institutions in meeting the recommendations of the *Guide*. These booklets are also available on their website.

3.3 Specific Controls on the Use of Chimpanzees

The *Chimpanzee Health Improvement, Maintenance, and Protection Act*, passed in 2000, established a national sanctuary system for federally owned or supported chimpanzees no longer needed for research. In 2013, the NIH announced its decision to reduce the use of chimpanzees in research to approximately 50 animals (NIH 2013). However, in 2015, the U.S. Fish and Wildlife Service announced its designation of captive chimpanzees as endangered. This resulted in a new requirement for any researcher proposing to use chimpanzees in research to obtain a permit. As no permits were subsequently sought by the research community, the NIH determined it would no longer maintain even the 50 chimpanzees they held and would retire them. However, there was concern that the relocation of the former research animals, many of whom were elderly or had chronic disease, to a sanctuary could have negative consequences for the more fragile animals. To address the increasing controversial question of relocation, the NIH convened a working group to evaluate what retirement location would be best for the animals' health and welfare. The working group concluded in a report published in May 2018 that the NIH-owned animals should be moved to retirement sanctuaries unless relocation was "extremely likely" to jeopardize the animal. Following a 60-day public comment period and review by the NIH, it was announced that a system of standardized approaches for assessing each chimpanzee based on health, behavior, social, and environmental requirements would be implemented. The goal is to transfer the remaining 180 NIH-owned chimpanzees into sanctuary, unless doing so "would severely or irreversibly accelerate deterioration of the chimpanzee's physical or behavioral health."

4 Other Countries

4.1 Australia

Animal research in Australia is regulated by the Australian Code of Practice (National Health and Medical Research Council 2013). The Code establishes governing principles; provides detailed descriptions of the responsibilities of institutions and animal ethics committees, as well as investigators and care staff; and establishes standards for animal care and management. These sections are designed to be applicable to all laboratory animals. The Code also contains specific reference to the use of nonhuman primates, noting that here is a requirement for "particular justification" for activities that may severely compromise primates' well-being and for which the tenets of the 3Rs (Russell and Burch 1959) cannot be applied. It should be noted that state and territory governments have regulatory responsibility for animal welfare, including that of nonhuman primates. Specific guidance for the care and use of nonhuman primates is provided in the "Principles and guidelines for the care and use of non-human primates for scientific purposes" (NHMRC 2016). The stated intent of this document is to describe basic principles and best practice guidance for the care and use of nonhuman primates and assist Animal Ethics Committees in reviewing applications for working with nonhuman primates. The funding agency, the National Health and Medical Research Council requires compliance with this guideline as part of its funding agreement.

4.2 Canada

The Canadian constitution precludes federal legislation pertaining to the use of animals in research, testing, or education because such use is under provincial jurisdiction. Eight provinces have established animal protection legislation and/or regulations that mention animals used in research in their animal protection legislation, five of which reference the Canadian Council on Animal Care (CCAC) guidelines and policies (see III.B.2.). The CCAC is an independent organization that establishes standards and certifies institutions that use animals in science in Canada. Contractors performing work for the Canadian government are required to adhere to CCAC guidelines, many of which have general applicability to nonhuman primates used in research, testing, or education. Specific guidance regarding the care and use of nonhuman primates is provided in Chapter 20 of the *Guide to the Care and Use of Experimental Animals, Volume 2* (under revision at the time of this writing).

4.3 China

The Ministry of Science and Technology (MOST) has issued regulations regarding the care of laboratory animals since 1988. Since that time, and certainly applicable to nonhuman primate research, is the requirement for adherence to standards pertaining to husbandry and transportation to promote animal welfare. MOST documents also reference the 3Rs (Kong and Qin 2010). In 2017, a new "GB" (Guobiao, Chinese for national standard) was issued, "Laboratory Animals - Codes of Welfare and Ethics" (currently only available in Chinese). These standards reference adherence to the 3Rs and Five Freedoms and provide details regarding the role of the Ethics Committee and the veterinarian, require an ethical review of the proposed research, require attention to humane endpoints and environmental enrichment, describe the importance of personnel training, and many other aspects of an animal care and use program. Although the standards are applicable to all laboratory animals, because China produces and uses a significant number of nonhuman primates (Cyranoski 2016), it is noteworthy that these new standards have national scope and will have a profound positive impact on the welfare of primates bred and used for research in China. In addition, a domestic accreditation system has been implemented that is based on national standards that encompass many of the same approaches as the *Guide*.

4.4 India

The cessation of the export of primates from India occurred in 1978. The Ministry of Environment, Forests, and Wildlife established the *Committee for the Purpose of the Control and Supervision of Experiment on Animals* (CPCSEA) under the 1960 Protection of Cruelty to Animals Act. As primates are regarded from a religious, cultural, and political aspect in India, there are particular controls on the use of such larger species, including avoiding prolonged restraint (e.g., chairing), considering windows in housing areas, social housing, access to a run for "free-ranging activities," and accommodating locomotion behavior (CPCSEA 2007). Any project using primates must first be approved by an Institutional Animal Ethics Committee. Plans to address the rehabilitation of the primates used in the study are an element of the protocol approval process.

4.5 Japan

In Japan, animal experimentation is regulated via (1) a number of laws, (2) amendments to these laws, and (3) ministerial guidelines. The Primate Research Institute of Kyoto University has issued very detailed guidelines for the care and use of nonhuman primates, including sections on facility design and equipment (2010). These guidelines, which are well known and used by Japanese scientists, are in good agreement with the European and American guidelines and provide useful information. Indeed, the Japanese guidelines often provide greater detail than the Western guidelines. They emphasize the importance of providing a captive environment in which the animals can perform their species-specific behavioral patterns at an optimal level, as determined by each individual's physiological, ecological, and behavioral characteristics, within a range that does not interfere with the objectives and methods of research. In sections on environmental enrichment, the guidelines state that if not all aspects can be improved adequately due to experimental and environmental limitations, improvement of selected aspects may be able to compensate for the loss of other aspects. For example, if there are necessary limitations to the social environment, efforts must be made to enrich the physical environments and increase human contact. This is in good agreement with the *Guide*, which emphasizes the importance of supplemental enrichment to compensate for situations in which an animal has to be singly housed for a period of its life. The Kyoto guidelines advocate that the animals' (functional) living space should be as large as possible, include novelty, manipulatable tools for foraging, and a suitable social environment. Interestingly, they advocate that if an adequate conspecific social environment cannot be provided for research reasons, then positive relationships with humans will, to some extent, compensate for the lack of conspecific social interaction.

4.6 Singapore

Singapore's National Advisory Committee for Laboratory Animal Research published the *Guidelines on the Care and Use of Animals for Scientific Purposes* in 2004 (NACLAR 2004) was adapted from the *Australian Code of Practice for the Care And Use of Animals for Scientific Purposes* (National Health and Medical Research Council, Australia); the *Guide to the Care and Use of Experimental Animals, Volume 1* (2nd Edition) (Canadian Council on Animal Care 1993); the *Good Practice Guide for the Use of Animals in Research, Testing, and Teaching* (Ministry of Agriculture and Forestry/National Animal Ethics Advisory Committee 2010, New Zealand); the *Guide for the Care and Use of Laboratory Animals* (National Academies Press, Washington, D.C., USA, 1996); the *Public Health Service Policy on Humane Care and Use of Laboratory Animals* (OLAW/NIH, revised 2015); and the ARENA/OLAW *Institutional Animal Care And Use Committee Guidebook* (2002). Appendix III of the NACLAR Guidelines contains specific recommendations regarding nonhuman primate care and use.

4.7 South America and Africa

To a lesser extent, nonhuman primate research is also being performed in South America and Africa. Although these continents lack supranational legislation or guidelines setting out rules for nonhuman primate research, some countries have implemented legislation that incorporates the 3Rs (Bayne et al. 2015), but information specific to nonhuman primates has not been included in legislative frameworks. An exception to this is the National Museums of Kenya's Institute of Primate Research (IPR), located in the Nairobi suburb of Karen. The IPR is perhaps the best known primate research center in Africa. Primate models, in particular vervet monkeys and baboons, are used in studies of tropical infectious diseases, human reproductive disorders, and conservation strategies. Research ethics and animal welfare issues are taken into account at IPR and according to their webpage "Ethical and animal welfare concerns form a strong component of the Department's animal husbandry and research activities." Before any experimental procedure is carried out, review committees evaluate all research proposals for scientific merit and welfare concerns. Housing conditions at IPR meet European standards. According to the Directive, nonhuman primates used for scientific research should be captive bred and reared on site to avoid transport stress, and where possible they should have access to outdoor enclosures. One of the advantages for the animals housed in nonhuman primate facilities in source countries, like Kenya, is that it is possible to house the animals in outdoor, seminatural enclosures in their natural habitat, preventing the animals from being subjected to long-distance transportation and the adjustments necessary to acclimatize to a new climate, and foreign biotic and abiotic environments (Lambeth et al. 2006). IPR is currently leading a multi-institutional effort to develop comprehensive guidelines for laboratory animals used in research and education in this part of the world. IPR has partnered with the National Council for Science and Technology, Kenya and the Consortium for National Health Research (a local funding agency for health research) for this task (Hau et al. 2014).

5   Professional Society Guidelines

The *Association of Primate Veterinarians* (APV) (http://www.primatevets.org/) is a professional organization of veterinarians concerned with the health, care, and welfare of nonhuman primates. Its objectives are: "To promote dissemination of information relating to the health, care, and welfare of nonhuman primates; to provide a mechanism by which primate veterinarians may speak collectively on matters regarding nonhuman primates; and to promote fellowship among primate veterinarians." The APV states that scientists, laboratory animal veterinarians, animal caregivers, and IACUCs/ethical review committees must work together to fully implement regulatory expectations to provide the most appropriate environment for captive nonhuman primates. The APV has published a formulary. Also on their website, the APV has additional relevant guidelines, including Guidelines for Use of Fluid Regulation in Biomedical Research, Laparoscopic Reproductive Manipulation of Female Nonhuman Primates Guidelines, Social Housing Guidelines, Food Restriction Guidelines, Jacket Use Guidelines, and Cranial Implant Care Guidelines.

The *IPS* [International Primatological Society] *International Guidelines for the Acquisition, Care, and Breeding of Nonhuman Primates*, published as the 2nd edition in 2007, represent one of the more detailed guidance documents and set of collated information. Subjects addressed include acquisition, transport, staff training, health control, staff safety, care, and husbandry that are specific to nonhuman primates. It serves as a useful source of information, especially where less common species are used, where national legislation is absent, or local guidance is not available. The IPS Guidelines are currently available in English, French, Spanish, Chinese, and Japanese.

The *National Center for the 3Rs* (NC3Rs) (http://www.nc3rs.org.uk) in the UK is a publicly and industry funded center to promote the utilization and development of the 3Rs. The NC3Rs website contains "Resource Hubs" that have information specific to nonhuman primates, such as blood sampling of marmosets; accommodations, care, and use guidelines; and chair restraint training. There is also guidance on the use of chronic implants, refining food, and fluid control and rehoming. The NC3Rs also supports "advances in nonhuman primate welfare through... research funding schemes, peer review service, and office led data sharing projects."

The *Institute for Laboratory Animal Research* (ILAR) (http://dels.nas.edu/ilar) was founded in 1952 under the auspices of the U.S. National Research Council, National Academies, and a congressionally chartered nongovernmental agency. The mission of ILAR is to evaluate and disseminate information on issues related to the scientific, technological, and ethical use of animals and related biological resources in research, testing, and education. The best-known report of the ILAR is the *Guide for the Care and Use of Animals* (NRC 2011). The *Guide* has been translated into several languages to facilitate its use as a reference internationally. Other reports that have applicability to nonhuman primates are: *Recognition and Alleviation of Pain in Laboratory Animals (2009); Recognition and Alleviation of Distress in Laboratory Animals (2008); Guidelines for the Humane Transportation of Research Animals (2006); Guidelines for the Care and Use of Mammals in Neuroscience and Behavioral Research (2003a); Occupational Health and Safety in the Care and Use of Nonhuman Primates (2003b); and The Psychological Well-Being of Nonhuman Primates (1998)*.

6 National Primate Societies

The *Primate Society of Great Britain* (http://www.psgb.org) promotes research in primate biology, conservation of primate populations and their habitats, and the welfare of primates. The organization offers grants for the conservation of primates and for their care in captivity. The organization publishes *Primate Eye*.

The *Primate Society of Japan* (PSJ) (https://primate-society.com/en/) with more than 600 members has as its mission the promotion of the development of primatology, organization of an annual meeting, publishing scientific information, primate conservation, and animal welfare. *Primates* is the official English language journal of PSJ while *Primate Research* is the official Japanese language journal.

The *China Laboratory Primates Breeding and Development Association* (CLPA) is a nonprofit and nongovernmental organization established in 1993 (http://clpa.org.cn). The CLPA is comprised of member companies and centers that conduct research on improving breeding and rearing of captive primates, as well as breed and supply animals. The CLPA is under the supervision of MOST and the State Forestry Administration of China (the capture and sale of wild caught primates are prohibited by the Chinese government).

6.1 Other National Primate Societies

Professional societies that are specific to nonhuman primates are found in several other countries. These include the American Society of Primatologists, Australasian Primate Society, Indonesian Primatological Association, Primate Specialist Group of Mammalian Society of China, Primatological Society of India, and numerous societies in Central and South America. Many country-specific primate societies are affiliated with the International Primatological Society (IPS).

The *Federation of Laboratory Animal Science Associations* (FELASA) (http://www.felasa.eu) represents the common interests of laboratory animal science associations of nations in Europe; there are currently 17 member associations representing more than 20 countries. FELASA organizes scientific meetings, issues policy statements, accredits LAS courses, is considered as the European specialist body in laboratory animal science by the EU and publishes recommendations and guidelines by expert working groups in all areas of laboratory animal science. Specific for nonhuman primates are guidelines on "Sanitary aspects of handling non-human primates during transport", and "Recommendations for health monitoring of non-human primate colonies" (http://www.felasa.eu/working-groups).

7 Networks

A range of networks provide useful information and standards for both researchers and laboratory animal professions.

Primate Info Net (http://pin.primate.wisc.edu) is based at the U.S. Wisconsin National Primate Research Center. It covers the broad field of primatology with original and linked resources and through email lists and other resources, Primate Info Net also supports an informal "primate information network" comprised of thousands of individuals around the world working with nonhuman primates.

The *European Primate Network* (EUPRIM-Net) (https://www.euprim-net.eu) provides specialized infrastructures and procedures for biological and biomedical research by bringing together the nine European primate centers that combine research and breeding to form a virtual European Primate Center. Its activities include defining a health control system for European primate centers, establishing standard operating procedures for nonhuman primate quarantine and maintenance at biosafety levels 2 and 3, standardizing procedures in common experiments, and providing access to tissue, serum, and gene banks. In the context of promoting animal welfare, the objectives of EUPRIM-Net are "the standardization of procedures and methods, the enhanced availability of primates, and training for those working with primates."

The *European Marmoset Research Group* (https://cordis.europa.eu/project/id/HPCF-CT-1999-00223) is a not-for-profit organization established in 1994 to facilitate interdisciplinary communication between institutions, both academic and commercial, conducting biological and/or biomedical research with marmosets and tamarins within Europe and beyond. It organizes workshops, produces publications on both care and use, and encourages communication between individuals and institutions.

The *Primate Specialist Group* (PSG) (http://www.primate-sg.org) of the International Union for Conservation of Nature (IUCN) is a network of scientists and conservationists focusing primarily on conservation. It is active across tropical world, working in dozens of nations in Africa, Asia, and Latin America promoting research on the ecology and conservation of hundreds of primate species. Their assessments contribute to the IUCN Red List, a comprehensive summary of threats to the world's biodiversity. Beyond these core functions, the PSG produces a journal or newsletter for each of the four global primate regions—Africa, Asia, Madagascar, and the Neotropics—as well as a journal, *Primate Conservation*.

The *South Asian Primate Network* (SAPN) (http://www.southasianprimatenetwork.org) is a component of the IUCN SSC Primate Specialist Group. South Asia is defined by the SAPN as consisting of Afghanistan, Bhutan, Bangladesh, India, Maldives, Nepal, Pakistan, and Sri Lanka. Although the focus of this information/dialogue group is South Asia, all primate researchers are eligible to join the network.

The *Universities Federation for Animal Welfare* (UFAW) (http://www.ufaw.org.uk) is a UK registered charity that works to develop and promote improvements in the welfare of all animals through scientific and educational activity worldwide. It has a strong interest in laboratory animal welfare, having commissioned the study that led to the promulgation of the principles of the 3Rs. It publishes the "The UFAW Handbook on the Care and Management of Laboratory and Other Research Animals," now in its 8th edition (2010), which is widely regarded as one of the definitive works on practical husbandry, breeding, laboratory procedures, and disease control for a wide variety of vertebrates from marine fish to nonhuman primates. It also publishes the journal, *"Animal Welfare"* and other reports that include information on nonhuman primates, such as Environmental Enrichment in Captive Primates: A Survey and Review (Dickie 1994).

8 Conclusion

Over the past 30 years, the way in which we house and care for nonhuman primates, and the rules and guidelines regulating their housing and care have changed dramatically. In the past, extensive single housing in small, barren, squeeze-back cages, sometimes in multiple tiers, was the norm, while currently social housing in large cages/enclosures equipped with furniture, fixtures, and toys is the default. Additionally, attention is now paid to the foraging and locomotor needs of the animals and management routines have been refined to provide the animals with opportunities to perform more natural behaviors. Effective behavioral management and educational programs for animal care staff, as for instance developed and implemented at American and European nonhuman primate centers—EUPrimNet (https://www.euprim-net.eu)—have resulted in widespread training programs for the animals, which now benefit from cognitive stimulation, reduced fearfulness, and the ability to voluntarily participate in a variety of husbandry, veterinary, and research procedures.

Captive care of nonhuman primates and behavioral management strategies is constantly evolving, and new refinements are continually being developed, discovered, and implemented by the forerunners in the field and followed quickly by nonhuman primate programs all over the world. Improving the ways in which we care for nonhuman primates will never finish and the next generation of guidelines will no doubt reflect this continuing process.

In the interests of medical progress and our responsibilities to future generations, nonhuman primates will continue to be vital as models for studies of debilitating diseases. It is thus of utmost importance that the care of the animals needed for this activity is continuously improved and resources allocated to studies that strive to ensure optimum physical and mental well-being of the animals in our care.

There is good agreement between the various guidelines (e.g., European, American, Japanese, and Australian), which have been developed and implemented in recent years. Together, they admirably capture (1) modern trends in housing and care of nonhuman primates and (2) the importance of a stimulating environment and training of the animals to lower stress associated with husbandry, veterinary, and research procedures. Adherence to the excellent guidance in the Directive and the *Guide* will ensure that laboratory nonhuman primates are housed and cared for in a manner that meets and exceeds today's best practice standards, standards based on analyses of the complex behavioral needs of nonhuman primates in a captive environment combined with the requirements of the research projects.

References

Animal Procedures Committee (2007) Consideration of policy concerning standards of animal housing and husbandry for animals from overseas non-designated sources. Available via http://webarchive.nationalarchives.gov.uk/20100816163757/http://apc.homeoffice.gov.uk/reference/2007-0404-web-version-standards.pdf. Accessed 30 Dec 2019

Applied Research Ethics National Association (ARENA) and Office of Laboratory Animal Welfare (OLAW) (2002) Institutional animal care and use Committee guidebook, 2nd edn. Available via https://olaw.nih.gov/sites/default/files/GuideBook.pdf. Accessed 30 Dec 2019

Bayne K (1998) Developing guidelines on the care and use of animals. Ann N Y Acad Sci 862:105–110. https://doi.org/10.1111/j.1749-6632.1998.tb09122.x

Bayne K, Morris TH (2012) Laws, regulations and policies relating to the care and use of nonhuman primates in biomedical research. In: Abee C, Mansfield K, Tardiff S, Morris T (eds) Nonhuman primates in biomedical research, Volume 1: Biology and management. Elsevier, New York, pp 35–56

Bayne K, Ramachandra GS, Rivera EA, Wang J (2015) The evolution of animal welfare and the 3Rs in Brazil, China and India. J Am Assoc Lab Anim Sci 54(2):181–191

Boccia ML, Laudenslager ML, Reite ML (1995) Individual differences in macaques' response to stressors based on social and physiological factors: implications for primate welfare and research outcomes. Lab Anim 29(3):250–257. https://doi.org/10.1258/002367795781088315

Canadian Council on Animal Care (1993) Guide to the care and use of experimental animals, Volume 1, 2nd edition. Available via https://www.ccac.ca/Documents/Standards/Guidelines/Experimental_Animals_Vol1.pdf. Accessed 30 Dec 2019

Carlsson HE, Schapiro SJ, Farah I, Hau J (2004) Use of primates in research: a global overview. Am J Primatol 63(4):225–237. https://doi.org/10.1002/ajp.20054

Committee for the Purpose of the Control and Supervision of Experiment on Animals (2007) Guidelines on the regulation of scientific experiments on animals. Available via http://cpcsea.nic.in/WriteReadData/userfiles/file/Compendium%20of%20CPCSEA.pdf. Accessed 30 Dec 2019

Council of the European Union (1992) Council Directive 92/65/EEC of 13 July 1992 laying down animal health requirements governing trade in and imports into the Community of animals, semen, ova and embryos not subject to animal health requirements laid down in specific Community rules referred to in Annex A (I) to Directive 90/425/EEC. https://publications.

europa.eu/en/publication-detail/-/publication/2de62d1d-48eb-43b1-97d6-f0b464deb5d1/language-en. Accessed 24 June 2021

Cyranoski D (2016) Monkey kingdom. Nature 532:21. https://doi.org/10.1038/532300a

Dickie LA (1994) Environmental enrichment in captive primates: a survey and review. Universities Federation for Animal Welfare, Herfordshire

EU Directive 2010/63/EU of the European Parliament and of the Council of 22 September 2010 On the protection of animals used for scientific purposes. Available via https://eur-lex.europa.eu/legal-content/EN/TXT/?uri=celex%3A32010L0063. Accessed 30 Dec 2019

European Convention for the Protection of Vertebrate Animals used for Experimental and Other Scientific Purposes (1986) Appendix A of the European Convention for the protection of vertebrate animals used for experimental and other scientific purposes (ETS NO. 123), Guidelines for accommodation and care of animals (Article 5 of the convention). Available via https://www.aaalac.org/pub/?id=E900CF34-9112-946E-C8A5-331F9E2897D9. Accessed 30 Dec 2019

Hau J, Bayne K (2017) Rules, regulations, guidelines, and directives. In: Schapiro S (ed) Handbook of primate behavioral management. CRC Press, Boca Raton, FL, pp 25–36

Hau J, Schapiro SJ (2004) The welfare of non-human primates. In: Kalisthe E (ed) The welfare of laboratory animals. Springer, Dordrecht, The Netherlands, pp 291–314

Hau AR, Guhad FA, Cooper ME, Farah IO, Souilem O, Hau J (2014) Animal experimentation in Africa: legislation and guidelines. In: Guillen J (ed) Laboratory animals: regulations and recommendations for global collaborative research. Elsevier, New York, pp 205–216

Honess P, Wolfensohn S (2010) A matrix of assessment and cumulative suffering in experimental animals. Alt Lab Anim 38(3):205–212. https://doi.org/10.1177/026119291003800304

Kong Q, Qin C (2010) Analysis of current laboratory animal science policies and administration in China. ILAR J 51(1):E1–E10. https://doi.org/10.1093/ilar.51.1.e1

Lambeth SP, Hau J, Perlman JE, Martino M, Schapiro SJ (2006) Positive reinforcement training affects hematologic and serum chemistry values in captive chimpanzees (*Pan troglodytes*). Am J Primatol 68(3):245–256. https://doi.org/10.1002/ajp.20148

Ministry of Agriculture and Forestry/National Animal Ethics Advisory Committee (2010) Good practice guide for the use of animals in research, testing and teaching. Available via https://www.agriculture.govt.nz/dmsdocument/33585-good-practice-guide-for-the-use-of-animals-in-research-testing-and-teaching. Accessed 30 Dec 2019

National Advisory Committee for Laboratory Animal Research (NACLAR) (2004) Guidelines on the care and use of animals for scientific purposes. Available via https://private.aaalac.org/intlRefs/IntRegs/Singapore/Singapore%20Animal%20Care%20Use%20%20Guide.pdf. Accessed 30 Dec 2019

National Health and Medical Research Council (2013) Australian code for the care and use of animals for scientific purposes, 8th edn. Available via https://www.nhmrc.gov.au/about-us/publications/australian-code-care-and-use-animals-scientific-purposes. Accessed 30 Dec 2019

National Health and Medical Research Council (2016) Principles and guidelines for the care and use of non-human primates for scientific purposes. Available via https://www.nhmrc.gov.au/about-us/publications/principles-and-guidelines-care-and-use-non-human-primates-scientific-purposes. Accessed 30 Dec 2019

National Institutes of Health (2013) Announcement of agency decision: recommendations on the use of chimpanzees in NIH-supported research. Available via http://dpcpsi.nih.gov/council/pdf/NIH_response_to_Council_of_Councils_recommendations_62513.pdf. Accessed 30 Dec 2019

National Institutes of Health (2017) Behavioral management of nonhuman primates. Available via https://oacu.oir.nih.gov/sites/default/files/uploads/arac-guidelines/d4b-2017_nhp_enrich-final_3-22-17.pdf. Accessed 30 Dec 2019

National Research Council (1998) The psychological well-being of nonhuman primates. National Academies Press, Washington, DC

National Research Council (2003a) Guidelines for the care and use of mammals in neuroscience and behavioral research. National Academies Press, Washington, DC

National Research Council (2003b) Occupational Health and safety in the care and use of nonhuman primates. National Academies Press, Washington, DC

National Research Council (2006) Guidelines for the humane transportation of research animals. National Academies Press, Washington, DC

National Research Council (2008) Recognition and alleviation of distress in laboratory animals. National Academies Press, Washington, DC

National Research Council (2009) Recognition and alleviation of pain in laboratory animals. National Academies Press, Washington, DC

National Research Council (2011) Guide for the care and use of laboratory animals. National Academies Press, Washington, DC

Office of Laboratory Animal Welfare, National Institutes of Health (2015) Public Health service policy on humane care and use of laboratory animals. Bethesda, MD. Available via https://grants.nih.gov/grants/olaw/references/phspol.htm. Accessed 30 Dec 2019

Primate Research Institute, Kyoto University (2010) Guidelines for care and use of nonhuman Primates, version 3. Available via https://www.pri.kyoto-u.ac.jp/research/sisin2010/Guidelines_for_Care_and_Use_of_Nonhuman_Primates20100609.pdf. Accessed 30 Dec 2019

Russell W, Burch R (1959) The principles of humane experimental technique. Methuen, London

Scientific Committee on Health Environmental and Emerging Risks (SCHEER) (2017) The need for non-human primates in biomedical research, production and testing of products and devices. Available via https://ec.europa.eu/health/scientific_committees/non-human-primates-testing-9-conclusions_en. Accessed 30 Dec 2019

United States Department of Agriculture (1991) Code of Federal Regulations, title 9, part 3, animal welfare; standards; final rule. Fed Reg 56(32):1–109

Universities Federation for Animal Welfare (2010) The UFAW handbook on the care and management of laboratory and other research animals, 8th Edition. R. Hubrecht and J. Kirkwood (eds). Wiley-Blackwell, Oxford, UK

van Vlissingen JMF (1997) Welfare implications in biomedical research. Primate Rep 49:81–85

Wolfensohn S, Sharpe S, Hall I, Lawrence S, Kitchen S, Dennis M (2015) Refinement of welfare through development of a quantitative system for assessment of lifetime experience. Anim Welf 24(2):139–149. https://doi.org/10.7120/09627286.24.2.139

Wood M, Smith MW (1999) Health and safety in laboratory animal facilities. Published for Laboratory Animals Ltd. by The Royal Society of Medicine Press, Ltd.

Index

A

Abnormal behavior, 135, 173
   age, 178–179
   extrinsic effects, 175
   function of, 179–180
      coping hypothesis, 180
      maladaptive, 180
   interventions for, 181–185
      drug therapy, 185
      inanimate enrichment, 181–183
      increased cage size, 183–184
      outdoor housing, 184
      positive reinforcement training, 184–185
      social housing, 181
   intrinsic effects, 177–178
   later environmental restriction, 176–177
   to measure welfare, 180–181
   pathological behavior, 173–174
   prevalence of, 174–175
   rearing, 175–176
   sex, 179
   species, 178
   stereotypies, 173
   stress, 177
Acclimation, 454–455
Acquired immunodeficiency syndrome (AIDS), 41
Adult male chimpanzees, 506, 508
   grooming equality, 511, 513
   medical care events of, 510
   repeated social association, 512
   risk of aggressions, 509
   social bond, 510–512
   social relationships, 509
   time allocation of behavioral repertoires, 511
Aerospace studies, chimpanzees in, 36–38
African green monkeys (*Chlorocebus sabaeus*), 330, 452, 454, 629
Allogrooming, 475, 476
Amazon species, 147
American Sign Language (AMESLAN), 38
American Veterinary Medical Association (AVMA), 134
Animal handling, 455
Animal husbandry, 464, 469, 490
   behavioral biology, 325
   overview, 324
   regulations, 325
   in research setting, 324
Animal–keeper interactions, 86
Animal models
   marmoset EAE model
      infections, 614–616
      predisposing genes, 613–614
      translational relevance of, 611–613
   in preclinical immunology research, 607–608
   specific pathogen-free (SPF)-bred laboratory mouse, 606
   for translational research, 606
   welfare aspects, 616–619
Animal rights, 634–636
Animal training programs
   facility-wide approach, 546
   personnel considerations, 546–548
   program structure, 545
   project-based approach, 545–546
   section-wide approach, 546
Animal–visitor interactions, 86
Animal welfare, 98, 522, 599, 633–634
   physical and psychological components, 414
Animal Welfare Act, 41, 44, 172, 647
Anthropoid Research Station, 33

Anxiety-related behavior, 185–186
  anxious phenotype, testing anxiety and identifying, 187–188
  behavioral measures of anxiety, 186–187
  to measure welfare using, 188–189
  and welfare-related studies, 189–190
Apes Seizure Database, 128
Applied behavior analysis (ABA)
  functional analysis, 543–544
  preference assessment, 544–545
Arnhem Zoo, 18
ARRIVE (Animal Research: Reporting of In Vivo Experiments) guidelines, 70
Association of Primate Veterinarians (APV), 61, 654
Atacicept, 620
Attention bias
  advantages, 223
  basics, 219
  description, 219
  limitations, 223–224
  Macaque Stimulus Set, 224
  measuring primate welfare
    dot-probe task, 220–222
    preferential-looking tasks, 220
    visual search task, 222
  phobia-related cues, 219
  scrambled stimulus, 224
Auditory enrichment, 487
Australia, 650–651

B
Baboons
  Guinea baboon (*Papio papio*), 16
  hamadryas baboons (*Papio hamadryas*), 4, 156
Behavioral disorders, 136
Behavioral management techniques, 36, 448, 451, 452, 455, 456
Behavioral needs
  activity, 288, 289
  social needs, 289–290
  space, 285
Biomedical research, chimpanzees in, 31–32
  in aerospace studies, 36–38
  captive research chimpanzee housing and care, 44–45
  ethical concerns, 45–48
  in experimental medicine and biology, 39–42
  large-scale sourcing of, 42–44

  in linguistic studies, 38–39
  under microscope, 48–49
  NIH decision and post-decision impact, 49–51
  to Yerkes Primate Research Center, 32–35
Biopharmaceuticals preponderance, 63
Blue monkeys (*Cercopithecus mitis*), 5
Bonobos (*Pan paniscus*), 8, 518
Boredom, 475
Bornean orangutans (*Pongo pygmaeus*), 88, 185
Brazilian Society of Zoos and Aquariums (SZB), 153
Brazil's National Primate Center, 152
Bristol Zoo, 12
Bronx Zoo, 24, 25, 31, 34
Brown capuchin monkeys (*Sapajus apella*), 83, 148
Bush fires, 521

C
Canada, 651
Canadian Council on Animal Care (CCAC), 651
Captive animals, 516
Captive chimpanzee welfare, 45–48
Captive nonhuman primates, in numbers, 148–154
  Brazil, 149–154
  Mexico, 154–155
  Uruguay, 156–157
Captive research chimpanzee housing and care, 44–45
Capuchin monkeys, 129, 148
  black capped capuchin (*Sapajus apella*), 83, 148
  blond capuchin monkey (*Sapajus flavius*), 158
  brown capuchin monkeys (*Sapajus apella*), 83, 148
  golden-bellied capuchin (*Sapajus xanthosternos*), 158
  tufted capuchin (*Cebus apella*), 83
  tufted capuchins (*Cebus apella margaritae*), 147
  wedge-capped capuchin (*Cebus olivaceus*), 147, 163
Cardiovascular system, 241–243
Caribbean countries, legal protection for, 158–162
Castration, 134

Chester Zoo, 23, 87
Chimpanzee Health Improvement,
  Maintenance, and Protection Act,
  650
Chimpanzees (*Pan troglodytes*), 10, 18, 80–81,
  506
  behavioral freedom of, 516, 522–523
  in biomedical research, 31–32
    in aerospace studies, 36–38
    captive research chimpanzee housing
      and care, 44–45
    ethical concerns, 45–48
    in experimental medicine and biology,
      39–42
    large-scale sourcing of, 42–44
    in linguistic studies, 38–39
    under microscope, 48–49
    NIH decision and post-decision impact,
      49–51
    to Yerkes Primate Research Center,
      32–35
  Bossou chimpanzees, 520, 522
  captive chimpanzees, 515
  cognitive competencies of, 523
  linguistic studies, 38–39
  in live-video footage, 513
  wild chimpanzees, 514
ChimpCare website, 503
China, 651–652
China Laboratory Primates Breeding and
  Development Association (CLPA),
  655
Chiropotes spp., 148
CN Bio Innovations, 575
Cognitive bias, 447
  animal welfare literature, 209
  design and application, 209
  human literature, 208
  nonhuman primate welfare, 209
  one-size-fits-all design, 225
  overview, 208
  valuable approach, 209
Cognitive challenges, 513, 515
Cognitive complexity, 100
Cognitive enrichment, 489–491, 503, 523
  behavioral management of NHPs, 286–287
  physical cognitive
    daily activity budget information, 279
    delay of gratification, 271–272
    memory and planning, 278, 280
    self-control, 271, 273, 275
    tool use and causal understanding,
      280–281
  social cognition
    behavioral testing, 284–285
    mirror self-recognition, 283–284
    social interaction, 281–282
    social learning, 282–283
    theory of mind, 282
Cognitive tasks, 516
Cold foods, 483, 486
Colony management program
  components of, 308
  fundamental principles, 308
  housing, 311–313
  social behavior, 313–315
  specific-pathogen-free (SPF), 308
  veterinary care, 309–311
  welfare and environment enrichment, 315
Common Marmoset Care site, 68
Compromised welfare, 446–448, 542
Connection cage system, 518
Conservation of Brazilian Primates (CPB), 150
Consul, 10, 11
Contraception, 91–92
Contract Research Organizations (CROs), 112
Convention on Biological Diversity (CBD),
  158
Convention on International Trade in
  Endangered Species (CITES), 42,
  647
Corticotropin-releasing hormone (CRH), 233
Coulston foundation, 41
Council Directive 92/65/EEC, 647
Counter-conditioning techniques, 541

D

Decision-making processes, 514, 515
Deep substrate, 481, 485
Deleterious effects, 414
Depressed monkeys, 450
Deslorelin, 92
Developing countries, primates under
  endangered New World monkeys, in
    captivity, 157–158
  history, 146–148
  in Latin America, 162–163
  Latin American and Caribbean countries,
    legal protection for, 158–162
  laws and documents consulted, 164–165
Diabetes, 131–133, 137, 309, 368, 370, 451,
  456, 473, 481, 591
Diana monkey (*Cercopithecus diana*), 84
Diazepam, 185
Dietary complexity, 502

Diet, as pets and entertainers primates, 131
Direct human-animal contact, 85–87
DNA-based technologies, 575
Documentation, 492
Domestication, 123–125
Drill (*Mandrillus leucophaeus*), 184
Drug Metabolism and Pharmacokinetics (DMPK), 104
Drug therapy, 185
Dusseldorf, 8

E
EAE progression mechanism, 613
East Javan langur (*Trachypithecus auratus*), 91
Emory University, 35
Emotion-mediated cognitive bias, 209
Enrichment
  aim, 465–466
  application, 464
  assessment, 491
    behavioral responses, 492
    documentation, 492
    indirect results, 491–492
    physiological responses, 491
    preference, 492
  definition, 464
  enrichment program
    designing, 471
    food-based enrichment, 466
    general principles of successful program, 467
    housing facility, 467
    questions and considerations for, 467–470
    S.P.I.D.E.R. framework, 467
    variability, 467
  food-based enrichment, 479–483
  physical enrichment
    durable items, 477, 479, 480
    environmental change, 477, 478
    indoor facilities, 476, 477
    outdoor facilities, 477, 478, 481
  preference testing, 465
  questions to achieve any goal, 466
  safety considerations, 471–475
  sensory-based enrichment
    auditory enrichment, 487
    olfactory enrichment, 487–488
    tactile enrichment, 486–487
    visual enrichment, 483–486
  social enrichment, 475–476
Entertainers, 122

Environmental enrichment, 450
Ethnographical exhibitions, 13
European Association of Zoos and Aquaria (EAZA), 7
European Endangered Species Programs (EEPs), 131
European Framework, 644–647
European Marmoset Research Group, 656
European Primate Network (EUPRIM-Net), 656
Exogenous retroviruses, 453
Experimental autoimmune encephalomyelitis (EAE), 606
Experimental Design Assistant (EDA), 70

F
Fearful behavior, in rhesus macaques, 541
Federation of Laboratory Animal Science Associations (FELASA), 656
Fission-fusion emulation (FFE), 506, 509, 514
Food-based enrichment, 479–483
Food or fluid control, 451
Franceville International Center for Medical Research (CIRMF), 51
Frozen enrichment items, 486
Furuvik Zoo, 24

G
Glucocorticoid desensitization, 454
Good primate welfare, 112
Good research practice, primates, 67–71
Gorillas
  Eastern gorilla (*Gorilla beringei*), 58
  mountain gorilla (*Gorilla beringei beringei*), 8, 89
  Western gorilla (*Gorilla gorilla*), 185
  Western lowland gorilla (*Gorilla gorilla gorilla*), 81
Great apes (Hominidae), 9
Green Corridor Project, 520–522

H
Harm-benefit assessment, 101
Health monitoring and disease prevention program
  daily health checks, 326–327
  disease prevention, 327–328
    pathogen surveillance, 328–329
    personal protective equipment (PPE), 328
    vaccination, 329

identification, 330–331
quarantine process, 329–330
routine health monitoring, 327
sedation/anesthesia, 330
Helsinki Declaration, 597
Hepatitis B virus, 40
Hepatitis C virus, 40
Herpes viruses, 453
Housing and husbandry for primates in zoos
balancing, 357–358
behavioral needs of, 358
diet and nutrition, 368–370
environmental enrichment, 368
exhibition needs, 358–359
indoor enclosures, 364–365
indoor vs. outdoor facilities, 362–363
materials and substrates, 366–367
mixed-species exhibits, 367–368
overview, 355–356
regulations, guidelines, and accreditation, 356
research needs, 360
retreat areas, 362
space, 360–362
three-dimensional structures, 365
Housing, as pets and entertainers primates, 129–130
Howler monkeys (*Alouatta* spp.), 88
Huddling and play, 475, 476
Human-animal bonds, 81
Human–animal interactions, 23–24, 80, 82
Human-animal relationships, 83–85
Human behavior change refinement, 111–113
Human care, in developing countries
endangered New World monkeys, in captivity, 157–158
history, 146–148
in Latin America, 162–163
Latin American and Caribbean countries, legal protection for, 158–162
laws and documents consulted, 164–165
Humane endpoint
definition of, 381
European regulations and the ethical perspective, 377–380
experimental endpoint, 380
incorporation management, 387–388
irreversible decline, 382
killing primates, 387
objective assessments of welfare, 382–384
outcome of welfare assessment, 384–385
using euthanasia, 386

Humane handling
animal management
appropriate rearing, 343
handling predictable and consistent, 344–345
reducing relocations, 343–344
human–animal interactions
positive reinforcement training, 346
positive relationships, 345–346
Humane Society of the United States (HSUS), 127
Human exceptionalism, 627
Human immunodeficiency virus (HIV), 40, 41
Humanized monoclonal antibody TGN1412, 59
Human socialization, 104
Husbandry
changes in, 21–23
and research procedures, 530
Hygiene/disinfectant era, 17
Hypothalamic–pituitary–adrenal (HPA) system
adrenocorticotropic hormone (ACTH), 233
advantages and disadvantages, 235
corticotropin-releasing hormone (CRH), 233
cortisol levels
blood and saliva, 236
hair follicle, 237–238
urine and feces, 236–237
exogenous ACTH, 239
measuring regulation, 238
neuroendocrine system, 233
pharmacological manipulation, 238
schematic diagram, 234
sympathetic–adrenal–medullary system, 234

I

IACUC-approved protocol, 648
Iconic animals, 10–12
Imaging techniques, 575
Immune system
adaptive immunity, 244
antibody responses to immunizations, 245
anti-inflammatory cytokine, 244
blood-borne measure, 247
blood sampling, 247
booster immunizations, 245
literature review, 246
lymphoid tissue, 243
natural killer (NK) cells, 244
neopterin, 248
overview of, 244

Immune system (*cont.*)
 psychological well-being, 246
 relocations and group formations, 246
Immunodeficiency virus vaccine research, 449
Immunosuppressive drugs, 453
Inanimate enrichment, 181–183
Inbreeding, 131
India, 652
Initiation mechanism, 613
Institute for Laboratory Animal Research (ILAR), 655
Institute of Medicine (IOM), 48–49
Intentional death, 105
International Union for the Conservation of Nature and Natural Resources (IUCN), 519
International Zoo Yearbook (IZY), 6, 22
Invasive approaches, 595
Invasive biomedical research, 503

J
Japan, 652–653
Javan langurs (*Trachypithecus auratus*), 91
Jello (jelly), 483
Jersey Zoo, 19
Judgment bias
 advantages, 217
 basics, 210
 description, 210
 limitations, 217–218
 measuring primate welfare
  active choice task: "Go-Go"—location, 216
  active choice task: "Go-Go"—size cues, 214–216
  active choice task: "Go-No Go"—various cue types, 213–214
  the classic task: "Go-No Go"—color cues, 213
  the classic task: "Go-No Go"—size cues, 211–213
Juvenile animals, 39, 184

K
Köhler, Wolfgang, 33
Kumamoto Sanctuary
 chimpanzees in Japan, 504
 medical care in, 509
 social enrichment, 504
 weekly schedule of enrichment program, 505
 wild chimpanzees, 504
Kyoto guidelines, 652

L
Laboratories
 and captive settings, 99
 primates in
  and captive settings, 99
  quality of scientific output, welfare for, 109–110
  refinement, human behavior change in, 111–113
  3Rs and welfare, ethical framework of, 101–109
  use in, 99–101
  welfare assessment, 110–111
Latin America
 human care for primates in, 162–163
 legal protection for, 158–162
Legal protection, Latin American and Caribbean countries, 158–162
Legislation, 645
 primates, as pets and entertainers, 126–127
Lemurs
 common brown lemur (*Eulemur fulvus*), 157
 crowned lemus (*Eulemur coronatus*), 86
 indri (*Indri indri*), 8
 ring-tailed lemur (*Lemur catta*), 6, 9
 ruffed lemurs (*Varecia* spp.), 90
London Zoo, 24
L-tryptophan, 185

M
Macaque monkeys
 Barbary macaque (*Macaca sylvanus*), 128
 cynomolgus macaques (*Macaca fasicularis*), 61, 107–108, 540
 cynomolgus monkey/long-tailed macaque (*Macaca fascicularis*), 61
 Japanese macaque (*Macaca fuscata*), 129
 lion-tailed macaque (*Macaca silenus*), 14, 82, 90
 rhesus macaque (*Macaca mulatta*), 61, 130, 595
 stump-tailed macaque (*Macaca arctoides*), 173
 Sulawesi macaque (*Macaca nigra*), 6, 88, 190
Macaque website, 68
Mangabeys
 white crowned mangabeys (*Cercocebus torquatus*), 90
 white-naped mangabey (*Cercocebus atys lunulatus*), 90
Marmelak, Adam, 597
Marmosets, 129

black-tufted marmosets (*Callithrix penicillata*), 150
common marmosets (*Callithrix jacchus*), 129, 150, 151
Goeldi's monkey (*Callimico goeldii*), 21
Pygmy marmoset (*Cebuella pygmaea*), 21
Measuring primate welfare
  attention bias
    dot-probe task, 220–222
    preferential-looking tasks, 220
    visual search task, 222
  judgment bias
    active choice task: "Go-Go"—location, 216
    active choice task: "Go-Go"—size cues, 214–216
    active choice task: "Go-No Go"—various cue types, 213–214
    the classic task: "Go-No Go"—color cues, 213
    the classic task: "Go-No Go"—size cues, 211–213
Mexico, captive nonhuman primates, in numbers, 154–155
Microbrains, 575
Modern zoo, 60
  development of, 4–6
Monkey house, 14, 15
Monoamine oxidase A (MAOA-LPR) genes, 187
Mother rearing, 176
Multiple sclerosis (MS), 608–610
Multiple stimuli, 465, 471

N
National Action Plans, 150
National Center for Chimpanzee Care, 451
National Centre for the Replacement, Reduction and Refinement of Animals in Research (NC3Rs), 61, 63
National Chimpanzee Mangement Program Advisory Committee regarding, 43
National Council for Animal Experimentation Control (CONCEA), 152
National Primate Societies, 655–656
National Research Council, 153
Naturalistic enclosures, 17
Naturalistic enrichment, 467
Negative reinforcement, 538–539
New World monkey, 148, 157–158
New York Blood Center (NYBC), 40, 46, 47
NIH Council of Councils Working Group, 48–49

Noise, 130–131
Nonaversive training techniques, 530
Non-depressed monkeys, 450
Nonhuman primates, 307, 560, 561
  biomedical research
    causality of, 590
    role of, 590–592
    scientific value of research, 593–599
    transparent science communication, 600–601
  captive nonhuman, 644
  disease modeling/basic research, 562
  drug/product testing, 561–562
  empirical evidence of benefits
    Alzheimer's disease, 566
    disease models/basic research, 564–565
    drug testing, 562–564
    HIV/AIDS, 565–566
    Parkinson's disease (PD), 567–568
    stroke, 566–567
  entertainment, 630–631
  ethical argument
    direct harms, 572–573
    indirect harms, 574
  genetic difference
    in drug tests, 570
    humans and, 569
    in neuroscience, 570–571
  laboratory research, 629–630
  pet trade, 632–633
  in science, 376–377
  stressed nonhuman primates, 571–572
  zoos, 631–632
Noninvasive approaches, 595
Nutrition
  developmental needs, 332–333
  food presentation, 332
  primates, as pets and entertainers, 131–132
  standard diet, 331

O
Odors, 488
Office of Laboratory Animal Welfare, 649–650
Olfactory enrichment, 487–488
Outdoor housing, 184

P
Paignton Zoo, 91
Pair housing, 191
Pan troglodytes, 80–81
Papio hamadryas, 4, 156
Parkinson's disease (PD), 103
Patas monkey (*Erythrocebus* spp.), 129

Pathogens, 452–453
Pathological behavior, 173–174
Pattern recognition receptors (PRR), 619
Peer-raised primates, 133
Personality, 396–397
    applications, 404–405
    future improvements, 405–406
    measurement, 397–398
        reliability, 398
        structure, 399
        validity, 398–399
    temperament, 178
    and welfare
        behavior, 402
        definitions of, 399
        health and physiology, 402–404
        ratings, 400–402
Pet, 122
Pharmaceutical companies, 112
Physical enrichment
    durable items, 477, 479, 480
    environmental change, 477, 478
    indoor facilities, 476, 477
    outdoor facilities, 477, 478, 481
Physical restraint and conventional handling method, 540
Physical safety concerns, 471
Pitheciidae
    bearded saki monkeys (*Chiropotes* spp.), 148
    white-faced saki monkey (*Pithecia pithecia*), 90
Plantation trees, 521
Portland Zoo, 22
Positive reinforcement training (PRT), 107–108, 184–185, 450–452
    animal welfare, 539–541
    biological sample collection, 534–535
    biomedical research, 530
    consistent and cooperative movements, 533
    counter-conditioning, 532
    habituation, 532
    operant conditioning, 531–532
    refinements, 530
    regulatory support for, 530
    research procedures, 535
    research settings, 541–542
    for restraint, 535–537
    scheduling training, 531
    for science improvement, 542–543
    and social housing, 534
    systematic desensitization, 532
    training cooperation, veterinary care, 537–538

Pre-Darwinian approach, 593
Preference testing, 465
Primate behavior, public understanding of, 125–126
Primate Info Net, 656
Primates
    captive nonhuman primates, in numbers, 148–149
        Brazil, 149–154
        Mexico, 154–155
        Uruguay, 156–157
    captive primates, 502
    cognitive abilities, 502
    cognitive challenges in, 502
    disciplines, species, origins and numbers, 61–63
    distinct behavioral and cognitive characteristics in, 502
    environmental enrichment, 503
    good research practice, 67–71
    history of, 4
        exhibition in zoo, 13–14
        housing, changes in, 14–21
        human–animal interaction, 23–24
        husbandry, changes in, 21–23
        iconic apes, 10–12
        modern zoo, development of, 4–6
        species kept by zoos, 6–10
        in zoos, 26
        zoos and primate research, 24–25
    under human care, in developing countries
        endangered New World monkeys, in captivity, 157–158
        history, 146–148
        in Latin America, 162–163
        Latin American and Caribbean countries, legal protection for, 158–162
        laws and documents consulted, 164–165
    in laboratories
        and captive settings, 99
        quality of scientific output, welfare for, 109–110
        refinement, human behavior change in, 111–113
        3Rs and welfare, ethical framework of, 101–109
        use in, 99–101
        welfare assessment, 110–111
    as pets and entertainers, 121–122
        definitions, 122–123
        diet and nutrition, 131–132
        domesticated or wild, 123–125
        history, 123

Index 669

housing, 129–130
humans and, transfer of disease, 136–137
inbreeding, 131
legislation, regulation and numbers, 126–127
noise, 130
pregnancy and weaning, 133
primate behavior, public understanding of, 125–126
rehabilitation issues, 137–138
skeletal disease, 132
social issues, 133–134
temperature, humidity, light, 130
trade, 128–129
"undesirable" behaviors and interventions, 134–136
research settings, 59–61
study, 58–59
survival and reproduction, 502
welfare, ethics and legislation, 63–66
in zoos, 80
   assessing human-animal relationships, 83–85
   direct human-animal contact, 85–87
   familiar humans, presence of, 80–82
   impact of visitors, reducing, 85
   social management (*see* Social management)
Primate Society of Japan (PSJ), 655
Primate Specialist Group (PSG), 656–657
Primate welfare
abnormal behavior, 173
   age, 178–179
   extrinsic effects, 175
   function of, 179–180
   interventions for, 181–185
   intrinsic effects, 177–178
   later environmental restriction, 176–177
   to measure welfare, 180–181
   pathological behavior, 173–174
   prevalence of, 174–175
   rearing, 175–176
   sex, 179
   species, 178
   stereotypies, 173
   stress, 177
   temperament, 178
anxiety-related behavior, 185–186
   anxious phenotype, testing anxiety and identifying, 187–188
   behavioral measures of anxiety, 186–187
   to measure welfare using, 188–189

and welfare-related studies, 189–190
good welfare, measures of, 190
species-appropriate behaviors, as benchmarks, 191–192
using behavior to assess, 172
Primate welfare questionnaires
behavioral observations, 256
with chimpanzees, 261
design and analysis, 257–258
quality of life (QoL), 262
reliable methodology, 259
types of, 260–262
WelfareTrak® system, 260
Primtrain, 451
Proboscis monkey (*Nasalis larvatus*), 89, 90
Proceedings of World Congresses on Alternatives, 574
Professional Society Guidelines, 654–655
Psychological stress, 135
Psychological well-being, 172
Purpose-bred primates, 100
Puzzle feeder, 479, 481, 482, 489, 505

R

Record keeping and pedigree management
demographic and genetic information, 318
demographic management, 316–317
genetic management plan, 317–318
population genetics and demography, 316
Reduction principle, 618
Refinement, 446, 448, 455, 456, 618
human behavior change in, 111–113
Reforestation, 522
Relocation-related manipulations, 454
Renaissance, 4
Replacement paradigm, 617
Reproducibility crisis, 599
Research veteran and/or aged nonhuman primates, 456
Retroviral infections, 453
Royal Society for the Prevention of Cruelty to Animals (RSPCA), 127

S

Saguinus spp. (tamarins), 19, 60, 163
Saki monkeys (*Pithecia* spp.), 9, 90, 148
Sanctuary ethos, 636–637
Sanwa Kagaku Kenkyusyo Co. Ltd., 503
Scatter feeding, 481
Scientific Committee on Health Environmental and Emerging Risk (SCHEER), 645
Self-injurious behavior, 173, 541

Sensory-based enrichment
  auditory enrichment, 487
  cognitive enrichment, 489–491
  olfactory enrichment, 487–488
  tactile enrichment, 486–487
  visual enrichment, 483–486
Serotonin transporter (5HTTLPR), 187
Shelter
  macroenvironment
    cleaning, 342
    lighting, 341
    noise, 341–342
    temperature, 340–341
  microenvironment
    cage complexity, 336–338
    caging, 334–335
    enclosure types, 335–336
    flooring and substrates, 338–339
Siamang (*Symphalangus syndactylus*), 6
Singapore, 653
Single-sex groups, 89–91
Snowflake (albino gorilla), 12
Social animals, 628
Social behaviors, 89
Social bonds and welfare
  affiliative interactions, 429, 432
  network position, 431
  stress physiology, 429
  strong social bonds, 429–430
  structure and welfare, 431
Social enrichment, 475–476, 503
Social grooming, 91
Social groupings, 90, 99, 449, 450, 534
Social housing, 176, 181
Socialization, 449–450
Social management, 87–89
  contraceptive methods, 91–92
  single-sex groups, 89–91
Social rehabilitation, 138
Social separations, 187, 236, 344, 455, 534
Social status, dominance relationships, and welfare
  aggression and trauma, 418–420
  aggressive interactions, 417–418
  high social status, 424–425
  potential costs, 416
  psychosocial stress and welfare, 420–424
  rank stability and certainty, 425–426
  resource competition, 417–418
  status-welfare relationship, 426–428
  welfare, 417–418
South America and Africa, 653–654
South Asian Primate Network (SAPN), 657
Speciesism, 627

Specific-pathogen-free (SPF), 308, 311, 313, 328, 329, 452, 453, 606, 607, 613, 615, 619, 620
S.P.I.D.E.R. framework, 467
Spider monkeys (*Ateles* spp.), 84
Squirrel monkeys (*Saimiri* spp.), 17, 85, 130
Stereotypies, 173
Stockmanship, 82
Stoic approach, 232
Stress, 454, 455
  abnormal behavior, 177
Support for African/Asian Great Apes, 503
Sympathetic–adrenal–medullary (SAM) system
  autonomic nervous system (ANS), 241
  HPA system, 240
  parasympathetic nervous system (PNS), 239
  sympathetic nervous system (SNS), 239, 240

T
Tacrolimus, 453
Tactile enrichment, 486–487
Tamarins
  golden lion tamarin (*Leontopithecus rosalia*), 60, 87, 99
Tantalus monkeys (*Chlorocebus tantalus*), 4
Temperament, 453–454
3R principle, 101, 105
3Rs-specific databases, 67
TissUse, 575
Training nonhuman primates, 451
Tuberculosis (*Mycobacterium tuberculosis*), 133

U
UK's Concordat on Openness on Animal Research, 600
United States Department of Agriculture, 647–649
Universities Federation for Animal Welfare (UFAW), 657
Uruguay captive nonhuman primates, in numbers, 156–157
U.S. Animal Welfare Act 1966 (USAW), 127
U.S. Public Health Policy, 44
Utilitarian approach, 101
UV radiation, 132

V
Validity, 448, 450
Visitor effect, 83, 84

Visual enrichment
  colors, 485–486
  mirrors, 485
  stimulus, 483
  televisions and videos, 485

W
Welfare and conservation, 522
Welfare assessment, 110–111
Welfare in captive primates
  acclimation and data quality, 454–455
  animal handling and data quality, 455
  behavioral abnormalities, 447
  behavioral management techniques, 448, 451, 452, 455, 456
  using behavior to assess primate welfare, 232, 256
  captive environments, 448
  enriched environments, 448
  environmental enrichment and data, 450
  multiple protocols, 455–456
  nonenriched environments, 448
  positive reinforcement training and data, 450–452
  re-use, 455–456
  socialization and data, 449–450
  subject characteristics and data
    genetic variation, 452
    pathogens, 452–453
    temperament, 453–454
Welfare-related studies, 189–190
Wild Futures, 135
WISH project, 517
Woolly monkeys
  common woolly monkey (*Lagothrix lagotricha*), 148
  silvery woolly monkey (*Lagothrix poeppigii*), 163
Working Group, 49

Y
Yerkes Primate Research Center, 32–35
Yerkes, Robert, 34

Z
Zoo Atlanta, 12
Zoo-based primate research, 65
Zoo-housed primates, 24
Zoos
  primates, 4
    exhibition in zoo, 13–14
    housing, changes in, 14–21
    human–animal interaction, 23–24
    husbandry, changes in, 21–23
    iconic apes, 10–12
    modern zoo, development of, 4–6
    species kept by zoos, 6–10
    in zoos, 26
    zoos and primate research, 24–25
  welfare of primates in, 80
    assessing human-animal relationships, 83–85
    direct human-animal contact, 85–87
    familiar humans, presence of, 80–82
    impact of visitors, reducing, 85
    social management, 80, 87–92
Zootierliste, 7

Printed in the United States
by Baker & Taylor Publisher Services